Global Warming: Impacts and Management

Global Warming: Impacts and Management

Editor: Vivian Moritz

STATES
ACADEMIC PRESS
www.statesacademicpress.com

States Academic Press,
109 South 5th Street,
Brooklyn, NY 11249, USA

Visit us on the World Wide Web at:
www.statesacademicpress.com

ISBN: 978-1-63989-241-9 (Hardback)

Cataloging-in-Publication Data

Global warming : impacts and management / edited by Vivian Moritz.
p. cm.
Includes bibliographical references and index.
ISBN 978-1-63989-241-9
1. Global warming. 2. Global temperature changes. 3. Greenhouse effect, Atmospheric. 4. Climatology. 5. Environmental impact analysis.
6. Environmental management. I. Moritz, Vivian.
QC981.8.G56 G56 2022
363.738 74--dc23

Table of Contents

Preface

The long term heating of the Earth's atmosphere due to anthropogenic activities is defined as global warming. These activities primarily include burning of fossil fuels, which is responsible for the excess release of heat trapping gases into the Earth's climate. Also known as greenhouse gases, methane, carbon dioxide and nitrous oxide are a few examples of greenhouse gases. Production of aerosols and deforestation are some of the other factors which increase global warming. It can have an adverse impact on the Earth's ecosystem. Melting of glaciers, rising sea levels, ocean acidification, heat waves and disease outbreaks are a few of the effects of global warming. Some of the techniques to mitigate the repercussions of global warming are energy conservation, use of renewable energy and afforestation. The topics included in this book on global warming are of utmost significance and bound to provide incredible insights to readers. From theories to research to practical applications, case studies related to all contemporary topics of relevance to this field have been included in it. This book is a vital tool for all researching or studying global warming as it gives incredible insights into emerging trends and concepts.

This book has been the outcome of endless efforts put in by authors and researchers on various issues and topics within the field. The book is a comprehensive collection of significant researches that are addressed in a variety of chapters. It will surely enhance the knowledge of the field among readers across the globe.

It gives us an immense pleasure to thank our researchers and authors for their efforts to submit their piece of writing before the deadlines. Finally in the end, I would like to thank my family and colleagues who have been a great source of inspiration and support.

Editor

1

Malaria Transmission in the African Highlands in a Changing Climate Situation: Perspective from Kenyan Highlands

Yaw A. Afrane[1,2], Andrew K. Githeko[1] and Guiyun Yan[3]
[1]*Climate and Human Health Research Unit,*
Centre for Global Health Research, Kenya Medical Research Institute;
[2]*School of Health Sciences, Bondo University College, Bondo;*
[3]*Program in Public Health, College of Health Sciences,*
University of California at Irvine, Irvine CA,
[1,2]*Kenya*
[3]*U.S.A*

1. Introduction

Vector-borne diseases are among the diseases that have been linked with climate change (IPCC. 2001). Malaria is probably the deadliest climate sensitive vector-borne disease (Githeko et al. 2000). About 90% of the 300-500 million cases of the reported malaria cases worldwide come from Africa. In the late 80s to the 90s, malaria epidemics occurred frequently in western Kenya highlands, often taking the population by surprise. The epidemics were caused by *Plasmodium falciparum* and transmitted by *Anopheles gambiae* and *Anopheles funestus*. Epidemics were associated with high morbidity and mortality in all age groups, with prevalence of the disease rising from about 20% to about 60%. The case mortality in functional health facilities were estimated at about 7.5% (Githeko and Ndegwa 2001). The malaria transmission system involves a complex interaction between humans, mosquitoes, the plasmodium parasite, climate and the physical environment.

Warming of the climate is expected to lead to latitudinal and altitudinal temperature increase. The temporal and spatial changes in temperature, precipitation and humidity that are expected to occur under different climate change scenarios will affect the biology and ecology of vectors and intermediate hosts and consequently the risk of disease transmission. The risk increases because, although arthropods can regulate their internal temperature by changing their behaviour, they cannot do so physiologically and are thus critically dependent on climate for their survival and development (Lindsay and Birley 1996).

Temperature is inversely related to altitude, thus the high-elevation areas in Africa, or highlands, generally exhibit low ambient temperature, which restricts the development of vectors and parasites. Because malaria transmission rate is temperature-dependent, any factor that alters the temperature in the highland would reduce the duration of parasite development, larval development and increase the mosquito biting rates, and subsequently increase malaria transmission in the highlands.

The greatest effect of climate change on transmission is likely to be observed at the extremes of the range of temperatures at which transmission occurs. For many diseases these lie in the range 14–18 °C at the lower end and 35–40 °C at the upper end. Warming in the lower range has a significant and non-linear impact on the extrinsic incubation period (Watts et al. 1987), and consequently disease transmission, while, at the upper end, transmission could cease. However, at around 30–32 °C, vectorial capacity can increase substantially owing to a reduction in the extrinsic incubation period, despite a reduction in the vector's survival rate (Githeko et al. 2000).

Climatic conditions affect malaria transmission in a number of ways. The development rate of immature mosquitoes is very much temperature dependent. Below 16°C, development of *Anopheles gambiae,* the main malaria vector in most parts of Africa, will completely stop, and the larvae will die in water temperature below 14°C. In low-temperature conditions, mosquito larval development is severely delayed and high mortality incurs. In the adult stage, increase in ambient temperature will accelerate the digestion of blood meals taken by mosquitoes leading to increased human biting frequency and malaria transmission. While female mosquitoes obtain blood meal to develop their eggs, they pick up malaria parasites which they develop and become infectious to the next human host that is subsequently fed upon. Increased biting frequency and faster blood meal digestion also means increased fecundity and better reproductive fitness. Increased temperature also shortens the development time for the malaria parasite in the mosquito (Githeko et al. 2000).

Mosquito species such as the *Anopheles gambiae* complex and *An. funestus* are responsible for transmission of most of the malaria in the continent of Africa and are sensitive to temperature changes as immature stages in the aquatic environment and as adults. If water temperature rises, the larvae take a shorter time to mature (Munga et al. 2007) and consequently there is a greater capacity to produce more offspring during the transmission period. In warmer climates, adult female mosquitoes digest blood faster and feed more frequently (Afrane et al. 2005), thus increasing transmission intensity. Increased biting frequency and faster blood meal digestion also means increased fecundity and better reproductive fitness (Afrane et al. 2006). Similarly, malaria parasites complete extrinsic incubation within the female mosquito in a shorter time as temperature rises (Afrane et al. 2008) thereby increasing the proportion of infective vectors. Warming above 34 °C generally has a negative impact on the survival of vectors and parasites (Rueda et al. 1990).

In addition to the direct influence of temperature on the biology of vectors and parasites, changing precipitation patterns can also have short- and long term effects on vector habitats. Increased precipitation has the potential to increase the number and quality of breeding sites for mosquitoes and the density of vegetation, affecting the availability of resting sites (Githeko and Ndegwa 2001).

In addition to warming of the climate, environmental changes, such as deforestation, could increase local temperatures in the highlands, which in turn accelerates vector and parasite development leading to increased transmission. Deforestation is a common event in many regions of Africa. For example, Malava forest, a tropical rain forest in Kakamega district, has shrunk from 150 km^2 in 1965 to 86 km^2 in 1997 (FAO 1993). In the East African highlands, 2.9 million hectares of forest were cleared between 1981 and 1990, representing an 8% reduction in forest cover in one decade (Lindblade 2000). Land use and land cover changes may modify the temperature and relative humidity of malaria vector habitats in the highlands. It was demonstrated that in the southwestern highlands of Uganda, maximum and minimum temperatures were significantly higher in communities bordering cultivated swamps than in those near natural swamps (Lindblade 2000). These changes in regional climate and microclimatic conditions of mosquito habitats causes abundant changes of the existing mosquito species, and may make some areas permissive to the proliferation of new species. For instance, deforestation caused the introduction of *An. gambiae* into a habitat that was previously dominated by *An. moucheti* in Cameroon (Manga et al. 1995). In northern Brazil, *An. marajoara,* a species previously of minor importance, has become the principal malaria vector following changes in land use (Conn et al. 2002).

2. Effect of microclimate changes due to deforestation on malaria transmission

In many parts of the world, including Africa and South America, changes in land use or land cover, especially deforestation, have been linked to changes in the microclimate of the area in question. These changes in microclimate due to changes in land uses and cover has been linked to changes in malaria transmission. The changes in land use affect temperature, humidity and rainfall or precipitation patterns of the area. These also, goes to affect the microhabitat of malaria transmitting mosquitoes and the parasite they transmit, thus affecting malaria transmission in the area. The link between changes in land use and land cover and malaria transmission has been clearly established, which are highlighted and discussed below.

2.1 Deforestation impact on malaria vectors: case studies from western Kenya highlands

We conducted studies in western Kenya to assess the possible effect of deforestation on the microclimate and the subsequent effect on malaria transmission indices in the area. We measured indoor and outdoor temperature and humidity in the forested and deforested areas in the highlands. Entomological parameters such as the duration of gonotrophic cycle and biting frequency, reproductive fitness, survivorship and development of mosquitoes were studied. The sporogonic development of malaria parasites in the mosquito was also studied. Vectorial capacity which measures the potential rate of contact between infectious vectors and susceptible hosts was calculated from these parameters. It is the product of the vector density in relation to man, the proportion that bite man twice, and the expectation of the infective life span of the vector (MacDonald 1957; Garrett-Jones and Shidrawi, 1969).

We found that deforestation increased the indoor mean temperature by 1.8 ^{0}C. Mean maximum and minimum temperatures were increased by 2.3^{0}C and 1.5 ^{0}C respectively. Outdoor temperatures were increased by 0.5^{0}C. The mean indoor relative humidity in the deforested area was about 22.6% lower than in the forest area during the dry season (79.88% vs. 57.29%) (Afrane et al. 2005; Afrane et al. 2006; Afrane et al. 2008; Afrane et al. 2007). Temperature and humidity were measured in houses in study sites that were originally forest but large portions of it have been deforested. Canopy coverage was over 60 % in the forested area and less than 10 % in the deforested area. Temperature and humidity were measured using HOBO® data loggers (Onset Computer Corporation, Bourne, Massachusetts, U.S.A.) which were placed inside all the human-habited experimental houses in the forested and deforested areas of the site. There were five experimental houses in each of the areas.

Since the general body functions of mosquitoes are temperature dependent, these small changes will have a biological significance on their life history characteristics. The mean minimum and maximum temperatures are very crucial for the development of mosquitoes and the parasite they transmit and by extension malaria transmission. Below 16°C, development of many malaria vectors in Africa will completely stop, and the larvae will die in water temperature below 14°C. The parasite in the mosquitoes has an upper and lower threshold of 32° C and 18°C respectively (Boyd 1949; MacDonald 1957).

The changes in the microclimate that were observed as a result of the deforestation caused a decrease in the duration of gonotrophic cycle of mosquitoes by 1.7 days (4.6 vs 2.9 days). The duration of gonotrophic cycle is the the period between the taking of a blood meal by a mosquito through digestion of the blood meal and oviposition or egg laying (Santos et al. 2002). The decreased duration of the gonotrophic cycles implies a decrease in human biting frequency from an average of once every 5 days to once every 3 days (Afrane et al. 2005). Increase in biting frequency means that the *An. gambiae* will feed more frequently on humans and in the process picks up infection, become infective and then transmit the parasite when they next bite another person.

The changes in the microclimate due to deforestation however, did not favour the survival of the adult *An. gambiae*. The effect of deforestation decreased median survival of *An. gambiae* by 5-7 days. The *Anopheles gambiae* mosquito rather prefers areas with high humidity. Since deforestation caused a decrease in the humidity, it rather caused a decrease in the survival of *An. gambiae*. However, despite the decreased survivorship of the mosquitoes due to the effects of deforestation, mosquitoes still had an enhanced reproductive fitness by 40%. Over the course of their life span, mosquitoes in deforested areas laid far more eggs and thus, had better fecundity than mosquitoes in the forested areas by over 65% (Afrane et al. 2006). The reproductive fitness of a mosquito is its ability to reproduce over its entire life span. Fecundity, measured by the number of offspring a female mosquito can produce, is a major fitness trait (Hard et al. 1989.). The implication of these findings is that *An. gambiae* could increase its population within a short time when breeding sites are available. This could potentially lead to an increase in malaria transmission when infected humans are available.

The microclimatic changes also caused the parasite in the *An. gambiae* to have a reduced sporogonic development time from an average of 14 to 12.6 days (Afrane et al. 2008). Both oocyst and sporozoite development times were reduced by 1 and 1.4 days respectively. Reduced parasite development time in mosquitoes implies that the parasite will take a shorter time to become infective inside the mosquito and therefore will be transmitted faster from the mosquito to humans.

Vectorial capacity of the malaria vector was then calculated using the parameters that were measured in the studies described above.

$$\text{Vectorial capacity} = \frac{ma^2pn}{(-\log_e p)}$$

Where m is the relative density of vectors in relation to human,

a is the average number of humans bitten by one mosquito in one day,

p is the proportion of vectors surviving per day, and

n is the duration of sporogony in days. (MacDonald 1957).

This formula postulates that the length of the sporogonic development of malaria parasites in mosquitoes, biting frequency and vector survivorship are the most important factors for vector-borne malaria transmission

We found that, deforestation substantially facilitated malaria transmission in the highland, as evidenced by two-fold increase in vectorial capacity (Afrane et al. 2008). These are shown in Table 1. Malaria is highly endemic in lowland areas. All the indices for malaria transmission were very much enhanced in the lowland site used for this study. This explains the huge difference between the lowland and two highland sites. In the highland sites, changes in the microclimate due to deforestation explain the changes in vectorial capacity of the two sites.

Site	Land use type	m	a	n	P	Vectorial capacity
Highland	Forested	3.05	0.198	13.9	0.927	0.54
	Deforested	4.64	0.233	12.8	0.917	0.96
Lowland	Deforested	7.85	0.465	11.7	0.923	8.30

Note.- m is the relative density of vectors in relation to human,
P is the proportion of vectors surviving per day,
a is the average number of men bitten by one mosquito in one day, and
n is the duration of sporogony in days.

Table 1. Estimated vectorial capacity of *A. gambiae* in forested and deforested areas in western Kenya highland and lowland.

2.2 Land use and land cover changes and malaria transmission

Land use and land cover changes have the potential to alter the micro-environment of malaria vectors which in turn affects their malaria transmission potential. In the highlands of Uganda, Lindblade and others (2000) compared mosquito density, biting rates, sporozoite rates and entomological inoculation rates between 8 villages located along natural papyrus swamps and 8 villages located along swamps that have been drained and cultivated. Since vegetation changes affect evapotranspiration patterns and, thus, local climate, they also investigated differences in temperature, humidity and saturation deficit between natural and cultivated swamps. They found that on average all malaria indices were higher near cultivated swamps. Maximum and minimum temperatures were significantly higher in communities bordering cultivated swamps. They found that the average minimum temperature of a village was significantly associated with the number of *Anopheles gambiae* s.l. per house. From their studies, it appears that replacement of natural swamp vegetation with agricultural crops led to increased temperatures, which may be responsible for elevated malaria transmission risk in cultivated areas.

To determine the effects of land cover types on mosquito productivity, Munga and others (2007) created semi-natural larval habitats within three land cover types (farmland, forest, and natural swamp) and filled them with *Anopheles gambiae* larvae. The mosquito pupation rate in farmland habitats was significantly greater than in swamp and forest habitats and larval-to pupal development times were significantly shorter. Land cover type may affect larval survivorship and adult productivity through its effects on water temperature and nutrients in the aquatic habitats. They found significantly higher water temperatures in farmland habitats than the other land cover types, which enhanced pupation rates and shortened development times. They also found a significant negative correlation of the occurrence of *An. gambiae* larvae with canopy cover and emergent plants in natural habitats. Their results suggest that deforestation and cultivation of natural swamps in the western Kenyan highland create conditions favorable for the survival of *An. gambiae* larvae, and consequently increase the risks of malaria transmission to the human population.

Other land cover changes such as swamp reclamation have been shown to have similar effects as deforestation on malaria vectors development. In other studies from western Kenya highlands, removal of swamp vegetation reduced shading and increased the breeding habitat water temperature by as much as 5 ^{o}C and whose effect reduced larval development time leading to increased vector productivity and abundance (Wamae et al. 2010). This has the potential to increase the population of mosquitoes and to which could also potentially cause an increase in malaria transmission if infected humans are available.

3. Proliferation of mosquito species to new areas

Deforestation has the potential to also change the microclimate of an area to become suitable to species that previously could not survive in the area. Each mosquito species has their own optimum survival conditions that allow them to survive or not to survive in an area. These survival conditions are mostly climate dependent. For instance, *An. gambiae* and *Anopheles arabiensis,* the most important vectors of human malaria in Africa are sibling species. *An. arabiensis* is either absent or shows a very low abundance in high-elevation areas where temperature is low and relative humidity and the amount of rainfall are high, but it is abundant in dry areas with less humidity which is the preferred habitat (Minakawa et al. 2002; Maharaj 2003). Studies were conducted by Afrane and others (2007) to investigate whether climate conditions in the western Kenya highlands were permissive to the development and survival of *An. arabiensis* and whether deforestation promoted *An. arabiensis* survivorship of immature and adult stages, using life-table analysis.

It was found that *An. arabiensis* larvae were able to survive in the deforested sites compared to the forested areas. Adult *An. arabiensis* placed in houses in the deforested area had an enhanced survival and better reproductive fitness than those in the forested area (Afrane et al. 2007). The implications of these findings are that, if the current trends of deforestation continue in the highlands, there could be a three vector system in the highlands, instead of the current two. This could potentially worsen the malaria transmission system in the highlands and could lead to the return of epidemics in the highlands. We expect that, environmental changes such as deforestation and global warming may facilitate the establishment of *An. arabiensis* populations in the highlands.

In the highland areas west to Mount Kenya which has an elevation of 1,720 - 1,921 m above sea level, malaria has been reported in the last 10 - 15 years. It was not clear whether these cases were introduced from the nearby lowland or resulted from local transmission because of no record of vector mosquitoes on the highlands. Chen and others (2006) reported finding mosquitoes breeding in the highlands. This means that the environment has become conducive for the proliferation of malaria vector species. Environmental changes and global warming could be a factor driving this emergence of malaria vectors to areas in mount Kenya region.

Manga and others (1995), working in an area that has been deforested to build a new airport in Cameroon, observed that deforestation caused the introduction of *An. gambiae* into a habitat that was previously predominated by *An. moucheti*. *An. gambiae* is the most ferocious of all the malaria vectors in Africa. It has been found to be the most efficient vector of *P.*

falciparum in Africa. If the effects of deforestation made it possible for *An. gambiae* to inhabit this new place, the implications are that malaria transmissions in this area will most likely increase.

In northern Brazil, Conn and others (2002) found that *An. marajoara,* a species previously of minor importance, has become the principal malaria vector following changes in land use. Malaria transmission in the area was previously dominated by *An. darlingi*. However, land use and land cover changes made *An. marajoara* to become the most abundant vector and the most infected with malaria parasites, with its anthropophilic behavior. Their discovery highlights one of the challenges of neotropical malaria control, namely that the targeting of specific vectors may be complicated by a changing mosaic of different locally important vectors and their interactions with human populations.

Since 2003, Afrane and others (unpublished) have been studying the spatial and temporal dynamics of adult mosquitoes in the highlands of western Kenya. Each month, mosquitoes are collected from over 40 houses using pyrethrum spray catches. Mosquitoes are collected and brought to the laboratiories of the Kenya Medical Research Institute and identified to species using the morphological keys of Gilles and Demellion (1968). Polymerase Chain Reactions (PCRs) are done to discriminate the sibling species of *An. gambiae* complex.

Species composition reflects a combination of environmental and historical intervention events at a site; hence, changes in species composition can provide a sensitive measure of ecological changes. Figure 1 shows the changes in species composition of the *An. gambiae* s.l. complex from 2003 to 2010 in the study site. Analysis of species composition illustrated that the proportion of *An. arabiensis* in the site was characterized by a significant increase in the proportion of *An. arabiensis* from < 1% in 2003 to 16% in 2010.

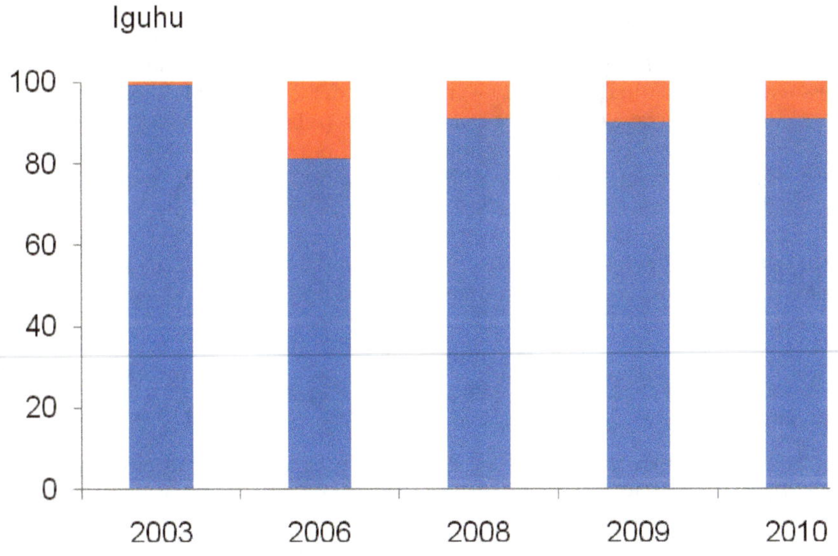

Fig. 1. Species composition of *Anopheles gambiae* (in blue) and *An. arabiensis (in red)* in Iguhu study site, Kakamega district in Kenya from 2003 to 2010.

4. Warming of the climate and malaria incidence in East Africa region

Deforestation in the western Kenya highlands has permitted the gradual colonisation of *An. arabiensis* into these areas that previously did not have this species. The implications are that with a new vector introduced in the area, the malaria transmission situation may worsen leading to resumption of epidemics in the site. The rate of malaria transmission is related to the abundance of malaria vectors while the stability of transmission is related to the diversity of species. Each species tends to dominate in the different seasons when its optimum survival conditions become available (Githeko et al. 2000).

There have been strong debates over the last decade by many authors on whether or not observed increases in malaria incidence in the Kenyan Highlands during the last thirty years are associated with co-varying changes in local temperature, possibly connected to global changes in climate. Studies using differing data sets and methodologies, produced conflicting results regarding the occurrence of temperature trends and their likelihood of being responsible, at least in part, for the increases in malaria incidence in the highlands of western Kenya. Omumbo and others (2011), used over 30 years of quality controlled daily observations of maximum, minimum and mean temperature in the analysis of trends at Kericho meteorological station, sited in Kenya's western highlands.

They found an upward trend of ≈0.2°C/decade in all three temperature variables. Mean temperature variations in the Kericho area were associated with large-scale climate variations including tropical Sea Surface Temperatures (SST). Local rainfall was found to have inverse effects on minimum and maximum temperature. They also used three versions of a spatially interpolated temperature data set, which showed markedly different trends when compared with each other and with their data.

Loevinsohn (1994) conducted a study that assessed the contribution of the climate to a malaria epidemic in Rwanda, focusing on the catchment area of one health centre where diagnosis was consistent and non-climatic variables well monitored. In late 1987 malaria incidence in the area increased by 337% over the 3 previous years. The increase was greatest in groups with little acquired immunity - children under 2 years (564%) and people in high-altitude areas (501%). Case-fatality rose significantly (relative risk = 4.85, $p < 0.001$). The same year also saw record high temperatures and rainfall. An autoregressive equation including lagged effects of these two variables explained 80% of the variance in monthly malaria incidence. Temperature (especially mean minimum) predicted incidence best at higher altitudes where malaria had increased most. Empirically derived relations were consistent with the estimated generation time of the disease and with the known sensitivity of the plasmodium parasite to temperature. These findings are most relevant to regions near the altitude or latitude limits of the disease, where several epidemics have lately been reported.

5. Commentary

Forests have been shown to stabilize local temperatures and this can annul the effects of climate variability that is related to malaria epidemic in the highlands. Deforestation accounts for nearly 20% of global annual carbon emissions. Thus this phenomenon is closely linked to global warming and subsequent increased spread of malaria in the highlands. Forests are carbon sinks as they absorb CO_2 and sequester it. Many policy makers are not familiar with the link between deforestation, climate change and their effect on malaria transmission.

Changes in temperature which are as a result of environmental changes have the potential to affect the vector and their parasites as well as the transmission of many parasite borne diseases including malaria. Environmental changes, either natural phenomenon or through human intervention, alter the ecological balance and context within which vectors and their parasites breed, develop, and transmit disease (Patz et al. 2000). Small increases in temperature affect the mosquitoes and the parasites they transmit (Bruce-Chwatt 1987).

Large-scale conversion of tropical forests for agricultural purposes can significantly change the surface properties, such as soil wetness and surface roughness, of an area and could lead to changes in the local climate. For example, radiation budget and energy balance could be affected (Nobre et al. 1991; (Bounoua 2002). According to Ramakrishna and Running (Ramakrishna and Running 1995), the geometry of the new surface and albedo, which is a measure of the reflectivity of a surface, affect the radiation budget whilst the energy balance is also changed. This leads to a change in the microclimate of the area concerned. Ambient and indoor temperatures as well as humidity are affected by a change in land use (Lindblade 2000). For instance in the Amazon forest of South America, extensive deforestation caused a significant increase in the mean surface temperature (by about 2.5°C) and a decrease in the annual evapo-transpiration (30% reduction) and precipitation (25% reduction) in the region, with the changes being more pronounced in the hot season (Nobre et al. 1991). The deforested area was associated with larger diurnal fluctuations of surface temperature and vapour pressure deficit. Land-cover changes that affect evaporation may also affect both the surface air temperature and the ground surface temperature in surrounding areas to a lesser extent (Skinner and Majorowicz 1999). In the tropics and subtropics, conversion warms canopy temperature by 0.8 °C all year round (Bounoua 2002.).

The data presented here have implications for understanding the effects of deforestation on an increased malaria transmission in the western Kenya and other African highlands. Malaria transmission force may be measured by the vectorial capacity of the vector. The duration of sporogony of *P. falciparum* in mosquitoes is exponentially related to the vectorial capacity (MacDonald 1957). If the daily survival and biting frequency of a vector are assumed to be constant, decreasing the duration of sporogony will lead to an increase in vectorial capacity. In this study deforestation led to a decrease in the duration of sporogony of *P. falciparum* by 1.1 days. This will translate into an exponential increase in vectorial capacity of *An. gambiae* which will lead to an increase in malaria transmission.

The studies presented here established the relationship between deforestation which is ongoing in the western Kenya highlands and the local climate as well as the effects of this altered local climate on malaria transmission indices in the study area.

The increase in indoor temperature due to land use and land cover changes enhanced digestion of blood meals by the mosquitoes in the deforested area leading to shorter gonotrophic cycles. Shorter gonotrophic cycles made mosquitoes to feed more frequently and thus increased their biting frequency. Frequent feeding by mosquitoes could increase the probability of picking infections from their hosts or infecting their hosts. Biting frequency affects the vectorial capacity and the Entomological Inoculation Rates (EIR), the two indices for measuring malaria transmission force in an area (Garrett-Jones and Shidrawi 1969). The vectorial capacity varies as the square of the daily man biting rates and EIR varies linearly with the daily man biting rates. An increase in the biting frequency of mosquitoes, thus, leads to an increase in vectorial capacity and EIR and hence to an increase in malaria transmission if there are infected individuals in the population.

The present study also showed that land use and land cover changes increased the population of mosquitoes. The increased temperature as well as the decrease in humidity in the deforested site led to a higher reproductive fitness for mosquitoes. The intrinsic rate of growth which is the number of progeny per unit of time and thus more accurately reflects maximal fitness was increased by the effects of deforestation. Generation time for mosquitoes was also decreased in the deforested areas. Increase in the number of mosquitoes has implication for malaria transmission in an area where the effects of deforestation lead to increase in biting frequency. In the case where there are infected individuals in the population, increase in mosquito population and an increase in biting frequency will cause an increase in the EIR leading to an increase in malaria transmission.

In the deforested area *An. arabiensis* had a shorter larva-to-adult development period. *An. gambiae* also exhibited the same phenomenon in a different study in the same area (Tuno et al. 2005). This renders the deforested area particularly vulnerable to mosquito proliferation and colonisation. The adults of *An. arabiensis* had a higher survival and reproductive fitness in the deforested area than in the forested areas. This could lead to proliferation in, and colonisation of the western Kenya highlands by *An. arabiensis*, a particularly crucial event since already *An. gambiae* and *An. funestus* are causing malaria epidemics in the highlands of western Kenya. Even though *An. arabiensis* is considered a zoophilic vector (prefers feeding on animals) rather than an anthropophilic vector (prefers feeding on humans), its efficiency to transmit malaria should not be underestimated. In many places in Africa such as the Kano plains and the lowland areas of western Kenya (White 1974; Githeko et al. 1996), Senegal (Robert et al. 1998) Ethiopia (Taye et al. 2005) and South Africa (Maharaj 2003), it is the main malaria vector. It is possible that it could become another vector in the highlands or it could also replace the current vectors in the highlands.

Deforestation had the effect of shortening the duration of the sporogony of *P. falciparum*. Time for the appearance of oocyst was shortened by 0.9 day and sporozoite appearance was also shortened by 1.1 days. This implies increased malaria transmission. The duration of sporogony of the malaria parasite in mosquitoes has an exponential relationship to the vectorial capacity of the vector. If all transmission variables remain the same, decreasing the duration of sporogony will cause an increase in the vectorial capacity. This translates into an increase in malaria transmission.

Besides their impacts on the duration of the gonotrophic cycle, biting frequency, duration of sporogony and subsequently on malaria transmission, deforestation and swamp reclamation also facilitate the creation of breeding habitats for *An. gambiae*. Those factors further enhance malaria transmission (Minakawa 2005; Munga et al. 2006). Mosquitoes will normally not breed in a natural swamp but will colonise a cultivated swamp (Munga et al. 2005). Deforestation opens up the forest, exposing the breeding habitats to direct sunlight which is a requirement for the maturation of the aquatic life stages of *An. gambiae* (Gilles and De Meillon 1968; Amerasinghe et al. 1997). As human populations increase in the highlands of western Kenya and elsewhere in the highlands of the East African region, deforestation and its subsequent effects on malaria transmission will continue.

The present study has shed some light on the possible impacts of increases in temperature in the highlands during periods of extreme events such as the El Niño events which are associated with anomalous warming and rainfall when they occur (Kovats et al. 2003; Rúa et al. 2005). In general a rise of 0.5°C in temperature in the highlands is equivalent to a decrease in altitude by 77m. This means highland climatic conditions move closer to lowland situations. The implication of this is that malaria transmission which is driven by temperature will also move closer to lowland situations.

Environmental changes such as deforestation and swamp reclamation have been shown to increase microclimate of an area. These changes in the microclimate have been shown to also affect the malaria mosquitoes and malaria transmission indices. Thus, if climate change should lead to altitudinal and latitudinal increases in temperature, this will go to affect malaria transmission especially in highaland areas of Africa.

The highlands of western Kenya are being impacted by three anthropogenic drivers of malaria transmission these being global warming, deforestation and swamp reclamation. Reforestation and swamp restoration can reverse increased local warming while providing, the local populations with ecological goods and services in a sustainable way. With regard to malaria such an intervention could potentially prevent the further spread of malaria to higher altitudes.

6. Acknowledgements

This study was supported by the National Institute of Health (R01 AI094580 , D43 TW01505 and R01 A150243).

7. References

Afrane, Y. A., B. W. Lawson, A. K. Githeko, and G. and Yan. 2005. Effects of Microclimatic Changes Due to Land use and Land Cover on the Duration of Gonotrophic Cycles of Anopheles gambiae Giles (Diptera: Culicidae) in Western Kenya Highlands. Journal of Medical Entomology 42: 974-980.

Afrane, Y. A., G. Zhou, B. W. Lawson, A. K. Githeko, and G. Yan. 2006. Effects of Microclimatic Changes Due to Deforestation on the Survivorship and Reproductive Fitness of Anopheles gambiae in Western Kenya Highlands. . Am. J., Trop Med Hyg. 74: 772-778.

Afrane, Y. A., T. Little, J., B. W. Lawson, A. K. Githeko, and G. Yan. 2008. Deforestation Increases the Vectorial Capacity of Anopheles gambiae Giles to Transmit Malaria in the Western Kenya Highlands. Emerg Infect Dis In Press.

Afrane, Y. A., G. Zhou, B. W. Lawson, A. K. Githeko, and G. Yan. 2007. Life-table analysis of Anopheles arabiensis in western Kenya highlands: effects of land covers on larval and adult survivorship. .Am J Trop Med Hyg. 77: 660-6.

Amerasinghe, F. P., F. Konradsen, K. T. Fonseka, and P. H. Amerasinghe. 1997. Anopheline (Diptera:Culicidae) breeding in a traditional tank-based village ecosystem in north central Sri Lanka. J Med Entomol. 34(290-7.

Bounoua, L., R. DeFries, G. J. Collatz, P. Sellers, and Khan, H. . 2002. Effects of land cover conversion on surface climate. Climatic Change 52: 29-64.

Boyd, M. F. 1949. Malariology. W. B. Saunders Co., Philadelphia.

Bruce-Chwatt, L. J. 1987. Malaria and its control: present situation and future prospects. Annu Rev Public Health 8: 75-110.

Chen, H., A. K. Githeko, G. Zhou, J. I. Githure, and G. Yan. 2006 New records of Anopheles arabiensis breeding on the Mount Kenya highlands indicate indigenous malaria transmission. Malar J. 7: 17.

Conn, J. E., R. C. Wilkerson, M. N. Segura, R. T. de Souza, C. D. Schlichting, R. A. Wirtz, and M. M. Povoa. 2002. Emergence of a new neotropical malaria vector facilitated by human migration and changes in land use. Am J Trop Med Hyg 66: 18-22.

FAO. 1993. Forest resources assessment, 1990: Tropical countries. FAO forestry paper No.112, Rome, Italy

Garrett-Jones, C., and G. R. Shidrawi. 1969. Malaria vectorial capacity of a population of Anopheles gambiae: An exercise of epidemiological entomology. Bulletin of the World Health Organization 40: 531 - 545.

Gilles, M. T., and B. De Meillon. 1968. The Anopheline of Africa South of the Sahara. Publication of the South African Institute for Medical research no. 54.

Githeko, A. K., and W. Ndegwa. 2001. Predicting malaria epidemics in the Kenya highlands using climate data: a tool for decision makers. Global Change and Human Health 2: 54-63.

Githeko, A. K., S. W. Lindsay, U. E. Confalonieri, and J. A. Patz. 2000. Climate change and vector-borne diseases: a regional analysis. Bull World Health Organ 78: 1136-47.

Githeko, A. K., N. I. Adungo, D. M. Karanja, W. A. Hawley, J. M. Vulule, I. K. Seroney, A. V. Ofulla, F. K. Atieli, S. O. Ondijo, I. O. Genga, P. K. Odada, P. A. Situbi, and J. A. Oloo. 1996. Some observations on the biting behavior of Anopheles gambiae s.s., Anopheles arabiensis, and Anopheles funestus and their implications for malaria control. Exp Parasitol 82: 306-15.

Hard, J. J., W. E. Bradshaw, and D. J. Malarkey. 1989. Resource- and density-dependent development in treehole mosquitoes. Oikos 54: 137-144.

IPCC., and C. C. 2001. 2001. Impacts, Adaptations and Vulnerability. Contribution of Working Group 2 to the Third Assessment Report of the Intergovernmental Panel on Climate Change; . Cambridge University Press: Cambridge, UK and New York, NY, USA, ;: 1-970.

Kovats, R. S., M. J. Bouma, S. Hajat, E. Worrall, and A. Haines. 2003. El Nino and health. Lancet 362: 1481-9.

Lindblade, K. A., Walker E. D, Onapa A.W, Katungu J, and Wilson M, . . : . 2000. Land use change alters malaria transmission parameters by modifying temperature in a highland area of Uganda. Trop. Med. Int. Health 5: 263-74.

Lindsay, S. W. and M. H. Birley. 1996. Climate change and malaria transmission. Ann Trop Med Parasitol 90: 573-88.

Loevinsohn, M. E. 1994. Climatic warming and increased malaria incidence in Rwanda. Lancet 343: 714-718.

MacDonald, G. 1957. The Epidemiology and Control of Malaria. Oxford University Press, Oxford.

Maharaj, R. 2003. Life Table Characteristics of Anopheles arabiensis (Diptera: Culicidae) Under Simulated Seasonal Conditions. J. Med. Entomol. 40: 737-742.

Manga, L., J. C. Toto, and P. Carnevale. 1995. Malaria vectors and transmission in an area deforested for a new international airport in southern Cameroon. Societes Belges Medicine Tropicale 75: 43-49.

Minakawa, N., G. Sonye, M. Mogi, A. Githeko, and G. Yan. 2002. The effects of climatic factors on the distribution and abundance of malaria vectors in Kenya. J Med Entomol 39: 833-41.

Minakawa, N., Munga, S., Atiel,.i F., Mushinzimana, E., Zhou, G, Githeko, A.K, Yan, G. 2005 Spatial distribution of anopheline larval habitats in Western kenyan highlands: effects of land cover types and topography. Am J Trop Med Hyg. 73: 157-65.

Munga, S., N. Minakawa, G. Zhou, A. K. Githeko, and G. Yan. 2007. Survivorship of immature stages of Anopheles gambiae s.l. (Diptera: Culicidae) in natural habitats in western Kenya highlands. J Med Entomol. 44: 758-64.

Munga, S., N. Minakawa, G. Zhou, E. Mushinzimana, O. J. Barrack, A. K. Githeko, and G. Yan. 2005. Association between land cover and habitat productivity of malaria vectors in western Kenya highlands. Am J Trop Med Hyg (in press).

Munga, S., N. Minakawa, G. Zhou, E. Mushinzimana, O. O. Barrack, A. K. Githeko, and Y. G. 2006. Association between land cover and habitat productivity of malaria vectors in western Kenyan highlands. Am J Trop Med Hyg 74: 69-75.

Nobre, C. A., P. J. Sellers, and J. Shukla. 1991. Amazonian deforestation and regional climate change. J. Clim 4: 957-988.

Omumbo, J. A., B. Lyon, S. M. Waweru, S. J. Connor, and M. C. Thomson. 2011. Raised temperatures over the Kericho tea estates: revisiting the climate in the East African highlands malaria debate. Malar J. 12.

Patz, J. A., T. K. Graczyk, N. Geller, and A. Y. Vittor. 2000. Effects of environmental change on emerging parasitic diseases. Int J Parasitol 30: 1395-405.

Ramakrishna, R., and S. W. Running. 1995. Satellite monitoring of global land cover changes and their impact on climate. Clim change 31: 315-493.

Robert, V., H. P. Awono-Ambene, and J. Thioulouse. 1998. Ecology of larval mosquitoes, with special reference to Anopheles arabiensis (Diptera: Culcidae) in market-garden wells in urban Dakar, Senegal. . J. Med Entomol. . 35: 948-55.

Rúa, G. L., M. L. Quiñones, I. D. Vélez, J. S. Zuluaga, W. Rojas, G. Poveda, and D. Ruiz. 2005. Laboratory estimation of the effects of increasing temperatures on the duration of gonotrophic cycle of Anopheles albimanus (Diptera: Culicidae). Mem Inst Oswaldo Cruz, Rio de Janeiro 100: 515-520,.

Rueda, L. M., l. K. J. Pate, R. C. Axtell, and R. E. Stinner, . 1990. Temperature-dependent development and survival rates of Culex quinquefasciatus and Aedes aegypti (Diptera: Culicidae). Journal of Medical Entomology 27: 892-898.

Santos, R. L., O. P. Forattini, and M. N. Burattini. 2002. Laboratory and field observations on duration of gonotrophic cycle of Anopheles albitarsis s.l. (Diptera: Culicidae) in southeastern Brazil. J Med Entomol. 39: 926-30. .

Skinner, W. R., and J. A. Majorowicz. 1999. regional climatic warming and associatd twentieth century land-cover changes in north western North America. Clim Res. 12: 39-52.

Taye, A., M. Hadis, N. Adugna, D. Tilahun, and R. A. Wirtz. 2005. Biting behavior and Plasmodium infection rates of Anopheles arabiensis from Sille, Ethiopia. . Acta Trop 16: ahead of print.

Tuno, N., W. Okeka, N. Minakawa, M. Takagi, and G. Yan. 2005. Survivorship of Anopheles gambiae sensu stricto (Diptera: Culicidae) larvae in western Kenya highland forest. J Med Entomol 42: 270-7.

Wamae, P. M., A. K. Githeko, D. M. Menya, and W. Takken. 2010. Shading by Napier Grass Reduces Malaria Vector Larvae in Natural Habitats in Western Kenya Highlands. Ecohealth PMID: 20602147

Watts, D. M., D. S. Burke, B. A. Harrison, R. E. Whitmire, and A. Nisalak, . 1987. Effect of temperature on the vector efficiency of Aedes aegypti for dengue 2 virus. American Journal of Tropical Medicine and Hygiene 36: 143-152.

White, G. B. 1974. Anopheles gambiae complex and disease transmission in Africa. Trans R Soc Trop Med Hyg 68: 278-301.

2

Crop Production and Global Warming

Masahumi Johkan[1], Masayuki Oda[2], Toru Maruo[1] and Yutaka Shinohara[1]
[1]*Graduate school of Horticulture, Chiba University*
[2]*Graduate school of Life Science and Environment, Osaka Prefecture University*
Japan

1. Introduction

Crop production will be affected by global warming, resulting in world-wide food shortages and starvation. Increased concentrations of carbon dioxide (CO_2), one of the main substances responsible for global warming, will promote plant growth through intensified photosynthesis. Some reports indicate that a rise in the levels of CO_2 would actually benefit plants, rather than harm them. The growth rates of C_3 plants increase in response to elevated concentrations of carbon dioxide. Thus, global warming might increase plant growth, because of higher temperatures and higher levels of atmospheric CO_2.

High atmospheric temperatures caused by elevated concentrations of CO_2 will induce heat injury and physiological disorders in some crops, which will decrease the incomes of farmers and agricultural countries. Photosynthesis is one of the most sensitive physiological processes to high temperature stress. Reproductive development is more sensitive than vegetative development to high temperatures, and heat-sensitivity differs among crops. In tomato, the optimal temperature for fruit set was reported as 21–24°C (Geisenberg and Stewart, 1986) or 22–25°C (Peet and Bartholomew, 1996), while pollen viability and release are adversely affected by high temperatures, and become major limiting factors for fruit set. Thus, global warming can have opposite effects on plant growth. From a long-term viewpoint, however, high atmospheric temperatures will drive the main sites of crop production further north, establishing new rules for the 'right crop for right land'.

Water shortages caused by global warming will be the greatest problem for crop production. Plants fundamentally rely on adequate fresh water, and agricultural water accounts for 70% of water use world-wide. As higher temperatures increase evaporation from water sources and decrease precipitation, arid regions will become further desertified. Particularly in semiarid regions, the cultivatable area will decrease because of drought, and this could result in famines and mass migration. As well, it is likely that there will be human conflicts over irrigation water and food.

Global warming is thought to be related to strong hurricanes, cyclones, and typhoons. These extreme weather events can seriously damage crop production, and destabilize farm management and the lives of consumers. However, these agricultural problems are most likely to occur in the medium and long-term future.

In this chapter, we summarize some of the agricultural problems and crop damage that result from global warming, and present some technical countermeasures (not political and administrative countermeasures) that could be used to ameliorate the effects of global warming on crop production. We also discuss parthenocarpy and grafting as potential countermeasures.

2. Effects of global warming on plant growth

2.1 Carbon dioxide

Carbon dioxide is a key molecule for photosynthesis. In plants, photosynthesis occurs mainly in the leaves. The chemical reaction driven by solar energy involves the reduction of CO_2 through water to create carbohydrates and release oxygen. The resulting carbohydrates are used for plant growth, and provide the energy source for living things. Under 'normal' conditions, the atmospheric CO_2 concentration is very low. The photosynthetic reactions under high temperatures and high light intensities are limited by the CO_2 concentration, and the photosynthetic rate does not exceed a certain value. Because CO_2 is the limiting factor, an increase in its concentration significantly affects photosynthesis. In general, increased CO_2 concentrations temporary promote photosynthesis; however, the photosynthetic rate decreases over an extended period of exposure to high CO_2 concentrations.

Photosynthesis is affected by various environmental conditions. In appropriate temperature conditions, the photosynthetic rate with the current CO_2 concentration of 370 ppm is strongly influenced by light intensity. Under low light conditions, the photosynthetic rate and the light saturation point are low, and so the CO_2 requirements are low. However, at higher light intensity, the photosynthetic rate and light saturation point are higher (Fig. 1). Thus, CO_2 is one of the major environmental factors affecting plant growth along with light, temperature, and water. However, unlike other environmental factors, the CO_2 concentration in the atmosphere does not fluctuate with climate change. Apart from anthropogenic sources, natural causes of changes in the atmospheric CO_2 concentration include plants, which at high density can change the CO_2 concentration by 25%, and respiration of soil microbes, which can change the CO_2 concentration by 30% near the soil surface.

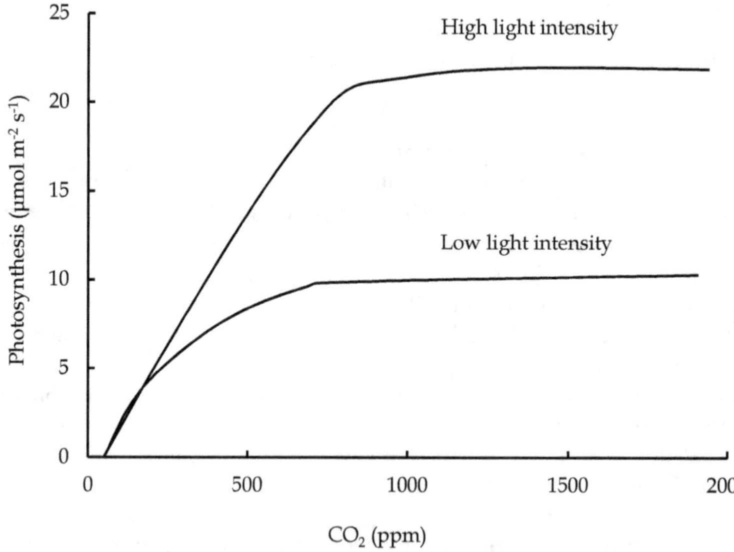

Fig. 1. Photosynthetic curves at high and low light intensities.

In plants, the promotion of photosynthesis under high CO_2 concentrations results in increased dry matter production. This is known as the "CO_2 fertilizer effect", and it is used to promote crop growth in greenhouses and plant factories. This effect is more pronounced in C_3 crops, such as wheat, soybean, and rice, but less so in C_4 crops, such as maize, millet, and sugarcane. This is because the current atmospheric CO_2 concentration limits photosynthesis in C_3 plants, but not in C_4 plants because of their ability to concentrate CO_2 in the cells (Fig. 2). If the concentration of CO_2 is doubled, the photosynthetic rate of many C_3 plants increases by 30–60%, but in most C_3 plants these promoted rates are temporary, and there are no promotion effects in the long term. Thus, it is thought that the changes in the photosynthetic apparatus or other photosynthetic factors that occur during short-term responses to increased CO_2 differ from those during long-term exposure to increased CO_2. This phenomenon often occurs when photosynthetic production exceeds plant growth (Nakano et al., 1997; Makino, 1994).

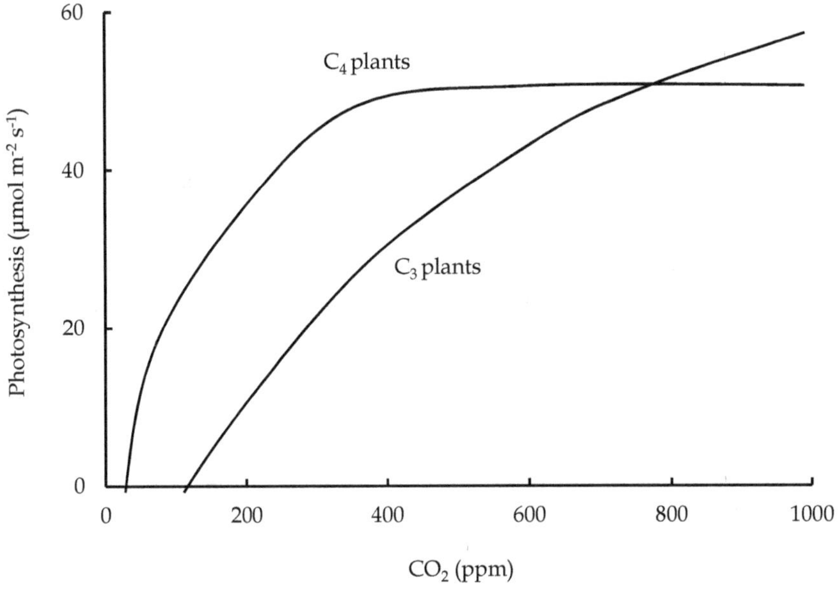

Fig. 2. Photosynthetic curves for C3 and C4 plants (Rogers &Runion, 1994).

The effects of CO_2 concentration on crop yield will become greater as the atmospheric CO_2 concentration increases. For many crops, the predicted yield increase in response to a 700 ppm CO_2 concentration is approximately 30%; specifically, a yield increase of 31% in wheat (Amthor, 2001), 29–35% in rice (Cock and Yoshida, 1973; Imai et al., 1985) 55% in soybean (Rogers et al., 1983), and 50% in maize (Rogers et al., 1983). However, the promoting effects will vary according to the specific nature of the plant and its sinks, e.g., seeds or bulbs. Even though photosynthesis in C_3-type crops is generally inhibited by long-term exposure to high CO_2 concentrations, those that have separate organs to accumulate photosynthetic products appear to be unaffected. For example, inhibition of photosynthesis in crops such as radish and potatoes has not been observed (Sage et al., 1989; Usuda and Simogawara, 1998). Also, if water and mineral nutrients become limiting factors, the promoting effects of increased CO_2

concentration on crop production are minor. This may because there is insufficient water and mineral nutrients to support an increase in photosynthesis, because of the balance of growth between the root and the shoot.

2.2 Temperature

Crop germination and growth rely on optimal temperatures during the period of greatest growth rate. Therefore, non-optimal temperatures slow the growth rate or stop growth altogether. The limiting temperatures for growth are minimum and maximum temperatures, and these vary among crops (Table 1), crop varieties (Table 2, Prasad et al., 2006), and among different growth stages in the same crop. In particular, temperature strongly affects crops during their reproductive period, from pollen formation to fertilization (Table 3). Low or high temperatures during this period can prevent crop fertilization and cause seed abortion.

Temperature type	Crops	Germination (°C)			Growth (°C)		
		Optimum	Min	Max	Optimum	Min	Max
High	Eggplant	20-25	15	33	22-28	17	33
	Pepper	20-30	15	35	25-30	15	35
	Water Melon	25-30	15	35	25-30	13	35
	Melon	25-30	15	40	25-30	18	35
	Soybean	25-30	10	35	25-30	10	35
	Maize	25-30	10	40	20-30	10	35
Midle	Wheat	25-30	2	40	20-25	4	30
	Tomato	15-27	11	30	21-26	5	35
	Cucumber	25-30	18	30	18-25	12	35
	Pumpkin	20-25	15	40	17-25	10	35
Low	Lettuce	15-20	4	25	15-20	8	25
	Carrot	15-25	11	30	16-20	5	28
	Potato[z]	-	-	-	15-24	10	29
	Strawberry[z]	-	-	-	17-23	10	23
	Cabbage	15-30	8	35	15-20	5	25
	Spinach	15-20	5	30	15-20	3	25

[z]: Potato and strawberry are mainly vegetatively propagated.

Table 1. Temperatures for germination and growth of crops (recasting and adding of Saitoh, 2008).

	Vegetative biomass (g $plant^{-1}$)			Grain yield (g $plant^{-1}$)		
	AT	HT	% decrease from AT	AT	HT	% decrease from AT
L-204	11.9	12.3	0.0	11.9	1.3	89.1
M-202	12.0	10.6	11.7	13.6	1.9	86.0
Labelle	22.7	12.3	45.8	21.8	4.5	79.4
Italica Livorma	8.7	11.6	0.0	12.5	1.4	88.8
S-102	9.2	9.6	0.0	15.3	1.6	89.5
Koshihikari	9.1	9.9	0.0	12.3	2.7	78.0
M-103	8.7	9.3	0.0	12.2	4.4	63.9
N-22	17.4	23.8	0.0	17.7	9.2	48.0
IR-8	17.7	11.2	36.7	8.2	6.2	24.4
IR-72	15.9	12.6	20.8	8.4	5.5	34.5

Table 2. Effects of ambient temperature (AT) and high temperature (ambient +5 ^{o}C, HT) on vegetative biomass and grain yield of different rice cultivars (Prasad et al., 2006).

Crop	High-temperature injyury
Wheat	Male serarity (over 30 ^{o}C)
Tomato	Male serarity (over 30 ^{o}C)
Cucumber	Male serarity (over 30 ^{o}C)
Pumpkin	Abnormal differentiation of male and femal flower (over 30°C)
Potato	Poor potatp formation (over 21°C), No potato formation (over 29°C)

Table 3. High temperature injury of crops (Yamazaki, 1985).

2.2.1 Effect of high temperatures on crop production in different areas

The increase in temperatures resulting from higher concentrations of CO_2 in the atmosphere may inhibit or promote crop growth. Naturally, such climatic changes will directly affect agricultural production. It is anticipated that for moderate global average temperature increases (1-3°C) there will be an overall increase in global food production. Additional temperature increases, however, would result in an overall decrease in food (IPCC, 2007). In low-latitude regions such as seasonally dry tropical areas, even moderate temperature increases (1–2°C) are likely to negatively affect major cereal crops, increasing the risk of famine. By the mid-21st century, crop yields in East and Southeast Asia may increase by up to 20%, while those in central and South Asia may decrease by up to 30% if the direct positive physiological effects of CO_2 are taken into account. Western New Zealand is likely to receive more rain, but large areas of mainland Australia and eastern New Zealand are likely to have less soil moisture. Therefore, the increased drought and frequency of fires could decrease crop production. In North America, it is predicted that there will be climate-related yield increases of 5–20% over the first decades of the century, with overall positive effects of climate persisting through much or all of the 21st century. However, crops that are currently growing in areas that are close to their climate thresholds (e.g., wine grapes in California) are likely to show decreases in yield and/or quality in response to even modest increases in temperature (Hayhoe et al., 2004; White et al., 2006). In drier areas of Latin America, climate change is likely to lead to salinization and desertification of agricultural lands, and therefore, food production and security will fluctuate. In some African countries, climate change could exacerbate erosion, and decreases in yields from rain-fed agriculture of up to 50% during the period from 2000 to 2020 (Agoumi, 2003).

2.2.2 Effect of high temperatures on crop growth

Rice

In rice, seed abortion occurs when plants are subjected to temperatures above 35°C. When the average temperature exceeds 26–27°C at 20 days after ear emergence, grain filling and grain quality are decreased, and there are more immature grains.

Wheat

The number of wheat tillers decreases in response to high temperatures, especially high night-time temperatures. In such conditions, shoot elongation is promoted but there are more immature grains and decreased yields because of dark respiration. Moreover, unusually early panicle initiation caused by warm winters can increase the risk of frost damage.

Soybean

The growth of soybean decreases above 38°C, but soybean is relatively heat-tolerant. However, pollen viability was lower at day/night temperature conditions of 37/27°C than at 27/27°C, and the rate of pod setting at 32/27°C was lower than that at 27/27°C (Kitano et al., 2006).

Tomato

At high temperatures, seedlings grow faster and the differentiation and development of flowers is also promoted. However, the rate of flower set decreases and flowers are of poor quality. High temperatures during flowering induce flower abscission, malformed flowers, and pollen sterility in tomato plants. High temperatures also result in poor flowering, poor fruit quality, and color disorders.

Cucumber

High temperatures result in faster seedling growth. However, the development of pistillate and bisexual flowers is suppressed, the node order of the flowers increases, and the number of flowers decreases. High temperatures inhibit flower differentiation and development, and result in smaller ovaries in pistillate and bisexual flowers.

Lettuce

Since lettuce is a cool-temperature crop, it is likely to be adversely affected by higher temperatures. Higher temperatures induce floral differentiation and flower stalk development in lettuce plants, both of which decrease yield and quality of the crop.

Carrot

Carrot is another cool-temperature crop, and so higher temperatures and drier conditions inhibit seed germination and plant growth. Moreover, high temperatures and wet conditions result in shorter, thicker roots. The major risk for carrots, however, is the predicted increase in disease damage in warmer, wet conditions.

Potato

The night-time temperature is important for formation of potato tubers. The optimum temperatures are 20°C days and 10–14°C nights, but almost no tubers form if night-time temperatures exceed 20°C. In warm, wet conditions, the stolon-formation period is shorter, the tubers are smaller, and the tuber quality is poor (e.g. some have hollow hearts).

Strawberry

Floral differentiation of strawberry is strongly affected by temperature and photoperiod. At low temperatures (5–10°C), flower buds form regardless of the photoperiod. At 11–15°C, flower buds can form, but only with increased light intensity under long-day conditions. At 16–26°C, flower buds form under a short-day photoperiod (less than 12-h days), and no flower buds form above 27°C. Therefore, high temperatures will delay floral formation and result in small fruits.

2.2.3 Weeds, harmful insects, and diseases

To achieve high yields, crops must be protected from pests, diseases, and weeds. Of the total cropping area worldwide, crop yield is decreased by 10–20% by weeds (Mirrabelli et al., 2005). Many C_4 weeds are found in arable C_3 crops and many C_3 weeds in arable C_4 crops. Therefore, under high temperatures and increased atmospheric CO_2 concentrations, weed damage to arable C_4 crops in tropical and subtropical semiarid areas is predicted to increase. It is likely that global warming will result in pests and diseases that are currently found in low-latitude regions spreading to high-latitude regions. Bacteria and fungi are the main causal agents of plant diseases, and their optimal growth temperatures are approximately 25°C and above 30°C, respectively. Temperature also affects the growth of insects, with higher temperatures increasing their growth rate. Therefore, continued warming is expected to increase damage to crops from bacteria, fungi, and insects.

3. Water shortages, desertification, and tropical cyclones

As global warming progress, there will be increased evaporation from the earth's surface and from plants. Even a 1°C temperature increase would increase the amount of evaporation from the earth. As well, increased temperatures result in more concentrated, heavy rainfall, and crops show decreased rainfall use efficiency in such circumstances. Snowfall, which results in stored water resources, also decreases with higher temperatures, and snow melts more rapidly. Together, these factors could combine to increase drought in the major agricultural regions—the mid-latitude continents. This could significantly restrict the world's food supply.

Desertification of dry areas is very likely to result from global warming. In dry areas, water moves from subsurface to surface layers of the soil. Therefore, when capillary water reaches the ground or irrigation water evaporates, the soluble salts that dissolved into the water in the lower layers are concentrated in the surface soil layer. As a result, salt accumulates at the soil surface, where it negatively affects or prevents plant growth. In dry regions, more water will evaporate as temperatures increase, turning these areas into deserts. Desertification decreases the amount of vegetation, reducing CO_2 absorption by plants, and further progressing global warming. The area of the world's deserts is expanding by 60,000 km^2 each year, and the desertification of humid and semi-arid areas and the dry sub-humid areas used to cultivate wheat and other grains could cause a serious food crisis.

Strong tropical storms, hurricanes, typhoons, and cyclones severely damage farmland. The numerous tropical storms in recent years are thought to be a result of global warming. Theoretically, the increase in the number of strong tropical storms is because of increased evaporation—the energy source for tropical storms—due to the rise in the temperature of the ocean's surface (Emanuel, 1987). Walsh (2004) indicated that the tropical cyclone intensity will increase by 5–10% around 2050, along with increased peak rainfall rates of approximately 25%. More recent climate model predictions are that the number of cyclones will decrease by approximately 30% under global warming, but their duration will be longer (Oouchi et al., 2005). Therefore, low-lying areas could be at high risk because of longer tropical storms and increased sea levels.

4. Countermeasures

Some strategies to ameliorate the effects of global warming on food production include the development and use of heat-tolerant varieties, appropriate nutrient and water management, coordination of growing periods, and control of pests and diseases. In particular, pest/disease control and the use of heat-tolerant crops are thought to be the most promising approaches. Soybean is sensitive to drought, and so adequate irrigation must be provided when its evapotranspiration rate increases in response to higher temperatures. Although wheat is relatively drought-tolerant, water shortages in dry areas are a concern, so efficient irrigation techniques, e.g. drip irrigation, should be used. The male sterility that results from high-temperature injury decreases the yields of wheat and maize, but treatment with the plant hormone auxin can restore fertility (Fig. 4; Sakata et al., 2010).

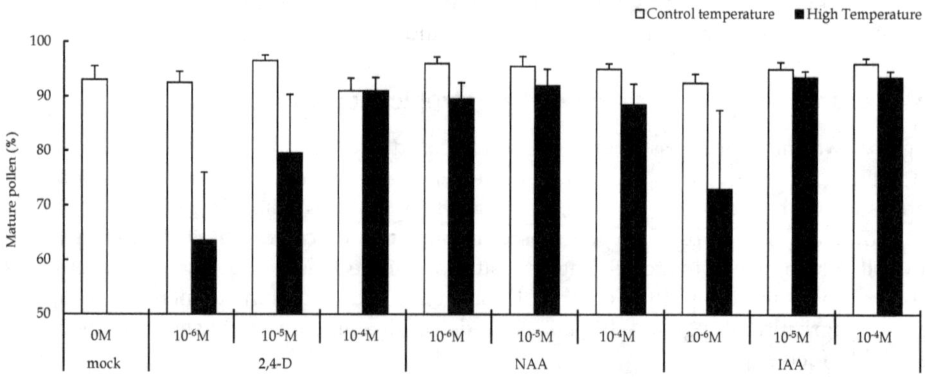

Fig. 4. Effects of exogenous auxin on high temperature injury of pollen developments in barley. A population of mature pollen grains were recovered from anthers after all auxin applications testeed. (Sakata et al., 2010).

For leafy crops, mulching and shading are useful techniques to prevent increases in the soil temperature. For fruit crops, useful techniques include the use of various materials to shade and heat-shield trees, and coordination of the growing period. A night-chilling treatment is a useful countermeasure to induce floral differentiation of strawberry. For crops in protected cultivation, the combination of heat pumps, evaporative cooling, air ventilation, and shading can achieve optimum growth conditions.

Rootstock	Desination	NaCl (mM)			
		0		50	
Moneymaker	M/M	201.2 ± 12.4	a	128.4 ± 10.1	b
Radja	M/R	242.9 ± 16.8	a	145.8 ± 24.3	b
Pera	M/P	238.2 ± 34.1	a	202.4 ± 24.4	a

Table 4. Shoot fresh weight (g plant−1) of self-grafted Moneymaker plants and Moneymaker grafted onto two rootstocks (Radja and Pera) after 35 days of salt treatment (Martinez-Rodriguez et al., 2008).

Results as means±S.E. (n = 6). Means within each column followed by different letters are significantly different according to the LSD test ($P \leq 0.05$).

4.1 Parthenocarpy

Fruit develops from the ovary, which itself develops in response to plant hormones that are produced by seeds formed after fertilization. Fertilization is the fusion of pollen (formed in the anther) and the ovule (formed in the ovary). However, because pollen is temperature-sensitive, high temperatures during pollen formation induce male sterility. If global warming proceeds at the current rate, fruit production could be affected in fruit and vegetables that require fertilization. Parthenocarpy is the natural or artificially induced production of fruit without fertilization of ovules, so this is a promising strategy for fruit production under global warming. Natural parthenocarpy involves using varieties that have parthenocarpic genes, while artificial parthenocarpy is induced by hormone treatment. Examples of parthenocarpy include the production of seedless grapes using GA treatment, and tomato, eggplant, and cucumber production using auxin treatment. Crops in which natural parthenocarpy is exploited include oranges, pineapple, banana, tomato, eggplant, and cucumber. In general, parthenocarpic fruit trees are propagated by vegetative propagation and parthenocarpic vegetables are propagated by F_1 seeds. Therefore, the use of parthenocarpic vegetables is limited because few seeds are formed in these vegetables. Recently, a method was reported to enhance seed production in parthenocarpic tomato fruits (Fig. 5; Johkan et al., 2010). This method can be applied to other crops, and therefore, it may allow more extensive use of parthenocarpic fruits in the future.

4.2 Grafting

Grafting is the vegetative propagation of fruit trees, but in Japan, it has also been used to avoid injury by continuous cropping of vegetables and soil pests and diseases. Vegetable fruits that can be grafted include cucurbits such as melons, watermelon, and cucumber, and

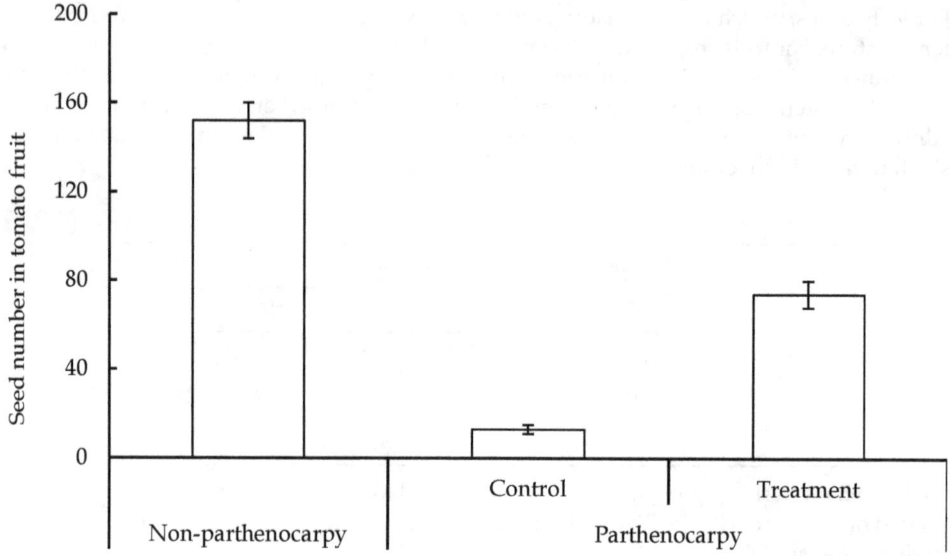

Fig. 5. Increase of seed formation in parthenocarpic tomato fruit (Johkan et al., 2010).

solanaceous species such as tomato and eggplant. Grafting can give the characteristics of the rootstock to the scion. Some rootstocks will be more tolerant to adverse environmental conditions. In particular, grafting scions to rootstock with high drought- or heat-tolerance can increase growth and yield under high temperatures (Fig. 6, Rivero et al., 2003). In addition, crops that are not salt tolerant can be cultivated in saline areas if grafted onto salt-tolerant rootstock. As mentioned above, grafting is an effective countermeasure for crop production without the need to breed lines that are tolerant to heat or drought.

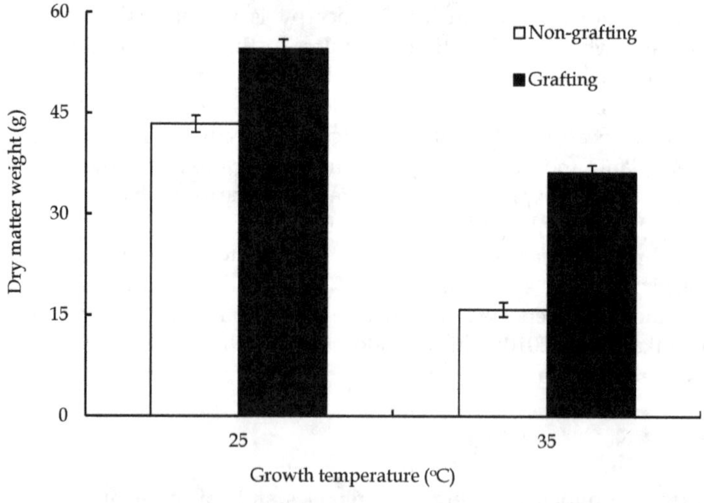

Fig. 6. Effect of grafting on growth of tomato shoots at high temperature (Rivero et al., 2003).

4.3 Vegetative propagation

The use heat-tolerant cultivars are an effective countermeasure against global warming. Breeding of heat-tolerant varieties is required if there are no suitable varieties for the local area. However, propagation by seeds requires genetic fixation, which is a lengthy process. In contrast, vegetative propagation does not require genetic fixation, so it can be used for rapid release of pathogen-resistant or desirable cultivars. One disadvantage of vegetative propagation is that it is less efficient than seed propagation. However, an *in vivo* mass propagation method was recently reported for tomato. This method markedly increased the multiplication rate (Johkan et al., 2008), and could be applied to other plants. As mentioned above, if immediate action is required to ameliorate the effects of global warming, vegetative propagation may be one effective method.

5. Conclusion

The effects of global warming on food production are complex, and are a combination of increased CO_2 concentrations in the atmosphere, higher temperatures, fluctuations in rainfall and solar radiation, and pests and diseases. The 'fertilizing effect' of elevated CO_2 concentrations on crop yields will decline slightly because of the negative effects of rising temperatures. Particularly in developing areas with undeveloped agricultural technologies, there is a high risk that a hotter, drier climate will negatively affect crop production. In mid- and high-latitude regions, even a moderate increase in the global average temperature (1–3°C) could increase food production, but additional temperature increases would cause an overall decline in food production. In low-latitude regions, especially seasonally dry semitropical areas, temperature increases of only 1 or 2°C are likely to negatively affect cereal production because of drought. Although the effect of global warming on food production will vary among regions and crop types, it is anticipated global food production will decrease, and the North-South gap in food production will become greater than it is at present. Therefore, it is important to reexamine existing varieties, to introduce new varieties with increased tolerance to high temperatures and pests/diseases, and to improve methods of pest and disease control.

Parthenocarpy, grafting, and vegetative propagation should also be considered as useful countermeasures against global warming. However, temperature strongly affects crop production, and the extent to which countermeasures can ameliorate the effects of global warming are limited. Global warming is a common problem for all people living on earth. There is no doubt that to address the issue of global warming together is the most effective strategy to guarantee stable crop production.

6. References

Agoumi, A. (2003). Vulnerability of North African countries to climatic changes: adaptation and implementation strategies for climatic change. In: *Developing Perspectives on Climate Change, Issues and Analysis from Developing Countries and Countries with Economies in Transition,* Climate Change Knowledge Network, (May, 2003), Available from http://www.cckn.net//pdf/north_africa.pdf.

Amthor, J. F. (2001). Effects of atmospheric CO_2 concentration on wheat yield: review of results from experiments using various approaches to control CO_2 concentration.

Field Crops Rearch, Vol.73, No.1, (October, 2001), pp. 1-34, ISSN 0378-4290

Aoki, M. & Yabuki, K. (1977). Studies on the carbon dioxide enrichment for plant growth. VII. Changes in dry matter, production and photosynthetic rates of cucumber during carbon dioxide enrichment. *Agricultural Meteorology*, Vol,18, No.6, (December, 1977), pp. 475-485, ISSN 0002-1571

Cock, J. H. & Yoshida, S. (1973). Changing sink and source relations in rice (*Oryza sativa* L.) using carbon dioxide enrichment in the field. *Soil Science of Plant Nutriton* Vol.19, No.3, (September, 1973), pp. 229-234, ISSN 0038-0768

Emanuel, K. A. (1987). The dependence of hurricane intensity. *Nature*, Vol. 326 No.6112, (April, 1987), pp. 483-485, ISSN. 0028-0836

Hayhoe, K.; Cayan, D.; Field, C.; Frumhoff, P.; Maurer, E.; Miller, N.; Moser, S.; Schneider, S.; Cahill, K.; Cleland, E.; Dale, L.; Drapek, R.; Hanemann, R. M.; Kalkstein, L.; Lenihan, J.; Lunch, C.; Neilson, R.; Sheridan S. & Verville J. (2004) Emissions pathways, climate change, and impacts on California. *Proceedings of the National Academy of Sciences*, Vol.101, No.34, (August, 2004), pp. 12422-12427, ISSN 0027-8424

Imai, K.; Coleman, D. F. & Yanagisawa, T. (1985). Increase in atmospheric partial pressure of carbon dioxide and growth and yield of rice (*Oryza sativa* L.). Japanese Journal of Crop Science, Vol.54, No.4, (December, 1985), pp. 413-418, ISSN 0011-1848 (In Japanese with English abstract)

Intergovernmental Panel on Climate Change (IPCC). (2007b). In: *Climate Change 2007: Impacts, Adaptation and Vulnerability*, 29.08.2007, Available from http://www.ipcc-wg2.org/

Johkan, M.; Mori, G.; Mitsukuri, K.; Mishiba, K.; Morikawa T. & Oda, M. (2008). In vivo shoot regeneration promoted by shading the cut surfaces of the stem in tomato plants. *HortScience*, Vol.43, No.1 (February, 2008), pp. 220-222, ISSN 0018-5345

Johkan, M.; Chiba, T.; Mitsukuri, K.; Yamasaki, S.; Tanaka, H.; Mishiba, K.; Morikawa, T.; Oda, M.; Yamamoto, C. & Ohkawa. H. (2010). Seed Production Enhanced by Antiauxin in the pat-2 Parthenocarpic Tomato Mutant. *Journal of the American Society for Horticultural Science*, No.135, Vol.1 (January, 2010), pp. 3-8, ISSN 003-1062

Kitano, M.; Saitoh, K. & Kuroda, T. (2006). Effect of high temperature on flowering and pod set in soybean. *Scientific Reports of the Faculty of Agriculture Okayama University*, No.95, Vol.1, (February, 2006), pp. 49-55. ISSN 0474-0254

Makino, A. (1994). Biochemistry of C_3-photosynthesis in High CO_2. *Journal of Plant Research*, Vol.107, No.1, (March,1994), pp. 79-84. ISSN 0918-9440

Martinez-Rodriguez, M. M.; Estan, M. T.; Moyano, E.; Garcia-Abellan, J. O.; Flores, F. B.; Campos, J. F.; Al-Azzawi, M, J,; Flowers, T.j. & Bolarin. M. C. (2008). The effectiveness of grafting to improve salt tolerance in tomato when an 'excluder' genotype is used as scion. *Environmental an Experimental Botany*, Vol.63, No.1-3, (May, 2008), pp. 392-401, ISSN 0098-8472

Millabelli, C.; Colla, G.; Fiorillo, A.; Cardarelli, M.; Rouphael, Y. & Paolini, R. (2005). The effect of mechanical weed control technique and irrigation method on yield, tuber quality and weed suppression in organic potato. Acta Horticulturae (ISHS), No.684, Vol.1, (October ,2005), pp. 127-134, ISSN 0567-7572

Nakano, H.; Makino, A. & Mae, T. (1997). The Effect of Elevated Partial Pressures of CO_2 on the Relationship between Photosynthetic Capacity and N Content in Rice Leaves. *Plant Physiology,* Vol.115, No.1 (February,1997), pp. 191-198, ISSN 0032-0889

Oouchi, K.; Yoshimura, J.; Yoshimura, H.; Mizuta, R.; Kusunoki, S. & Noda, A. (2006). Tropical cyclone climatology in global-warming climate as simulated in a 20 th km-mesh global atmospheric model. *Journal of the Meteorological Society of Japan,* Vol.84, No.2, (April, 2006), pp. 259-276. ISSN 0026-1165

Prasad, P. V. V.; Boote, K. J.; Allen Jr., L. H.; Sheehy, J. E. & Thomas, J. M. G. (2006). Speicies, ecotype and cultivar differeces in spikelet fertility and harvest index of rice in response to high temperature stress. Field Crop Research, Vol.95 No.2-3, (February, 2006), pp. 398-411, ISSN 0378-4290

Rivero, R. M.; Ruis, J. M.; Sanchez, E. & Romero, L. (2003). Does grafting provide tomato plants an advantage against H2O2 production under conditions of thermal shock? *Physiologya Plantarum,* Vol.117, No.1, (January, 2003), pp. 44-50, ISSN 0031-9317

Rogers, H. H. (1983). Responses of selected plant species to elevated carbon dioxide in the field. *Journal of Environnemtal Quality,* Vol.12, No.4 , (December, 1983), pp. 569-574. ISSN 0047-2425

Rogers, H. H. and Runion, G. B. (1994). Plant responses to atmospheric CO2 enrichment with emphasis on roots and the rhizosphere. *Environ. Pollu.* Vol.83, No.1-2, (October,1994), pp. 155-189. ISSN 0269-7491

Sage, R.F.; Sharkey, T. D. & Seemann, J. R. (1989). The acclimation of photosynthesis to elevated CO_2 in five C_3 species. *Plant Physiology,* Vol.89, No.2, (February, 1989), pp. 590-596, ISSN 0032-0889

Saitoh, H. (Ed.). (2008). *Ecological and physiology of vegetable,* Nousangyoson Bunka Kyoukai, ISBN 978-4-540-08114-9, Tokyo, Japan (In Japanese)

Sakata, T.; Oshino, T.; Miura, S.; Tomabechi, M.; Tsunaga, Y.; Higashitani, N.; Miyazawa, Y.; Takahashi, H.; Watanabe, M. & Higashitani, A. (2010). Auxins reverse plant male sterility caused by high temperatures. *Proceedings of the National Academy of Sciences,* Vol.107, No.19, (May, 2010), pp. 8569-8574, ISSN 0027-8424

Usuda, H. & Shimogawara, K. (1998). The Effects of Increased Atmospheric Carbon Dioxide on Growth, Carbohydrates, and Photosynthesis in Radish, Raphanus sativus. *Plant Cell Physiology,* Vol.39, No.1, (January, 1998), pp. 1-7, ISSN 0032-0781

Walsh, K. (2004). Tropical cyclones and climate change: unresolved issues. *Climate Research* Vol.27, No.1, (January, 2004), pp. 77-83, ISSN 0936-577X

White, M. A.; Diffenbaugh, N. S.; Jones, G. V.; Pal, J. S. & Giorgi, F. (2006). Extreme heat reduces and shifts United States premium wine production in the 21st century.

Proceedings of the National Academy of Sciences, Vol.103, No.30, (July, 2006), pp. 11217-11222, ISSN 0027-8424

Yamazaki, K. (1985). Growth and physiology of vegetable, In *Vegetable Horticultural Dictionary*, S. Shimizu, (Ed.) pp. 245-252, Yokendo, Tokyo, Japan (In Japanese)

3

Decreasing of Population Size of *Imperata cylindrica* Mangrove Ecotype and Sea-Level Rising

Ping Kao[1], Ting-Ying Wu[1], Chia-Lun Chang[2], Chang-Hung Chou[3] and Ing-Feng Chang[1,2]
[1]*Department of Life Science, National Taiwan University, Taipei,*
[2]*Institute of Plant Biology, National Taiwan University, Taipei,*
[3]*Graduate Institute of Ecology and Evolutionary Biology, China Medical University, Taichung, Taiwan*

1. Introduction

A very recent on-going trend worldwide is that sea level is found rising up due to global warming (Cazenave and Llovel, 2010). With one to two degrees temperature elevated, the polar iceberg will melt which eventually will lead to significant sea-level rise. There is increasing interest in knowing how mangrove population changes under sea-level rising. Mangrove forest (mangrove) is usually termed "kidney of the earth". The morphology and anatomy of these plants have unique structure, such as breathing roots, knee roots. Viviparous seedlings of the Rhizophoraceae plants further drift with the sea which results in implantation and reproduction. The origin of the name mangrove is derived from plants of a kind of mangrove tree *Rhizophora mucronata*. The plant's timber, tree trunks, branches, flowers are red. These trees grow in tropical and subtropical regions of rivers and coastal salt marshes, evergreen shrub or tree forest.

According to the literature, mangrove forests are mainly distributed in the western coastal estuary in Taiwan (Hsueh and Lee, 2000). Taiwan's largest and most well-known mangrove protected areas are in Danshui district, including Chuwei mangrove conservation area. These areas are along the Danshui River. The main object of the protected areas is *Kandelia* mangrove forests. *Kandelia candel* (L.) Druce is grouped into Rhizophoraceae *Kandelia* genera. Kandelia is the dominant species and turned out to be a precious national treasure along the Danshui River in Taipei, Taiwan. In addition to *Kandelia*, there are a wide range of plants associated with its surrounding area. These plants have also evolved special morphological and physiological characteristics to adapt to flooding and high salinity environment. These halophytes include *Phragmites austrakis* (reed), *Sporobolus virginicus* (salt sage), and *Imperata cylindrica* (cogon grass). In 1982 Lue carried out an ecological survey of Chuwei mangrove forest, and found the population of *Imperata cylindrica* (Lue, 1982).

Imperata cylindrica, a perennial herb, is a C4-type plant. It has 10 pairs of chromosomes. A single *Imperata* individual can generate many seeds. Seed dispersal can be assisted by wind. *Imperata* can also use their underground rhizome for asexual reproduction. It is adapted to a wide range of environmental factors and confers a high degree of stress tolerance. The underground rhizome is a commonly used medicinal herb. There are about 10 species of *Imperata* worldwide. In Taiwan there is only one genus one species, namely *Imperata cylindrica* (L.) Beauv. var. *major* (Nees) Hubb. (Hsu, 1975). Cheng & Chou (1997a) examined the leaves of *I. cylindrica* from Chuwei using scanning electron microscope (SEM) and found that Chuwei population differs from others in anatomy and morphology. The stele was empty instead of solid. The lower stem was surfaced with white wax instead of trichomes. In addition, polymorphism among populations was also analyzed by rapid amplify polymorphic DNA (RAPD) (Cheng & Chou, 1997b) and restriction fragment length polymorphism (RFLP) (Chou & Tsai, 1999) on ribosomal DNA (rDNA) (Chiang *et al.*, 1998; Tsai & Chou, 1999). In 2006 Chang found that Chuwei population is salt and flood tolerant (Chang & Chou, 2006). In 2008 Chang utilized proteomics approach and identified differentially expressed protein among Chuwei, Sarlun and Neihu Imperata populations (Chang, 2008). All these results revealed that Chuwei population was found to be a distinct ecotype.

Under salt pressure, plants accumulated various organic compounds *i.e.* proline, glycinebetaine, choline, glycerol, and sorbitol as compatible solutes in their bodies. Barley (Stewart & Michelle, 1983), tobacco (Binzel *et al.*, 1987), eggplant (Jain *et al.*, 1987), and spinach (Coughlan & Wyn Jones, 1980) accumulated large amounts of proline in response to salt treatment. The amount of proline increased with increased salinity. In addition, the expression of proline was regulated by abscisic acid (ABA) (Stewart, 1980; Stewart & Voeberg, 1985). Proline played diverse roles in stress physiology. Stewart & Lee (1974) pointed out that the accumulation of proline in plants correlated its tolerance to salt stress. They also indicated that high level proline would protect many N metabolism-related enzymes from harm. Moreover, large amount of proline would inhibit ACC from converting to ethylene so that plant would not hurt by ethylene (Chrominski *et al.*, 1988; Chrominski *et al.*, 1989). Proline could also be osmo-protectant which functions to maintain osmotic potential balance (Jain, 1987). It is suggested that proline in *I. cylindrica* from Chuwei may function as an osmo-regulator against salt stress.

Mangrove swamp environment is very unique. It is not only a salty land but is also subjected to periodic flooding which is very detrimental to plant growth. It has been discussed that sea-level rise may impact the survival of mangrove forests (Feller *et al.*, 2010). The rises of sea water may impact the survival of mangrove trees in the world. We suspect that the rise of sea water may be one of the causes of the population size decrease of Chuwei mangrove population of *Imperata*. A field work survey discovered that the population size of Chuwei ecotype is decreasing. The proline content of Chuwei population in the field before and after flooding was measured. In present study, we tried to investigate the relationship between the decrease of population size of *Imperata* Chuwei ecotype and proline content changes. We hypothesized that the dynamics of proline content may be related to the population size decrease. The study of how flooding affects proline accumulation may elucidate the possible impact of sea-level rise using Imperata mangrove population as an example.

2. Materials and methods

2.1 Sampling sites and plant materials

Imperata cylindrica (L.) Beauv. var. *major* (Nees) Hubb, Cogon grass, was used as plant materials. Chuwei mangrove salt marsh wetland (Hwang & Chen, 1995) where periodical flooding pressure threatens plant survival was chosen as a sampling site (Figure 1, Figure 2). Sarlun sandy beach was chosen as a control site (Figure 1, Figure 3) (Figure 1, 2, and 3 were adapted and modified as previously described (Chang, 1996)). These two sampling sites are close to each other in Danshui area, Taipei as shown in Figure 1. Plant leaves from Chuwei were harvested between July and October in 1995 and 2011 for proline content assay. Plant samples were collected on spring-tide days (high tide and flooding; Jul 14, Aug 13, Sep 11, Sep 28 and Oct 12 in 1995; Feb 20 in 2011). During harvesting, each leaf sample was excised by sterilized scissors and stored in zip-block in ice bucket with dry ice to keep it fresh and brought back to lab immediately for study the same day. Plant leaves harvested from each site were assay for water proline. Plants rhizomes collected from the field were washed by sterilized water and cultured in pots (60× 20 ×20 cm^3) in greenhouse for two weeks, then transplanted to Kimura's culture solution (Chang & Chou, 2006) to grow plantlet, the vegetative shoot. The culture solution was aerated with air pump without break. The culture solution was changed every week. These plantlets were used for differential display assay.

2.2 Hypotheses to be tested

We have two hypotheses for the interpretation of possible reasons for the population size decrease of Chuwei ecotype. The first hypothesis is a synecology hypothesis. This hypothesis address that the overpopulated *Kandelia* may threaten the survival of Chuwei Imperata. The second hypothesis is an autoecology hypothesis. This hypothesis address the local flooding may affect the survival of Chuwei Imperata. We hypothesized that Chuwei ecotype may respond to local flooding in a manner different from *Kandelia*. We suspected that local flooding may impact the survival of Chuwei ecotype even though it is flooding tolerant (Chang and Chou, 2006).

2.3 Proline content measurement

L-proline content was measured by the method as previously described (Bates *et al.*, 1973). Half a gram of plant fresh leaf tissues were chopped into pieces and quick-frozen in liquid nitrogen in 5 g sea sand and 5mL extraction buffer containing 3 % (w/v) 5-sulfosalicylic acid for grinding and homogenization. The homogenates were centrifugated at 538×g (Sigma, Model 2K15, Taiwan) for 10 minutes. The supernatant was obtained as proline crude extract. The reaction mixture contained 2 ml acid-ninhydrin solution, 2ml proline crude extract. Acid-ninhydrin solution contained a mixture of 0.14 M ninhydrin, 60 % (v/v) acetic acid, 2.4 M phosphatic acid and 2 ml acetic acid. The reaction took place at 100°C for 1 hour. Samples were then put into -20°C refrigerator to stop the reaction immediately. Four ml methane was mixed with samples and vortex viguously. The solution was divided into two layers, the upper methane layer and the lower water layer. Three ml upper ethane layer solution was placed in a cuvette. Proline content was analyzed spectrophotometrically at wavelength 520nm using spectrophotometer (Beckman, Model DU-50, Taiwan). Ten replicas were included for each data set. Data were analyzed by Student□s pairwise t-test in SPSS statistical package.

2.4 RNA extraction

RNA extraction was conducted by use of pine tree method as previously described (Chang *et al.*, 1993). Leaves of Sarlun and Chuwei population of *Imperata cylindrica* were ground in liquid nitrogen using mortar and pestle. About 2 grams of leaf powder were added into 15 mL extraction buffer containing 100 mM Tris-HCl pH 8.0, 25 mM EDTA, 2 M NaCl, 0.5 g/L

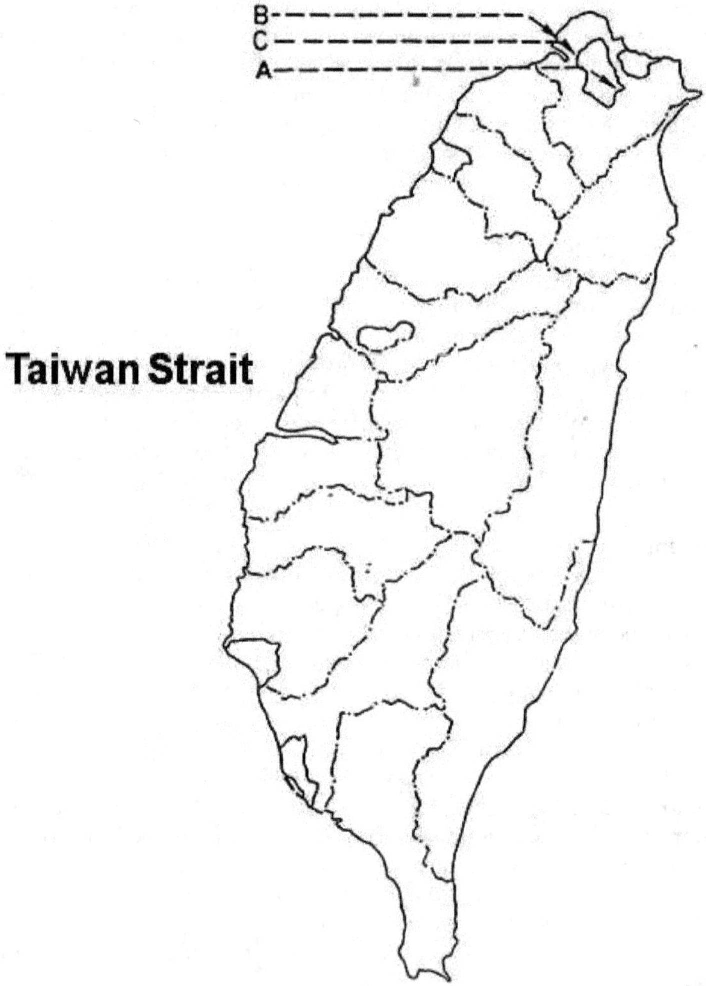

Fig. 1. Sampling sites of *Imperata cylindrica* in Danshui area, Taipei, Taiwan. *Imperata cylindrica* sampling sites were labeled as arrows indicated. A: Neihu; B: Sarlun; C: Chuwei.

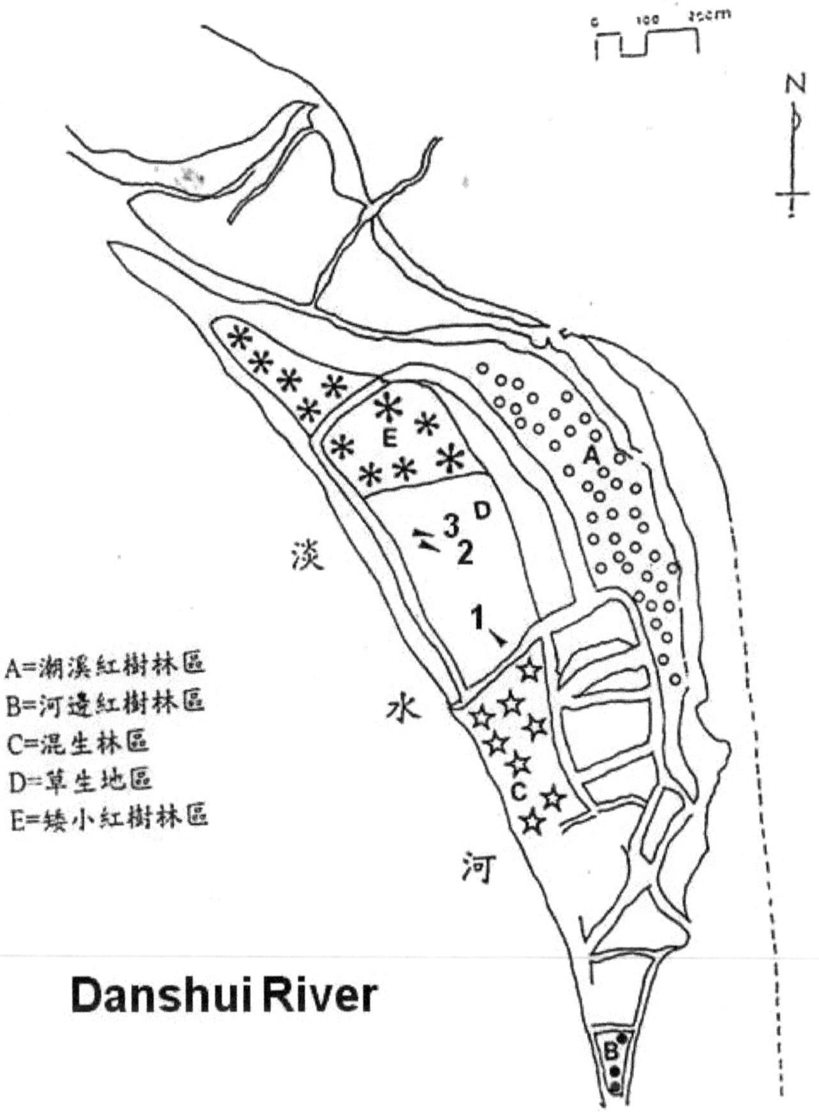

Fig. 2. Sampling sites of *Imperata cylindrica* in Chuwei, Taipei, Taiwan. *Imperata cylindrica* is populated in area D. In area D, sampling site 1, 2, and 3 were labeled as arrows indicated. Site 2 and 3 were lost just very recently for unknown reasons.

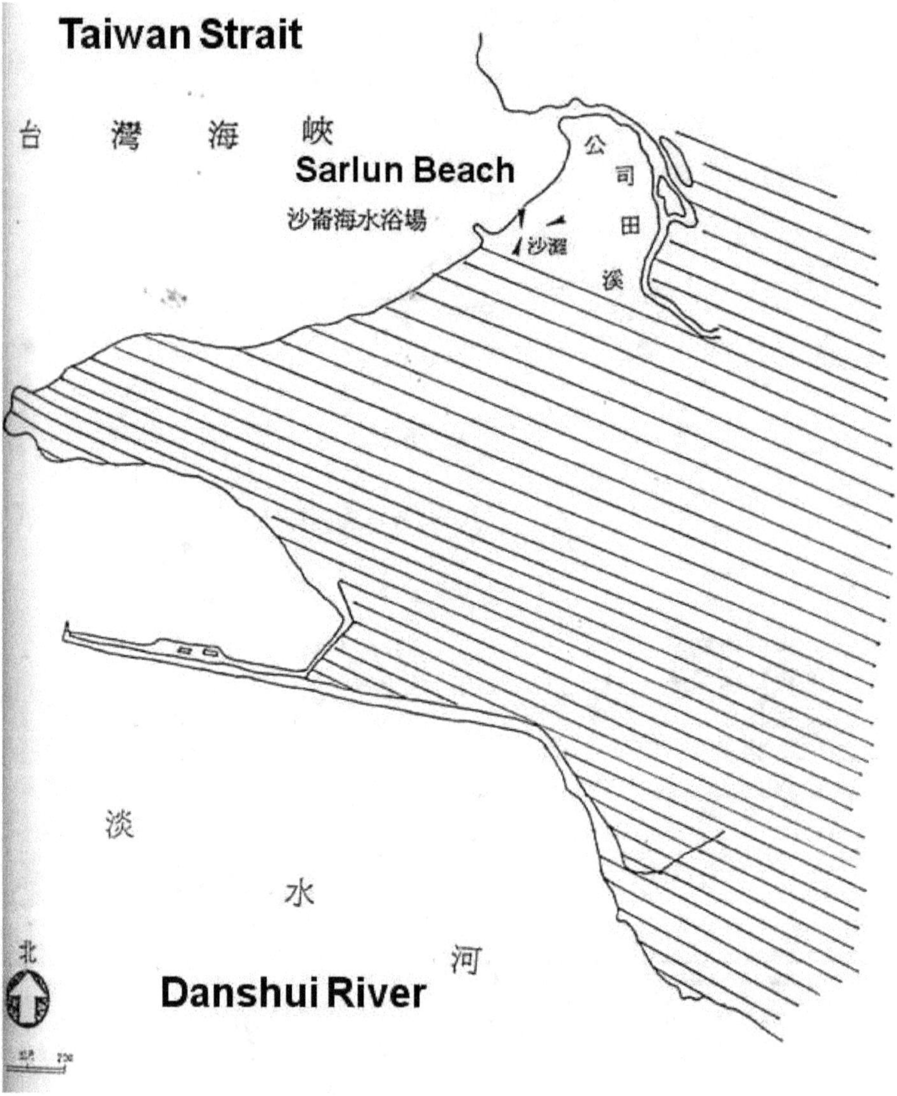

Fig. 3. Sampling sites of *Imperata cylindrica* in Sarlun, Taipei, Taiwan. Sampling sites were labeled as arrows indicated.

spermidine, 2% beta-mercaptoethanol, 2% CTAB, and 2% PVP. The homogenate was mixed well and subjected to water bath at 65°C for 10 minutes. The mixture was then centrifuged at 13000 rpm for 15 minutes at room temperature. The upper layer was recovered to a new centrifuge tube. Equal volume of chloroform:isoamyl alcohol (24:1) was added. After thorough mixing, centrifugation was carried out at 13000 rpm for 15 minutes at room temperature. This step was repeated twice in order to remove proteins and metabolites. The upper water layer was transferred to a new centrifuge tube. One third volumes of 10 M LiCl was added, mix well and subjected to incubation under 4°C overnight. The solution was centrifuged at 13000 rpm for 30 minutes at 4°C. The pellet was washed by 1mL 70% (v/v) ethanol. The solution was transferred to a sterilized microcentrifuge tube for centrifugation at 13000 rpm for 15 minutes at 4°C. The wash step was repeated twice to remove salt. The pellet was air dried. One hundred μL sterilized ddH_2O was added to resuspend RNA. RNA concentration was measured by NanoDrop spectrometer. The RNA integrity was checked by 2% agarose gel electrophoresis, and the solution was used as templates for reverse-transcription immediately followed by storage under -80°C.

2.5 Differential display assay

Differential display and the subsequent silver stain were carried out by following the manufacturer instruction (RNAimage Kit 5, GenHunter Corporation, Cat. No. G505; SILVER SEQUENCE™ Staining Reagents, Promega, Cat. No. Q4132). After reverse-transcription, cDNAs were used as templates to perform another PCR with AP primers provided by the manufacturer. PCR products were loaded into denature acryl-amide gel for electrophoresis at 1700V for 4 hours. The gel was stained by silver stain immediately after the electrophoresis was completed. The Air-dried gel was scanned by scanner (EPSON PERFECTION V750 PRO, EPSON). The electrophoretic bands were compared to identify gel bands specific to either Chuwei or Sarlun population.

2.6 Isolation of electrophoretic bands for DNA cloning and sequencing

The electrophoretic bands were excised from the acrylamide gel by sterilized dissection knife, and collected in 1.5 mL eppendorfs. The recovery step was conducted by use of QIAGEN II Gel Extraction Kit (QIAGEN, Cat. No. 28704). The DNA concentrations were detected by NanoDrop spectrometer. The recovered DNA was used for TA cloning. Cloning was conducted using TA Cloning Vector (RBC, # 200-09205) as cloning vector. T4 DNA ligase was used for ligation between TA vector and DNA. *E. coli* (DH5α) was used as hosts. The ligated plasmids were transformed into DH5α competent cells by heat-shock, and incubated under 37°C on LB medium (with ampicillin) for 14 hours. Five to eight colonies were selected from each plate to perform colony PCR and subsequent sequencing.

2.7 Database searching

Sequences acquired were searched against NCBI database using BLAST algorithm (http://blast.ncbi.nlm.nih.gov/Blast.cgi) (Altschul *et al.*, 1990). Plant Genome Database (PlantGDB, http://www.plantgdb.org/), the Gene Ontology (http://www.geneontology.org/), the Arabidopsis Information Resource (TAIR, http://www.arabidopsis.org/), and MIPS *Oryza sativa* database (MOsDB, http://mips.helmholtz-muenchen.de/plant/rice/) were also used. In nucleotide sequence searching, the cloned sequences were compared against databases by BLASTn program and

BLASTtx program for matches. In protein sequence searching, the acquired sequences were first translated by ExPASy Translate Tool (http://au.expasy.org/tools/dna.html), and then screen for possible protein sequences manually. The translated protein sequences were compared against databases by BLASTp program.

3. Results and discussion

3.1 Population size of Chuwei ecotype is much smaller than before

Imperata cylindrica (L.) Beauv. var. *major* (Nees) Hubb., a top-ten weed (Holm *et al.*, 1977), was found in Taiwan all around the island (Hsu, 1975). In America, *Imperata cylindrica* was found as a non-endemic species dominating Florida and neighboring states in both dry and wetlands (King & Grace, 2000). Interestingly, it was found growing in Chuwei mangrove salt marsh wet land as well. To track the dynamics of population size of Chuwei ecotype, we carried out field survey since 2008. Surprising, we observed decrease of population size of Chuwei ecotype. Two of the sampling sites (Chuwei and Sarlun) we selected for the field work (site 2 and site 3 as shown in Figure 2) before year 2000 were found missing since the beginning of 2009. Only sampling site 1 (around 30 m^2) still existed so far (Figure 2). In addition, the size of sampling site 1 is only less than half of the size as before. Therefore, we concluded that Chuwei ecotype is now decreasing its population and is endangered. However, the reason why the population size decreased is unknown.

3.2 Proline content in leaves of Chuwei ecotype decreased dramatically after flooding in the field

We have two hypotheses for the interpretation of possible reasons for the population size decrease of Chuwei ecotype. The first hypothesis is supported by the evidence that more and more *Kandelia* trees are invading the habitat of Chuwei Imperata. These *Kandelia* trees may threaten the survival of Imperata. It is expected that in ten years, this area will be dominated by *Kandelia* trees and Imperata Chuwei ecotype will eventually be extinct. In order to test the second hypothesis, we compare proline content in leaves of *I. cylindricain* in the field before and after local flooding on the spring tide (high tide) date. Samples were collected at monthly interval from August to December in 1995. For proline content analysis, leaves sampling was performed with 20 individuals within a day. To test if proline content changed in response to flooding, same leaf samples were collected before and after flooding (1 hr difference) on spring-tide days. Proline content analysis showed that proline content dropped dramatically after flooding. On July 14 in 1995, proline content decreased 53 % after flooding. On August 13, proline content decreased 46 % after flooding. On October 12, proline content decreased 67 % after flooding. Moreover, on February 20 in 2011, proline content decreased 40% after flooding (Table 1).

The decrease of proline content in response to local flooding in Chuwei supported our second hypothesis. In the salt marsh wetland, high level of proline content is helpful in maintaining the osmotic equilibrium in the cells. When local flooding occurred, proline dropped rapidly, which is harmful to plants since osmotic equilibrium was not maintained in the cells. This might eventually lead to cell death. In order to determine if proline content in leaves of plantlet increases in responses to flooding, flooding treatment was performed three months in the greenhouse to see long-term effect on proline. Leaf samples were collected and analyzed. Results showed that the accumulation of proline was not found in leaves of flooding-treated plantlet in the greenhouse (data not shown).

Sampling date	Before flooding	After flooding
07.14.1995	138.99± 7.53	64.62± 1.58[1]
08.13.1995	130.94 ± 8.29	70.49± 4.84[1]
10.12.1995	156.82± 14.84	52.16± 2.57[1]
02.20.2011	150.45 ± 2.37	89.97± 3.68[1]

[1] Values showed significant difference in Student's t-test at 5% level. S.E. was at 5% level. Proline count unit: micro g gfw^{-1}

Table 1. Average proline content in leaves of *I. cylindrica* from Chuwei population before and after flooding on spring-tide (high tide) days. Sampling was performed in 1995 and 2011.

3.3 Proline serves more than one function in *Imperata cylindrica* Chuwei ecotype? - A cross-talk between flooding and salt stress

In a previous study, *I. cylindrica* from Chuwei was tolerant to 1% salt treatment. Therefore, it was grouped as a salt-tolerant ecotype (Chang & Chou, 2006). The tolerance ability was even stronger than *Kandelia candel* (tolerant to 0.8 % salt), a dominant mangrove forest species in Chuwei salt marsh wetland (Hwang & Chen, 1995). Since Chuwei sampling site is a mangrove salt marsh land, how plants survived in such a high salt land is poorly known. According to our study, proline content of *I. cylindrica* from Chuwei was much higher than those from others in the field (Chang & Chou, 2006). They were also higher under salt treatment. In the greenhouse experiments, proline of *I. cylindrica* from Chuwei was highly accumulated in response to salt treatment.

However, does proline serve only single role in plants? It has been reported that the regulation of ion uptake and production of organic solutes (*i.e.* methyl proline) is related to combined salt and flood stress tolerance in some wetland plants (Carter *et al.*, 2006). In addition, mechanism of how some halophytes tolerate flooding is reviewed (Colmer and Flowers, 2008). In present study, proline content dropped dramatically (almost 50 %) after local flooding (only in one hour) in Chuwei ecotype, suggesting that proline may serve as a signal molecule in response to flooding. The dynamic of proline in leaves of Chuwei *Imperata* was suggested which might not only be involved in salt stress, but also in flooding stress response.

The decrease of proline content in response to flooding can be explained by two mechanisms. The first mechanism is that proline acts as a direct response to flooding. It has been reported the accumulation of proline was regulated by abscisic acid (ABA) (Stewart, 1980; Stewart & Voeberg, 1985). Alcohol dehydrogenase *ADH* gene expression was also known regulated by ABA (de Bruxelles *et al.*, 1996) through G-box, a promoter region (Walker *et al.*, 1987; Lu *et al.*, 1996). It is possible that *ADH* gene expression and proline accumulation cross-talks, and is regulated through ABA. Since proline accumulated in response to salt stress in Chuwei ecotype, decrease in proline content may be an easier way for cell to recognize to activate fermentation pathway, including activation of *ADH* gene. An alternative mechanism is an indirect response of proline to flooding. The accumulated proline transported downward toward root (shoot to root) to protect root from salt damage, and anti-transpirated with a flooding signal molecule from root to shoot (Else *et al.*, 1996) for stomata closes in response to flooding. This pathway might overlap with water-deficit pathway. Besides, a long distance transport of proline as a signal molecule was required. However, we did not confirm any of the possibility because it was very complicated to simulate a physiology experiment with two stresses acting at the same time. Timing makes the experiment even difficult to resolve. Maybe an easier way to keep tract on proline dynamic was to label proline using isotope as a tracer. Although the role of proline as signal molecule in response to salt stress has been proposed (Maggio *et al.*, 2002; Hare & Cress, 1997), the involvement of proline in flooding response has not been discovered before. Anyway, due to the fact that proline functioned in response to both flood and salt stress in present study, a crosstalk between two distinct pathways, salt and flooding response, was suggested.

3.4 Differential display reverse transcription assay revealed five genes which were differentially expressed in Chuwei ecotype as compared to Sarlun population

Differential display was first introduced In 1992, by Liang and Pardee as an easy application for comparison, identification and isolation of genes expressed as mRNA in various cells under designated conditions (Liang and Pardee, 1995). The basic steps of differential display include isolation of intact RNA from sample, reverse transcription of the RNA using an arbitrary primer to produce ssDNA, several rounds of polymerase chain reactions using the same primer to amplify the corresponding cDNAs, and display of the cDNAs as bands on a matrix gel, such as a DNA sequencing gel. In order to identify differentially expressed genes in Chuwei ecotype, differential display (DD) was introduced. The differences of gene expression between Chuwei ecotype and Sarlun population were compared. By comparing and collecting the bands uniquely appeared to either Chuwei or Sarlun populations. Population-specific ESTs was isolated and identified.

Three biological replicates were included. RNAs extracted from Chuwei ecotype and Sarlun population were subjection to QC. To check RNA integrity, RNAs extracted were used to perform electrophoresis in 2% agarose gel at 120V for 15 minutes. Differential display was conducted by three replicates with one of the replicate. Bands with high reproducibility were indicated and excised from the gel for subsequent cloning procedure. A total of twenty-six bands specific to Chuwei ecotype were detected, whereas seventeen bands were specific to Sarlun population.

Detail information about Chuwei-specific bands is listed in Table 2 and described as follow. BLAST search result of band C3 nucleotide sequence showed that part of the sequence of *Zea mays* chloroplast genome which coding for 16s rRNA was identified. This matched sequence was also found to be present in *Arabidopsis thaliana* chloroplast, *Zea perennis* mitochondrion and uncultured environmental bacterium. (Takeuchi *et al.*, 2005), whereas others had no significant matches. For BLAST search result of band C12 nucleotide sequence, a candidate uncharacterized mRNA sequence of *Triticum aestivum* was identified (Kawaura *et al.*, 2009). This cDNA had similar sequence matches in other plant species, including *Stellaria longipes*, *Sorghum bicolor*, *Oryza sativa*, and *Arabidopsis thaliana*. No hits were found in animals, which suggested it's a plant-specific gene. It appeared to be a putative disease resistance gene (Yang *et al.*, 2008, Sequencing Project International Rice, 2005, Tanaka *et al.*, 2008, Ohyanagi *et al.*, 2006), but remained to be a hypothetical one (Paterson *et al.*, 2009). For BLAST search result of band C12 nucleotide sequence, most ID matched hypothetical disease-resistance related genes in plants, which supported C12 to be an EST sequence of a plant-specific gene for a hypothetical protein. For BLAST search result of band C7 nucleotide and protein sequence, nevertheless no significant matches were found in NCBI, PlantGDB, TAIR and MOsDB databases.

Detail information about Sarlun-specific bands is listed in Table 3 and described as follow. BLAST search result of band S4 nucleotide sequence showed that an ATPase III subunit in *Dendrocalamus latiflorus* chloroplast, *Sorghum bicolor* chloroplast and *Zea mays* chloroplast

(Strittmauer and Kossel, 1984, Alexandrov *et al.*, 2009, Wu *et al.*, 2009) was identified. Moreover, BLAST search result of band S4 protein sequence showed that ATP synthase CF0 subunit III of many eudicotyledons such as *Geranium carolinianum* and *Erodium texanum* (Guisinger *et al.*, 2008) was identified. For BLAST search result of band S8 nucleotide sequence, the most likely candidate was a hypothetical protein of *Sorghum bicolor* (Paterson *et al.*, 2009). Similar sequences were also found in *Zea mays* and *Oryza sativa*, still hypothetical proteins (Yu *et al.*, 2005, Lai *et al.*, 2004). Moreover, for BLAST search result of band S8 protein sequence, eight possible queries out of six reading frames were blast against NCBI database. Five of these queries had no significant similarities, the others matched to hypothetical proteins of *Sorghum bicolor, Zea mays* and *Oryza sativa.*

For band S9 nucleotide BLAST search, a putative casein kinase of *Oryza sativa* was found. This sequence was also similar to another putative serine/threonine protein kinase of *Oryza sativa* (Sasaki *et al.*, 2002). For band S9 protein BLAST search, eight possible queries out of six reading frames were submitted to NCBI database. Two of them had matches indicating serine/threonine protein kinase of *Oryza sativa*; four were found to share similarities with hypothetical proteins among various vascular plants; the rest had no significant similarities found. For band S6 nucleotide BLAST search, no significant similarities were found among all databases mentioned above. For band S6 protein BLAST search, four possible queries were submitted. No significant similarities were found.

Band #	Size (bp)	Accession #	Identity	Annotation of matched gene
C3	776	AY928077	99%	*Zea mays* chloroplast, 16S rRNA
C12	553	AK333595	82%	*Triticum aestivum*, hypothetical

Table 2. Differential display bands specific to Chuwei ecotype.

Band #	Size (bp)	Accession #	Identity	Annotation of matched gene
S4	404	AY928077	98%	*Zea mays*, ATP synthase C chain
S8	468	AK333595	96%	*Sorghum bicolor*, hypothetical
S9	383	AAT44311	79%	*Oryza sativa*, casein kinase

Table 3. Differential display bands specific to Sarlun population.

3.5 Adaptational divergence of *Imperata cylindrica* Chuwei ecotype

Imperata Chuwei ecotype encountered both flooding and salt pressure. In order to survive, plants must possess adaptation mechanism against flooding and salt stress. The increase of

ADH activity and decrease of proline content may be a mechanism against flooding stress, and accumulation of proline may be a mechanism against salt stress (Chang & Chou, 2006). The working model for flooding stress tolerance of Chuwei ecotype was proposed. Before flooding, proline content accumulated in leaves of Chuwei ecotype in response to salt stress. After flooding, leaf ADH activity increased. By contrast, leaf proline content decreased after flooding. These changes can be reached in only one hour. In addition, we identified several differentially expressed genes in Chuwei ecotype and Sarlun population in present study. All these biochemical, physiological changes and genes may contribute to the tolerance of Chuwei ecotype to both flooding and salt stresses, and therefore drove this population diverged form other populations as a distinct ecotype through evolution. However, what was the signal transduction relationship among ADH gene activation and proline, and what was the crosstalk link between flooding and salt stress response remained unknown. More studies on crosstalk of flooding and salt stress physiology of Chuwei ecotype on the signal transduction level are needed.

3.6 Sea-level rise as a potential threat to the survival of *Imperata cylindrica* ?

In present study, we observed a decrease of population size of *Imperata* Chuwei ecotype. In the proline content assay, we also observed dramatic drop of proline content after local flooding in the field in Chuwei ecotype. The drop of proline may be detrimental to plants. These results suggested that sea-level rise can be a potential threat to the survival of *Imperata* Chuwei ecotype. However, it is still not sure that the decrease of population size of *Imperata* Chuwei ecotype is directly or indirectly related to sea-level rise. What can be expected is that in the near future the population may be totally wiped out due to sea-level rise. Since Chuwei ecotype may possess many unknown and valuable salt-tolerant genes, we urge to preserve Chuwei ecotype for conservation.

3.7 Conclusions

The results from present study showed that proline content dropped dramatically after the local flooding at Chuwei. The results supported our second hypothesis that local flooding-mediated proline drop may threaten the survival of *Imperata*. However, the first hypothesis was also supported by the fact of overpopulation of *Kandelia* trees in Chuwei. A concurrent contribution of the two is suggested. In addition, differential display identified Chuwei ecotype or Sarlun population-specific transcripts in the greenhouse. These differentially expressed genes can be important in salt or flooding stress tolerance. However, further validation and confirmation of gene expression using other molecular tools is needed.

4. Acknowledgments

We appreciated National Science Council, Taiwan for funding support (NSC# 98-2815-C-002-146-B, NSC# 98-2621-B-002-008-MY3, and NSC # 99-2815-C-002-041-B). We thank Technology Commons, College of Life Science, National Taiwan University for the equipment support. We appreciated technical support from Wuu-Yang Chen, Prof. Chan-Pin Lin, and Prof. Kuo-Chieh Ho at National Taiwan University.

5. References

Alexandrov, N.; Brover, V.; Freidin, S.; Troukhan, M.; Tatarinova, T.; Zhang, H.; Swaller, T.; Lu, Y.P.; Bouck, J.; Flavell, R. & Feldmann, K. (2009). Insights into corn genes derived from large-scale cDNA sequencing. *Plant Molecular Biology*, 69, pp. 179-194.

Altschul, S.F.; Gish, W.; Miller, W.; Myers, E.W. & Lipman, D.J. (1990). Basic local alignment search tool. *J. Mol. Biol.*, 215, pp. 403-410

Bates, L.S.; Walden, R.P. & Teare, I.D. (1973). Rapid determination of free proline for water-stress. *Plant & Soil*, 39, pp. 205-207.

Binzel, M.L.; Hasegawa, P.M.; Rhode, D.; Handa, S.; Handa, A.K. & Bressan, R.A. (1987). Solute accumulation in tobacco cells adapted to NaCl. *Plant Physiology*, 84, pp. 1408-1415.

Carter, J.L.; Colmer, T.D. & Veneklaas, E.J. (2006). Variable tolerance of wetland tree species to combined salinity and waterlogging is related to regulation of ion uptake and production of organic solutes. *New Phytologist*, 169, pp. 123-134.

Cazenave, A. & Llovel, W. (2010). Contemporary sea level rise. *Ann. Rev. Mar. Sci.*, 2, pp. 145-173.

Chang, I.F. (1996). Physio-ecological responses of *Imperata cylindrica* var. *major* from different habitats to saline and flooding. Master thesis, pp. 9-12.

Chang, I.F. (2008). Ecotypic variation of a medicinal plant Imperata cylindrica populations in Taiwan: mass spectrometry-based proteomic evidence. *J. Med. Plant Res.*, 2, pp. 71-76.

Chang, I.F. & C.H. Chou. (2006). Ecotypic variation of Imperata cylindrica populations in Taiwan: II. Physiological and biochemical evidence. *Botanical Studies*, 47, pp. 175-184.

Chang, S.; Puryear, J. & Cairney, J. (1993). A Simple and Efficient Method forIsolating RNA from Pine Trees. *Plant Mol. Biol.* Rep., 11, pp. 113-116.

Cheng, K.T. & Chou, C.H. (1997a). Ecotypic variation of *Imperata cylindrica* populations in Taiwan: I. Morphological and molecular evidences. *Bot. Bull.Acad. Sin.*, 38, pp. 215-223.

Cheng, K.T. & Chou, C.H. (1997b). Specific RAPD Markers of *Imperata cylindrica* populations in Taiwan. *J. Genet. Mol. Biol.*, 8, pp. 41-54.

Chiang, Y.C.; Chiang, T.Y.; Chang, I.F. & Chou, C.H. (1998). Sequence announcement: rDNA IGS of *Imperata cylindria* (L.) Beauv. Var. *major* (Ness) C E Hubb. *Plant Mol. Biol.*, 39, pp. 391-392.

Chou, C.H. & Tsai, C.C. (1999). Genetic variation in intergenic spacer of ribosomal DNA of *Imperata cylindrical* (L.) Beauv. Var. *major* (Cogograss) population in Taiwan. *Bot. Bull. Acad. Sin.*, 40, pp. 319-327.

Chrominski, A.; Bhat, R.B.; Weber, D.J. & Smith, B.N. (1988). Osmotic stress dependent conversion of ACC to ethylene in the halophyte, *Allentolfea occidentals*. *Environmental & Experimental Botany*, 28, pp. 171-174.

Chrominski, A.; Halls, S.; Weber, D.J. & Smith, B.N. (1989). Proline affects ACC to ethylene conversion under salt and water stress in the halophyte, *Allentolfea occidentals*. Environmental & Experimental Botany, 29, pp. 359-363.

Colmer, T.D. & Flowers, T.J. (2008). Flooding tolerance in halophytes. *New Phytologist*, 179, pp. 964-974.

Coughlan, S.J. & Wyn Jones, R.G. (1980). Some response of *Spinaceae oleraceae* to salt stress. *Journal of Experimental Botany*, 31, pp. 883-893.

de Bruxelles, G.L.; Peacock, W.J.; Dennis, E.S. & Dolferus, R. (1996). Abscisic acid induces the alcohol dehydrogenase gene in Arabidopsis. *Plant Physiology*, 111, pp. 381-391.

Else, M.A.; Tiekstra, A.E.; Croker, S.J.; Davies, W.J. & Jackson, M.B. (1996). Stomata closure in flooded tomato plants involved abscisic acid and a chemically unidentified anti-transpirant in xylem sap. *Plant Physiology*, 112, pp. 239-247.

Feller, I.C.; Lovelock, C.E.; Berger, U.; McKee, K.L.; Joye, S.B. & Ball, M.C. (2010). Biocomplexity in mangrove ecosystems. *Ann. Rev. Mar. Sci.*, 2, pp. 395-417.

Guisinger, M.M.; Kuehl, J.V.; Boore, J.L. & Jansen, R.K. (2008). Genome-wide analyses of Geraniaceae plastid DNA reveal unprecedented patterns of increased nucleotide substitutions. *Proceedings of the National Academy of Sciences* 105, pp. 18424-18429.

Hare, P.D. & Cress, W.A. (1997). Metabolic implications of stress-induced proline accumulation in plants. *Plant Growth Regulation*, 21, pp. 79-102.

Hsu, C.C. (1975). Gramineae. *In Flora of Taiwan*, Vol.6, Li, H.L.; Liu, T.C.; Huang, T.C.; Koyama, T. & Devel, C.E. (eds.). Taipei. pp.661-662.

Hsueh, M.L. & Lee, H.H. (2000). Diversity and distribution of the mangrove forests in Taiwan. *Wetlands Ecology and Management*, 8, pp. 233-242.

Holm, L.; Plucknett, J.P. & Herberger, J. (1977). *The world's worst weeds: distribution and biology*, University Hawaii Press, Honolulu. Xii + pp. 609 .

Hwang, Y.H. & Chen, S.C. (1995). Salt tolerance in seedlings of the mangrove *Kandelia candel* (L.) Druce, Rhizophoraceae. *Bot. Bull. Acad. Sin.*, 36, pp. 25-31.

Jain, P.K.; Dhawan, R.S.; Sharma, D.R. & Chowdhury, J.B. (1987). Salt-tolerance and proline accumulation: a comparative study in salt-tolerance and wild type cultered cells of eggplant. *Plant Cell Report*, 6, pp. 382-384.

Kawaura, K.; Mochida, K.; Enju, A.; Totoki, Y.; Totoda, A.; Sakaki, Y.; Kai, C.; Kawai, J.; Hayashizaki, Y.; Seki, M.; Shinozaki, K. & Ogihara, Y. (2009). Assessment of adaptive evolution between wheat and rice as deduced from full-length common wheat cDNA sequence data and expression patterns. *BMC Genomics*, 10, pp. 271.

Lai, J.; Dey, N.; Kim, C.S.; Bharti, A.K.; Rudd, S.; Mayer, K.F.X.; Larkins, B.A.; Becraft, P. & Messing, J. (2004). Characterization of the Maize Endosperm Transcriptome and Its Comparison to the Rice Genome. *Genome Research*, 14, pp. 1932-1937.

Liang, P. & Pardee, A.B. (1995). Recent Advances in Differential Display. *Current Opinion in Immunology*, 7, pp. 274-280.

Lu, G.; Paul, A.L.; McCarty, D.R.; & Ferl, R.J. (1996). Transcription factor veracity: is GBF3 responsible for ABA-regulated expression of Arabidopsis *Adh*? *The Plant Cell* 8, pp. 847-857.

Lue, K.Y. (1982). The ecological study on Chu-wei mangrove swamp. *Quarterly Journal of Chinese Forestry* 15, pp. 69-76.

Ma, J.F.; Goto, S.; Tamai, K. & Ichii, M. (2001). Role of Root Hairs and Lateral Roots in Silicon Uptake by Rice. *Plant Physiology*, 127, pp. 1773-1780.

Maggio, A.; Miyazaki, S.; Veronese, P.; Fujita, T.; Ibeas, J.I.; Damsz, B.; Narasimhan, M.L.; Hasegawa, P.M. & Bressan, R.A. (2002). Does proline accumulation play an role in stress-induced growth reduction? *The Plant Journal* 31, pp. 699-712.

Ohyanagi, H.; Tanaka, T.; Sakai, H.; Shigemoto, Y.; Yamaguchi, K.; Habara, T.; Fujii, Y.; Antonio, B.A.; Nagamura, Y.; Imanishi, T.; Ikeo, K.; Itoh, T.; Gojobori, T. & Sasaki, T. (2006). The Rice Annotation Project Database (RAP-DB): hub for *Oryza sativa* ssp. japonica genome information. *Nucleic Acids Research*, 34, pp. D741-4.

Paterson, A.H. et al., (2009). The Sorghum bicolor genome and the diversification of grasses. Nature, 457, pp. 551-556.

Sasaki, T. et al., (2002). The genome sequence and structure of rice chromosome 1. *Nature*, 420, pp. 312-316.

Stewart, C.R. (1980) The mechanism of abscisic acid-induced proline accumulation in barley leaves. *Plant Physiology*, 66, pp. 230-233.

Stewart, G.R. & Lee, J.A. (1974). The role of proline accumulation in halophytes. *Planta*, 120, pp. 279-289.

Stewart, C.R. & Michelle, B.B. (1983). Effect of NaCl on proline synthesis and utilation in excised barley leaves. *Plant Physiology*, 72, pp. 664-667.

Stewart, C.R. & Voeberg, G. (1985). Relationship between stress-induced ABA and proline accumulation and ABA and proline accumulation in excised barley leaves. *Plant Physiology*, 79, pp. 24-27.

Strittmauer, G. & Kossel, H. (1984). Cotranscription and processing of 23S, 4.5S and 5S rRNA in chloroplasts from *Zea mays*. *Nucl. Acids Res.*, 12, 7633-7647.

Takeuchi, F.; Watanabe, S.; Baba, T.; Yuzawa, H.; Ito, T.; Morimoto, Y.; Kuroda, M.; Cui, L.; Takahashi, M.; Ankai, A.; Baba, S.I.; Fukul, S.; Lee, J.C. & Hiramatsu, K. (2005). Whole-Genome Sequencing of Staphylococcus haemolyticus Uncovers the Extreme Plasticity of Its Genome and the Evolution of Human-Colonizing Staphylococcal Species. *J. Bacteriol.*, 187, pp. 7292-7308.

Tanaka, T.; (2008). The Rice Annotation Project Database (RAP-DB): 2008 update. *Nucleic Acids Research*, 36, pp. D1028-33.

The Rice Chromosome 11 and 12 Sequencing Consortia. (2005). The sequence of rice chromosomes 11 and 12, rich in disease resistance genes and recent gene duplications. *BMC Biology*, 3, pp. 20.

Tsai, C.C. & C.H. Chou. (1999). Sequence of a 5.8S rRNA gene and internal transcribed spacer (ITS) from *Imperata cylindrica* var. *major*. *Plant Mol. Biol.*, 39, pp. 1313.

Walker, J.C.; Howard, E.A.; Dennis, E.S. & Peacock, W.J. (1987). DNA sequences required for anaerobic expression of the maize alcohol dehydrogenase 1 gene. *Proceedings of National Academy of Science*, 84, pp. 6624-6628.

Wu, F.H.; Kan, D.P.; Lee, S.B.; Daniell, H.; Lee, Y.W.; Lin, C.C.; Lin, N.S. & Lin, C.S. (2009). Complete nucleotide sequence of Dendrocalamus latiflorus and Bambusa oldhamii chloroplast genomes. Tree Physiol, 29, pp. 847-856.

Yang, S.; Gu, T.; Pan, C.; Feng, Z.; Ding, J.; Hang, Y.; Chen, J.Q. & Tian, D. (2008). Genetic variation of NBS-LRR class resistance genes in rice lines. Theor. Appl. Genet., 116, pp. 165-77.

Yu, J. *et al.* (2005). The Genomes of *Oryza sativa*: A History of Duplications. *PLoS Biol.*, 3, pp. e38.

4

Climate Change Adaptation using Agroforestry Practices: A Case Study from Costa Rica

Maren Oelbermann and Carolyn E. Smith
University of Waterloo
Canada

1. Introduction

Conventional agricultural practices in tropical latitudes, using modern plant breeding techniques, fertilizers, and irrigation, have resulted in an increased grain yield (Huxley, 1999). However, these agricultural practices have also played a major role in increasing the global total area of marginal land that is now substandard for the long-term production of food and livestock (FAO, 1990), and has contributed significantly to the accumulation of greenhouse gases in the atmosphere (IPCC, 2007). Such activities are most notable in developing countries which are strongly affected by climate change due to their lower capacity to adapt compared to developed nations (IPPC, 2007). Comparatively, developed nations have the infrastructure and financial capabilities to cope more effectively to a changing climate.

As a result of the inherent environmental degradation and poverty associated with deforestation and conventional agriculture, there is a resurgence in the implementation of sustainable land-use practices including agroforestry (Kandji et al., 2006). In this paper agroforestry is defined as the deliberate integration of woody species with agricultural crops and/or pastures on the same land-unit resulting in the integration of economical and ecological interactions between components (Young, 2002). The cultivation of trees in combination with agricultural crops was a common practice dating back to the beginning of plant and animal domestication. Since then, several models of various agroforesry practices, from Asia, Africa and Europe to North and South America, have been developed (King, 1987). In tropical latitudes, farmers imitated vertical forest structure and diversity by planting a variety of crops with different growth habits. It was not unusual to plant up to 24 species on a plot one-tenth of a hectare with each layer corresponding to the natural stratification of a tropical forest (King, 1987). Until the 1860s, the focus of agroforestry practices was on tree production. It was not until 1975, when the International Development Research Centre (IDRC) in Canada concluded that priority in research should be given to combined agroforestry production systems in tropical regions in order to optimize land-use, establish food security, and address the increasing problem of environmental degradation (King, 1987). Since then, agroforestry practices were promoted as a sustainable land-use management system in developed and developing countries. For example, agroforestry practices range from low-input systems such as alley cropping and short-term improved fallow with leguminous shrubs to shade-grown coffee (*Coffea arabica* L.) in tropical regions and high-input cereal-legume systems and riparian plantings in temperate biomes (Nair, 1993; Gordon & Newman, 1997).

In Central America, agroforestry practices were also diverse and commonly included a combination of trees with food crops, trees with livestock (pasture for fodder) or all elements combined (Budowski, 1987). In this region, pre-Columbian people mixed trees with food crops in a system similar to shifting cultivation. Other land-use techniques included home gardens and the integration of trees and crops along banks. Agroforestry practices in Central America were first described by O.F. Cook in a 1901 report to the U.S. Department of Agriculture, noting that coffee plants were growing under shade trees. Cook also noted the beneficial effects of leguminous shade trees on soil fertility. Since then there has been a plethora of information published on Central American, particularly Costa Rican agroforestry practices. For example, Holdridge reported on the practice of planting *Alnus acuminate* Kunth. in pastures in the Costa Rican highlands (Holdridge, et al. 1977). Budowski (1987) noted the use of *Cupressus lusitanica* L. windbreaks in the highlands and the use of *Cordial alliodora* L. in lowland pastures in Costa Rica. Other research from Costa Rica reported secondary forest management on abandoned coffee plantations (Budowski, 1987), and use of life-fences and increased biodiversity (Harvey et al., 2005). Alley cropping was established in Costa Rica as a new agroforestry practice in the early 1980s, where crops were grown between rows of trees (Kass et al. 1995). The aim of this modern agroforestry practice was to minimize the need for external nutrient inputs from fertilizers or manure, which are difficult or not economically feasible for small producers. Instead this agroforestry practice relied on the biannual input of organic material derived from leguminous plants, high in nutrient content, of completely shoot pruned of trees. The application of the organic material to the soil as mulch was synchronized with the seeding of crops, resulting in the maximum release of nutrients from the decomposing organic material to the growing crops. Over the past decade the potential of agroforestry systems to sequester carbon (Oelbermann et al., 2004; Oelbermann & Voroney, 2011), and their role in providing ecosystem services (Pagiola et al., 2008) has become the forefront of research as a result of global climate change.

Agroforestry systems are unique because they are a land management practice that simultaneously addresses biophysical, economical, and socio-ecological components. Such diversity and interactions leads to a greater functional and structural complexity compared to conventional agroeosystems (Nair, 1993). Leaky & Simmons (1996) suggested that over the long-term agroforestry systems become successively similar to natural systems, where biodiversity increases with each stage in the development of this succession. Such functional and structural complexity has led to the great diversity currently observed in modern agroforestry practices.

However, the adoption of agroforestry practices is based on socio-cultural and sociecological considerations including land tenure and land rights, in addition to providing enhanced biodiversity and ecological services (McGinty et al. 2008; Sood & Mitchell 2009). Currently, agroforestry practices are increasingly adopted due to their potential of providing ecosystem services. Ecosystem services not only provide an economic incentive but are also of value to society by maintaining environmental sustainability (FAO, 2007). The benefits derived from agroforestry practices occur over a range of spatial and temporal scales (Jose, 2009). For example, when comparing the large-scale removal of trees due to deforestation and its resultant high levels of erosion, agroforestry systems provide ecological services by decreasing soil erosion and preventing the sedimentation of waterways; ultimately protecting downstream fisheries (Pearce & Mourato 2004). Thus, ecological services provided by agroforestry systems are derived at the farm- and / or landscape-scale but are enjoyed by society at regional- and / or global-scales (Jose, 2009).

Verchot et al. (2007) suggested that agroforestry practices, apart from providing ecosystem services, are a means of diversifying agricultural production and increase food security for smallholder agricultural producers, especially under current climate change scenarios. For example, recent observations on inter-annual changes in temperature and precipitation along Costa Rica's Pacific coast have shown more intense and longer dry seasons, coinciding with changes predicted by the IPCC in this region (IPCC, 2007). Although maintaining agricultural productivity under changing climatic conditions will be challenging (Watson et al., 2000), Sanchez (2000) suggested that structurally and functionally complex agroecosystem management systems, including agroforestry practices, may show a greater resilience to changing environmental conditions. Although several studies have been published on the adoption of agroforestry practices (Pattanayak et al., 2003; Mercer, 2004), few studies have investigated the efficacy of agroforestry as an agroecosystem management practice with greater resilience to climate change compared to less diversified (conventional) agroecosystems. Verchot et al. (2007) noted that the contribution of agroforestry practices to buffer against climate shifts is currently not well understood. Such knowledge is of great importance because agroforestry is also considered as a key player in achieving the Millennium Development Goals as described in the United Nations annual report (Verchot et al., 2007). The goal of this research was to present a case study from Costa Rica that evaluated the knowledge and perception of agricultural producers on climate change. This study also evaluated the implementation of agroforestry practices as a strategy to climate change adaptation.

2. Study site

The research area and agricultural producers described in this paper corresponded to those currently under evaluation by the GEF-Silvopastoral Study (Pagiola et al., 2007; Yamamoto et al., 2007; Pagiola et al., 2008; Hänsela et al., 2009). The Silvopastoral Approaches for Ecosystem Management, or Global Environment Facility-Silvopastoral Study (GEF-Silvopastoral Study), is a pilot project that was initiated in 2002 in rural Nicaragua, Costa Rica and Colombia where degraded pastures were prevalent. Land owners participating in the project received payment for the ecological services they generated including the integration of trees on their land, improvement of pasture management, maintaining habitat for biodiversity conservation, enhancing soil and water conservation, and increasing carbon sequestration. In association with the Global Environment Facility, the World Bank, the Regional Integrated Silvopastoral Management Project (RISEMAP) and researchers from the Tropical Agricultural Research and Higher Education Center (CATIE) in Costa Rica, each participating agricultural producer was surveyed to determine baseline data and the type of already existing or new agroforestry practices, that ultimately provided ecological services, they were interested in implementing. Financial incentives were received over a four-year period, and were based on the type and increment of ecological services they implemented relative to the previously collected baseline data (Pagiola et al., 2008).

The GEF Silvopastoral study and our study was conducted surrounding the area of Esparza (09°59′N, 84°38′W), located in the Central Pacific Coastal region of Costa Rica. The study area surrounding Esparza encompassed 432 km^2, with land-use consisting predominantly of mixed-agricultural production under conventional practices with beef, dairy, rice, sugarcane, and fruit. The majority of agricultural producers in this area have a beef-dairy production (62%), followed by beef (20%), mixed crop-livestock (14%), and beef/dairy breeding stock production (8%).

The area surrounding Esparza was characterized as a sub-humid tropical forest zone. The average annual temperature was 27.9°C and the mean annual daytime temperature was 32°C and the mean annual nighttime temperature was 22°C (Figure 1). The average annual precipitation of this region was ~1599 mm, with a distinct dry season from December to April and a distinct wet season from May to November. The mean number of sunshine hours in this region was 10 h/day from November to April, and 6 h/day from May to October. This area had a relative humidity ranging from 65% during the dry season and 80% in the wet season. This region is also influenced by large-scale inter-annual fluctuations such as the El Niño or La Niña. The elevation above sea level ranges from 50 m near the Pacific coast to 1000 m in inland areas. The geology of this region was based upon volcanic origin and alluvial sedimentation. This led to the development of Nitisols, with a predominantly loamy texture. The soils of the region were dark red, brown or yellow in color with the presence of an agrillic B horizon, but lack ferric and vertic properties.

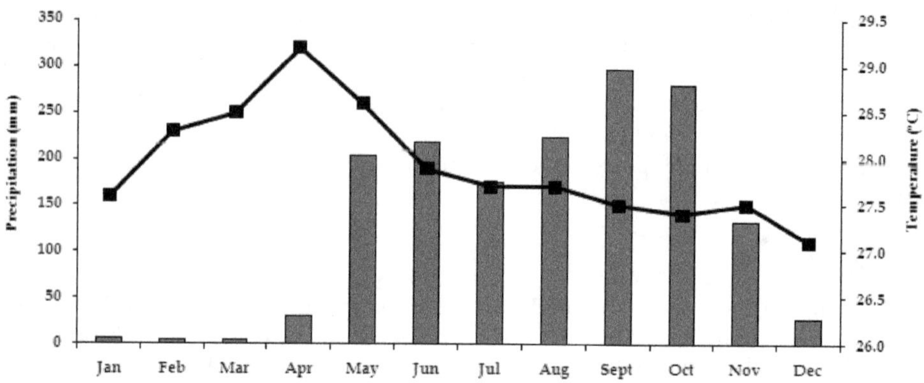

Fig. 1. Mean monthly maximum and minimum temperature (line graph) over 41 years and mean monthly precipitation (bar graph) over 64 years for Esparza, Costa Rica.

3. Sampling methods and statistical analysis

Agricultural producers that participated in this study were chosen because of their existing participation in the GEF-Silvopastoral Study. As such, the participating agricultural producers were already informed about climate change, ecological services, and payment for such services through interaction with CATIE researchers and extension personnel. In addition to conducting interviews with agricultural producers in association with CATIE rural extension officers and the Costa Rican Ministry of Agriculture, further information for this study was obtained from an existing database collected by the GEF-Silvopastoral Study. Information from this database, such as land biophysical characteristics and the type of agroforestry practices implemented provided more depth to the interviews conducted in our study.

In 2007, a total of 50 out of 105 agricultural producers were randomly chosen for our study; and were asked questions based on the following themes: 1) background and farming experience based on information not already collected by the GEF-Silvopastoral Study; 2) knowledge and perceptions about climate change; 3) knowledge on the effects of climate change on current agricultural production; 4) the type of agroforestry practice implemented and its outcome; and 5) why agroforestry practices were adopted. In order to evaluate the adaptive capacity of agricultural producers to climate change, and to contextualize how landowners perceived climate change, the survey included questions on changes in crop and livestock productivity, diseases and pests; changes in the amount and intensity of ambient rainfall and temperature, onset of rainy and dry seasons, frequency of droughts and how agroforestry practices helped to manage droughts. Additionally, landowners were also asked what strategies they were currently using to adapt to climate change and what their incentives (e.g. environmental, financial, extension support) were for implementing agroforestry practices. The semi-structured interviews lasted between 1-3 hours and each interview was recorded.

The surveys were developed and conducted in association with a representative from the Costa Rican Ministry of Agriculture as well as rural extension officers and researchers from CATIE. Details of the survey questions are presented in Smith (2008). Textual data included information provided by agricultural producers that elaborated on the questions of the survey. Recorded interview data was fully transcribed and organized manually into the different themes. In order to understand how the different themes are interconnected, each individual theme was analyzed using QSR NVivo® software (QSR International, Melbourne, Australia). This was done by counting who said what within each theme. Nodes and memos were created using NVIVO to mark relevant concepts, topics and themes. The model explorer tool in NVIVO was used to diagrammatically map out how the themes related to each other.

4. Results and discussion

4.1 Producer perception and understanding of climate change

Results from this study only incorporated agricultural producers with the greatest experience (more than 30 years) in farming because these participants were more likely to notice changes in climate. The majority of these producers (90%) understood the concept of climate change and its cause. Approximately 80% of those surveyed understood that climate change is currently impacting and will impact their land, their family and livelihoods in the future. Similar results were reported by Maddison (2007) who surveyed agricultural producers from 11 different African countries. He found that agriculturalists with the greatest amount of farming experience reported changes in climate. Our results found that many agricultural producers within the study region had knowledge on climate change likely as a consequence of their involvement with the GEF-Silvopastoral Study, and the social networks they established with other agricultural producers, the Costa Rican Ministry of Agriculture, and with researchers and extension personnel from CATIE. Agricultural producers in our survey noted that their primary source of information on climate change was derived from CATIE (56%) and from the Costa Rican Ministry of Agriculture (34%). Information from CATIE and the Costa Rican Ministry of Agriculture was gained through the participation of agricultural producers in organized workshops, seminars and / or on-farm visits.

Agricultural producers interviewed in this study observed several changes in climate, particularly relating to increased ambient temperatures and a prolonged dry season, in the Esparza region over the past 30 years and how this may potentially have affected agricultural productivity (Table 1). Approximately 90% of the agricultural producers noticed increased temperatures while 70% observed a more intense rainy season, but the overall amount of rainfall was lower. An additional 78% of the agricultural producers observed negative impacts on their land, crops, and livestock as a result of a prolonged and more intense dry season including a decrease in crop yield and water availability to livestock and their family. These observations coincided with those predicted by the IPCC (2007), who suggested that the Pacific region of Costa Rica is expected to be drier and hotter. A temperature increase of 0.1°C per decade was reported between the years 1901 and 2005 (IPPC, 2007). According to the IPPC, carbon dioxide (CO_2) started to have an effect on Costa Rica's temperature in the 1970s (IPCC, 2007). Walther et al. (2002) noted that Central America is one of the global regions where climate change impacts on the environment will be pronounced and the loss of species with climate change has already been identified. In the future, IPPC Climate models predict a minimum increase in temperature of 1.5°C and a maximum increase of 5.6°C, from current conditions for this region of Central America (IPCC, 2007). Predictions by the IPCC for changes in precipitation range from a decrease by 11 to 38% during the dry season and an increase of 7 to 21% during the wet season from current conditions. Agricultural producers in our study observed that changes in climate may have potentially led to a decrease in livestock weight, lower milk production, and a decrease in yield of their harvestable crops and pasture grasses. They also noted that the overall decrease in farm productivity affected food availability for their families and livestock. A prolonged dry season resulted in drought leading to a reduction of the farm's carrying capacity, thereby affecting the number of animals, harvestable crops, and the quantity and quality of milk produced. A mixed response, approximately 38% of the agricultural producers, also related climate change to increased pest outbreaks such as locusts (*Schistocera* spp.), ultimately affecting their crop productivity. Others, especially those at the higher elevations, noted an increase in snakes that are commonly found in lowland areas of the Pacific coast. These observations were similar to those reported by Raxworthy et al. (2008) who found that amphibians were moving upslope as a result of a warmer and drier climate at lower elevations.

Climate Characteristics	Potential Changes in Farm Production Quantity and / or Quality	Landowner Observation (%)
Increased Ambient Temperatures (IPCC, 2007)	Overall decrease in farm productivity	72
	Decrease in food production	76
	Livestock weight loss	80
	Decreased milk production	72
	Loss of harvestable products	38
Prolonged Dry Season (IPCC, 2007)	Decreased forage production	72
	Loss of livestock	76
	Loss of harvestable products	46
	Decreased milk production	74
	Decrease in water levels	64
	Overall decrease in water availability for livestock, crops, and people	54

Table 1. Agricultural producers observation of changes in quantity and / or quality of available food, feed and water potentially linked to the increase of temperatures and prolonged drought over the past 30 years in Esparza, Costa Rica (n=50).

4.2 Adaptation and implementation of agroforestry systems

As a result of the payment for ecological services (PES) program through the GEF-Silvopastoral Study, agricultural producers participating in our study had implemented a variety of agroforestry practices, including silvopasture with *Brachiaria brizantha* (Hochst. Ex A. Rich) Stapf. grass, fodderbanks, integration of trees on farmland or farmyards, and the establishment of live fences, that aided their adaptation to climate change. They were also aware that such activities helped in mitigating methane emissions from cattle and significantly increased the amount of carbon captured and stored for the long-term in trees and in soil. A total of 96% of the agricultural producers surveyed integrated trees on their land (Table 2). Specifically, agricultural producers adopted different strategies in order to cope with a longer and more intense dry season.

Agroforestry Land management Practice as a Climate Change Adaptation Strategy	Implementation by Landowners (%)
Silvipasture with *Brachiaria brizantha*	100
Maintenance/implementation of trees on farmland/farmyard	96
Fodder banks	56
Live fences	80
Shade for pasture grass and livestock	98
Enhanced nutrient cycling	98
Soil and water conservation	96
Fuelwood	100
Fruit production	69

Table 2. Percentage of agricultural producers implementing agroforestry land management practices as an adaptation strategy to a more intense dry season, and indicated other benefits (ecological services) as a result of implementing trees on their land in Esparza, Costa Rica (n=50).

For example, 100% of the agricultural producers in our study adapted *B. brizantha* as an improved pasture grass species when they initiated agroforestry silvopastural practices. *Brachiaria brizantha* is native to the African continent but it has been integrated in other tropical areas outside of its native range for use in permanent pasture, cutting for fresh feed, and as a ground cover for erosion control (Barnes et al. 2007). This grass species grows on a wide variety of soil types ranging from light to heavy soil textures and can tolerate a wide range in soil pH (4 to 8). *Brachiaria brizantha* is well adapted to a dry season ranging from 3 to 6 months in length during which time its leaves remain green while native species become dry, brown and unproductive. Additionally, *B. brizantha* can also endure wet soils and grows at an altitude up to 2000 m and thus can tolerate light frost (Barnes et al., 2007). Agricultural producers in our study found many advantages from growing *B. brizantha* on their land (Table 3).

Producers in our study observed an increase in milk production during the dry season by up to 70% as a result of establishing pastures with *B. brizantha*. Similar results were reported by Holmann et al. (2004) who found that the overall production of milk in Costa Rica increased by 55.5% as a result of establishing pastures or fodder banks with *B. brizantha*. This increase is likely due to the higher protein content in *B. brizantha* compared to other grass species traditionally used in Costa Rica (Cook et al., 2005). Results from our study showed that replacing conventional pasture grass with *B. brizantha* resulted in an improved management practice of grasslands, and also lowered grazing pressure and increased livestock carrying capacity, while at the same time provided a strategy to adapt to a changing climate. For example, Cameron (2002) observed that the implementation of *B. brizantha* grass into pastures resulted in soil and water conservation through a decrease in soil compaction and a reduction in external nutrient inputs.

Benefits Observed from *B. brizantha*	**Landowner Observation (%)**
Reduced erosion	96
Greater drought resistance	100
Increased forage production during dry season	98
Increased number of livestock per unit area	100
Livestock weight maintenance	92
Increased livestock meat and / or milk production	92
Improved calf health	96

Table 3. Benefits observed by agricultural producers by integrating silvopasture with *B. brizantha* grass in Esparza, Costa Rica (n=50).

Fig. 2. Fodderbank with *Gliricidia sepium* (Jacq.) Walp. The trees in the foreground show they were recently completely shoot pruned as fodder for livestock, and are already re-sprouting. Photo used with permission from M. Oelbermann.

Fodderbanks were another adaptive measure to climate change implemented by agricultural producers in our study (Figure 2). For example, a total of 56% of the agricultural producers surveyed used fodderbanks with commonly used tree species such as *Gliricidia sepium* (Jacq.) Walp. Pagiola et al. (2008) observed that adaptation of fodderbanks surrounding Esparza was much larger than that compared to other areas of Costa Rica, where they were virtually non-existent. This suggested that the integration of fodderbanks as an additional management practice was due to the close interaction between the agricultural producers, the Costa Rican Ministry of Agriculture, and CATIE researchers and extension officers taking part in the GEF-Silvopastoral Study. Agricultural producers found that *B. brizantha* fodderbanks helped to maintain livestock productivity during years with lower forage biomass and during drought or extended dry periods. A large proportion

(96%) of agricultural producers surveyed in our study implemented and maintained trees on their farmland or in the farmyard. Tree species most commonly established on farmland or in farm yards in our study included *Ficus hartwegii* (Miq.) Miq., *Guazuma ulmifolia* Lam., *Dipteryx panamensis* (Pittier) Rec. & Mell, *Enterolobium cyclocarpum* Jacq. Griseb., *Mangifera indica* L., *Byrsonima crassifolia* (L.) Kunth, *Psidium friedrichsthalianum* (O. Berg) Nied., *Gliricidia sepium* (Jacq.) Walp., and *Acrocomia aculeate* (Jacq.) Lodd. Ex Mart. Agricultural producers found that these trees served multiple functions such as the production of fruit for livestock and human consumption (69%), and the production of leaves for livestock fodder. They also expressed the hardiness of these trees during the dry season and as such did not shed their leaves. Other valuable services these trees provided included fuel wood (100%), shade for pasture grass and livestock, and enhanced nutrient cycling (98%), and soil and water conservation (96%).

Another agroforestry practice commonly implemented by participants in the GEF-Silvopastoral Study and widely used in Costa Rica included the use of live fences. In our study, all (100%) farmers surveyed implemented live fences and in a previous study by Harvey et al. (2005) over 80% of the agricultural producers surveyed in Costa Rica implemented live fences. In our study, the most common use of live fences was to provide an economical enclosure for livestock, and separating pastures from roads, forested areas, and waterways. Our study participants also observed that live fences provided shade for livestock and pasture grasses. In addition, many of the agricultural producers interviewed believed that live fences provided an ecological service by conserving biodiversity in addition to preventing soil erosion and protecting streams when they are used to separate pastureland from waterways. Live fences also provided foliage for cattle fodder. However, in our study, none of the agricultural producers surveyed discussed the use of live fences for this purpose. Similarly, Harvey et al. (2005) noted in their study that only 10% of Costa Rican agricultural producers used live fences for the production of fodder in addition to their use as an enclosure for pasture lands. This suggested that agricultural producers in Costa Rica were not aware of all possible benefits of live fences, but those that participated in the GEF-Silvopastoral Study had a greater knowledge on the diverse utility of live fences.

Other studies also revealed that the conversion of farmland previously containing few or no trees into areas dominated by agroforestry decreased the amount of degraded land by 15% over a 3-year period (Zamora-Lopez, 2006). Zamora-Lopez (2006) also found that the poorest agricultural producers were able to reduce pastureland degradation by introducing improved pasture with trees, by adopting fodder banks, and live fences. As such, a diversification in agricultural production systems included the implementation of a significant tree component that may buffer against income risks associated with current and future shifts in climate (Verchot et al., 2007). Kandji et al. (2006) also pointed out that agroforestry systems have the ability to improve the microclimate which in turn improves the adaptive capacity of land owners to climate change. Additionally, Kandji et al. (2006) suggested that the presence of trees in agricultural croplands can provide agricultural producers with an additional source of income that helps to strengthen their socioeconomic resilience. For example, tree products including fodder, resins, timber, and fruits provide a higher source of income than corps, and can also buffer against crop failure.

4.3 Producer contingency plans and adaptive capacity

According to the IPCC (2007) introducing a contingency plan to cope with unexpected changes in climate will increase food security and reduce financial losses. Our study showed that 28% of the agricultural producers researched the issue of climate change in detail and were able to explain what climate change is and how it will affect food security, their livelihood and families. These agricultural producers were able to link changes caused in the Earth's atmosphere to deforestation and industrial processes, and therefore had implemented a climate change contingency plan. These agricultural producers suggested, and were aware of, that those without a climate change contingency plan may face many additional challenges compared to those with a plan. An additional 50% of the agricultural producers in our study were seeking information about climate change contingency plans. This was because of their interaction with the GEF-Silvopastoral Study and the knowledge gained through this interaction. These agricultural producers expected that climate change will force them to make major adjustments to their land management practices in order to ensure food security and livelihoods for their family in the future. Approximately 56% of these agricultural producers relied on CATIE and 34% on the Costa Rican Ministry of Agriculture to obtain information on climate change and strategies for adaptation. The remaining agricultural producers from our study expressed that the different adaptation strategies they had already implemented as a result of their association with the GEF-Silvopastoral Study provided sufficient protection from negative impacts of climate change on their land.

Adger (2001) suggested that the capacity to adapt to climate change may be enhanced through the improvement of social networks that provide information about climate change. Such social networks, comprised of institutions and organizations fluent in issues surrounding climate change, should work in close association with agricultural producers to assist them in developing strategies to enhance their capacity to adapt to climate change (Gallopin, 2006). Our results showed that 84% of the agricultural producers sought assistance from CATIE and the Ministry of Agriculture to increase their capacity to adapt to climate change. Through the GEF-Silvopastoral Study, CATIE has played a key role in enhancing social networks in Esparza by connecting agricultural producers with the Costa Rican Ministry of Agriculture. A network specifically established for the agricultural producers in this region provided opportunities to connect with each other to learn about new climate change adaptation strategies through the implementation of agroforestry practices.

Nhemachena & Hassan (2007) observed that widespread information on climate change to agricultural producers enhanced their capacity to adapt to climate change, and led to the development and implementation of long-term contingency plans. Füssel (2007) also found that if agricultural producers understood the inherent changes expected as a result of climate change, then they improved their ability to adapt and decreased their overall cost of adaptation. Agricultural producers decided which land management practices to implement based on profit, grain yield and/or livestock productivity (Adger, 2001). Approximately 70% of agricultural producers in our study stated that market incentives either through credits or PES will increase their capacity to adapt to climate change. Based on the qualitative discussions with agricultural producers and the quantitative data collected in the surveys, several factors influenced their capacity and motivation to implement adaptation strategies (Figure 3). Our results showed that the producer's ability and motivation to implement climate change adaptation strategies was strongly influenced by financial incentives through a PES system which would also cover costs of tree planting and other expenses related to the adoption of agroforestry practices. Increased market incentives (e.g. eco-labelling) or other incentives to increase the marketability of their products, and non-financial assistance to help increase their capacity to adapt to climate change were requested by those surveyed in our study. Maddison (2007) pointed out that differences in the capacity to adapt to climate change may be due to the underlying differences in the perception of climate change. Some of the producers in his study expressed a lack of knowledge about adaptation strategies, rationing of key resources such as water, lack of appropriate seed, lack of market access and insecure property rights. Pagiola et al. (2008) also noted that the initial costs of implementing fodderbanks using *Brachiaria* grass species and the time lag before receiving any benefit from such activity in Costa Rica was a primary determinant of this management practice. Agricultural producers in our study, who thought climate change is not an environmental issue, were encouraged to implement environmental services for financial gain. The incentives offered by the GEF-Silvopastoral Study through a PES system and their increasing knowledge on the value of environmental services allowed them to understand and benefit financially through implementing these services.

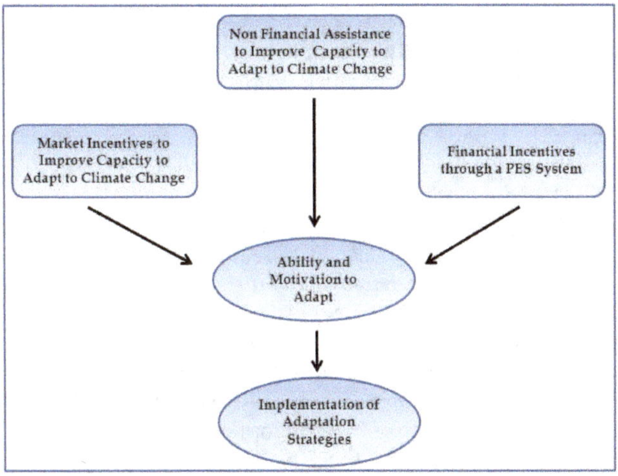

Fig. 3. Factors influencing the ability and motivation of agricultural producers to implement agroforestry systems as a climate change adaptation strategy, Esparza, Costa Rica (n=50).

5. Conclusions

Results from our study showed that the majority of agricultural producers had already implemented specific management practices to help cope with climate change. Our results also suggested that producers implemented specific agroforestry practices under the direction and guidance of CATIE extension personnel, researchers and the Costa Rican Ministry of Agriculture as part of the GEF-Silvopastoral Study. Through this interaction, agricultural producers became aware that the adoption of agroforestry practices also provided an ecological service, and therefore helped them to adapt to current and future climate change, for which they received a financial incentive. Only a small number of agricultural producers had a contingency plan in place which would help them cope with imminent climate change and reduce financial losses while maintaining food security and livelihoods for their families. However, a large proportion of agricultural producers were seeking further information on climate change contingency plans and relied on social networks built through their interaction with the GEF-Silvopastoral Study. A large number of agricultural producers in our study would readily adopt agroforestry practices as part of an incentive program including a PES system and market incentives, and receive further advice (non-financial assistance) from extension personnel on increasing their capacity to adapt to climate change.

6. Acknowledgements

Financial support was provided by Inter-American Institute for Co-operation in Agriculture, Canada, the Global Environment Facility Silvopastoral Study, and the University of Waterloo. We also thank Dr. M.A. Ibrahim and C. Cepulveda for assistance with research logistics and access to the GEF-Silvopastoral Study data base. We thank Francisco Casasola, and the Costa Rican Ministry of Agriculture for assistance in the field.

7. References

Adger, W.N. (2001). Scales of governance and environmental justice for adaptation and mitigation of climate change. *Journal of International Development* 16:238-281, ISSN 1099-1328.

Barnes, R.F., Nelson, C.J., Moore, K.J., Collins, M. (2007). *Forages: The Science of Grassland Agriculture*. Wiley-Blackwell, ISBN 0813806801, NJ, USA.

Budowski, G. (1987). The development of agroforestry in Central America. In: *Agroforestry: A Decade of Development* H.A. Steppler, Nair, P.K.R., pp. 69-88. International Council for Research in Agroforestry (ICRAF), ISBN 92 9059 036, Nairobi, Kenya.

Cameron, A.G. (2002). *The value of improved pastures*. Agnote 297, No. E32. Agdex 131. Australian Agronomy Society. ISSN No. 0157-8243.

Cook, B.G., Pengelly, B.C., Brown, S.D., Donnelly, J.L., Eagles, D.A., Franco, M.A., Hanson, J., Mullen, B.F., Partridge, I.J., Peters, M. and Schultze-Kraft, R. (2005). *Tropical Forages: An Interactive Selection Tool*, [CD-ROM], CSIRO, DPI&F (Qld), CIAT and ILRI, Brisbane, Australia, http://www.tropicalforages.info/.

Food and Agriculture Organization of the UN (FAO). (1990). *Production Yearbook*. ISBN 9789250030678, Rome, Italy.

Food and Agriculture Organization of the UN (FAO). (2007). *State of Food and Agriculture Report*. FAO Economic and Social Development Department, Corporate Document Repository. http://www.fao.org/docrep/010/a1200e/a1200e00.htm.

Füssel, H.M. (2007). Adaptation planning for climate change: Concepts, assessment approaches, and key lessons. *Sustainability Science* 2:265-275, ISSN 1862-4057.

Gallopin, G.C. (2006). Linkages between vulnerability, resilience, and adaptive capacity. *Global Environmental Change* 16:293-303, ISSN 0959-3780.

Gordon, A.M., Newman, S.M. (1997). *Temperate Agroforestry Systems*. CAB International, ISBN 0851991475,Wallingford, U.K.

Hänsela, G, Ibrahim, M.A., Villanueva, C, Andradec, H.J. (2009). Exploiting synergies between silvopastoral system components for carbon sequestration and an increase cattle productivity: Experience from Costa Rica and Nicaragua. *Proceedings of the XIII World Forestry Congress*, Buenos Aires, Argentina 18-23rd October, 2009, http://www.cfm2009.org/es/programapost/trabajos/exploiting_synergies_fd.pdf

Harvey, C.A., Villanueva, C., Villacis, J., Chacon, M., Monuz, D., Lopez, M., Ibrahim, M., Gomez, R., Taylor, R., Martinez, J., Navasa, A., Saenz, J., Sanchez, D., Medina, A., Vilchez, S., Hernandez, B., Pereza, A., Ruiz, R., Lopez, F., Lang, I., Sinclair, F.L. (2005). Contribution of live fences to the ecological integrity of agricultural landscapes. *Agriculture, Ecosystems and Environment* 111:200-230, ISSN 0167-8809.

Holdridge, L.R., Greneke, W.C., Hatheway, W.H., Liang, T., Tosi Jr., J.A. (1977). *Forest Environments in Tropical Life Zones*. Pergamon Press, ISBN 0080163408, Oxford, U.K.

Holmann, F., Rivas, L, Angel, P.J., Pérez, E. (2004). Impacts of adoption of *Brachiaria* grasses: Central America and Mexico. *Livestock Research for Rural Development* 16:1-9, ISSN 02123784.

Huxley, P. (1999). *Tropical Agroforestry*. Blackwell, ISBN 978-0-632-04047-6, Oxford, UK.

IPCC. (2007). Climate Change 2007 - Impacts, Adaptation and Vulnerability. *IPCC Fourth Assessment Report*.http://www.ipcc.ch/publications_and_data/publications_and_data_reports.shtml.

Jose, S. (2009). Agroforestry for ecosystem services and environmental benefits: An overview. *Agroforestry Systems* 76:1-10, ISSN 1572-9680.

Kandji, S.T., Verchot, L.V., Mackensen, J., Boye, a., Van Noordwijk, M., Tomich, C.K., Ong, C.K., Albrecht, A., Palm, C.A. (2006). Opportunities for linking climate change adaptation and mitigation through agroforestry systems. In: *World Agroforestry into the Future*, Garrity, D.P., Okono, A., Grayson, M., Parrott, S., pp. 113-121. World Agroforestry Centre (ICRAF), ISBN 9290591846, Nairobi, Kenya.

Kass, D.L.C., Araya, J.S., Sanchez, J.O., Pinto, L.S., Rerreia, P. (1995). Ten years of experience with alley farming in Central America. *Proceedings of the International Alley Farming Conference*, IITA, Ibadan, Nigeria, pp. 393-402.

King, K.F.S. (1987). The history of agroforestry. In: *Agroforestry: A Decade of Development* H.A. Steppler, Nair, P.K.R., pp. 3-13. International Council for Research in Agroforestry (ICRAF), ISBN 92 9059 036, Nairobi, Kenya.

Leakey, R.R.B., Simons, A.J. (1996). The domestication and commercialization of indigenous trees in agroforestry for the alleviation of poverty. *Agroforestry Systems* 38:165-176.

McGinty, M.M., Swisher, M.E. Alvalapati, J. (2008). Agroforestry adoption and maintenance: Self-efficacy, attitudes and socio-economic factors. *Agroforestry Systems* 73:99-108, ISSN 1572-9680.

Maddison, D. 2007. The perception of and adaptation to climate change in Africa. *Policy Research Working Paper WPS4308*. The World Bank, Development Research Group, Sustainable and Urban Development Team, http://papers.ssrn.com/sol3/papers.cfm?abstract_id=1005547.

Nair, P.K.R. (1993). *An Introduction to Agroforestry*. Kluwer Academic Publishers, ISBN 0792321340, Dordrecht, NL.

Nhemachena, C., Hassan, R. (2007). Micro-level analysis of farmer's adaptation to climate change in Southern Africa. *International Food Policy and Research Institute Discussion Paper 00714*. Environment and Production Technology Division, http://www.ifpri.org/sites/default/files/publications/ifpridp00714.pdf.

Mercer, D.E. (2004). Adoption of agroforestry innovations in the tropics: a review. *Agroforestry Systems* 74:311-328, ISSN 1572-9680.

Oelbermann, M., Voroney, R.P., Gordon, A.M. (2004). Carbon sequestration in tropical and temperate agroforestry systems: A review from Costa Rica and southern Canada. *Agriculture, Ecosystems and Environment* 104:359-377, ISSN 0167-8809.

Oelbermann, M., Voroney, R.P. (2011). An evaluation of the Century model to predict soil organic carbon in tropical and temperate agroforestry systems. *Agroforestry Systems* 81: DOI 10.1007/s10457-010-9351-6, ISSN 1572-9680.

Pagiola, S. (2007). Payments for environmental services in Costa Rica. *Ecological Economics* 6: 712-724, ISSN 09218009.

Pagiola, S., Rios, A.R., Arcenas, A. (2008). Can the poor participate in payments for environmental services? Lessons from the Silvopastoral Project in Nicaragua. *Environment and Development Economics* 13:299-325, ISSN 1355770.

Pattanayack, S.K., Mercer, D.E., Sills, E., Yang, J.C. (2003). Taking stock of agroforestry adoption studies. *Agroforestry Systems* 57:186-203, ISSN 1572-9680.

Pearce, D., Mourato, S. (2004). The economic valuation of agroforestry's environmental services. In: *Agroforestry ad Biodiversity Conservation in Tropical Landscapes*, Schroth, G., da Fonseca, G.A.B., Harvey, C.A., Gascon, C., Vasconcelos, H.L., Izac, A.M.N., pp. 67-86. Island Press, ISBN 1559633565, Washington, USA.

Raxworthy, C.J., Pearson, R.G., Rabibisoa, N., Rakotondrazafy, A.M., Ramanamanjato, J.B., Raselimanana, A.P., Wu, S., Nussbaum, R.A., Stone, D.A. (2008). Extinction vulnerability of tropical montane endemism from warming and upslope displacement. *Global Change Biology* 14:1703-1720, ISSN 1354-1013.

Sanchez, P.A. (2000). Linking climate change research with food security and poverty reduction in the tropics. *Agriculture, Ecosystems and Environment* 82:371-383, ISSN 0167-8809.

Smith, C.E. (2008). Encouraging climate change adaptation though the payment for environmental services: Case studies in the Pacific region of Costa Rica. MES Thesis, Department of Environment and Resource Studies, University of Waterloo, Canada.

Sood, K.K, Mitchell, C.P. (2009). Identifying important biophysical and social determinants of on-farm tree growing in subsistence-based traditional agroforestry systems. *Agforestry Systems* 75:175-187, ISSN 1572-9680.

Verchot, L.V., van Noordwijk, M., Kandji, S., Tomich, T., Ong, C., Albrecht, A., Mackensen, J., Bantilan, C. Palm, C. (2007). Climate change: Linking adaptation and mitigation through agroforestry. *Mitigation and Adaptation Strategies for Global Change* 12:902-918, ISSN 1381-2386.

Walther, G.R., Post, E., Convey, P., Menzel, A., Parmesan, C., Beebee, T.J.C., Formentin, J.M., Hoegh-Gugdberg, O., Bairlein, F. (2002) Ecological responses to recent climate change. *Nature* 416:389-395.

Watson, R.T., Noble, I.R., Bolin, B., Ravindranath, N.H., Verardo, D.J., Dokken, D.J. (2000). *IPCC Special Report on Land Use, Land-use Change and Forestry.* http://www.grida.no/climate/ipcc/land use/

Yamamoto, W, Dewi, I.A., Ibrahim, M.A. (2007). Effects of silvopastoral areas on milk production at dual-purpose cattle farms at the semi-humid old agricultural frontier in central Nicaragua. *Agricultural Systems* 94:368-375, ISSN 0308-521.

Young, A. (2002). *Agroforestry for Soil Management*. CAB International, ISBN 0851991890, Wallingford, UK.

Zamora-Lopez, S.E. (2006). Efecto de los pagos por servicios ambientales en la estructura, composición, conectividad y el stock de carbono presente en el paisaje ganadero de Esparza, Costa Rica. M.Sc. Thesis Centro Agronómico Tropical de Investicación y Enseñanza (CATIE), Turrialba, Costa Rica.

5

Global Warming and Its Economic Effects on the Anchovy Fishery and Tourism Sector in North-Western Spain

M. Dolores Garza-Gil, M. Xosé Vázquez-Rodríguez,
Albino Prada-Blanco and Manuel Varela-Lafuente
University of Vigo
Spain

1. Introduction

In recent years much evidence has been gathered on climate change and its impact on different sectors and systems. Global warming is one of the main threats to sustainable development and, consequently, one of the most significant environmental challenges in the last decades affecting the economy, health and social welfare. It is necessary, therefore, to identify evidence of the impact of global warming on biodiversity and carry out an economic evaluation. In the specific case of marine ecosystems, changes in rainfall frequency and intensity, acidity, water temperature, wind, dissolved CO_2 and salinity, combined with anthropogenic nutrient and toxin contamination, can affect water quality both in coastal regions as well as in the open sea. All of this will consequently affect the productivity of the marine environment. And given that fishing is one of the economic activities which critically depend on natural conditions or characteristics, the influence of environmental changes on fishing is notably higher than that which might occur in other primary activities. Furthermore, climate has a vital impact on the tourism and recreation sector and, therefore, this sector will be affected by any changes in climate.

In this chapter, we assess the possible economic effects (losses or gains) of global warming on some of the main economic activities in north-western Spain. The economy of this region specialises in products derived from fishing and aquaculture as well as tourism, among others (IGE - Galician Statistics Institute-, 2010), and both activities are extremely sensitive to environmental conditions. It is highly probable that global warming will alter the intensity and conformation of ocean currents, affect marine organisms and generate coastal alterations (IPCC, 2007). Such environmental changes will have important repercussions on these economic activities.

A considerable number of studies have been carried out internationally which have aimed to assess the economic effects of climate change on these activities. Among other references in the case of fishing, we would underline the following: Arnason (2005) evaluates the possible impact of climate change on Iceland's fishery production, proposing different scenarios involving temperature increase; along the same lines, Eide (2005) analyses the possible impact on the Barents Sea fisheries; Gallagher (2005) makes an application to the New Zealand cod fishery, differentiating between zone and fishing method; Röckmann

(2005) analyses the effects of possible changes in salinity on the Baltic cod fishery; Sissener & Bjorndal (2005) study the effect of climate change on the migratory patterns of the Norwegian herring; Arnason (2006) proposes a theoretical model into which he introduces the risks deriving from global warming; Briones et al (2006) develops a model applied to small pelagic species in fisheries in India, the Philippines and Thailand; Hannesson (2006) analyses the effects of warming on the Norwegian herring fishery; Herrick et al (2006) make an application to the sardine fishery in north-American Pacific waters; and Garza-Gil et al (2011) evaluate the effects on the Ibero-atlantic sardine fishery.

In relation with the references which study the complex relationship between climate change and tourism, most of them point out that not only will the volume of tourism flows change but the destinations as well (Smith, 1990, 1993); Viner & Agner, 1999; Maddison, 2001; Lise & Tol, 2002; Scott *et al.*, 2004, 2005; Hamilton *et al.*, 2005a, 2005b; Gómez Martín, 2005; Bigano *et al.*, 2006, among others. In Europe, we would underline the PESETA study (European Commission, 2007) to evaluate the impacts forecast in Europe for the time periods 2011-2040 and 2071-2100, and which predicts that summer climate conditions will change radically and that the zone that currently has excellent conditions (the Mediterranean) will be displaced northwards (figure 1). In the case of Spain, the Spanish Climate Change Office reaches similar conclusions, predicting a "Mediterraneanisation" of the north of the peninsula and the aridization of the south (Ministry of the Environment, 2005). This result is particularly relevant for the research carried out, as we will see that for Galicia tourism flows come directly from the rest of Spain and this Mediterraneanisation of the north might generate a change in direction to the north from the current flows to the south. In this respect, the analysis by Bigano *et al.* (2004), which points out that approximately 86% of total tourism is the domestic or internal tourism of each country, is confirmed when we analyse the tourism flows from the rest of Spain to the north-west of the country. It should also be pointed out that, given the volume of domestic tourism, the analysis of the effects of climate change has in general dealt with international tourism, obviating internal or domestic tourism (Seddighi & Shearing, 1997; Coenen & Van Eekeren, 2003).

For this chapter and in relation with fishing activity, the study case chosen is the anchovy (*Engraulis encrasicholus*) fishery, it being a hugely important pelagic species for the fisheries sector in the region analysed. The pelagic species would be among those most affected by the impacts of climate change on seas and oceans due to the high level of instability and sensitivity to environmental impacts. These species are especially sensitive to temperature changes and the upwelling of nutrients in the marine environment. Therefore, any water temperature variation will have repercussions to a greater or lesser extent on these species' reproduction levels. In particular, we assess the possible economic effects of global warming on this fishery and sea surface temperature management is introduced into the management problem. This variable allows us to gather evidence of climate change and its repercussion on ecosystems and marine species, which are the bases of fish reproduction functions (McGowan *et al.*, 1998; Levitus *et al.*, 2000; IPCC, 2001; Stenevik and Sundby, 2007). Other variables, such as the frequency and intensity of rainfall, acidity, dissolved carbon and salinity, are also prone to experience environmental changes; however there is a high level of correlation between all of these variables. In this study, we describe the evolution of fish biomass, based on the sensitivity of the species' growth function with respect to fluctuations in oceanographic conditions (through the sea temperature), and will analyse impacts on the economic yield of the fishery deriving from a possible change in the temperature conditions

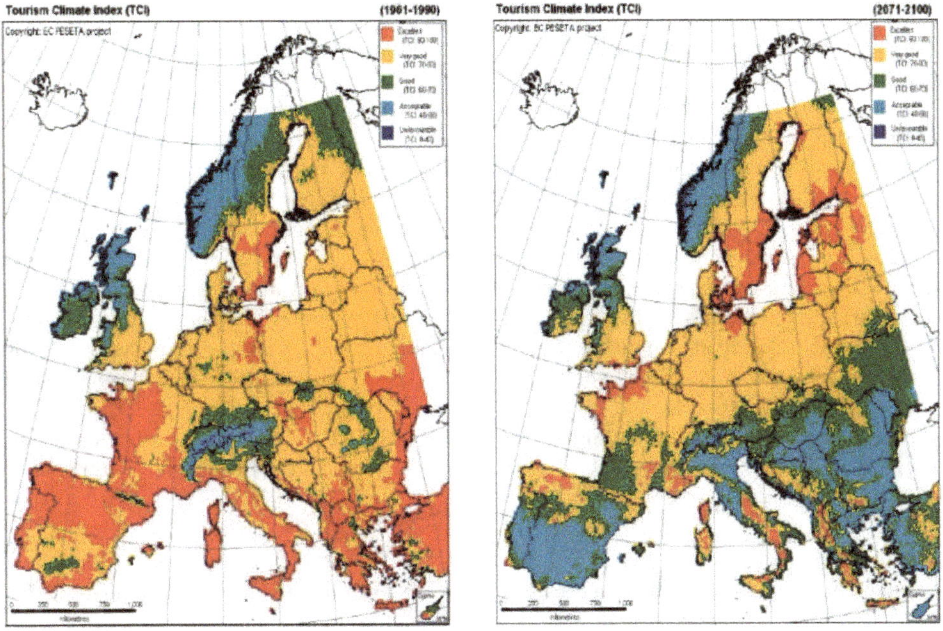

Source: PESETA Project (www.jrc.es/docs/Tourism.html) cited in EEA (2007).

Fig. 1. Simulations for summer tourism en Europa for 1961-1990 (left) and 2071-2100 (right), based on high emissions scenario of IPCC A2.

of the marine ecosystem. We will apply a bio-economic model for evaluating these effects on the fishing profits.

With relation to tourist activity, the physical changes in the coastal landscape, in the availability of certain resources or basic provisions (water, energy, food, etc.), in risks to health (new illnesses, effects of extreme temperatures) and the risk of catastrophic events (droughts, floods, storms and extreme weather conditions) are some of the consequences of climate change that will directly influence tourism. We assess the possible effects of global warming on tourist visits to the north-west of Spain. The analysis is based on prior qualitative information gathered on expected changes in the climate at different times of the year in Spain, the main source of tourist demand in the region studied. A prior description or zero scenario of tourism flows to the region is carried out. On this basis, changes in tourist preferences to estimate how the new climate scenario will affect tourist travel to the region are analysed. The methodology used consists of a field study based on direct methods using questionnaires presented to a representative sample at the source market. The objective market chosen was Madrid, the main tourist source market for the north-west of Spain. Climate is hugely important in the satisfaction or dissatisfaction of tourist visits at present and, in consequence, the choice of destination.

The structure of the chapter is as follows: in section 2, we will analyse the effects of climate change on the anchovy fishery, while section 3 deals with effects on tourist activity. Lastly, in section 4, we will sum up the study's main conclusions.

2. Effects of climate change on the anchovy fishery

Given that fishing activity is one of the economic activities which most critically depends on natural conditions or characteristics, the influence of environmental changes on fishing is notably higher than that which might occur in other primary activities (Hannesson 2006). It is assumed with some degree of probability that global warming will alter the intensity and disposition of ocean currents, and the effect of this, among others, will mean an increase in ocean temperatures, variations in salinity levels and changes in upwellings (ACIA 2004; IPCC 2007). The impacts will differ according to ecosystems and coastal or ocean zones, and will affect different groups of organisms, from phytoplankton and zooplankton to fish and algae (Ministry of the Environment 2006). Among these organisms, pelagic species (and, in particular, small pelagic species), on account of their high level of instability and sensitivity to environmental impacts, will be among those most affected by the impacts of climate change on seas and oceans. The small pelagic species are target species for the majority of the Spanish fleet, in general, and for the north-west in particular; and, of these, the anchovy is important on account of its high commercial value. For this reason, any significant modification in anchovy biomass levels can affect fishermen's net profits. We will begin by describing the situation of this fishery.

2.1 The situation of the fishery

The Spanish purse seine fishery in the Atlantic is made up of 491 vessels, of which 346 fish in north-western Cantabrian waters and the remaining 146 in the waters of the Canary Islands and the Bay of Cadiz (MAPA, 2008). It is one of the fleets with the most vessels operating in these waters, second only to the artisanal fleet, and it targets small pelagic species, among which are the sardine, horse mackerel, mackerel and the anchovy (MAPA, 2008; Ibermix, 2007). The Spanish purse seine fleet in the North Atlantic is made up of relatively homogenous vessels insofar as their technical characteristics are concerned, with an average capacity of 34.2 GT, power of 151.8 Kw per vessel and a length of 21m. The average life of the fleet is 20 years, with a crew of 8 per vessel. All of the vessels use nets made from synthetic materials, they are equipped with hydraulic haulers and electronic fish detectors.

Regarding the management, in general there are no specific regulations from the European Union. Given the poor situation in which the anchovy has found itself over the last years, a precautionary TAC was implemented for several years and a moratorium for this species has been in place in the Bay of Biscay since 2005. The Spanish government uses input control (area restrictions, entrance restrictions and gear regulation) and, in addition, some regional governments implement output control for some species (maximum catches per fishing day in the case of the anchovy).

This fleet targets pelagic species, in this case, the sardine and the anchovy. The anchovy has been subject to a precautionary TAC in area VIIIc for several years, but since the collapse of this species in 2005, the European Union established a moratorium (ICES, 2008). Stock biomass has been low because recruitment has been low since 2002, and the fishery has been closed since 2005. There are no indications how long the low recruitment will last and whether a continued low SSB will reduce future recruitments. Biologists consider it likely that the closure of the fishery over the last few years has led to an increase in the abundance of older anchovy, and because of the very low recruitment in 2008, the contribution of older fish to spawning in 2009 will be crucial. The fishery was opened again in 2010 and the EC implemented a TAC of 7,000 tons. In any case, the harvest control rules for anchovy are currently under development. Under the rules being considered, a TAC is set on the basis of the estimated spawning biomass and may be set for the whole period July–June, with or without a provision to revise it at the beginning of the year based on the results of juvenile surveys. The criterion for accepting these rules as precautionary would be that the rule implies a low risk of reducing the SSB to a level which may imply further reduction in recruitment (ICES, 2008). Other supplementary measures (area closures, minimum landing size) may be considered in addition to TACs.

Year	**Sea Surface Temperature (°C)**	**Anchovy Biomass (tons)**	**Catches in Spain (tons)**
1987	16.08	34128	15308
1988	16.13	51754	20302
1989	16.62	30193	16558
1990	16.61	81330	40759
1991	15.86	43675	25556
1992	16.02	18286	41051
1993	15.88	117981	42377
1994	16.08	79257	38019
1995	16.45	97017	43071
1996	16.32	78517	38968
1997	16.76	71573	27632
1998	16.64	124322	42579
1999	16.50	102308	34668
2000	16.41	132228	39496
2001	16.48	113063	49247
2002	16.24	46391	26313
2003	16.62	41317	15864
2004	16.36	52016	22205
2005	16.49	24413	5643
2006	16.51	37141	6244
2007	16.54	49579	6595
2008	16.56	36532	6457

Source: Own compilation from ICES (2008) and, for the temperature, Spanish Centre for Higher Scientific Research.

Table 1. The fishery.

The landings of this species were on the increase up to the year 2000, after which they dropped until the temporary closure of the fishery was imposed on account of poor stock recruitment in preceding years (table 1). For application purposes, we will not bear in mind the data corresponding to the period 2005-2008 due to the moratorium established by the European Commission for this stock, given that it is not the usual situation in the fishery. In figure 2, we can observe the variability that exists in the evolution of the biomass. However, in general, and except in the middle years of the period, as the sea surface temperature in this zone increased, the biomass dropped in the same way as catches, especially in the last years of the period analysed. And in table 2 we can observe the correlation matrix between these three variables, from which it can be gathered that variations in the sea temperature generate variations of the opposite kind in biomass and catches.

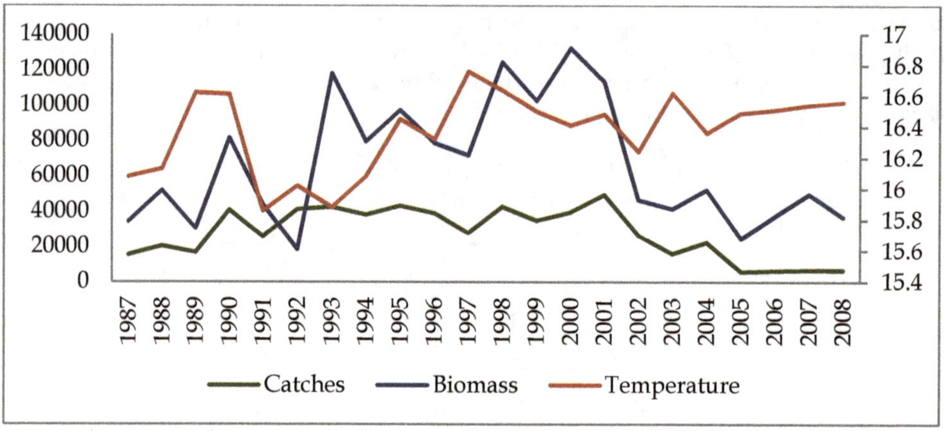

Fig. 2. Evolution of fishery. 1987-2008.

With regard to the economic parameters, we have average data for the period 1995-2006 from the European Commission (2006) and from direct interviews with fishery sector representatives. On this basis, the price per unit of catches landed (p) stands at 3950.01 €/ton (constant euros for 2008); the cost per unit of catch (c) stands at 982.7 €/ton (constant euros for 2008); and the discount rate (δ) is approximately 5%.

	Biomass	Biomass t+1	Catches	Temperature
Biomass	1.0000			
Biomass t+1	0.7923	1.0000		
Catches	0.7141	0.7428	1.0000	
Temperature	-0.4659	-0.6366	-0.6781	1.0000

Table 2. Correlation matrix among the variables biomass, catch and temperature

On the other hand, the north-west zone represents 36% of the total fishery insofar as anchovy landings for the period 1995-2006 are concerned. This will be the data we will use as a reference to estimate the variations in net profits in the face of the new climate scenario.

2.2 Methodology

Bio-economic modeling enables us to incorporate natural, environmental and institutional contributing factors, as well as typically economic factors, into a single analytical body. For the specific case of fisheries, the aim is to control the size of the fish population by limiting catches (or fishing effort) in such a way as to maximise the present value of the flow of net profits generated by the fishery over a specific time horizon and bearing in mind the dynamics of said natural resource. In this way, we can determine the extent to which society can invest (or disinvest) in the natural resource and what the appropriate extraction rate would be over time, allowing for the sustainable exploitation of said resource.

Given that we do not have data on fishing effort for this fishery, we can represent the bio-economic problem in simplified form as follows:

$$Max_h \int_0^{\infty} \pi(X,h)e^{-\delta t}\, dt \tag{1}$$

$$s.a.\ \dot{X} = \frac{dX}{dt} = F(X) - h(t)$$

where $\pi(X, h) = [p - c]\, h(t)$ represents the net profits generated by fishing in the instant t, X the fish population, h the catch rate, p the price per unit of fish, c the cost per unit of fishing, δ the social discount rate and F the natural dynamics of the fish population (or the natural growth function of the stock without considering human activity). The fundamental problem (Kamien & Swartz, 1991) consists of determining the optimal feasible control, $h(t)=h^*(t)$ con $t \geq 0$, which maximises the objective function while satisfying the problem's conditions in the new climate context.

In order to resolve the problem (1), we need to previously define marine resource dynamics $\dot{X}$. This function is statistically tested on the basis of the data that exists on biomass, catches and, in our case, the sea surface temperature, generally using Ordinary Least Squares. From the data in table 1, and once the correlation matrix shown in table 2 has been obtained, the function which presents the best economic results is as follows:

$$\dot{X} = \alpha X_t + \beta X_t^2 + \gamma T_t - h_t \tag{2}$$

The equation (2) corresponds to the logistic model, widely used in fisheries economics literature, where α, β and γ are parameters containing biological information on anchovy stock, and T denotes the sea surface temperature. The concrete results of the econometric estimation of (2) are shown in table 3. Therefore, the concrete form of this function is given by the following expression:

$$\dot{X} = 1.98\, X_t - 0.7E^{-5}\, X_t^2 - 0.03648\, T_t - h_t \tag{3}$$

	$X_{t+1} + h_t = \alpha X_t + \beta X_t^2 + \gamma T_t$
α	1.98 (4.86)
β	-0.7 E^{-5} (-2.28)
γ	-0.03648 (-2.14)
Jb	0.3564
Q-Stat	0.0883
LM (ARCH)	0.8153
R^2	0.7982
R^2 adjusted	0.7684

Note: t-ratio between brackets. Jb is the Jarque-Bera statistic of the normality test; Q-Stat is the Ljung-Box statistic used in the correlation test; LM (Lagrange multiplier) is the one used in the heteroscedasticity test; and AIC is the statistic used in the prediction error model.

Table 3. Econometric results for growth function of anchovy biomass.

In this way, the *Hamiltonian* function in usual terms (current moment *t*) associated with problem (1) is given by the following expression:

$$H\ (X, h, t;\ \mu) = (p-c)h_t + \mu\ (\alpha X_t - \beta X_t^2 - \gamma T_t - h_t) \tag{4}$$

where μ denotes the shadow price of the marine resource in current terms.
The conditions necessary (Kamien & Swartz, 1991) to solve the problem are given by the expressions:

$$\frac{dH\ (X,h,t;\ \mu)}{dh} = 0 \Longrightarrow (p-c) - \mu = 0 \tag{5}$$

$$\dot{\mu} - \delta\mu = -\frac{dH\ (X,h,t;\ \mu)}{dX} = -\mu(\alpha - 2\beta X) \tag{6}$$

$$\frac{dX}{dt} = 0 \Longrightarrow \alpha X - \beta X^2 - \gamma T - h = 0 \tag{7}$$

By using expressions (5)-(7) the catch level is obtained $h^*(t)$, which will depend on the level of sea surface temperature, as will stock dynamics (expression (2)):

$$h^* = \alpha\left(\frac{\alpha-\delta}{2\beta}\right) - \beta\left(\frac{\alpha-\delta}{2\beta}\right)^2 - \gamma T \tag{8}$$

Once this level is known, we can obtain the losses or profits associated with the new climate scenario in relation with the fishery's present situation through the net profit function for the fishery.

2.3 Results

Declines in the abundance of the most important commercial fish species have often considered as a result of overfishing and occasionally from a combination of environmental effect and fishing pressure (IPCC, 2001). The impacts of climate variations have been shown to have substantial effects on decreases as well as increases in stock abundance, and the success of the future fish stock assessment depends to a large extent on the ability to predict the impacts of climate change on the dynamics of marine ecosystems.

The temperature in the North Atlantic has shown, in general, an increasing trend over the recent three decades. It may be an indication of the climate change caused by emission of different greenhouse gases. However, there is natural variability in the climate in addition to long term climate change induced by anthropogenic activity. And the difficulty in obtain many highly confident outcomes is why the term "climate scenarios" has been adopted in most impact assessment (IPCC, 2001). The climate scenarios should be regarded as internally consistent patterns of plausible future climates, and not predictions based in probabilities. Since most climate models focus on the atmosphere, the climate change scenarios for the ocean are particularly prone to uncertainty (Stenevik & Sundby, 2007). It is, however, concluded that the global warming will affect the ocean through changes in the sea temperature, among other variables (IPCC, 2001). And it has been shown that there has been a general warming of a large part of the world oceans during the past fifty years (Levitus *et al.*, 2000). For the Ibero-Atlantic waters, the sea surface temperature is expected to increase by approximately the same amount as in the last decades (Rosón, 2008), this is 0.027°C per year ("Current warming" scenario in table 4).

In the table 4, the results obtained in the face of sea surface temperature increases in these fishing grounds are shown for different climate scenarios. For the current warming scenario, we can see that as the sea surface temperature increases, both the catch level as well as the net profits fall for the overall fishery. In the specific case of the net profits, the decrease is estimated to stand at 1.28% using the year 2040 as the time horizon, a medium-term time horizon established in the European Commission's Peseta Report (EEA, 2007).

Current Warming			Higher Warming		
Temperature (°C)	**Catches (tons)**	**Net Benefits (euros)**	**Temperature (°C)**	**Catches (tons)**	**Net Benefits (euros)**
16.6790	13991.89542	41519550.5	16.6979	13849.31678	41096462.6
16.7060	13991.89444	41519547.6	16.7276	13849.31570	41096459.4
16.7330	13991.89346	41519544.7	16.7573	13849.31462	41096456.2
16.7600	13991.89248	41519541.8	16.7870	13849.31355	41096453.0
16.7870	13991.89150	41519538.9	16.8167	13849.31247	41096449.8
16.8140	13991.89053	41519535.9	16.8464	13849.31139	41096446.6
16.8410	13991.88955	41519533.0	16.8761	13849.31032	41096443.4
16.8680	13991.88857	41519530.1	16.9058	13849.30924	41096440.2
16.8950	13991.88759	41519527.2	16.9355	13849.30816	41096437.0
16.9220	13991.88661	41519524.3	16.9652	13849.30709	41096433.8
16.9490	13991.88563	41519521.4	16.9949	13849.30601	41096430.7
16.9760	13991.88465	41519518.5	17.0246	13849.30493	41096427.5
17.0030	13991.88368	41519515.6	17.0543	13849.30386	41096424.3
17.0300	13991.88270	41519512.7	17.0840	13849.30278	41096421.1
17.0570	13991.88172	41519509.8	17.1137	13849.30170	41096417.9
17.0840	13991.88074	41519506.9	17.1434	13849.30063	41096414.7
17.1110	13991.87976	41519504.0	17.1731	13849.29955	41096411.5
17.1380	13991.87878	41519501.1	17.2028	13849.29847	41096408.3
17.1650	13991.87780	41519498.2	17.2325	13849.29740	41096405.1
17.1920	13991.87682	41519495,3	17.2622	13849.29632	41096401.9
17.2190	13991.87585	41519492.4	17.2919	13849.29524	41096398.7
17.2460	13991.87487	41519489.5	17.3216	13849.29417	41096395.5
17.2730	13991.87389	41519486.6	17.3513	13849.29309	41096392.3
17.3000	13991.87291	41519483.7	17.3810	13849.29201	41096389.1
17.3270	13991.87193	41519480.8	17.4107	13849.29094	41096385.9
17.3540	13991.87095	41519477.9	17.4404	13849.28986	41096382.7
17.3810	13991.86997	41519475.0	17.4701	13849.28878	41096379.5
17.4080	13991.86899	41519472.1	17.4998	13849.28771	41096376.3

Table 4. Results of estimations.

Table 4, on the other hand, also shows the results for these variables when faced with an even greater increase in sea surface temperature in these fishing grounds. As was foreseeable, there is an even sharper decrease in both variables. It should be noted that in any of the thermal oscillation scenarios, the level of catches obtained is lower than the average for the period 1987-2008.

3. Effects of climate change on tourism

Given that the aim is to carry out a preliminary estimation of changes in tourism flows to the north-west coast from one of the visitors' main source markets at present, it is necessary to begin by clarifying the basic terminology and, consequently, by delimiting the work done. In this sense, the World Tourism Organisation establishes that if an activity is to be considered a tourist activity, it must involve at least one overnight stay away from the permanent residence. Thus, for example, Evans *et al.* (2000) differentiate between these activities which involve overnight stays and a certain prior planning of recreational activities away from home in general which would be included in the concept of recreation or recreational activities. In the study described below, we will look exclusively at tourism flows according to the internationally-accepted definition, which does not cover recreational activities. This differentiation is also significant for the study's aims, as tourist trips are usually strongly conditioned by climate (when choosing destination and date), while recreation depends more on short-term weather forecasts. In this respect, Limb & Spellman (2001) identify the climatic factors that most influence tourists' decisions and observe that they are temperature, sun and rain. We will take into consideration the direct effect of climate change on tourism demand, and not look at indirect effects (through changes to the landscape, natural resources, etc.). Firstly, we need to identify the tourism destinations and uses of Madrid residents at present and the variables which affect their choice of destination. Secondly, we will reveal how preferences are modified under a climate change scenario, initially in their own region, but then also in different coastal destinations currently preferential for Madrid residents (especially the Mediterranean and the north-west). We will begin by presenting a description of current tourism demand in Galicia. Then we will briefly describe the methodology used to estimate changes in future demand and present the main results obtained.

3.1 Analysis of tourism demand under the baseline scenario

From the data presented in table 5, the relative decisive importance of domestic tourism in comparison with foreign tourism can be observed when we analyse tourist visits to the north-west of Spain (Galicia) 2008. This would explain why a single domestic market (Madrid) makes up half of the total figure for tourism from outside Spain to Galicia. This is also true in terms of the value of increased consumption, although the source used would seem to overestimate the former and underestimate the tourism that comes from the rest of Spain (36% of overnight stays generates only 18% of consumption). This said, it should be underlined that the estimation of tourism consumption received from the rest of Spain (in which Madrid is fundamental) is very similar to the importance of foreign tourism consumption (907 and 1187 million, respectively).

Of the 46 million overnight stays[1] by domestic tourists, 18.3 million (40%) are explicitly identified as summer holidays. To these we should add, among others, visits to family and friends, which can also be highly seasonalised. Per type of accommodation, the distribution of these overnight stays in visitors' own homes and/or those belonging to family and friends (which exceed 53% of the total) will be closely related to the data on access to a second home. It is significant that with regard to Galicia, in 85% of cases the location of this

[1] The number of overnight stays in hotel accommodation is an "...indicator that in terms of volume it is much more significant than the mere arrival of tourists" (Xunta de Galicia 2005: 13).

	Overnight stays (people)	%	Consumption (millions of euros)	%
Domestic total	46,166,152	79.5	3,929	76.8
- Residents	24,889,152	42.7	3,022	59.1
- Rest of Spain	21,277,140	36.5	907	17.7
Madrid[1]	6,497,107	11.2	n.a	n.a
Rest[2]	14,730,033	25.3	n.a	n.a
From outside Spain	11,922,500	20.5	1,187	23.2
Total	58,088,652	100.0	5,187	100.0

[1]Makes up 5% of overnight stays, where the feeder market is Madrid, in the rest of Spain.
[2]Cataluña comes after Madrid, with 3,190,147 visits. Source: www.iet.tourspain.es e www.exceltur.org

Table 5. Analysis of visits and visitors.

second home is in Galicia, whereas with regard to Madrid in 84% of cases the second home is situated outside the region, that is, completely the opposite.

Non-regulated establishments are defined as homes and apartments not legalised as lodgings for business purposes, and therefore reliable statistics in this regard do not exist. For this reason, in order to assess their importance within the tourism sector, we need to resort to indicators that relate to their effective use, such as the overnight stays in non-regulated accommodation by Spanish and foreign tourists. However, we do not have disaggregated data at local level, which obliges us to make *ad hoc* estimations. The data for the three Galician provinces with a coastline is as shown in the Table 6.

NW coastal provinces	Regulated	Non-regulated	Total	% Non-regulated	% prov./GAL
A Coruña	3,857,573	11,267,254	15,124,827	74.5%	27.8%
Lugo	925,826	10,032,247	10,958,072	91.5%	20.1%
Pontevedra	3,798,427	12,425,654	16,224,081	76.5%	29.8%
Galicia (GAL)	9,191,098	45,099,554	54,290,652	83.0%	3.4%
Spain	344,022,030	1,228,798,432	1,572,820,462	78.1%	

Source: Compiled by the authors based on data from *Anuario económico de La Caixa.*

Table 6. Regulated and non-regulated overnight stays in the Galician coastal provinces.

As we can see in the table, overnight stays in non-regulated establishments account for more than three-quarters of the total in Galicia. Such a volume of non-regulated tourism business stands at a slightly higher rate than the average for Spain (5% more), making it especially appropriate to evaluate the real behaviour (under the baseline scenario) or potential behaviour (under a climate change scenario) from the point of view of tourism demand (in our case, the source market of Madrid), as to do so from the point of view of regulated supply[2] would involve only looking at a minor part of tourism uses associated with the use of the coast. On the other hand, overnight stays in coastal areas are especially decisive for

[2] By its very nature, it would be difficult for non-regulated supply to be the subject of a statistical analysis in fieldwork that could be considered rigorous.

the region as a whole, as the main reason for approximately 44% of domestic trips is to visit and enjoy its beaches (Familitur, 2006).

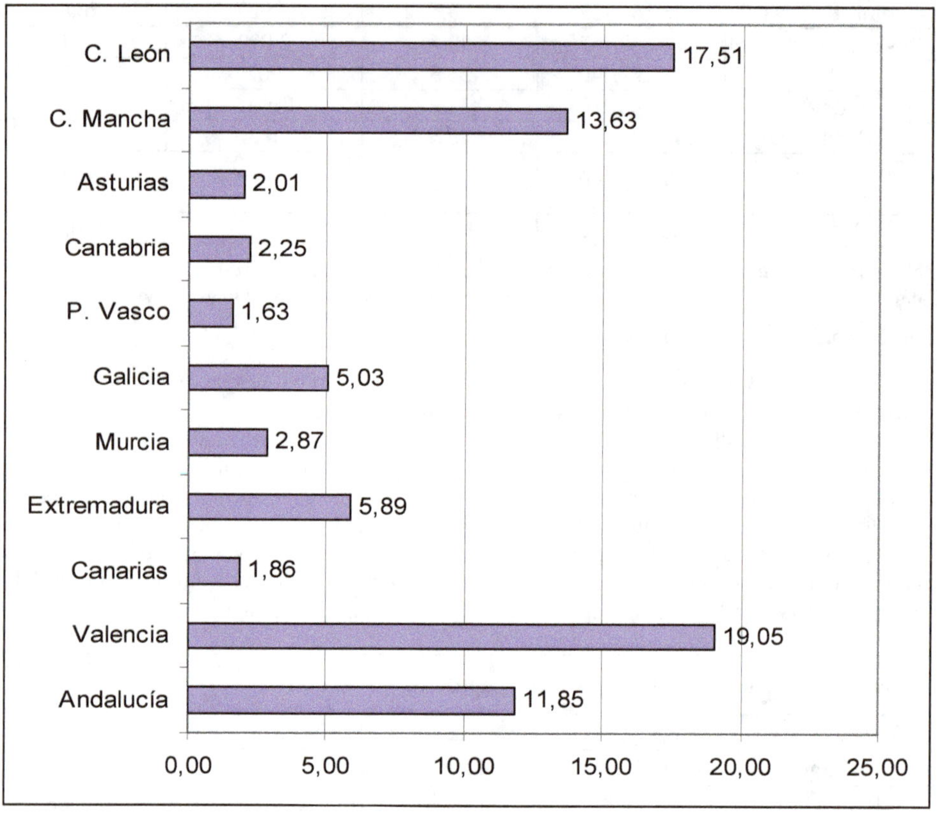

Source: Compiled by the authors based on data from Familitur, 2006.

Fig. 3. Distribution of tourism destinations visited by Madrid residents. Overnight stays with domestic destinations in Spain.

With regard to domestic tourism, the case of Madrid is notable. Therefore, as the main tourism feeder market for Galicia and in relation with the eventual effects of climate change on destination preferences, it is worth noting the following distribution under the baseline scenario (figure 3), in order to contextualise that 5% which translates into 6.5 million overnight stays in Galicia by Madrid residents.

And in the case of foreign tourism, in the minority in the north-west of Spain, it makes up 2.1% of the Spanish total. This fact implies the existence of a potential margin for other currently dominant destinations but concentrated in the south (Andalusia with 14.5%, the Canary Islands with 16.4% or Valencia with 9.2%) to move their preferences towards the north of Spain under a scenario we will evaluate herein. We believe this could be feasible in feeder markets from central and northern Europe, from where tourists mainly travel to spend their holidays (78.8%), usually in the summer.

3.2 Method

The methodology used to analyse changes in tourists' preferences towards coastal destinations under a climate change scenario is based on the economic valuation of stated preferences developed within the scope of environmental economics (Bateman *et al.*, 2002). Economic valuation with stated preference techniques is applied to projects or measures which cause changes to social welfare, and its aim can be to quantify in economic terms the magnitude of such impacts or forecast modifications in the behaviour of individuals in certain simulated scenarios, to quantify such changes and subsequently convert them into economic dimensions. In this study, we will adopt this latter approach within the stated-preference techniques. The method used is a variation of the so-called contingent behaviour analysis. This method, as with the other techniques based on stated preferences, is based on a questionnaire in which a representative sample of the target population takes part. The questionnaire is defined following internationally-accepted protocols and it attempts to place the individual in a hypothetical future situation in which he will have to decide what his behaviour as a consumer would be or, implicitly, how his consumption behaviour would vary in quantity and type in this new situation.

In order to carry out the estimation, we drew up a questionnaire to be completed by Madrid residents who might have visited a Spanish coastal destination in the last year (June 2007-June 2008). The questionnaire was divided up into four sections:

1. The first section aims to find out how many visits were made to the north-west of Spain during the reference period, in order to obtain specific data on each visit: the month in which it took place, the type of accommodation availed of, the approximate number of nights the visit lasted, the means of transport used to go to Galicia and the main reason for the trip. With sole regard to visits to the coast, the respondent was asked to point out the trips he thought were especially agreeable and satisfactory as well as those which did not live up to his expectations, and to specify the reasons why in both cases.
2. In the second section a similar process was undertaken, but relating to visits made to the Spanish coast (excluding Galicia), requesting specific data on each visit and for respondents to point out those that were particularly agreeable and disagreeable, as well as the reasons why.
3. Having completed the analysis of current tourist behaviour, the third section covers the importance of climate variables when deciding a holiday destination. Below we describe the hypothetical future scenario (in a context of climate change) which is accompanied by the corresponding contingent behaviour questions. It begins with a general line of questioning on how the respondent thinks tourist trips to Galicia and the number of overnight stays might change in the new context, if he personally would modify his trips to Galicia and the number of overnight stays on the coast. Finally, a question was posed which began by reminding the respondent of how many overnight stays he had made in the last year on the Galician coast and the rest of the Spanish coast, the aim of which was to quantify the behavioural change. He was then asked to estimate what trips he would make in the future bearing in mind the changes described. Respondents were also asked if they had a second home. Those who did were asked where, and if they would change the location of the second home in the new context. If they did not have a second home, they were asked where they would like to have one in such a future situation.

4. The aim of the fourth and final section of the questionnaire was to find out the socio-economic characteristics of respondents, such as their age, academic background, career situation, income, etc.

The text depicting the hypothetical future scenario described for the respondents to determine their contingent behaviour was as follows:

Access by road to the Spanish coast, including the north-west, has improved considerably, and in the coming years new high-speed rail links will make it possible to travel from Madrid to Galicia in approximately two hours.

Furthermore, experts consider that the following climate changes are already occurring:

- Precipitation will be concentrated in the autumn and will be lower throughout the rest of the year.
- Temperatures will undergo a general increase, especially in spring and summer. Furthermore, in summer and autumn the nights will be warmer.
- The sea temperature will rise.
- The sea level will rise.

With regard to Madrid, as well as the Mediterranean coast and the islands, being inland and further to the south, respectively, these changes will be more intensive than in Galicia".

This scenario was defined on the basis of prior information provided by biologists and other scientists. We chose to show qualitative information in order to facilitate the understanding of the information as well as the simplicity and brevity necessary for a telephone questionnaire.

With this methodology the individual has to choose between two alternatives: acquiring the product (tourism) with the usual characteristics, z_0 (in the current climate scenario); and acquiring the product with an additional attribute (the new climate scenario), at a specific cost for respondents (option z_1). We assume that the researcher does not know the true preferences of the respondents; therefore the social welfare function will be defined by the following expression:

$$V(z_j) = v(z_j) + \varepsilon_j, \qquad j = 0,1$$

where ε is the error term or the part not observed by the researcher, and it is considered to be a identically distributed independent random variable with mean zero. Through the valuation exercise, the researcher offers the respondent the possibility of acquiring products at a price A. The researcher can only assume that the reply is a random variable with a probability function,

$$\begin{aligned} \Pr\ (if\ /\ A) &= \Pr\ \{V(z_1) > V(z_0)\} \\ &= \Pr\ \{v(z_1) + \varepsilon_1 > v(z_0) + \varepsilon_0\} \\ &= \Pr\ \{v(z_1) - v(z_0) > (\varepsilon_0 - \varepsilon_{1i})\} \\ &= \Pr\{\Delta v > (\varepsilon_0 - \varepsilon_{1i})\} \end{aligned} \tag{9}$$

Where *Pr(if/A)* is the probability the respondent will accept the exchange at the proposed price and *P(not/A)=1- P (if/A)*. $\eta=\varepsilon_0-\varepsilon_1$ and $F\eta$ will be its cumulative distribution function. If this distribution function is specified as logistic, we are facing a logit model,

$$\Pr(if / A) = F_\eta(\Delta v) = \frac{1}{1+e^{-\Delta v}} \tag{10}$$

whereby, if we also assume that the utility function is linear, the utility differential can be expressed as follows:

$$\Delta v = \alpha + \gamma A, \qquad \alpha = \alpha_0 - \alpha \tag{11}$$

where γ is the coefficient associated with the price attribute. By substituting in the probability expression, the following is obtained:

$$\Pr(if / A) = F_\eta(\Delta v) = \frac{1}{1+e^{-(\alpha+\gamma A_i)}} \tag{12}$$

the resulting regression equation will be

$$\ln\left(\frac{\Pr(if / A)}{1-\Pr(if / A)}\right) = \alpha + \gamma A \tag{13}$$

On the other hand, if the individual is prepared to pay the quantity *A*, then the willingness to pay (*D*) is greater than *A*. Formally,

$$F_\eta(\Delta v) = \Pr(\Delta v > \eta) = \Pr(D > A) = 1 - G_D(A) \tag{14}$$

where $G_D(A)$ is the cumulative distribution of the individual's willingness to pay. That is, the probability that the individual's social welfare will be greater if he accepts the exchange is directly related to the probability that his true willingness to pay is higher than the price.

3.3 Results

The questionnaires designed in order to analyse tourist preferences and their changes under a climate change scenario were carried out by professional pollsters in the month of July 2008 at Madrid City Council. The questionnaires were administered over the telephone and the numbers were dialled at random with quotas in accordance with age. 1495 people were contacted, of which 745 agreed to complete the questionnaire, which is a response rate of 49.8%, usual in this type of study. Of the 745 people who agreed to take part, 131 (17.6%) had visited Galicia in the last year (17.6%). (Therefore, 614 or 82.4% had not visited Galicia). This percentage exceeds to some degree the data provided by Familitur on the source Madrid and destination Galicia put at 5.03% on the baseline scenario. 73 people declared themselves to be visitors to the Galician coast (55.7% of those who had visited Galicia); 38 were inland visitors (29%) and 20 had visited both the coast and an inland region (15.3%). Therefore, 93 people (71% of those who had visited Galicia) had visited the Galician coast. From the initial sample 396 (53.2%) had visited the Spanish coast (excluding the Galician coast) during the last year (349 or 43.8% had not visited it). Considering the initial sample, we can see that 349 people had not visited either the Galician or the Spanish coast (46.84%);

59 had visited both (8%); 34 only the Galician coast (4.5%) and 337 only the Spanish coast (45.23%).

	First trips		Second trips	
	Frequency	Percentage	Frequency	Percentage
Work	8	7.0		
To visit family and/or friends	18	15.7	4	33.3
To enjoy the beaches	13	11.3	1	8.3
To go walking or trekking	6	5.2	0	0
Mountain sports	1	0.9	0	0
To visit historical monuments	1	0.9	1	8.3
To visit charming towns and villages	10	8.7	1	8.3
To enjoy the gastronomy	10	8.7	1	8.3
Others	48	41.7	1	33.3
Total	115	100.0	12	100.0

Table 7. Reason for the trip (to the Galician coast).

The sample size finally used for the statistical analysis is obtained after applying a sample filter, and the useful sample is constituted by integrating the visitors to the Spanish or Galician coast in the last year. With the selection, the final sample is made up of 430 people (who visited either the Galician or Spanish coast). We excluded those who visited inland Galicia and did not visit the coast (16 people are not current users of the coast). The end distribution of the sample according to their use of the coast remains as follows: of the 430 visits to the Spanish coast (excluding the Galician coast), 78.37% only visited the Spanish coast, 13.72% visited the Spanish as well as the Galician coast, and 7.9% visited the Galician coast but not the Spanish coast.

Table 7 shows the main reasons for the trips made to Galicia in the last year. 41% did not choose any of the reasons specified and gave a combination of reasons which were equally important for their decisions. It can be seen that visits to family and friends are the principle cause for those who chose some main reason (15.7% of the sample), the second (11.3%) being to enjoy the beaches and the third (8.7%) for its gastronomy.

When the reasons for especially satisfactory trip or trips are analysed, the main reason most of the respondents underlined (table 8) was the green landscape and woodland (46.51%), followed by the agreeable climate (11.63%). Secondary reasons included firstly the gastronomy (39%), followed by the green landscape and woodland (19.4%) and the climate (6%). Of those who state a third reason, 32% specify the amicability of the people and 13% the climate.

As we can see, the climate features in the three most-mentioned reasons, be it in first, second or third place. The people who mentioned more than one reason were asked what they considered the most important to be, and in the distribution of responses 40% replied the green landscape and woodland, 16% the gastronomy, 15% the amicability of the people, 12% the climate and 10% the quiet and non-overcrowding. In the case of trips to Galicia, only five people mentioned a trip that did not meet their expectations, four for reasons relating to the poor quality of the accommodation and one for the unpleasantness of the people.

In relation with trips to the Spanish coast, when the respondents were asked to specify the main reason for their trip (table 9), and in the same way as in the case of trips to Galicia, a significant number pointed to a combination of factors all of which were equally important. Of the individuals who stated a specific reason, beach-based recreation was the most frequent reason (with 28% of first trips and 20% of second ones), followed by visits to family or friends (14% and 18%, respectively).

	Reason 1		Reason 2		Reason 3	
	Frequency	%	Frequency	%	Frequency	%
Green landscape and woodland	40	46.51	13	19.40	3	6.38
Gastronomy	7	8.14	26	38.81	5	10.64
Amicability of the people	7	8.14	7	10.45	15	31.91
Pleasant climate	10	11.63	4	5.97	6	12.77
Cultural and historical heritage	2	2.33	3	4.48	2	4.26
Quality and proximity of the beaches	2	2.33	3	4.48	1	2.13
Quiet and non-overcrowding	4	4.65	3	4.48	4	8.51
Other	14	16.28	8	11.94	11	23.40
Total	86	100.00	67	100.00	47	100.00

Table 8. Reasons why the trip (to the Galician coastline) was especially pleasant or satisfactory.

	First trips		Second trips	
	Frequency	Percentage	Frequency	Percentage
Work	14	3.5	7	6.7
To visit family and/or friends	56	14.1	19	18.3
To enjoy the beaches	112	28.3	21	20.2
To go walking or trekking	25	6.3	8	7.7
Mountain sports	1	0.3	1	1.0
To visit historical monuments	3	0.8	0	0
To visit charming towns and villages	11	2.8	4	3.8
To enjoy the gastronomy	15	3.8	1	1.0
Others	159	40.2	43	41.3
Total	396	100.0	104	100.0

Table 9. Reason for the trip (to the Spanish coast).

In the total sample, 306 people were of the opinion that their trips were especially satisfactory, 4 that none was satisfactory and 86 mentioned an especially pleasant trip. In total, 392 had at least one especially satisfactory trip to the Spanish coast. Table 10 shows the main reasons for their opinions. Of the reasons they mention in first place, landscape is the most frequent (32.4% of the sample), followed by climate (21%). Of those mentioned in second place, the most important are gastronomy and climate (25% and 12%, respectively), and those who give a third reason point to the amicability of the people (21%) and the climate (14%). As we can see, the climate, in this case, is also quoted as one of the first three reasons relating to the satisfactory nature of the trip.

	Reason 1		**Reason 2**		**Reason 3**	
	Frequency	**%**	**Frequency**	**%**	**Frequency**	**%**
	127	32.4	26	11.02	14	9.59
Gastronomy	15	3.8	60	25.42	13	8.90
Amicability of the people	36	9.2	37	15.68	31	21.23
Pleasant climate	83	21.2	28	11.86	21	14.38
Historical and cultural heritage	4	1.0	9	3.81	5	3.42
Quality and proximity of the beaches	25	6.4	26	11.02	17	11.64
Quiet and non-overcrowding	37	9.4	22	9.32	16	10.96
Others	65	16.6	28	11.86	29	19.86
Total	392	100.0	236	100.00	146	100.00

Table 10. Reasons why the trip (to the Spanish coast) was especially pleasant or satisfactory.

	Positively		**Negatively**	
	Frequency	**Percentage**	**Frequency**	**Percentage**
Mild temperatures	22	61.1%	16	13.8%
Rainfall	7	19.4%	81	69.8%
Wind	1	2.8%	2	1.7%
Cloud/fog	0	0%	3	2.6%
Colder sea temperature	4	11.1%	12	10.3%
Waves	0	0	0	0
Snow	0	0	0	0
Others	2	5.6%	2	1.7%
Total	36	100.0	116	100.0

Table 11. Influence of climate on your decision to visit Galicia.

In relation with the contingent behavior analysis, when questioned whether the climate influenced their decision to visit Galicia or not, 35% of the respondents (152 individuals) answered that it did, 60% (258) that it did not and 4.7% (20) did not reply. Of those who answered in the affirmative, 23.7% (36 replies) indicated that the climate influenced them positively, and 76.3% (116) that it influenced them negatively. The climate variable that has the most positive influence is the mild temperature, mentioned by 61% of respondents. The most important negative climate variable is the risk of rain (table 11).

Once the expected changes for the years to come have been described, both with regard to improvements in infrastructures as well climate (in Galicia and Madrid), the respondents were asked if they thought that under this scenario tourist visits to Galicia would increase or decrease. 67% of those who visited the coast (287 people) considered that the affluence of tourists to Galicia would increase; 18% (79 people) thought that it would not change; and only 3% thought that visits to Galicia would decrease in the new context; 12% pointed out that they could not venture to make such a prediction.

Under the new scenario and when asked whether Galicia would be more attractive to them as a destination, they stated the following. For approximately 50%, Galicia's attractiveness as a destination would increase in the new situation; 36% would not modify their current opinion and, finally, 5% would consider Galicia as a less attractive destination. 9% did not venture an opinion on this point.

When asked whether their overnight stays on the Galician coast would increase or decrease, the replies were distributed in the following manner: 48% stated that under the new scenario they would make more overnight stays on the Galician coast; approximately 35% said that they would maintain their current number of stays; and only 4% said that they would make less overnight stays. 13% stated no opinion on this point.

The respondents were then asked to quantify changes in overnight stays. We will begin by analysing the variation in total overnight stays on the Spanish coast (including the Galician coast) to subsequently describe how future overnight stays are divided up between the Galician coast and the remaining Spanish coast.

We start off from the information obtained from the sample itself which indicates that 430 of the respondents visited the Spanish coast (including the Galician coast) and made 5339 overnight stays. Of the 428 respondents who answered the question on quantifying the changes, 76 (17.76%) would reduce their overnight stays on the coast under the new scenario, 157 (36.68%) would maintain the same number and 195 (or 45.46%) would increase their overnight stays.

Having presented the variations in overnight stays, it is deduced that they would increase by 2384 and drop by 652, with a final positive result in the form of a net increase of 1732 overnight stays (per year). The number of overnight stays would rise to 7071 (as opposed to the current 5339), with an increase in percentage terms of 32.44.

However, these global figures hide a very different situation between the changes observed in demand if we consider the Spanish coast and focus on the Galician coast. Beginning with the Spanish coast, currently 337 of the people polled visited the Spanish coast (excluding Galicia) and made 4421 overnight stays in the last year. In this case, of the 428 people who answered the question on changes, 161 (37.61%) would maintain their overnight stays, 153 (35.74%) would reduce them and 82 (19.16%) would increase them.

When presenting the variation in overnight stays, it can be observed that they would increase by 725 and drop by 1194, with a negative net result of -469 overnight stays in a year. The prediction, therefore, would be that the number of overnight stays per year in the sample on the Spanish coast, excluding the Galician coast, would drop from 4421 to 3952, with an approximate decrease in percentage terms of 11.

Analysing the variations in visits to the Galician coast, the data from the sample shows that in the last year, 93 people visited the Galician coast and made 918 overnight stays. Of the 428 people who answered the question on the quantification of changes, 22 (5.14%) would reduce their overnight stays on the Galician coast in the new situation, 139 (32.48%) would maintain them and 267 (62.38%) would increase the number of overnight stays they make.

Overnight stays would increase by 2226 and fall by 207, with a positive final result in the form of a net increase of 2019 overnight stays (per year). The number of overnight stays would rise to 2937, therefore (as opposed to 918 at present), with an approximate increase of 220%.

We will now describe the changes in preferences associated with a second home in a future climate change context. In the final sample, 146 people (34%) currently have a second home, a figure which closely resembles that obtained in the baseline scenario for Madrid (36.9% according to Familitur). Of these, 1.9% (8 individuals) has a second home in Galicia; 31.6% (136 people) on the Spanish coast and 2 people have a second home in another country. Only 17 people (11.6%) would change their residence under the new scenario, which indicates more rigid tourist behavior associated with second homes. The places to which they would change are presented in table 12, with a greater preference for Andalusia (23.53%), followed by Galicia and the Autonomous Community of Valencia (both with 17.65%).

Of the 284 individuals from the sample who do not currently have a second residence (66% of visitors to the coast), 33% would not know where to buy a second home if they had the opportunity to do so (79 people), 39% (111 people) would prefer to buy it on the Mediterranean, 22.5% in northern Spain (64 people), 4% in inland Spain (12 people) and 4 people (1.4%) chose other places (other countries).

	Frequency	Percentage
Asturias	1	5.88
Andalusia	4	23.53
Galicia	3	17.65
Avila	2	11.76
A.C. of Valencia	3	17.65
North	1	5.88
The Coast	1	5.88
Zamora	1	5.88
Ibiza	1	5.88
Total	17	100.00

Table 12. Place to which they would change their second home under the new climate scenario.

Lastly, in relation with the characteristics of the sample, 405 people were in their habitual residence, 94.2% of the sample. 18 was the average number of years that they had been living in that residence. The average age of those polled was 45 (with a minimum of 25 and a maximum of 75), which is representative of the target population. 63% of the sample was women. This percentage is higher than that of the population because in this case quotas per sex were not applied given that the interest lay in the questionnaire being answered by a family member who made the decision on holiday destinations. The average number of family members was 3.13 (median 4). In 65% of the homes that participated in the sample there were no under-19s, in 17% there was one and in 15% there were two. There were more than two in the remaining 10%. In 79% of the homes there were no over-65s, there was one in 14% and two in 6%. Only in 0.9% were there more than two over-65s. Only 4.2% stated they had a family member born in Galicia. 96%had heating in their homes (62% had natural gas) and 61% had air conditioning. 86% of respondents had a car and, of these, 98% had heating-ventilation-air conditioning. 138 personas (32.2% of respondents) had studied to high-school level and 227 people (53%) affirmed that they were the main breadwinner in the family.

4. Conclusions

In this chapter we have analysed the effects of climate change on the anchovy fishery and tourism flows in the north-west of Spain. To do so, we have taken into consideration possible future scenarios involving sea surface temperature increases and possible coastal and climate modifications in the region.

In relation with the effects on the anchovy fishery, we have used a bio-economic model applied to this fishery, which has allowed us to gauge the environmental effects as well as natural and economic factors. Climate change was gauged on the basis of sea surface temperature. The results obtained indicate a reduction both in future fish biomass as well as catch levels. This would be so both if the trend forecast for rises in sea temperatures follows past trends (foreseeable according to oceanographers; Roson, 2008) as well as if an even greater warming occurred as a result of global warming.

Given the poor situation of anchovy biomass as a consequence of low recruitment in recent years, the European Commission implemented a moratorium in the fishery for the period 2005-2009, both inclusive. Consequently, we assumed as a baseline scenario that which corresponded to the last year of fishing activity before the moratorium, that is, 2004. As a summery, table 13 shows the net profits that correspond to the baseline scenario and those that would be obtained on the 2040 time horizon, and for both scenarios with regard to sea surface temperature oscillation (this horizon corresponds to the intermediate period used in the EC Peseta Report).

Scenarios	Value (x10^6 Euros)
Baseline scenario	23,721
Climate scenario (1)	14,947
Climate scenario (2)	14,795
Variation rate for scenario (1)	-1.274 %
Variation rate for scenario (2)	-1.303 %

Table 13. Summary of estimated losses in the anchovy fishery (NW zone).

Therefore, the effects of sea surface temperature increases, assuming that the rest of the parameters remain constant, generate a reduction in net profits of 1.27% and 1.30%, respectively, if the sea temperature increases according to the trend over recent years, or if it increases at a greater rate than that observed to date.

With respect to tourist activity, a questionnaire was drawn up to reveal the tourist behavior, present and future, of the main tourism market in north-western Spain (Madrid). The questionnaire was applied over the telephone and a useful sample (of visitors to the Spanish coast over the previous year) of 430 people was obtained. Of these, 78% had visited the Spanish coast in that period, 8% had visited the Galician coast and 14% had visited both. Of those who had visited Galicia, more than half had made their trips in the summer, at an average of 10 overnight stays per visitor per year. Visitors to Galicia stayed mainly in hotels (42%) and rented apartments or flats (18%). They had travelled mainly in their own car (75%) and the main reason for the trip was to holiday, with a mixture of aims (beach, gastronomy, etc.). In general, they stated they were very satisfied with their trips, mainly motivated by the landscape, climate being among the three main reasons mentioned as agreeable factors.

Visitors to the Spanish coast spent an average of 11 nights per year and also made their trips mainly in the summer (62%). 33% stayed in hotels, followed by 25% who stayed in rented apartments or flats. Once again, their own cars were the main means of travel from Madrid to the coast (72% used their own car to travel) and the reason for the trip was also to holiday, in general. The respondents also affirmed that they were highly satisfied with their trips, the main reasons for this being the landscape and the climate, although 10% stated that one of the trips did not live up to their expectations, the reasons being the quality of the accommodation, the climate or noise and congestion.

It can be observed that the climate appears as a key reason to satisfaction or dissatisfaction with trips throughout the analysis. When asked about the extent to which the climate influences their decisions to visit the north-west, 35% admit that this variable did indeed influence their decisions. For 76% of those who admitted to being influenced by the climate, it was negative or off-putting, mainly due to the risk of rain, low temperatures and the low temperature of the sea.

With regard to the future climate change scenario, both in the north-west as well as the source (Madrid), and in the rest of the Spanish coast, most of the respondents (67%) were of the opinion that the tourism flow to the north-west would increase in the new context. For 50%, the attractiveness of the region would increase, and 48% affirm that they would increase their number of overnight stays in the region. The quantification of this increase in overnight stays indicates that only 5.4% would reduce their current number of stays, 32% would maintain them and 62% would increase them. The net increase would be +2.019 overnight stays (per year), which would involve a 220% increase with regard to the current number of stays in the sample analysed. This increase would occur to the detriment of trips to other areas of the Spanish coast (not the north-west). Thus, 38% of visitors to other parts of the Spanish coast would maintain the number of overnight stays in those destinations, 19% would increase them and 36% would reduce them. The total number of overnight stays would undergo a net reduction of approximately 11% with respect to the situation as it stands at present.

If we assume that tourist demand for the north-west of Spain originating in the rest of Spain adapts to the preferences of the source market of Madrid for the climate change scenario proposed, the consequences could be simulated (table 14) in two alternative scenarios (minimum with a 50% increase; and 100% or duplication maximum). Therefore, we would have the following differential impact on the GDP and in respect of the baseline scenario described in various paragraphs above:

Scenarios	Value (x10^6 Euros)
Baseline scenario	907.00
Minimum scenario (1)	1,360.00
Maximum scenario (2)	1,814.00
Effect on the GDP (1)	+1%
Effect on the GDP (2)	+2%

Table 14. Summary of estimated gains in tourism (NW zone).

To sum up, and considering only these two economic activities, the effect of climate change on the north-west of Spain would generate estimated gains of approximately 1300 million euros on account of the significant importance of the tourist sector in the region and the foreseeably greater number of tourists who would visit it from other regions which would be harmed by the temperature increase forecast (the South and East of Spain).

In summary, in this chapter we assess the possible economic effects of global warming on some of the main economic activities in north-western Spain: the anchovy fishery and tourist activity. With regard to fishing activity, the results showed that if the sea surface temperature trend in the Ibero-Atlantic fishing grounds continues to show warming, both the biomass and the expected profits will drop. With relation to tourist activity, the results showed that the demand for trips and overnight stays would increase in the new climate scenario and the north-west, which would become a more attractive destination for current and potential visitors. The increase in demand would come about as a result of a drop in demand for current destinations as well as a net increase in the number of tourist trips regarding the market analysed.

5. References

ACIA (2004): *Artic climate impacts assessment: Overview Repo*rt. Cambridge University Press, Cambridge.

Arnason, R. (2005): Climate change and fisheries: assessing the economic impact in Iceland and Greenland. *Workshop on economic effects of climate change on fisheries*, Bergen.

Arnason, R. (2006): Global warming, small pelagic fisheries and risk. En R. Hannesson, M. Barange, S. Herrick (Eds): *Climate change and the economics of the world's fisheries.* Edward Elgar Publishing.

Bateman, I. J.; Carson, R. T.; Day, B.; Hanemann, M.; Hanley, N.; Hett, T.; Jones-Le, M.; Loomes, G.; Mourato, S.; Ozdemiroglu, E.; Pearce, D. W.; Sugden, R.; Swanson, J. (2002), *Economic Valuation with Stated Preference Techniques: A Manual*. Edwar Elgar, Cheltenham, UK.

Bigano, A.; Hamilton, J. M.; Lau, M.; Tol, R. S. J. e Zhou, Y. (2004). *A Global Database of Domestic and International Tourist Numbers at National and Subnational Level.* Fondazione Eni R. Mattei. Working Papers, n 305, Milano, Italy.

Bigano, A.; Hamilton, J. M.; Maddison, D. J. e Tol, R. S. J. (2006). Predicting tourism flows under climate change. *Climatic Change*, 79, 175-180.

Briones, R., Garces, L., Ahmed, M. (2006): Climate change and small pelagic in developing Asia: the economic impact on fish producers and consumers. En R. Hannesson, M. Barange, S. Herrick (Eds): *Climate change and the economics of the world's fisheries.* Edward Elgar Publishing.

Coenen & Van Eekeren. (2003). A study of demand for domestic tourism by Swedish households using a two-staged budgeting model. *Scandinavian Journal of Hospitality and Tourism*, 3(2), 114-133.

EEA (2007). *Climate change: the cost of inaction and the cost of adaptation.* European Environmental Agency Technical Report 13/2007.

Eide, A. (2005): Economic impactas of global warming. The case of the Barents Sea fisheries. *Workshop on economic effects of climate change on fisheries*, Bergen.

European Commission (2006): Economic Performance of important segments fleets. Brussels.

European Commission (2007). *Limiting Global Climate Change to 2 degrees Celsius. The way ahead for 2020 and beyond.* Communication from the Commission to the Council, the European Parliament, the European Economic and Social Committee and the Committee of the Regions.

Evans *et al.* (2000). *Planning for leisure and tourism.* Report prepared to DETR Planning Research Programme, DETR, London.

Gallagher, Ch. (2005): Variable abundance and fishery movements in New Zeland squid fisheries: Preliminary findings from global and regional investigations. *Workshop on economic effects of climate change on fisheries*, Bergen.

Garza-Gil, M.D., Torralba-Cano, J., Varela-Lafuente, M. (2011): Evaluating the economic effects of climate change on the European sardine fishery. *Regional Environmental Change* 11, 87-95.

Gómez Martín, M. B. (2005). Weather, Climate and Tourism: a geographical perspective. *Annals of Tourism Research*, 32(3), 571-591.

Hamilton, J. M.; Maddison, D. J. e Tol, R. S. J. (2005a). Climate Change and International tourism: a simulation study. *Global Environmental Change*, 15, 253-266.

Hamilton, J. M.; Maddison, D. J. e Tol, R. S. J. (2005b). The effects of Climate Change on international tourism. *Climate Research*, 29, 255-268.

Hannesson, R. (2006): Sharing the herring: fish migrations, strategic advantage and climate change. En R. Hannesson, M. Barange, S. Herrick (Eds): *Climate change and the economics of the world's fisheries.* Edward Elgar Publishing.

Herrick, S., Hill, K., Reiss, C. (2006): An optimal harvest policy for the recently renewed United States Pacific sardine fishery. En R. Hannesson, M. Barange, S. Herrick (Eds): *Climate change and the economics of the world's fisheries.* Edward Elgar Publishing.

Ibermix (2007). *Identification and segmentation of mixed-species fisheries operating in the Atlantic Iberian Peninsula waters.* EC Project, FISH/2004/03-33.

ICES (2008). *Report of the Working Group on the Assessment of Mackerel, Horse Mackerel, Sardine and Anchovy.* Headquarters, ICES CM 2007/ACFM; 31.

IET (2007). *Familitur-Comunidad de Madrid, Año 2006.* www.iet.tourspain.es

IET (2007). *Movimientos turísticos de los españoles (Familitur) 2006.* www.iet.tourspain.es

IET (2007). *Movimientos turísticos en fronteras (Frontur) 2006.* www.iet.tourspain.es

IGE (2010): *A economía galega a través do Marco Input-Output de Galicia 2005.* Instituto Gallego de Estatística, Santiago de Compostela.

IPCC (2001): *Climate Change 2001: impacts, adaptation, and vulnerability. Contribution of Working Group II to the third assessment report of the Intergovernmental Panel on Climate Change.* New York, USA: Cambridge University Press.

IPCC (2007): *4th Report of International Panel for Climate Change*. www.ipcc.ch.

Kamien, M. & Schwartz, N. (1991). *Dynamic Optimization. The Calculus of Variations and Optimal Control in Economics and Management*. New York: North-Holland Ed.

Levitus, S., Antonov, J.I., Boyer, T.P., Stephens, C. (2000): Warming of the world ocean. *Sciences*, 287, 2225-2229.

Limb, M. & Spellman, G. (2001). *Evaluating domestic tourists' attitudes to British weather. A qualitative approach*. Proceedings of the First International Workshop on climate, tourism and recreation. Halkidiki, Greece.

Lise, W. & Tol, R. S. J. (2002). Impact of Climate Change on Tourist Demand. *Climatic Change*, 55, 429-449.

Maddison, D. (2001). In search of warmer climates?. The impact of Climate Change on Flows of British Tourists. *Climatic Change*, 49, 103-208.

MAPA (2008). *La Agricultura, la Pesca y la Alimentación en España, Informe Anual*. Ministerio de Agricultura, Pesca y Alimentación, Madrid.

McGowan, J.A., Cayan, D.R., Dorman, L.M. (1998): Climate-ocean variability and ecosystem response in North Pacific. *Science*, 281, 201-217.

Ministerio de Medio Ambiente (2005). *Principales conclusiones de la evaluación preliminar de los impactos en España por efecto del cambio climático*. Oficina Española de Cambio Climático (OECC), Ministerio de Medio Ambiente.

Ministerio de Medio Ambiente (2006): *Plan Nacional de Adaptación al Cambio Climático*, Madrid.

Naranjo, L. & Pérez, V. (2006). *A variabilidade natural do clima en Galicia*. Fundación CaixaGalicia e Xunta de Galicia.

Perry, A. (2001). More heat and drought.Can Mediterranean tourism survive and prosper? In A., Matzarakis and C.R. de Freitas, eds., *Procedings of the First International Workshop on Climate, Tourism and Recreation*, International Society of Biometereology, Comission on Climate Tourism and Recreation, pp. 35-40.

Röckmann, C. (2005): Rebuilding the Eastern Baltic cod stock under environmental change. *Workshop on economic effects of climate change on fisheries*, Bergen.

Rosón, G. (2008): Índices climáticos y su impacto en la hidrografía y dinámica marina, Workshop *Evidencias del cambio climático en Galicia*, Santiago de Compostela.

Santos, F. D. & Miranda, P. (2006). *Alteraçoes climáticas em Portugal: cenarios, impactos e medidas de adaptaçao*. Gradiva, Lisboa.

Scott, D.; McBoyle, G. & Schwarzentruber, M. (2004). Climate Change and the distribution of climatic resources for tourism in North America. *Climate Research*, 27, 105-117.

Scott, D.; Wall, G. & McBoyle, G. (2005). The evolution of the climate change issue in the Tourism sector. En: Hall, C. M.; Higham, J. (Eds.). *Tourism, recreation and climate change*. Channelview, London.

Seddigi, H. R. & Shearing, D. F. (1997). The demand for tourism in North East England with special reference to Northumbria: an empirical analysis. *Tourism Management*, 18(8), 499-511.

Sissener, E.H. & Bjorndal, T. (2005): Climate change and the migratory pattern for Norwegian spring-spawning herring: implications for management. Marine Policy 29, 299-309.

Smith, K. (1990). Tourism and Climate Change. *Land Use Policy*, 40, 371-389.

Smith, K. (1993). The influence of weather and climate on recreation and tourism. *Weather*, 48, 398-404.

Stenevik, E.K., Sundby, S. (2007): Impacts of climate change on commercial fish stocks in Norwegian waters. *Marine Policy*, 31, 19-31.

Viner. D. & Agner, M. (1999). *Climate Change and its impacts on tourism*. Climatic Research Unit, University of East Anglia, Norwich, U.K.

VV.AA. (2007): *El cambio climático en España. Estado de situación*. Ministerio de la Presidencia, Madrid.

World Tourism Organization (2006). *Overview of International and European Tourism: 2005 results and short term outlook*. www.world-tourism.org.

Xunta de Galicia-Exceltur (2005). *Galicia-Impacto del turismo sobre la economía y el empleo. Estudio Impactur*. www.exceltur.org

6

Vulnerability of South American Pinnipeds Under El Niño Southern Oscillation Events

Larissa Rosa de Oliveira[1,2,3]
[1]*Laboratório de Ecologia de Mamíferos, Universidade do Vale do Rio dos Sinos (UNISINOS), São Leopoldo, RS,*
[2]*Grupo de Estudos de Mamíferos Aquáticos do Rio Grande Sul (GEMARS), Porto Alegre, RS,*
[3]*Centro para la Sostenibilidad Ambiental, Universidad Peruana Cayetano Heredia (UPCH), Miraflores, Lima 18,*
[1,2]*Brazil*
[3]*Peru*

1. Introduction

The fur seal and sea lions, as well as other marine vertebrates along the Peruvian coast, were truly the eyewitness of cyclic events of global changes in the Pacific Ocean during *El Niño* Southern Oscillation (ENSO) events. In November 1997, the Punta San Juan Reserve (15°22'S, 75°12'W) concentrated with the National Reserve of *Paracas* (13°54'S) and Punta San Bernardo (15°09'S), over 90% of the Peruvian fur seal and sea lion populations (Majluf & Trillmich, 1981). The ENSO event developed unexpectedly and rapidly early in February of that year and became more intense than anticipated, with sea surface temperatures reaching up to 9°C in some locations along the Peruvian coast (Glantz, 2001). The 1997-98 ENSO was considered the "El Niño of the twentieth century", challenging the 1982-83 event in terms of intensity, marine mortality and economic loss.

The ENSO is a climatological phenomenon characterized by anomalous conditions in the atmosphere and ocean that are mainly related to warming of the sea surface temperature (SST) from 2°C to 9°C along the coast of Ecuador and Peru (Glantz, 1996). At intervals of two to seven years (Cane, 1983), the Humboldt Current upwelling system, which is the richest in the world (Cushing, 1982; Idyll, 1973), is affected by ENSO, with increased SST and reduced primary productivity which directly influences the depth distribution and abundance of Peruvian anchovy, *Engraulis ringens*, the most important prey of the South American sea lion, *Otaria flavescens*, and South American fur seal, *Arctocephalus australis*, in Peru (Arias-Schreiber, 2003; Idyll, 1973; Majluf, 1992). The South American sea lion and South American fur seal are distributed along the Atlantic and Pacific coasts of South America (from southern Brazil to northern Peru (Capozzo & Perrin, 2008; Rosas et al., 1994; Vaz-Ferreira, 1981, 1982). Nevertheless, the distribution of *O. flavescens* is larger and continuous, while that of *A. australis* has gaps mainly in the central coast of Chile and along the Argentinean coast (Guerra and Torres, 1987; Oliveira et al., 2008; Repenning et al., 1971). In the Peruvian territory these species are distributed among breeding colonies and haul-out areas on both the continent and islands. *O. flavescens* is the most abundant pinniped species with 118,220 individuals, followed by *A. australis* with 15,317 individuals (Imarpe, 2006). Both have a polygynic breeding system with very few males mating with many females (Capozzo & Perrin, 2008; Majluf 1987). These two species of South American seals were hunted intensively for several centuries, and Uruguay became the last country to prohibit hunting in 1991 (Vaz-Ferreira &Bianco, 1998). Nowadays the most important threat to the conservation of these species are the interactions with fishery activities (mainly in the case of *O. flavescens*), and mortality during ENSO events on the Pacific coast.

Starting in February instead of the usual December, this 1997-98 ENSO hit South American fur seals and sea lions at the time when they normally replenish their reserves for the upcoming breeding season in November-December and January-February, respectively (Majluf, 1998). Most of the female sea lions gave birth prematurely in December 1997 and their pups died within hours of birth, probably as a result of a bacterial infection. At the Punta San Juan colony over 2,000 sea lion pups died in a single week. In the same area all the few fur seal pups that were born died within a few weeks. Their mothers had to spend very long periods at the sea foraging (10-20 days) and the pups died of starvation. Adult fur seals and sea lions did not die significantly until January-February 1998, when sea temperature anomalies peaked at 7-8° above normal (Majluf, 1998). As a final outcome, the Peruvian population of both species, sea lions and fur seals, declined roughly 81% and 72%, respectively (Arias-Schreiber & Rivas, 1998; Arias-Schreiber, 2000), as a result of low food availability due to the replacement of cold and nutrient rich waters of the upwelling system with warm, nutrient-deficient and low productivity waters (Majluf & Trillmich, 1981).

Throughout my field work at Peru I recovered hundreds of skulls and skeletons (Figure 1) of marine mammals in order to prepare a scientific collection for the project. All of the collected animals were a consequence of the remarkable mortality caused by this ENSO, and consisted mainly of adult fur seals of both sexes and adult female sea lions. In fact, the national census conducted by *Instituto del Mar del Perú* (IMARPE) along the Peruvian coast from 1996 to 1997 indicated that the sea lion population declined from 144,087 (Arias-Schreiber & Rivas, 1998) to 27,991 individuals in December 1998 (Imarpe, 2006) and the fur seals from 24,481 in December 1996 (Arias-Schereiber & Rivas, 1998) to 8,223 individuals in December 1999 (Arias-Schereiber, 2000). Due to the drastic population declines, the sea lions were classified as vulnerable (Decreto Supremo No. 034-2004-AG) and the fur seals as in danger of extinction along the Peruvian coast (Decreto Supremo No. 013-99-AG).

Trillmich & Ono (1991) comprehensively documented the consequences of the 1982-1983 ENSO event on many species of pinnipeds along the coasts of North, Central and South America in the book "Pinnipeds and El Niño – responses to Environmental stress." The biology of fur seals, sea lions and also elephant seals were deeply affected by this ENSO event. As a result, authors have reported an increase in pup and adult mortality, an increase in foraging trips, changes in migration movements, establishment of new breeding colonies and diminished reproductive success. Most of these conclusions were based on counting numbers before and after ENSO events.

In fact, the census population size (*N*) is usually the only information available for most threatened species. However, for evolutionary matters, the effective population size, not the census number, is the prime concern. The effective population size (*Ne*) is envisioned as the size of an ideal population that has the same rate of increase in homozygosity or gene frequency change as the actual population under consideration (Wright, 1931).

Fig. 1. South American sea lion (*Otaria flavescens*) breeding colony at Punta San Juan, Peru. **A.** during a non-ENSO period (Photo: Larissa Oliveira). **B.** during 1997-98 ENSO (Photo: Patricia Majluf).

An important application of N_e in conservation biology is the estimation of the minimum viable effective population size, particularly in cases like the Peruvian fur seals and sea lions that suffered significant declines after an environmental change due to ENSO. Adequate assessment of viability requires, in part, determining whether the population is large enough to avoid inbreeding or to maintain adaptive genetic variation (Vucetich & Waite, 1998). A population with a high N_e retains high levels of genetic diversity and reduces the probability of effects of inbreeding depression. In contrast, a population with a very low N_e is more susceptible to genetic drift and less able to respond to selection. This is because in small populations there is less genetic variation for natural selection to act upon, and there is a higher probability that beneficial alleles will not be maintained by selection and will instead be lost from the population because of random drift effects (Willi et al., 2007). Furthermore, the estimate of N_e reflects the number of individuals responsible for maintaining the genetic diversity of the species as well as its evolutionary potential. Since the goal is the conservation of species as dynamic entities capable of evolving to cope with environmental change, it is important that the species' evolutionary potential must be retained in order to respond to the current unpredictable climate change scenario (Frankham et al., 2002). The calculation of N_e numbers for the Peruvian coast fur seals and sea lions provide us with a quick insight into the ENSO effects and the short-term consequences of rapid changes in an ecosystem of large vertebrate species. It is important to mention that besides the ENSO effects, the Peruvian sea lions and fur seals also suffered local extinctions in the region due to indiscriminate commercial sealing for fur, meat and oil until 1946, when the hunting of both species of seals (*O. flavescens* and *A. australis*) was prohibited (Piazza, 1969; Tovar & Fuentes, 1984). The sealing was totally banned only in 1959 (Grimwood, 1969). Nevertheless, despite all the legal protection, seal poaching still occurs in order to supply the Asian aphrodisiac market (Lama, 2010).

Another important conservation problem that affects the Peruvian sea lions is the mortality due to interactions with fishery activities (Arias-Schreiber, 1993; Majluf et al., 2002), which have been documented throughout the species' range (Aguayo & Maturana, 1973; Dans et al., 2003; George-Nascimiento et al., 1985; Koen Alonso et al., 1999; Sepúlveda & Oliva, 2005; Szteren & Paez, 2002). The South American sea lion is an opportunistic predator that forages on the most abundant prey (Oliveira et al., 2008), including many benthic and pelagic fish that usually are economically important (Jefferson et al., 2008). As a result, many sea lions were incidentally captured or even intentionally killed by fishery and fish farming operations throughout their range (see Crespo et al., 2009 for a review).

The interactions with the fishery activities could be intensified during ENSO events and increase the mortality of sea lions during this period. In this sense the estimated N_e combined with the current ENSO events and fishery interactions are causes for great concern for the survival of the species. These concerns should be taken into account in future management plans in order to ensure the conservation and protection of this species on the Peruvian coast.

This chapter presents the estimated N_e of the Peruvian population of the South American sea lions and fur seals based on the effects of species polygyny and oscillations in population size in different generations, which includes the fluctuations caused by the most severe ENSO of the century (1997-1998). Also discussed is the importance of this value for the conservation of a population considered to be vulnerable and which faces environmental changes like ENSO events.

2. Effective population size of fur seals and sea lions

Estimated N_e is obtained by genetic (reviewed by Neigel, 1996; Nunney, 2002) and demographic methods (reviewed by Caballero, 1994; Husband & Barrett, 1992; Nunney, 1995; Oliveira et al., 2006; Traill et al., 2007, 2010). Perhaps the most important influence on N_e and a difficult factor to obtain is the temporal oscillation in population size based on long-term census (Vucetich et al., 1997). According to Nunney & Elam (1994), estimates based on data collected during a single season ignore the influence of temporal fluctuation and thus may represent gross overestimates of N_e. Vucetich & Waite (1998) highlight the importance of long-term counts in order to improve the accuracy of the estimates of N_e. Traill et al. (2010) also reinforce that conservationists working within developing nations rarely have the resources available to collect the long-term demographic and other data necessary to model viability for specific species or taxa. Fortunately, due to the systematic efforts of IMARPE for more than 25 years, results from a long-term census size are available for fur seals and sea lions along the Peruvian coast. Therefore I present and discuss data compiled and published by IMARPE staff (Arias-Schereiber, 2000; Oliveira et al., 2006 and Imarpe, 2006), taking into account the differences in the number of breeding males and females of South American fur seals and sea lions according to N_e calculations proposed by Hedrick (2000) (see below). I also comment on the consequences of this value for the conservation of bottlenecked populations.

For *O. flavescens* I used data from eight census (1984, 1993, 2000, 2002, 2003, 2004, 2005 and 2006 - IMARPE, 2006) (Table 1) and six censuses for *A. australis* (1984, 1992, 1996, 1999, 2005 and 2006 - Arias-Schereiber, 2000; IMARPE, 2006) (Table 2), both activities conducted by IMARPE including the census during 1999 (just after the 1997-1998 ENSO event, Arias-Schereiber, 2000) which covered 71 breeding colonies from South American sea lion and 46 from South American fur seal, from Los Órganos (04°10'S, 81°07'W) to Morro Sama (18°00'S, 70°53'W) and included the most important reproductive colonies of both species on the Peruvian coast (Isla Brava 11°22'S, 77°45'W, Islas Chincha 13°38'S 76°24'W, Punta Arquillo 13°54'S 76°19'W, Morro Quemado 14°20'S, 76°07'W, San Fernando 15°04'S 75°21'W, Tres Hermanas 15°26'S 75°04'W, Punta Atico 16°14'S 73°41'W and Punta Coles 17°42'S 71°23'W, Figure 2).

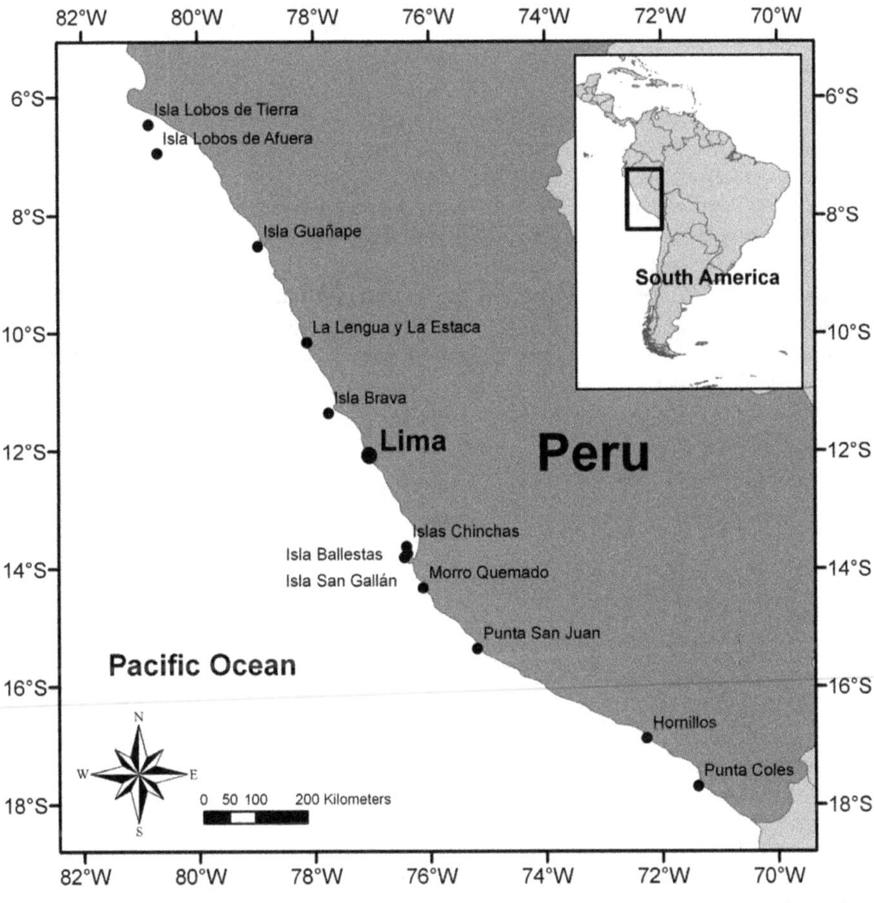

Fig. 2. Most important breeding sites and haul-out areas of South American sea lions (*Otaria flavescens*) and South American fur seals (*Arctocephalus australis*) along the Peruvian coast. These areas were censused between 1984 and 2006 by IMARPE (see text).

In order to estimates N_e we only used censuses that clearly identified adult breeding males and females (Oliveira et al., 2006). Breeding males, also called "territorial males", are considered all the males that copulate on the beach and assemble harems, while other males that are not in reproductive age and do not mate with females are counted as "subadult" males in the census (Imarpe, 2006).

The South American sea lion presents a polygynic breeding system with very few males being able to mate with many females (Capozzo & Perrin, 2008). The South American fur seal also presents a polygynic breeding system. However, it was modified from lekking to a territorial reproductive system on the Peruvian coast, following changes in population density and availability of space for male territories (Majluf 1987). In both systems, a small number of males are able to mate with many females, so the effective population size should be smaller than the actual population size (Crow and Kimura, 1970). The N_e equation that accounts for the effects of unequal sex-ratio is:

$$N_e = \frac{4N_f N_m}{N_f + N_m}$$

where N_f is the number of breeding females and Nm is the number of breeding males. In order to calculate the effect of changes in population size over time we estimated effective population sizes before and after the ENSO event for both species. To calculate the effective population size prior to the ENSO event for fur seals we used data collected for 1996-1997 by IMARPE (Arias-Schreiber & Rivas, 1998), with a census size of 24,481 individuals, among which 2,903 were reproductive males and 10,720 were reproductive females. For sea lions the census size was 76,349 individuals for the year 1993 (Imarpe, 2006), among which 6,435 were reproductive males and 45,080 were reproductive females.

The estimate of N_e for both species before and after the ENSO of 1997-1998 was then used to estimate an overall effective population size, which accounts for the variation in population size in different generations. The N_e for a population that varies in size over generations is given by the harmonic mean of the N_e in each generation (Hedrick, 2000):

$$N_e = \frac{t}{\sum 1/N_i}$$

where N_i is the effective population size in the ith generation and t is the number of generations considered.

The data obtained here was compared to the minimum viable population size (MVP), calculated by Reed et al. (2003), for 102 vertebrate species, including three pinnipeds. The MVP calculated was on average approximately 5,000 breeding age adults. The MVP can be defined as the smallest size required for a population or species to have a predetermined probability of persistence for a given length of time (Shaffer, 1981). All MVP estimates in the study of Reed et al. (2003) are for a 99% probability of persistence for 40 generations and the models for this estimate are comprehensive and include age-structure, the effect of demographic stochasticity, environmental stochasticity, and inbreeding depression (MacCarthy et al., 2001). The authors estimated MVP using three different criteria: the mean carrying capacity required for a 99% probability of persistence for 40 generations (MVPK), the minimum viable adult population size (MVPA), calculated by the software Vortex (Miller and Lacy, 1999) and the effective population size (MVPN_e). These three measures of

MVP all correlated very strongly with each other (r >0.93 in all comparisons) and the choice of measure does not qualitatively change the conclusions reached in any of the analysis.

3. Results and implications for conservation

The oscillations in census size and its respective effective population size by each year censused for *O. flavescens* and for *A. australis* are presented in tables 1 and 2, respectively. The N_e value that accounts for the effects of unequal sex-ratio for the Peruvian population of South American fur seal, *A. australis,* was 1,220 (year 1999) and the N_e prior to the 1997-98 ENSO was 9,138 for 1996. This yielded an overall effective population size of 2,153 which accounts for both the effects of the mating system and variation in population size. For South American sea lions, *O. flavescens,* the estimate N_e that accounts for the effects of unequal sex-ratio was 22,525 (year 1993) and 2,135 for the year 2000, which means that the N_e prior to the 1997-1998 ENSO was 10 times bigger than the N_e for the year 2000. Nevertheless, the overall effective population size given by the harmonic mean of the N_e in each generation was 7,715, which accounts for both the effects of the mating system and variation in population size. In all, the N_e values for the Peruvian population of *A. australis* and *O. flavescens* are 2,135 and 7,715 specimens respectively.

Year census	N	N_{em}	N_{ef}	N_e year
1984	33,816	5,887	18,872	17,948.94
1993	76,349	6,435	45,080	22,524.69
2000	48,088	558	12,323	2,135.31*
2002	62,840	3,166	29,676	11,443.18
2003	75,158	1,942	27,315	7,252.38
2004	59,399	2,665	19,508	9,378.76
2005	100,256	4,314	38,630	15,522.52
2006	118,220	3,786	40,737	13,856.23

*effective population size declined after 1997-1998 ENSO.

Table 1. Census size and effective population size for *Otaria flavescens* along the Peruvian coast.

Year census	N	N_{em}	N_{ef}	N_e year
1984	15,369	1,827	9,596	6,139.16
1992	27,219	2,557	15,670	8,793.15
1996	24,481	2,903	10,720	9,137.54
1999	8,223	3,215	337	1,220.11*
2005	14,320	716	4,725	2,487.12
2006	15,137	919	4,135	3,007.57

Table 2. Census size and effective population size for *Arctocephalus australis* along the Peruvian coast. *effective population size declined after 1997-1998 ENSO.

These results showed that even under the same conditions of ENSO and suffering similar selective pressures, both species presented highly different results, which is probably due to differences in original census size.

The South American sea lion is the most abundant pinniped species along the Peruvian coast, the last census from 2006 counted 118,220 individuals distributed on breeding colonies and haul-out areas on the continent as well as on islands (Imarpe, 2006). On the other hand, the South American fur seal population was estimated 15,317 individuals for the same year, after a fast recovery from the 6,257 remaining fur seals after 1997-98 ENSO (Imarpe, 2006). It is obvious that the census size population declined significantly for both species. However, are these remaining N_e critical for the conversation of both species along the Peruvian coast?

The calculated N_e for the Peruvian population of *A. australis* (N_e = 2,135) seems extremely large when compared with another pinniped species (14 Guadalupe fur seals (*Arctocephalus townsendi*) in 1954 (Hubbs, 1956); 20 Northern elephant seals (*Mirounga angustirostris*) in 1985 (Hedrick, 1985) and another 596 in 2003 (Reed et al., 2003); 3,249 Southern elephant seals (*M. leonina*); 76 Hawaiian monk seals (*Monachus schauinslandi*) and 2,344 grey seals, *Halichoerus grypus*, (Reed et al., 2003)). However, according to Reed et al. (2003), for 102 vertebrate species, including three pinnipeds, the minimum viable population size (MVP) calculated was on average approximately 5,000 breeding age adults. The mean N_e estimated for 102 vertebrate species was 1,752 individuals (SD = 156) (Reed et al., 2003), which is a very close value to the 2,153 individuals estimated for the Peruvian population of *A. australis*. In this 23 sense the estimated N_e for the Peruvian population of fur seal represents a critical value, because it is significantly lower than the average MVP and close to the N_e values estimated for vertebrates.

Moreover, recently Oliveira et al. (2009) found that this population decline affected significantly the N_e, causing a genetic bottleneck. A direct consequence of short periods of small N_e (bottlenecks) or continued small population sizes is typically the loss of genetic diversity. The loss of genetic diversity in a bottlenecked population is a matter of concern because decreased polymorphism may reduce the evolutionary potential of a population to respond to a changing environment (Fisher, 1930). Furthermore, numerous studies have shown a direct correlation between genetic diversity and measures of fitness (*e.g.* Leary et al., 1983; Kohen et al., 1988; Soule, 1979), mainly related to inbreeding depression. The bottleneck found in the nuclear genes of South American fur seals could be an indirect result of the synergic effect between the repetitive ENSO events and the human impact in the past (sealing and overfishing) on the Pacific population of South American fur seals.

It is important to mention that a population bottleneck does not necessarily doom a species to immediate extinction (*e.g.* Northern elephant seal - Frankham et al., 2002; LeBoeuf & Bonnell, 1980). However, the loss of genetic diversity is likely to make it more prone to extinction due to the emergence of new diseases or environmental changes. Further, the population will be partially inbred, and is likely to have reduced reproductive fitness as a consequence. An important feature of such bottleneck events is the large chance element in the outcome. Some situations will be relatively harmless if few deleterious mutations are, by chance, present in the remaining population. In other cases, populations are not so lucky; deleterious mutations are fixed and the species may decline to extinction.

The estimated N_e of 2,153 combined with the apparently current increase in frequency of ENSO events are reasons of great concern for the survival of the species on the Peruvian coast. According to Majluf (1998), the ENSO event from 1997-1998 caused a high juvenile and pup mortality, as well as a high mortality of reproductive females, due to the need to spend very long periods foraging (10-20 days), suffering subsequently from physical stress and even starvation. Although the South American fur seal population on the Pacific side of its distribution most likely is well adapted to recurrent ENSO events, the already depleted current population may not be able to survive more events similar in magnitude to the 1997-1998 ENSO. Indeed, the continued viability of *A. australis* on the Peruvian coast may depend primarily on nongenetic factors, such as local availability of food resources during breeding seasons and its consequent effects on pup growth and survival. Global warming models predict stronger and more frequent ENSOs in the future (NCDC-NOOA, 2004). If a strong one were to hit the surviving adults in the near future, there would be few juveniles to replace them and the future generations may be compromised and in a much greater risk than ever imagined (Majluf, 1998).

On the other hand, I believe that the estimated N_e of 7,715 individuals for the Peruvian population of *O. flavescens* is not a critical value because it is higher than the mean minimum viable population for vertebrates (7000 breeding age adults, Reed et al., 2003). However, the viability of *O. flavescens* as well of *A. australis* on the Peruvian coast may depend primarily on local availability of food resources and its effects on pup growth and survival (Soto et al., 2004). The species faces the productive but unpredictable Peruvian upwelling ecosystem (Ryther, 1969), and is directly exposed to interannual and highly stochastic fluctuations in the distribution and abundance of its principal prey, *E. ringens* (Arias-Schreiber, 2003). According to Soto et al. (2006) there are strong linkages between maternal attendance patterns and the abundance of prey and oceanographic features close to the rookeries. Acute prey shortage during ENSO resulted in females increasing the length of their foraging trips and decreasing the time they spent onshore with their pups, which died due to starvation (Soto et al., 2004). Thus, stochastic fluctuations in the marine environment should directly affect the maternal behavior and possibly the reproductive success of this species.

ENSO year	Intensity of the event (based on SST)	Gap between events (years)
1877 - 1878	Strong	10
1888 - 1889	Moderate	7
1896 - 1897	Strong	2
1899	Strong	3
1902 - 1903	Strong	2
1905 - 1906	Strong	5
1911 - 1912	Strong	1
1913 - 1914	Moderate	4
1918 - 1919	Strong	4
1923	Moderate	2
1925 - 1926	Strong	6
1932	Moderate	7
1939 - 1941	Strong	5
1946 - 1947	Moderate	4
1951	Weak	2
1953	Weak	4
1957 - 1959	Strong	4
1963	Weak	2
1965 - 1966	Moderate	2
1968 - 1970	Moderate	2
1972 - 1973	Strong	3
1976 - 1977	Weak	1
1977 - 1978	Weak	1
1979 - 1980	Weak	2
1982 - 1983	Strong	3
1986 - 1988	Moderate	2
1990 - 1993	Strong	1
1994 - 1995	Moderate	2
1997 - 1998	Strong	4
2002 - 2003	Moderate	1
2004 - 2005	Weak	1
2006 - 2007	Weak	1

Table 3. ENSO events from 1877 - 2006 based on sea surface temperature (SST) reconstructions (Smith & Reynolds, 2004).

In this sense, the estimated N_e of 7,715 for *O. flavescens* should be considered as a value to be maintained in order to keep the population large enough to avoid inbreeding or to retain adaptive genetic variation to survive to future ENSO events. Moreover, this N_e result has an important application related to management decisions for the conservation of sea lions in Peruvian waters. This N_e value must be taken into account mainly during periods of suggestion of culling, based on the increasing competition between fishery activity and sea lions during ENSO events. In 1997 Peruvian fishermen called for a cull of sea lions to protect fisheries and the Peruvian Fisheries Ministry was considering a pilot program to kill up to 60 sea lions. However, as a result of the 1997-1998 ENSO event the numbers of sea lions onshore were drastically reduced and the program was abandoned (Seal Conservation Society, 2010; Lama, 2010).

The most important drawback of the estimate of N_e is the necessity of a data set from long term censuses that include temporal oscillations in population size (~20 years) (Vucetich et al., 1997). Nunney & Elam (1994) suggested that estimates based on data collected during a single season ignores the influence of temporal fluctuation and thus may represent gross overestimates of N_e. This is why the data presented here are so important and unique in comparison to the information available on pinniped populations from other countries throughout South America. The long-term census size available for fur seals and sea lions along the Peruvian coast are results of the systematic efforts of IMARPE for more than 25 years.

Vucetich & Waite (1998) highlight the importance of long-term counts in order to improve the accuracy of the estimates of N_e. Traill et al. (2010) also reinforce that conservationists working within developing nations rarely have the resources available to collect the long term demographic and other data necessary to model viability for specific species or taxa.

Global warming models predict stronger and more frequent ENSOs in the future (see table 3) (NCDC-NOOA, 2004). I recommend that conservation planners consider the estimated N_e for both species in future management strategies to ensure the conservation of the South American sea lions and fur seals on the Peruvian coast.

4. Acknowledgments

The author would like to thank Patricia Majluf for opening the doors of the Punta San Juan Project and providing all of the knowledge about the Peruvian fur seals and sea lions. To the Instituto del Mar del Perú (IMARPE), that kindly provided information related to Peruvian studies. To the research team of the Punta San Juan Project (1997 and 2003): Nora Rueda, Gabriela Battistini, Rosana Paredes, Carlos Zavalaga, Diana and Juan Cervantes-Sanchez, Milena Roca, Susana Cárdenas, Armando Valdés, Marco Cardeña, Walter Diaz, Manuel Apaza and Pedro Llerena for their assistance in the field work in Peru. To Nathalia dos Santos Freitas for supply important references, to Adriano Duarte, who kindly prepared Figure 2. This study was Gemars contribution n. 37.

5. References

Aguayo, A. & Maturana, R. (1973). Presencia del lobo marino común *Otaria flavescens* en el litoral chileno. *Biol. Pesq.* 6, 49-75.

Arias-Schreiber, M. (1993). *Interacciones entre lobos marinos Otaria byronia y la pesquería artesanal en el puerto San Juan de Marcona, Perú*. Bachelor thesis, Univesidad Nacional Agraria La Molina, Lima, Perú.

Arias-Schreiber, M. & Rivas, C., 1998. Distribuición, tamaño y estructura de las poblaciones de lobos marinos *Arctocephalus australis* y *Otaria byronia* en el litoral Peruano, en Noviembre 1996 y Marzo 1997. Inf. Progres. Inst. Mar Perú 73, 17-32.

Arias-Schereiber, M., 2000. Distribuición, tamaño y estructura de las poblaciones de lobos marinos *Arctocephalus australis* y *Otaria byronia* en el litoral Peruano durante 1999. Informe Anual 1999. Instituto del Mar del Perú. Informe Interno. Instituto Del Mar Del Perú (IMARPE), Esq. Gamarra y Gral. Valle s/n, Chucuito, Callao, Peru.

Arias-Schreiber, M. (2003). *Prey spectrum and feeding behavior of two sympatric pinnipeds (Arctocephalus australis and Otaria flavescens) in relation to the 1997–98 ENSO in southern Peru*. Master thesis, University of Bremen, Germany.

Caballero, A. (1994). Developments in the prediction of effective population size. *Heredity* 73, 657–679.

Campagna, C., Le Boeuf, B.J. & Cappozzo, H.L. (1988). Group raids: a mating strategy of male Southern sea lions. *Behaviour* 105, 224-249.

Cane, M.A. (1983). Oceanographic events during El Niño. *Science* 222, 1189-1195.

Capozzo L.H. & Perrin, W.F. (2008). South American Sea Lion, *Otaria flavescens*. In *Encyclopedia of Marine Mammals*: 1076-1079. Perrin, W.F., Würsig, B., Thewissen, J.G.M. (Eds.). Amsterdam: Academic Press.

Crespo, E.A., Oliva, D., Dans, S.L. & Sepúlveda, M. (2009). Estado de situación del lobo marino común *Otaria flavescens* en su área de distribución. Informe del Taller de Trabajo, Valparaíso, Chile, 15-17 June 2009.

Crow, J.F. & Kimura, M. (1970). *An Introduction to Population Genetics Theory*. New York: Harper and Row.

Cushing, D.H. (1982). *Climate and fisheries*. London: Academic Press.

Dans, S., Alonso, M., Crespo, E., Pedraza & S., García, N. (2003). Interactions between marine mammals and high seas fisheries in Patagonia: An integrated approach. In *Marine mammals: fisheries, tourism and management issues*: 100-115. Gales, I., Nicholas J., Hindell, M.A. (Eds.). Collingwood: CSIRO Publishing.

Frankham, R., Balou, J.D. & Briscoe, D.A. (2002). *Introduction to conservation genetics*. Cambridge: Cambridge University Press.

Fisher, R. A., 1930. *The genetical theory of natural selection*. Claredon Press, Oxford.

George-Nascimiento, M., Bustamante, R. & Oyarzún, C. (1985). Feeding ecology of the South American sea lion *Otaria flavescens*: food contents and food selectivity. *Mar. Ecol. Prog.* Ser. 21, 135-143.

Glantz, M.H., (1996). *Currents of Change - El Niño's impact on climate and society*. Cambridge: Cambridge University Press.

Grimwood, I.R. (1969). Notes on the distribution and status of some Peruvian mammals. American Commitee for International Wild Life Protection and New York Zoological Society, Special Publication 21, 1-86.

Guerra, C.C. & Torres, D.N. (1987). Presence of South American fur seal, *Arctcephalus australis*, in northern Chile. In: Croxall JP, Gentry RL (eds) Status, Biology and Ecology of Fur Seals, Proceedings of an International Symposium and Workshop, United Kingdom, 23-27 April 1984, 169-176.

Hedrick, P.W. (1985). Elephant seals and the estimation of a population bottleneck. *Journal of Heredity* 86: 232-235.

Hedrick, P.W. (2000). *Genetics of Populations*. 2nd edn. Sudbury: Jones and Barlett.

Hubbs, C.L. (1956). Back from oblivion. Guadalupe fur seal: still a living species. *Pacific Discovery* 9: 14-51.

Husband, B.C. & Barrett, S.C.H. (1992). Effective population-size and genetic drift in tristylous *Eichhornia paniculata* (*Pontederiaceae*). *Evolution* 46, 1875–1890.

Idyll, C.P. (1973). The anchovy crisis. *Sci. Am.* 228, 22-29.

IMARPE (2006). Censo Nacional de Lobo Chusco (*Otaria flavescens*) (2006). UBI Depredadores Superiores. Inf. Int..

Jefferson, T.A., Webber, M.W. & Pitman, R.L. (2008). *Marine mammals of the world a comprehensive guide to their identification*. Amsterdam: Academic Press.

Kohen, R.K., Diehl, W.J. & Scott , T.M. (1988). The differential contribution by individual enzymes of glycosis and protein catabolism to the relationship between heterozygosity and growth rate in the coot clam. *Genetics* 118: 121-130.

Koen-Alonso, M., Crespo, E.A., Pedraza, S.N., Garcia, N.A. & Dans, S.L. (1999). Food consumption by the southern sea lion, *Otaria flavescens*, population in northern and central Patagonia. Trabajo presentado en la 13th Annual Conference of the European Cetacean Society, Valencia, España.

Lama, A. (2010). Aphrodisiac market fuels killing of sea lions. Available at http://www.tierramerica.info/nota.php?lang=engandidnews=1715.

Leary, R.F., Allendorf, F. W. & Knudson, K.L (1983). Developmental stability and enzyme heterozygosity in rainbow trout. *Nature* 301: 71-72.

LeBoeuf, B.J. & Bonnell, M. (1980). Pinnipeds on the California islands: abundance and distribution. In: D. Power (Ed.) Proceedings of multidisciplinary symposium. *Santa Barbara Museum of Natural History*, 475-493.

MacCarthy, M.A., Lindenmayer, D.B. & Possingham, H.P. (2001). Assessing spatial PVA models of arboreal marsupials using significance tests and Bayesian statistics. *Biological Conservation* 98: 191-200.

Majluf, P. (1987). *Reproductive ecology of female South American fur seals at Punta San Juan*. Ph.D. thesis M-22, University of Cambridge, 127p.

Majluf, P. (1992). Timing of births and juvenile mortality in the South American fur seal in Peru. *Journal of Zoology of London* 227, 367-383.

Majluf, P. (1998). Effects of the 1997/1998 El Niño on pinnipeds in Peru. In Abstracts of 8a. Reunião de Trabalho de Especialistas em Mamíferos Aquáticos da América do Sul e 2°. Congresso da Sociedade Latinoamericana de Especialistas em Mamíferos Aquáticos (SOLAMAC), Olinda, Brazil, 25-29 October 1998. p. 120.

Majluf, P. & Trillmich, F. (1981). Distribution and abundance of sea lions (*Otaria byronia*) and fur seals (*Arctocephalus australis*) in Peru. *Z. Säugetierkunde* 46, 384-393.

Majluf, P., Babcock, E.A., Riveros, J.C., Schreiber, M.A. & Alderete, W. (2002). Catch and bycatch of sea birds and marine mammals in the small-scale fishery of Punta San Juan, Peru. *Conservation Biology* 16, 1333-1343.

NCDC-NOOA (2004). The top 10 El Niño Events of the 20th Century. Available at http://www.ncdc.noaa.gov/oa/climate/research/1998/enso/10elnino.html.

Neigel, J.E. (1996). Estimation of effective population size and migration parameters from genetic data. In: *Molecular genetic approaches in conservation*: 329–346. Smith, T.B., Wayne, R.K. (Eds.). Oxford: Oxford University Press.

Nunney, L. (1995). Measuring the ratio of effective population size to adult numbers using genetic and ecological data. *Evolution* 49, 389–392.

Nunney, L. (2002). The effective size of annual plant populations: The interaction of a seed bank with fluctuating population size in maintaining genetic variation. *The American Naturalist* 160, 195–204.

Nunney, L. & Elam, D.R. (1994). Estimating the effective population size of conserved populations. *Conservation Biology* 8, 175–184.

Oliveira, L.R., Arias-Schreiber, M., Meyer, D. & Morgante, J.S. (2006). Effective population size in a bottlenecked fur seal population. *Biological Conservation* 131, 505-509.

Oliveira, L.R., Ott, P.H. & Malabarba, L.R. (2008). Ecologia alimentar dos pinípedes do Sul do Brasil e uma avaliação de suas interações com atividades pesqueiras. Ecologia de Mamíferos. Reis, N.R., Peracci, A.L., Santos G.A.S.D. (Eds). En: Technical Booksed pp. 97-116.

Oliveira, L.R., Meyer, D., Hoffman, J.I., Majluf, P. & Morgante, J.S. (2009). Evidence of a genetic bottleneck in an *El Niño* affected population of South American fur seals, *Arctocephalus australis. Journal of the Marine Biological Association of the United Kingdom* 89, 1717–1725.

Piazza, A. (1969). Los lobos marinos. *Pesca y caza* 9, 1-29.

Reed, D.H., Grady, J.J.O., Brook, B.W., Ballou, J.D. & Frankham, R. (2003). Estimates of minimum viable population sizes for vertebrates and factors influencing those estimates. *Biological Conservation* 113, 23-24.

Repenning, C.A., Peterson, R.S., Hubbs, C.L. (1971) Contributions to the systematics of the southern fur seals, with particular reference to the Juan Fernández and Guadalupe species. In: Burt WH (ed), "Antarctic Pinnipedia", Antarctic Research, 18:1-34. American Geophysical Union.

Riedman, M.L. (1990). *The Pinnipeds. Seals, sea lions and walruses.* Berkeley: University of California Press.

Rosas, F.C.W.; Pinedo, M.C.; Marmontel, M.& Haimovici, M. (1994). Seasonal movements of the South American sea lion (Otaria flavescens, Shaw) off the Rio Grande do Sul coast, Brazil, *Mammalia* 58: 51-59.

Ryther, J.H. (1969). Photosynthesis and fish production in the sea. *Science* 166, 72–76.

Seal Conservation Society (2010). South American Sea Lion. Available at http://www.greenchannel.com/tec/species/samslion.html.

Sepúlveda, M. & Oliva, D. (2005). Interactions between South American sea lions *Otaria flavescens* (Shaw) and salmon farms in southern Chile. *Aquaculture Research* 36, 1062-1068.

Shaffer, M.L. (1981). Minimum population sizes for species conservation. *BioScience* 31, 131-134.

Smith, T. M., Reynolds, R. W. (2004). Improved extended reconstruction of SST (1854-1997). *Journal of Climate* 7, 2466-2477.

Soto, K.H., Trites, A.W. & Arias-Schreiber, M. (2004). The effects of prey availability on pup mortality and the timing of birth of South American sea lions (*Otaria flavescens*) in Peru. *The Zoological Society of London* 264, 419-428.

Soto, K.H., Trites, A.W. & Arias-Schreiber, M. (2006). Changes in diet and maternal attendance of South American sea lions indicate changes in the marine environment and prey abundance. *Marine ecology progress series* 312, 277–290.

Soule, M.E. (1979). Heterozygosity and developmental stability: another look. Evolution 33, 396-401.

Szteren, D. & Paez, E. (2002). Predation by southern sea lions (*Otaria flavescens)* on artisanal fishing catches in Uruguay. *Marine and Freshwater Research* 53, 1161-1167.

Tovar, H. & Fuentes, H. (1984). Magnitud poblacional de lobos marinos en el litoral peruano en marzo de 1984. Informe Instituto del Mar del Perú 88.

Traill, L.W., Bradshaw, C.J.A. & Brook, B.W. (2007). Minimum viable population size: a meta-analysis of 30 years of published estimates. *Biological Conservation* 139, 159–166.

Traill, L.W., Brook, B.W., Frankham, R.R. & Bradshaw, C.J.A. (2010). Pragmatic population viability targets in a rapidly changing world. *Biological Conservation* 143, 28-34.

Trillmich, F. & Ono, K.A. (1991). *Pinnipeds and El Niño: Responses to environmental stress.* Springer-Verlag, Berlin, Pp. 66-74.

Vaz-Ferreira, R. South American sea lion, *Otaria flavescens* (Shaw, 1800) (1981). *In: Handbook of Marine Mammals,* 39-66. Academic Press, Londres.

Vaz-Ferreira R. (1982) *Arctocephalus australis* Zimmerman, South American fur seal. In *Mammals in the seas, small cetaceans, seals, sirenians and otters* FAO Fisheries series 4, 497-508.

Vaz-Ferreira, R. & Bianco, J. (1998). Explotación, sobrevivencia y preservación de los otariideos en el Uruguay. Paper presented at the 8ª Reunião de Trabalho de Especialistas em Mamíferos Aquáticos da América do Sul e 2º Congresso da Sociedade Latinoamericana de Especialistas em Mamíferos Aquáticos de América do Sul, Olinda, 25-29 October 1998, p 221.

Vucetich, J.A., Waite, T.A. & Nunney, L. (1997). Fluctuating population size and the rao of effective to census population size (*Ne*/*N*). *Evolution* 51, 2015–2019.

Vucetich, J.A. & Waite, T.A. (1998). Number of censuses required for demographic estimation of effective population size. *Conservation Biology* 12, 1023–1030.

7

Effects of Temperature and Light Conditions on Growth of Current-Year Seedlings of Warm-Temperate Evergreen Tree Species and Cool-Temperate Deciduous Tree Species

Koichi Takahashi[1], Hiroyuki Kobori[1] and Tatsuyuki Seino[2]
[1]Department of Biology, Faculty of Science,
Shinshu University, Matsumoto, Nagano,
[2]Yatsugatake Forest, Agricultural and Forestry Research Center,
Graduate School of Life and Environmental Sciences,
University of Tsukuba, Minamimaki, Nagano,
Japan

1. Introduction

It is suggested that global warming affects plant distribution along latitudinal and altitudinal gradients because vegetation changes with thermal conditions. Simulation studies predicted that global warming largely affects plant distribution (e.g., Morin et al., 2008). Actually, vegetation change during several decades has been observed (Penuelas et al., 2007; Lenoir et al., 2008). By contrast, some other studies did not observe vegetation changes (Holtmeier & Broll, 2007; Harsch et al., 2009). Interpretation of results of simulation models also needs caution because simulation results are different according to modeling methods even for same species (Thuiller, 2003). Therefore, there is still uncertainty of effects of global warming on plant distribution.

Plant distribution is determined by integrated demographic processes such as seed dispersal, seed germination, growth and survival of individual plants. Since early demographic phase such as seedling establishment is more susceptible to environmental conditions than the adult phase (Kullman, 2002), it is important to clarify effects of temperature on seedling growth to predict effects of global warming on plant distribution. There are many experimental studies that examined effects of temperature on growth of tree seedlings (Danby & Hik, 2007; Hoch & Körner, 2009; Munier et al., 2010). For example, Yin et al. (2008) reported that seedling growth of *Betula albo-sinensis* increased in the warm condition with 0.51°C higher than the ambient air condition. Many experimental studies that examined effects of temperature were conducted at bright conditions (e.g. Danby & Hik, 2007; Way & Sage, 2008). However, most seedlings distribute in dark closed-canopy conditions in forests. Therefore, it is necessary to examine effects of temperature on seedling growth not only in bright conditions but also in dark conditions.

Plants plastically change morphology according to light conditions. For example, relative biomass allocation to leaves is greater in dark conditions than in bright conditions, accompanied with reduction of leaf mass per area (LMA) (Ellsworth & Reich, 1992; Gould, 1993; Niinemets et al., 1999; Takahashi et al., 2005). These morphological changes increase light capture per plant, which is an adaptive strategy in dark conditions. Thus, plants adapt to changing environments through morphological plasticity. This study compared growth responses to temperature and light conditions among three species with different climatically distribution ranges. This comparison is important to clarify effects of global warming on distribution shift of vegetation.

Central Japan is a latitudinal vegetation ecotone between warm-temperate evergreen broad-leaved forests and cool-temperate deciduous broad-leaved forests. Ishigami et al. (2003) simulated net primary production (NPP) of forests by the modified model of BIOME3 (Haxeltine & Prentice, 1996). They predicted that NPP of warm-temperate evergreen broad-leaved forests is greater than that of cool-temperate deciduous broad-leaved forests above the current northern distribution limit of warm-temperate evergreen broad-leaved forests if air temperature is increased by global warming. According to the simulation result of NPP, Ishigami et al. (2003) suggested that the northern distribution limit of warm-temperate evergreen broad-leaved forests will move to the north. However, the northern distribution limit would not move to the north easily because of competition with existing northern vegetation (Kohyama & Shigesada, 1995). Therefore, it is important to compare growth responses of seedlings to temperature and light conditions between warm-temperate evergreen broad-leaved species and cool-temperate deciduous broad-leaved species to clarify vegetation changes due to global warming.

It is considered that warm-temperate evergreen broad-leaved species cannot distribute in the cool-temperate zone due to low winter temperature, not due to low summer temperature for the growth (Kira, 1949). Cold winter temperature decreases photochemical efficiency of leaves in warm-temperate evergreen broad-leaved species (Aranda et al., 2005; Taneda & Tateno, 2005). Reduction of photochemical efficiency due to cold temperature is more conspicuous in bright conditions than dark conditions (Matsuki et al., 2003). Therefore, it is necessary to examine growth responses to temperature and light conditions not only during growth period but also during dormant period (winter) to clarify how temperature affects growth of warm-temperate evergreen broad-leaved species.

This study examined effects of temperature and light conditions on seedling growth of a warm-temperate evergreen broad-leaved species and two cool-temperate deciduous broad-leaved species to answer the following two questions.

(1) Do the warm-temperate evergreen and cool-temperate deciduous broad-leaved species grow more in higher temperature and light conditions?

(2) Do high temperature conditions mitigate reduction of photochemical efficiency in winter for the warm-temperate evergreen broad-leaved species?

2. Materials and methods

2.1 Study site

This study was conducted at the two sites in Nagano Prefecture, central Japan: the campus of Shinshu University (N36°15′04″, E137°58′41″, 630 m above sea level) in Matsumoto, and the Yatsugatake experimental forest (N35°56′40″, E138°28′11″, 1350 m a.s.l.) of Tsukuba University in Nobeyama (Fig. 1). The mean annual precipitation of Matsumoto was 1035.7 mm during 2000–2009. The mean annual temperature was 12.2°C. The mean monthly

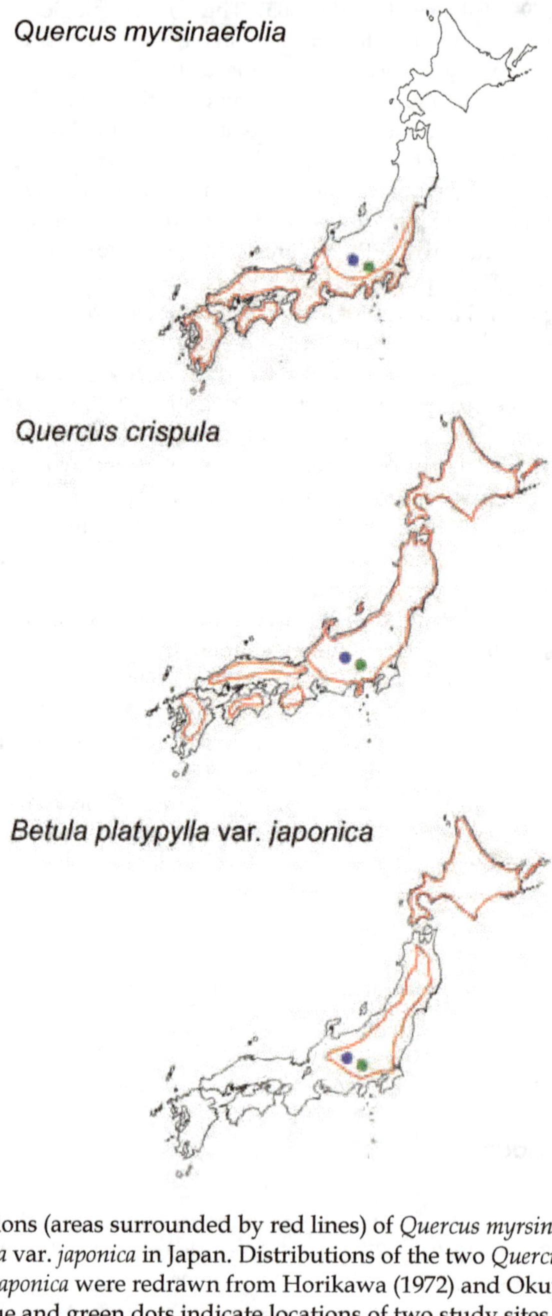

Fig. 1. Distributions (areas surrounded by red lines) of *Quercus myrsinaefolia*, *Q. crispula* and *Betula platyphylla* var. *japonica* in Japan. Distributions of the two *Quercus* species and *B. platyphylla* var. *japonica* were redrawn from Horikawa (1972) and Okuyama (1982), respectively. Blue and green dots indicate locations of two study sites, Matsumoto and Nobeyama, respectively.

temperatures in January and in August were -0.3 and 24.8°C, respectively. The mean annual precipitation of Nobeyama was 1435.3 mm during 2000–2009. The mean annual temperature was 7.2°C. The mean monthly temperatures in January and in August were -5.4 and 19.2°C, respectively.

2.2 Study species

This study examined a warm-temperate evergreen broad-leaved species (*Quercus myrsinaefolia* Blume) and two cool-temperate deciduous broad-leaved species (*Betula platyphylla* var. *japonica* Hara and *Quercus crispula* Blume) (Fig. 1). *Q. myrsinaefolia* is a shade-tolerant tall tree species that distributes in the northern part of warm-temperate zone (Horikawa, 1972). *B. platyphylla* var. *japonica* and *Q. crispula* are shade-intolerant and mid-shade-tolerant tall tree species, respectively (Samejima, 1979; Koike, 1988; Masaki et al., 1992). *B. platyphylla* var. *japonica* and *Q. crispula* often form pure stands after disturbances such as strong wind, forest fire and clear cutting (Samejima, 1979; Kamitani, 1993; Namikawa et al., 1997).

Q. myrsinaefolia distributes below 700 m a.s.l. in the warm-temperate zone, and the northern distribution limit is about N38° (Horikawa, 1972). Distribution of *Q. myrsinaefolia* is limited to below 500 m a.s.l. in the north of N36°. Although the southern part of Nagano Prefecture was the northern distribution limit of *Q. myrsinaefolia* (Horikawa, 1972), the distribution of *Q. myrsinaefolia* is recently observed in central part (Matsumoto area) of Nagano Prefecture probably because of escapes of seeds from planted *Q. myrsinaefolia* trees (Otsuka & Ozeki, 2008). *B. platyphylla* var. *japonica* distributes in central and northern Japan (Okuyama, 1981). *B. platyphylla* var. *japonica* distributes mainly at about 1000 m and 1300 m a.s.l. in Matsumoto and Nobeyama, respectively (Editorial Board of Flora of Nagano Prefecture, 1997). *Q. crispula* is one of the dominant species in cool-temperate deciduous broad-leaved forests, and widely distributes in the cool-temperate zone of Japan (Horikawa, 1972). *Q. crispula* distributes mainly at about 1000 m and 1300 m a.s.l. in Matsumoto and Nobeyama, respectively, like *B. platyphylla* var. *japonica*. Therefore, Matsumoto is the northern distribution limit for the warm-temperate evergreen species *Q. myrsinaefolia*, and Nobeyama is optimal thermal conditions for the two cool-temperate deciduous species *B. platyphylla* var. *japonica* and *Q. crispula*.

2.3 Seedling growth experiments

Growth experiments were conducted at Matsumoto and Nobeyama to examine effects of temperature and light conditions on the current-year seedling growth of *Q. myrsinaefolia*, *B. platyphylla* var. *japonica* and *Q. crispula*. A greenhouse experiment was also conducted at Matsumoto to make optimal temperature conditions for the warm-temperate evergreen *Q. myrsinaefolia* because Matsumoto is the northern distribution limit of this species. Windows of the greenhouse were opened in summer to avoid extreme rising of air temperature inside the greenhouse. Increase of air temperature in the greenhouse was 3°C on the average. Therefore, air temperature increased in the order of Nobeyama, Matsumoto and the greenhouse at Matsumoto. Nobeyama, Matsumoto and the greenhouse at Matsumoto are referred to T1, T2 and T3, respectively, in this study. Two light conditions (20% and 54% light) were set at each temperature condition by using shade cloth. 20% and 54% light conditions are referred to L1 and L2, respectively.

Species	Variables	T	L	T × L	*n*
Qm	Stem height	56.1***	< 0.1	0.3	99
	Stem diameter	16.1***	12.5***	0.2	46
	Dry mass	24.2***	15.1***	1.5	46
	LMA	3.5*	7.0*	1.2	47
Qc	Stem height	46.2***	3.1	1.3	94
	Stem diameter	18.0***	36.2***	1.6	94
	Dry mass	11.6***	14.2***	0.6	94
	LMA	32.4***	67.4***	0.4	93
Bp	Stem height	517.5***	0.4	1.0	120
	Stem diameter	768.0***	11.5***	1.3	120
	Dry mass	595.9***	49.5***	8.3***	117
	LMA	2.3	52.9***	2.3	109

*: $p < 0.05$, **: $p < 0.01$, ***: $p < 0.001$.

Table 1. ANOVA results of stem height, stem diameter, seedling dry mass and leaf mass per area (LMA) for *Quercus myrsinaefolia* (Qm), *Quercus crispula* (Qc) and *Betula platyphylla* var. *japonica* (Bp) to test three temperature conditions (T), two light conditions (L) and the interaction (T × L) on the four variables. *F*-values and the number of samples (*n*) are shown.

Seeds of *Q. myrsinaefolia*, *B. platyphylla* var. *japonica* and *Q. crispula* were collected in 2008. Seeds were sown in plastic pots (diameter 15 cm × height 13 cm) filled with horticulture soil and leaf mold. The number of pots was 20 for each light condition (L1 and L2) at each temperature condition (T1, T2 and T3) for each species. Seed sowing was conducted between late March and early April in 2009. However, seeds of *B. platyphylla* var. *japonica* were sown again at T1 in early June because seeds did not germinate. Pots were watered once a week or according to the need during the experiment period. All pots were randomly moved within each treatment once a week to minimize effects of pot position on seedling growth.

Seedlings of the three species were excavated and washed carefully in early October in 2009 after the cease of the current-year growth. In terms of *Q. myrsinaefolia*, only the half number of seedlings was excavated at each light and temperature condition, and the other seedlings were remained for the measurement of photochemical efficiency (see next paragraph). Stem diameter at the base and stem height were measured at the harvest for the three species. Seedlings were divided into stem, root and leaf, and then all leaves were scanned by using the free graphic software ImageJ (http://rsbweb.nih.gov/ij/) to measure total leaf area per seedling. Each organ was oven-dried at 80°C for two days, and was weighed.

The ratio of light-induced variable to maximum fluorescence of chlorophyll (Fv/Fm) of *Q. myrsinaefolia* was measured as a surrogate for photochemical efficiency of PSII (Demmig-Adams et al., 1989). Photochemical efficiency was measured around the noon once a month from summer of 2009 to the next summer by using a chlorophyll fluorometer OS-30p (Opti-Science, NH, USA). The measurement was conducted after 30 min of dark acclimation.

Two-way ANOVA was done to examine effects of temperature and light conditions on stem height, stem diameter, LMA and seedling dry mass (including root) for the three species (*Q. myrsinaefolia*, *B. platyphylla* var. *japonica* and *Q. crispula*).

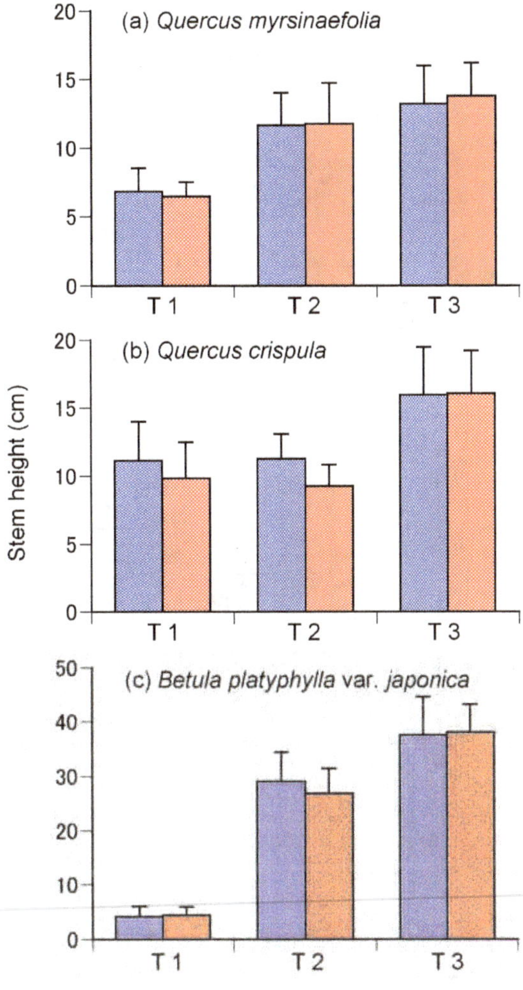

Fig. 2. Stem height for the three species at the three temperature and two light conditions. Blue and red bars indicate L1 and L2 light conditions, respectively. Mean values with positive standard deviations are shown.

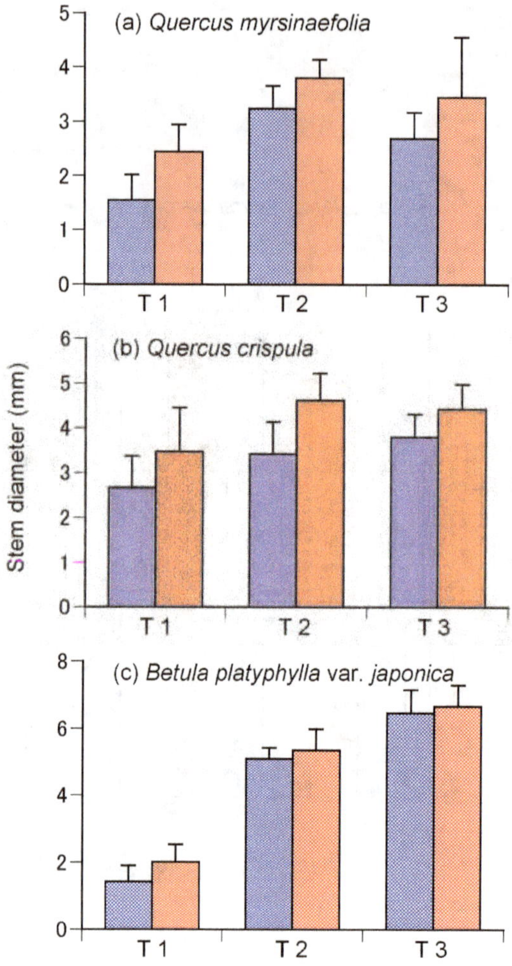

Fig. 3. Stem diameter for the three species at the three temperature and two light conditions. Blue and red bars indicate L1 and L2 light conditions, respectively. Mean values with positive standard deviations are shown.

3. Results

3.1 Growth of the three species

Effect of temperature on stem height was significant for the three species (Table 1), i.e., stem height was greater in higher temperature conditions, irrespective of light conditions (Fig. 2). However, stem height of *Q. crispula* was not different between T1 and T2 because of the shoot re-growth from the base after the shoot die-back in late May at T2. The difference in stem height of *Q. myrsinaefolia* between T2 and T3 was also not large (Fig. 2). Light conditions did not affect stem height for the three species (Table 1). By contrast, stem diameter of the three species was different not only among the three temperature conditions but also between the two light conditions (Table 1). Stem diameter of the three species was greater at L2 than L1 for each temperature condition (Fig. 3). Stem diameter was greater at T2 and T3 than T1 for *Q. crispula*. Stem diameter of *B. platyphylla* var. *japonica* increased in the order of T1, T2 and T3. However, stem diameter of *Q. myrsinaefolia* was slightly smaller at T3 than T2 for the two light conditions.

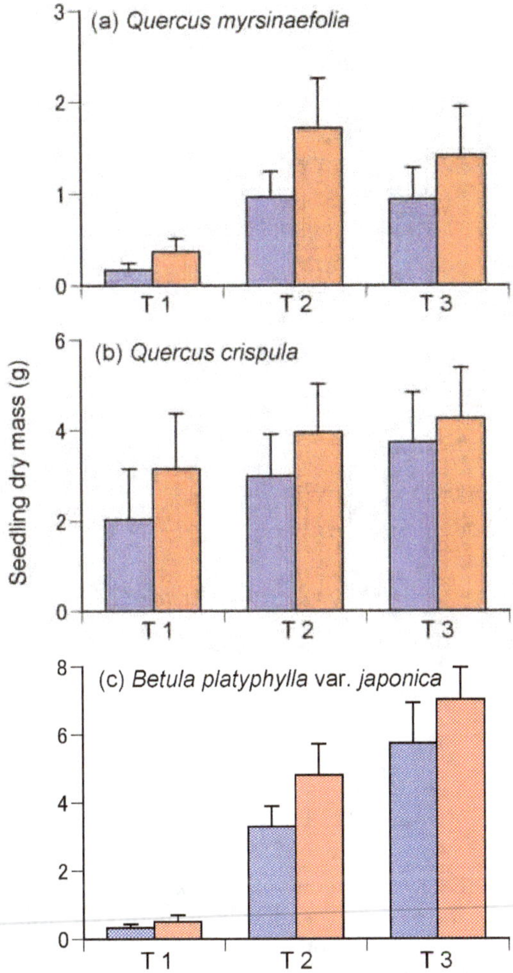

Fig. 4. Seedling dry mass for the three species at the three temperature and two light conditions. Blue and red bars indicate L1 and L2 light conditions, respectively. Mean values with positive standard deviations are shown.

Effects of temperature and light conditions on seedling dry mass were significant for the three species (Table 1), i.e., seedling dry mass was greater at higher temperature and brighter light conditions (Fig. 4). However, seedling dry masses of *Q. myrsinaefolia* at T3 were similar to and smaller than those at T2 for L1 and L2, respectively.

LMA was greater at L2 than L1 for the three species (Fig. 5, Table 1). LMA of *Q. myrsinaefolia* and *Q. crispula* was significantly different among the three temperature conditions (Table 1). Although LMA of the two species was greater at T2 than T1 and T3, the difference was not large (Fig. 5).

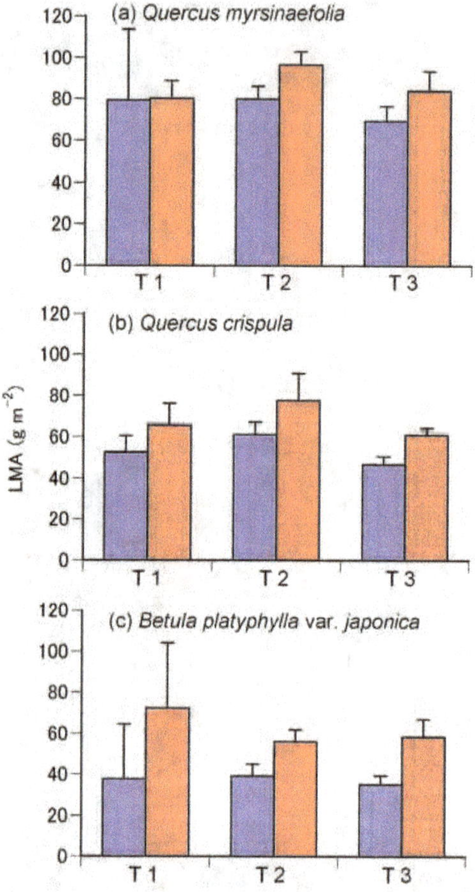

Fig. 5. Leaf mass per area (LMA) for the three species at the three temperature and two light conditions. Blue and red bars indicate L1 and L2 light conditions, respectively. Mean values with positive standard deviations are shown.

3.2 Photochemical efficiency of *Q. myrsinaefolia*

Values of photochemical efficiency (Fv/Fm) differed, according to temperature and light conditions and season (Fig. 6). In August of 2009, Fv/Fm values were similar between the two light conditions for the three temperature conditions. Although Fv/Fm value was about 0.8 at T3 in August, Fv/Fm values slightly decreased as temperature decreased from T3 to T1. Fv/Fm value was about 0.7 at T1 in August of 2009. Fv/Fm value decreased from autumn to winter in each temperature condition. This tendency was more conspicuous at L2 than L1 in each temperature condition. In addition, Fv/Fm value decreased to almost zero in February at T1, and all seedlings died by March (Fig. 7). By contrast, although Fv/Fm values decreased in winter at T2 and T3, these values increased again from spring to summer (Fig. 6) and the seedlings grew well by autumn (Fig. 8).

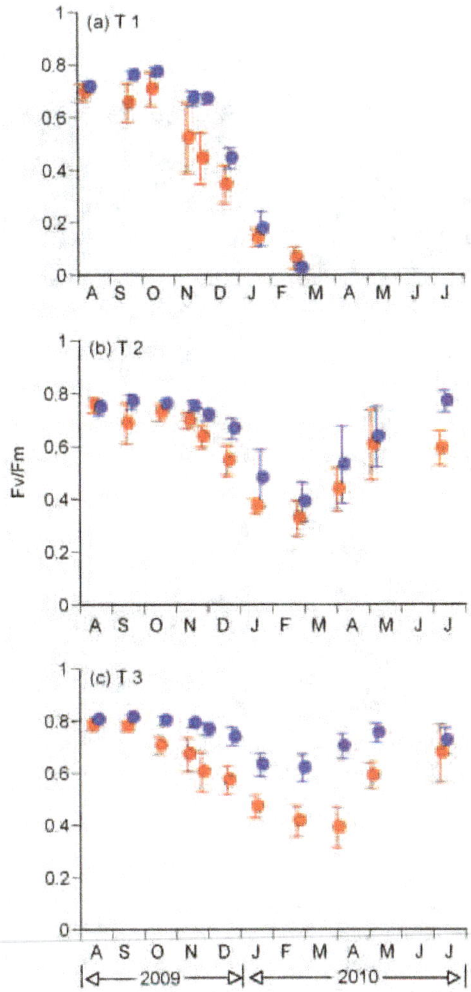

Fig. 6. Seasonal change of Fv/Fm of *Quercus myrsinaefolia* at three temperature and two light conditions. Blue and red circles indicate L1 and L2 light conditions, respectively. Mean values with standard deviations are shown.

4. Discussion

This study showed (1) that high temperature and light conditions increased the current-year seedling growth of the warm-temperate evergreen and the two cool-temperate deciduous species, except for the effect of light on stem height, and (2) that high temperature mitigated reduction of photochemical efficiency (Fv/Fm) of *Q. myrsinaefolia* in winter. Furthermore, winter temperature was so cold at T1 for *Q. myrsinaefolia* that all seedlings of this species died in winter. Therefore, this study showed that high temperature and light conditions increased the seedling growth of the three species and survival of *Q. myrsinaefolia* in winter.

Fig. 7. A current-year seedling of *Quercus myrsinaefolia* at T1 with L1 light condition on February 16 and March 11, 2010. These photos were taken for a same seedling. Die-back was observed on March 11.

Light conditions did not affect stem height of the three species. Seiwa & Kikuzawa (1989) also reported that stem height of current-year seedlings was not different between bright and dark conditions for deciduous broad-leaved species with small seeds, while stem height was smaller in bright conditions than dark conditions for those of large seeds. Thus, stem height is not always taller in brighter conditions for current-year seedlings. This may reflect a survival strategy in forests. Litter accumulation on forest floor is a factor reducing seed germination and seedling survival (Goldberg & Werner, 1983). Therefore, the current-year seedlings germinated below litter layer have to grow over the litter layer. Furthermore, high growth of stem height is also advantageous for survival because of competition with other plants. Therefore, seedlings of the three species would preferentially grow stem height more than stem diameter in dark conditions as compared with bright conditions.

LMA was smaller at darker conditions for the three species at the three temperature conditions, which is advantageous for light capture efficiency per unit leaf mass. LMA of *Q. myrsinaefolia* and *Q. crispula* was also affected by temperature. Although LMA of the two species tended to be greater at T2 than T1 and T3, these differences were not large. By contrast, Woodward (1979) observed that low temperature increased LMA of *Phleum bertolonii* and *P. alpinum* because of increase of cell diameter. Unfortunately, this study did not examine cell diameter of leaves at the three temperature conditions. However, it is possible that light dominates the control of LMA of the current-year seedlings of the three species, compared with temperature.

Fig. 8. A 1-year old seedling of *Quercus myrsiaefolia* at T3 with L1 light condition on October 14, 2010.

Many previous studies about effects of temperatures on plant growth have been conducted so far in cold environments such as alpine region and tundra (Chapin & Shaver, 1985; Molau, 1997; Arft et al., 1999; Hollister & Webber, 2000). It is often reported that high temperature increases plant growth in such cold environments (Chapin & Shaver, 1996; Takahashi, 2005), meaning that cold temperature limits plant growth. Increase of temperature enhances photosynthetic rates in cold environments because of temperature dependency of photosynthetic rates (DeLucia & Smith, 1987). Furthermore, high temperature throughout a year expands the growth period of plants (Chmielewski & Rotzer, 2001). Nevertheless, seedling dry mass, stem height and diameter of *Q. myrsinaefolia* at T3 were similar to or smaller than those at T2 (the temperature conditions of the northern distribution limit). This suggests that temperature during the growth period is not a major factor limiting the growth of *Q. myrsinaefolia*. Northern distribution limits of warm-temperate evergreen broad-leaved forests tend to be determined by coldness in winter, not by warmth during the growth period (Kira, 1991). Thus, winter temperature may be important for the distribution of *Q. myrsinaefolia* along latitudinal gradients.

Increase of temperature clearly mitigated reduction of photochemical efficiency (Fv/Fm) of *Q. myrsinaefolia* during winter. This tendency was more conspicuous at brighter conditions. In forests, seedlings hardly grow in understory dark conditions, and grow vigorously in canopy gaps (Takahashi et al., 2001; Takahashi & Rustandi, 2006). Therefore, it is suggested that increase of photochemical efficiency of *Q. myrsinaefolia* by increase of temperature would enhance the growth in bright conditions. This effect may be large in the northern distribution limit of *Q. myrsinaefolia* because *Q. myrsinaefolia* co-dominates with many deciduous broad-leaved species there (Otsuka et al., 2004). Much light penetrates into forest floor in winter because of leaf fall of deciduous broad-leaved species. Understory seedlings and saplings of *Q. myrsinaefolia* can assimilate during winter after leaf fall of deciduous broad-leaved species, and the photosynthetic production during winter increases the growth and survival of *Q. myrsinaefolia* in the understory (Takenaka, 1986). Therefore, increase of winter temperature would enhance the growth and photochemical efficiency of *Q. myrsinaefolia* in the northern distribution limit.

The growth of the two cool-temperate deciduous broad-leaved species (*Q. crispula* and *B. platyphylla* var. *japonica*) was greater at T2 and T3 conditions than T1 (the optimal temperature conditions). Many studies showed that tree growth is greater nearer the southern and lower distribution limits along latitudinal and altitudinal gradients, respectively (Persson, 1998; Mäkinen et al., 2000; Li et al., 2003; Takahashi & Yoshida, 2009). Although the results of the previous studies were conducted in natural distribution ranges, the result of this study showed that tree growth increases in the temperature conditions warmer than the natural distribution range. Drought stress increases near southern and lower distribution limits (Buckley et al., 1997; Takahashi et al., 2003a; Adams & Kolb, 2005; Hart et al., 2010; Lebourgeois et al., 2010). On the contrary, increase of temperature prolongs the growth period (Sparks et al., 2000; Chmielewski & Rotzer, 2001; Fujimoto, 2008). Although increase of temperature induces drought stress to some extent, it would increase annual growth of plants by increasing annual carbon gain through increase of growth period.

The high growth of *Q. crispula* and *B. platyphylla* var. *japonica* at the two temperature conditions (T2 and T3) warmer than the current distribution range suggests that the natural distribution ranges of the two species are not determined by optimal temperature conditions alone (i.e., the concept of ecological niche). Although *Q. crispula* is a common species in old-growth deciduous broad-leaved forests (Masaki et al., 1992; Takahashi et al., 2003b), *Q. crispula* is recognized as a gap-dependent species (Yamamoto, 1989). It is reported that saplings of deciduous broad-leaved species distributed mainly in canopy gaps in an evergreen broad-leaved forest (Miura et al., 2001). In deciduous broad-leaved forests, understory saplings increases carbon gain by leafing before leaf expansion of canopy trees (Seiwa & Kikuzawa, 1996). However, understory of evergreen broad-leaved forests is dark conditions throughout a year, which have restricted saplings of deciduous broad-leaved species into canopy gaps. Therefore, actual distribution ranges are largely affected not only by temperature conditions but also by competition with other species (Takahashi, 2003).

5. Conclusion

This study predicted that high winter temperature mitigates the reduction of photochemical efficiency and reduces winter mortality of *Q. myrsinaefolia*. Furthermore, this study showed that increase of temperature enhances the growth of not only *Q. myrsinaefolia* but also *Q. crispula* and *B. platyphylla* var. *japonica*. It is possible that global warming intensifies competition between warm-temperate and cool-temperate broad-leaved species at their latitudinal ecotones. Therefore, it is unclear whether *Q. myrsinaefolia* will gain parts of the habitat now covered by *B. platyphylla* var. *japonica* and *Q. crispula* under global warming. This study examined only one warm-temperate broad-leaved species and two cool-temperate broad-leaved species, and the results of this study are not enough to show general responses of the two life forms to temperature and light conditions because of large variations in responses to temperature and light conditions within a life form (Parsons et al., 1994; Takahashi et al., 2003a; van der Werf et al., 2007). Therefore, it is necessary to analyze many species for each life form to clarify general growth responses to temperature and light conditions.

6. Acknowledgment

This study was partially supported by grants from the Ministry of Education, Culture, Sports, Science and Technology, Japan (Nos. 19580168, 22580160). We thank Dr. H. Kobayashi for constructive comments.

7. References

Adams, H. D. & Kolb, T. E. (2005) Tree growth response to drought and temperature in a mountain landscape in northern Arizona, USA. *Journal of Biogeography*, 32, 1629-1640.

Aranda, I., Castro, L., Alia, R., Pardos, J. A. & Gil, L. (2005) Low temperature during winter elicits differential responses among populations of the Mediterranean evergreen cork oak (*Quercus suber*). *Tree Physiology*, 25, 1085-1090.

Arft, A. M., Walker, M. D., Gurevitch, J., Alatalo, J. M., Bret-Harte, M. S., Dale, M., Diemer, M., Gugerli, F., Henry, G. H. R., Jones, M. H., Hollister, R. D., Jónsdóttir, I. S., Laine, K., Lévesque, E., Marion, G. M., Molau, U., Mølgaard, P., Nordenhäll, U., Raszhivin, V., Robinson, C. H., Starr, G., Stenström, A., Totland, Ø., Turner, P. L., Walker, L. J., Webber, P. J., Welker, J. M. & Wookey, P. A. (1999) Responses of tundra plants to experimental warming: meta-analysis of the international tundra experiment. *Ecological Monographs*, 69, 491-511.

Buckley, B. M., Cook, E. R., Peterson, M. J. & Barbetti, M. (1997) A changing temperature response with elevation for *Lagarostrobos franklinii* in Tasmania, Australia. *Climatic Change*, 36, 477-498.

Chapin, F. S., III & Shaver, G. R. (1985) Individualistic growth response of tundra plant species to environmental manipulations in the field. *Ecology*, 66, 564-576.

Chmielewski, F. & Rotzer, T. (2001) Response of tree phenology to climate change across Europe. *Agricultural and Forest Meteorology*, 108, 101-112.

Danby, R. K. & Hik, D. S. (2007) Responses of white spruce (*Picea glauca*) to experimental warming at a subarctic alpine treeline. *Global Change Biology*, 13, 437-451.

DeLucia, E. H. & Smith, W. K. (1987) Air and soil temperature limitations on photosynthesis in Engelmann spruce during summer. *Canadian Journal of Forest Research*, 17, 527-533.

Demmig-Adams, B., Adams, W.I., Winter, K., Meyer, A., Schreiber, U., Pereira, J. S., Krüger, A., Czgan, F. C. & Lange, O. L. (1989) Photochemical efficiency of photosystem II, photon yield of O_2 evolution, photosynthetic capacity, and carotenoid composition during the midday depression of net CO_2 uptake in *Arbutus unedo* growing in Portugal. *Planta*, 177, 377-387.

Editorial Board of Flora of Nagano Prefecture (1997) Flora of Nagano Prefecture. Shinano Mainichi Shinbunsha, Nagano, Japan (in Japanese).

Ellsworth, D. S. & Reich, P. B. (1992) Leaf mass per area, nitrogen content and photosynthetic carbon gain in *Acer saccharum* seedlings in contrasting forest light environments. *Functional Ecology*, 6, 423-435.

Fujimoto, S. (2008) Estimating the impact of thermal change on broad-leaved tree leaf phenology in the warm temperate zone. *Japanese Journal of Conservation Ecology*, 13, 75-87 (in Japanese).

Goldberg, D. E. & Werner, P. A. (1983) The effects of size of opening in vegetation and litter cover on seedling establishment of goldenrods (*Solidatgo* spp.). *Oecologia*, 60, 149-155.

Gould, K. S. (1993) Leaf heteroblasty in *Pseudopanax crassifolius*: functional significance of leaf morphology and anatomy. *Annals of Botany*, 71, 61-70.

Harsch, M. A., Hulme, P. E., McGlone, M. & Duncan, R. P. (2009) Are treelines advancing? A global meta-analysis of treeline response to climate warming. *Ecology Letters*, 12, 1040-1049.

Hart, J. L., Gevel, S. L., Sakulich, J. & Grissino-Mayer, H. D. (2010) Influence of climate and disturbance on the growth of *Tsuga canadensis* at its southern limit in eastern North America. *Trees*, 24, 621-633.

Haxeltine, A. & Prentice, I. C. (1996) BIOME3: An equilibrium terrestrial biosphere model based on ecophysiological constraints, resource availability, and competition among plant functional types. *Global Biogeochemical Cycles*, 10, 693-709.

Hoch, G. & Körner, C. (2009) Growth and carbon relations of tree line forming conifers at constant vs. variable low temperatures. *Journal of Ecology*, 97, 57-66.

Hollister, R. D. & Webber, P. J. (2000) Biotic validation of small open-top chambers in a tundra ecosystem. *Global Change Biology*, 6, 835-842.

Holtmeier, F.-K. & Broll, G. (2007) Treeline advance -driving processes and adverse factors. *Landscape Online*, 1, 1-33.

Horikawa, Y. (1972) Atlas of the Japanese flora. Gakken, Tokyo, Japan.

Ishigami, Y., Shimizu, Y. & Omasa, K. (2003) Projection of climatic change effects on potential natural vegetation distribution in Japan. *Journal of Agricultural Meteorology*, 59, 269-276 (in Japanese).

Kamitani, T. (1993) Ecological studies on regeneration of beech coppice forests in heavy snowfall region. *Memoirs of the Faculty of Agriculture, Niigata University*, 30, 1-108 (in Japanese).

Kira, T. (1949) Forest zones of Japan. Ringyo Gizyutu Kyokai, Tokyo, Japan (in Japanese).

Kira, T. (1991) Forest ecosystems of east and southeast Asia in a global perspective. *Ecological Research*, 6, 185-200.

Kohyama, T. & Shigesada, N. (1995) A size-distribution-based model of forest dynamics along a latitudinal environmental gradient. *Vegetatio*, 121, 117-126.

Koike, T. (1988) Leaf structure and photosynthetic performance as related to the forest succession of deciduous broad-leaved trees. *Plant Species Biology*, 3, 77-87.

Kullman, L. (2002) Rapid recent range-margin rise of tree and shrub species in the Swedish Scandes. *Journal of Ecology*, 90, 68-77.

Lebourgeois, F., Rathgeber, C. B. K. & Ulrich, E. (2010) Sensitivity of French temperate coniferous forests to climate variability and extreme events (*Abies alba*, *Picea abies* and *Pinus sylvestris*). *Journal of Vegetation Science*, 21, 364-376.

Lenoir, J., Gegout, J. C., Marquet, P. A., de Ruffray, P. & Brisse, H. (2008) A significant upward shift in plant species optimum elevation during the 20th century. *Science*, 320, 1768-1771.

Li, M. H., Yang, J. & Krauchi, N. (2003) Growth responses of *Picea abies* and *Larix decidua* to elevation in subalpine areas of Tyrol, Austria. *Canadian Journal of Forest Research*, 33, 653-662.

Mäkinen, H., Nöjd, P. & Mielikäinen, K. (2000) Climatic signal in annual growth variation of Norway spruce (*Picea abies*) along a transect from central Finland to the Arctic timberline. *Canadian Journal of Forest Research*, 30, 769-777.

Masaki, T., Suzuki, W., Niiyama, K., Iida, S., Tanaka, H. & Nakashizuka, T. (1992) Community structure of a species-rich temperate forest, Ogawa Forest Reserve, central Japan. *Vegetatio*, 98, 97-111.

Matsuki, S., Ogawa, K., Tanaka, A. & Hara, T. (2003) Morphological and photosynthetic responses of *Quercus crispula* seedlings to high-light conditions. *Tree Physiology*, 23, 769-775.

Miura, M., Manabe, T., Nishimura, N. & Yamamoto, S. (2001) Forest canopy and community dynamics in a temperate old-growth evergreen broad-leaved forest, south-western Japan: a 7-year study of a 4-ha plot. *Journal of Ecology*, 89, 841-849.

Molau, U. (1997) Responses to natural climatic variation and experimental warming in two tundra plant species with contrasting life forms: *Cassiope tetragona* and *Ranunculus nivalis*. *Global Change Biology*, 3 (Suppl. 1), 97-107.

Morin, X., Viner, D. & Chuine, I. (2008) Tree species range shifts at a continental scale: new predictive insights from a process-based model. *Journal of Ecology*, 96, 784-794.

Munier, A., Hermanutz, L., Jacobs, D. M. & Lewis, K. (2010) The interacting effects of temperature, ground disturbance, and herbivory on seedling establishment: implications for treeline advance with climate warming. *Plant Ecology*, 210, 19-30.

Namikawa, K., Ishikawa, Y. & Sano, J. (1997) Stand dynamics during a 12-year period in a second-growth stand in a cool temperate forest in northern Japan. *Ecological Research*, 12, 277-287.

Niinemets, Ü., Kull, O., Tenhunen, J. D. (1999) Variability in leaf morphology and chemical composition as a function of canopy light environment in coexisting deciduous trees. *International Journal of Plant Science*, 160, 837-848.

Okuyama, S. (1982) *Wild plants in Japan (spring to early summer)*, Seibundo-Shinkosha Publishling Co., Tokyo, Japan (in Japanese).

Otsuka, K. & Ozeki, M. (2008) Monitoring of warm-temperate evergreen broad-leaved species *Quercus myrsinaefolia* in Nagano Prefecture, In: *Investigation report for the actual conditions of global warming at Nagano Prefecture*, 35-41, Nagano Environmental Conservation Research Institute, Nagano, Japan (in Japanese).

Otsuka, K., Ozeki, M. & Maekawa, M. (2004) Distribution of *Quercus myrsinaefolia* Blume (Fagaceae), escaped evergreen broad-leaved tree, in mid and downstream region of the Chikuma River, northern Nagano Prefecture. *Bulletin of Nagano Environmental Conservation Research Institute*, 7, 17-22 (in Japanese).

Parsons, A. N., Welker, J. M., Wookey, P. A., Press, M. C., Callaghan, T. V. & Lee, J. A. (1994) Growth responses of four sub-Arctic dwarf shrubs to simulated environmental change. *Journal of Ecology*, 82, 307-318.

Penuelas, J., Ogaya, R., Boada, M. & Jump, A. S. (2007) Migration, invasion and decline: changes in recruitment and forest structure in a warming-linked shift of European beech forest in Catalonia (NE Spain). *Ecography*, 30, 829-837.

Persson, B. (1998) Will climate change affect the optimal choice of *Pinus sylvestris* provenances? *Silva Fennica*, 32, 121-128.

Samejima, J. (1979) Useful hard-wood in Hokkaido and its characters for improvement (II). *Forest Tree Breeding of Hokkaido*, 22, 6-14 (in Japanese).

Seiwa, K. & Kikuzawa, K. (1989) Seasonal growth patterns of seedling height in relation to seed mass in deciduous broad-leaved tree species. *Japanese Journal of Ecology*, 39, 5-15 (in Japanese).

Seiwa, K. & Kikuzawa, K. (1996) Importance of seed size for the establishment of seedlings of five deciduous broad-leaved tree species. *Vegetatio*, 123, 51-64.

Sparks, T. H., Jeffree, E. P. & Jeffree, C. E. (2000) An examination of the relationship between flowering times and temperature at the national scale using long-term phenological records from the UK. *International Journal of Biometeorology*, 44, 82-87.

Takahashi, K. (2003) Effects of climatic conditions on shoot elongation of alpine dwarf pine (*Pinus pumila*) at its upper and lower altitudinal limits in central Japan. *Arctic, Antarctic, and Alpine Research*, 35, 1-7.

Takahashi, K. (2005) Effects of artificial warming on shoot elongation of alpine dwarf pine (*Pinus pumila*) on Mt. Shogigashira, central Japan. *Arctic, Antarctic, and Alpine Research*, 37, 620-625.

Takahashi, K., Azuma, H. & Yasue, K. (2003a) Effects of climate on the radial growth of tree species in the upper and lower distribution limits of an altitudinal ecotone on Mt. Norikura, central Japan. *Ecological Research*, 18, 549-558.

Takahashi, K., Mitsuishi, D., Uemura, S., Suzuki, J. & Hara, T. (2003b) Stand structure and dynamics during a 16-year period in a sub-boreal conifer-hardwood mixed forest, northern Japan. *Forest Ecology and Management*, 174, 39-50.

Takahashi, K. & Rustandi, A. (2006) Responses of crown development to canopy openings by saplings of eight tropical submontane forest tree species in Indonesia: a comparison with cool temperate trees. *Annals of Botany*, 97, 559-569.

Takahashi, K., Seino, T. & Kohyama, T. (2001) Responses to canopy openings in architectural development of saplings in eight deciduous broad-leaved tree species. *Canadian Journal of Forest Research*, 31, 1336-1347.

Takahashi, K., Seino, T. & Kohyama, T. (2005) Plastic changes of leaf mass per area and leaf nitrogen content in response to canopy openings in saplings of eight deciduous broad-leaved tree species. *Ecological Research*, 20, 17–23.

Takahashi, K. & Yoshida, S. (2009) How the scrub height of *Pinus pumila* decreases at the treeline. *Ecological Research*, 24, 847-854.

Takenaka, A. (1986) Comparative ecophysiology of two representative *Quercus* species appearing in different stages of succession. *Ecological Research*, 1, 129-140.

Taneda, H. & Tateno, M. (2005) Hydraulic conductivity, photosynthesis and leaf water balance in six evergreen woody species from fall to winter. *Tree Physiology*, 25, 299-306.

Thuiller, W. (2003) BIOMOD - optimizing predictions of species distributions and projecting potential future shifts under global change. *Global Change Biology*, 9, 1353-1362.

van der Werf, G. W., Sass-Klaassen, U. G. W. & Mohren, G. M. J. (2007) The impact of the 2003 summer drought on the intra-annual growth pattern of beech (*Fagus sylvatica* L.) and oak (*Quercus robur* L.) on a dry site in the Netherlands. *Dendrochronologia*, 25, 103-112.

Way, D. A. & Sage, R. F. (2008) Elevated growth temperatures reduce the carbon gain of black spruce [*Picea mariana* (Mill.) B.S.P.]. *Global Change Biology*, 14, 624-636.

Woodward, F. I. (1979) The differential temperature responses of the growth of certain plant species from different altitudes. II. Analyses of the control and morphology of leaf extension and specific leaf area of *Phleum bertolonii* D.C. and *P. alpinum* L. *New Phytologist*, 82, 397-405.

Yamamoto, S. (1989) Gap dynamics in climax *Fagus crenata* forests. *Botanical Magazine, Tokyo*, 102, 93-114.

Yin, H., Lai, T., Cheng, X., Jiang, X. & Liu, Q. (2008) Warning effect on growth and physiology of seedling of *Betula albo-sinensis* and *Abies faxoniana* under two contrasting light conditions in subalpine coniferous forests of western Sichuan, China. *Journal of Plant Ecology*, 32, 1072-1083.

8

The Influence of Climate Change on Rice in China from 1961 to 2009

Yanling Song, Bo Liu and Guoli Tang
National Climate Centre, China Meteorological Administration, Beijing, China

1. Introduction

At present, 40% of the Earth's land surface is managed for cropland and pasture (Foley et al., 2005). In developing countries, nearly 70% of people live in rural areas where agriculture is the largest supporter of livelihoods. For China, as such large developing country, the total population accounted to 13 billion with 73% of them pursuing agriculture, and the cropland was 1.3 billion ha in 2006 (CAY, 2007). About 20% of the global population lives in China supported only by 7% of the world's cultivated land. Although Chinese agriculture has undergone tremendous structural changes over the last decades, the average staple crop productivity has doubled in 25 years while the population increased by 25 % (CSY, 2003), until now, agriculture is still the most important industry for China which can affect food supplies of 13 billion populations.

Rice is one of China's most important staple food crops, and it's planting area amounted to 27.4% of all food crop area, and it's production was 37.1% of total food production in China (CAY, 2007). Meanwhile, the production of rice was influenced strongly by climate (Olesen, 2002), so it's confirmed that the agriculture will be affected greatly by climate change.

2. Indexes

2.1 Rice planting regions

The triple-cropping rice can be planted in southern China when the accumulated temperature amounted to 7000 °C (daily temperature >10 °C), while when the accumulated temperature changed during 5300 – 7000 °C accumulated temperature amounted to, the double-cropping rice can growth (Cheng and Li, 2007).

2.2 The growing season length

Cheng and Li (2007) gave the indexes of the growing length of rice in southern China, the start of growing length was defied by the daily temperature more than 10 °C stably, and the end of growing season was defied by the daily temperature more than 20 °C.

2.3 High temperature

The production of rice was influenced strongly when the daily temperature was more than 35 °C in southern China, so 35 °C was defied as high temperature.

3. Results

3.1 Air temperature and precipitation variation in China, 1961-2009

The high quality data collected by 511 weather stations (selected from 748 CMA baseline or basic stations in line with the criteria of continuity and integrity) were employed to analyze and understand the temperature and precipitation variation trends in China during the period of 1961-2009. Anomalies were calculated based on the benchmark values of meteorological elements. Meanwhile, temperature and precipitation anomalies were calculated in grids (5 ° × 5 °, longitude × latitude). The averaged nationwide series were obtained through an area-weighted process. The benchmark values of meteorological elements were derived from the averages of 1971-2000.

3.1.1 Temperature variation trends, 1961-2009

Fig. 1 depicts the averaged temperature anomalies in China during the period of 1961-2009. It is apparent that in a period of 49 years, China has witnessed an ascending mean temperature trend across the country, or 1.5°C up for the period of 1961-2009. The mean temperature in most years has registered a positive anomaly since the 1990s, with 2007 being the warmest.

Most parts of China enjoyed an ascending trend of air temperature, as shown by the spatial distribution of air temperature variation in Fig. 2, with the northern part sitting in a noticeably warming range between 0.3-0.6 °C/10a. The northeast part of the northeast region, the central part of Inner Mongolia, and the east part of Xinjiang also saw a temperature rise up to 0.6-0.9 °C/10a. Most part of the Yellow River in the south came up with a universal small rise of temperature, basically under 0.3 °C/10a.

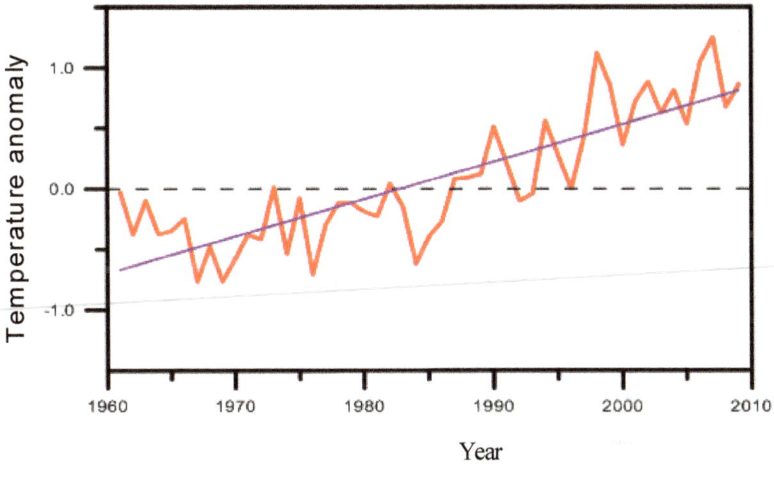

Fig. 1. Averaged temperature anomalies, 1961-2009 (°C).

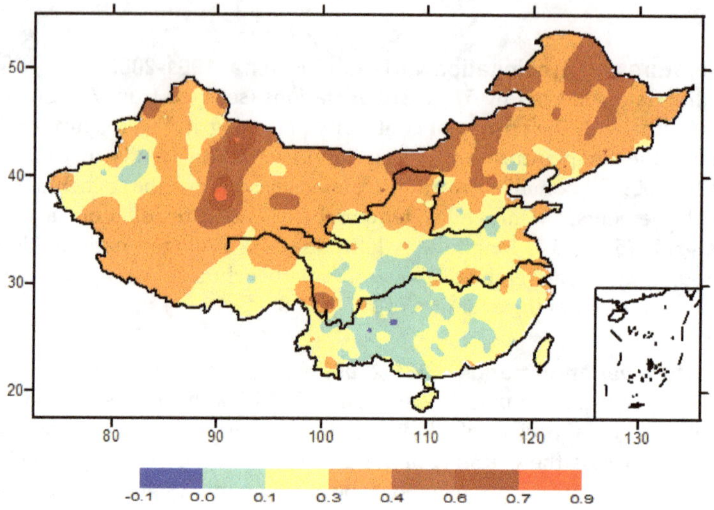

Fig. 2. Spatial distribution of temperature variation across the country, 1961-2009 (°C/10a).

3.1.2 Precipitation variation, 1961-2009

China had not registered a noticeable trendy variation of averaged precipitation during the period of 1961-2009, with a slight increase of 12.86mm (Fig. 3), though the curve went down in the latest 10 years. Fig. 4 shows the spatial distribution of precipitation trends across the country, suggesting a decline in most parts of East China, noticeably in most parts of North China, the Pan-Bohai Sea region, and the central and east part of the southwest region. During the same period of time, the lower and middle streams of the Yangtze River, the southeast coastal areas, and Hainan saw increased rainfall. On the other hand, most parts of China west of 105 °E tended to have an increased precipitation, noticeably in the southeast part of Tibet.

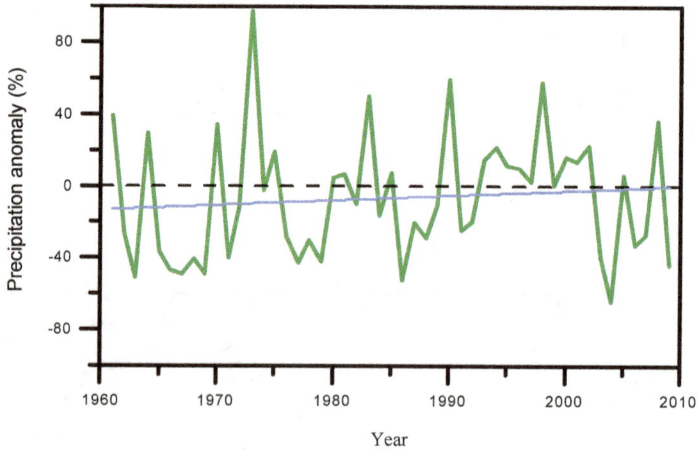

Fig. 3. Averaged precipitation anomalies, 1961-2009 (mm).

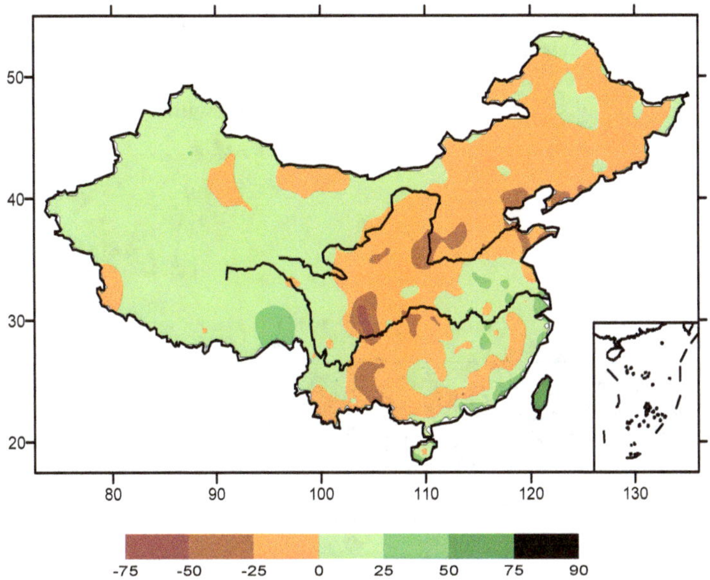

Fig. 4. Spatial distribution of precipitation variation, 1961-2009 (mm/10a).

3.2. Climate change impact on the northward shift of double rice cropping in South China

3.2.1 Active accumulated temperature variation across the southern rice growing areas

Paddy rice is a short-day plant which needs the warm and wet environment. Rice growing activities have to be defined in line with a range of ecological and environmental elements, including heat, growing season, water, sunshine, elevation, soil, production condition, and cropping system (Cheng, 2007). Heat resource makes a pivotal element in defining the distribution of rice growing areas. Theoretically, the heat resource defined for paddy rice is technically expressed in an active accumulated temperature≥10°C. Active accumulated temperature is the sum total of the temperatures that are above the lowest biological temperature a plant has had during the entire growth season in a year, or the sum total of the averaged daily temperatures that are above a given threshold value. In southern China, triple cropping rice can be planted when the active accumulated temperature was more than 7000°C, and double cropping rice can grow when the active accumulated temperature was more than 5300°C (Cheng, 2007).

Previous studies show that most parts of China have registered an increased heat resource against the background of global warming, which extends crops' growing season and allows thermophilic crops to move further north, changing the pattern of cropping systems. During the period of 1961-2009, the southern rice growing areas witnessed a noticeably increased active accumulated temperature (≥ 10 °C), with an averaged active accumulated temperature at 5712.8 °C in the 60's, and 5686.3 °C and 5677.0 °C in the 70's and 80's, respectively. The same indicator rose to 5809.8 °C in the 1990's, or 97.0 °C up compared with the 60's. During the period of 2001-2009, the averaged active accumulated temperature hit 5980.8°C, or 268.0 °C higher compared with the 60's, or 303.8 °C up against the 80's. During the 19 years running from 1991 to 2009, 16 years reported an active accumulated temperature that was higher than it from 1971 to2000, with 1998 being the warmest at 6144.9 °C, or 420.6 °C higher than what would have been in a normal year. The same indicator appeared at 6011.1 °C in 2009, or 286.8 °C higher compared with a normal year (Fig. 5).

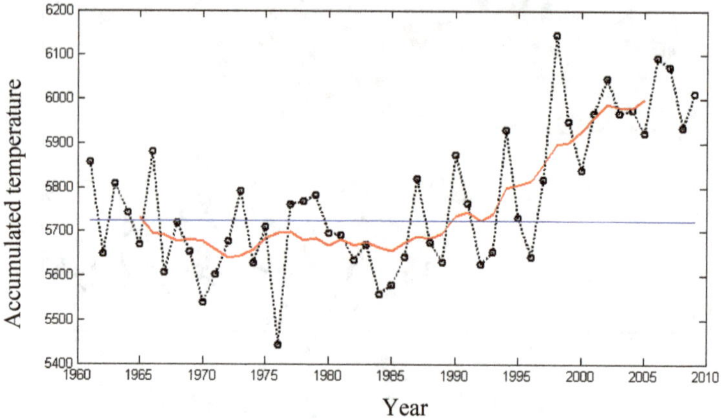

Fig. 5. Active accumulated temperature variation in the southern rice growing areas, 1961-2009 (°C).

Rice is a staple food crop in China. In 2000, the rice sown area reached 29.962 million hectares, accounting for 27.6% of the grain sown area in the country, with a yield of 187.91 million tons, or 40.7% of the total food crop yield in the country (CSY, 2001). Heat resource defines the distribution and cropping of rice growing. For example, an area having an accumulated temperature at 2000-5300°C (≥ 10 °C) is desirable for the one-cropping system, while the one at 5300-7000°C may enjoy a double cropping practice. An area with an accumulated temperature at 7000°C or above is allowed to grow three cropping.

During the period of the 1960's-1980's, China saw limited variations in heat distribution. The areas having an annual active accumulated temperature (≥ 10 °C) at 7000°C or above sat mostly in the central and southern part of South China, practicing a three cropping system. The areas with an annual active accumulated temperature (≥ 10 °C) between 5300°C and 7000°C have included most parts of Hunan, Jiangxi, the central and southern part of Zhejiang, most parts of Fujian, the southeast part of Hubei, some parts of Sichuan and Chongqing, and the southern part of Yunnan (Figs. 6).

During the period from the 1990s to 2009, the safe threshold for double cropping (5300°C) and triple cropping (7000°C) moved further north. For example, the northern boundary of double cropping moved up for more than 2 degrees, or some 300 km (Figs. 6). As a result, a range of areas, including the east part of Sichuan, most parts of Chongqing, the east part of Guizhou, most parts of Hubei, most parts of Anhui, and the southern part of Jiangsu, became desirable for double cropping.

China National Climate Center predicts that China will see a further warming climate during the period of 2011-2020 (Ding et al., 2007). Most parts of the country will have a sustained rise of active accumulated temperature. The trend may further increase the heat resource for the newly added double cropping areas. As a result, the northern boundary of double cropping may move further north (Fig.6). These areas apparently need to readjust the cropping system in a scientific manner, in line with the local conditions, adapting to the changed climate resource and environment, and enhancing the utilization of climate resources.

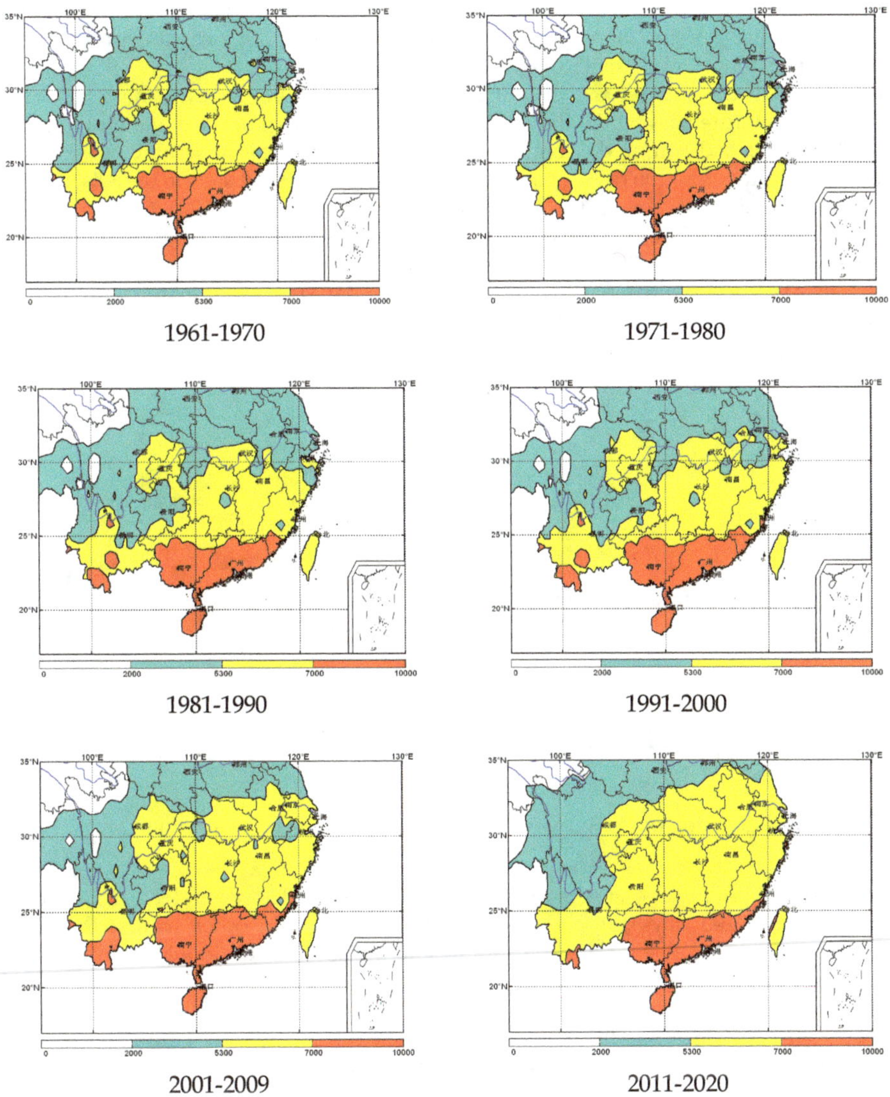

(the data during 1961-2009 from observed meteorological stations, and the data during 2011-2020 projected by RegCM3)

Fig. 6. Distribution of active accumulated temperature (≥ 10 °C) in the southern rice growing areas. Note: the red color represents the areas desirable for triple cropping, and the yellow one for double cropping.

3.2.2 Changed duration of the growing season

The growing season duration is also an element that defines the distribution of rice growing areas. For example, early rice is sown in the late February in the southern part of China, with late rice being harvested in the early November. The sowing could be later, starting from the mid-March in the south of the Yangtze River, with late rice becoming ripened in the mid-October. In southern China, triple cropping rice can be planted when the growing season length was longer than 270 days, and double cropping rice can grow when the growing season length was longer than 200 days (Cheng, 2007).

During the period of 1961-2009, the duration of rice growing season (the days with an averaged temperature passing the threshold of ≥ 10 °C in the beginning of the growing season and ≥ 20 °C at the end of it) in the southern part of the country was noticeably prolonged (Fig. 7). For example, the duration of rice growing season lasted for 214.9 days in the 60's, with 221.9 days and 211.7 days in the 1970's and 1980's, respectively. In the 90's, an averaged rice growing season would last for 220.3 days, a bit longer compared with the 60's and 80's, though slightly shorter against the 70's. During the period of 2000-2009, the rice growing season rose to 231.4 days in the southern rice growing areas, or 16.5 days longer compared with the 60's, or 19.7 days longer against the 80's. Apparently, the duration of rice growing season has become noticeably longer in the southern part of China since the 21st century.

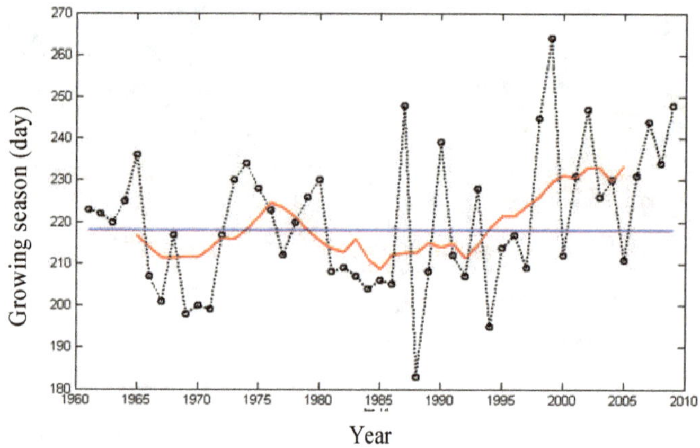

Fig. 7. Changed duration of rice growing season in the southern rice growing areas, 1961-2009 (day).

In China, the safe rice growing season would last for some 270 days under a triple cropping system, or some 200 days for a double cropping system. In other words, the 270-day growing season is desirable for triple cropping, and the 200-day for double cropping.

Figs. 8 show that as far as the safe rice growing season is concerned, climate warming has imposed no significant impacts on the scope of the areas desirable for triple cropping. However, climate warming have noticeably pushed the northern boundary of double cropping further north. Such northward shift has become increasingly noticeable since the 21st century. In the period from the 60s to the 90s, the double cropping system was mainly practiced in the south of the Yangtze River Valley. However, the northern boundary of double cropping has gone further to the north of the Yangtze River since the 21st century, including the northeast part of Sichuan, the east part of Guizhou, Chongqing, most parts of Hubei, the central and southern part of Anhui, and the southern part of Jiangsu. Apparently, the northern boundary of double cropping has gone further north, compared with the 90's.

According to the analysis and prediction made by China National Climate Center, the period of 2011-2020 will see a further intensified climate warming trend in China, with a prolonged rice growing season in the southern part of the country, and a further northward shift of the double cropping boundary. The conclusion agrees with the northward shift of the double cropping boundary depicted by the results of active accumulated temperature calculation.

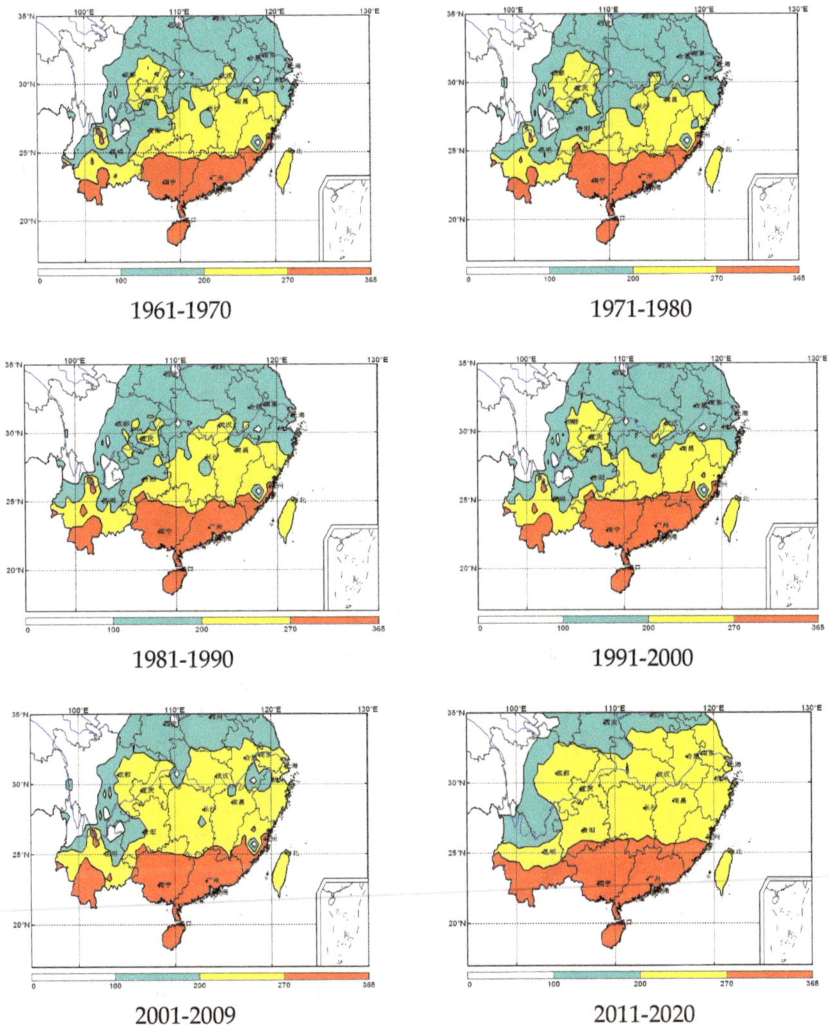

(the data during 1961-2009 from observed meteorological stations, and the data during 2011-2020 projected by RegCM3)

Fig. 8. Changed duration of rice growing season in the southern rice growing areas (day). Note: the red color represents the areas desirable for triple cropping, and the yellow one for double cropping.

3.2.3 Annual precipitation variation

Precipitation is also in a position to affect the distribution of rice growing areas. In China, both double and triple cropping need an annual rainfall no less than 1,000 mm. Fig. 19 shows that climate warming has produced limited impacts on the rainfall in the southern part of China. In the 60's, the southern part of the country where paddy rice was grown registered an averaged annual precipitation of 1379mm. The 1970's, 1980's, and 1990's recorded a sustained increase of precipitation at 1412 mm, 1435mm, and 1491 mm respectively (Fig. 9) . China, however, has seen a slightly decreased precipitation to 1431 mm starting from the 21st century.

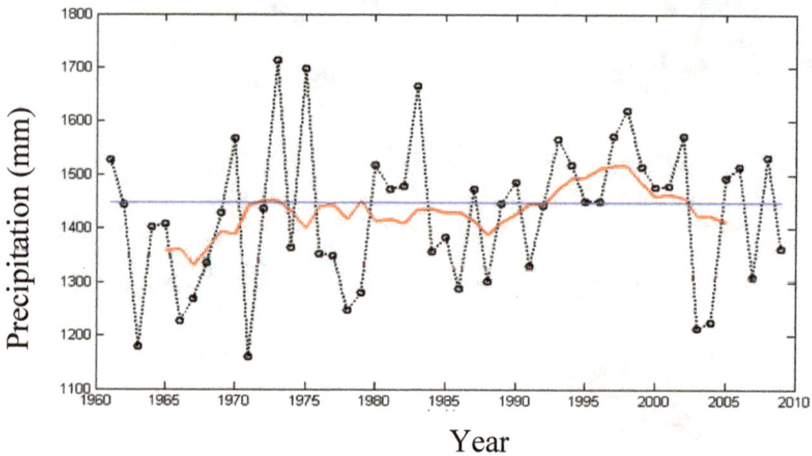

Fig. 9. Annual precipitation variation in the southern rice growing areas, 1961-2009 (mm).

In addition, most parts of Hainan Province, the east part of Guangxi, Guangdong, Fujian, and Jiangxi reported an annual precipitation of 1,500 mm or above in the 60's, with the rest of rice growing areas in the south sitting between 1,000mm and 1,500 mm (Fig. 10). In the 70's, the areas enjoying an annual precipitation of 1,500 mm or above did not see much change in number, though the areas sitting between 1,000mm and 1,500 mm in both Sichuan and Jiangsu reported a slight drop in number. In the 80's, the areas having an annual precipitation of 1,500 mm or above in Guangxi saw a slight drop in number, with little change of the areas sitting between ,1000mm and 1,500 mm. In the 90's, the areas having an annual precipitation of 1,500 mm or above in Guangxi reported some increase in number, though the areas sitting between 1,000mm and 1,500 mm in Sichuan suffered a noticeable drop in number. Zhejiang, Anhui, and Jiangxi have reported a noticeably declined number of the areas having an annual precipitation of 1,500 mm or above since the 21st century, with a slight drop of the areas sitting between 1,000mm and 1,500 mm in both Sichuan and Yunnan.

Overall speaking, the southern paddy rice growing areas in China didn't see much change in precipitation during the period of 1961-2009, though a number of provinces, including Jiangsu, Hubei, and Sichuan, had reported a declined number of the areas sitting between 1,000mm and 1,500 mm since the 21st century, which deserves further attention.

China National Climate Center predicts that the southern part of China will not see a large variation in precipitation during the period of 2011-2020 (Ding et al., 2007). As a result, precipitation will produce no significant impacts on the southern rice growing areas in the coming decade if the precipitation change slightly.

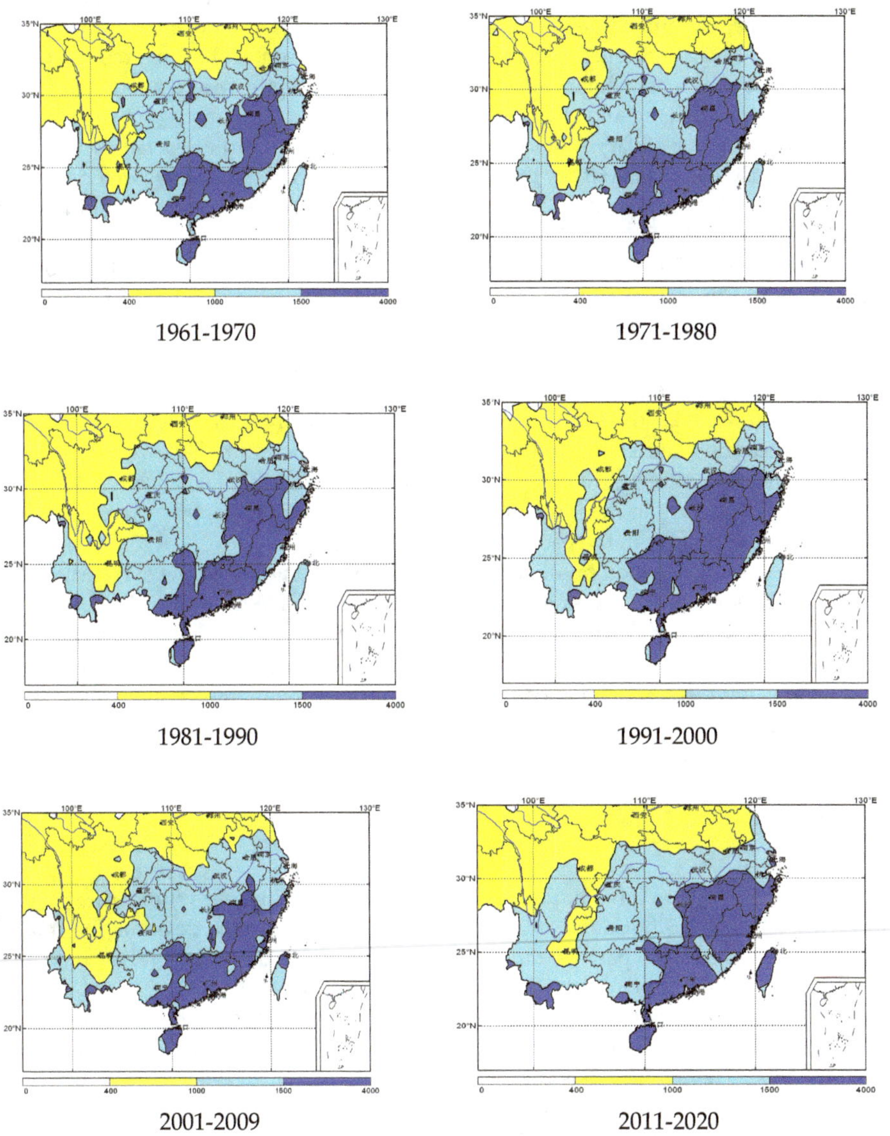

(the data during 1961-2009 from observed meteorological stations, and the data during 2011-2020 projected by RegCM3)

Fig. 10. Distribution of annual precipitation in the southern rice growing areas (mm).

3.2.4 High temperature day trends in the southern rice growing areas

The Jiangnan region

In the most part of the Jiangnan of China, rice would enter the heading, flowering, and filling stages in June and July. In the Jiangnan, the first cropping rice enters the heading and flowering stage in August. This period of time is featured with the positive dominance of subtropical highs, claiming more high temperature days. The abnormal frequency of hot days would result in a shortened period for heading, flowering, and filling, reducing the leaf assimilation on the one hand, and causing the surge of blighted rice and a reduced yield and quality on the other (Zhang and Wang, 1998).

In June, hot weather (≥ 35 °C) starts to affect the lower and middle reaches of the Yangtze River. In the early June, this part of the Jiangnan would have 0.3 hot days in a normal year, with 0.5 hot days for the period of 2001-2009. In the mid-June, 0.4 hot days would appear in a normal year, with 0.8 days in the period of 2001-2009. In the late June, 0.6 hot days is a common place, with 1.7 days for the period of 2001-2009. It is noteworthy that the lower and middle reaches of the Jiangnan has witnessed a noticeably raised number of hot days at 2.1 days during the period of 2005-2009 since 2005 (Table 1).

July is a month featured with a noticeably raised number of hot days in the lower and middle reaches of the Yangtze River. In the early July, hot days would reach 1.9 in number in a normal year, with 2.6 days for the years since 2000. The mid-July would produce 2.7 hot days, though 3.5 days since 2000. In the late July, high temperature days would go up to 3.2 days on an average, with 4.5 days for the period of 2000-2009. Technically, the late July claims the most hot days in the lower and middle reaches of the Jiangnan with a noticeably raised number of hot days in the past 20 years (Table 1).

August starts to see a declined number of hot days in the Jiangnan. In the early August, high temperature days reach 2.9 on an average, though with a slightly descending trend in the past 10 or 20 years. The mid-August would have 1.5 hot days, expecting no ascending trend. In the late August, high temperature days are averaged at 1.2 days. Technically, the early August would have a slightly increased number of high temperature days, and the mid-August a steadily decreased number of hot days since that time on.

Overall, the Jiangnan would have 14.7 hot days for the summer (June-August). The number of high temperature days has witnessed a noticeable rise, especially in the late June and late July, since 2000.

South China

The South China starts to have hot days (≥ 35 °C) in June. In the early June, this part of the country would have 0.5 hot days on an average (Table 1), with 0.6 days for the mid-June, and 0.8 days for the late part of the month. This part of the country has registered a noticeably increased number of hot days in the late June since 2000.

July is featured with a noticeably raised number of hot days, with 1.5 hot days for the early July (Table 1). The mid-July would have 1.9 hot days on an average, though 2-5 days for the years since 2000. The late July has registered a pattern similar to the mid-July, with 2.2 hot days on an average, and a noticeably increased number of hot days to 2-5 days since 2000.

In August, the lower and middle reaches of the Yangtze River start to see fewer hot days. The early August is averaged with 1.7 hot days, witnessed a noticeably increased number of hot days in a 10-year period. The mid-August would have 1.4 hot days, and the late August 1.2 hot days. Overall speaking, the early August has slightly more high temperature days, with an ascending trend in the past decade. The number of hot days steadily declines starting from the mid-August (Table 1).

		1961-1970	1971-1980	1981-1990	1991-2000	2001-2009
Jiangnan	Jun 1-10	0.3	0.2	0.4	0.3	0.3
	Jun 11-20	0.5	0.6	0.4	0.3	0.8
	Jun 21-30	0.8	0.7	0.6	0.7	1.7
	July 1-10	1.4	2.5	1.8	1.4	2.8
	July 11-20	3.6	2.1	3.3	2.8	3.4
	July 21-31	4.3	3.1	2.5	3.8	4.5
	Aug 1-10	3.1	3.3	2.6	2.8	2.9
	Aug 11-20	2.6	1.4	1.8	1.4	1.8
	Aug 21-31	2.7	1.4	1.0	1.1	2.1
South China	Jun 1-10	0.3	0.2	0.3	0.9	0.4
	Jun 11-20	0.4	0.7	0.6	0.5	0.5
	Jun 21-30	0.8	0.7	1.0	0.7	1.5
	July 1-10	1.1	1.7	1.5	1.2	2.1
	July 11-20	1.8	1.9	2.2	1.6	3.0
	July 21-31	2.5	1.9	2.1	2.4	3.1
	Aug 1-10	1.4	1.2	2.1	1.9	2.7
	Aug 11-20	1.4	0.9	1.9	1.4	1.8
	Aug 21-31	1.7	0.8	1.2	1.5	1.8

Table 1. The high temperature days(≥35 °C)for every 10days over Jiangnan and South China regions during 1961-2009.

In summary, the South China has, on an average, 11.8 hot days in the summer, though with a noticeably raised number of hot days since 2000, especially from the late June to the early August. In the past decade, the southern rice growing areas have witnessed a noticeably increased number of hot days in the summer, which may cause heat injury to the rice plants in their filling stage, compromising both the yield and quality of rice. In this context, one has to breed heat resistant rice varieties and develop technical measures to ease the damages caused by high temperature.

4. Conclusion

Rice is one of China's most important staple food crops, and the climate change have already influenced the growth of rice in China. All the above studies showed that the climate change were influenced the rice planting in southern China. For recent two decades, as the active accumulated temperature increases in the southern China, the rice growth period extended, with the double-cropping rice growing boundaries being apparently pushed northward, even showing a slight northward shit of triple-cropping rice boundaries. From 1960s to 1980s, the double-cropping rice only grew in the areas to the south of the Yangtze River. However, for the recent 10 years, its growth areas had shifted to those to the north of the Yangtze, with their northern boundaries having been pushed approximately 300 km to the north, extending to northeastern part of the Sichuan, eastern Guizhou, Chongqing, most Hubei, central and southern Anhui, as well as southern Jiangsu (Fig.11). Although climate warming increased the thermal resources for rice growth areas over the region, creating favorable conditions for making use the thermal resources, and for increasing the multi-cropping index, farmers may face challenges as how to better address the issues related to availability of additional farming machinery required for the enlarged double-cropping rice production. On the other side, as climate will continue to warm, the early rice may be negtively affected by high temperature (≥35 °C) on filling stage of rice, and therefore it is imperative to develop and use heat-resistant rice varieties to mitigate the high temperature-induced crop damages.

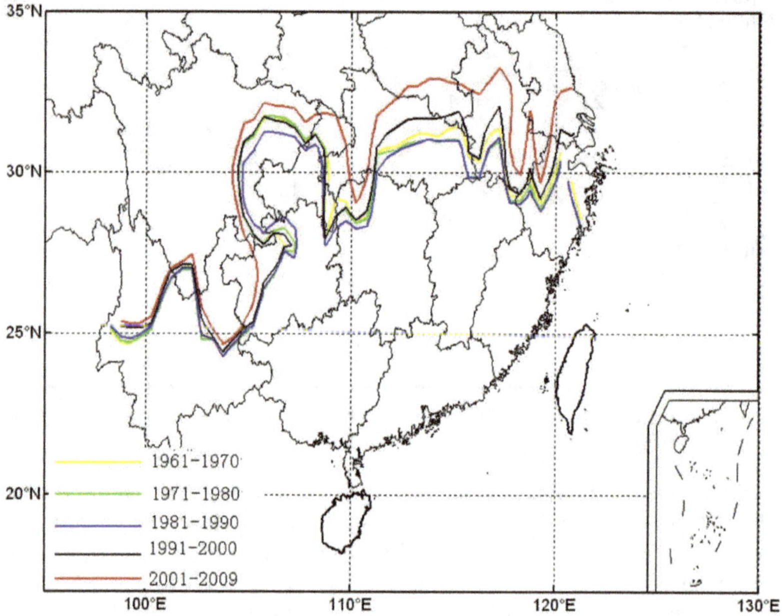

Fig. 11. The influence of climate change on growing northern boundaries of double-cropping rice in China from 1961 to 2009.

5. References

CAY, China Agriculture Yearbook, 2007: China Agriculture Publishing House, 140. (In Chinese)

Cheng S.H., Jian Li, 2007: Rice in China. Jindun Publishing, 90-93

CSY. China Statistical Yearbook. 2001: National Bureau of Statistics of China. China Statistics Press (In Chinese)

CSY. China Statistical Yearbook. 2003: National Bureau of Statistics of China. China Statistics Press (In Chinese)

Ding Y., Ren, G., Shi, G., Gong, P., Zheng, D., Zai, P., Zhang, D., Zhao, C., Wang, S., Wang, H., Luo, Y., Chen, D., Gao, X., Dai, X., 2007. China's National Assessment Report on Climate Change (I): Climate change in China and the future trend. Advances in Climate Change Research 3, 1-5.

Foley, J.A., R. DeFries, G.P. Asner, C. Barford, G. Bonan, S.R. Carpenter, F.S. Chapin, M.T. Coe and Co-author, 2005: Global consequences of land use. Science, 309,570-574

Olesen, J.E., M. Bindi · Consequences of climate change for European agriculture productivity, land use and policy. Eur.J.Agron , 2002 , 16: 239-262

Zhang Y., F T Wang, 1998: On the possible impacts of climate warming on rice production in China. Acta Meteorologica Sinica, 56(3), 369-376

9

Effects of High Night Temperature on Crop Physiology and Productivity: Plant Growth Regulators Provide a Management Option

Abdul Razack Mohammed and Lee Tarpley
Texas AgriLife Research and Extension Center, Beaumont, Texas USA

1. Introduction

Crop production is experiencing both increases in the frequency and intensity of high night temperatures (HNT) (IPCC, 2007). The HNT threatens the sustainability of crop production both currently and in the future. Recent meteorological data indicated faster increases in night temperatures (NT) than day temperatures (Alward et al., 1999). Experimental evidence also indicated the importance of assessing the effects of NT separately from the day temperature, as a smaller increase in NT can show a drastic decrease in crop production (Peng et al., 2004). Studies have indicated decreased crop yields as a result of HNT (Hall, 1992; Mohammed & Tarpley, 2009a). Apart from HNT, climate change will also increase a variety of environmental stresses affecting crop production, namely heat, drought, salinity, and relative humidity, hence farming in the future climate will have to be better adapted to a range of abiotic and biotic stresses. A long-term approach to negate the effects of abiotic stress is to develop stress-tolerant cultivars. The short-term approach includes the use of agrochemicals, especially those with plant growth regulator (PGR) capabilities, for the prevention and/or amelioration of various environmental stresses, including heat stress. Studies focusing on day temperature stress and, more recently, on NT stress have increased, but, studies looking at the effects of PGR under heat stress are rare (Ashraf & Foolad, 2007). This chapter primarily focuses on the effects of HNT on crop production and the beneficial effects of PGR (glycine betaine [GB] and salicylic acid [SA]) application against heat stress on crop production with special emphasis on rice (*Oryza sativa* L.) production.

2. Night temperature and its impact on crop production

Global climate change is a dynamic process affecting global air temperature, oceanic temperature, rainfall, wind and quality of incoming solar radiation. Global circulation models predict 1.4 to 5.8 °C rise in global temperature because of projected increase in the concentrations of all greenhouse gases by the end of the 21st century (Intergovernmental Panel on Climate Change [IPCC], 2007). Much of this increase in average daily temperature is projected to be due to an increase in NT. Night temperatures are expected to increase at a faster rate than day temperatures due to less radiant heat loss because of increased cloudiness (Alward et al., 1999). In some cases, increases in NT have been documented. For example, the scientists at the International Rice Research Institute (IRRI) reported an increase of 1.13 °C in NT over a period of 25 years (1979-2003) in the Philippines (Peng et al., 2004). The HNT increased at a rate of 0.18 °C per decade over a period of 45 years (1950-1995) in Libya (Jones et al., 1999).

Long- and short-term episodes of heat stress are predicted to occur more frequently as a result of global warming, affecting many aspects of crop growth and development, reducing crop yield and decreasing crop quality. Many studies have reported the impacts of long- and short-term temperature stresses on crop production. However, most of the studies assume no difference in the influence of day versus night temperature (Peng et al., 2004). High night temperature is known to decrease yield in several crops like cereals [rice, wheat (*Triticum aestivum* L.) and corn (*Zea Mays* L.)], legumes [cowpea (*Vigna unguiculata* L. Walp.)], oil seeds [soybean (*Glycine max* L. Merr.) and sunflower (*Helianthus annuus L.*), vegetables [tomato (*Solanum lycopersicum* L.)], fruits [apples (*Malus domestica* L.) and pineapple (*Ananas comosus* L.)] and fibers [cotton (*Gossypium hirsutum* L.)] (Chang, 1981; Gibson & Mullen, 1996; Izquierdo et al., 2002; Kondo & Takahashi, 1989; Loka & Oosterhuis, 2010; Mohammed & Tarpley, 2009a; Neales et al., 1980; Peters et al., 1971; Seddigh & Jolliff, 1984; Warrag & Hall, 1984; Willits & Peet, 1998).

High night temperature decreases crop production by decreasing photosynthetic function, sugar and starch content (Loka & Oosterhuis, 2010; Turnbull et al., 2002), increasing respiration rate (Mohammed & Tarpley, 2009b), suppressing floral bud development (Ahmed & Hall, 1993), causing male sterility and low pollen viability and hastening crop maturity (Mohammed & Tarpley, 2009a; Seddigh and Jolliff, 1984). Another effect of HNT that can contribute to reduced crop yield is decreased antioxidant capacity of the plants. Under normal physiological conditions, the toxic effects of reactive oxygen species (ROS) are minimized by enzymatic and non-enzymatic antioxidants. Under stress conditions, oxidant levels can overwhelm the antioxidant levels leading to cell damage. The increased production of ROS [oxide radical (O_2^-), H_2O_2, and the hydroxyl radical (-OH)], or the plant's decreased ability to neutralize ROS, as a result of heat stress negatively affects many physiological processes in plants, thus decreasing yield.

2.1 Methodology for determining rice plant response to high night temperature stress

Plant height and the numbers of tillers and leaves were recorded at harvest. Daily observations were made for the appearance of panicles. At harvest, leaves were separated from the stems and leaf area was measured using a CI-251 area meter (CID Inc., Camas, Washington, USA). During the rice plant vegetative stage, leaf photosynthetic rate, internal CO_2 concentration and respiration rate were measured using LI-6400 portable photosynthesis system (LI-COR Inc., Lincoln, Nebraska, USA). Leaf nitrogen content was determined using a FP-528 Nitrogen/Protein analyzer (LECO Corporation, St. Joseph, Michigan, USA) and was expressed as percentage (%; w/w). Total chlorophyll content, pollen germination and spikelet fertility were determined using the procedures from Mohammed & Tarpley (2011a), and relative injury to the membrane and total antioxidant capacity were determined using procedures from Mohammed & Tarpley (2009b).

2.2 Effects of high night temperature on plant morphology

Plant morphology deals with plant development, form, and structure. In a narrow sense, plant morphology refers to the external appearance of the plant. For example plant height and the numbers of tillers, leaves, panicles and grains are referred as rice morphological parameters. The effects of HNT on plant morphology vary from negative to positive, including no effect. Seddigh & Jolliff (1984) reported that HNT promoted early vegetative growth and hastened physiological maturity, but did not affect morphological characteristics such as plant height, number of auxiliary branches and number of nodes in soybean. The HNT has no effect on rice (cultivar Cocodrie which is a japonica cultivar) plant height and number of tillers but can increase number of leaves and leaf area (Fig. 1). However, Cheng et al. (2009) and Yoshida et al. (1981) reported reduction in plant height, number of tillers and total biomass for rice cultivar 'IR 72' (indica cultivar) as a result of HNT. Similar results were seen in redstem stork's bill (*Erodium cicutarium* L.) for which HNT decreased number of branches, leaf area and plant dry matter (Blackshaw & Entz, 1995). The decreased production of plant dry matter was directly correlated with decreased production of photosynthates.

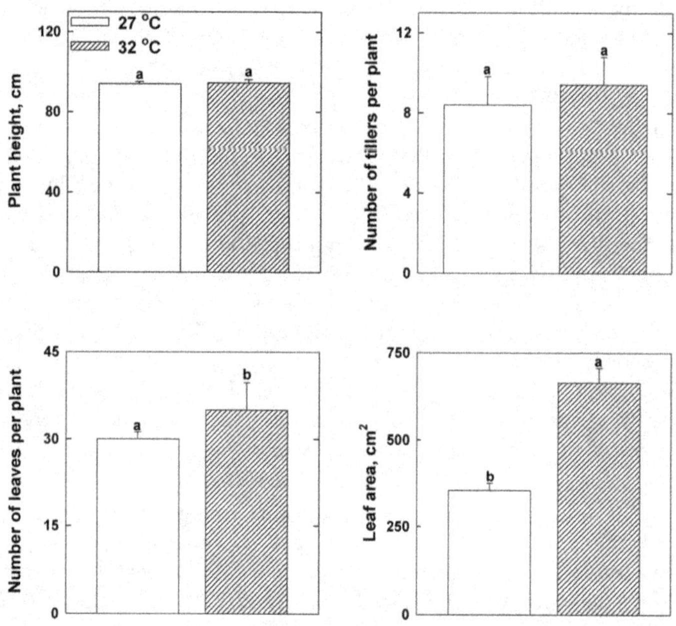

Fig. 1. Effects of high night temperature on rice morphology. Each bar represents average ± S.E. Different letters indicate means are significantly different at the P<0.05 level.

2.3 Effects of high night temperature on plant physiology

Plant physiology deals with the functioning of plants and is closely related with plant morphology. Fundamental processes such as photosynthesis, respiration, plant water and nutrition status, plant hormone functions and translocation of photosynthates are encompassed in plant physiology. Photosynthesis is a complex process and leaf photosynthetic rates depend upon leaf chlorophyll and nitrogen content, photosystems, stomatal characteristics and enzyme activities. The environmental stresses have a direct effect on leaf photosynthetic rates and an indirect effect through their effects on leaf chlorophyll and nitrogen content, stomatal characteristics and enzyme activities. The HNT had positive and negative effects, including no effect, on the following day's photosynthetic rates (Frantz et al., 2004; Mohammed & Tarpley 2009a). Rice leaf photosynthetic rates showed negative correlation with HNT (Fig. 2). The decrease in photosynthetic rates as a result of HNT was accompanied with decrease in leaf chlorophyll and leaf nitrogen content in rice (Fig. 2). Premature loss of chlorophyll as a result of heat stress has been noticed in many crop species (Guo et al., 2006; Reynolds et al., 1994). Damage to the photosynthetic electron transport particularly at the site of photosystem II (Havaux and Tardy, 1996) and inhibition of photosynthetic CO_2 fixation (Sayed et al., 1989; Yamane et al., 1997) as a result of heat stress can also reduce photosynthetic rates, thus leading to decrease in yield. Apart from decreases in photosynthetic rates, increased respiration rates can also decrease the crop yield.

Plant respiration is one of the important processes in terms of understanding plant growth and development when subjected to stress. Increase in respiration from climate warming is of serious concern as respiratory processes can consume a larger portion of total photosynthates (Paembonan et al., 1992). On average, the carbon lost from respiratory metabolism within an individual plant ranges between 30 and 70% of the carbon gained through photosynthesis (Peterson & Zelitch, 1982). Respiration is typically partitioned into the functional components of construction (growth), maintenance and ion uptake to facilitate our understanding of the impact of the environment on respiratory processes (Amthor, 1986; Farrar, 1985; Lambers, 1985). Maintenance respiration is mainly associated with turnover of proteins and lipids and maintenance of ion concentrations across membranes (Penning de Vries, 1975). It is the most responsive to environmental changes among the functional components of respiration (Ryan, 1991). At high temperatures, the cost of maintenance increases to support protein turnover and to maintain active ion fluxes across the membranes (Penning de Vries, 1975), thereby increasing maintenance respiration. An increase in NT from 27 °C to 32 °C increased respiration rates by 40% in rice leaves (Fig. 2). Previous studies on rice, cotton, lettuce (*Lactuca sativa* L.), tomato and soybean indicated an increase in maintenance respiration with warmer nights (Frantz et al., 2004; Loka & Oosterhuis, 2010, Mohammed & Tarpley, 2009b). Hence, HNT can stimulate respiration rates, thereby negatively affecting the yield (Zheng et al., 2002). The other consequence of increased respiration is increased production of ROS.

The production of ROS is an unavoidable consequence of aerobic respiration, with the majority of the ROS produced in photosynthetic tissue (mostly leaves) in the dark by mitochondrial electron transport chain activity (miETC) (McDonald & Vanlerberghe, 2005). The production of ROS by mitochondria has been shown to increase in many plants as a result of biotic and abiotic stresses (McDonald & Vanlerberghe, 2005; Moller, 2001). An increase in NT increases maintenance respiration and thus increases production of ROS. Physiological injury due to heat stress has been associated with increases in oxidative damage to the membrane in plant species (Larkindale & Knight, 2002). Plants increase maintenance respiration to support repair mechanisms of the membranes due to oxidative damage (Amthor & McCree, 1990). Thus, an increase in respiration occurs 1) with an increase in temperature (Huang et al., 1998) and 2) to support repair mechanisms of the membranes due to oxidative damage (Amthor & McCree, 1990). In addition, ROS interferes with photosynthesis and respiration by disrupting water, ion, and organic solute movement across plant membranes by affecting membrane stability (Christiansen, 1978).

Destabilized membranes are leaky membranes, thus interpreting the amount of electrolytic leakage from the membrane can be used as an indicator of cell membrane stability (Sullivan & Ross, 1979). Many studies have examined stability loss of the membranes subjected to environmental stresses by measuring electrolytic leakage from the membranes (Blum & Ebrercon, 1981; Dexter, 1956; Ibrahim and Quick, 2001; Ismail and Hall, 1999; Sullivan, 1972). An increase in NT from 27 °C to 32 °C increased injury to the membrane by 60% in rice leaves (Fig. 2). Similar results of increased injury to the membrane as a result of heat stress were seen in many crop species (Ibrahim & Quick, 2001; Ismail & Hall, 1999;, Martineau et al., 1979; Mohammed & Tarpley, 2009b). The properties of the photosynthetic system, including key enzymes and thylakoid membrane activities depend on the stability of membranes (Björkman et al., 1980). Moreover, it is well known that a functional cell-membrane system is central to crop productivity and acclimation of plants to high temperature (Raison et al., 1980). Mohammed & Tarpley (2009b) and Reynolds et al. (1994) positively associated membrane stability with yield performance under heat-stressed conditions in rice and wheat.

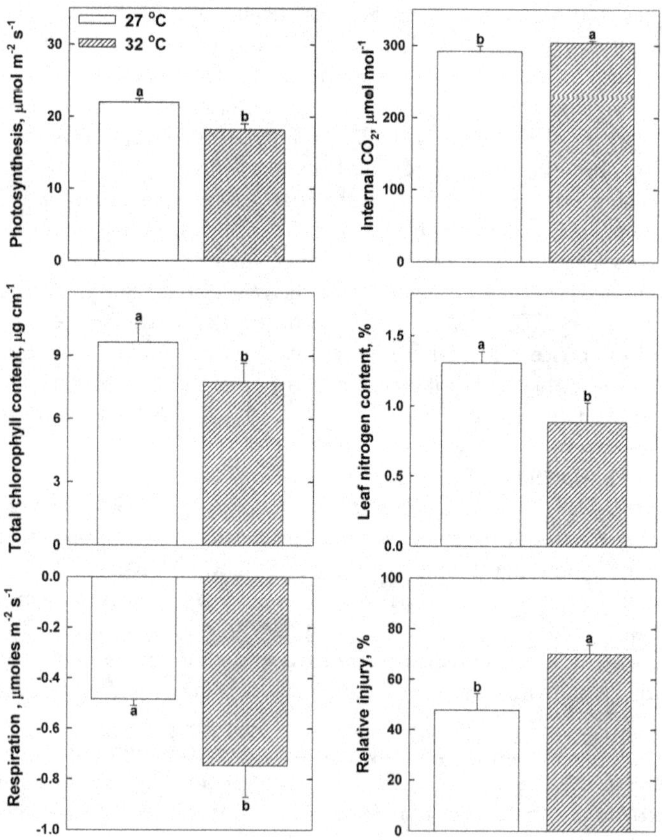

Fig. 2. Effects of high night temperature on rice physiology. Each bar represents average ± S.E. Different letters indicate means are significantly different at the P<0.05 level.

2.4 Effects of high night temperature on yield and yield-related parameters

The reproductive phase is relatively more sensitive than the vegetative phase to heat stress in many crop species (Hall, 1992). Moreover, differential temperature sensitivity for reproductive and vegetative growth has been reported in rice (Baker et al., 1992), soybean (Baker et al., 1989), wheat (Mitchell et al., 1993) and many other crops. Warrag & Hall (1984) reported that HNT induced male sterility and excessive floral abscissions in cowpea, whereas high day temperature had no adverse effect. The HNT can decrease crop yields by decreasing crop growth duration, suppressing floral bud development and decreasing pollen production and viability (Ahmed & Hall, 1993; Mohammed & Tarpley, 2009a; Prasad et al., 1999; Prasad et al., 2006). Increase in NT from 27 °C to 32 °C decreased crop growth duration by 2 days, as indicated by emergence of 1st panicle, in rice (Fig. 3). Decreased crop duration due to HNT decreases the time for carbohydrate accumulation, thus decreasing the yield (Cantarero et al., 1999). Badu-Apraku et al. (1983) attributed dramatic yield loss under high temperatures during the period of grain filling to sucrose availability (Afuakwa et al., 1984) and activity levels of enzymes involved in starch and sugar metabolism (Singletary et al., 1994). Moreover, the suppression of floral buds and flowering under HNT was also attributed to a shortage of photosynthetic assimilates supplied to the floral buds (Guinn, 1974), and/or an inability of floral buds to mobilize carbohydrates under heat stress (Dinar & Rudich, 1985). Decreased pollen germination due to heat stress has been noticed in many crop species (Hall, 1992; Matsui et al., 2001; Mohammed & Tarpley, 2009a). This decrease in pollen germination at high temperatures is due to poor anther dehiscence and pollen reception (Prasad et al., 2006), reduced pollen swelling and decreased anther pore size (Matsui & Kagata, 2003).

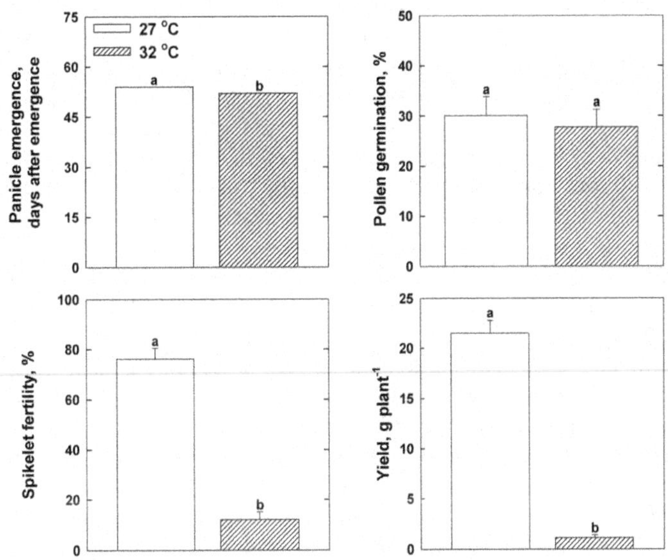

Fig. 3. Effects of high night temperature on rice yield-related parameters. Each bar represents average ± S.E. Different letters indicate means are significantly different at the P<0.05 level.

In most cereal crops, spikelet fertility, described as the ratio between filled grains and total grains, decreased with decrease in pollen germination. Increase in NT from 27 °C to 32 °C decreased spikelet fertility by 70% in rice (Fig. 3). Apart from pollen germination, hormonal balance in the sink (Micheal & Beringer, 1980) and/or availability and transportability of photosynthates to the sink from the source (Afuakwa et al., 1984) and/or inability of floral buds to mobilize carbohydrates (Dinar & Rudich, 1985) and/or altered activities of starch and sugar biosynthesis enzymes (Keeling et al., 1994; Singletary et al., 1994) can govern the spikelet fertility at high temperatures. The capacity to survive heat stress varies with plant species, genotype and developmental stage. The PGR provide an option to protect the plants from heat stress, thus providing sustainability for crop productivity.

3. Plant growth regulators

Genetic improvement and the modification of agronomic practices can be beneficial for crop adaptation to future climate conditions. However, these approaches are more applicable for long-term episodes of heat stress. The use of agrochemicals, especially those with PGR capabilities, for the prevention and/or amelioration of various environmental stresses, such as those resulting from global climatic change, is a viable approach to make crop production more resilient to short-term episodes of heat stress. Plant growth regulators are natural or synthetic chemical compounds that are used to promote or inhibit plant growth and development or alter specific physiology or metabolic factors. Like plant hormones, small concentrations of PGRs are enough to influence plant growth and development. In recent years, numerous PGRs have been developed to promote crop production under stress conditions. Glycine betaine, SA, vitamin E, proline and choline are some of the agrochemicals which can induce stress-tolerance (thermotolerance, drought tolerance, cold tolerance and/or salinity tolerance) in various crop plants. In this chapter, we focus on the beneficial effects of GB and SA application against heat stress on crop production.

Oxygen is potentially toxic to all organisms because metabolism and environmental stresses generate ROS from the oxygen. Under normal physiological conditions, the toxic effects of ROS are minimized by enzymatic and non-enzymatic antioxidants (Kreiner et al., 2002). Under stress conditions, oxidant levels overwhelm the antioxidant levels leading to cell damage (Kreiner et al., 2002). Plants exposed to environmental stress, such as heat, cold, drought and salinity produce ROS, which damage macromolecules and cell membranes (Zhang & Kirkham, 1996). Angiosperms possess several enzymatic and non-enzymatic scavenging systems to minimize deleterious effects of ROS. These include lipid-soluble antioxidants (e.g. α-tocopherol and β-carotene), water-soluble reactants (e.g. ascorbic acid and glutathione), and enzymatic antioxidants (e.g. superoxide dismutase, catalase and enzymes of the ascorbate and glutathione cycle) (Zhang & Kirkham, 1996). Glycine betaine and SA are synthesized in the plants and play important but different, roles in preventing oxidative damage to the membranes (Bowler et al., 1992; Demiral & Turkan, 2004) and inducing theromotolerance in plants subjected to stresses (Ashraf & Foolad, 2007; Caldas et al., 1999; Farooq et al., 2008a, 2008b; Larkindale & Huang, 2004; Mohammed & Tarpley, 2009b; Pan et al., 2006; Raskin, 1992). The commercial availability of GB and SA provides potential crop-management options to reduce crop heat stress events.

Glycine betaine, an important osmoprotectant is an amino-acid derivative that is naturally synthesized and accumulates under stress conditions in some plants (Bohnert & Jensen, 1996; Demiral & Turkan, 2004). It enhances stress tolerance by protecting enzymes (Paleg et al., 1981), photosystem II (Allakhverdiev et al., 1996), membrane integrity and antioxidant status of the plant (Mohammed & Tarpley, 2009b). Glycine betaine may also protect plants from environmental stresses indirectly via its role in signal transduction and ion homeostasis (John, 2002; Yilmaz, 2004). It regulates signal transduction and ion homeostasis by affecting MAP kinases, and expression of many genes related to signal transduction (John, 2002). Exogenous application of GB improved growth, survival and induced stress tolerance in many crop species (Diaz-Zorita et al., 2001; Harinasut et al., 1996; Rajasekaran et al., 1997) and also increased total antioxidant capacity and antioxidant enzyme activities (Farooq et al., 2008a; Mohammed & Tarpley, 2009b).

Salicylic acid is an endogenous plant growth regulator of phenolic nature and is considered to be a potent plant hormone because of its diverse regulatory roles in plant metabolism (Raskin, 1992). It plays an important role in the regulation of plant growth, development and responses to abiotic and biotic stresses (Raskin, 1992; Senaratna et al., 2000; Yalpani et al., 1994) by affecting photosynthetic rate, stomatal conductance, transpiration, ion uptake and transport, respiration rates, membrane stability, antioxidant capacity and antioxidant enzyme activities (Farooq et al., 2008a; Harper & Balke, 1981; Khan et al., 2003; Mohammed & Tarpley, 2009b). In addition, SA is considered to be an important signaling molecule involved in stress tolerance. It also promotes pathogenisis-related (PR) gene expression and induces systemic acquired resistance (SAR) in plants (Ryals et al., 1994). Besides providing disease resistance to the plants, SA can modulate plant responses to a wide range of oxidative stresses (Shirasu et al., 1997). It plays an essential role in preventing oxidative damage in plants by detoxifying superoxide radicals (Bowler et al., 1992) and by altering antioxidant capacity and antioxidant enzyme activities in plants (Chen et al., 1997; Dat et al., 1998, Fodor et al., 1997; Larkindale & Huang, 2004; Mohammed & Tarpley, 2009b; Rao et al., 1997; Wang & Li, 2006). Plants pre-treated with SA have shown increased thermotolerance (Larkindale & Knight, 2002).

3.1 Effects of plant growth regulators on plant morphology and physiology under high night temperature

Exogenous application of PGRs on plants under high night temperatures have shown positive, negative or no effect on plant morphology. Exogenous foliar application of GB or SA did not affect plant height in rice (Fig. 4). However, Farooq et al., (2008a, 2008b) state that seed priming of maize (*Zea mays* L.) with GB or SA increased shoot length, root length and biomass production. Increase in plant height as a result of SA application is due to increased cell enlargement, endoreduplication and/or cell division (Kang et al., 2007; Vanacker et al., 2007).

Glycine betaine and SA have profound effects on crop production through their effects on crop physiology. Foliar application of GB or SA increases leaf photosynthetic rates and decreases internal CO_2 concentration, respiration rates and membrane injury in rice (Fig. 5). Similar results were seen in other crop species and growth conditions where application of GB or SA increased photosynthetic rates and decreased respiration rates and injury to the membranes (Farooq et al., 2008a; Khan et al., 2010; Zhou et al., 1999).

In rice, application of GB or SA did not affect leaf chlorophyll or nitrogen content (Fig. 5). However, studies have shown that exogenous application of GB stabilizes pigments and prevents water oxidation and photooxidation (Cha-um et al., 2006, Sakamoto et al. 1998). Stabilization of pigments (chlorophyll and carotenoids) aids in light energy capture as required for photosynthesis. Several studies have reported that application of GB aids in protecting the photosynthetic machinery of the plant organelles by stabilizing the ultrastructure of the chloroplast, photosystem-II reaction centers and maintaining the oxygen-evolving machineries (Allakhverdiev et al., 2003; Busheva & Apostolova, 1997; Lee et al., 1997; Makela et al., 2000; Pospisil & Dau 2000; Sakamoto & Murata, 2000). On the other hand SA increases photosynthetic efficiency by increasing pigment concentrations and protecting photosystem function via activation of antioxidant defense systems as a signaling molecule (Dat et al., 1998; Ervin et al., 2005; Senaratna et al., 2001). The other physiological parameters such as respiration rates, transpiration rate and injury to leaf membrane decreased with application of GB or SA (Mohammed & Tarpley, 2011b). Studies have reported antitranspiration effects of GB and SA applications in crop species suggesting that GB and SA play an important role in controlling stomatal opening and closing under stress conditions (Cha-um et al., 2006; Khan et al., 2010; Larque-Saavedra, 1978). The most important effect of PGR's on crop production is through their ability to maintain the integrity of the membranes. The functional cell-membrane system is central to crop productivity. Stable membranes facilitate free movement of ions and water across the membrane and provide a favorable environment for enzymes and thylakoid membrane activities (Björkman et al., 1980). Several studies have reported that GB or SA application can maintain membrane integrity under stress conditions (Bohnert & Jensen, 1996; Farooq et al., 2008a, 2008b; Hussain et al., 2008; Mohammed & Tarpley, 2011b; Pan et al., 2006) by up regulating antioxidant enzyme activities and total antioxidant capacity (Demiral & Türkan, 2004; Farooq et al., 2008a, 2008b; He et al., 2005; Mohammed & Tarpley, 2009b).

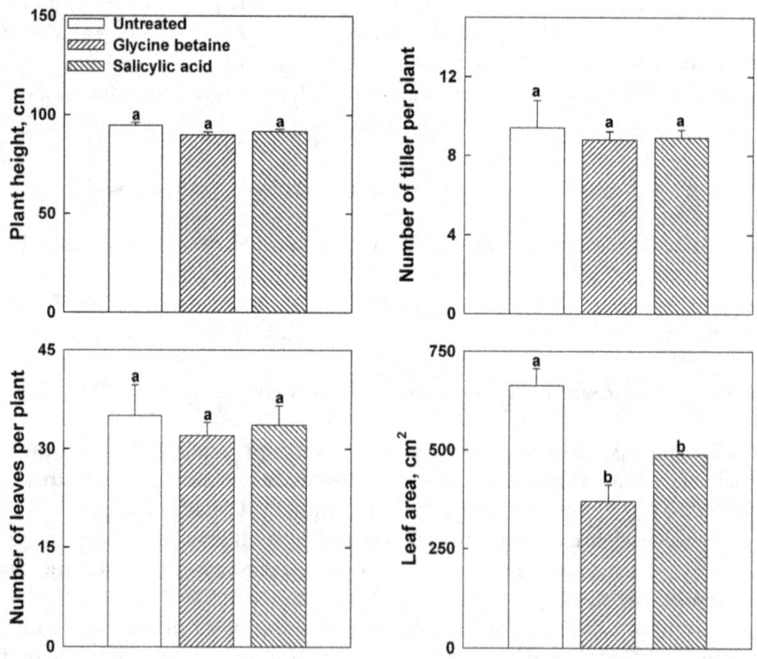

Fig. 4. Effects of glycine betaine and salicylic acid on rice morphology under high night temperature. Each bar represents average ± S.E. Different letters indicate means are significantly different at the P<0.05 level.

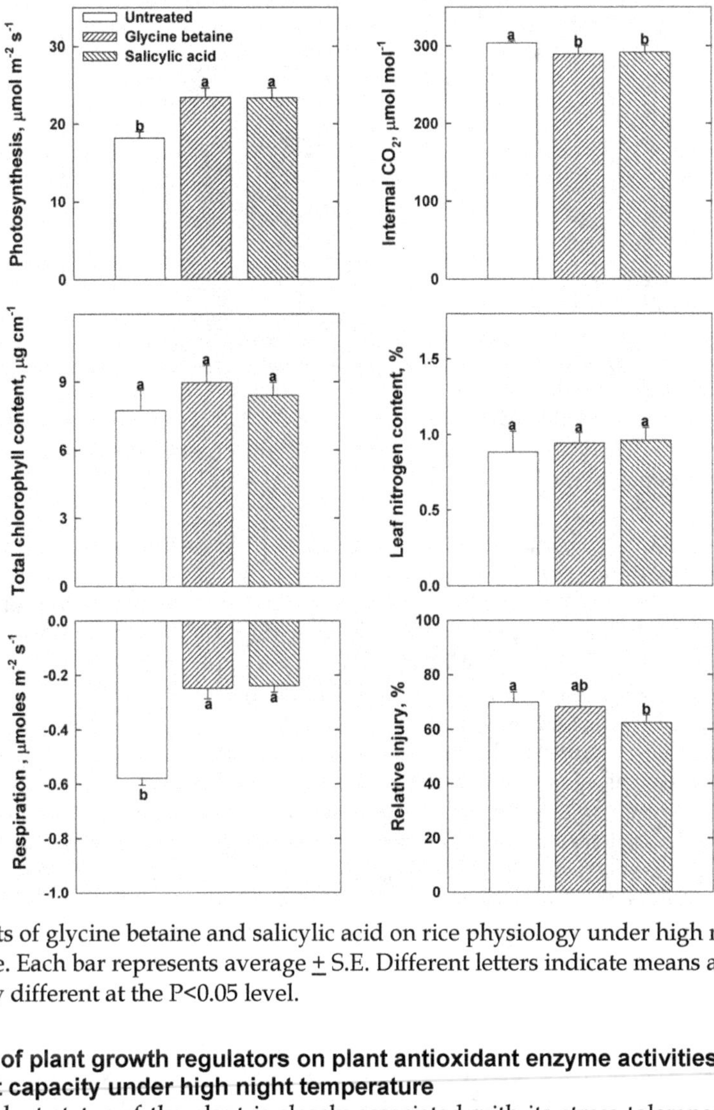

Fig. 5. Effects of glycine betaine and salicylic acid on rice physiology under high night temperature. Each bar represents average ± S.E. Different letters indicate means are significantly different at the P<0.05 level.

3.2 Effects of plant growth regulators on plant antioxidant enzyme activities and total antioxidant capacity under high night temperature

The antioxidant status of the plant is closely associated with its stress tolerance (Smirnoff, 1995). The severity of ROS-induced damage depends upon the antioxidant status of the plant and activities of antioxidant enzymes such as superoxide dismutase (SOD), ascorbate peroxidase (AP), catalase (CAT) and glutathione redutase (GR). Many studies have shown an increase in endogenous antioxidant levels as a result of exogenous application of GB or SA (Chen et. al., 1997; Diaz-Zorita et. al., 2001; Fang et al., 2009; Mohammed & Tarpley, 2009b). In rice, exogenous application of GB or SA increased total antioxidant capacity under high night temperature (Fig. 6). These increases in the levels of endogenous antioxidant are due to increases in antioxidant enzyme activities (Demiral & Türkan, 2004; Farooq et al., 2008a, 2008b; He et al., 2005; Khan et al., 2010).

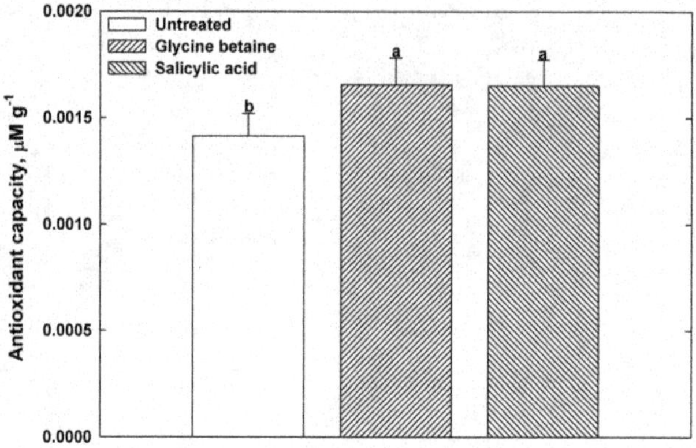

Fig. 6. Effects of glycine betaine and salicylic acid on total antioxidant capacity under high night temperature. Each bar represents average ± S.E. Different letters indicate means are significantly different at the $P<0.05$ level.

3.3 Effects of plant growth regulators on yield and yield-related parameters under high night temperature

Rice phenology under high night temperature was not altered by exogenous application of GB or SA as indicated by 1st panicle emergence date (Fig. 7). However, previous studies by Oota (1975) and Datta & Nanda (1985) showed that SA application induces early flowering in duckweeds (*Lemna gibba* L.) and chenna millet (*Panicum miliaceum* L.). Exogenous application of GB or SA increased spikelet fertility in rice (Fig. 7). The increase in spikelet fertility was due to increased photosynthesis, and decreased respiration and injury to the membrane stability as a result of GB or SA application. Exogenous application of GB or SA increases stability of the membranes as a result of increased antioxidant capacity of the plants. The enhanced membrane stability is responsible for increased photosynthesis and decreased respiration, which in turn increases crop production. In a nutshell, exogenous applications of GB or SA have been shown to increase stress tolerance, and often enhance plant growth and final crop yield under stress conditions, including HNT.

4. Conclusion

Yield, the final manifestation of all the physiological processes, decreased due to high night temperatures. The decrease in crop yields as a result of HNT was due to increased respiration and decreased photosynthesis and membrane stability. The decrease in photosynthesis was associated with decreases in photosynthesis-governing parameters such as stomatal conductance, internal CO_2 concentration, leaf chlorophyll content and leaf nitrogen, as a result of HNT. Moreover, decreased crop growth duration, pollen production and viability also contributed to decreased yields under HNT. Exogenous application of PGRs such as GB and SA increased production of photosynthates and decreased consumption of photosynthates and injury to the membrane, thereby increasing crop productivity under HNT. The exogenous application of GB and SA increases antioxidant levels, thereby protecting the membranes and enzymes against heat-induced ROS-mediated degradation, thus increasing crop productivity. However, there is a significant variation in

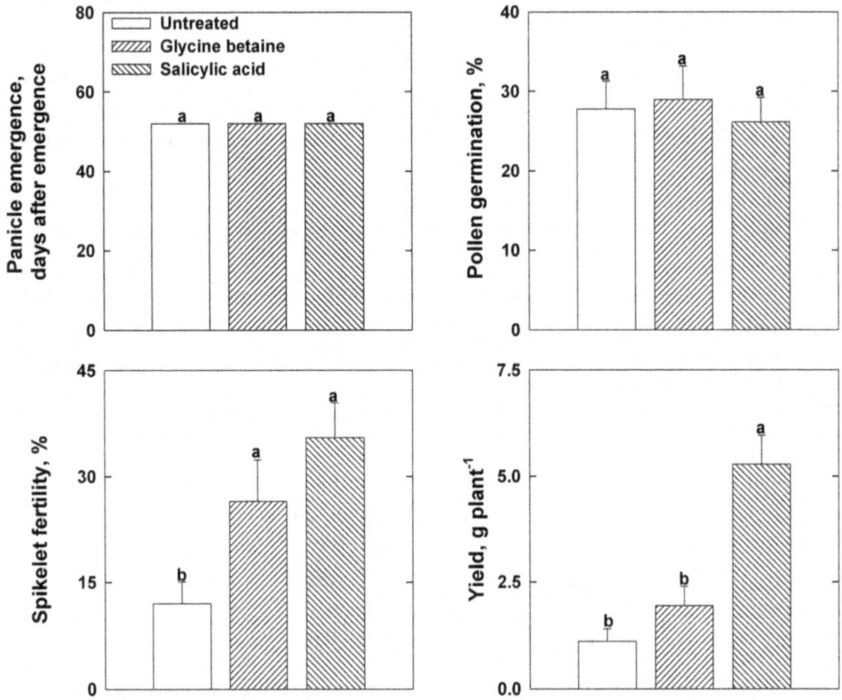

Fig. 7. Effects of glycine betaine and salicylic acid on rice yield-related parameters under high night temperature. Each bar represents average ± S.E. Different letters indicate means are significantly different at the P<0.05 level.

response to exogenous application of PGRs such as GB and SA among different plant species. This varied response among the plant species is due to 1) plant quantitative response to exogenous application of PGRs such as GB and SA at different stages of plant development, 2) plant exhibition of different responses to PGRs such as GB and SA at different stages of plant development and 3) variation in the effective concentration of a particular PGR among species. The PGRs such as GB and SA provide an option to protect the plants from different stresses like high day temperatures, high night temperature, drought and salinity, however, over-application of the PGRs such as GB and SA may be toxic to some plants resulting in growth inhibition or yield reduction.

5. Acknowledgment

We thank Texas Rice Belt Warehouse and Texas Rice Research Foundation for their support throughout this work.

6. References

Afuakwa, J.J.; Crookston, R.K. & Jones, R.J. (1984). Effect of Temperature and Sucrose Availability on Kernel Black Layer Development in Maize. *Crop Science*, Vol. 24, pp. 285-288, ISSN 0011-183X.

Ahmed, F.E. & Hall, A.E. (1993). Heat Injury during Early Floral Bud Development in Cowpea. *Crop Science*, Vol. 33, pp. 764-767, ISSN 0011-183X.

Allakhverdiev, S.I.; Feyziev, Y.M.; Ahmed, A.; Hayashi, H.; Aliev, J.A.; Kimlov, V.V.; Murata, N. & Carpentier, R. (1996). Stabilization of Oxygen Evolution and Primary Electron Transport Reactions in Photosystem II Against Heat Stress with Glycinebetaine and Sucrose. *Journal of Photochemistry and Photobiology*, Vol. 34, pp. 149-157, ISSN 0031-8655.

Alward, R.D.; Detling, J.K. & Milchunas, D.G. (1999). Grassland Vegetation Changes and Nocturnal Global Warming. *Science*, Vol. 283, pp. 229-231, ISSN 0036-8075.

Amthor, J.S. & McCree, K.J. (1990). Carbon Balance of Stress Plants: A Conceptual Model for Integrating Research Results, In: *Stress Response in Plants: Adaptation and Acclimation Mechanism*, R.G. Alscher & J.R. Cumming (Eds.), 1-15, Alan R. Liss, New York, USA.

Amthor, J.S. (1986). Evolution and Applicability of a Whole Plant Respiration Model. *Journal of Theoretical Biology*, Vol. 122, pp. 473-490, ISSN 0022-5193.

Ashraf, M. & Foolad, M.R. (2007). Roles of Glycine Betaine and Proline in Improving Plant Abiotic Stress Resistance. *Environmental and Experimental Botany*, Vol. 59, pp. 206-216, ISSN 0098-8472.

Badu-Apraku, B.; Hunter, R.B., & Tollenaar, M. (1983). Effect of Temperature During Grain Filling on Whole Plant and Grain Yield in Maize (*Zea mays* L.). *Canadian Journal of Plant Science*, Vol. 63, pp. 357-363, ISSN 0008-4220.

Baker, J.T.; Allen, L.H. Jr. & Boote K.J. (1992). Response of Rice to Carbon Dioxide and Temperature. *Agricultural and Forest Meteorology*, Vol. 60, pp. 153-166, 0168-1923.

Baker, J.T.; Allen, L.H., Jr.; Boote, K.J.; Jones, P. & Jones, J.W. (1989). Response of Soybean to Air Temperature and Carbon Dioxide Concentration. *Crop Science*, Vol. 29, pp. 98-105, ISSN 0011-183X.

Blackshaw, R.E. & Entz, T. (1995). Day and Night Temperature Effects on Vegetative Growth of *Erodium cicutarium*. *Weed Research*, Vol. 35, pp. 471-476, ISSN 0043-1737.

Blum, A. & Ebercon, A. (1981). Cell Membrane Stability as a Measure of Drought and Heat Tolerance in Wheat. *Crop Science*, Vol. 21 pp. 43-47, ISSN 0011-183X.

Bohnert, H.J. & Jensen, R.G. (1996). Strategies for Engineering Water-stress Tolerance in Plants. *Trends in Biotechnology*, Vol. 14, pp. 89-97, ISSN 0167-7799.

Bowler, C.; Montagu, M.V. & Inze, D. (1992). Superoxide Dismutase and Stress Tolerance. *Annual Review of Plant Physiology and Plant Molecular Biology*, Vol. 43, pp. 83-116, ISSN 1040-2519.

Busheva, M. & Apostolova, E. (1997). Influence of Saccharides and Glycinebetaine on Freezing of Photosystem II-enriched Particles: a Chlorophyll Fluorescence Study. *Photosynthetica*, Vol. 34, pp. 591-594, ISSN 0300-3604.

Caldas, T.; Demont-Caulet, N.; Ghazi, A. & Richarme, G. (1999). Thermoprotection by Glycine Betaine and Choline. *Microbiology*, Vol. 145, pp. 2543–2548, ISSN 1350-0872.

Cantarero, M.G.; Cirilo, A.G. & Andrade, F.H. (1999). Night Temperature at Silking Affects Kernel Set in Maize. *Crop Science*, Vol. 39, pp. 703-710, ISSN 0011-183X.

Cha-um, S.; Supaibulwatana, K. & Kirdmanee, C. (2006). Water Relation, Photosynthetic Ability and Growth of Thai Jasmine Rice (*Oryza sativa* L. ssp. indica cv. KDML105) to Salt Stress by Application of Exogenous Glycinebetaine and Choline. *Journal of Agronomy and Crop Science*, Vol. 192, pp. 25-36, ISSN 0931-2250.

Chen, Z.; Iyer, S.; Caplan, A.; Klessig, D.F. & Fan, B. (1997). Differential Accumulation of Salicylic Acid and Salicylic Acid-sensitive Catalase in Different Rice Tissues. *Plant Physiology*, Vol. 114, pp. 193-201, ISSN 0032-0889.

Cheng, W.; Sakai, H.; Yagi, K. & Hasegawa, T. (2009). Interactions of Elevated [CO2] and Night Temperature on Rice Growth and Yield. *Agricultural and Forest Meteorology*, Vol. 149, pp. 51-58, ISSN 0168-1923.

Christiansen, M.N. (1978). The Physiology of Plant Tolerance to Temperature Extremes. In: *Crop Tolerance to Suboptimal Land Conditions*, G.A. Jung (Ed.), 173–191, American Society of Agronomy, Madison, Wisconsin, USA.

Dat, J.F.; Lopez-Delgado, H.; Foyer, C.H. & Scott, I.M. (1998). Parallel Changes in H_2O_2 and Catalase During Thermotolerance Induced by Salicylic Acid or Heat Acclimation in Mustard Seedlings. *Plant Physiology*, Vol. 116, pp. 1351-1357, ISSN 0032-0889.

Datta K.S. & Nanda K.K. (1985). Effect of Some Phenolic Compounds and Gibberellic Acid on Growth and Development of Cheena Millet (*Panicum miliaceum* L.). *Indian Journal of Plant Physiology*, Vol. 28, pp. 298-302, ISSN 0019-5502.

Demiral, T. & Türkan, I. (2004). Does Exogenous Glycinebetaine Affect Antioxidative System of Rice Seedlings Under NaCl Treatment? *Journal of Plant Physiology*, Vol. 161, pp. 1089-1100, ISSN 0176-1617.

Dexter, S.T. (1956). Evaluation of Crop Plants for Winter Hardiness. *Advances in Agronomy*, Vol. 8, pp. 203-209, ISSN 0065-2113.

Diaz-Zorita, M.; Fernandez-Canigia, M.V. & Grosso, G.A. (2001). Application of Foliar Fertilizers Containing Glycinebetaine Improved Wheat Yields. *Journal of Agronomy and Crop Science*, Vol. 186, pp. 209-215, ISSN 0931-2250.

Dinar, M. & Rudich, J. (1985). Effect of Heat Stress on Assimilate Partitioning in Tomato. *Annals of Botany*, Vol. 56, pp. 239-249, ISSN 0305-7364.

Ervin, E.H.; Zhang, X. & Schmidt, R.E. (2005). Exogenous Salicylic Acid Enhances Post-Transplant Success of Heated Kentucky Bluegrass and Tall Fescue Sod. *Crop Science*, Vol. 45, pp. 240-244, ISSN 0011-183X.

Fang, C-X.; Xiong, J.; Qiu, L.; Wang, H-B.; Song, B-Q.; He, H-B.; Lin, R-Y. & Lin, W-X. (2009). Analysis of Gene Expressions Associated with Increased Allelopathy in Rice (*Oryza sativa* L.) Induced by Exogenous Salicylic Acid. *Plant Growth Regulation*, Vol. 57, pp. 163-172, ISSN 0167-6903.

Farooq, M.; Aziz, T.; Hussain, M.; Rehman, H.; Jabran, K. & Khan, M.B. (2008a). Glycinebetaine Improves Chilling Tolerance in Hybrid Maize. *Journal of Agronomy and Crop Science*, Vol. 194, pp. 152-160, ISSN 0931-2250.

Farooq, M.; Aziz, T.; Basra, S.M.A.; Cheema, M.A. & Rehman, H. (2008b). Chilling Tolerance in Hybrid Maize Induced by Seed Priming with Salicylic Acid. *Journal of Agronomy and Crop Science*, Vol. 194, pp. 161-168, ISSN 0931-2250.

Farrar, J.F. (1985). The Respiratory Source of CO_2. *Plant Cell and Environment*, Vol. 8, pp. 427-438, ISSN 0140-7791.

Fodor, J.; Gullner, G.; Adam, A.L.; Barna, B.; Komives, T. & Kiraly, Z. (1997). Local and Systemic Responses of Antioxidants to Tobacco Mosaic Virus Infection and to Salicylic Acid in Tobacco. *Plant Physiology*, Vol. 114, pp. 1443-1451, ISSN 0032-0889.

Frantz, J.M.; Cometti, N.N. & Bugbee, B. (2004). Night Temperature has a Minimal Effect on Respiration and Growth in Rapidly Growing Plants. *Annals of Botany*, Vol. 94, pp. 155-166, ISSN 0305-7364.

Gibson, L.R., & Mullen, R.E. (1996). Influence of Day and Night Temperature on Soybean Seed Yield. *Crop Science*, Vol. 36, pp. 98-104, ISSN 0011-183X.

Guinn, G. (1974). Abscission of Cotton Floral Buds as Influenced by Factors Affecting Photosynthesis and Respiration. *Crop Science*, Vol. 14, pp. 291-293, ISSN 0011-183X.

Guo, Y.P.; Zhou, H.F. & Zhang, L.C. (2006). Photosynthetic Characteristics and Productive Mechanism against Photooxidation during High Temperature Stress in Two Citrus Species. *Scientia Horticulturae*, Vol. 108, pp. 260-267, ISSN 0304-4238.

Hall, A.E. (1992). Breeding for Heat Tolerance. *Plant Breeding Reviews*, Vol. 10, pp.129-168, ISSN 0730-2207.

Harinasut, P.; Tsutsui, K.; Takabe, T.; Nomura, M.; Takabe, T. & Kishitani, S. (1996). Exogenous Glycinebetaine Accumulation and Increased Salt-tolerance in Rice Seedlings. *Bioscience Biotechnology and Biochemistry*, Vol. 60, pp. 366-368, ISSN 0916-8451.

Harper, J .R. & Balke, N.E. (1981). Characterization of the Inhibition of K+ Absorption in Oats Roots by Salicylic Acid. *Plant Physiology*, Vol. 68, pp. 1349-1353, ISSN 0032-0889.

Havaux, M. & Tardy, F. (1996). Temperature-dependent Adjustment of the Thermal Stability of Photosystem 2 in vivo: Possible Involvement of Xanthophylls-Cycle Pigments. *Planta*, Vol. 198, pp. 324-333, ISSN 0032-0935.

He, Y.; Liu, Y.; Cao, W.; Huai, M.; Xu, B. & Huang, B. (2005). Effects of Salicylic Acid on Heat Tolerance Associated with Antioxidant Metabolism in Kentucky Bluegrass. *Crop Science*, Vol. 45, pp. 988-995, ISSN 0011-183X.

Huang, B., Liu, X. & Fry J.D. 1998. Shoot Physiological Responses of Two Bentgrass Cultivars to High Temperature and Poor Soil Aeration. *Crop Science*, Vol. 38, pp. 1219-1224, ISSN 0011-183X.

Ibrahim, A.M.H. & Quick, J.S. (2001). Heritability of Heat Tolerance in Winter and Spring Wheat. *Crop Science*, Vol. 41, pp. 1401-1405, ISSN 0011-183X.

IPCC (Intergovernmental Panel on Climate Change) (2007). Climate Change 2007: Impacts, Adaptation and Vulnerability. In: *Contribution of Working Group II to Fourth Assessment Report of the Intergovernmental Panel on Climate Change*, M.L. Parry; O.F. Canziani; J.P. Palutikof; P.J. van der Linden; & C.E. Hanson (Eds.), 1000, Cambridge University Press, Cambridge, UK.

Ismail, A.M. & Hall, A.E. (1999). Reproductive-stage Heat Tolerance, Leaf Membrane Thermostability and Plant Morphology in Cowpea. *Crop Science*, Vol. 39, pp. 1762-1768, ISSN 0011-183X.

Izquierdo, N.; Aguirrezábal, L.; Andrade, F. & Pereyra. V. (2002). Night Temperature Affects Fatty Acid Composition in Sunflower Oil Depending on the Hybrid and the Phenological Stage. *Field Crops Research*, Vol. 77, pp. 115-126, ISSN 0378-4290.

John, E. (2002). Candidate Effector and Regulator Genes Activated by Glycine Betaine in Arabidopsis. Available from http://abstracts.aspb.org/pb2002/public/P63/0153.html.

Jones, P.D.; New, M.; Parker, D.E.; Martin, S. & Rigor, I.G. (1999). Surface Air Temperature and its Changes over the Past 150 Years. *Reviews of Geophysics*, Vol. 37, pp. 173–199.

Kang, G.Z., Z.X. Wang, K.F. Xia, and G.C. Sun, 2007: Protection of Ultrastructure in Chilling-stressed Banana Leaves by Salicylic Acid. *Journal of Zhejiang University Science*, Vol. 8, pp. 277-282, ISSN 1009-3095.

Keeling, P.L.; Banisadr, R.; Barone, L.; Wasserman, B.P. & Singletary, G.W. (1994). Effect of Temperature on Enzymes in the Pathway of Starch Biosynthesis in Developing Wheat and Maize Grain. *Australian Journal of Plant Physiology*, Vol. 21, pp. 807-827, ISSN 0310-7841.

Khan, N.A.; Syeed, S.; Masood, A.; Nazar, R. & Iqbal, N. (2010). Application of Salicylic Acid Increases Contents of Nutrients and Antioxidative Metabolism in Mungbean and Alleviates Adverse Effects of Salinity Stress. *International Journal of Plant Biology*, Vol. 1, pp. e1, ISSN 2037-0156.

Khan, W.; Prithiviraj, B. & Smith, A. (2003). Photosynthetic Responses of Corn and Soybean to Foliar Application of Salicylates. *Journal of Plant Physiology*, Vol. 160, pp. 485-492, ISSN 0176-1617.

Kondo, S. & Takahashi, Y. (1989). Relation between Early Drop of Apple Fruit and Ethylene Evolution under High Night-temperature Conditions. *Journal of Japanese Society for Horticultural Science*, Vol. 58, pp. 1-8, ISSN 1882-3351.

Kreiner, M.; Harvey, L.M. & McNeil, B. (2002). Oxidative Stress Response of a Recombinant *Aspergillus niger* to Exogenous Menadione and H_2O_2 Addition. *Enzyme and Microbial Technology*, Vol. 30, pp. 346-353, ISSN 0141-0229.

Lambers, H. (1985). Respiration in Tack Plants and Tissues: Its Regulation and Dependence on Environmental Factors, Metabolism and Invaded Organism. In: *Higher Plant Cell Respiration. Encyclopedia of Plant Physiology*, R. Douce & D.A. Day (Eds.), New series, Vol. 18, Springer-Verlag, Berlin, Gremany.

Larkindale, J. & Huang, B. (2004). Thermotolerance and Antioxidant Systems in *Agrostis stolonifera*: Involvement of Salicylic Acid, Abscisic Acid, Calcium, Hydrogen Peroxide and Ethylene. *Journal of Plant Physiology*, Vol. 161 pp. 405-413, ISSN 0176-1617.

Larkindale, J. & Knight, M.R. (2002). Protection Against Heat Stress-induced Oxidative Damage in Arabidopsis Involves Calcium, Abscisic Acid, Ethylene, and Salicylic Acid. *Plant Physiology*, Vol. 128, pp. 682-695, ISSN 0032-0889.

Larque-Saavedra, A. (1978). The Antitranspirant Effect of Acetylsalicylic Acid on *Phaseolus vulgaris*. *Plant Physiology*, Vol. 43, pp. 126-128, ISSN 0032-0889.

Lee, C.B.; Hayashi, H. & Moon, B.Y. (1997). Stabilization by Glycinebetaine of Photosynthetic Oxygen Evolution by Thylakoid Membranes from Synechococcus PCC7002. *Molecular Cell*, Vol. 30, pp. 296-299, ISSN 1097-2765.

Loka, D.A., Oosterhuis, D.M., 2010. Effect of High Night Temperatures on Cotton Respiration. ATP Levels, and Carbohydrate Content. *Environmental and Experimental Botany*, Vol. 68, pp. 258-263, ISSN 0098-8472.

Mäkelä, P.; Karkkainen, J. & Somersalo, S. (2000). Effect of Glycine Betaine on Chloroplast Ultrastructure, Chlorophyll and Protein Content, and RUBPCO Activities in Tomato Grown under Drought or Salinity. *Biologia Plantarum*, Vol. 3, pp. 471-475, ISSN 0006-3134.

Martineau, J.R.; Specht, J.E.; Williams, J.H. & Sullivan, C.Y. (1979). Temperature Tolerance in Soybeans. I. Evaluation of a Technique for Assessing Cellular Membrane Thermostability. *Crop Science*, Vol. 19, pp. 75-78, ISSN 0011-183X.

Matsui, T. & Kagata, H. (2003). Characteristics of Floral Organs Related to Reliable Self Pollination in Rice (*Oryza sativa* L.). *Annals of Botany*, Vol. 91, pp. 473-477, ISSN 0305-7364.

McDonald, A.E. & Vanlerberghe, G.C. (2005). Alternative Oxidase and Plastoquinol Terminal Oxidase in Marine Prokaryotes of the Sargasso Sea. *Gene*, Vol. 349, pp. 15-24, ISSN 0378-1119.

Micheal, G. & Beringer, H. (1980). The Role of Hormones in Yield Formation. In: *Physiological Aspects of Crop Productivity*, G. Micheal & H. Beringer (Eds.), 85-116, Proceedings of the 15th Colloquium of the International Potash Institute, Bern, Switzerland.

Mitchell, R.A.C.; Mitchell, V.J.; Driscoll, S.P.; Franklin, J. & Lawlor, D.W. (1993). Effects of Increased CO_2 Concentration and Temperature on Growth and Yield of Winter Wheat at Two Levels of Nitrogen Application. *Plant, Cell and Environment*, Vol. 16, pp. 521-529, ISSN 0140-7791.

Mohammed, A.R. & Tarpley, L. (2011a). Morphological and physiological responses of nine southern U.S. rice cultivars differing in their tolerance to enhanced ultraviolet-B radiation. *Environmental and Experimental Botany*, Vol. 70, pp. 174-184, ISSN 0098-8472.

Mohammed, A.R. & Tarpley, L. (2011b). Characterization of Rice (*Oryza sativa* L.) Physiological Responses to α-Tocopherol, Glycine Betaine or Salicylic Acid Application. *Journal of Agriculture Science*, Vol. 3, pp. 3-13, ISSN 0021-8596.

Mohammed, A.R. & Tarpley, L. (2009a). High Nighttime Temperatures Affect Rice Productivity through Altered Pollen Germination and Spikelet Fertility. *Agricultural and Forest Meteorology*, Vol. 149, pp. 999-1008, ISSN 0168-1923.

Mohammed, A.R. & Tarpley, L. (2009b). Impact of High Nighttime Temperature on Respiration, Membrane Stability, Antioxidant Capacity, and Yield of Rice Plants. *Crop Science*, Vol. 49, pp. 313-322, ISSN 0011-183X.

Moller, I.M. (2001). Plant Mitochondria and Oxidative Stress: Electron Transport, NADPH turnover, and Metabolism of Reactive Oxygen Species. *Annual Review of Plant Physiology and Plant Molecular Biology*, Vol. 52, pp. 561-591, ISSN 1040-2519.

Neales, T.F.; Sale, P.J.M. & Meyer, C.P. (1980). Carbon Dioxide Assimilation by Pineapple Plants, *Ananas comosus* (L.) Merr. II. Effects of Variation of the Day/Night Temperature Regime. *Australian Journal of Plant Physiology*, Vol. 7, pp. 375-385, ISSN 0310-7841.

Oota, Y. (1975). Short-day Flowering of *Lemna gibba* G3 Induced by Salicylic Acid. *Plant Cell and Physiology*, Vol. 16, pp. 1131-1135, ISSN 0032-0781.

Paembonan, S.A.; Hagihara, A. & Hozumi, K. (1992). Long-term Respiration in Relation to Growth and Maintenance Processes of the Aboveground Parts of a Hinoki Forest Tress. *Tree Physiology*, Vol. 10, pp. 101-110, ISSN 0829-318X.

Paleg, L.G.; Douglas, T.J.; van Daal, A. & Keech, D.B. (1981). Proline, Betaine and Other Organic Solutes Protect Enzymes Against Heat Inactivation. *Australian Journal of Plant Physiology*, Vol. 8, pp. 107-114, ISSN 0310-7841.

Pan, Q.; Zhan, J.; Liu, H.; Zhang, J.; Chen, J.; Wen, P. & Huang, W. (2006) Salicylic Acid Synthesized by Benzoic Acid 2 Hydroxylase Participates in the Development of Thermotolerance in Pea Plants. *Plant Science*, Vol. 171, pp. 226-233, ISSN 0168-9452.

Peng, S.; Huang, J.; Sheehy, J.E.; Laza, R.C.; Visperas, R.M.; Zhong, X.; Centeno, G.S.; Khush, G.S. & Cassman, K.G. (2004). Rice Yields Decline with Higher Night Temperature from Global Warming. *Proceedings of the National Academy of Sciences of the United States of America* Vol. 101, pp. 9971-9975, ISSN 0027-8424.

Penning de Vries, F.W.T. (1975). The Cost of Maintenance Process in the Cell. *Annals of Botany*, Vol. 39, pp. 77-92, ISSN 0305-7364.

Peterson, R.B. & Zelitch, I. (1982). Relationship between Net CO_2 Assimilation and Dry Weight Accumulation in Field-grown Tobacco. *Plant Physiology*, Vol. 70, pp. 677-685, ISSN 0032-0889.

Pospisil, P. & Dau, H. (2000). Chlorophyll Fluorescence Transients of Photosystem II Membrane Particles as a Tool for Studying Photosynthetic Oxygen Evolution. *Photosynthesis Research*, Vol. 65, pp. 41-52, ISSN 0166-8595.

Prasad, P.V.V.; Boote, K.J.; Allen Jr, L.H.; Sheehy, J.E. & Thomas, J.M.G. (2006). Species, Ecotype and Cultivar Differences in Spikelet Fertility and Harvest Index of Rice in Response to High Temperature Stress. *Field Crops Research*, Vol. 95, pp. 398-411, ISSN 0378-4290.

Prasad, P.V.V; Craufurd, P.Q. & Summerfield, R.J. (1999). Fruit Number in Relation to Pollen Production and Viability in Groundnut Exposed to Short Episodes of Heat Stress. *Annals of Botany*, Vol. 84, pp. 381-386, ISSN 0305-7364.

Rajasekaran, L.R.; Kriedemann, P.E.; Aspinall, D. & Paleg, L.G. (1997). Physiological Significance of Proline and Glycinebetaine: Maintaining Photosynthesis during NaCl Stress in Wheat. *Photosynthetica*, Vol. 34, pp. 357-366, ISSN 0300-3604.

Rao, M.V.; Paliyath, G.; Ormrod, P.; Murr, D.P. & Watkins, C.B. (1997). Influence of Salicylic Acid on $H_2O_2^{\bullet}$ Production, Oxidative Stress, and H_2O_2-metabolizing Enzymes. *Plant Physiology*, Vol. 115, pp. 137-149, ISSN 0032-0889.

Raskin, I. (1992). Role of Salicylic Acid in Plants. *Annual Review of Plant Physiology and Plant Molecular Biology*, Vol. 43, pp. 439-463, ISSN 1040-2519.

Reynolds, M.P.; Balota, M.; Delgado, M.I.B.; Amani, I. & Fisher, R.A. (1994). Physiological and morphological traits associated with spring wheat yield under hot irrigated conditions. *Australian Journal of Plant Physiology*, Vol. 21, pp. 717–730, ISSN 0310-7841.

Ryals, J.; Uknes, S. & Ward, E. (1994). Systematic Acquired Resistance. *Plant Physiology*, Vol. 104, pp. 1109-1112, ISSN 0032-0889.

Ryan, M.G. (1991). Effects of Climate Change on Plant Respiration. *Ecological Applications*, Vol. 1, pp. 157-167, ISSN 1051-0761.

Sakamoto, H. & Murata, N. (1998). Metabolic Engineering of Rice Leading to Biosynthesis of Glycinebetaine and Tolerance to Salt and Cold. *Plant Molecular Biology*, Vol. 38, pp. 1011-1019, ISSN 0167-4412.

Sayed, O.H.; Earnshaw, M.J. & Emes, M.J. (1989). Photosynthetic Response of Different Varieties of Wheat to High Temperature. *Journal of Experimental Botany*, Vol. 40, pp. 633-638, ISSN 0022-0957.

Seddigh, M. & Jolliff, G.D. (1984). Night Temperature Effects on Morphology, Phenology, Yield and Yield Components of Indeterminate Field-grown Soybean. *Agronomy Journal*, Vol. 76, pp. 824-828, ISSN 0002-1962.

Senaratna, T.; Touchell, D.; Bunn, T. & Dixon, K. (2000). Acetyl salicylic acid (aspirin) and salicylic acid induce multiple stress tolerance in bean and tomato plants. *Plant Growth Regulation*, Vol. 30, pp. 157-161, ISSN 0167-6903.

Shirasu K.; Nakajima H. & Rajashekar K. (1997). Salicylic Acid Potentiates an Agonist-dependent Gain Control that Amplifies Pathogen Signal in the Activation of Defense Mechanisms. *Plant Cell*, Vol. 9, pp. 261-270, ISSN 1040-4651.

Singletary, G.W.; Banisadr, R. & Keeling, P.L. (1994). Heat Stress during Grain Filling in Maize: Effects on Carbohydrate Storage and Metabolism. *Australian Journal of Plant Physiology*, Vol. 21, pp. 829-841, ISSN 0310-7841.

Smirnoff, N. (1995). Antioxidant Systems and Plant Response to the Environment. In: *Environment and Plant Metabolism: Flexibility and Acclimation*, N. Smirnoff (Ed.), 217-243, Bios Scientific Publishers, Oxford, UK.

Sullivan, C.Y. (1972). Mechanisms of Heat and Drought Resistance in Grain Sorghum and Methods of Measurement. In: *Sorghum in the Seventies*, N.G.P. Rao & L.R. House (Eds.), 247-264, Oxford and IPH Publishing Company, New Delhi, India.

Sullivan, C.Y. & Ross, W.M. (1979). Selecting for Drought and Heat Resistance in Grain Sorghum. In: *Stress Physiology in Crop Plants*, H. Mussell, H. & R.C. Staples (Eds.), 263-281, John Wiley & Sons, New York, USA.

Turnbull, M.H.; Murthy, R. & Griffin, K.L. (2002). The Relative Impacts of Daytime and Night-time Warming on Photosynthetic Capacity in *Populus detoides*. *Plant Cell and Environment*, Vol. 25, pp. 1729-1737, ISSN 0140-7791.

Vanacker, H.; Lu, H.; Rate, D.N. & Greenberg, J.T. (2001). A Role for Salicylic Acid and NPR1 in Regulating Cell Growth in Arabidopsis. *Plant Journal*, Vol. 28, pp. 209-216, ISSN 0960-7412.

Warrag, M.O.A. & Hall, A.E. (1984). Reproductive Responses of Cowpea (*Vigna unguiculata* (L.) Walp.) to Heat Stress. II. Responses to Night Air Temperature. *Field Crops Research*, Vol. 8, pp. 17-33, ISSN 0378-4290.

Willits, D.H. & Peet, M. (1998). The Effect of Night Temperature on Greenhouse Grown Tomato Yields in Warm Climates. *Agricultural and Forest Meteorology*, Vol. 92, pp. 191-202, ISSN 0168-1923.

Yalpani, N.; Enyedi, A.J.; Leon, J. & Raskin, I. (1994). Ultraviolet Light and Ozone Stimulate Accumulation of Salicylic Acid, Pathogen-related Proteins and Virus Resistance in Tobacco. *Planta*, Vol. 193, pp. 372-376, ISSN 0032-0935.

Yamane, Y.; Kashino, Y. & Koike, H. (1997). Increase in Fluorescence F_0 Level and Reversible Inhibition of Photosystem 2 Reaction by High-temperature Treatments in Higher Plants. *Photosynthetic Research*, Vol. 52, pp. 57-64.

Yilmaz, J.L. (2004). Abiotic Stress Tolerance: Metabolic and Physiological Effects of Compatible Solutes and Calmodulin on *E. coli* and Tobacco. In: *Pure and Applied Biochemistry*, J.L. Yilmaz (Ed), 1-40, Lund University, Lund, Sweden.

Yoshida, S. (1981). Physiological Analysis of Rice Yield. In: *Fundamentals of Rice Crop Science*, S. Yoshida (Ed.), 231-251. International Rice Research Institute, Los Banos, Philippines.

Zhang, J. & Kirkham, M.B. (1996). Antioxidant Responses to Drought in Sunflower and Sorghum Seedlings. *New Phytologist*, Vol. 132, pp. 361-373, ISSN 0028-646X.

Zheng, S.H.; Nakamoto, H.; Yoshikawa, K.; Furuya, T. & Fukuyama, M. (2002). Influence of High Night Temperature on Flowering and Pod Setting in Soybean. *Plant Production Science*, Vol. 5, pp. 215-218, ISSN 1343-943X.

Zhou, X.M., Mackenzie, A.F., Madramootoo, C.A., & Smith, D.L.J. (1999). Effects of Some Injected Plant Growth Regulators, With or Without Sucrose, on Grain Production, Biomass and Photosynthetic Activity of Field-grown Corn Plants. *Journal of Agronomy and Crop Science*, Vol. 183, pp. 103-110, ISSN 0931-2250.

10

A Review of Sea-Level Rise Effect on Mangrove Forest Species: Anatomical and Morphological Modifications

Laura Yáñez-Espinosa[1] and Joel Flores[2]
[1]*Instituto de Investigación de Zonas Desérticas; Programas Multidisciplinarios de Posgrado en Ciencias Ambientales; Universidad Autónoma de San Luis Potosí*
[2]*División de Ciencias Ambientales; Instituto Potosino de Investigación Científica y Tecnológica, A.C.*
México

1. Introduction

Mangrove forests are among the most productive and biologically important ecosystems of the world, because they provide important and unique ecosystem goods and services to human society and coastal and marine systems, as stabilizing shorelines and reducing the devastating impact of natural disasters, providing breeding and nursing grounds for marine species, and food, medicine, fuel and building materials (Tomlinson, 1986; Giri et al., 2011).

Mangroves are taxonomically diverse trees and shrubs that have evolved independently through convergence (Hogarth, 1999). The principal genera are *Avicennia* (Avicenniaceae), *Laguncularia* and *Lumnitzera* (Combretaceae), *Nypa* (Palmae), *Bruguiera, Ceriops, Kandelia* and *Rhizophora* (Rhizophoraceae), and *Sonneratia* (Sonneratiaceae) (Tomlinson, 1986). These plants have developed complex physiological, morphological and anatomical adaptations allowing survival and success in the high stress habitat where they inhabit (Hogarth, 1999). They can tolerate the stress of waterlogging and salinity prevailing in coastal environments influenced by tides and have adapted to wide salinity levels, may be influenced by local hydrology and episodic disturbance events (Doyle, 2003). The maximum concentration of soil water salinity that mangrove species can tolerate is suggested up to 155 ‰ with annual averages of 100 ‰ (Tomlinson, 1986).

Mangrove forests are generally distributed along tropical coastlines of America, Africa and Asia between 25° N and 25° S, although this range extends beyond due to the movement of unusually warm waters from the equator, including the east coast of Africa, Australia, and New Zealand (Hogarth, 1999; McLeod & Salm, 2006).

Nevertheless, sea-level rise in the future could be the biggest threat to mangrove ecosystems as climate change consequence (Giri et al., 2011). In the last century, sea-level has risen 10-20 cm mainly due to thermal expansion of the oceans and melting of glacial ice caused by global warming, with climate models predicting an accelerated rate of sea-level rise over coming decades from 0.09 to 0.88 m (McLeod & Salm, 2006). This will generate salinity concentration, along with rising CO_2, and temperature, determining future species distributions, abundances, and viability (Kareiva et al., 1993; Yáñez-Arancibia et al., 1998), concerning worldwide from both scientific and policy perspectives (Alongi, 2008; Gilman et al., 2008; Semeniuk, 1994).

Increased surface temperature is expected to affect mangrove forests (Field, 1995). However, it is uncertain the effect of rising temperature when the interactive effects of changes in precipitation and other weather-related factors are considered (Alongi, 2008). In addition, temperature change in the tropics may not be as great as in boreal and temperate regions (Beaumont et al., 2011; Solomon et al., 2007). Responses of mangroves to rising temperature depend on reactions of individual plants, but such responses have not been addressed experimentally.

The chapter approach involve a review of worldwide studies regarding the association of mangroves morphological and anatomical traits variation with local and regional environmental factors, analyzing their stem, leaves, roots and growth, allowing to infer the effect of potential impacts of sea-level rise on their structure.

2. Morphological and anatomical adaptations of mangrove forest species to environmental factors

Mangrove species anatomical and morphological adaptations are relevant from ecological point of view, due to the particular environment where they inhabit, frequently flooded and highly saline (Tomlinson, 1986) (Fig. 1).

Fig. 1. Mangrove *Avicennia germinans* (L.) Stearn (Avicenniaceae) trees in the coastal lagoon "La Mancha", Veracruz, in the Gulf of Mexico (photo by Jorge López-Portillo).

There is in addition a group of species described as mangrove associates that comprises a large number of species typically occurring on the landward margin of the forest or non-mangrove habitats like salt marsh or lowland fresh water swamps that are completely or partially flood tolerant (Hogarth, 1999) (Fig. 2). There is a long list of mangrove associates like *Hibiscus* and *Pavonia* (Malvaceae), *Amoora* (Meliaceae), *Ardisia* and *Myrsine* (Myrsinaceae), *Calamus* and *Phoenix* (Arecaeae), *Pandanus* (Pandanaceae), *Pouteria* (Sapotaceae), *Pachira* (Bombacaceae), *Ficus* (Moraceae), *Tabebuia* (Bignoniaceae), *Achrosticum* (Pteridaceae), *Annona* (Annonaceae), and others (Tomlinson, 1986).

The most typical adaptations of mangrove species are the aerial roots, stilt-roots, pneumatophores, root knees, and plank roots that have a higher proportion of gas space when waterlogged, mechanisms of salt exclusion by the roots, tolerance of high tissue salt concentrations and excretion of salt excess from leaves, vivipary or the development of the embryo in seed plants before they are dispersed, and seeds and propagules have different morphological adaptations that facilitate flotation (Baskin & Baskin, 2001; Hogarth, 1999; Tomlinson, 1986).

Fig. 2. Mangrove associate *Annona glabra* L. (Annonaceae) trees showing massive root system in the Everglades, Florida.

Trees vascular system has strong influence on leaves structure and function, carbon gain, nutrient use efficiency and growth rate. Diverse studies have revealed that mangroves show great plasticity in wood, bark and leaves structure as an adaptation to a wide gradient of water salinity, flooding level and waterlogging period. These modifications are important to account and understand the possible effects of the predicted sea-level rise as a consequence of global warming, because as all mangrove forests occur between high and low tide marks it is evident that they will be drastically influenced by any changes in sea-level.

3. Modifications of wood structure in response to environmental factors

Several studies have demonstrated that mangroves like *Bruguiera, Rhizophora, Laguncularia* and *Avicennia* trees stem vascular system, particularly wood anatomical traits as vessel density, grouping, diameter and length, as well as fiber wall thickness, are affected by variations on salinity and flooding level. (Table 1).

Anatomical traits	Species	Environmental factors	Reference
Vessel density	*Annona glabra* *Laguncularia racemosa* *Rhizophora mangle*	Salinity Flooding level Soil texture	Yáñez-Espinosa et al., 2001
	Laguncularia racemosa	Salinity Flooding level	Yáñez-Espinosa et al., 2004
	Rhizophora mucronata	Salinity	Schmitz et al., 2006
	Avicennia marina *Rhizophora mucronata*	Salinity	Robert et al., 2009
	Avicennia germinans	Salinity Flooding period	Yáñez-Espinosa et al., 2009
	Bruguiera gymnorrhiza	Flooding period	Xiao et al., 2010
Vessel diameter	*Rhizophora mangle*	Salinity Flooding level	Yáñez-Espinosa et al., 2001
	Avicennia marina *Rhizophora mucronata*	Salinity	Robert et al., 2009
	Avicennia marina	Flooding period	Xiao et al., 2009
	Rhizophora mucronata	Salinity	Schmitz et al., 2006
Vessel grouping	*Annona glabra*	Salinity Flooding level Soil texture	Yáñez-Espinosa et al., 2001
	Avicennia marina *Rhizophora mucronata*	Salinity	Robert et al., 2009
Vessel length	*Avicennia marina* *Rhizophora mucronata*	Salinity	Robert et al., 2009
	Laguncularia racemosa	Salinity Flooding level	Yáñez-Espinosa et al., 2004
Fiber wall thickness	*Laguncularia racemosa*	Salinity	Yáñez-Espinosa et al., 2004
	Avicennia germinans	Salinity Flooding period	Yáñez-Espinosa et al., 2009
	Bruguiera gymnorrhiza	Flooding period	Xiao et al., 2010

Table 1. Mangrove and mangrove associate species wood anatomical characters affected by different environmental variables.

Vessel density increases from low- to high-salinity areas in all species, from high- to low-flooding level in most of them, and from temporal- to prolonged flooding period. Increasing vessel density with salinity can be explained by an interference with nutrient uptake and auxin physiology with higher soil water salinity (Schmitz et al., 2006).

Numerous vessels promote protection against cavitation in stressed environments, avoiding cavitation damage when flooding level is higher, because a sporadic drought could occur (Yáñez-Espinosa et al., 2004).

The concentration of salinity effects on vessel density could be observed in the relationship of *Rhizophora mucronata* 23–30 vessels mm^{-2}, and *Avicennia marina* 78-68 vessels mm^{-2}, ranging from 26.4–49.2‰ in Gazi Bay, Kenya (Robert et al., 2009; Schmitz et al., 2006); *Avicennia germinans* 30 - 52 vessels mm^{-2}, ranging from 3.2 - 30‰ of soil water in La Mancha lagoon, México (Yáñez-Espinosa et al., 2009); *Laguncularia racemosa* 7-16 vessels mm$^{-2,}$ ranging from 6-9‰ in an estuarine river in Nayarit, México (Yáñez-Espinosa et al., 2004).

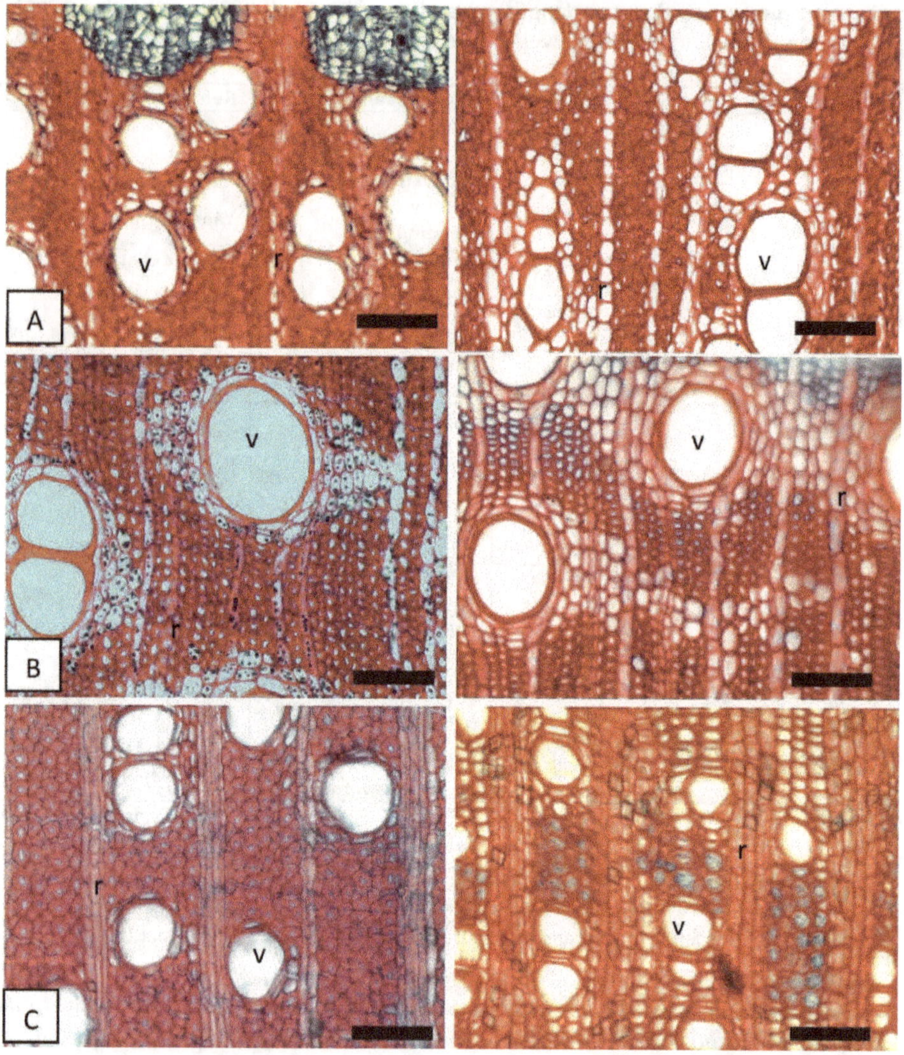

Fig. 3. *Avicennia germinans* (A), *Laguncularia racemosa* (B), and (C) *Rhizophora mangle* wood of trees growing in sites with lower (left) and higher (right) salinity. Differences in vessel density and diameter are present. Light microscope micrograph. Transverse section. v = vessel; r = wood ray; bar = 100 μm.

The stem water transport system consisting of a high vessel density, a high vessel grouping, small vessel diameters and short vessel elements in *Avicennia* and *Laguncularia* species, is contrasting with the water transport system of *Rhizophora* species, suggesting to be safer based on a physiological interpretation of the observed characteristics (Fig. 3). Structure modifications of *Bruguiera gymnorrhiza* seedlings of fiber wall thickness and length, and vessel diameter were reduced as flooding prolonged, suggesting distinct strategies for maintaining a balance between growth, conductive capacity, conductive safety and mechanical strength (Xiao et al., 2010). Lovelock et al. (2006a) suggest that over a range of species and sites, hydraulic conductivity in mangroves is constrained by salinity, with a variable relationship due to species differences (*Avicennia germinans* [2.13 kg m^{-1} s^{-1} $MPa^{-1}\cdot10^{-4}$]; *L. racemosa* [0.81 kg m^{-1} s^{-1} $MPa^{-1}\cdot10^{-4}$]), and to variation in nutrient availability.

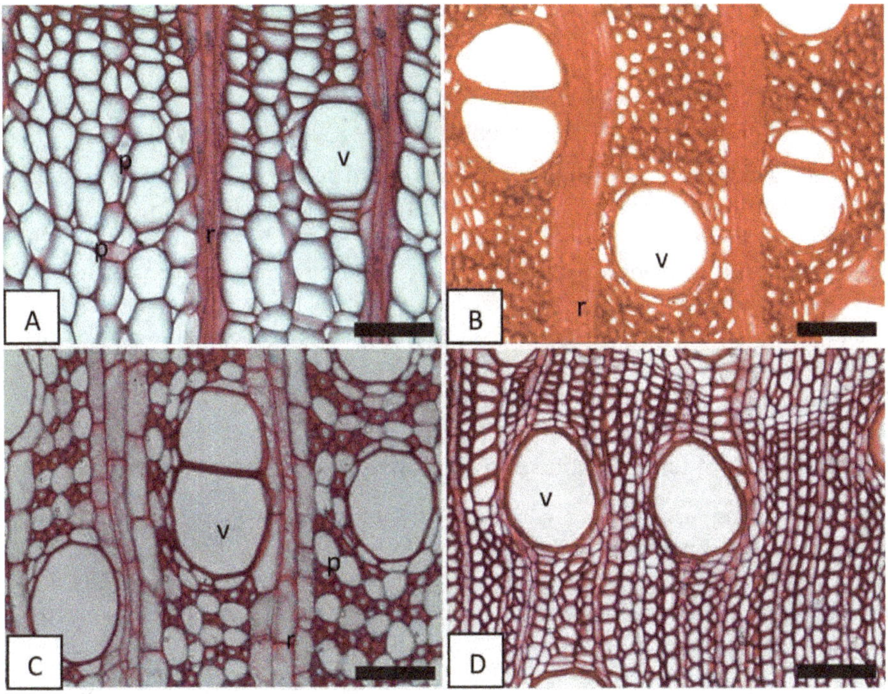

Fig. 4. The mangrove associate species *Annona glabra* (A), *Hibiscus tiliaceus* (B), *Pachira aquatica* (C), and *Rhabdadenia biflora* (D), showing vessels with wider tangential diameter than mangrove species. Light microscope micrograph. Transverse section. v = vessel; p = axial parenchyma; r = wood ray; bar = 100 μm.

Numerous mangrove associate species are more susceptible to modifications in flooding regime and salinity concentration, increasing vessel density like *Annona glabra,* from 4 to 7 vessels mm^{-2}. In general, vessel tangential diameter of mangrove associates is wider than those of mangrove species, suggesting an increment in vulnerability based on a physiological interpretation (Yáñez-Espinosa & Terrazas, 2001) (Fig. 4) (Table 2).

Nevertheless, abundant axial and radial parenchyma cells may be storing water in the stem, like *Annona, Pachira* and *Hibiscus* species. This mechanism would compensate the effects of increased axial resistance on leaf water status, extracting water from storage in the stem and then sap flow would increase more slowly during the morning (Stratton et al., 2000).The mangrove associate vine, *Rhabdadenia biflora,* growing on *Avicennia germinans* mangrove tree has scarce axial and radial parenchyma cells, however fiber diameter is wider like the other associate mangrove species.

Mangrove	Vessel tangential diameter (μm)	Mangrove associate	Vessel tangential diameter (μm)
Avicennia germinans	70	*Annona glabra*	93
Laguncularia racemosa	96	*Hibiscus tiliaceus*	122
Rhizophora mangle	73	*Pachira aquatica*	130
Conocarpus erectus	68	*Phyllanthus elsiae*	74
		Rhabdadenia biflora	93

Table 2. Comparison of average vessel tangential diameter from mangrove and mangrove associate species growing in Mexican mangrove forests (unpublished data).

Analysis of wood traits plasticity in some studies has identified hydraulic properties of trees as more plastic than those of leaf structural and physiological characters, concluding that hydraulic properties explain growth control.

The wide tolerance of mangroves wood traits to environmental gradient of salinity, soil texture and flooding conditions suggests that even if increases in relative sea level will eventually raise saturation and salinity conditions at ecotonal boundaries, mangroves are capable to advance or invade inland into freshwater marsh and swamp habitats (Doyle, 2003), modifying distribution and composition of the mangrove.

4. Modifications of bark structure in response to environmental factors

Prolonged flooding also affects bark anatomy of mangrove species, modifying the secondary phloem, rhytidome and periderm tissues, suggesting that prolonged flooding modifies vascular cambium and phellogen differently (Yáñez-Espinosa et al., 2008). Typical structural responses include formation of hypertrophied lenticels and adventitious root (Fig. 5), and increased aerenchyma development in the bark (Table 3).

Species	Rhytidome	Aerenchyma in rhytidome	Aerenchyma in collapsed phloem	Ray dilatation	Hypertro-phied lenticels	Adventitious roots
*Annona glabra**	++	++	++	++	++	++
*Hibiscus tiliaceus**	++	++	++	++	++	--
*Phyllanthus elsiae**	--	--	++	++	++	++
*Avicennia germinans**	--	--	--	--	++	--
*Laguncularia racemosa**	++	++	--	±±	++	±±
Conocarpus erectus•	++	++	--	±±	++	--
Pachira aquatica•	--	--	++	++	++	++

Table 3. Mangrove and mangrove associate species bark anatomical characters. ++ = present; -- = abscent; ±± = slightly present. * Data from Yáñez-Espinosa et al., 2008; • = Unpublished data.

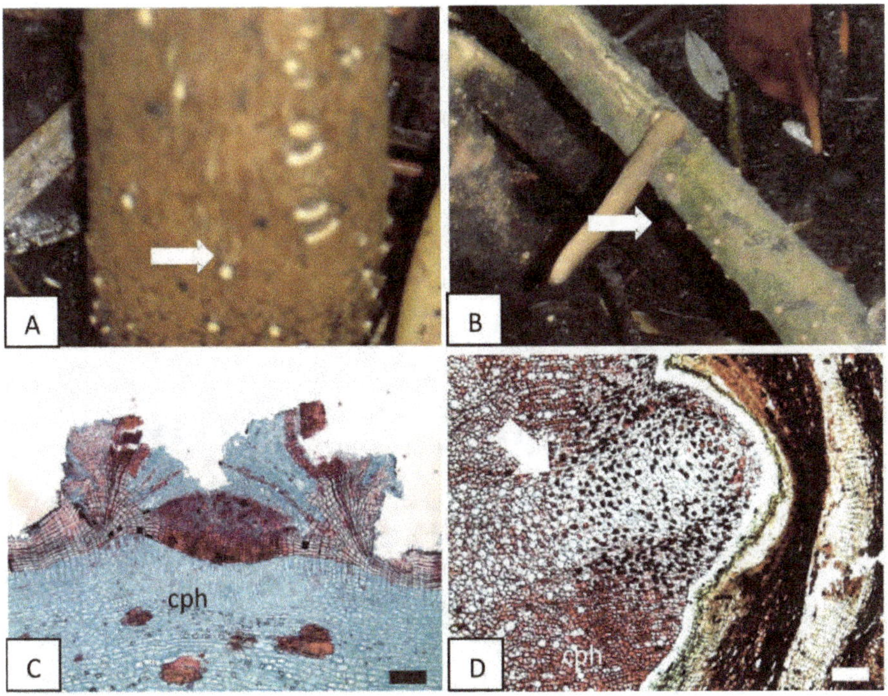

Fig. 5. Hypertrophied lenticels in the immersed portion of the stem of *Avicennia agerminans* (arrow) (A); and in the immersed stilt root of *Rhizophora mangle* (arrow) (B) macroscopic appearance; Light microscope micrograph. Transverse section. *Phyllanthus elsieae* hypertrophied lenticel in the immersed portion of the stem (C); *Annona glabra* adventitious root (arrow) in the immersed portion of the stem (D). p = periderm; cph = collapsed phloem; f = filling tissue; bar = 100 μm.

Mangrove species develop hypertrophied lenticels in the immersed portion of the stem, as well as mangrove associates tolerant to flooding, like *Annona glabra* (Mielke et al., 2005), excepting *Phyllanthus elsiae* (Euphorbiaceae), which develops hypertrophied lenticels above and below flooding level (Yáñez-Espinosa et al., 2008). Adventitious roots were present in the mangrove associate species *Annona glabra, Hibiscus tiliaceus, Pachira aquatica* and *Phyllanthus elsiae* (Mielke et al., 2005; Yáñez Espinosa et al., 2008). These species have not specialized roots like stilt roots or pneumatophores, then aeration depends on abundant hypertrophied lenticels and adventitious roots to tolerate prolonged or permanent flooding.

The importance of aerenchyma in the development of a continuous interconnected series of intercellular spaces is the most important adaptation to flooding in plants (Lambers et al., 1998). Trees without cork aerenchyma in their barks might develop larger air spaces in the phloem parenchyma, phelloderm or primary cortex (Roth, 1981), like *Rhizophora mangle* develops aerenchyma in phloem axial and radial parenchyma of collapsed region (Yáñez-Espinosa, unpublished data).

Although studies realized on bark rays show no statistical differences from short- to long-flooding period, mangrove species *Laguncularia racemosa, Avicennia germinans* and *Rhizophora mangle*, present shorter phloem rays under longer flooding period, may be associated with the effect of prolonged flooding similar to that of drought. But mangrove associate *Annona glabra* tend to have lower rays under short-flooding period, may be due to anoxic conditions caused by prolonged flooding (Yáñez- Espinosa et al., 2001).

Increments in rays are related to enlargement of cells that could mobilize a higher oxygen volume and more photosynthates (Kozlowski et al., 1991). The intercellular space system of ray tissue is essential, because ray cells have indefinite viability and require gas exchange for metabolic activity (Carlquist, 1988) (Fig. 6).

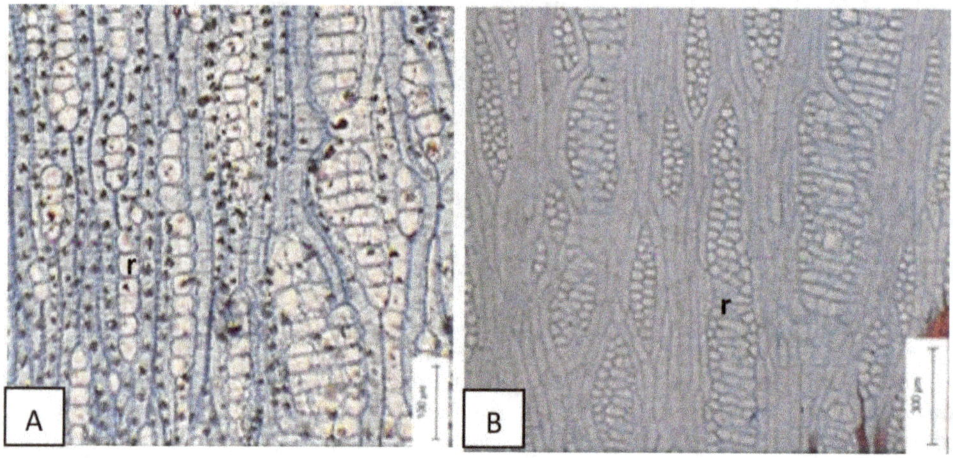

Fig. 6. Secondary phloem rays of *Laguncularia racemosa* (A) and *Annona glabra* (B) close to vascular cambium, showing a slightly dilatation. Light microscope micrograph. Tangential section. r = phloem ray.

Annona glabra, Pachira aquatica and *Hibiscus tiliaceus* develop aerenchyma in the rhytidome, phelloderm, and phloem parenchyma, which suggests a greater volume for aeration, particularly in the immersed portion of the stem. *Laguncularia racemosa* and *Conocarpus erectus* (Combretaceae) axial and radial parenchyma cells in the collapsed phloem region dilate slightly close to periderm, but develop abundant aerenchyma in the rhytidome. *Avicennia germinans* only develops rhytidome in the immersed portion of the stem and presents an aerenchymatous cork that may be the path connecting the aerial atmosphere with the immersed tissues (Fig. 7).

Even though aerenchyma and hypertrophy are the most common events related to flooding, each type of tissue responded differently, depending on the species. The mangrove and mangrove associate species respond to different flooding periods and then cohabit on a wide environmental gradient, suggesting that accelerated sea-level rise caused by global warming, would affect distribution of the species according to their particular tolerance.

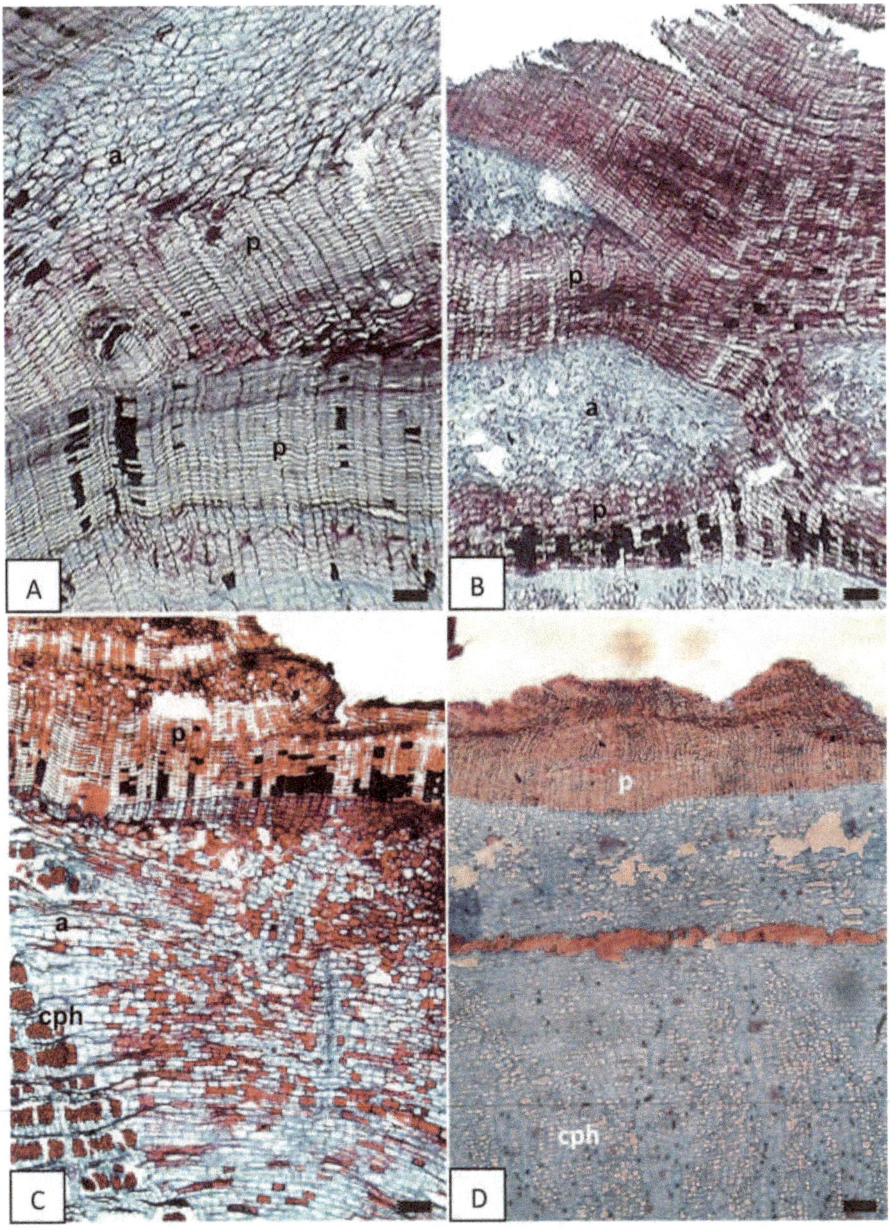

Fig. 7. *Annona glabra* aerenchyma in rhytidome of immersed portion of the stem (A), *Avicennia germinans* aerenchyma in rhytidome of immersed portion of the stem (B), bar = 200 µm; *Hibiscus tiliaceus* axial and radial parenchyma dilatation close to periderm (C), *Rhizophora mangle* aerenchyma close to periderm and dilatation of axial and radial parenchyma in collapsed phloem region (D). Light microscope micrograph. Transverse section. a = aerenchyma; p = periderm; cph = collapsed phloem; bar = 100 µm.

5. Modifications of leaf structure in response to environmental factors

Mangroves show many xerophytic adaptations that are evident in their leaves. Leaves are dorsiventral (Fig. 8), excepting *Lumnitzera racemosa,* where they are isobilateral and amphistomatic; the cells of the lower epidermis are larger than those of the upper surface, and water storage tissue is present in all the species, except *Aegialitis rotundifolia* (Seshavatharam & Srivalli, 1989; Tomlinson, 1986).

Experimental studies point to mangrove species-specific responses to increases in atmospheric CO_2 concentration. Increase in atmospheric CO_2 can be expected to improve mangrove tree growth, *i.e.* leaf area ratio increased with a decrease in humidity when *R. stylosa* was grown under elevated CO_2 (700 cm^3 m^{-3}) (Ball et al. 1997). However, elevated CO_2 (until 700 $\mu l l^{-1}$) did not affect leaf area of *Rhizophora mangle* seedlings (Farnsworth *et al.* 1996). In addition, increased CO_2 until 700 cm^3 m^{-3} did not affect leaf area of *Rhizophora apiculata* limited by high salinity, but did enhance leaf area when the plants were limited by humidity (Ball et al. 1997).

Responses of mangrove to sea-levels rise depend on reactions of individual plants, and some evidence exists on modifications of leaf structure in response to increasing flooding or waterlogging. *Bruguiera gymnorrhiza* and *Rhizophora stylosa* seedlings decreased leaf area intensely with decreasing tidal elevation (from 30-40 cm) in Yingluo Bay, a core zone within the Shankou Mangrove Reserve of Guangxi, China. Inversely, *Aegiceras corniculatum* and *Avicennia marina* seedlings increased leaf area with decreasing tidal elevation, at 10, 0, and - 30 cm (He et al. 2007).

Fig. 8. Leaves with xerophytic adaptations of *Avicennia germinans* (A), *Bruguiera* sp. (B), *Rhizophora mangle* (C), *Laguncularia racemosa* (D).

In addition, leaf area of adult individuals from *Laguncularia racemosa, Rhizophora mangle, R. racemosa, R. harrisonii,* and *Avicennia germinans* was generally significantly larger for species growing in wet (dry season lesser than 1 month) than in dry sites (dry season of seven months) in the Caribbean coast of Venezuela (Medina & Francisco, 1997). Similarly, *Rhizophora mangle* seedlings in Belize decreased leaf area in plants under water (Ellison & Farnsworth, 1997). However, there are some species with leaf area unaffected by tidal flooding, like *Xylocarpus granatum* seedlings in the Federated States of Micronesia (Allen et al. 2003), as well as *Aegiceras corniculatum* and *Avicenia marina* seedlings in Yingluo Bay, China (He et al. 2007), and *Annona glabra* after 56 days of flooding showed no symptoms of stress, and there were no changes in total leaf area (2.20 m^2) (Mielke et al., 2005).
Mangrove leaf anatomy is very uniform matched by a suite of features common to most mangrove species, including the colorless "water storage" tissue, the short tracheids terminating vein endings, the marked absence of sclerotic vein sheaths, presence of sclereids, thickened outer epidermal wall strongly cutinized, thick mesophyll including the adaxial hypodermis layers, palisade parenchyma layers and spongy parenchyma (Tomlinson, 1986) (Fig. 9).

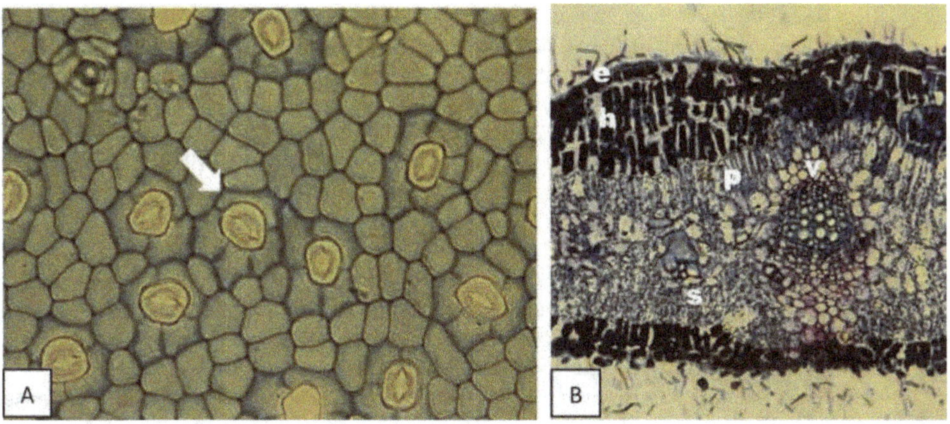

Fig. 9. *Conocarpus erectus L. (Combretaceae)* leaf epidermis showing stomata (arrow) (A), and showing dorsiventral structure: epidermis, adaxial hypodermis with two distinct layers, assimilating tissue below, and a vascular bundle. Light microscope micrograph .Paradermal (A) and transverse (B) section, by Guillermo Angeles. e = epidermis; h = hypodermis; p = palisade parenchyma; s = spongy parenchyma; v = vascular bundle.

Nevertheless, studies of modifications on leaf anatomical features under experimental conditions are scarce. Changes of *Avicennia marina* seedlings under experimental conditions were expressed in the progressive increase of upper and lower epidermis thickness to leaf thickness ratio and mesophyll to leaf thickness ratio with prolonged waterlogging duration, but prolonged waterlogging duration have negative effects on hypodermis thickness, leaf thickness, mesophyll thickness, palisade parenchyma thickness, palisade-spongy ratio, tangential vessel diameter and vessel wall thickness in leaf tissue (Xiao et al., 2009), contrary to *Bruguiera gymnorrhiza* seedlings grown under experimental conditions of simulated semidiurnal tides with salinities of 15‰, after 12 h treatments, that showed declination in leaf thickness, palisade parenchyma thickness, spongy parenchyma thickness, xylem length of the vascular system and number of vessels and vessel lines, up to 67.1%, compared with the 0 h treatments. However, the upper and lower epidermis and stomatal density of 12 h treatments showed increases of up to 104.3% over the 0 h treatments. The cuticle and percentage of intercellular spaces in spongy tissue decreased significantly with waterlogging duration at first and then increased. These results suggest that modifications in the anatomical features of *B. gymnorrhiza* as a result of periods of immersion longer than 2 h, would result in a reduction of photosynthesis and water transport (Wang et al., 2007). However, anatomical characters of *Rhizophora mangle* leaves did not differ when Ellison & Farnsworth (1997) simulated a 16 cm decrease and increase of sea-level.

There are many studies indicating the importance of salinity for mangrove species as well as evidence that various mangroves may have different tolerances and optimal salinity (Ball et al., 1997; Ball, 2002). The physiology of their ability to survive in saline environment may shed light on the evolution of mangroves from terrestrial species (Parani et al. 1998).

Leaf structure traits reflect the long-term adaptive strategy successfully (Cunningham et al., 1999; Wang et al,. 2003; Wright et al., 2004). Salinity affects ion accumulation in leaves, thereby membrane permeability and chlorophyll synthesis (Cram et al., 2002). In addition, with increase in salinity, carbon allocation to roots increases at the expense of leaf area (Ball 1988). Leaf area decreased at higher salinity concentrations (47-92‰) for *Laguncularia racemosa* in the Sontecomapan lagoon in Veracruz, Mexico (Medina et al., 1995), and for *Sonneratia alba* at 100‰ sea water and *S. lanceolata* seedlings at 50‰ sea water, in northern Australia (Ball & Pidsley 1995). The average leaf area of *Rhizophora apiculata* and *R. stylosa* propagules in the Northern Territory of Australia decreased with an increase in salinity from 125 to 350 mol m^{-3} NaCl, with *R. apiculata* being the more sensitive (Ball et al. 1997). For *Bruguiera parviflora* in Bhitarakanika mangrove forest, Orissa, India, leaf area was significantly less at 400 mM NaCl as compared to 0 mM NaCl (Parida et al., 2004), for *Laguncularia racemosa* in Sepetiba Bay, Rio de Janeiro, Brazil, which showed significant lower leaf area in salt marsh plants with high soil salinity and low nutrient input, than in riverside plants with daily tidal fluxes and consequently high nutrient input, that can be associated with environmental restrictions of their habitat (Lira-Medeiros et al., 2010), and for *Rhizophora mucronata* seedlings in the 45 PSU in South Africa (Hoppe-Speer et al., 2011).

In some species, leaf area is unaffected by salinity concentrations, *i.e. Rhizophora mangle*, which had similar leaf area in plants from El Real (at 47-92 ‰) and from La Boca (at 0-22 ‰) in the Sontecomapan lagoon in Veracruz, Mexico (Medina et al., 1995), and *Xylocarpus granatum* seedlings at 0.5 and 23 ‰ salinity in the Federated States of Micronesia (Allen et al,. 2003).

Increasing flood level and salinity due to sea-level rise, may result in mangrove leaves morphology and anatomy modifications, considering that extant studies show perceptible modifications, although not all species are affected (i.e. *Rhizophora* and *Xylocarpus)*. Modifications include increasing stomatal density, reduction of leaf area and thickness, palisade parenchyma thickness, number of vessels and vessel diameter in leaf tissue, whose may generate the reduction of transpiration, photosynthesis and water transport.

6. Modifications of root structure in response to environmental factors

The anatomy of mangrove roots has been extensively studied in relation to development and function (Tomlinson, 1986). Prolonged flooding alters soil physical and chemical properties, as well as bacterial composition, directly affecting species roots (Rajaniemia &

Allison, 2009). Root vascular system affects stem vascular system and leaves structure of mangrove species, even when there are present modified roots like pneumatophores, cable, knee and stilt roots (Visser & Pierik, 2007).

Pneumatophores arise vertically from cable roots and have evolved independently in at least five mangrove families and genera: *Laguncularia* (Combretaceae), *Avicennia* (Avicenniaceae), *Bruguiera* (Rhizophoraceae), *Xylocarpus* (Meliaceae), and *Sonneratia* (Sonneratiaceae) (Tomlinson 1986). They have abundant lenticels, and aerenchyma may account for up to 70% of root volume (Hogarth, 1999) (Fig. 10).

The normal pneumatophores of *Sonneratia* and *Avicennia* are slender and cone shaped, standing erect and aligned on the cable root, of 1-20 m or more in length, which spread horizontally in the soil. *Sonneratia alba* trees in Ranong and Phang-nga Bay, Thailand, produce pneumatophores 25-35 cm long, 4–6 cm thick at the base, with a regular conical shape. However, pneumatophores of *S. caseolaris* may become 1.2 m tall and 12 cm thick at the base. The cable roots morphology and length are more related to soil structure and tidal regime than tree variables, as well as pneumatophores that may be deformed due to massive waves, sediments deposition, and other environmental changes (Nakamura et al., 2004).

Avicennia marina in Kenya has the ability to adapt its pneumatophores to micro-topographical irregularities in the regularly sloping intertidal zone. Significantly higher pneumatophore densities and total pneumatophore lengths are present in the center of the landward depression, and significantly lower lengths in the center of the seaward depression (Dahdouh-Guebas et al., 2007), suggesting that cable roots and pneumatophores morphology, density and length may be affected by sea-level rise. *Rhizophhora mangle* underground roots are in permanently hypoxic or anoxic environment, and then vertical roots have the role of supplying oxygen to underground roots, reaching adequate gas exchange at low tide. Air passes through aerenchyma tissue, constituted by air spaces running longitudinally the root axis (Hogarth, 1999) (Fig. 10).

Experiments with *Avicennia marina* have demonstrated that oxygen concentrations decrease from 16 to less than 2% within one hour after exposure to hypoxic conditions. The roots increased then the capacity for alcoholic fermentation, although ethanol concentration was low in intact and hypoxic roots, but may have diffused from the roots into the surrounding substrate (McKee & Mendelssohn, 1987). *Avicennia germinans* roots, commonly found in anaerobic substrate, create oxidized rhizospheres substantially larger, but when air cannot enter the root system through the pneumatophores, the rhizospheres become as reduced as nearby non-vegetated soil. *Rhizophora mangle* have no effect on the oxidation state of surrounding anaerobic soils (Thibodeau & Nickerson, 1986).

Studies realized in *Laguncularia racemosa* cable roots and pneumatophores showed that cable roots are very efficient for water transport, and that this occurs mainly through the external vessels of the secondary xylem (Angeles et al., 2002). The development of a massive root system under soil flooding conditions in *Annona glabra* can be a mechanism to compensate for the high resistance to water flow in the roots (Mielke et al., 2005). However, more studies on mangrove root water transport are necessary to understand hydraulic properties of the whole plant system.

Sea-level rise caused by global warming may affect mangrove roots in different ways, depending of different functional root types (McLeod & Salm, 2006). Lenticels in the aerial roots have the role of supplying oxygen to underground roots, then if a rapid increment of sea-level occurs, pneumatophores of *Sonneratia* and *Avicennia* may have not the capacity of increase pneumatophore density and length fast enough to avoid oxygen concentration decrease, producing death of the plant. However, *Rhizophora* aerial roots could not be affected, because lenticels would remain above sea-level.

Fig. 10. *Rhizophora mangle* stilt roots (A); Light microscope micrograph of aerenchyma tissue close to periderm in a *R. mangle* aerial root (B); *Avicennia germinans* pneumatophores (C); Light microscope micrograph of dilated parenchyma tissue in *Laguncularia racemosa* pneumatophore (D). p = periderm; a = aerenchyma; x =secondary xylem; ph = secondary phloem; bar = 100 μm.

7. Modifications of growth in response to environmental factors

Growth of a plant is the increment in dry mass, volume, area or length as a result of division, expansion and differentiation of cells, determined by the interaction of photosynthesis, long-distance transport, respiration, water relations, and mineral nutrition (Lambers et al., 1998). Total above-ground biomass varies widely being highest at low latitude and declining northward and southwards from the equator. Undisturbed *Rhizophora* forests in northern Australia may reach up to 700 t ha^{-1} dry weight, but in old mangrove forests of South-east Asia range from 500 to 550 t ha^{-1} dry weight (Hogarth, 1999). However, the growth rate of mangroves is critically related to the availability of water to the trees and this is reflected in the soil water content and soil salinity (Field, 1995).

7.1 Biomass

The relative growth rate of *Bruguiera gymnorrhiza* decreased significantly with waterlogged time, with the highest value found for drained plants and the lowest in plants less than 12 weeks waterlogging. On the contrary, no significant difference was found between waterlogged and drained *Kandelia candel* plants. The shoot to root biomass ratio of *K. candel* increased when subjected to 8 or 12 weeks waterlogging but little change was recorded in *B. gymnorrhiza*, indicating a shift in biomass allocation from roots to shoots in *K. candel* under prolonged waterlogging but not in *B. gymnorrhiza*. Chlorophyll contents of *K. candel* increased more rapidly in response to waterlogging than *B. gymnorrhiza*. These physiological indicators supported the hypothesis that *K. candel* is more tolerant to waterlogging than *B. gymnorrhiza* (Ye et al., 2003). In mangrove associate *Annona glabra* seedlings, the high survival and growth rates are directly related to the capacity to develop a massive root system and maintain a high stomatal conductance and net photosynthesis under soil flooding. In most flood-tolerant species the initial decrease in stomatal conductance is followed by recovery, like in *A. glabra* seedlings under flooding treatment, where stomatal conductance was reduced to 65% of control levels with four days of flooding, but reached 152% at day 56. Significant increases in root, stem and total plant biomass and in collar diameter were observed. These results suggest that *A. glabra* is highly adapted to both periodic and permanent flooding (Mielke et al., 2005).

The effect of salinity on the growth of *Avicennia germinans* propagules collected in Cuman´a, Venezuela, was experimentally determined comparing the relative growth rates (RGR) after 27 weeks, reaching a maximum 10.4 mg g^{-1} d^{-1} in 170 mol m^{-3} NaCl, followed by the 0 and 430 mol m^{-3} NaCl treatments (8.4 and 7.9 mg g^{-1} d^{-1} respectively). At the same period the RGR of plants grown in 680 and 940 mol m^{-3} NaCl were 47 and 44% lower than that of plants grown in 170 mol m^{-3} NaCl. Higher salt tolerance of *A. germinans* in relation to other mangrove species might be due to low relative growth rates and high water-use efficiency over a wide salinity range. The decline in RGR with increasing salinity was clearly associated with a decrease in net assimilation rate (Suárez & Medina, 2005).

Also nutrient deficiency is relevant, when comparing dwarf and taller fringing tree forms of *Rhizophora mangle* in Belize, allow identifying trait plasticity in hydraulic properties of trees as more plastic than those of leaf structural and physiological characteristics, implying that hydraulic properties are significant in controlling growth in mangroves. Improvement of P deficiency reduced the structural and functional distinctions between dwarf and taller fringing tree forms, releasing trees from hydraulic limitations (Lovelock et al., 2006b).

Topography would be relevant if tidal effect would change. In one experiment, seven mangrove species *Rhizophora mucronata, R. apiculata, Bruguiera cylindrica, Ceriops tagal, Sonneratia alba, Avicennia officinalis* and *Xylocarpus granatum*, were planted at various topographic sites in an intertidal zone of Phang Nga, Thailand. The experimental plots were on a slope and showed a maximal elevation difference of 1.8 m, submerged with 2-3% saline water twice a day. Survival and growth performance of seedlings were measured differing and showing increasing tolerance to higher tidal inundations in the order: *R. mucronata, S. alba, R. apiculata, A. officinalis, C. tagal, B. cylindrica* and *X. granatum* (Kitaya et al., 2002).

7.2 Growth rings

Secondary growth, or secondary tissues added lateral to stems and roots increasing their thickness, are produced by the vascular cambium. In secondary xylem or wood, growth rings may be produced during one season, and usually they are delimited by growth boundaries (Beck, 2010) (Fig. 11). In mangroves, the highly dynamic intertidal environment and the overriding ecological drivers difficult the existence of growth rings. Nevertheless, climatic conditions that result in a range of soil water salinity experienced over the year are a prerequisite for the formation of growth ring (Robert et al., 2011).

Recently, annual growth rings were discovered in *Rhizophora mucronata* in Kenya, but the ring boundaries are indistinct and growth ring consists of a low vessel density earlywood, produced during the rainy seasons, and a high vessel density latewood, produced during the dry season. Intra-annual differences in the vessel features revealed a trade-off between hydraulic efficiency (large vessels) during the rainy season and hydraulic safety (small vessels) during the dry season (Verheyden et al., 2005).

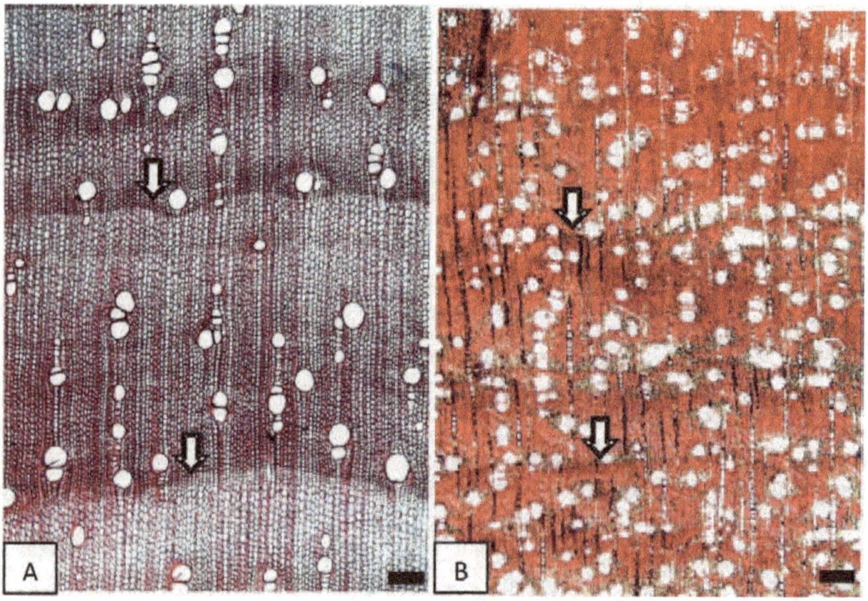

Fig. 11. Light microscope micrograph growth boundaries (arrows) distinct in *Pachira aquatica* (A), slightly defined in *Conocarpus erectus* (B). bar = 100 μm.

Laguncularia racemosa from Rio de Janeiro, Brazil, also present distinguishable growth rings macroscopically, with alternating early wood formed between spring and autumn and consists of layers with abundant paratracheal axial parenchyma, while late wood is formed in winter and consists of narrower dark layers which are basically fibrous zones with scarce paratracheal axial parenchyma. The occurrence of a dry season in winter could be the seasonal factor determining the formation of annual rings (Duque Estrada et al., 2008). These results are relevant because a factor that may be important for the future of mangrove ecosystems is any change in precipitation (Field, 1995).

Sea-level rise might reduce survival and growth performance of mangroves seedlings, expressed as biomass decrement due to variation in flooding and salinity. However, initial decrease is followed by recovery, suggesting that mangrove species are adapted to both periodic and permanent flooding, and high salinity levels. Secondary growth apparently is determined by seasonal precipitation, expressed by the formation of growth rings. Considering that precipitation rates are predicted to increase by about 25% by 2050 in response to global warming (McLeod & Salm, 2006), changes in precipitation patterns may affects at local scale the patterns of freshwater inflow, diminishing salinity concentration in consequence, and affecting mangroves growth and their distribution.

8. Conclusions

Modification of anatomical and morphological patterns of wood, bark, leaf and root of mangrove and mangrove associate species are predictable as result of sea-level rise caused by global warming. Sea-level rise will alter environmental conditions, increasing salinity, flooding level, and altering the rate of deposition of sediment. These species show a wide plasticity in structural and physiological patterns, allowing them to survive and prosper in this ecosystem exposed to a dynamic environment, and occasionally subjected to extreme events, like hurricanes and massive waves. General tendencies suggest a strong association between structure patterns and environmental gradient (salinity, flooding level and periodicity, and soil structure). Cell and organ size decrease while salinity and flooding or waterlogging increases.

Nevertheless, each species has its specific tolerance range, allowing to success along different zones of the mangrove ecosystem with particular geomorphology and environmental conditions, causing a change in forest species composition and dominance, allowing expanding distribution of more tolerant species and limiting or suppressing those less tolerant, particularly in places where sedimentation rates are low. Some studies suggest that mangroves can adapt to sea-level rise if it occurs slowly, there are sufficient space for distribution expansion, and the other environmental factors are tolerable. They may adapt growing inland on more elevated areas, probably increasing their distribution, but if these not occur fast, mangroves adaptation to stressing factors will result in smaller trees until disappearing.

9. Acknowledgments

We thank the financial support of CONACYT (L.Y.E. scholarship no. 112170), Colegio de Postgraduados (M.Sc.), and Instituto de Ecología, A.C. (postdoctoral research), to develop the research projects on mangrove forests from the Pacific and Gulf of Mexico; National Tropical Botanical Garden (Kenan Fellowship for College Professors) to visit and photograph the mangrove forests in Florida, USA; Idea Wild (2007) and PROMEP 103.5/07/2574 (2008) for the acquisition of the digital camera and the light microscope. Thanks to Teresa Terrazas (Universidad Nacional Autónoma de México), Guillermo Ángeles and Jorge López-Portillo (Instituto de Ecología, A.C.), for the time we have dedicated to understand the ecoanatomy of mangroves.

10. References

Allen, J.A.; Krauss, K.W. & Hauff, R.D. (2003). Factors Limiting the Intertidal Distribution of the Mangrove Species *Xylocarpus granatum*. *Oecologia*, Vol.135, No. 1, (March 2003), pp. 110-121, ISSN 0029-8549

Alongi, D.M. (2008). Mangrove Forests: Resilience, Protection from Tsunamis, and Responses to Global Climate Chang. *Estuarine, Coastal and Shelf Science*, Vol. 76, No. 1, (January 2008), pp. 1-13, ISSN 0272-7714

Angeles, G.; López-Portillo, J. & Ortega-Escalona, F. (2002). Functional Anatomy of the Secondary Xylem of Roots of the Mangrove *Laguncularia racemosa* (L.) Gaertn. (Combretaceae). *Trees – Structure and Function*, Vol. 16, No.4-5, (May 2002), pp. 338-345, ISSN 0931-1890

Ball, M.C. & Pidsley, S.M. (1995). Growth Responses to Salinity in Relation to Distribution of Two Mangrove Species, *Sonneratia alba* and *S. lanceolata*, in Northern Australia. *Functional Ecology*, Vol. 9, No. 1, (February1995), pp. 77-85, ISSN 0269-8463

Ball, M.C. (1988). Ecophysiology of Mangroves. *Trees – Structure and Function*, Vol. 2, No. 3, (November 1988), pp. 129-142, ISSN 0931-1890

Ball, M.C. (2002). Interactive Effects of Salinity and Irradiance on Growth: Implications for Mangrove Forest Structure along Salinity Gradients. *Trees – Structure and Function*, Vol. 16, No. 2-3, (March 2002), pp. 126-39, ISSN 0931-1890

Ball, M.C.; Cochrane, M.J. & Rawson, H.M. (1997). Growth and Water Use of the Mangroves *Rhizophora apiculata* and *R. stylosa* in Response to Salinity and Humidity under Ambient and Elevated Concentrations of Atmospheric CO_2. *Plant, Cell and Environment* Vol. 20, No. 9, (September 1997), pp. 1158-1166, ISSN 0140-7791

Baskin, C.C. & Baskin, J.M. (2001). Seeds: Ecology, Biogeography, and Evolution of Dormancy and Germination. Academic Press, ISBN 0-12-080260-0, London, United Kingdom.

Beaumont, L.J.; Pitman, A.; Perkinsc, S.; Zimmermann, N.E.; Yoccoz, N.G. & Thuiller, W. (2011). Impacts of Climate Change on the World's Most Exceptional Ecoregions. *Proceedings of the National Academy of Sciences*, Vol. 108, No. 6, (February 2011), pp. 2306–2311, ISSN 0027-8424

Beck, C.B. (2010). *An Introduction to Plant Structure and Development: Plant Anatomy for the Twenty-First Century*, Cambridge University Press, ISBN 978-0-521-51805-5, New York, USA

Carlquist, S. (1988). *Comparative Wood Anatomy: Systematic, Ecological, and Evolutionary Aspects of Dicotyledon Wood*, Springer-Verlag, ISBN 978-3540411734, New York, USA

Cram, J.W.; Torr, P.G. & Ross, D.A. (2002). Salt Allocation During Leaf Development and Leaf Fail in Mangroves. *Trees – Structure and Function*, Vol. 16, No. 2-3, (May 2002), pp. 112-19, ISSN 0931-1890

Cunningham, S.A.; Summerhayes, B. & Westoby, M. (1999). Evolutionary Divergences in Leaf Structure and Chemistry, Comparing Rainfall and Soil nutrient Gradients. *Ecological Monographs*, Vol. 69, No. 4, (December 1999), pp. 569-588, ISSN 00129615

Dahdouh-Guebas, F.; Kairo, J.G; De Bondt, R. & Koedam, N. (2007). Pneumatophore Height and Density in Relation to Micro-Topography in the Grey Mangrove *Avicennia marina*. *Belgian Journal of Botany*, Volume 140, Number 2, (December 2007), pp. 213-221, ISSN 0778-4031

Doyle, T.W., 2003, Predicting Future Mangrove Forest Migration in the Everglades Under Rising Sea Level: U.S. Geological Survey Fact Sheet 030-03, 2 p. Available from http://www.nwrc.usgs.gov/factshts/030-03.pdf

Duque Estrada, G.C.; Henriques Callado, C.; Gomes Soares, M.L. & Lisi, C.S. (2008). Annual Growth Rings in the Mangrove *Laguncularia racemosa* (Combretaceae). *Trees – Structure and Function*, Vol. 22, No. 5, (October 2008), pp. 663–670, ISSN 0931-1890

Ellison, A.M. & Farnsworth, E.J. (1997). Simulated Sea Level Change Alters Anatomy, Physiology, Growth, and Reproduction of Red Mangrove (*Rhizophora mangle* L.). *Oecologia*, Vol. 112, No. 4, (November 1997), pp. 435-446, ISSN 0029-8549

Farnsworth, E.J.; Ellison, A.M. & Gong, W.K. (1996). Elevated CO_2 Alters Anatomy, Physiology, Growth, and Reproduction of Red Mangrove (*Rhizophora mangle* L.). *Oecologia*, Vol. 108, No. 4, (November 1996), pp. 599-609, ISSN 0029-8549

Field, C.D. (1995). Impact of Expected Climate Change on Mangroves. *Hydrobiologia*, Vol. 295, No. 1-3; (January 1995), pp. 75-81, ISSN 0018-8158

Gilman, E.L.; Ellison, J.; Duke, N.C. & Field, C. (2008). Threats to Mangroves from Climate Change and Adaptation Options. *Aquatic Botany*,Vol. 89, No. 2, (August 2008), pp. 237-250, ISSN 0304-3770

Giri, C.; Ochieng, E.; Tieszen, L. L.; Zhu, Z.; Singh, A.; Loveland, T.; Masek, J. & Duke, N. (2011). Status and Distribution of Mangrove Forests of the World using Earth Observation Satellite Data. *Global Ecology and Biogeography*, Vol. 20, No. 1, (January 2011), pp. 154-159, ISSN 1466-822X

He, B.; Lai, T.; Fan, H.; Wang, W. & Zheng, H. (2007). Comparison of Flooding-Tolerance in four Mangrove Species in a Diurnal Tidal Zone in the Beibu Gulf. *Estuarine, Coastal and Shelf Science*, Vol. 74, No. 1-2, (August 2007), pp. 254-262, ISSN 0272-7714

Hogarth, P.J. (1999). *The Biology of Mangroves*, Oxford University Press, ISBN 0-19-850222-2, , New York, USA

Hoppe-Speer, S.C.L.; Adams, J.B.; Rajkaran, A. & Bailey, D. (2011). The Response of the Red Mangrove *Rhizophora mucronata* Lam. to Salinity and Inundation in South Africa. *Aquatic Botany*, Vol. 95, No. 2, (August 2011), pp. 71-76, ISSN 0304-3770

Kareiva, P.M.; Kingsolver, J.G. & Huey, R.B. (1993). *Biotic Interactions and Global Change*, Sinauer Associates Inc., ISBN 0–87893–430–8, Sunderland, USA

Kitaya, Y.; Jintana, V.; Piriyayotha, S.; Jaijing, D.; Yabuki,K.;Izutani, S.; Nishimiya, A. & Iwasaki, M. (2002). Early Growth of Seven Mangrove Species Planted at Different Elevations in a Thai Estuary. *Trees – Structure and Function*, Vol. 16, No. 2-3, (March 2002), pp. 150-154, ISSN 0931-1890

Kozlowski, T.T.; Kramer, P.J. & Pallardy, S.G. (1991). *The Physiological Ecology of Woody Plants*, Academic Press, ISBN 0-12-424160-3, London, England

Lambers, H.; Chapin, F.S. & Pons, T.L. (1998). *Plant Physiological Ecology*, Springer-Verlag, ISBN 0-387-98326-0, New York, USA

Lira-Medeiros, C.F.; Parisod, C.; Fernandes, R.A.; Mata, C.S.; Cardoso, M.A. & Ferreira, P.C.G. (2010). Epigenetic Variation in Mangrove Plants Occurring in Contrasting Natural Environment. In: *PloS One,* Vol. 5, No. 4, e10326. doi:10.1371/journal.pone.0010326

Lovelock, C.E.; Ball, C.E.; Feller, I.C.; Engelbrecht, B.M.J. & Ewe, M.L. (2006a). Variation in hydraulic conductivity of mangroves: influence of species, salinity, and nitrogen and phosphorus availability. *Physiologia Plantarum,* Vol. 127, No. 3, (July 2006), pp. 457–464, ISSN 0031-9317

Lovelock, C.E.; Ball, M.C.; Choat, B.; Engelbrecht, B.J.; Holbrook, N.M. & Feller, I.C. (2006b) Linking physiological processes with mangrove forest structure: phosphorus deficiency limits canopy development, hydraulic conductivity and photosynthetic carbon gain in dwarf *Rhizophora mangle*. *Plant, Cell and Environment* Vol. 29, No. 5, (May 2006), pp. 793-802, ISSN ISSN 0140-7791

McKee, K.L. & Mendelssohn, I.A. (1987). Root Metabolism in the Black Mangrove (*Avicennia germinans* (L.)L): Response to Hypoxia. *Environmental and Experimental Botany*, Vol. 27, No. 2, (April 1987), pp. 147-156, ISSN 0098-8472

McLeod, E. & Salm, R.V. (2006). *Managing Mangroves for Resilience to Climate Change,* IUCN, ISBN-10: 2-8317-0953-9, Gland, Switzerland

Medina, E. & Francisco, M. (1997). Osmolality and $\delta^{13}C$ of Leaf Tissues of Mangrove Species from Environments of Contrasting Rainfall and Salinity. *Estuarine, Coastal and Shelf Science*, Vol. 45, No. 3, (September 1997), pp. 337-344, ISSN 0272-7714

Medina, E.; Lugo, A.E. & Novelo, A. (1995). Contenido Mineral del Tejido Foliar de Especies de Manglar de la Laguna de Sontecomapan (Veracruz, Mexico) y su Relación con la Salinidad. *Biotropica,* Vol. 27, No. 3, (September 1995), pp. 327-323, ISSN 0006-3606

Mielke, M.S.; Matos, E.M.; Couto, V.B.; de Almeida, A.F.; Gomes, F.P. & Oliveira Mangabeira, P.A. (2005). Some Photosynthetic and Growth Responses of *Annona glabra* L. Seedlings to Soil Flooding. *Acta Botanica Brasilica*, Vol. 19, No. 4, (October 2005), pp. 905-911, ISSN 0102-3306

Nakamura, T., Minagawa, R. & Havanond, S. (2004). Some Ecological Aspects of the Morphology of Pneumatophores of *Sonneratia alba* and *Avicennia officinalis.* In: *Mangrove Management and Conservation: Present and Future,* M. Vannucci, (Ed.), 39-44, United Nations University Press, ISBN 9789280810844, Tokyo, Japan

Parani, M.; Lakshmi, M.; Senthilkumar, P.; Ram, N. & Parida, A. (1998). Molecular Phylogeny of Mangroves V. Analysis of Genome Relationships in Mangrove Species using RAPD and RFLP Markers. *Theoretical and Applied Genetics,* Vol. 97, No. 4, (September 1998), pp. 617-25, ISSN *0040-5752*

Parida, A.K.; Das, A.B. & Mittra, B. (2004). Effects of Salt on Growth, Ion Accumulation, Photosynthesis and Leaf Anatomy of the Mangrove. *Bruguiera parviflora. Trees – Structure and Function,* Vol. 18, No. 2, (March 2004), pp. 167-174, ISSN 0931-1890

Rajaniemia,T.K. & Allison, V.J. (2009). Abiotic Conditions and Plant Cover Differentially Affect Microbial Biomass and Community Composition on Dune Gradients. *Soil Biology and Biochemistry*, Vol. 41, No.1, (January 2009), pp. 102-109, ISSN 0038-0717

Robert, E.M.R.; Koedam, N.; Beeckman, H. & Schmitz, N. (2009). A Safe Hydraulic Architecture as Wood Anatomical Explanation for the Difference in Distribution of the Mangroves *Avicennia* and *Rhizophora. Functional Ecology*, Vol. 23, No. 4, (August 2009), pp. 649-657, ISSN 0269-8463

Robert, E.M.R.; Schmitz, N.; Auma Okello, J.; Boeren, I.; Beeckman, H. & Koedam, N. (2011). Mangrove Growth Rings: Fact or Fiction? *Trees – Structure and Function*, Vol. 25, No. 1, (February 2011), pp. 49-58, ISSN 0931-1890

Roth, I. (1981) *Structural Patterns of Tropical Barks*, Encyclopedia of Plant Anatomy Band IX 3, Borntraeger, ISBN 978-3-443-14012-0, Berlin, Germany

Schmitz, N.; Verheyden, A.; Beeckman, H.; Gitundu Kairo, J. & Koedam, N. (2006) Influence of Salinity Gradient on the Vessel Characters of the Mangrove Species *Rhizophora mucronata. Annals of Botany*, Vol. 98, No. 6, (December 2006), pp. 1321-1330, ISSN 0305-7364

Semeniuk, V. (1994). Predicting the Effect of Sea-level Rise on Mangroves in Northwestern Australia. *Journal of Coastal Research*, Vol. 10, No. 4, (Autumn 1994), pp. 1050-1076, ISSN 0749-0208

Seshavatharam, V. & Srivalli, M. (1989). Systematic Leaf Anatomy of Some Indian Mangroves. *Proceedings: Plant Sciences*, Vol. 99, No. 6, (December 1989), pp. 557-565, ISSN 0306-4484

Solomon, S.; Qin, D.; Manning, M.; Alley, R.B.; Berntsen, T.; Bindoff, N.L.; Chen, Z.; Chidthaisong, A.; Gregory, J.M.; Hegerl, G.C.; Heimann, M.; Hewitson, B.; Hoskins, B.J.; Joos, F.; Jouzel, J.; Kattsov, V.; Lohmann, U.; Matsuno, T.; Molina, M.; Nicholls, N.; Overpeck, J.; Raga, G.; Ramaswamy, V.; Ren, J.; Rusticucci, M.; Somerville, R.; Stocker, T.F.; Whetton, P.; Wood, R.A. & Wratt, D. (2007). Contribution of Working Group I to the Fourth Assessment Report of the Intergovernmental Panel on Climate Change. In: *Climate Change 2007, The Physical Science Basis*. S. Solomon, D. Qin, M. Manning, Z. Chen, M. Marquis, K.B. Averyt, M. Tignor, H.L. Miller (Eds.) Cambridge University Press, ISBN 978-0-521-70596-7, Cambridge, United Kingdom

Stratton, L.; Goldstein, G. & Meinzer, F.C. (2000). Stem Water Storage Capacity and Efficiency of Water Transport: Their Functional Significance in Hawaiian Dry Forest. *Plant, Cell and Environment*, Vol. 23, No. 1, (January 2000), pp. 99-106, ISSN ISSN 0140-7791

Suárez, N. & Medina, E. (2005). Salinity Effect on Plant Growth and Leaf Demography of the Mangrove, *Avicennia germinans* L. *Trees – Structure and Function*, Vol. 19, No. 6, (November 2005), pp. 721-727, ISSN 0931-1890

Thibodeau, F.R. & Nickerson, N.H. (1986). Differential Oxidation of Mangrove Substrate by *Avicennia germinans* and *Rhizophora mangle*, *American Journal of Botany*, Vol. 73, No. 4, (April 1986), pp. 512-516, ISNN 0002-9122

Tomlinson, P.B. (1986). *The Botany of Mangroves*, Cambridge University Press, ISBN 0-521-25567-8, New York, USA

Verheyden, A.; de Ridder, F.; Schmitz, N.; Beeckman, H. & Koedam, N. (2005). High-resolution Time Series of Vessel Density in Kenyan Mangrove Trees Reveal a Link with Climate. *New Phytologist*, Vol. 167, No. 2, (August 2005), pp. 425-435, ISSN 0028-646X

Visser, E.J. & Pierik, R. (2007). Inhibition of Root Elongation by Ethylene in Wetland and Non-Wetland Plant Species and the Impact of Longitudinal Ventilation. *Plant, Cell and Environment,* Vol. 30, No. 1, (January 2007), pp. 31-38, ISSN 0140-7791

Wang WQ, Wang M, Lin P. (2003). Seasonal Changes in Element Contents in Mangrove Element Retranslocation During Leaf Senescence. *Plant and Soil,* Vol. 252, No. 2, (May 2003), pp. 187-193, ISSN *0032-079X*

Wang, W.; Xiao, Y.; Chen, L. & Lin, P. (2007). Leaf Anatomical Responses to Periodical Waterlogging in Simulated Semidiurnal Tides in Mangrove *Bruguiera gymnorrhiza* Seedlings. *Aquatic Botany,* Vol. 86, No. 3, (April 2007), pp. 223-228, ISSN 0304-3770

Wright, I.J.; Reich, P.B.; Westoby, M.; , Ackerly, D.D.; Baruch, Z.; Bongers, F.; Cavender-Bares, J.; Chapin, T.; Cornelissen, J.H.C.; Diemer, M.; Flexas, J.; Garnier, E.; Groom, P.K.; Gulias, J.; Hikosaka, K.; Lamont, B.B.; Lee, T.; Lee, W.; Lusk, C.; Midgley, J.J.; Navas, M.L.; Niinemets, Ü.; Oleksyn, J.; Osada, N.; Poorter, H.; Poot, P.; Prior, L.; Pyankov, V.I.; Roumet, C.; Thomas, S.C.; Tjoelker, M.G., Veneklaas, E.J. & Villar, R. (2004). The Worldwide Leaf Economics Spectrum. *Nature,* Vol. 428, No. 6985, (April 2004), pp. 821-827, ISSN 0028-0836

Xiao, Y.; Wang, W. & Chen, L. (2010). Stem Anatomical Variations in Seedlings of the Mangrove *Bruguiera gymnorrhiza* Grown under Periodical Waterlogging. *Flora - Morphology, Distribution, Functional Ecology of Plants,* Vol. 205, No. 8, (January 2010), pp. 499-505, ISSN 0367-2530

Xiao, Y.; Zuliang, J.; Wang, M.; Lin, G. & Wang, W. (2009). Leaf and Stem Anatomical Responses to Periodical Waterlogging in Simulated Tidal floods in Mangrove *Avicennia marina* Seedlings. *Aquatic Botany,* Vol. 91, No. 3, (October 2009), pp. 231-237, ISSN 0304-3770

Yáñez-Arancibia, A.; Twillry, R.R. & Lara-Domínguez, A.L. (1998). Los Ecosistemas de Manglar frente al Cambio Climático Global. *Madera y Bosques,* Vol. 4, No. 2, (July 1998), pp. 3-19, ISSN 1405-0471

Yáñez-Espinosa, L. & Terrazas, L. (2001). Wood and Bark Anatomy Variation of *Annona glabra* L. Under Flooding. *Agrociencia,* Vol. 35, No. 1, (January-February 2001), pp. 51-63, ISSN 1405-3195

Yáñez-Espinosa, L.; Angeles, G.; López-Portillo, J. & Barrales, S. (2009). Variación Anatómica de la Madera de *Avicennia germinans* en la Laguna de La Mancha, Veracruz, México. *Boletín de la Sociedad Botánica de México,* No. 85, (December 2009), pp. 7-15, ISSN 0366-2128

Yáñez-Espinosa, L.; Terrazas, L. & Angeles, G. (2008). The Effect of Prolonged Flooding on the Bark of Mangrove. *Trees – Structure and Function,* Vol. 22, No. 1, (Febraury 2008), pp. 77-86, ISSN0931-1890

Yáñez-Espinosa, L.; Terrazas, L. & López-Mata, L. (2001). Effects of Flooding on Wood and Bark Anatomy of Four Species in a Mangrove Forest Community. *Trees – Structure and Function,* Vol. 15, No. 2, (Febraury 2001), pp. 91-97, ISSN0931-1890

Yáñez-Espinosa, L.; Terrazas, L.; López-Mata, L. & Valdez-Hernández, J.I. (2004). Wood Variation in *Laguncularia racemosa* and its Effect on Fibre Quality. *Wood Science and Technology*, Vol. 38, No. 3, (June 2004), pp. 217-226, ISSN *0043-7719*

Ye, Y.; Tam, N.F.Y.; Wong, Y.S. & Lu, C.Y. (2003). Growth and Physiological Responses of Two Mangrove Species (*Bruguiera gymnorrhiza* and *Kandelia candel*) to Waterlogging. *Environmental and Experimental botany*, Vol. 49, No. 3, (June 2003), pp. 209-221, ISSN 0098-8472

11

Developing Urban Adaptation Strategies for Global Warming by Using Data Mining Techniques: A Case Study of Major Metropolitan Areas in Japan

Yu-Chi Weng
Division of Environmental Engineering, Faculty of Engineering, Hokkaido University
Japan

1. Introduction

Modern life and high population density is the main characteristics of cities. Urbanization diffuses all over the world in recent centuries. Metropolitan areas are formed to provide more industrial production as well as business communication. Furthermore, economies of scale arising from spatial concentration of activity within industries in metropolitan area, i.e. industry agglomeration (Rosenthal & Strange, 2001); population, resources, capital concentrate into the cities. Industry agglomeration makes cities more adaptive to uncertainty of business environment (Strange et al., 2006). Hence the most profit seems to be created in metropolitan area in this globalization age. Nonetheless, natural resources are rapidly produced into goods, consumed and transformed into side-products, e.g., solid waste and a variety of pollutants, into the environment. Besides, more and more infrastructures are developed in the progress of civilization, but only a few natural habitats are preserved, leading to the decreases of biodiversity (van Bohemen, 1998; Mcdonald et al, 2008). Though people have convenient life in metropolitan area, however, the urbanization is bringing about a great deal of critical global environmental changes. In particular, the enhanced global warming, due to the urban growth, is threatening the human security and the possibly irreversible changes on ecological systems in a variety of dimensions and scales (Chung et al., 2009; Firman et al., 2011; Kataoka et al., 2009; Khasnis & Nettleman, 2005; Robert & Cory, 2003). Thereby, it is imperative for the municipalities to develop adaptation strategies for metropolitans in the context of human security, urban sustainability and urban growth. In the progress of making urban growth policies, important factors should be taken into consideration from the socio-economic, environmental, cultural, public health and ecological perspectives regarding the potential threats of global warming.

Several types of environmental indicators are developed to diagnose the current situation and formulate adaptation strategies against the global warming and associated issues of urban sustainability. Normally, recent studies argued that the urban sustainability should take the composite system ecology into consideration, implying that the interactions of the stakeholders in the city have to be considered in the evaluation simultaneously (Mistch, 2003; Roseland, 1997). Also, several evaluation bases could be used to examine the performance of the environment system, e.g. the energy, monetary, material bases (Weng & Fujiwara, 2011). A variety of environmental impact assessment tools, e.g., life-cycle assessment (LCA), life-cycle cost (LCC) assessment and cost-benefit analysis (CBA), could serve as efficient evaluation methods in the evaluation of countermeasures against the global warming, given that credible parameters are available (Jeong & Lee, 2009). In fact urban environment is a socio-economic-natural composite ecological system such that public policies should consider all the dimensions simultaneously, reorganizing the urban system into ecological network both for human system and natural system. Some attempts have developed the Environmental Kuznets Curve (EKC) to analyze the relationships between the economic and environmental quality, regarding driving factors of income, consumption and policy interventions (Arrow et al., 1995; Azomahou et al, 2006; Magnani, 2000; Weng et al., 2010). Although more and more evidences showed that the EKCs explain the relationships of several environmental pollutants, including the CO_2 emission, some methodological issues of EKC have to be dealt with from the perspective of statistics (Müller-Fürstenbergera & Wagner, 2007). One reason is that the available socio-economic and environmental data is not sufficient. Meanwhile, as for the conventional environmental evaluation methods, the data availability would lead to the uncertainty of the quantification outcomes, limiting the credibility of the interpretations. Hence, the dimensionless composite environmental indicators could be an alternative tool for environmental evaluation. For this purpose, the pressure-status-response (PSR) framework was established by the Organisation for Economic Co-operation and Development [OECD] in 1993 to serve as an environmental policy evaluation tool (OECD, 1993) and afterward, the driving forces are adopted as driving forces-state-response (DSR) framework in 1996 (OECD, 1996). Also, Kessler and Van Dorp (1998) proposed the adoption of environmental indicators under a strategic environment assessment (SEA) framework, in which socio-economic and environmental variables should be considered simultaneously. Subsequently, a driving forces-pressure-state-impact-response (DPSIR) framework was proposed by European Environmental Agency [EEA] as an extension of the PSR (EEA, 1999). Based on the aforementioned frameworks, Hu and Wang (1998) argued the urban environment should be considered as a socio-economic-natural composite ecological system; they adopted economic, cultural, environmental, and infrastructure variables to simulate the linkages among the functional modules in the evaluation of the performance of eco-reconstructing of eco-villages in China. Button (2002) proposed an analytical indicator framework composed evaluating urban environmental system; the indicators are composed of economic, environmental, social, cultural, and political variables. Jago-on et al. (2009) analyzed the critical urban environmental issues by using the DPSIR framework and proposed countermeasures for Asian cities. Regarding global warming, Omann et al. (2009) conducted a DPSIR analysis discussing the impacts of the climate changes on the biodiversity conservation. The outcomes indicated that the driving forces from the modern human society, e.g., the socio-economic and cultural attributes and the energy demands of transportation, bring about significant negative impacts on the global warming and the related challenges of biodiversity conservation. Also, Rounsevell et al. (2010) proposed a conceptual DPSIR framework qualitatively analyzing the apparent driving factors of socio-ecological indicators and their influences on the eco-services. A normal evaluation scheme is proposed in the study arguing that the attributes and the influencing time-spatial scales of the driving factors should be clarified, and thus possible quantitative impact analyses and efficient countermeasures could be subsequently implemented.

Regarding the quantification of indicator systems, Song et al. (2004) developed a hierarchical indicator system to evaluate the ecological sustainability of inshore cities in China. In their study, many socio-economic and environmental variables are categorized into three functional groups: structure, function, and coordination. Variables were linked as an ecological network, and the sustainability of each city was calculated and compared by a composite evaluating indicator. In addition, Srebotnjak (2007) presented a quantitative indicator system developing an integrated environmental performance index. She also particularly argued the importance of the development and the utilization of credible environmental statistics while the data availability is the main problem in environmental evaluations. Furthermore, Pan & Kao (2009) developed an inter-generation equity indicator (IGEI) to quantify the sustainability among generations in recent decades at a world scale, under a pressure-state-response (PSR) framework.

In fact, local actions are of particular importance in the enhancement of sustainability regarding the urbanization. In this sense, this study aims at developing a precautionary indicator system (PIS), which is in aid of the formation of adaptation strategies of the global warming mitigation. After the World War II, Japan made great economic development in the past six decades. Some regionally nuclear cities have been formed all over Japan. However, the current deficiencies of policies and future adaptation strategies are expected to deal with facing the critical challenges of sustainable development, e.g., the urban growth management and the global warming mitigation. In this study, by evaluating current social, economic, and environmental system of urban area, the precautionary indicators are developed to diagnose the current situation. Finally, strategies for developing a sustainable city are proposed in the final part.

2. Research approach

In order to deal with the information hidden in an enormous amount of statistics, data mining approaches are rapidly developed in recent decades. Mainly, data mining approaches are based on the integration of statistical theories for pattern recognition, causal relationships development, behavior analysis and system control & forecasting. In particular, several methods show superiority for the purpose of data mining, including regression analysis, multivariate analysis, indicator system techniques, artificial neural network, and data envelope analysis (Chen et al., 2010; Ngai et al., 2009). Moreover, the developing information system technologies, e.g., the geographic information systems and the remote sensing technology, provide high quality data and platforms for data analysis and integration.

The aforementioned approaches could be applied to establish the precautionary indicator system of global warming from a variety of perspectives. In developing a precautionary indicator system of global warming, several principles should be taken into consideration in constructing a practicable and informative indicator system (Duke & Aull-Hyde, 2002; Niemeijer, 2002; Solnes, 2003; Valentin & Spangenberg; 2000; Verdoodt & Van Ranst, 2006):

- The indicator system should have sufficient rational theoretical bases;
- The boundary of the indicator system should be identified clearly for the purpose of nature hazard mitigation;
- Credible and consistent databases should be available to support the calculation of the indicator system;

- The interpretations of the indicator system should be direct and informative for all the stakeholders, e.g. policy-makers and citizens.

Some PSR- or DPISR-like indicators have been established in the aforementioned literature. Each indicator system has its specific application purpose and respective data requirement at different scales. Based on the above principles, a modified DPSIR framework is considered in this study to develop a representative precautionary indicator system. Thereby the research flow of this study is represented in Fig. 1.

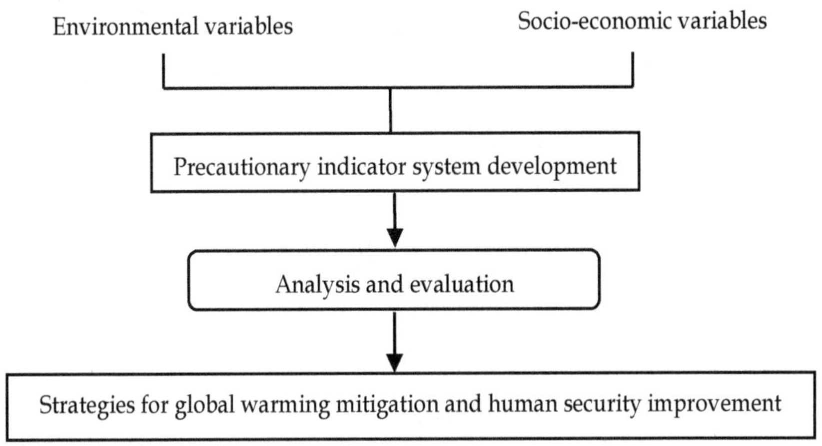

Fig. 1. The research flow diagram of this study.

To meet the goal of reflecting the urban adaptation mechanisms in terms of global warming, a three-layer hierarchical indictor system is induced to develop an evaluating indicator framework. A composite urban system is divided into three levels of indicators, and finally an overall score is estimated. The levels are decided by considering to the urban socio-ecological networks using socio-economic and environmental variables, which serve as the fundmental components of the indicator system. The variables are linked as a composite socio-economic and ecological network, representing the levels of the urban growth and the sustainability. According to the definition of Tanguay et al. (2008), the terms in the statistics are identified as variables, and an indicator means composite information obtained from specific variables.

The PSR analytical framework in this study is composed of three categories- "Pressure and State," "Function" and "Coordination." In addition, the driving factors and the responses are emphasized within the flows among the components. By such manipulation, a composite urban social-economic-natural system could be described by the intrinsic relationships among the components and the flows, on the basis of system ecology, as shown in Fig. 2.

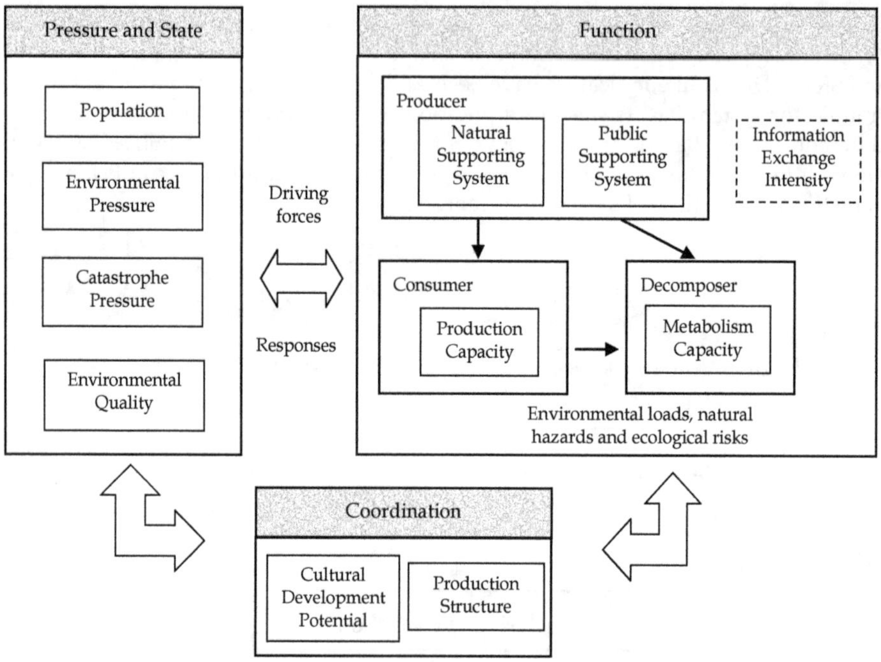

Fig. 2. The conceptual framework of the precautionary indicator system towards a sustainable city.

The characteristics of the intrinsic relationships of the precautionary indicator system are designed on the basis of the system ecology thinking and the research purpose. Some points of view are assumed in the framework as follows:

- The indicator system is to represent the intrinsic interactions of the socio-economic-environment systems, implying that all the sub-systems are integrated by rationally choosing the variables and determining the indicators' weights.
- Each 1st-level indicator is composed of several 2nd-level indicators, which represent the intrinsic properties. Each 2nd -level indicator is composed of some significant 3rd-level indicators, which are specifically selected to represent the characteristics of the upper level indicator, considering the data availability.
- The first 1st-level indicator, "Pressure and State," reflects the current pressure levels on the social, environmental and ecological states. This category reflects the quality and pressures of an urban system. Also, some available natural hazard variables would be highlighted and linked to other common environmental variables.
- The second 1st-level indicator, "Function", denotes the main socio-ecological network in an urban system. Like biological food-web, "Producer" in an urban system is assumed as "Natural Supporting System" and "Public Supporting System". They provide the resource for both social society and biosphere. "Consumer" in an urban system is assumed as "Production Capacity" representing the efficiency of the production sectors. "Decomposer" in an urban system is assumed as the "Metabolism Capacity", which means the efficiency and performance of environmental loads treatment. Finally, the linkages of socio-environmental-ecological networks are

assumed to be highly affected by the "Information Exchange Intensity" in this study. In each 2nd -level indicator, some representative variables are selected as well.

- In principle, the stronger the producer and decomposer are, the more robust an ecological system would be. That is, the production sectors could be transformed into a sustainable manner and the environmental quality could thus be self-regulated at an optimal level through the intrinsic mechanisms. Thereby, a powerful natural supporting sub-system and metabolism sub-system means that the potential of self-resiliency would be large, and the environmental quality and human security can be promoted. In this category, the social security indicators are particularly emphasized regarding the human security.
- The third 1st-level indicator, "Coordination," denotes the future potential indicator of ecologically sustainable city. The importance of the culture and the industrial structure are stressed herein.
- It is assumed that the driving forces are imposed and the responses are exchanged interactively among the main parts.
- Each 3rd-level indicator is given an attribute in the calculation procedure. A larger value of the 3rd-level indicator represents that the city has a better sustainable quality with regard to this indicator, and thus the attribute is positive; otherwise, it is negative. This attribute is determined by the domain knowledge of social and natural science.

In this study, the aggregation procedures of the precautionary indicator system are proposed as follows:

1. Normalize the values of each 3rd-level indicator using the following equations.
    - If the variable is a positive driving indicator (i.e. the larger value of original variable denotes a higher level of the urban sustainability):

$$Q_{i_{3rd-level},j} = \frac{C_{i_{3rd-level},j} - \overline{C}_{i_{3rd-level}}}{S_{C_{i_{3rd-level}}}} \quad \forall i,j \tag{1}$$

where $Q_{i_{3rd-level},j}$ is the normalized score of the i item of the 3rd-level variable in j region; $C_{i_{3rd-level},j}$, $\overline{C}_{i_{3rd-level}}$ and $S_{C_{i_{3rd-level}}}$ are the original value of the i item of the 3rd-level variable in j region, its average and its standard error, respectively.

- If the variable is a negative driving indicator (i.e. the larger value of the original variable denotes a lower level of the urban sustainability):

$$Q_{i_{3rd-level},j} = -\left(\frac{C_{i_{3rd-level},j} - \overline{C}_{i_{3rd-level}}}{S_{C_{i_{3rd-level}}}}\right) \quad \forall i,j \tag{2}$$

2. Aggregate the upper level indicator by the weighted aggregation, the formula is:

$$Q_{m,j} = \sum_i W_i \times Q_{i,j} \quad \forall m,i,j \tag{3}$$

where W_i is the weight of each lower level indicator; $Q_{m,j}$ denotes the value of the m upper-level indicator of in j region, and $Q_{i,j}$ is the value of the i lower-level indicator in j region.

After implementing the normalization transformation, the values of each 3rd-level indicator would obey a standard normal distribution, with a zero average and an unity standard error. Thus, all the indicators could be compared at the same scale. After the articulation, a larger normalized value of the indicator would imply the region is relatively more ecologically sustainable among the regions of interest with respect to the indicator.

The aggregation indicator is calculated by weighted summation. In terms of the determination of the weight of each indicator, expert consultation (i.e., the Delphi Method), analytical hierarchical procedure (AHP), analytical network procedure (ANP) and principal component analysis (PCA), could be considered (Hsu et al., 2010; Solnes, 2003; Song et al., 2004; Srebotnjak, 2007). Moreover, the nonlinear fuzzy theory could be introduced into the aforementioned methods. In this study the weight of each indicator is determined based on the outcomes of Song et al. (2004) and adjusted by the author according to the research purpose. In this sense, some variables associated with the global warming would be highlighted in the precautionary indicator system. In principle, the larger weight means the indicator with the higher degree of global warming.

3. Data and results

Japan has experienced the processes of extensive industrialization and urbanization, and now attempts to construct a sustainable society with particular focuses on the development of low-carbon society, material-cycle society and symbiotic society. In order to achieve the aforementioned objectives and to deal with the potential deficiencies, the adaptation strategies of the metropolitan areas are necessary. For this reason, 18 major cities in Japan are selected as the study areas to prepare required and feasible policy measures from a comprehensive perspective. In this sense, the official statistics of socio-economic and environmental variables in 2008 are utilized to develop the precautionary indicator system (Yokohama City Government, 2011). The structure of the precautionary indicator system is constructed according to the principles discussed in Section 2 and data availability. 30 representive variables are particularly selected in the indicator system, and the detailed definitions and the values of the variables are provided in Table 1 and Table 2, respectively.

In fact, it is difficult and tricky to select appropriate variables. In addition, the measurement bases, such as the measure of area or capita, would have important implications. In this study, normally, per capita variables are often used in constructing the socio-economic relationships while area-specific variables are preferred in establishing the natural-ecological-environmental relationships. Still, some particular considerations are imposed for several variables considering the complex interactions. For instance, this study adopts the spatial density of college students to reflect the degree of the potential on research and regional development. Furthermore, no variables are selected as the adequate industrial structure while the attributes and expectations may vary by regions.

Using the cross-section data in 2008 as shown in the Table 2, the calculation of the indicator system is implemented hierarchically. First, each 3rd-level indicators are calculated using Eq. (1) and Eq. (2), and the results are also shown in Table 2; secondly, the 2nd-level indicators and the 1st-level ones are estimated by Eq. (3), sequentially. Consequently, the overall score of each city can be obtained. A comparative analysis is performed based on the outcomes.

The 1st-level indicator	Weight	The 2nd-level indicator	Weight	The 3rd-level indicator	Description	Primative unit	Weight	Attribute of the original variable
Pressure and State	0.6	Population Scale	0.15	The population density	The indicator is regarded as the driving force of urban resources consumption and environment pollution.	10^3 capita/km^2	0.4	-
				The population natural growth rate	The indicator reflects the population sustainability.	‰	0.3	+
				The household density	The indicator is regarded as the driving force of urban resources consumption and environment pollution.	household/km^2	0.3	-
		Environmental Pressure	0.2	Per capita electricity consumed	The indicator is regarded as the environmental pressure in the urban system.	kWh/day	0.2	-
				Per capita water consumed	The indicator is regarded as the environmental pressure in the urban system.	m^3/yr	0.2	-
				Per capita natural gas consumed	The indicator is regarded as the environmental pressure in the urban system.	10^5 MJ/yr	0.15	-
				Per capita general waste generated	The indicator is regarded as the environmental pressure in the urban system.	kg/day	0.15	-
				The spatial car density	The indicator is regarded as the environmental pressure in the urban system.	number/km^2	0.15	-
				The ratio of road in total area	The indicator is regarded as the heat island effect and ecological habitat fragment in the urban system.	%	0.15	-
		Catastrophe Pressure	0.5	The maximum daily rainfall	The indicator is regarded as the risk indicator for the climate change due to global warming.	mm/day	0.5	-
				The maximum daily temperature	The indicator is regarded as the risk indicator for the climate change due to global warming.	℃	0.15	-
				Crimes incidents intensity	The indicator attempts to reflect the social safety pressure in the urban system.	cases/10^3capita	0.15	-
				Fire accident intensity	The indicator attempts to reflect the social safety pressure in the urban system.	cases/10^3capita	0.2	-
		Environmental Quality	0.15	Ambient Photochemical oxidant concentration	The indicator is regarded as the environmental response to the human activities and the regarded as the current environmental quality.	ppm	0.33	-
				Ambient NOx concentration	The indicator is regarded as the environmental response to the human activities and the regarded as the current environmental quality.	ppm	0.33	
				Ambient PM concentration	The indicator is regarded as the environmental response to the human activities and the regarded as the current environmental quality.	ppm	0.33	-

Table 1. The structure of the precautionary indicator system in this study.
Note: The unit denotes that of the original variable.

The 1st-level indicator	Weight	The 2nd-level indicator	Weight	The 3rd-level indicator	Description	Primative unit	Weight	Attribute of the original variable
Function	0.3	Production Capacity	0.1	Per capita income	The indicator is regarded as the intensity of production and consumption in urban system though only the one year lag data is available.	10^4¥(2008 prices) /yr	1	+
		Natural Supporting System	0.35	The annual rainfall	The indicator is regarded as the resource abundance for the natural ecological system.	mm	0.2	+
				The ratio of natural surfaces within the total area	Natural surfaces denote urban parks, forests, vegetation fields and lakes. The indicator is regarded as the resource abundance for urban environment metabolism, ecological habitats development and natural functions operation.	%	0.8	+
		Public Supporting System	0.15	The spatial availability of hospital bed	The indicator represents the capacity of public health care.	number/km²	0.15	+
				The spatial availability of doctors	The indicator represents the capacity of public health care.	number/km²	0.3	+
				Crime clearance rate	The indicator implies the efficiency of social safety system	%	0.25	+
				The spatial availability of disaster assistance	The indicator is regarded as the efficiency of social safety system.	cases/km²	0.3	+
		Information Exchange Intensity	0.2	The spatial intensity of fixed phone and mobile phone services registered	The indicator is regarded as the efficiency of information exchange flow.	number/km²	0.5	+
				The spatial intensity of internet service registered	The indicator reflects the efficiency of information exchange flow.	number/km²	0.5	+
		Metabolism Capacity	0.2	Sewage prevailing rate	The indicator is regarded as the capacity of urban environment metabolism.	%	0.5	+
				The recycling ratio of general waste	The indicator is regarded as the capacity of urban environment metabolism.	%	0.5	+
Coordination	0.1	Civilization Development Potential	1	The ratio in the municipal expenditure on education	The indicator is regarded as the administrative support for civic education and cultural preservation.	%	0.5	+
				The density of college students	The indicator is regarded as the potential for regional development.	number/km²	0.3	+
				The spatial intensity of museums, libraries, natural and cultural heritage appointed	The indicator is regarded as the resource abundance for cultural development.	number/km²	0.2	+

Table 1. The structure of the precautionary indicator system in this study. (*conti.*)
Note: The unit denotes that of the original variable.

Variable \ City	Sapporo	Sendai	Saitama	Chiba	Tokyo (special wards)	Kawasaki	Yokohama	Niigata	Shizuoka	Hamamatsu	Nagoya	Kyoto	Osaka	Sakai	Kobe	Hiroshima	Kitakyushu	Fukuoka
The population density (10^3 capita/km^2)	1. 693 (0.709)	1 .308 (0.803)	5.521 (-0.226)	3 .481 (0.272)	14.046 (-2.310)	9.631 (-1.231)	8.394 (-0.929)	1.118 (0.850)	0.511 (0.998)	0.538 (0.991)	6.886 (-0.560)	1.772 (0.690)	1.193 (-1.793)	5.574 (-0.239)	2. 773 (0.445)	1.289 (0.808)	2.019 (0.629)	4.215 (0.093)
The population natural growth rate (‰)	0.30 (-0.431)	2.50 (0.992)	2.40 (0.928)	2.00 (0.669)	0.40 (-0.367)	4.10 (2.028)	1.90 (0.604)	-0.50 (-0.949)	-1.3 (-1.467)	0.80 (-0.108)	0.90 (-0.043)	- 0.70 (-1.079)	- 0.50 (-0.949)	1.10 (0.086)	- 0.30 (-0.820)	2.30 (0.863)	- 1.00 (-1.273)	3.00 (1.316)
The household density (number/km^2)	783.45 (0.659)	575.14 (0.764)	2,284.94 (-0.095)	1,448.92 (0.325)	7,008.77 (-2.469)	4,438.23 (-1.177)	3,582.28 (-0.747)	421.50 (0.841)	195.82 (0.954)	209.58 (0.947)	3,062.58 (-0.486)	810.80 (0.645)	5,800.22 (-1.862)	2,427.58 (-0.167)	1,207.35 (0.446)	572.92 (0.765)	872.69 (0.614)	2,007.32 (0.044)
Per capita electricity consumed (kWh/day)	4,854.18 (0.520)	6,248.44 (0.114)	6,855.80 (-0.063)	10,603.46 (-1.154)	7,161.74 (-0.152)	6,594.90 (0.013)	6,558.47 (0.024)	5,727.29 (0.266)	7,425.71 (-0.229)	6,725.81 (-0.025)	17,011.74 (-3.020)	2,796.31 (1.119)	2,951.33 (1.074)	2,417.32 (1.230)	2,512.08 (1.202)	5,779.90 (0.251)	8,737.47 (-0.610)	8,574.22 (-0.563)
Per capita water consumed (m^3/yr)	93.03 (1.532)	110.18 (0.328)	106.29 (0.602)	100.50 (1.008)	126.21 (-0.797)	110.11 (0.334)	109.59 (0.370)	125.25 (-0.729)	123.60 (-0.613)	102.62 (0.859)	123.28 (-0.591)	121.28 (-0.451)	155.19 (-2.831)	113.06 (0.126)	118.81 (-0.277)	122.29 (-0.521)	108.99 (0.412)	97.20 (1.240)
Per capita natural gas consumed (10^5 MJ/yr)	69.97 (1.026)	110.26 (0.648)	89.10 (0.847)	455.50 (-2.596)	235.034 (-0.525)	379.868 (-1.886)	138.11 (0.386)	153.46 (0.242)	133.56 (0.429)	55.15 (1.166)	159.84 (0.182)	170.54 (0.081)	259.76 (-0.757)	272.70 (-0.879)	190.24 (-0.104)	138.51 (0.382)	118.55 (0.570)	95.34 (0.788)
Per capita general waste generated (kg/day)	1.08 (-0.096)	1.03 (0.191)	1.03 (0.208)	1.08 (-0.123)	1.11 (-0.326)	0.90 (1.083)	0.83 (1.550)	1.13 (-0.441)	1.08 (-0.096)	0.96 (0.682)	0.91 (1.028)	1.09 (-0.173)	1.43 (-2.434)	1.06 (-0.016)	1.09 (-0.184)	0.88 (1.194)	1.07 (-0.029)	1.37 (-2.018)
The spatial car density (number/km^2)	884.33 (0.769)	757.59 (0.865)	2,549.93 (-0.489)	1,746.95 (0.118)	4,221.22 (-1.752)	3,229.92 (-1.003)	3,337.07 (-1.084)	779.73 (0.849)	339.66 (1.181)	403.52 (1.133)	3,838.11 (-1.462)	728.04 (0.888)	3,850.18 (-1.471)	2,518.65 (-0.465)	1,142.98 (0.574)	710.12 (0.901)	1,175.30 (0.550)	2,037.04 (-0.101)
The ratio of roads in total area (%)	0.0033 (-0.597)	0.0028 (-0.150)	0.0019 (0.541)	0.0025 (0.092)	0.0011 (1.154)	0.0012 (1.112)	0.0015 (0.836)	0.0051 (-2.060)	0.0031 (-0.430)	0.0057 (-2.555)	0.0025 (0.093)	0.0017 (0.715)	0.0015 (0.851)	0.0020 (0.449)	0.0023 (0.201)	0.0026 (0.002)	0.0034 (-0.664)	0.0021 (0.409)
The maximum daily rainfall (mm)	37.5 (1.581)	87.5 (-0.157)	111.5 (-0.992)	83.5 (-0.018)	111.5 (-0.992)	78.5 (0.156)	78.5 (0.156)	65.5 (0.608)	125.5 (-1.479)	103 (-0.697)	133.5 (-1.757)	118 (-1.218)	57 (0.903)	57 (0.903)	47 (1.251)	50.5 (1.129)	65.5 (0.608)	82.5 (0.016)
The maximum daily temperature (℃)	31.4 (2.641)	33.6 (1.277)	37.3 (-1.016)	35.3 (0.224)	35.3 (0.224)	35 (0.410)	35 (0.410)	34.8 (0.534)	36 (-0.210)	36.6 (-0.582)	37.9 (-1.388)	37.7 (-1.264)	36.4 (-0.458)	36.4 (-0.458)	34.7 (0.596)	36.7 (-0.644)	34.5 (0.720)	37.3 (-1.016)
Crimes incidents intensity (cases/10^3capita)	15.34 (0.481)	14.21 (0.689)	17.79 (0.026)	19.59 (-0.305)	17.56 (0.069)	12.69 (0.970)	12.28 (1.046)	13.40 (0.839)	11.94 (1.109)	12.21 (1.058)	26.27 (-1.541)	22.46 (-0.837)	31.14 (-2.441)	21.37 (-0.636)	18.69 (-0.140)	13.17 (0.880)	20.34 (-0.445)	22.39 (-0.823)
Fire accident intensity (cases/10^3capita)	0.38 (-0.179)	0.35 (0.115)	0.34 (0.294)	0.33 (0.395)	0.48 (-1.180)	0.33 (0.366)	0.30 (0.698)	0.19 (1.764)	0.33 (0.385)	0.40 (-0.362)	0.47 (-1.020)	0.13 (2.342)	0.49 (-1.265)	0.40 (-0.323)	0.49 (-1.297)	0.43 (-0.622)	0.44 (-0.762)	0.30 (0.653)
Ambient NOx concentration (ppm)	0.032 (0.488)	0.015 (1.233)	0.029 (-0.853)	0.024 (-0.108)	0.032 (-1.300)	0.032 (-1.300)	0.030 (-1.002)	0.010 (1.979)	0.019 (0.637)	0.015 (1.233)	0.027 (-0.555)	0.021 (0.339)	0.034 (-1.598)	0.026 (-0.406)	0.024 (-0.108)	0.02 (0.488)	0.024 (-0.108)	0.017 (0.935)
Ambient Photochemical oxidant concentration (ppm)	0.030 (0.304)	0.033 (-0.560)	0.030 (0.304)	0.030 (0.304)	0.028 (0.880)	0.028 (0.880)	0.027 (1.168)	0.036 (-1.424)	0.023 (2.320)	0.036 (-1.424)	0.031 (0.016)	0.030 (0.304)	0.031 (0.016)	0.031 (0.016)	0.037 (-1.712)	0.033 (-0.560)	0.031 (0.016)	0.034 (-0.848)
Ambient PM concentration (ppm)	0.013 (2.842)	0.021 (0.806)	0.027 (-0.721)	0.023 (0.297)	0.024 (0.042)	0.025 (-0.212)	0.028 (-0.976)	0.022 (0.551)	0.023 (0.297)	0.022 (0.551)	0.029 (-1.230)	0.020 (1.060)	0.027 (-0.721)	0.026 (-0.467)	0.023 (0.297)	0.028 (-0.976)	0.025 (-0.212)	0.029 (-1.230)

Table 2. The original values of entering variables of the cities of interest.
Note: The number in the parentheses denotes the normalized value of each variable, i.e. the value of the 3rd-level indicator, in a specific category.

Variable \ City	Sapporo	Sendai	Saitama	Chiba	Tokyo (special wards)	Kawasaki	Yokohama	Niigata	Shizuoka	Hamamatsu	Nagoya	Kyoto	Osaka	Sakai	Kobe	Hiroshima	Kitakyushu	Fukuoka
Per capita income (10⁴¥(2008 prices)/yr)	345.53 (-0.437)	322.62 (-1.147)	381.64 (0.682)	338.35 (-0.659)	400.48 (1.266)	395.05 (1.098)	437.29 (2.407)	372.08 (0.386)	369.29 (0.300)	361.56 (0.060)	345.32 (-0.443)	353.12 (-0.201)	338.79 (-0.646)	323.91 (-1.107)	361.78 (0.067)	379.98 (0.631)	345.08 (-0.450)	301.29 (-1.808)
The annual rainfall(mm)	843.0 (-2.032)	1,349.0 -0.496)	1,392.5 (-0.364)	1,639.0 (0.384)	1,857.5 (1.047)	1,919.0 (1.233)	1,919.0 (1.233)	1,530.0 (0.053)	1,955.5 (1.344)	1,869.5 (1.083)	1,579.5 (0.203)	1,430.5 -0.249)	1,262.5 (-0.759)	1,262.5 (-0.759)	1,041.0 (-1.431)	1,148.50 (-1.105)	1,447.0 (-0.199)	1,780.5 (0.813)
The ratio of natural surfaces within the total area (%)	61.55 (0.393)	70.27 (0.785)	50.20 (-0.118)	56.09 (0.147)	17.28 (-1.598)	27.50 (-1.139)	32.60 (-0.909)	76.48 (1.064)	85.41 (1.465)	78.11 (1.137)	30.89 (-0.986)	71.18 (0.825)	13.61 (-1.763)	33.80 (-0.855)	68.97 (0.726)	74.07 (0.956)	50.38 (-0.110)	52.38 (-0.020)
The spatial availability of hospital bed (number/km²)	36.04 (-0.408)	17.46 (-0.850)	39.61 (-0.323)	35.38 (-0.424)	133.95 (1.922)	73.85 (0.492)	66.56 (0.318)	16.07 (-0.883)	6.06 (-1.121)	6.82 (-1.103)	84.18 (0.738)	29.55 (-0.562)	155.47 (2.434)	87.41 (0.814)	35.45 (-0.422)	18.93 (-0.815)	43.16 (-0.239)	71.41 (0.434)
The spatial availability of doctors (number/km²)	50.73 (-0.535)	30.17 (-0.767)	78.63 (-0.219)	69.02 (-0.328)	316.36 (2.472)	140.24 (0.479)	139.33 (0.468)	26.92 (-0.804)	9.51 (-1.001)	11.78 (-0.976)	152.04 (0.612)	46.87 (-0.578)	297.21 (2.255)	117.45 (0.221)	63.37 (-0.392)	30.98 (-0.758)	61.83 (-0.409)	120.88 (0.259)
Crime crack rate (%)	40.99 (0.729)	33.61 (-0.347)	35.00 (-0.144)	36.13 (0.021)	41.95 (0.870)	43.87 (1.150)	45.51 (1.389)	32.57 (-0.499)	36.75 (0.111)	31.38 (-0.673)	27.98 (-1.168)	27.94 (-1.174)	26.53 (-1.380)	22.91 (-1.909)	35.83 (-0.023)	45.24 (1.350)	40.74 (0.693)	42.88 (1.005)
The spatial availability of disaster assistance (case/km²)	62.33 (-0.654)	47.60 (-0.713)	224.25 (0.002)	165.18 (-0.237)	783.25 (2.266)	383.32 (0.646)	334.14 (0.447)	39.94 (-0.744)	18.61 (-0.831)	19.60 (-0.827)	294.39 (0.286)	83.57 (-0.568)	869.85 (2.617)	277.70 (0.218)	114.38 (-0.443)	49.88 (-0.704)	92.74 (-0.531)	165.91 (-0.230)
The spatial intensity of fixed phone and mobile phone services registered (number/km²)	513.06 (-0.583)	432.79 -0.622)	1,508.39 (-0.100)	1,041.03 (-0.327)	8341.10 (3.215)	2,858.91 (0.555)	2,622.62 (0.440)	353.54 (-0.661)	227.08 (-0.722)	181.19 (-0.744)	2,481.46 (0.372)	612.89 -0.535)	1,955.58 (1.573)	1,405.99 (-0.150)	871.38 (-0.409)	434.04 (-0.622)	559.96 (-0.561)	1,471.64 (-0.118)
The spatial intensity of internet service registered (number/km²)	3,723.07 (-0.763)	2,425.76 (-0.840)	26,359.83 (0.589)	18,244.63 (0.104)	26,962.28 (0.625)	52,788.36 (2.167)	17,421.92 (0.055)	2,260.09 (-0.850)	2,099.09 (-0.860)	1,961.09 (-0.868)	1,9133.48 (0.157)	2,649.58 (-0.827)	35,864.38 (1.156)	53,154.55 (2.189)	8,122.09 (-0.500)	2,588.69 (-0.831)	8,724.48 (-0.464)	12,478.38 (-0.240)
Sewage prevailing rate (%)	99.70 (0.642)	97.60 (0.422)	85.00 (-0.896)	97.10 (0.370)	100.00 (0.673)	99.30 (0.600)	99.80 (0.652)	73.42 (-2.107)	75.7 (-1.868)	74.1 (-2.036)	98.60 (0.527)	99.20 (0.590)	99.90 (0.663)	93.9 (0.035)	98.60 (0.527)	92.90 (-0.069)	99.80 (0.652)	99.50 (0.621)
The recycling ratio of general waste (%)	9.54 (-0.026)	10.17 (0.062)	22.03 (1.688)	20.85 (1.525)	12.28 (0.350)	6.22 (-0.481)	13.24 (0.483)	25.68 (2.187)	6.94 (-0.381)	10.67 (0.130)	0.60 (-1.251)	5.65 (-0.559)	3.07 (-0.912)	3.15 (-0.901)	4.23 (-0.754)	14.77 (0.692)	3.36 (-0.873)	2.59 (-0.979)
The ratio in the municipal expenditure on education (%)	7.91 (-0.938)	11.27 (0.801)	10.76 (0.540)	10.36 (0.330)	13.11 (1.756)	8.07 (-0.860)	8.37 (-0.700)	10.62 (0.466)	9.27 (-0.234)	12.27 (1.321)	8.44 (-0.668)	7.68 (-1.059)	7.43 (-1.190)	9.62 (-0.054)	9.65 (-0.040)	13.81 (2.120)	8.11 (-0.834)	8.27 (-0.755)
The density of college students (number/km²)	45.36 (-0.554)	61.71 (-0.459)	91.91 (-0.284)	95.23 (-0.264)	758.28 (3.588)	212.62 (0.418)	181.74 (0.238)	27.12 (-0.660)	10.35 (-0.757)	7.24 (-0.775)	272.31 (0.764)	159.46 (0.109)	125.84 (-0.086)	79.48 (-0.356)	115.48 (-0.147)	33.01 (-0.626)	44.72 (-0.558)	210.95 (0.408)
The spatial intensity of museums, libraries, natural and cultural heritage appointed (number/km²)	0.074 (-0.491)	0.077 (-0.488)	0.221 (-0.363)	0.107 (-0.463)	4.510 (3.368)	0.277 (-0.314)	0.334 (-0.265)	0.088 (-0.479)	0.055 (-0.508)	0.039 (-0.521)	0.573 (-0.057)	2.578 (1.687)	1.345 (0.615)	0.393 (-0.213)	0.355 (-0.247)	0.064 (-0.500)	0.066 (-0.498)	0.337 (-0.262)

Table 2. The original values of entering variables of the cities of interest. (*conti.*)
Note: The number in the parentheses denotes the normalized value of each variable, i.e. the value of the 3rd-level indicator, in a specific category.

Table 3 shows the points of overall score and the 1st-level indicator of each city. Based on the situation in 2008, Top 3 cities in terms of the global warming adaptation are Sapporo, Hiroshima and Niigata; on the other hand, the last three ones are Fukuoka, Osaka and Nagoya. The overall score and the 1st-level indicators give a comprehensive image of urban system performance. Furthermore, the 2nd-level and the 3rd-level indicators can provide information to make substantial strategy for global warming adaptation on a city level. Table 4 gives the results of the 2nd-level and original indicators for each city.

City	Overall Score	The 1st-level indicator		
		Pressure and State	Function	Coordination
Sapporo	0.452 (1)	0.959 (1)	-0.184 (15)	-0.689 (18)
Sendai	0.196 (4)	0.384 (3)	-0.126 (12)	0.036 (5)
Saitama	-0.129 (6)	-0.277 (15)	0.115 (6)	0.022 (6)
Chiba	0.052 (9)	0.033 (10)	0.135 (5)	-0.086 (8)
Tokyo (special wards)	0.011 (10)	-0.710 (18)	0.528 (1)	2.789 (1)
Kawasaki	0.112 (7)	0.105 (7)	0.266 (3)	-0.313 (14)
Yokohama	0.181 (5)	0.182 (6)	0.336 (2)	-0.288 (13)
Niigata	0.294 (3)	0.471 (2)	0.089 (7)	-0.155 (10)
Shizuoka	-0.073 (14)	-0.067 (14)	0.048 (8)	-0.473 (16)
Hamamatsu	-0.019 (11)	-0.013 (11)	-0.083 (11)	0.139 (3)
Nagoya	-0.732 (18)	-1.055 (17)	-0.312 (17)	-0.055 (7)
Kyoto	-0.033 (12)	-0.050 (12)	-0.048 (9)	0.115 (4)
Osaka	-0.465 (17)	-0.651 (16)	-0.141 (13)	-0.318 (15)
Sakai	-0.068 (13)	0.078 (9)	-0.319 (18)	-0.192 (11)
Kobe	0.093 (8)	0.203 (5)	-0.052 (10)	-0.134 (9)
Hiroshima	0.315 (2)	0.371 (4)	0.137 (4)	0.510 (2)
Kitakyushu	-0.079 (15)	0.094 (8)	-0.236 (16)	-0.651 (17)
Fukuoka	-0.107 (16)	-0.059 (13)	-0.152 (14)	-0.258 (12)

Table 3. The overall score and the 1st-level indicators of the interested city.
Note: The number in the parentheses denotes the rank of each indicator in a specific category.

City \ Indicator	Pressure and State				Function					Coordination
	Population Scale	Environmental Pressure	Catastrophe Pressure	Environmental Quality	Production Capacity	Natural Supporting System	Public Supporting System	Information Exchange Intensity	Metabolism Capacity	Civilization Development Potential
Sapporo	0.352(6)	0.576 (1)	1.223 (1)	1.199 (1)	-0.437 (11)	-0.092 (10)	-0.235 (12)	-0.673 (12)	0.308 (6)	-0.689 (18)
Sendai	0.848 (1)	0.322 (4)	0.239 (7)	0.488 (4)	-1.147 (17)	0.529 (6)	-0.658 (14)	-0.731 (15)	0.242 (7)	0.036 (5)
Saitama	0.159 (9)	0.274 (6)	-0.586 (16)	-0.419 (15)	0.682 (4)	-0.167 (12)	-0.150 (8)	0.244 (7)	0.396 (4)	0.022 (6)
Chiba	0.407 (5)	-0.406 (15)	0.058 (10)	0.163 (6)	-0.659 (15)	0.194 (8)	-0.228 (11)	-0.111 (8)	0.948 (1)	-0.086 (8)
Tokyo (Special Wards)	-1.775 (18)	-0.407 (16)	-0.688 (17)	-0.124 (9)	1.266 (2)	-1.069 (17)	1.927 (1)	1.920 (1)	0.512 (3)	2.789 (1)
Kawasaki	-0.237 (14)	-0.035 (13)	0.358 (6)	-0.208 (10)	1.098 (3)	-0.664 (14)	0.699 (3)	1.361 (3)	0.060 (8)	-0.313 (14)
Yokohama	-0.414 (16)	0.332 (3)	0.436 (4)	-0.267 (11)	2.407 (1)	-0.481 (13)	0.670 (4)	0.248 (6)	0.567 (2)	-0.288 (13)
Niigata	0.307 (7)	-0.304 (14)	0.863 (2)	0.365 (5)	0.386 (6)	0.862 (3)	-0.722 (16)	-0.755 (16)	0.040 (9)	-0.155 (10)
Shizuoka	0.245 (8)	-0.006 (12)	-0.528 (15)	1.074 (2)	0.300 (7)	1.441 (1)	-0.690 (15)	-0.791 (17)	-1.125 (18)	-0.473 (16)
Hamamatsu	0.648 (3)	0.231 (8)	-0.349 (13)	0.119 (7)	0.060 (8)	1.126 (2)	-0.875 (18)	-0.806 (18)	-0.953 (17)	0.139 (3)
Nagoya	-0.383 (15)	-0.746 (17)	-1.522 (18)	-0.584 (17)	-0.443 (12)	-0.748 (15)	0.088 (6)	0.265 (5)	-0.362 (16)	-0.055 (7)
Kyoto	0.146 (10)	0.360 (2)	-0.456 (14)	0.562 (3)	-0.201 (10)	0.611 (4)	-0.722 (16)	-0.681 (13)	0.015 (10)	0.115 (4)
Osaka	-1.560 (17)	-0.923 (18)	-0.236 (12)	-0.760 (18)	-0.646 (14)	-1.562 (18)	1.482 (2)	1.364 (2)	-0.125 (13)	-0.318 (15)
Sakai	-0.120 (13)	0.135 (9)	0.223 (8)	-0.283 (12)	-1.107 (16)	-0.836 (16)	-0.223 (9)	1.019 (4)	-0.433 (15)	-0.192 (11)
Kobe	0.066 (11)	0.258 (7)	0.434 (5)	-0.503 (16)	0.067 (9)	0.295 (7)	-0.319 (13)	-0.455 (10)	-0.113 (12)	-0.134 (9)
Hiroshima	0.811 (2)	0.318 (5)	0.476 (3)	-0.346 (13)	0.631 (5)	0.544 (5)	-0.223 (9)	-0.726 (14)	0.311 (5)	0.510 (2)
Kitakyushu	0.054 (12)	0.024 (10)	0.193 (9)	-0.100 (8)	-0.450 (13)	-0.128 (11)	-0.145 (7)	-0.512 (11)	-0.110 (11)	-0.651 (17)
Fukuoka	0.445 (4)	-0.003 (11)	-0.137 (11)	-0.377 (14)	-1.808 (18)	0.147 (9)	0.325 (5)	-0.179 (9)	-0.179 (14)	-0.258 (12)

Table 4. The 2nd-level indicators of the interested cities.

Note: The number in the parentheses is the rank of each 2nd-level indicator in a specific category.

From Table 2-4 the comparative advantages and disadvantages of each city in 2008, in terms of global warming adaptation and human security, could be observed under the assumed intrinsic relationships of the precautionary indicator system. The upper-level cities in the overall ranking primarily obtain high scores on "Pressure and State" while this perspective is highlighted for the precautionary purpose in this study. Even the top 1 city, Sapporo, has low rankings in the other two 1st-level indicators, the differences compared with other cities are not too large. This is also one reason that the cities with high 1st-level scores would occupy high rankings. Still, some important policy implications for the development of adaptation strategies could be found by examining the scoring structure in Table 2-4. In principle, some implications could be observed as the following:

Pressure and State

Sapporo, Niigata and Sendai obtain the highest scores in this category, implying that the current pressures on global warming and human security are less than other cities in 2008. Regarding global warming, cities in north areas, e.g., Sapporo, Sandai and Niigata, might face less threatens of heat waves and the extreme weather than the sourth cities. Such outcome implies the geological condition would be an influencing factor on the regional climate stability. In view of the catastrophe pressures and climate change adaptation, meteorologically, the adequate latitude condition of a city might shift to a higher level than that of most current major cities in Japan. However, variables related to some sorts of natural hazards, e.g., earthquakes and tsunami, are not included herein while it is difficult to obtain appropriate quantitative precautionary variables in the existing statistics. On the other hand, Nagoya, Tokyo (special wards) and Osaka get relatively low scores while the current megacity scale is not suitable to relieve the pressures on global warming and human security.

Regarding the "Population Scale" indicator, for lower-scoring cities, the control of high population and household densities could be considered, while the later term seems to be a characteristic of modern society (Weng et al., 2009). As for the "Environmental Pressure" indicator, countermeasures on eliminating the resources and energy consumption and controlling the heat island effect could be taken into consideration. Also, incentives for using public transportation could be further designed. In fact, the public environmental education should be promoted as well so as to change the unsustainable lifestyle. In terms of the estimates of the "Catastrophe Pressure" and "Environmental Quality" indicators, as mentioned earlier, the urban heat island effect should be noticed and controlled by increasing green area inside the city while this argument is also highlighted in the "Natural Supporting System" indicator. In addition, the promotion of the green roof at buildings could be a multi-functional measure on pollution mitigation, stormwater runoff control, energy saving and heat island effect mitigation (Cartera & Keeler, 2008). Meanwhile, the restriction of private cars, incentives for using public transportation and the effective management on the pollution sources would reduce the air pollutant emissions associated with global warming and improve the urban environmental quality. In particular, the spatial car density data shows a strong positive relation to the ambient NOx concentration (the correlation coefficient is 0.84 for the original series). Economic instruments, e.g., car taxes and pollutant emission credits, would be potentially effective policy measures. Some strict emission standards of greenhouse gas could be suggested as well. Regarding the human security in urban area, municipal governors should eliminate the occurrence of crime and fire acccidents by launching strict regulations and frequent inspections.

Function

In fact, except for the "Natural Supporting System" indicator, the cities with high population density may have sufficient financial supports to enhance the capacities on this part. Therefore, Tokyo (special wards), Yokohama and Kawasaki have high scores on the "Function" indicator. On the other hand, Kitakyushu, Nagoya and Sakai obtain relatively low scores, implying that the natural service and the public governances should be promoted to cope with essential tasks of adapting global warming and improving human security.

In terms of the "Production Capacity," this study chooses per capita income only as a reference index. Some additional influencing factors might exist in this category, e.g., the technology innovation towards global warming mitigation. However, some quantitative variables might be difficult to be obtained. In addition, since the main purpose of this study is to developing the precautionary indicator for global warming, the weight of this category is not significant. In this sense, a moderate representative index might be sufficient. However, the representativeness of this index might be biased sine many employees are not inhabitants in the area where they work. As for the "Natural Supporting System" indicator, mainly, an open natural space inside the city is expected to enhance the potential of the service of natural systems. In addition, there is high potential to implement a variety of ecological engineering approaches in urban area to increase the natural supporting capacity (Chapman & Blockley, 2009). In addition, as aforementioned in the interpretation of the "Pressure and State" indicator, constructing new urban parks and green roof (or roof gardening) are effective measures. Regarding the "Public Supporting System" indicator, concerning about the human security capacity, the development of the public health service and the social security enhancement should be improved although the tasks might be low priority in conventional urban governances. In order to distribute in-time emergent alarms and implement reaction plans, the enhancement of the information change systems is required. In addition, the practice of the emergency reaction plans should be performed routinely. Though not strongly related, the metabolism capacity in large cities should be promoted considering the improvement of all the perspectives of urban sustainability.

Coordination

In this category, Tokyo (special wards), Hiroshima and Kobe have high scores while Shizuoka, Kitakyushu and Sapporo get relatively low scores on the "Civilization Development Potential" indicator. Examining the detailed scoring structure in Table 1, the low ratio of the public expenditure on education seems to be a main reason for this result. Meanwhile, more public education facilities are expected in the low-score cities to provide more opportunities for adult education. The public education facilities indeed could serve as platforms for promoting the education for sustainable development (ESD), which highlights issues covering global warming mitigation and human security improvement (UNESCO, 2011). Still, some additional indicators are required to take the sustainable industrial structure into consideration in this part.

Although the evaluation outcomes come out only by using the data in 2008, the results are informative for the decision-makers and citizens. The related stakeholders could find potential directions to improve upon the current situation in the comparative analysis. Based on the previous observations, concrete adaptation strategies of global warming mitigation and human security improvement for each city are proposed in Table 5.

City	Proposal	
Sapporo	Pressure and State	Improve the greening of roads
	Function	Build water storage facilities and facilitate the water resources management against potential droughts; Improve the spatial homogeneity of the resources of public health;
	Coordination	Improve the public investment on education; Improve the regional research capacity
Sendai	Pressure and State	Improve the greening of roads and promote the utilization of public transportation systems
	Function	Increase the resource (doctors and facilities) of public health care; Improve the capacity of crime crack by increasing the manpower; Establish efficient alarm networks for natural hazards and human security
	Coordination	Improve the regional research capacity; Improve the resources and infrastructures of public education
Saitama	Pressure and State	Manage the utilization of car by imposing a higher car tax for instance; Promote roof gardening, improve the drainage systems, and establish detention ponds to prevent from potential stormwater; Create natural surfaces to eliminate potential heat waves, by roof gardening for example
	Function	Increase the natural surfaces and promote green roof or green buildings by revising built environment regulations; Create urban parks; Increase the resource of public health care; Improve the capacity of crime crack and hazard security, by increasing the manpower for instance
	Coordination	Improve the regional research capacity; Improve the resources and infrastructures of public education
Chiba	Pressure and State	Encourage energy/natural gas saving by using economic incentives and adjusting the prices for instance
	Function	Improve the regional research capacity
	Coordination	Improve the regional research capacity; Improve the resources and infrastructures of public education

Table 5. Proposals towards sustainable city for interested cities.

City	Proposal	
Tokyo (special wards)	Pressure and State	Control the population/household density; Encourage water/natural gas saving by using economic incentives and adjusting the price levels for instance; Promote roof gardening, improve the drainage systems, and establish detention ponds to prevent from potential stormwater; Control NOx pollution: implement more strict pollution countermeasures such as making more strict emission standards and eliminating pollutant emission sources.
	Function	Increase the natural surfaces and promote green roof or green buildings by revising built environment regulations; Create urban parks;
	Coordination	Under stable development
Yokohama	Pressure and State	Manage the utilization of car by imposing a higher car tax for instance
	Function	Increase the natural surfaces and promote green roof or green buildings by revising built environment regulations; Create urban parks
	Coordination	At an appropriate level
Niigata	Pressure and State	Improve the greening of roads and promote the utilization of public transportation systems; Encourage the water saving by using economic incentives and adjusting the price levels for instance; Improve the greening of roads
	Function	Establish efficient alarm networks for natural hazards and human security
	Coordination	At an appropriate level;
Shizuoka	Pressure and State	Encourage the increase of newborn babies; Promote roof gardening, improve the drainage systems, and establish detention ponds to prevent from potential stormwater
	Function	Increase the resource (doctors and facilities) of public health care; Improve the capacity of hazard security by increasing the manpower; Improve the prevailing of sewage services; Establish efficient alarm networks for natural hazards and human security
	Coordination	Improve the regional research capacity; Improve the resources and infrastructures of public education

Table 5. Proposals towards sustainable city for interested cities. (*conti.*)

City	Proposal	
Hamamatsu	Pressure and State	Improve the greening of roads and promote the utilization of public transportation systems; Prevent from Oxidant pollution: survey the possible precursors' emission and eliminate the sources.
	Function	Increase the resource (doctors and facilities) of public health care; Improve the capacity of crime crack and hazard security, by increasing the manpower for instance; Improve the capacity of hazard security by increasing the manpower; Improve the prevailing of sewage services; Establish efficient alarm networks for natural hazards and human security
	Coordination	Improve the regional research capacity; Improve the resources and infrastructures of public education
Nagoya	Pressure and State	Encourage energy saving by using economic incentives and adjusting the price levels for instance; Manage the utilization of car, e.g. by imposing a higher car tax; Create natural surfaces to eliminate potential heat waves, by roof gardening for example; Prevent crime and fire accidents
	Function	Increase the natural surfaces and promote green roof or green buildings by revising built environment regulations; Create urban parks; Improve the capacity of crime crack and hazard security, by increasing the manpower for instance; Improve the recycling and reuse of general waste (the original data might be questionable due to the natural hazard occurred in the year)
	Coordination	At an appropriate level
Kyoto	Pressure and State	Encourage the increase of newborn babies; Promote roof gardening, improve the drainage systems, and establish detention ponds to prevent from potential stormwater; Create natural surfaces to eliminate potential heat waves, by roof gardening for example
	Function	Improve the capacity of crime crack and hazard security, by increasing the manpower for instance; Establish efficient alarm networks for natural hazards and human security
	Coordination	Improve the public investment on education

Table 5. Proposals towards sustainable city for interested cities. (*conti.*)

City		Proposal
Osaka	Pressure and State	Control the population/household density; Encourage water/natural gas saving by using economic incentives and adjusting the price levels for instance; Promote the reduction of general waste generation; Manage the utilization of car, e.g. by imposing a higher car tax; Prevent crime; Control NOx pollution: implement more strict pollution countermeasures such as making more strict emission standards and eliminating pollutant emission sources.
	Function	Increase the natural surfaces and promote green roof or green buildings by revising built environment regulations; Create urban parks; Improve the capacity of crime crack and hazard security, by increasing the manpower for instance; Improve the recycling and reuse of general waste
	Coordination	Improve the public investment on education
Sakai	Pressure and State	Encourage natural gas saving by using economic incentives and adjusting the price levels for instance; Manage the utilization of car, e.g. by imposing a higher car tax
	Function	Improve the capacity of crime crack and hazard security, by increasing the manpower for instance; Improve the recycling and reuse of general waste
	Coordination	At an appropriate level
Kobe	Pressure and State	Prevent crime and fire accidents; Prevent from Oxidant pollution: survey the possible precursors' emission and eliminate the sources.
	Function	Improve the recycling and reuse of general waste
	Coordination	At an appropriate level
Hiroshima	Pressure and State	Create natural surfaces to eliminate potential heat waves, by roof gardening for example
	Function	Increase the resource (doctors and facilities) of public health care; Establish efficient alarm networks for natural hazards and human security
	Coordination	At an appropriate level
Kitakyushu	Pressure and State	Encourage the increase of newborn babies; Improve the greening of roads and promote the utilization of public transportation systems
	Function	At an appropriate level
	Coordination	At an appropriate level
Fukuoka	Pressure and State	Create natural surfaces to eliminate potential heat waves, by roof gardening for example
	Function	Improve the recycling and reuse of general waste
	Coordination	At an appropriate level

Table 5. Proposals towards sustainable city for interested cities. (*conti.*)

In addition, applying the ideas of the EKC, a quadratic equation is preliminarily established to fit the possible relationship between the overall scores and the area of the regions. The ordinary least squares (OLS) method is used to estimate the parameters, and the results are statistically diagnosed to confirm the statistic credibility of the model. As shown in Fig.3, the results of the F test and the t test denote that most of the parameter estimators are significant; the Durbin-Watson (DW) statistic shows that no significant serial correlation problem exist in the error term. However, only about 17% of the model variances are explained by the current one; one reason is that the equation only considers the influencing factor of area, on the basis of the concept of EKC. Still, the developed equation could provide a preliminary estimation for the optimal area of a sustainable city for the research purpose. In Fig. 3, the results imply that 987.26 km^2 would be the optimal area according to the methodological assumptions and data in 2008. Besides, the area of the top 2 cities in the ranking is close to this optimal value, supporting the current outcome. On the other hand, satisfactory results are not obtained in developing the function of the overall scores and the population scales of the interested regions. Given that more rational weights of the indicators, the result suggests that the coupling of the indicator system and the EKC could be an informative tool to analyse the optimal development scale from a variety of perspectives with the current available statistics, and thus this methodology could aid future urban design towards sustainable development.

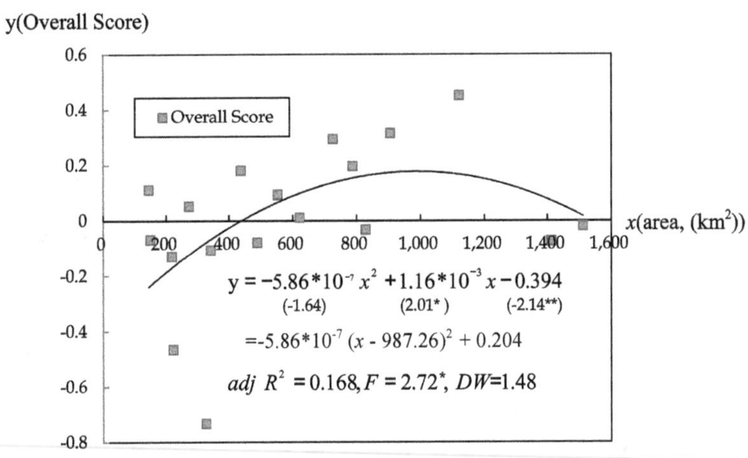

Fig. 3. The relationship between the overall scores and areas for the study regions. Note: Values in the parentheses are t statistics for the parameter estimator; * and ** denote significance at the 10% and 5% levels, respectively.

In fact, some deficiencies in this study are expected to be improved at the next step. First, an additional analysis on the weights in the current precautionary indicator system is required. Second, the aggregation process in the precautionary indicator system actually assumes that the intrinsic relationships among the variables are linearly. In this sense, the weights of the indicators serve as the coefficients in a linear equation in each category. Therefore, the potential collinearity among the indicators in the current indicator system has to be further examined given that a more credible spatial-temporal panel data. Third, regarding the cooperation and competition among the cities, some neighboring cities may have strong interactions to form a regional economy. For example, in the eastern Japan, several major cities comprise of the informal Kando regional economy. In this sense, the current methodology of the development of indicator systems should take the interactions among stakeholders into consideration by adopting import/export flows. Still, the current methodology could provide adaptation strategies to a certain extent with a small data requirement.

4. Conclusion and future prospect

Considering the composite socio-economic-environmental relationships, this study makes an attempt to develop a precautionary indicator system to quantify the degree of the risks with regard to global warming and human security in urban areas. An normalization procedure considering the attributs of impacts is proposed to develop the indicator system. Regarding the improvement of the current methodology, two parts could be further studied. Firstly, considering the manipulability of the calculation, the behaviors of the variables are assumed to obey the standard normal distribution in this study while some might not. Further distribution tests could be implemented for the recognition of the behaviour of each index. Secondly, the weights within the hierarchies could be re-structured by using the AHP or ANP-like methods, e.g., AHP, fuzzy-AHP, or ANP methods, so that all the stakeholders' view points could be taken into consideration.

Using the official data in 2008, the preliminary empirical results provide concrete adaptation strategies for the major cities in Japan. Furthermore, every city could find some solutions to improve their capacity regarding the global warming and human security. In addition, concerning the catastrophe pressures and the climate change adaptation, the adequate latitude condition of a city might shift to a higher level than that of most major cities in Japan. Furthermore, using the overall scores of the precautionary indicator systems, a preliminary EKC is established to discover the optimal size of a city, showing that a city in the area of around 987.26 km^2 would be adequate for the purpose of global warming adaptation and human security improvement with regard to city governances. Consequently, as shown in the demonstrated EMC example, more attributes on the city scales could be exploited based on the established indicator sytem using a comprehensive data. Based on the outcomes of this study, the precautionary indicator system and the proposed adaptation strategies regarding the global warming mitigation and human security improvement would contribute to the facilitation of the current urban planning for sustainable development.

5. References

Arrow, K, Bolin, B., Costanza, R., Dasgupta, P., Folke, C., Holling, C. S., Jansson, B. O., Levin, S., Miiler, K. G., Perrings, C., Pimentel, D. (1995). Economic growth, carrying capacity, and the environment. *Science*, Vol.268, No.5210, (April 1995), pp. 520-521, ISSN 1095-9203

Azomahou, T., Laisney, F., Nguyen Van, P. (2006). Economic development and CO2 emissions: A nonparametric panel approach. *Journal of Public Economics*, Vol.90, No.6-7, (August 2006), pp. 1347–1363, ISSN 0047-2727

Button, K. (2002). City management and urban environmental indicators. *Ecological Economics*, Vol.40, No.2, (February 2002), pp. 217-233, ISSN 1470-160X

Cartera, T., Keeler, A. (2008). Life-cycle cost–benefit analysis of extensive vegetated roof systems. *Journal of Environmental Management*, Vol.87, No.3, (May 2008), pp. 350-363, ISSN 0301-4797

Chapman, M. G., Blockley, D. J. (2009). Engineering novel habitats on urban infrastructure to increase intertidal biodiversity. *Oecologia*, Vol.161, No.3, (September 2009), pp. 625-635, ISSN 1432-1939

Chen, H. W., Chang, N. B., Chen, J. C., Tsai, S. J. (2010). Environmental performance evaluation of large-scale municipal solid waste incinerators using data envelopment analysis. *Waste Management*, Vol.30, No. 7, (July 2010), pp. 1371-1381, ISSN 0956-053X

Chung, J. Y., Honda, Y., Hong, Y. C., Pan, X. C., Guo, Y. L., Kim, H. (2009). Ambient temperature and mortality: An international study in four capital cities of East Asia. *Science of the Total Environment*, Vol.408, No.2, (December 2009), pp. 390-396, ISSN 0048- 9697

Duke, J. M., Aull-Hyde, R. (2002). Identifying public preferences for land preservation using the analytic hierarchy process. *Ecological Economics*, Vol.42, No.1-2, (August 2002), pp. 131-145, ISSN 1470-160X

European Environment Agency [EEA]. (1999). Environmental Indicators: Typology and Overview, Report No. 25. Copenhagen: EEA, Retrived from < http://www.eea.europa.eu/publications/TEC25 >.

Firman, T., Surbakti, I. M., Idroes, I. C., Simarmata, H. A. (2011). Potential climate-change related vulnerabilities in Jakarta: Challenges and current status. *Habitat International*, Vol.35, No.2, (April 2011), pp. 372-378, ISSN 0197-3975

Hsu, Y. L., Lee, C. H. Kreng, V. B. (2010). The application of Fuzzy Delphi Method and Fuzzy AHP in lubricant regenerative technology selection. *Expert Systems with Applications*, Vol.37, No.1, (January 2010), pp. 419-425, ISSN 0957- 4174

Hu, D., Wang, R. (1998). Exploring eco-construction for local sustainability: An eco-village case study in China. *Ecological Engineering*, Vol.11, No.1-4, (October 1998), pp. 167-176

Jago-on, K. A. B., Kaneko, S., Fujikura, R., Fujiwara, A., Imai, T., Matsumoto, T., Zhang, J., Tanikawa, H., Tanaka, K., Lee, B., Taniguchi, M. (2009). Urbanization and subsurface environmental issues: An attempt at DPSIR model application in Asian cities. *Science of the Total Environment*, Vol.407, No.9, (15 April 2009), pp. 3089-3104, ISSN 0048- 9697

Jeong, I. J., Lee, K. M. (2009). Assessment of the ecodesign improvement options using the global warming and economic performance indicators. *Journal of Cleaner Production*, 17, 13, (September 2009), pp. 1206-1213, ISSN 0959-6526

Kataoka, K., Matsumoto, F., Ichinose, T., Taniguchi, M. (2009). Urban warming trends in several large Asian cities over the last 100 years. *Science of the Total Environment*, Vol.407, No.9, (April 2009), pp. 3112-3119, ISSN 0048- 9697

Kessler, J. J., Van Dorp, M. (1998). Structural adjustment and the environment: the need for an analytical methodology. *Ecological Economics*, Vol.27, No.3, (December 1998), pp. 267-281, ISSN 1470-160X

Khasnis, A. A., Nettleman, M. D. (2005). Global warming and infectious disease. *Archives of Medical Research*, Vol.36, No.6, (November-December 2005), pp. 689-696, ISSN 0188-0128

Magnani, E. (2000). The Environmental Kuznets Curve, environmental protection policy and income distribution. *Ecological Economics*, Vol.32, No.3, (March 2000), pp. 431-443, ISSN 1470-160X

Mcdonald, R. I., Kareiva, P., Forman, R. T. T. (2008). The implications of current and future urbanization for global protected areas and biodiversity conservation, *Biological Conservation*, Vol.141, No.6, (June 2008), pp. 1695-1703, ISSN 0006-3207

Mistch, W. J. (2003). Ecology, ecological engineering, and the Odum brothers. *Ecological Engineering*, Vol.20, No.5, (October 2003), pp. 331-338, ISSN 0925-8574

Müller-Fürstenbergera, G., Wagner, M. (2007). Exploring the environmental Kuznets hypothesis: Theoretical and econometric problems. *Ecological Economics*, Vol.62, No.3-4, (May 2007), pp. 648-660, ISSN 1470-160X

Niemeijer, D. (2002). Developing indicators for environmental policy: data-driven and theory-driven approaches examined by example. *Environmental Science and Policy*, Vol.5, No.2, (April 2002), pp. 91-103, ISSN 1462-9011

Ngai, E. W. T., Xiu, L., Chau, D. C. K., (2009). Application of data mining techniques in customer relationship management: A literature review and classification. *Expert Systems with Applications*, Vol.36, No.2, (March 2009), pp. 2592–2602, ISSN 0957-4174

Organization for Economic Cooperation and Development [OECD]. (1993). OECD Core Set of Indicators for Environmental Performance Reviews. OECD Publishing, Retrieved from <http://fao.org/ag/againfo/programmes/en/lead/toolbox/Refer/gd93179.pdf>

OECD, (July 30, 1996). Developing OECD Agri-Environmental Indicators, Mimeograph. Paris, France.

Omann, I., Stocker, A., Jäger J. (2009). Climate change as a threat to biodiversity: An application of the DPSIR approach. *Ecological Economics*, Vol.69, No.1, (November 2009), pp. 24-31, ISSN 1470-160X

Pan, T. C., Kao, J. J. (2009). Inter-generational equity index for assessing environmental sustainability: An example on global warming. *Ecological Indicators*, Vol.9, No.4, (July 2009), pp. 725-731, ISSN 1470-160X

Robert, B. E, Cory, R. (2003). Socio-economic indicators and integrated coastal management, *Ocean and Coastal Management*, (2003), Vol.46, pp. 299-312, ISSN 0964-5691

Roseland, M. 1997. Dimensions of the eco-city. Cities, Vol.14, No.4, (August 1997), pp. 197-202.

Rosenthal, S. S., Strange, W. C. (2001). The determinants of agglomeration. *Journal of Urban Economics*, Vol.50, No.2, (September 2001), pp. 191-229, ISSN 0094-1190

Rounsevell, M. D. A., Dawson, T. P., Harrison, P. A. (2010). A conceptual framework to assess the effects of environmental change on ecosystem services. *Biodiversity and Conservation*, Vol.19, No.10, (September 2010), pp. 2823-2842, ISSN 0960-3115

Solnes, J. (2003). Environmental quality indexing of large industrial development alternatives using AHP. *Environmental Impact Assessment Review*, Vol. 23, No.3, (May 2003), pp. 283–303, ISSN 0195-9255

Song, D., Xiao, D., Shen, Y. (2004). Assessment for the level of ecological city in inshore regions in China. *Progress in Geography*, Vol.23, (July 2004), pp. 80-85, ISSN 1007-6301 (in Chinese)

Srebotnjak, T. (2007). The role of environmental statisticians in environmental policy: the case of performance measurement. *Environmental Science and Policy*, Vol.10, No.5, (August 2007), pp. 405-418, ISSN 1462-9011

Strange, W., Hejazi, W., Tang, J. (2006). The uncertain city: Competitive instability, skills, innovation and the strategy of agglomeration. *Journal of Urban Economics*, Vol.59, No.3, (May 2006), pp. 331-351, ISSN 0094-1190

Tanguay, G. A., Rajaonson, J., Lefebvre, J.-F., Lanoie, P. (2010). Measuring the sustainability of cities: An analysis of the use of local indicators. *Ecological Indicators*, Vol.10, No.2, (March 2010), pp. 407–418, ISSN 1470-160X

UNESCO [United Nations Educational, Scientific and Cultural Organization], (2011). Education for Sustainable Development (ESD). April 1, 2011, Available from: <http://www.unesco.org/new/en/education/themes/leading-the-international-agenda/education-for-sustainable-development>.

Valentin, A., Spangenberg, J. H. (2000). A guide to community sustainability indicators. *Environmental Impact Assessment Review*, Vol.20, No.3, (June 2000), pp. 381-392, ISSN 0195-9255

van Bohemen, H. D. (1998). Habitat fragmentation, infrastructure and ecological engineering, *Ecological Engineering*, Vol.11, No.1-4, (October 1998), pp. 199-207, ISSN 0925-8574

Verdoodt, A., Van Ranst, E. (2006). Environmental assessment tools for multi-scale land resources information systems: A case study of Rwanda. *Agriculture, Ecosystems and Environment*, Vol.114, No.2-4, (June 2006), pp. 170-184, ISSN 0167-8809

Weng, Y. C., Fujiwara, T., Matsuoka, Y. (2009). Municipal solid waste management and short-term projection of the waste discard levels in Taiwan. *Journal of Material Cycles and Waste Management*, Vol.11, No.2, (May 2009), pp. 110-122, ISSN 1438-4957

Weng, Y. C., Fujiwara, T., Matsuoka, Y. (2010). An analysis of municipal solid waste discards in Taiwan based on consumption expenditure and policy interventions. *Waste Management and Research*, Vol.28, No.3, (March 2010), pp. 245-255, ISSN 0734-242X

Weng, Y. C., Fujiwara, T. (2011). Examining the effectiveness of municipal solid waste management systems: An integrated cost-Benefit analysis perspective with a financial cost modeling in Taiwan. *Waste Management*, Vol.31, No.6, (June 2011), pp. 1393-1406, ISSN 0956-053X

Werner, P., Werner, G., Kristina, V. (2006). Survey of environmental informatics in Europe. *Environmental Modelling and Software*, Vol.21, No.11, (November 2006), pp. 1519-1527.

Yokohama City Government. (2011). Comparative Statistics in Japanese Major Cities. March 1, 2011, Available from: <http://www.city.yokohama.lg.jp/ex/stat/daitoshi/>.

12

Change in Species Composition and Distribution of Algae in the Coastal Waters of Western Japan

Satoshi Nagai[1], Goro Yoshida[2] and Kenji Tarutani[2]
[1]*Fisheries Research Agency, National Research Institute of Fisheries and Environment of Inland Sea, Research Center for Environmental Conservation*
[2]*Fisheries Research Agency, National Research Institute of Fisheries and Environment of Inland Sea, Coastal Fisheries and Environment Division*
Japan

1. Introduction

Estuarine and coastal ecosystems are sensitive to human activities that lead to nutrient pollution, excess or insufficient sedimentation, and other types of pollution. The effects of climate change are particularly noticeable in these shallow ecosystems, and increases in non-native species and gelatinous zooplankton such as jellyfish can have harmful effects on fisheries. With climate change related to human activities well under way (IPCC, 2007), there is a need to assess the impacts of these changes on estuaries and coastal waters. Growing awareness of climate change has stimulated considerable recent interest in the possible shifting of biological events in marine ecosystems and the ecological consequences of these shifts. It has been suggested that enhanced stratification caused by an increase in surface water temperature decreases nutrient supply, leading to a decrease in primary productivity throughout a large area of the North Pacific Ocean (Ono et al., 2002; Watanabe et al., 2005). It has also been reported that the timing and magnitude of phytoplankton blooms have changed markedly, affecting the dynamics of zooplankton and fisheries resources (e.g., Edwards & Richardson, 2004; Chiba et al., 2008). Impacts on estuarine and coastal ecosystems remain unclear, however, because of the difficulty in distinguishing the effects of global warming from those of other anthropogenic factors. Although phytoplankton are major primary producers in most aquatic ecosystems, benthic primary production, such as macroalgae, can be often substantial and even dominant in shallow waters. *Sargassum* algal beds are highly productive and offer habitats for numerous marine animals. The decline and qualitative change of *Sargassum* beds in western Japan, which may be caused by global warming, and their impact on coastal ecosystems and fisheries that rely on them is of great concern. In last two decades, macroalgal beds composed of *Sargassum* and kelp (mainly *Ecklonia*) have been degrading or declining along the coast of the Kyushu and Shikoku areas in western Japan, which has southern temperate climate characteristics. It is estimated that the mass disappearance of local kelp and *Sargassum* beds in these regions totals several thousand hectares and more than 40% of the total macroalgal beds surveyed in the early 1990s. Degradation of macroalgal beds in these areas has coincided with increased water temperatures and grazing pressures by herbivorous animals (fish and sea urchins). Shifts of flora from temperate to subtropical components, accompanied by quantitative changes in macroalgal beds, have also been observed in many areas of western Japan.

Subtropical *Sargassum* species such as *S. duplicatum* and *S. carpophyllum* have begun to appear in many *Sargassum* beds in the Kyushu and Shikoku regions, sometimes becoming more dominant by replacing temperate species. These subtropical species exhibit an annual life cycle trait and develop their stocks only between winter and spring. This seasonality is attributable to the growth characteristics of the species in which shoot growth is triggered by cooler temperatures and is advantageous for avoiding active grazing by herbivorous animals during warmer seasons. Subtropical *Sargassum* species also have high regenerative ability at their holdfasts, which allows them to grow in severely turbulent conditions during which grazing activity is restricted. These eco-physiological characteristics of subtropical *Sargassum* species contrast greatly with those of their temperate counterparts and are well adapted to an environment with high grazing pressure from herbivorous animals. The quantitative and qualitative changes in macroalgal beds will extend to cooler regions in Japan if coastal water temperatures continue to increase. Monitoring and continuous surveying of coastal areas are needed in addition to countermeasures to protect biological and fisheries production against the coastal environmental changes induced by global warming.

Several species of dinoflagellates produce toxins responsible for paralytic shellfish poisoning (PSP). When these dinoflagellates bloom, their toxins accumulate in mussels and oysters that ingest them. Outbreaks of PSP in mussels and oysters caused by *Alexandrium*, *Gymnodinium catenatum* Graham, and *Pyrodinium bahamense* Plate var. *compressum* (Böhm) Steidinger, Tester & Taylor are currently a serious problem from both an economic and a food hygiene point of view. The occurrence of toxic *Alexandrium* blooms has increased in frequency, and these species are widely distributed in the coastal waters of Europe, North America, Japan, the Philippines, Australasia, India, and Argentina (Hallegraeff, 1995). Extensive taxonomic, physiological, and ecological studies have resulted in the recognition of some 10 species of the genus *Alexandrium* that are toxic or presumed toxic (Balech, 1995; Yoshida & Fukuyo, 2000). In Japan, *A. catenella* (Whedon & Kofoid) Balech, and *A. tamarense* (Lebour) Balech are the major *Alexandrium* species causing PSP. Appearances of the toxic tropical dinoflagellate species *A. tamiyavanichii* have been confirmed in the Seto Inland Sea since the 1990s, although it remains unclear whether global warming is directly responsible. Rising winter water temperatures in the Seto Inland Sea in Japan have occurred in response to progressing global warming caused by the massive exhaustion of carbon dioxide and the decomposition of the ozone layer related to civilization and industrialization (Yamamoto, 2003; Takahashi & Seiki, 2004; Wanishi, 2004; Tarutani, 2007).

Our purpose is to present and summarize some of the evidence of the impact of global warming on the coastal waters of western Japan. In this chapter, we discuss three topics that are closely related to global warming: 1) recent trends in environmental changes—mainly in surface water temperature (Section 2-1), 2) the degradation of macroalgal beds composed of algal genus *Sargassum* and kelp and their replacement by subtropical *Sargassum* species (Section 2-2), and 3) the new invasion of tropical species that cause harmful microalgal blooms and its settlement in the introduced area (Section 2-3).

2. Potential climate change effects on environmental characteristics in the coastal ecosystems of western Japan

2.1 Long-term changes in water temperature in the coastal waters of western Japan

According to the *Global Warming Projection*, published by the Japan Meteorological Agency (2008), the sea surface temperature (SST) around Japan has increased 0.8–1.7°C over the past 100 years (1900–2007). The SST around Japan is projected to increase, and the long-term SST trends for the 21st century predicted using the North Pacific Ocean General Circulation Model (Sato et al., 2006) are increases of 2.0–3.1°C and 0.6–2.1°C per century under scenarios A1B and B1, respectively. The findings for near-shore areas were not included in this report, however, and the impacts of global warming on environmental factors in these waters remain unclear. In this section, we take the Seto Inland Sea as an example of the coastal waters of western Japan and discuss recent trends in water temperature.

2.1.1 Site description

The Seto Inland Sea is the largest semi-enclosed coastal sea in Japan, surrounded by Honshu (the main island of Japan), Shikoku, and Kyushu Islands (Fig. 1). It is 500 km long, has an average depth of 30 m, and contains more than 700 small islands. The region of the Seto Inland Sea is meteorologically referred to as the Seto Inland Climate Zone. The annual average temperature and precipitation are 15°C and 1,000–1,600 mm, respectively. The watershed land covers 68,000 km^2, which corresponds to about 18% of the total land area of Japan. This region is one of the most industrialized areas in Japan. At present, approximately 35 million people—about 18% of the total population of Japan—live within the watershed, and the marine environment has been significantly affected by their activities.

Although the Seto Inland Sea is enclosed, phenomena in the Pacific Ocean can strongly influence its marine environment thorough two boundary regions, the Kii and Bungo Channels. Off the southeastern coasts of Kyushu, Shikoku, and Honshu, the Kuroshio Current flows northeastward. This current corresponds to the northeastern part of the subtropical ocean gyre that circulates clockwise in the North Pacific and transports a large amount of water and heat to the coast of Japan. The current path shows significant interannual variations off the southern coast of Japan (e.g., Kawabe, 2005; see Fig. 1) and strongly influences the thermal conditions and nutrient environment in the Seto Inland Sea, especially around the Kii and Bungo Channels.

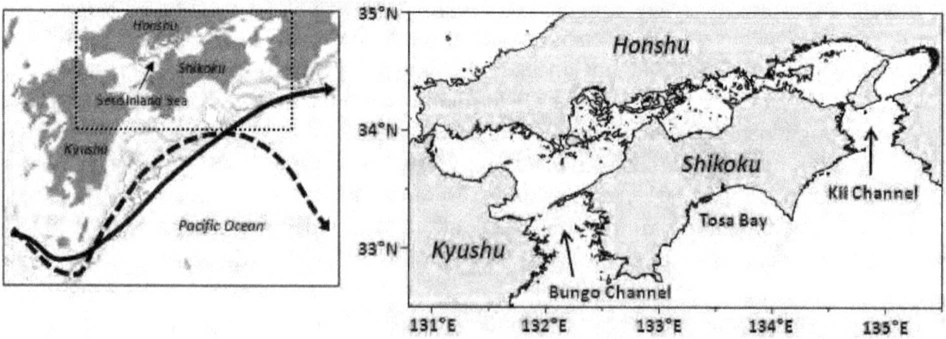

Fig. 1. Maps of the Seto Inland Sea and the typical Kuroshio Current path off the southern coast of Japan (left). The dashed and solid lines in the left panel represent the typical offshore (large meandering) and near-shore (non-large meandering) path, respectively.

The Seto Inland Sea has extremely high productivity per unit area of fishery products, which is the largest among comparable semi-enclosed coastal seas worldwide (e.g., Chesapeake Bay, Baltic Sea, North Sea, Mediterranean Sea; Takeoka, 2002). This high productivity is related to the extremely complex and semi-enclosed geography of the sea and the presence of essential nutrients supplied from the many rivers around the sea in sufficient amounts for the growth of phytoplankton. In recent decades, decreasing fishery production in the Seto Inland Sea has become recognized as a serious problem (Yamamoto 2003; Tarutani et al., 2009).

2.1.2 Recent trends in water temperature

The mean seasonal pattern for water temperature in the Seto Inland Sea has a maximum value in late summer (August to September) and a minimum value in late winter (February to March). The pattern differs somewhat spatially: the winter temperature is higher in the areas near the Kii and Bungo Channels, whereas the summer temperature is higher in the western part of the sea. Fig. 2 shows the long-term variations in mean water temperature throughout the Seto Inland Sea, and these variations were extracted from data gathered by the Fisheries Monitoring Program since 1973. After 1989–1990, the mean water temperature at a depth of 10 m generally remained higher than that measured during the 1980s (Tarutani, 2007). Considerable interannual variability is superimposed on long-term warming—for example, in Hiroshima Bay, located in the central part of the Seto Inland Sea, the annual mean temperature of water at 10 m of depth has warmed by about 1.0°C since 1973. The water temperature fluctuations differed seasonally, however. The warming trend was noticeable in autumn-winter, but was weak in spring-summer (Fig. 3). Such warming phenomena have occurred in several of the western coastal waters of Japan (e.g., Tosa Bay, located in the southern part of Shikoku Island; Setou et al., 2004).

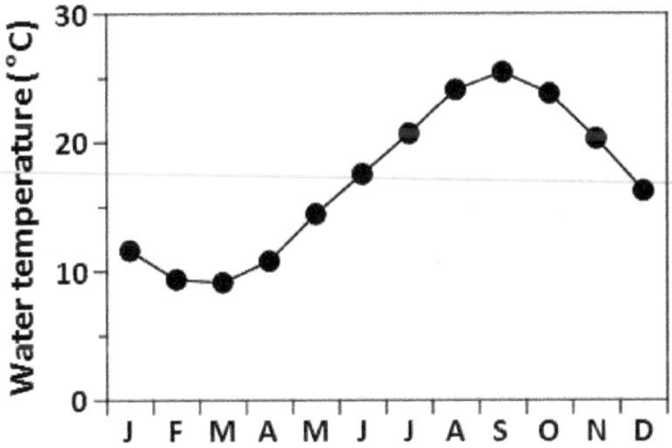

Fig. 2. Seasonal variations in water temperature in Osaka Bay, in the eastern part of the Seto Inland Sea (the mean from 1973 to 2002).

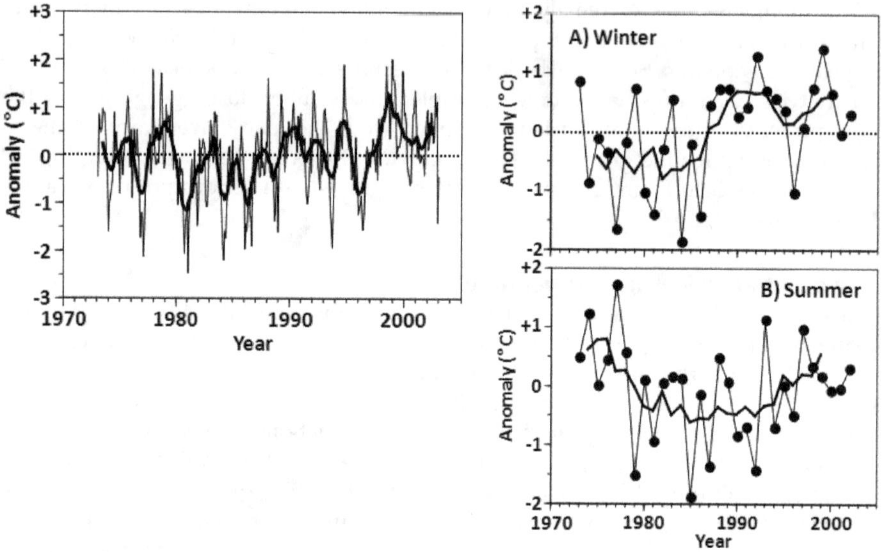

Fig. 3. Long-term variations in water temperature at 10 m depth in the Seto Inland Sea from 1973 to 2002. The thick line in the left panel indicates the running mean from 13 months of data. The right panels are (a) winter and (b) summer data, and the thick lines indicate the running means of 3 years of data.

The seasonality and future trends of water temperatures in the Seto Inland Sea are extremely important because they provide insight into the phenological changes in phytoplankton and macroalgae and may also play a role in regulating nutrient delivery to the sea from the Pacific Ocean. A hydrological modeling study revealed that the rising water temperature in the Seto Inland Sea was caused not only by changes in heat flux at the sea surface but also by changes in open oceanic temperature (Takahashi & Seiki, 2004). This discovery suggests that further detailed investigations of the mechanisms determining both regional atmospheric and oceanic temperature are required to accurately assess the impacts of climate change on hydrology in the Seto Inland Sea.

2.2 Changes in macroalgal vegetation on Japanese coasts

2.2.1 Degradation of macroalgal beds in southern temperate Japan

As mentioned in the introduction, coastal water temperatures have been increasing in Japan since the 1970s. The temperature increase is exerting a large influence on coastal ecosystems both directly and indirectly. Luxuriant macroalgal beds with diverse flora along the Japanese coast reflect variable climatic characteristics (cold and warm temperate to subtropical). These beds, called *Moba* in Japanese, were estimated to cover approximately 201,000 ha during the last research survey conducted between 1989 and 1991 (Environmental Agency & Marine Parks Center of Japan, 1994). In temperate Japan, large brown algae of the genus *Sargassum* and kelps (*Eisenia* and *Ecklonia)* dominate rocky shores and exhibit high productivity (Fig. 4). Along with the seagrass *Zostera* on sandy shores, these algae play important ecological roles as habitats for many marine animals. A recent report has shown that these brown algal beds have become extremely degraded or have disappeared during the last 20 years, especially in the southern part of temperate Japan (Fig. 5). In the southern area of the Miyazaki Prefecture in Kyusyu, for example, nearly 90% of local *Sargassum* beds have been lost (Aratake et al., 2007). In another case in the Kochi Prefecture, most kelp beds along the coast were lost after 1990 (Hiraoka et al., 2005).

The mass disappearance of local kelp and *Sargassum* beds in these regions is estimated to encompass several thousand hectares and 44.7% of the macroalgal beds reinvestigated after the survey by the Environmental Agency in 1989-1991 (Fujita et al., 2010). In most cases, the disappearance of macroalgal beds has seriously affected local coastal fisheries such as abalone fisheries (Serisawa et al., 2004). The scenery of the seabed after the decrease or loss of these large brown algae is called *Isoyake* (Fig. 6a). *Isoyake* has been known to occur occasionally since the 19th century with the approach of the Kuroshio Current offshore in the Pacific region (Fujita et al., 2010). The recent appearance of *Isoyake* is more frequent, however, and the phenomenon has extended to the entire Japanese coast. In the most remarkable case of *Isoyake*, the only dominant algae are crustose coralline algae on rocks or boulders that frondose macroalgae once occupied. Another common aspect among *Isoyake* regions is increased grazing pressure on algae by herbivorous animals (mainly fish and sea urchins; Fujita et al., 2010). Although it is unconfirmed whether these grazers are the cause of *Isoyake*, they prevent the recovery of frondose macroalgae and surely contribute to the continuity of *Isoyake* conditions.

Fig. 4. Brown algal beds in southern temperate Japan. A *Sargassum* bed (a) in Hiroshima Bay, the Seto Inland Sea, and a kelp (*Ecklonia*) bed (b) at Sata-Misaki Peninsula in the Bungo Channel.

2.2.2 Changes in *Sargassum* flora in *Isoyake* areas

Concurrent with *Isoyake* during and after the 1990s, changes in the species composition of *Sargassum* beds that remain in *Isoyake* areas have also been reported in southern temperate Japan (e.g., Fujita et al., 2010). Along the coast of the Kochi Prefecture in the Pacific region, temperate species such as *S. micracanthum* and *S. okamurae* that dominated during the 1970s have been decreasing, and the subtropical species *S. duplicatum* (Fig. 6b) has become a major component of the flora since the 1990s (Hiraoka et al., 2005). In the Miyazaki Prefecture, *Sargassum* flora composed mainly of temperate species in 1970s has contained many subtropical species since 2000 (Aratake et al., 2007). The appearance of subtropical *Sargassum* has also been reported in the *Isoyake* area of Nagasaki, in the western Kyushu region (Fujita et al., 2010). Subtropical *Sargassum* exhibits annual characteristics in its life cycle that contrast greatly with the perennial characteristics of most temperate species (Fig. 7). Although the perennial species sustain shoots year round, subtropical species in Japan display shoot growth only in winter and spring. After reproduction in early summer, their shoots become senescent, and no visible thalli are observed in autumn (Shimabukuro et al., 2007).

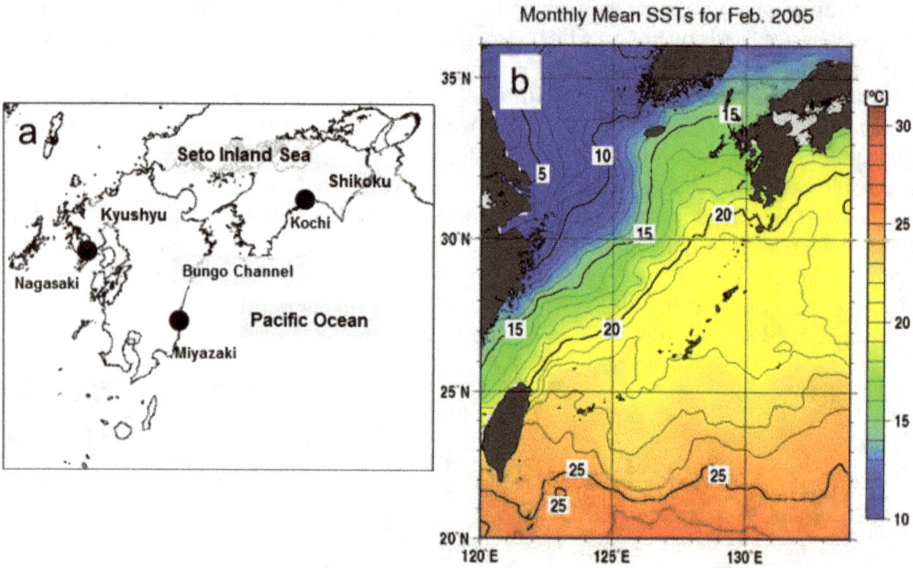

Fig. 5. Places cited in the text (a) and surface water temperature in winter (Feb. 2005) in the southern sea area of Japan (b) from the website of the Japan Meteorological Agency (2011).

To understand the physiological mechanism of subtropical *Sargassum* seasonality, we examined growth characteristics against variable temperatures for several common subtropical *Sargassum* species under experimental culture conditions. The results showed that the optimum temperature range for shoot growth in these species is 16–21°C, a range that coincides with winter coastal water temperatures in southern temperate Japan (see Fig. 5). Shoot development was inhibited above 24°C (Fig. 8), which is the average summer-autumn temperature in the field (Yoshida et al., 2009). This finding suggests that a drop in winter water temperature triggers shoot growth in subtropical *Sargassum*. On the contrary, the optimum temperature range for shoot growth is wider for temperate species. Even at summer temperatures (25°C), the shoots of many temperate species exhibit smooth growth (Haraguchi et al., 2005; Yoshida et al., 2008). These data seemed contradictory—a warmer temperature suppresses the growth of subtropical species more than it suppresses the growth of temperate species. Such a growth characteristic seems to be advantageous, however, because it allows subtropical species to escape grazing pressure by herbivorous animals. For example, the feeding activity of rabbit fish (*Siganus fuscescens*), a common herbivorous fish in warm temperate Japan, is temperature dependant. It grazes brown algae intensively under warmer temperatures (>25°C), whereas grazing activity is low below 20°C and almost stops at 15°C (Yamauchi et al., 2006). Owing to their growth characteristics, subtropical species can develop shoots during winter when the herbivorous fish is inactive. On the contrary, temperate species are defenseless and vulnerable to grazing by these fish owing to the year-round growth pattern of their stands.

Fig. 6. *Isoyake* (a) and subtropical *Sargassum duplicatum* with co-occurring corals (b). Both were photographed at Ainan, Ehime Prefecture, in the Bungo Channel.

Another advantageous characteristic of subtropical species revealed by our study is their high regenerative ability, which enables them to develop stands under high grazing pressure. A high frequency of new budding production from holdfasts is observed in several subtropical species (e.g., *S. carpophyllum*) when the holdfasts are experimentally excised from their thalli (Fig. 9a). On the contrary, less or no budding occurs in the excised holdfasts of temperate species—e.g., *S. yamamotoi*, one of major species in warm temperate regions. Many subtropical *Sargassum* can develop stands in specific environments with relatively severe physical turbulence—for example, strong surge or frequent sand movement (Fig. 9b). Holdfasts of subtropical *S. carpophyllum* can endure being buried in sand for several months, and after reappearing, they can generate new thalli. Less grazing occurs in such severe environments, and even when grazing does occur, the high regenerative ability of subtropical species allows them to grow despite physical damage.

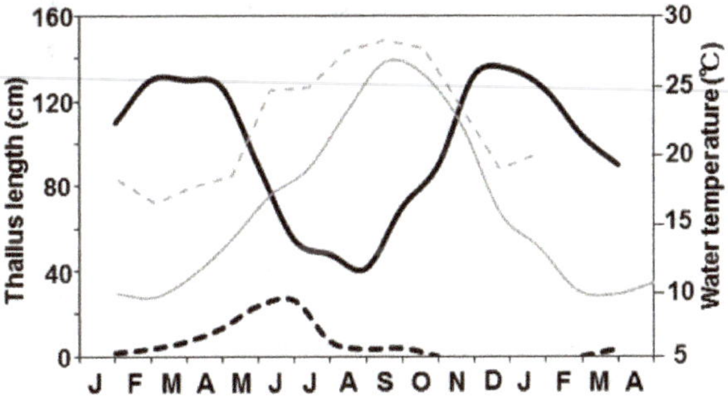

Fig. 7. Schemes of seasonality of temperate and subtropical *Sargassum*. The bold, solid line indicates seasonal growth of temperate *S. fulvellum* (from Yoshida et al., 2008), and the bold dotted line indicates subtropical *S. duplicatum* (from Shimabukuro et al., 2007). The thin solid and dotted lines indicate water temperature in their habitats, respectively.

2.2.3 Mechanisms of the changes and human countermeasures

Although many factors (e.g., nutrient depletion, change in food web structure related to overfishing of predators) are possible contributors to *Isoyake*, the increase in coastal water temperature is one of the most plausible (Fujita et al., 2010). In the past few decades, the mean yearly surface temperature along the warm temperate Japanese coast has increased 0.5°C to 0.7°C (e.g., Serisawa et al., 2004; Shimabukuro et al., 2007). Although extreme rises in summer temperature could cause physiological damage to macroalgae, long-term monitoring also indicates that the temperature increase has been larger in winter than in summer (Hiraoka et al., 2005; Shimabukuro et al., 2007). Rising water temperature in autumn and winter allows herbivorous animals to be active for longer periods. An increase in grazing by these animals may be the most potent cause for the ousting of temperate large brown algae in warm temperate regions. It has been reported that when grazing pressure increases, shallow coastal seabeds shift from being macroalgal dominated to being coral- or coralline algal dominated in tropical and subtropical regions (Lapointe et al., 1997). Are subtropical *Sargassum* extending their distribution in temperate regions of Japan? Some past records indicate that the subtropical species are minor but original members of the coastal flora once dominated by temperate species in southern temperate Japan (Aratake et al., 2007). As the result of environmental change over several decades, subtropical *Sargassum* ecologically adapted to high grazing pressure and has been becoming conspicuous, replacing major temperate species that are more sensitive to grazing (Yoshida et al., 2009).

Fig. 8. Growth of subtropical *Sargassum duplicatum* cultured in a laboratory under (a) 24°C, and (b) <18°C.

The mass disappearance of macroalgae in warm temperate coastal Japan has had serious effects on coastal biological productivity, and countermeasures to sustain coastal fisheries are needed. It is well known that physical defense of *Moba* against grazing by using nets or cages and removing herbivores is often effective for recovering macroalgal beds in *Isoyake* areas (Fujita et al., 2010). Another possible countermeasure that protects fisheries production is the exploitation of productivity of the new vegetation of subtropical *Sargassum*. Because subtropical *Sargassum* species have characteristics that differ from those of temperate ones—e.g., seasonality, habitat, stand structure—the ecological roles they could play in the coastal ecosystem should be evaluated (Fujita et al., 2010).

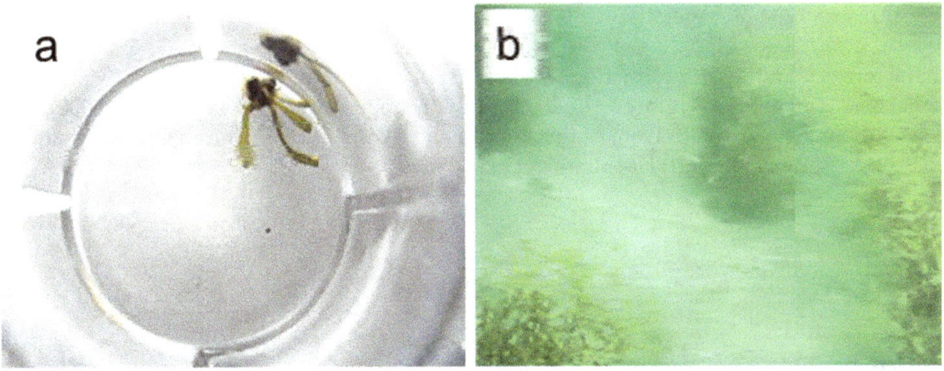

Fig. 9. (a) Regeneration of new buddings from an excised holdfast of *Sargassum carpophyllum*, and (b) a *Sargassum* habitat buried in sand (Kushima, Miyazaki Prefecture).

If coastal water temperature continues to increase as predicted, quantitative and qualitative change in macroalgal beds will extend to cooler regions in Japan. To recognize these changes as soon as possible, fixed survey stations around the Japanese coast have been established for a long-term national monitoring project (e.g., Biodiversity Center of Japan, 2011). One of the most interesting survey points is in the channels between the Pacific regions and the Seto Inland Sea in southern Japan. In the Bungo Channel, a gradient in winter seawater temperature ranges between 6°C and 8°C within ca. 200 km. Within this distance, both luxuriant brown algal beds and *Isoyake* with subtropical *Sargassum* and corals are observed (Yoshida et al, 2009), as shown in Figs. 4 and 6. Ecological and oceanological surveys in this transition zone, or ecotone, between temperate and subtropical areas will provide valuable information about the mechanisms of global warming effects on coastal ecosystems.

2.3 Appearance of a tropical harmful algal bloom species in temperate Japanese coastal waters

The first appearance of *A. tamiyavanichii* was confirmed in a towing plankton net sample at Aburatsubo in Sagami Bay, Japan, in 1988 (Ogata et al., 1990). After that, *A. tamiyavanichii* bloomed for the first time at Shioya Bay in Okinawa Prefecture, Japan, in 1997 and again in 1998, and an outbreak of PSP was confirmed in green muscles fed *A. tamiyavanichii* (Koja et al., 2001). This record is the first of toxification of shellfish caused by this species in Japan. The appearance of *A. tamiyavanichii* was also confirmed in the Seto Inland Sea in 1997 when 6.6 MU g^{-1} of PSP from the edible part of Japanese oysters in Uchinoumi, Tokushima Prefecture (Sato et al., 2001) and 8.8 MU g^{-1} of PSP from muscles and 5.0 MU g^{-1} of PSP from arch shells at Tsuta-cho in Kagawa Prefecture were detected in 1999 (Yoshimatsu et al., 2000)—the first record of PSP outbreaks attributed to this species in the Seto Inland Sea. In 2001, *A. tamiyavanichii* blooms were confirmed in almost all areas of the Seto Inland Sea, and 2.4 MU g^{-1} of PSP was detected in manila clams (Tanada et al., 2003).

The blooming of this species has often been observed not only the Seto Inland Sea but also around the Tsushima, Gotou, and Amami Islands (Fig. 10). This species has also been collected in the Gulf of Mexico, the Straits of Malacca in Malaysia (Usup et al., 2002), the Gulf of Thailand (Fukuyo et al., 1988), and Manila Bay in the Philippines (S. Sakamoto, personal communication). Research has provided some information on the morphology and the ultrastructure of vegetative cells (Phanichyakarn et al., 1993; Wisessang et al., 1993; Balech 1994; Nagai et al., 2003), life cycle (Nagai et al., 2003; Nagai et al., 2008), growth characteristics and toxin production (Kodama et al., 1988; Fukuyo et al., 1989; Ogata et al., 1990; Nagai et al., 2008; Oh et al., 2009), phylogenetic position (Usup et al., 2002), and distribution (Fukuyo et al., 1988) of this species and toxin accumulation in mussels that feed on it (Wisessang et al., 1991; Hashimoto et al., 2002).

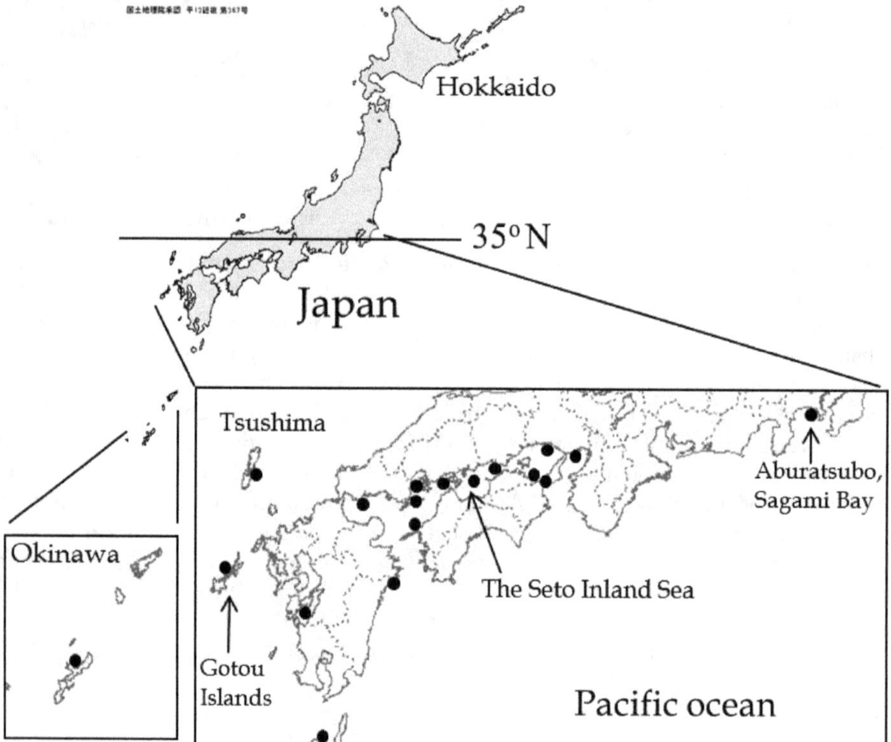

Fig. 10. Occurrences of the toxic dinoflagellate *Alexandrium tamiyavanichii* in Japanese coastal waters. Note that until now, the distribution has been limited in areas south of the northern latitude of 35°C.

item	vegetative cells (μm)	planozygotes (μm)	item	cysts (μm)
length	30-60	50-70	long diameter *	45-75
width	35-65	50-75	short diameter*	35-60
			height	40-60

Table 1. Measurements of cell sizes of *Alexandrium tamiyavanichii* in culture (n=50). *, They were measured from the upper side.

2.3.1 Morphological characteristics of *A. tamiyavanichii*

Vegetative cells of *A. tamiyavanichii* are yellow-brown and contain numerous chloroplasts (Nagai et al., 2003). This species forms long chains that often contain more than 200 cells during the early and middle exponential growth phases under laboratory conditions (Fig. 11a). Cell chains separate into smaller chains or even single cells in the stationary phase. Vegetative cells are 30–60-μm long and 35–65-μm wide, and the sizes drastically change in response to culturing (Table 1). The cell is round to subpentagonal in the ventral view and usually more wide than long. The first apical plate has a ventral pore at the posterior portion of the right anterior margin with the fourth apical plates (Fig. 11b). The most typical characteristic of this species is that the sulcal anterior plate has a triangular or rectangular anterior expansion, and the anterior and posterior portions are delimited by a transverse rib, but the two parts are inseparable (Balech, 1994; see Fig. 11b).

2.3.2 Toxin composition in *A. tamiyavanichii*

Vegetative cells of *A. tamiyavanichii* collected from the Tokushima Prefecture (eastern Seto Inland Sea) in 1999 showed PSP toxicity, of which 83 mol% was accounted for by gonyautoxin (GTX) 2, GTX3, and GTX4 with fluorescent high-performance liquid chromatography analysis. Its specific toxicity was 112.5 fmol cell^{-1}, and 1 MU was equivalent to 7,200 cells (Hashimoto et al., 2002). The toxic component of the Aburatsubo strains from Sagami Bay, Japan, was dominated by the highly potent carbamate toxin GTX4 (37–45 mol%; Ogata et al., 1990). The toxic composition observed in samples from the Tokushima Prefecture was closely related to that of strains isolated from the Gulf of Thailand (Ogata et al., 1990; Wisessang, et al., 1991). The toxic components of *A. tamiyavanichii* clonal strains, isolated from the Seto Inland Sea, showed that *N*-sulfocarbamoyl (C-) 2 and GTX4 were the main toxins, and C3+4, GTX2+3, GTX5, neo-saxitoxin (neo-STX), and STX were minor components (Oh et al., 2009). The percentages of the main toxins (C2 and GTX4) are closely related to those of *A. tamarense* and *A. catenella*. Potent toxic components including GTX2+3 and STX existed at higher concentrations than was the case with *A. tamarense* and *A. catenella*, however. Additionally, the average toxicity of *A. tamiyavanichii* was higher than that of *A. tamarense* and *A. catenella*. Toxic components detected from several shellfish, including the mussel *Mytilus galloprovincialis*, the Pacific oyster *Crassostrea gigas*, and the ark shell *Scapharca broughtonii* after the large-scale bloom of *A. tamiyavanichii* in the eastern Seto Inland Sea in early December 1999 were similar to those detected in *A. tamiyavanichii* plankton cells (Hashimoto et al., 2002), and the PSP toxicity in the shellfish that were higher than regulation levels (4 MU g^{-1}) were seen when the abundance of *A. tamiyavanichii* in the bloom was greater than 30 cells mL^{-1} (Yoshimatsu et al., 2000; Sato et al., 2001; Tanada et al., 2003).

2.3.3 Sexual reproduction in *A. tamiyavanichii* under laboratory conditions

In one study, encystment through sexual reproduction was observed in 54 pairs (39.7%) of 136 pairs of *A. tamiyavanichii*, which included 16 that were self-crossing, using 16 clonal strains under laboratory conditions (Nagai et al., 2003). Neither planozygote formation nor encystment was confirmed in any of the self-crossings. These results clearly show that *A. tamiyavanichii* is heterothallic—i.e., compatible strains of an opposite mating type are required for sexual reproduction. Crossing experiments have been carried out in a modified f/2 medium (made up with 1/6 nitrate and phosphate, 1/3 metals and 1/10 vitamins; see

Nagai et al., 2004) using a sexually compatible mating pair under the same conditions used in growth experiments. Vegetative cells in the mixed culture at 25°C grew exponentially, and the growth reached a peak (2,400 ± 200 cells mL^{-1}) on day 8, with density thereafter decreasing gradually owing to an increase in the number of planozygotes and resting cysts (Fig. 12). Suspected gametes began to appear in the culture 2–3 days after the crossing experiment started (see Fig. 12). These cells were morphologically indistinguishable from gametes (hologamy), but the swarming of single vegetative cells, which make an association of sexual mating behaviour as seen in *A. tamarense*, were often observed in the early and middle exponential growth phases.

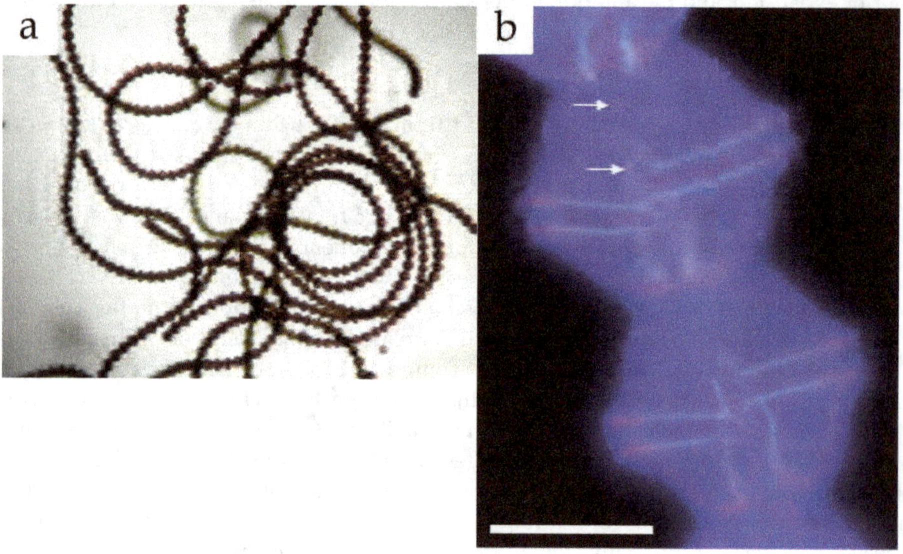

Fig. 11. Vegetative cells of *Alexandrium tamiyavanichii*. (a) A long chain consisting of more than 200 cells. (b) Vegetative cells stained with calcofluor-white. Arrows indicate a ventral pore and a transverse rib. Scale bar is 25 µm. Observation in (a) was made using a Nikon TE-300 light microscope (Tokyo, Japan) and in (b) using an epifluorescence microscope under ultraviolet and blue light excitation.

Pairs of fusing gametes were isolated in the wells of a microplate with 1 mL of the culture medium to observe the sequence of zygotes. The gametes swam rapidly and made contact with each other bilaterally (gametes remained side by side; Fig. 13a), and the couplets swam helicoidally with rotation due to the almost perpendicular orientation of their longitudinal flagella, as reported in *A. pseudogonyaulax* (Biecheler) Horiguchi ex Yuki et Fukuyo by Montresor (1995). The length of this phase was variable, and a couplet was observed swimming for more than 1 hour before conjugation, whereas others started fusing sooner and formed a planozygote (Fig. 13b, c). Conjugation took place through the progressive engulfment of one gamete by the other, which consequently increased in size (see Fig. 13c). Sexual reproduction was by conjugation of isogametes, although the cell length of one mating gamete was often up to 1.2–1.3 times longer than that of the other. Plasmogamy was completed 60–80 min after conjugation started and produced a planozygote with one transverse and two longitudinal flagella (Fig. 13d). There was no trace of thecal plates near the newly formed zygotes, and only the thecal plates of the gamete that engulfed the other gamete appeared to be maintained as the zygotic thecal plates.

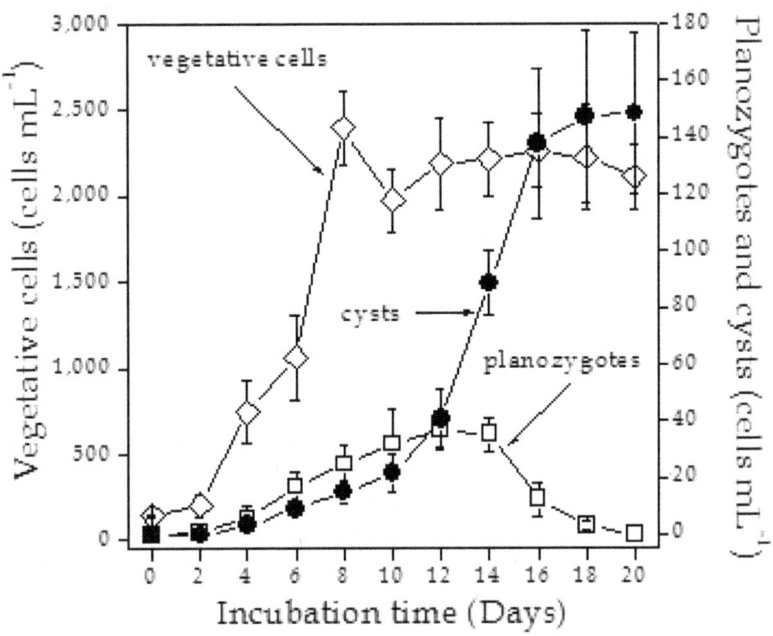

Fig. 12. Encystment of an *Alexandrium tamiyavanichii* batch culture at 25°C. Time-course measurements were carried out to observe the sexuality and the encystment process in 1.9 mL of modified f/2 medium in Iwaki 24-well microplates (Chiba, Japan) under an irradiance of 100–150 µmol $m^{-2} s^{-1}$ provided by cool-white fluorescent lamps with a 12:12 h light:dark cycle (Nagai et al., 2003, 2008). The culture medium contained 1/6 nitrate and phosphate, 1/3 metals, and 1/10 vitamins. Two compatible strains of AT0112T06 and AT0112F06 were mixed and incubated for 20 days. A portion of the *Alexandrium* culture (100 µL) was sampled every 2 days, and the numbers of vegetative cells, planozygotic cells, and resting cysts were counted using a microscope.

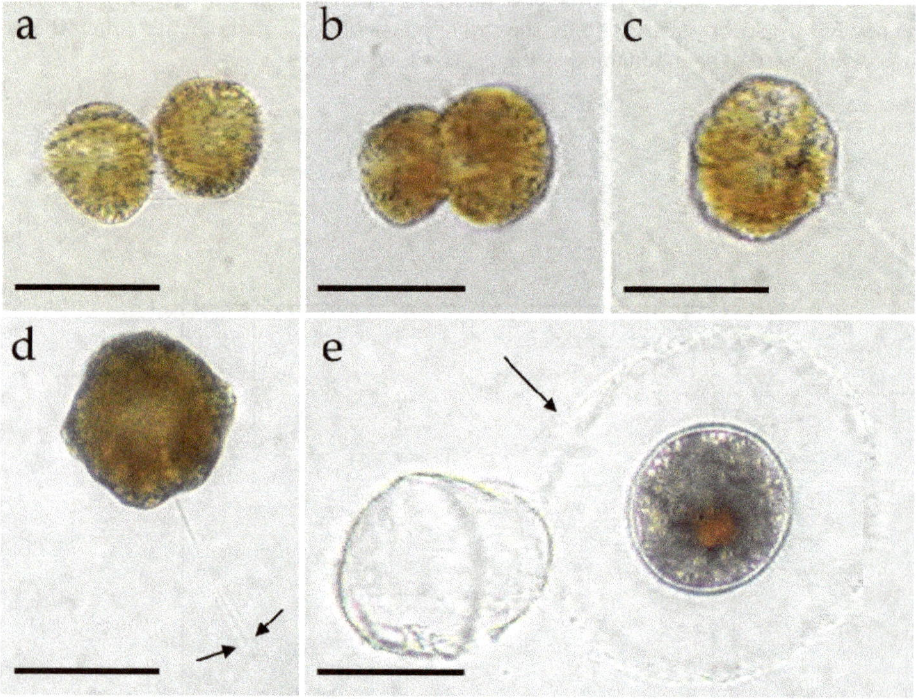

Fig. 13. Life cycle stages of *Alexandrium tamiyavanichii* obtained in the sexual reproduction experiment described in Fig. 12 (Nagai et al., 2003). Scale bars are 50 μm (all at the same magnification). (a) Mating pair of isogametes entwined by their flagella (initial contact phase). (b) Conjugating gametes. (c) Completion of conjugation in which one gamete engulfs the other. (d) Planozygote with two longitudinal flagella indicated by arrows. (e) Threshing during encystment and resting cysts covered with transparent mucilaginous matter indicated by arrows. Observation was made using a Nikon TE-300 light microscope (Tokyo, Japan).

Planozygotes were typically a more blackish color compared with vegetative cells and tended to swim more slowly; therefore, they were readily distinguishable. Planozygotes were 50–70-μm long and 50–75-μm wide and larger than vegetative cells (see Table 1). In the transformation process of planozygotes to hypnozygotic cysts, a small number of planozygotes failed to thresh, but most succeeded and after 2–3 days the cysts were covered with transparent mucilaginous matter as seen in *A. tamarense* and *A. catenella* reported by Turpin et al. (1978), Yoshimatsu (1981), and Nagai et al. (2004) (Fig. 13e). Cysts were vertically compressed (Fig. 14a, b) or spherical (Fig. 14c, d) and 45–75-μm long, 35–60-μm wide, and 40–60-μm high. Scanning electron microscopy clearly showed that the surface of *A. tamiyavanichii* cysts was smooth and lacked paratabulation (see Fig. 14). The density of planozygotes increased gradually until day 12 (74 ± 13 cells mL^{-1}) and then decreased on day 14 owing to transformation into cysts. The period from planozygote formation to transformation into cysts depended on the individual planozygote and varied from 2 to approximately 10 days. Cyst formation was observed on day 4 and encystment continued exponentially until day 16, with density slightly increasing until the end of the incubation (150 ± 30 cysts mL^{-1}; see Fig. 12).

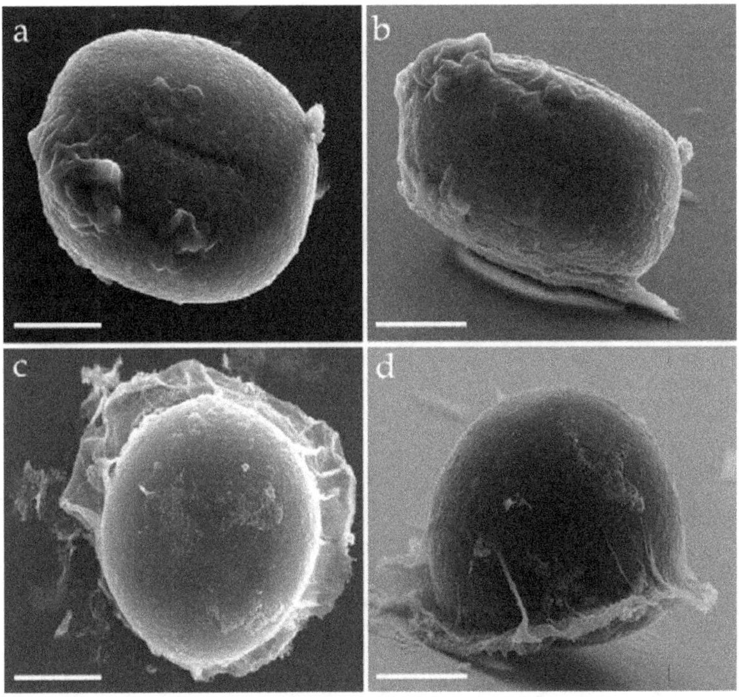

Fig. 14. Scanning electron microscopic observations of resting cysts in *Alexandrium tamiyavanichii* (Nagai et al., 2003). Scale bars are 20 μm (all at the same magnification). (a) Vertically compressed cyst without any paratabulation on the surface (view from above). (b) Side view of the cyst shown in (a). (c) Spherical cyst covered with transparent mucilaginous material (viewed from above). (d) Side view of the cyst shown in (c).

2.3.4 Effect of temperature on vegetative growth and sexual reproduction in *A. tamiyavanichii* and *A. tamarense* under laboratory conditions

Vegetative growth in *A. tamiyavanichii* has been observed at temperatures ranging from 20°C to 32.5°C, and the culture strains survived briefly at 17.5°C and 35°C but died quickly at 15°C (Fig. 15a). Optimum growth was seen around 27.5–30°C, and the growth rates were 0.71–0.76 divisions day^{-1} (Fig. 15b). The maximum yield of vegetative cells ranged from 300 ± 41 cells mL^{-1} at 20°C to 3,513 ± 452 cells mL^{-1} at 30°C, and the yields were highest at 27.5–30°C (see Fig. 15a). The effect of temperature on growth rate tended to be very similar to that for the maximum yield of vegetative cells. Encystment was observed at temperatures between 20°Cand 32.5°C and cyst yields ranged from 7.0 ± 2.6 cysts mL^{-1} at 20°C to 148 ± 28 cysts mL^{-1} at 25°C (Fig. 15c). The cyst yield was highest at 25°C and significantly higher than that at temperatures between 20°C and 32.5°C (*t*-test, $P < 0.005$–0.001). Cyst formation (cyst yield/maximum yield of vegetative cells × 100 [%]) ranged from 0.1 ± 0.1% at 32.5°C to 6.2 ± 1.1% at 25°C (data not shown).

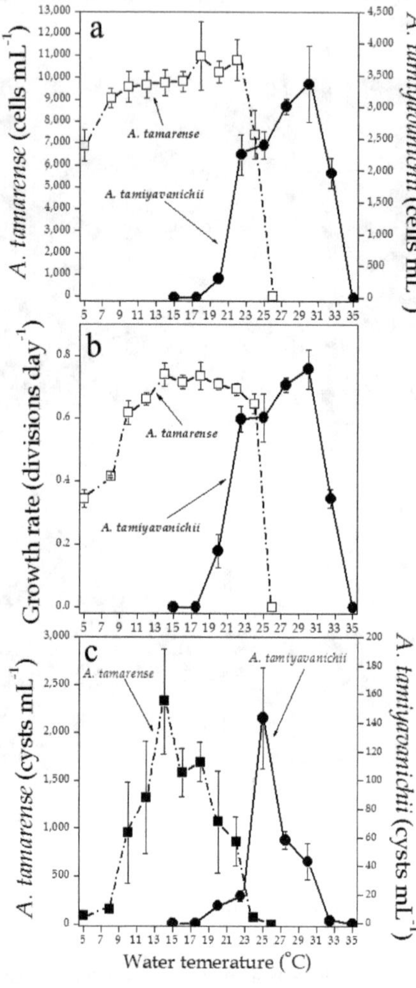

Fig. 15. Effect of temperature on vegetative growth and resting cyst formation in *Alexandrium tamiyavanichii* and *A. tamarense* under laboratory conditions. (a) Maximum yields of vegetative cells at each temperature. (b) Growth rates of vegetative cells at each temperature. (c) Maximum yields of resting cysts at each temperature. Effect of temperature on vegetative growth was investigated at nine temperatures between 15°C and 35°C in *A. tamiyavanichii*, and 11 temperatures between 5°C and 26°C in *A. tamarense* under the conditions described in Fig. 12 (except temperature). Two compatible strains of AT0104H15 and AT0104H26 were used to study *A. tamarense*. After pre-incubation, both strains were mixed at each temperature setting and incubated for 20 days (*A. tamiyavanichii*) and 30 days (*A. tamarense*). Five wells of each were used for counting motile cells and cysts, and the maximum yields of vegetative cells and cysts. The number of vegetative cells counted using the microscope, and the growth rates (divisions day^{-1}) in the exponential growth phase were calculated using the method of Guillard (1973).

In Japanese coastal waters, the toxic dinoflagellate *A. tamarense* was reported mostly along the northern Pacific coast, such as in the Hokkaido and Tohoku regions, until the 1980s (Uchida et al., 1980; Fukuyo, 1982; 1985; Sekiguchi et al., 1986). Since the 1990s, however, *A. tamarense* has become a conspicuous species in southwestern Japan (Yamamoto & Yamasaki, 1996; Kotani et al., 1998), especially in Hiroshima Bay (Asakawa et al., 1995; Yamaguchi et al., 1995; Itakura et al., 2002; Yamaguchi et al., 2002). In laboratory cultures of *A. tamarense* isolated from Hiroshima Bay, vegetative growth was observed at temperatures ranging from 5°C to 24°C, and the maximum yields of motile cells ranged from 6,880 ± 716 cells ml^{-1} at 5°C to 10,980 ± 1,553 cells ml^{-1} at 18°C. No growth was seen at 26°C (see Fig. 15a, b). The growth rate ranged from 0.34 ± 0.03 divisions day^{-1} at 5°C to 0.74 ± 0.04 divisions day^{-1} at 14°C and 18°C (see Fig. 15b), and the rates at 5°C and 8°C were remarkably lower than those at the other temperatures. The range of growth rates between 10°C and 24°C was 0.62–0.74 divisions day^{-1}. Encystment was observed at temperatures ranging from 5°C to 24°C using compatible strains with different mating types, and cyst yields ranged from 68 ± 34 cysts ml^{-1} at 24°C to 2,326 ± 551 cysts ml^{-1} at 14°C (see Fig. 15c). The cyst yields at 5°C, 8°C, and 24°C were remarkably lower than those at the other temperatures. The cyst yield at 14°C was significantly higher than that at temperatures between 10°C and 22°C (*t*-test, $P < 0.05$–0.001). Cyst formation (cyst yield/maximum yield of vegetative cells × 100 [%]) ranged from 0.9 ± 0.5% at 24°C to 23.8 ± 5.1% at 14°C. The effect of temperature on cyst formation tended to be similar to that for cyst yield. Cyst formation at 14°C was also significantly higher than that at temperatures between 10°C and 22°C (*t*-test, $P < 0.05$–0.001; data not shown, but see Nagai et al., 2004). Therefore, the optimal temperature for encystment of *A. tamarense* strains from Hiroshima Bay is approximately 14°C.

2.3.5 Effect of temperature on germination of resting cysts in *A. tamiyavanichii* under laboratory conditions

Experiments have been conducted to examine the germination physiology of resting cysts obtained from laboratory cultures of *A. tamiyavanichii* under laboratory conditions after storage at nine temperatures ranging from 5°C to 30°C for several months in darkness. The experiments were conducted at 25°C or at each temperature at which the resting cysts had been stored. In the experiment at 25°C, the first excystment (germination) was confirmed on day 7 because the cysts had been stored in darkness at 17.5°C to 30°C (3.3–10.8% germination success—i.e. germinated cysts/total cysts × 100 [%]), and germination was confirmed at all temperatures except 5°C on day 15 (4.2–91.2%). Germination success was higher than 60% at temperatures between 10°C and 17.5°C (Fig. 16a). After day 30, germination success at 10°C continued to be approximately 50%, but >80% at 15°C to 27.5°C. Germination success at 30°C increased after day 51, and high germination success (>80%) was observed after that. No germination was confirmed at 5°C throughout the experiment, however. At 7.5°C, germination was seen only on day 15 (4.2%) and day 30 (9.0%). In the germination experiment at each temperature at which cysts were stored in darkness, successful germination was observed at ≥17.5°C, but not at <15°C (Fig. 16b). High germination success (79.4–96.6%) was observed at 20°C, 27.5°C, and 30°C after day 64. At 17.5°C, germination was confirmed after day 79, but no vegetative growth was seen after germination. Accordingly, the optimum temperature for storage in darkness and germination in resting cysts in *A. tamiyavanichii* is around 25°C, and this species has no dormancy period, or if it does, it would be within 1 week. A germination experiment was carried out at 25°C to examine the germination potential of resting cysts after 3 months of storage in darkness at each temperature, and germination was observed on day 2, although germination success was low (0.8–8.8%). After that, active germination was confirmed in the resting cysts stored at 15–25°C, with germination success >70% on day 6 (Fig. 16c).

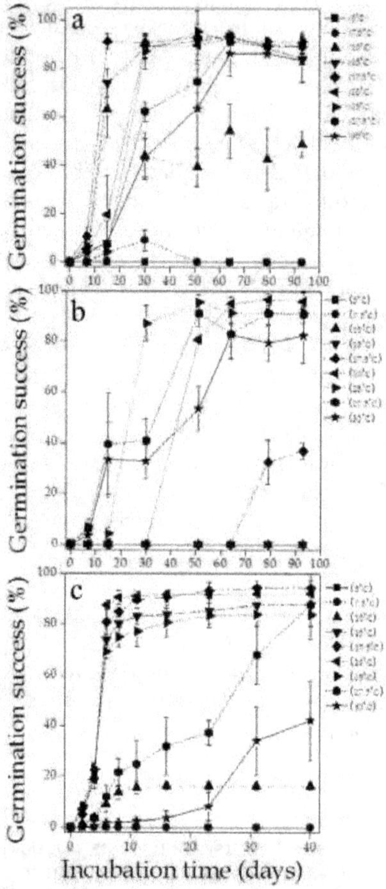

Fig. 16. Effect of temperature on germination in resting cysts of *Alexandrium tamiyavanichii* under laboratory conditions. Cysts were produced under appropriate incubation conditions at 25°C by mixing two compatible strains of AT0112T06 and AT0112F06 as described in Fig. 12. Resting cysts were harvested 3 weeks after the start of the encystment experiment. Cysts were stored in darkness at nine temperatures between 5°C and 30°C to investigate the period of mandatory dormancy of cysts. Some of the cysts (14–65 cysts) were picked up on days 7, 15, 30, 51, 64, 79, and 93 days and incubated at 25°C (a) and at the same temperature at which they had been stored (b). One month after incubation under the light conditions described in Fig. 12, the number of germinated cysts, dead cysts, and total cysts were measured under an inverted microscope. The percentage of germination success was calculated (germinated cysts/total cysts × 100 [%]).Resting cysts stored at the nine temperatures for 90 days were incubated under appropriate light conditions at 25°C (c). Germination success was calculated every 2–9 days for 40 days.

2.3.6 Difference in life cycle strategies of *A. tamiyavanichii* and *A. tamarense*

A. tamiyavanichii grows at temperatures ranging from 20°C to 32.5°C and the optimum temperatures for growth are 27.5–30°C (see Fig. 15). *A. tamiyavanichii* cannot survive below 15°C, however, demonstrating the growth physiology of a typical tropical microalgal species (Ogata et al., 1990; Nagai et al., 2008). Actively swimming vegetative chains of this species have been confirmed around 15°C in the Seto Inland Sea at the end of blooms in the middle of December (Yoshimatsu et al., 2000; Tanada et al. 2003; Nagai et al. 2008), suggesting a higher tolerance for low water temperatures in natural populations than in laboratory cultures. Anderson & Wall (1978), Anderson & Morel (1979), and Itakura & Yamaguchi (2001) have emphasized the importance of *Alexandrium* cysts as the seed population for overwintering, bloom initiation, and cyst dispersal within estuaries and near-coastal waters—i.e., the geographical spread of the population. Hallegraeff (1998) has also pointed out the role of dinoflagellate cysts in settling and spreading over a wider distribution through transportation either in ship's ballast water or associated with the movement of shellfish stocks from one area to another. *A. tamiyavanichii* produces resting cysts at temperatures of 20–32.5°C in laboratory cultures, and the optimal temperature for encystment is 25°C (see Fig. 15). Interestingly, this temperature is slightly lower than the optimal temperature for vegetative growth (see Fig. 15). Conversely, resting cyst production in *A. tamarense* is observed at temperatures of 5–24°C in laboratory cultures (see Fig. 15). The optimal temperature for encystment is around 14°C, and cyst formation is ca. 25%. Accordingly, these data clearly indicate that *A. tamiyavanichii* produces resting cysts at remarkably higher temperatures and does not produce resting cysts at high density compared with *A. tamarense*. In the coastal areas around Hokkaido in Japan, *A. tamarense* usually appears in early spring when the water temperature is 1–3°C (start of blooming) but appears when waters temperatures reach 8–10°C in western Japan. *A. tamarense* has a vegetative stage that usually lasts for <4 months in the water column (Itakura et al. 2002; Nagai et al. 2007), although the duration of the vegetative stage varies with the water temperature as well as with the pattern of annual water temperature fluctuation. This species disappears in the water columns when the water temperature exceeds ca. 15–16°C, clearly indicating an adaptation to cold waters (Itakura et al., 2002; Nagai et al., 2007; Shimada et al., 1996, 2010).

It is assumed that in the annual life cycle of *A. tamarense*, to appear for a limited time as vegetative populations in the water column in a year and to endure the most unsuitable season (high water temperatures), the organism has developed a survival strategy that produces high amounts of resting cysts at the end of blooms (Itakura et al., 2002; Nagai et al., 2007). Conversely, *A. tamiyavanichii* is a typical tropical species that blooms in stable environmental conditions—i.e., little fluctuation in annual water temperature. Therefore, no preparation is necessary to survive adverse environmental conditions (e.g., unsuitably cold winter season) through the production of high amounts of resting cysts as seen in *A. tamarense*. Regarding germination, *A. tamiyavanichii* also cannot germinate at temperatures below 20°C, and although occasional germination was observed at this temperature, the organism did not grow. *A. tamiyavanichii* loses the ability to germinate when kept at the temperatures below 10°C in darkness even for a week or so, even though the resting cysts look alive. Conversely, resting cysts of *A. tamiyavanichii* have high potential for germination, even when kept in darkness for more than 3 months at relatively high temperatures of 20–30°C (see Fig. 16). Thus, we can conclude from the results of these laboratory experiments that *A. tamiyavanichii* is the species with the life cycle strategy that adapts to tropical environmental conditions, thereby resulting in limited distribution, until now, in areas south of a northern latitude of 35 degrees.

2.3.7 Environmental conditions that allow the appearance of *A. tamiyavanichii* in the Seto Inland Sea

In the Seto Inland Sea, the first appearance of *A. tamiyavanichii* has been observed in middle and late August when water temperature rises to around 28°C, and the species has grown up to 30 cells mL^{-1} by late autumn (Yoshimatsu et al., 2000). Tropical species cannot survive at water temperatures lower than 15°C, and they die when temperatures drop below 15°C (Ogata et al., 1990; Nagai et al., 2008). Vegetative populations of this species are not found in middle December, therefore (Fig. 17). We speculate that this species produces durable resting cysts when growing in water columns as vegetative populations, perhaps in November-December when its abundance reaches its maximum, and resting cysts overwinters in surface sediments as seed populations between December and early June, during which it cannot survive as vegetative populations (<15°C). Accordingly, this life cycle characteristic enables this species to appear for 6–7 months a year in the water column. *A. tamiyavanichii* begins to germinate next in the middle of June, when the bottom water temperature exceeds 20°C, and swims out as a vegetative population in water columns. These characteristics strongly suggest that cysts play an important role in prolonged overwintering in surface sediments as a seed population in temperate Japanese coastal waters. Also, the fact that planozygotes have been observed in sea water samples collected from the Seto Inland Sea during *A. tamiyavanichii* blooms in 2001 and that appearances of this species have since been observed often not only in the Seto Inland Sea but also around Tsushima and Gotou Islands imply that they may have gained a niche and become settled in this coastal area. Moreover, *A. tamiyavanichii* might be transferred to other areas of Japan by natural and human-assisted dispersals, expanding its distribution.

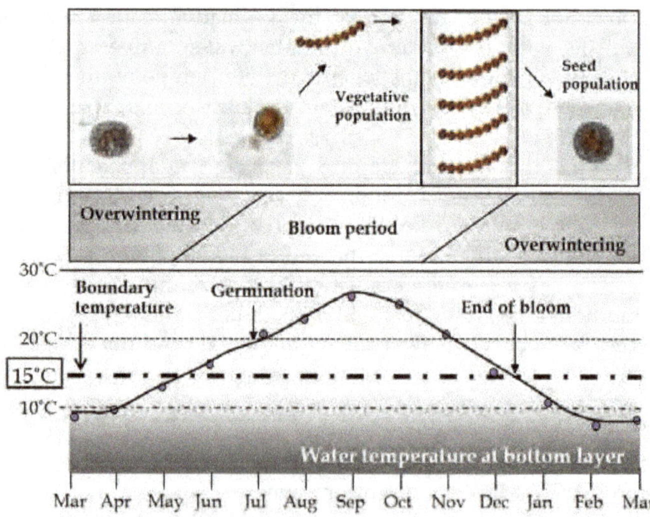

Fig. 17. A conceptual diagram showing the relationship between bottom water temperature and bloom mechanism in *Alexandrium tamiyavanichii* in the Seto Inland Sea. The solid line shows the change in water temperature in Hiroshima Bay (see Fig. 10).

But why does *A. tamiyavanichii* bloom during September–December and not during August–September in western Japan? We have investigated cyst abundance in the surface sediments collected from several locations in the Tokushima Prefecture region of the eastern Seto Inland Sea just after the dense bloom and carried out a germination experiment using 10 g of wet sediments with seawater under appropriate incubation conditions for germination and vegetative growth. No germinated cells were detected, however. We speculate, therefore, that *A. tamiyavanichii* is undetectable in July or August if the resting cysts germinate and begin blooming when the water temperature exceeds 20°C in mid-June (see Fig. 17), because cyst abundance is too low to grow to detectable levels in July-August. Occurrences of bloom in this species, resulting in PSP outbreaks in shellfish, are recognized as a new biological threat in western Japan. We have begun investigating the origin of the Japanese population of this species using highly polymorphic genetic markers (Nishitani et al., 2009).

3. Conclusion

The increase in atmospheric carbon dioxide concentrations caused by human activities has influenced the change of world's ocean environments such as increasing water temperature, decreasing sea ice, and causing acidification (Tadokoro et al., 2008). It has been suggested that enhanced stratification, caused by the increases in the surface layer temperature, is decreasing the nutrient supply from the subsurface layer hindering the primary productivity over a large area of the North Pacific Ocean, resulting in the changes of marine ecosystems in primary productivities and seasonal succession in phytoplankton (Ono et al., 2002), zooplanktons (Aoyama et al., 2008), fish (Tian et al., 2006; Shida et al., 2007; Masuda, 2008) and coral (Nojima & Okamoto, 2008). In this chapter, we demonstrated the influence of water temperatures on the changes of habitat of microalgal species or the Isoyake phenomena and the introduction of a toxic phytoplankton from tropical regions to Japanese coastal waters. Thus, effects of global warming on marine ecology have been studied with high research activities. In future, it is highly possible that the influence of global warming on marine ecosystems will be more prominent, therefore, continuous efforts for environmental and biological monitoring would be crucial.

4. Acknowledgement

We are grateful to Dr. T. Kamiyama, National Research Institute of Fisheries and Environment of Inland Sea, Fisheries Research Agency of Japan, for his useful suggestions and encouragement during this study. This work was supported in part by a grant from the Fisheries Research Agency of Japan.

5. References

Anderson, D.M. & Morel, F.M.M. (1979). The seeding of two red tide blooms by the germination of benthic *Gonyaulax tamarensis* hypnocysts. *Estuarine and Coastal Marine Science*, Vol.8, No.3, pp. 279-293, ISSN 0272-7714

Anderson, D.M. & Wall, D. (1978). Potential importance of benthic cysts of *Gonyaulax tamarensis* and *G. excavata* in initiating toxic dinoflagellate blooms. *Journal of Phycology*, Vol.14, No.6, pp. 224-234, ISSN 0022-3646.

Aoyama, M., Goto, H., Kamiya, H., Kaneko, I., Kwae, S., Kodama, H., Konishi, Y., Kusumoto, K., Miura, H., Moriyama, E., Murakami, K., Nakano, T., Nozaki, F., Sasano, D., Shimizu, T., Suzuki, H., Takathuki, Y. & Toriyama, A. (2008). Marine biochemical response to a rapid warming in the main stream of the Kuroshio in the western North Pacific. *Fisheries Oceanography,* Vol.17, No. 3, pp.206-218. ISSN 10546006

Aratake, H., Shimizu, H., Watanabe, K. & Yoshida, G. (2007). Long-term change in *Sargassum*-bed distribution along the coast of Kushima-city, southern part of Miyazaki Prefecture, Japan. *Bulletin of the Miyazaki Prefectural Fisheries Experimental Station,* Vol. 11, (March 2007), pp. 1-13.

Asakawa, M., Miyazawa, K. & Noguchi, T. (1995). Studies on paralytic shellfish poison (PSP) toxification of bivalves, in association with appearance of *Alexandrium tamarense,* in Hiroshima Bay, Hiroshima Prefecture. *Journal of Food Hygiene Society of Japan,* Vol.34, Vol.1, pp. 50-54, ISSN 00156426.

Balech, E. (1994). Three new species of the genus *Alexandrium* (Dinoflagellate). *Transactions of the American Microscopical Society,* Vol.113, pp. 216-220, ISSN 0003-0023.

Balech, E. (1995). *The Genus* Alexandrium *Halim (Dinoflagellate).* Sherkin Island Marine Station, Sherkin Island, County Cork, Ireland.

Biodiversity Center of Japan (2011). In: *Monitoring Sites 1000.* Available from:<www.biodic.go.Jp/moni1000/index.html>.

Chiba, S., Aita, M.N., Tadokoro, K. Saino, T., Sugisaki, H. & Nakata, K. (2008). From climate regime shifts to lower-trophic level phenology: Synthesis of recent progress in retrospective studies of the Western North Pacific. *Progress in Oceanography,* Vol.77, No.2, (May 2008), pp. 112-126, ISSN 0079-6611.

Edwards, M. & Richardson, A.J. (2004). Impact of climate change on marine pelagic phenology and trophic mismatch. *Nature,* Vol.430, No.7002, (August 2004), pp. 881-884, ISSN 0028-0836.

Environmental Agency & Marine Parks Center of Japan (1994). *The report of the marine biotic environment survey in the 4th national survey on the natural environment. Vol.2. Algal and sea-grass beds,* Tokyo, March 1994.

Fujita, D., Murase, N. & Kuwahara, H. (Eds.) (2010). *Monitoring and Maintenance of Seaweed Beds.* ISBN 978-4-425-88501-5, Seizando-Shoten, Tokyo.

Fukuyo, Y. 1982. Taxonomical and ecological studies on *Protogonyaulax* occurring in Japanese coastal waters. Ph.D. thesis. University of Tokyo (in Japanese).

Fukuyo, Y. 1985. Morphology of *Protogonyaulax tamarensis* (Lebour) Taylor and *Protogonyaulax catenella* (Whedon and Kofoid) Taylor from Japanese coastal waters. *Bulletin of Marine Science,* Vol.37, Vol.2, pp. 529-537, ISSN 0007-4977.

Fukuyo, Y., Pholpunthin, P. & Yoshida, K. (1988). *Protogonyaulax* (Dinophyceae) in the Gulf of Thailand. *Bulletin of Plankton Society of Japan,* Vol.35, Vol.1, pp. 9-20, ISSN 0387-8961.

Fukuyo, Y., Yoshida, K., Ogata, T., Ishimaru, T., Kodama, M., Pholpunthin, P., Wisessang, S., Phanichyakarn, V. & Piyakarnchana, T. (1989). In: *Red Tides: Biology, Environmental Science and Toxicology,* Okaichi, T., Anderson D.M. & Nemoto, T (Ed.), pp. 403-406, Elsevier, New York.

Guillard, R.R.L. (1973). Division rates. In: *Handbook of Phycological Methods: Culture Methods and Growth Measurements,* Stein J.R. (Ed.), pp. 289-311, ISBN 0-521-20049-0, Cambridge University Press, Cambridge.

Hallegraeff, G.M. (1995). Harmful algal blooms: A global overview. In: *Manual on Harmful Marine Microalgae,* Hallegraeff, G.M. Anderson, D.M. & Cembella, A.D. (Eds.), pp. 1-22, IOC Manuals and Guides No. 33, UNESCO, Paris.

Hallegraeff, G.M. (1998). Transport of toxic dinoflagellates via ships' ballast water: bioeconomic risk assessment and efficacy of possible ballast water management strategies. *Marine Ecology Progress Series,* Vol.168, pp. 297-309, ISSN 0171-8630.

Haraguchi, H., Murase, N., Misukami, Y., Noda, M., Yoshida, G. & Terawaki, T. (2005). The optimal and maximum critical temperatures of nine species of Sargassaceae in the coastal water of Yamaguchi Prefecture, Japan. *Japanese Journal of Phycology,* Vol. 53, No.1, (March 2005), pp. 7-13, ISSN 0038-1578.

Hashimoto, T., Matsuoka, S., Yoshimatsu, S., Miki, K., Nishibori, N., Nishio, S. & Noguchi, T. (2002). First paralytic shellfish poison (PSP) infestation of bivalves due to toxic dinoflagellate *Alexandrium tamiyavanichii,* in the southeast coasts of the Seto Inland Sea, Japan. *Journal of Food Hygiene Society of Japan,* Vol.43, No.1, pp. 1-5, ISSN 0022-5304.

Hiraoka, M., Ura, Y. & Haraguchi, H. (2005). Relationship between seaweed beds and seawater temperature in the Tosa Bay. *Aquabiology,* Vol. 27, No. 5, (October 2005), pp. 485-493, ISSN 0285-4376

IPCC (2007). *Climate Change 2007: Physical Sciences Basis.* Cambridge University Press, New York, 996 pp.

Itakura, S. & Yamaguchi, M. (2001). Germination characteristics of naturally occurring cysts of *Alexandrium tamarense* (Dinophyceae) in Hiroshima Bay, Inland Sea of Japan. *Phycologia,* Vol.40, No.3, pp. 263-267, ISSN 0031-8884.

Itakura, S. Yamaguchi, M. Yoshida, M. Fukuyo, Y. (2002). The seasonal occurrence of *Alexandrium tamarense* (Dinophyceae) vegetative cells in Hiroshima Bay, Japan. *Fisheries Science,* Vol.68, Vol.1, (January 2002), pp. 77-86 ISSN 0919-9268.

Japan Meteorological Agency (2008). *Global Warming Projection,* Vol.7, (March 2008), 59 pp.

Japan Meteorological Agency (2011). Monthly mean surface water temperature in *Kyushu* and *Okinawa.* Available from: <www.data.kishou.go.jp/db/nagasaki/monthly/sst_n/html>

Kawabe, M. (2005). Variations of the Kuroshio in the Southern Region of Japan: Conditions for large meander of the Kuroshio. *Journal of Oceanography,* Vol.61, No.6, (December 2005), pp. 529-537, ISSN 0916-8370.

Kotani, Y., Koyama, A., Yamaguchi, M. & Imai, I. (1998). Distribution of resting cysts of the toxic dinoflagellate *Alexandrium catenella* and/or *A. tamarense* in the coastal areas of western Shikoku and Kyushu, Japan. *Bulletin of the Japanese Society of Fisheries Oceanography,* Vol.62, No.2, (May 1998), pp. 104-111 ISSN 09161562 (in Japanese with English abstract).

Kodama, M., Ogata, T. Fukuyo, Y., Ishimaru, T., Wisessang, S., Saitanu, K., Phanichyakarn, V. & Piyakarnchana, T. (1988). *Protogonyaulax cohorticula,* a toxic dinoflagellate found in the Gulf of Thailand. *Toxicon,* Vol.26, No.8, pp. 707-712, ISSN 0041-0101.

Koja, A., Tamanaha, K., Abe, Y., Oshiro, N. & Teruya, N. (2001). Studies on paralytic shellfish poisons in Okinawa prefecture II. Annual report of Okinawa Prefectural

Institute of Health and Environment, Vol.35, (April 2001), pp. 59-61, ISSN 1341-0636 (in Japanese).

Lapointe, B.R., Littler, M.M. & Littler, D. S. (1997). Macroalgal overgrowth of fringing coral reefs at Discovery Bay, Jamaica: Bottom-up versus top-down control. *Proceedings of the 8th International Coral Reef Symposium 1*, pp. 927-932, (June 1996), ISBN 0935868909, Panama.

Matsuda, R. (2008). Seasonal and interanual variation of suctidal fish assemblages in Wakasa Bay with referenced to the warming trend in the Sea of Japan. *Environmental Biology of Fishes*, Vol.82, pp. 387-399.

Montresor, M. (1995). The life history of *Alexandrium pseudogonyaulax* (Gonyaulacales, Dinophyceae). *Phycologia*, Vol.34, No.6, pp. 444-448, ISSN 0031-8884.

Nagai, S., Itakura, S., Matsuyama, Y. & Kotani, Y (2003). Encystment under laboratory conditions of the toxic dinoflagellate *Alexandrium tamiyavanichii* (Dinophyceae) isolated from the Seto Inland Sea, Japan. *Phycologia*, Vol.42, No.6, pp. 646-653, ISSN 0031-8884.

Nagai, S., Matsuyama, Y., Oh, S.J. & Itakura, S. (2004). Effect of nutrients and temperature on encystment of the toxic dinoflagellate *Alexandrium tamarense* (Dinophyceae) isolated from Hiroshima Bay, Japan. *Plankton Biology & Ecology*, Vol.51, No.2, pp. 103-109, ISSN 1343-0874.

Nagai, S., Lian, C.L., Suzuki, M., Hamaguchi, M., Matsuyama, Y., Itakura, S., Shimada, H., Kaga, S., Yamauchi, H., Sonda, Y., Nishikawa, T., Kim, C.H. & Hogetsu, T. (2007). Microsatellite markers reveal population genetic structure of the toxic dinoflagellate *Alexandrium tamarense* (Dinophyceae) in Japanese coastal waters, *Journal of Phycology*, Vol.43, No.1, pp. 43-54, ISSN 0022-3646.

Nagai, S., Kotani, Y. & Itakura, S. (2008). New appearance of toxic dinoflagellates and toxification of shellfish in the Seto Inland Sea, Japan. *Nippon Suisan Gakkaishi*, Vol.74, pp. 880-883, ISSN 0021-5392 (in Japanese).

Nishitani, G., Nagai, S., & Kotani, Y. (2009). Development of compound microsatellite markers in the marine phytoplankton *Alexandrium tamiyavanichii* (Dinophyceae). *Conservation Genetics*, Vol.10, No.5, pp. 1561-1563, ISSN 1566-0621.

Nojima, A. & Okamoto, M. (2008). Enlargement of habitats of scleractinian corals to north and coral bleaching events. *Nippon Suisan Gakkaishi*, Vol.74, No.5, (May 2008), pp. 884-888, ISSN 00215392 (in Japanese with English abstract)

Ogata, T., Pholpunthin, P., Fukuyo, Y. & Kodama, M (1990). Occurrence of *Alexandrium cohorticula* in Japanese coastal water. *Journal of Applied Phycology*, Vol.2, No.4, pp.351-356, ISSN 0921-8971.

Oh, S.J., Matsuyama, Y., Nagai, S., Itakura, S., Yoon, Y.H. & Yang, H.S. (2009). Comparative study on the PSP component and toxicity produced by *Alexandrium tamiyavanichii* (Dinophyceae) strains occurring in Japanese coastal water. *Harmful Algae*, Vol.8, No.2, pp. 362-368, ISSN 1568-9883.

Ono, T., Tadokoro, K., Midorikawa, T., Nishioka, J. & Saino, T. (2002). Multi-decadal decrease of net community production in Western Subarctic North Pacific. *Geophysical Research Letters*, Vo.29, No.8, (April 2002), pp. 1186, DOI 10.1029/2001GL014332, ISSN 0094-8276.

Phanichyakarn, V., Wisessang, S., Piyakarnchana, T., Fukuyo, Y., Ishimaru, T., Kodama, M. & Ogata, T. (1993). Ultrastructure of *Alexandrium cohorticula* found in the Gulf of

Thailand. In: *Toxic Phytoplankton Blooms in the Sea*, Smayda, T.J. & Shimizu, Y. (Eds.), pp. 165-168, Elsevier Scientific Publications, Amsterdam.

Sato, K., Tenshin, M. & Miyata, T. (2001). Environmental monitoring-survey of fishing ground in Uchinoumi. Research project report of Tokushima Agriculture, Forestry and Fisheries Technology Support Center, Fisheries Research Institute in Fiscal 1999, (March 2001), pp. 85-87.

Sato, Y., Yukimoto, S., Tsujino, H., Ishizaki, H. & Noda, A. (2006). Response of North Pacific Ocean circulation in a Kuroshio-resolving ocean model to an Arctic oscillation (AO)-like change in Northern Hemisphere Atmospheric circulation due to greenhouse-gas forcing. *Journal of the Meteorological Society of Japan*, Vol.84, No.2, (May 2006), pp.295-309, ISSN 0026-1165.

Sekiguchi, K., Watanabe, S., Shimizu, M. & Saito, S. (1986). Occurrence of *Protogonyaulax tamarensis* on the coast of Iwate Prefecture, in relation to toxification of the scallop *Patinopecten yessonensis*. *Bulletin of Tohoku Regional Fisheries Research Laboratory*, Vol.48, (January 1986), pp. 115-123, ISSN 0049-402X.

Setou, S., Akiyama, H. & Saito, T. (2004). Interannual variability of the upper ocean temperature in Tosa Bay. *Umi To Sora*, Vol.79, No.4, (March 2004), pp. 97-106, ISSN 0503-1567 (in Japanese with English abstract).

Serisawa, Y., Imoto, Z., Ishikawa, T. & Ohno, M. (2004). Decline of the *Ecklonia cava* population associated with increased seawater temperatures in Tosa Bay, southern Japan. *Fisheries Science*, Vol. 70, No.1, (February 2004), pp. 189-191, ISSN 0919-9268.

Shida, O., Hamatsu, T., Nishimura, A., Suzakiic, A., Yamamoto, J. Miyashita, K. & Sakurai, Y. (2007). Interannual fluctuations in recruitment of walleye Pollock in Oyashio region related to environmental changes. *Deep-Sea Research II*, Vol.54, 2822-2831, ISSN 0967-0645

Shimabukuro, H., Terada, R., Sotobayashi, J., Nishihara, G.N. & Noro, T. (2007). Phenology of *Sargassum duplicatum* (Fucales, Phaeophyta) from the southern coast of Satsuma Peninsula, Kagoshima, Japan. *Nippon Suisan Gakkaishi*, Vol.73, No. 3, (May 2007), pp. 454-460, ISSN 0021-5392.

Shimada, H., Hayashi, T. & Mizushima, T. (1996). Spatial distribution of *Alexandrium tamarense* in Funka Bay, Southwestern Hokkaido, Japan, In: *Harmful and Toxic Algal Blooms*, Yasumoto, Y., Oshima, Y. & Fukuyo, Y. (Eds.), pp. 219-221, International Oceanographic Commission of UNESCO, Paris.

Shimada, H., Sawada, M., Kuribayashi, T., Nakata, A., Miyazono, A. & Asami, H. (2010). Spatial distribution of the toxic dinoflagellate *Alexandrium tamarense* in summer in the Okhotsk Sea off Hokkaido, Japan. *Plankton Benthos Research*, Vol.5, No.1, pp. 1-10, ISSN 1880-8247.

Tadokoro, K., Sugimoto, T. & Kishi, M.J. (2008). The effect of anthropogenic global warming on the marine ecosystem. *Oceanography in Japan*, Vol.17, No.6, pp. 404-420, ISSN 09168362 (in Japanese with English abstract)

Takahashi, S. & Seiki, S. (2004). Long-term change of water temperature in the Seto Inland Sea. *Umi To Sora*, Vol.80, No.2, (September 2004), pp.69-74, ISSN 0503-1567 (in Japanese with English abstract).

Takeoka, H. (2002). Progress in Seto Inland Sea Research. *Journal of Oceanography*, Vol.58, No.2, (April 2002), pp. 93-107, ISSN 0916-8370.

Tanada, N., Tenshin, M., Yamazoe, Y. & Miyata, T. (2003). Monitoring survey of toxic phytoplankton. Research project report of Tokushima Agriculture, Forestry and Fisheries Technology Support Center, *Fisheries Research Institute in Fiscal 2001*, (March 2003), pp. 98-99 (in Japanese).

Tarutani, K. (2007). Long-term variations in water environments in the Seto Inland Sea of Japan during 1973 to 2002 based on data from the Fisheries Monitoring Program. *Japanese Journal of Benthology*, Vol.62, No.1, pp. 52-56, ISSN 1345-112X (in Japanese with English abstract).

Tarutani, K., Nakajima, M., Harada, K., Kudoh, T. & Wanishi, A. (2009). Possible ecological deterioration due to the changes in anthropogenic nutrient loading in the Seto Inland Sea, Japan. *Kaiyo Monthly*, Vol.41, No.9, (September 2009), pp. 508-513, ISSN 0916-2011 (in Japanese).

Tian, Y., Kidokoro, H. & Watanabe, T. (2006). Long-term changes in the fish community structure from the Tsushima warm current region of the Japan/East Sea with an emphasis on the impacts of fishing and climate regime shift over the last four decades. Progress in Oceanography, Vol.68, (March 2006), pp. 217-237. ISSN 0079-6611

Turpin, D.H., Dobell, P.E.R. & Taylor, F.J.R. (1978). Sexuality and cyst formation in pacific strains of the toxic dinoflagellate *Gonyaulax tamarensis*. *Journal of Phycology*, Vol.14, No.2, pp. 235-238, ISSN 0022-3646.

Uchida, T., Kawamata, K. & Nishihama, Y. (1980). Vertical distribution of paralytic toxin-producing species, *Protogonyaulax* sp. in Funka Bay, Hokkaido. *Japanese Journal of Phycology*, Vol.25, pp. 133-139, ISSN 0038-1578 (in Japanese with English abstract).

Usup, G., Pin, L.C., Ahmad, A. & Teen, L.P. (2002). Phylogenetic relationship of *Alexandrium tamiyavanichii* (Dinophyceae) to other *Alexandrium* species based on ribosomal RNA gene sequences. *Harmful Algae*, Vo.1, No.1, (March 2002), pp. 59-68. ISSN 1568-9883.

Wanishi, A. (2004). Variations of water temperature during the recent 30 years in the Suo-Nada region off Yamaguchi Prefecture in the western Seto Inland Sea. *Bulletin of Yamaguchi Prefectural Fisheries Research Center*, Vol.2, (March 2004), pp. 1-6 ISSN 1347-2003.

Watanabe, Y., Ishida, H., Nakano, T. & Nagai, N. (2005). Spatiotemporal decreases of nutrients and chlorophyll-a in the surface mixed layer of the Western North Pacific from 1971 to 2000. *Journal of Oceanography*, Vol.61, No.6 (December 2005), pp. 1011-1016, ISSN 0916-8370.

Wisessang, S., Ogata, T., Kodama, M., Fukuyo, Y., Ishimaru, T., Saitanu, K., Yongvanich, T. & Piyakarnchana, T. (1991). Accumulation of paralytic shellfish toxins by green mussel *Perna viridis* by feeding on cultured cells of *Alexandrium cohorticula* isolated from the Gulf of Thailand. *Nippon Suisan Gakkaishi*, Vol.57, No.1, pp. 127-131, ISSN 0021-5392.

Wisessang, S., Phanichyakarn, V., Piyakarnchana, T., Fukuyo, Y., Ishimaru, T., Kodama, M. & Ogata, T. (1993). Scanning electron microscope study of *Alexandrium cohorticula*, a toxic dinoflagellate from the Gulf of Thailand. In: *Toxic Phytoplankton Blooms in the Sea*, Smayda T.J. & Shimizu, Y. (Eds.), pp. 191-195. Elsevier Scientific Publications, Amsterdam.

Yamaguchi, M., Itakura, S. & Imai, I. (1995). Vertical and horizontal distribution and abundance of resting cysts of the toxic dinoflagellate *Alexandrium tamarense* and

Alexandrium catenella in sediments of Hiroshima Bay, the Seto Inland, Sea, Japan. *Nippon Suisan Gakkaishi,* Vol.61, No.5, pp. 700-706, ISSN 0021-5392.

Yamaguchi, M., Itakura, S., Nagasaki, K. & Kotani, Y. (2002). Distribution and abundance of resting cysts of the toxic *Alexandrium* spp. (Dinophyceae) in sediments of the western Seto Inland Sea, Japan. *Fisheries Science,* Vol.68, No.5, pp. 1012-1019, ISSN 0919-9268.

Yamauchi, M., Kimura, H. & Fujita, D. (2006). Seasonal and diurnal feeding patterns of the herbivorous fish Siganus fuscescens and scaring by optic and auditory stimuli. Fisheries Engineering, Vol. 43, No.1, (July 2006), pp. 65-68, ISSN 0916-7617.

Yamamoto, M. (2003). The long-term variations in water temperature and salinity in Bisan-Seto, the Central Seto Inland Sea. *Bulletin of the Japanese Society of Fisheries Oceanography,* Vol.67, pp. 136-167, ISSN 0916-1562.

Yamamoto, M. & Yamasaki, M. (1996). Japanese monitoring system on shellfish toxins. In: *Harmful and Toxic Algal Blooms,* Yasumoto, T. Oshima, Y. & Fukuyo, Y (Eds.), pp. 19-22, International Oceanographic Commission of UNESCO, Paris.

Yoshida, M. & Fukuyo, Y. (2000). Taxonomy of armored dinoflagellate *Alexandrium* Halim based on morphology. *Bulletin of Plankton Society of Japan,* Vol.47, No.1, pp. 34-43, ISSN 1343-0874.

Yoshida, G., Yatsuya, K. & Terawaki, T. (2008). Growth pattern of the large brown alga *Sargassum fulvellum* in the natural habitat and outdoor culture. *Japanese Journal of Phycology,* Vol. 56, No.1, (March 2008), pp. 1-8, ISSN 0038-1578.

Yoshida, G., Terawaki, T.& Yoshimura, T. (2009). Deserted coast- changes in macroalgal beds and global warming. In: *Chikyu ondanka to sakana,* Fisheries Research Agency, pp. 121-137, ISBN 978-4-425-88471-1, Seizando-Shoten, Tokyo.

Yoshimatsu, S. (1981). Sexual reproduction of *Protogonyaulax catenella* in culture I. Heterothallism. *Bulletin of Plankton Society of Japan,* Vol.28, No.2, pp. 131-139, ISSN 0387-8961.

Yoshimatsu, S., Matsuoka, S., Ochi, H., Nishio, S. T., Hashimoto, T., Nishibori, N., Yoshida, M. & Fukuyo, Y. (2000). Toxification of shellfish by the PSP causing dinoflagellate *Alexandrium tamiyavanichii* in southern Harima-Nada, I. Occurrences of *A. tamiyavanichii.* Abstract for the meeting of the Japanese Society of Fisheries Science, 45, (April 2000), Tokyo, Japan.

Permissions

All chapters in this book were first published in GWICSEHHUNE, by InTech Open; hereby published with permission under the Creative Commons Attribution License or equivalent. Every chapter published in this book has been scrutinized by our experts. Their significance has been extensively debated. The topics covered herein carry significant findings which will fuel the growth of the discipline. They may even be implemented as practical applications or may be referred to as a beginning point for another development.

The contributors of this book come from diverse backgrounds, making this book a truly international effort. This book will bring forth new frontiers with its revolutionizing research information and detailed analysis of the nascent developments around the world.

We would like to thank all the contributing authors for lending their expertise to make the book truly unique. They have played a crucial role in the development of this book. Without their invaluable contributions this book wouldn't have been possible. They have made vital efforts to compile up to date information on the varied aspects of this subject to make this book a valuable addition to the collection of many professionals and students.

This book was conceptualized with the vision of imparting up-to-date information and advanced data in this field. To ensure the same, a matchless editorial board was set up. Every individual on the board went through rigorous rounds of assessment to prove their worth. After which they invested a large part of their time researching and compiling the most relevant data for our readers.

The editorial board has been involved in producing this book since its inception. They have spent rigorous hours researching and exploring the diverse topics which have resulted in the successful publishing of this book. They have passed on their knowledge of decades through this book. To expedite this challenging task, the publisher supported the team at every step. A small team of assistant editors was also appointed to further simplify the editing procedure and attain best results for the readers.

Apart from the editorial board, the designing team has also invested a significant amount of their time in understanding the subject and creating the most relevant covers. They scrutinized every image to scout for the most suitable representation of the subject and create an appropriate cover for the book.

The publishing team has been an ardent support to the editorial, designing and production team. Their endless efforts to recruit the best for this project, has resulted in the accomplishment of this book. They are a veteran in the field of academics and their pool of knowledge is as vast as their experience in printing. Their expertise and guidance has proved useful at every step. Their uncompromising quality standards have made this book an exceptional effort. Their encouragement from time to time has been an inspiration for everyone.

The publisher and the editorial board hope that this book will prove to be a valuable piece of knowledge for researchers, students, practitioners and scholars across the globe.

List of Contributors

Yaw A. Afrane
Climate and Human Health Research Unit, Centre for Global Health Research, Kenya Medical Research Institute, Kenya
School of Health Sciences, Bondo University College, Bondo, Kenya

Andrew K. Githeko
Climate and Human Health Research Unit, Centre for Global Health Research, Kenya Medical Research Institute; Kenya

Guiyun Yan
Program in Public Health, College of Health Sciences, University of California at Irvine, Irvine CA, U.S.A

Masahumi Johkan, Toru Maruo and Yutaka Shinohara
Graduate school of Horticulture, Chiba University, Japan

Masayuki Oda
Graduate school of Life Science and Environment, Osaka Prefecture University, Japan

Ping Kao and Ting-Ying Wu
Department of Life Science, National Taiwan University, Taipei, Taiwan

Chia-Lun Chang and Chang-Hung Chou
Institute of Plant Biology, National Taiwan University, Taipei, Taiwan

Ing-Feng Chang
Department of Life Science, National Taiwan University, Taipei, Taiwan
Institute of Plant Biology, National Taiwan University, Taipei, Taiwan

Maren Oelbermann and Carolyn E. Smith
University of Waterloo, Canada

M. Dolores Garza-Gil, M. Xosé Vázquez-Rodríguez, Albino Prada-Blanco and Manuel Varela-Lafuente
University of Vigo, Spain

Larissa Rosa de Oliveira
Laboratório de Ecologia de Mamíferos, Universidade do Vale do Rio dos Sinos (UNISINOS), São Leopoldo, RS, Brazil
Grupo de Estudos de Mamíferos Aquáticos do Rio Grande Sul (GEMARS), Porto Alegre, RS, Brazil
Centro para la Sostenibilidad Ambiental, Universidad Peruana Cayetano Heredia (UPCH), Miraflores, Lima 18, Peru

Koichi Takahashi and Hiroyuki Kobori
Department of Biology, Faculty of Science, Shinshu University, Matsumoto, Nagano, Japan

Tatsuyuki Seino
Yatsugatake Forest, Agricultural and Forestry Research Center, Graduate School of Life and Environmental Sciences, University of Tsukuba, Minamimaki, Nagano, Japan

Yanling Song, Bo Liu and Guoli Tang
National Climate Centre, China Meteorological Administration, Beijing, China

Abdul Razack Mohammed and Lee Tarpley
Texas AgriLife Research and Extension Center, Beaumont, Texas, USA

Laura Yáñez-Espinosa
Instituto de Investigación de Zonas Desérticas, Programas Multidisciplinarios de Posgrado en Ciencias Ambientales, Universidad Autónoma de San Luis Potosí, México

Joel Flores
División de Ciencias Ambientales, Instituto Potosino de Investigación, Científica y Tecnológica, A.C. México

Yu-Chi Weng
Division of Environmental Engineering, Faculty of Engineering, Hokkaido University, Japan

Satoshi Nagai
Fisheries Research Agency, National Research Institute of Fisheries and Environment of Inland Sea, Research Center for Environmental Conservation, Japan

Goro Yoshida and Kenji Tarutani
Fisheries Research Agency, National Research Institute of Fisheries and Environment of Inland Sea, Coastal Fisheries and Environment Division, Japan

Index

www.ingramcontent.com/pod-product-compliance
Lightning Source LLC
LaVergne TN
LVHW051046220926
848485LV00001BA/24

Ann Salandre

Die Rückeroberung des Lebens

Einblicke in eine Therapie nach sexueller Gewalt in der Familie

Ich widme dieses Buch
in tiefer Dankbarkeit all jenen,
die mich bis heute begleitet
und unterstützt haben.

Ann Salandre

Die Rückeroberung des Lebens

Einblicke in eine Therapie nach sexueller Gewalt in der Familie

Asanger Verlag • Kröning

Die Publikation dieses Buches wurde durch erhaltene Leistungen aus dem *Fonds Sexueller Missbrauch* (FSM) zur kreativen Aufarbeitung sexueller Gewalt finanziert.

Layout: Wolfgang Wohlers, einsatz.berlin
Druck: PBtisk, a.s., Czech Republic

Bibliographische Informationen der Deutschen Nationalbibliothek:
Die Deutsche Nationalbibliothek verzeichnet diese Publikation in der Deutschen Nationalbibliographie; detaillierte bibliographische Daten sind im Internet über http://dnb.d-nb.de abrufbar.

ISBN 978-3-89334-657-8

…, dass eine verkörperte Erfahrung von Verletzlichkeit eine Sprache finden muss. Oder, anders ausgedrückt, traumatisches Wissen verlangt nach einer Geschichte, da es ohne irgendeinen Ausdruck unerträglich ist.
Elisabeth Bronfen

Wenn Gefühle auf Worte treffen
Siri Hustvedt im Gespräch mit Elisabeth Bronfen

Prolog

Im Frühjahr 1977 übernachtete ich mit meinem Mann in unserem Campingbus. Ich erwachte in der Nacht mit starken Schmerzen im unteren Bauch. Es war stockfinster. Ich konnte mich nicht rühren, fühlte mich wie gelähmt. Und spürte immer mehr Angst, starke Angst. Auch Angst zu sterben. Schmerzen und Schrecken und Tod in mir! Ich lag weiter völlig starr, weckte auch nicht meinen Mann. Ich hätte ihn doch wecken und mit ihm reden können! Aber in der Nacht war ich nicht auf diese Idee gekommen. Ich lag da, es war dunkel, die Schmerzen blieben und auch meine Angst.

Irgendwie überstand ich diese Nacht – mit wenig und nur leichtem Schlaf. Am nächsten Tag hatte ich permanent schnelleres Herzklopfen. Das blieb so, tagelang, und verunsicherte mich sehr.

Im Nachhinein sehe ich diese Nacht als Beginn meines „Absturzes“, denn in der Folgezeit erlebte ich immer stärkere Herzattacken der verschiedensten Art, obwohl mein Herz organisch gesund war, wie sich bei einer Untersuchung herausstellen sollte. Hinzu kamen bald Angst- und Panikattacken. Angst zu sterben, aber auch einfach nur Angst, ohne dass ich sagen konnte wovor. Ängste und Panik traten anfangs besonders nachts auf. Manchmal schlotterten mir die Beine oder mein ganzer Körper schlotterte. Ich erlebte die Angst wie ein Fallen ins Bodenlose: Niemand und nichts kann mich halten, mich retten! Ich selbst schon gar nicht!

Was ist denn nur los mit mir? Mir geht es doch eigentlich gut! Warum habe ich denn Angst? Und wovor? Ich wusste es einfach nicht. Ich fragte mich sogar, ob ich meine Hausfrauen-Pflichten nicht ernster nehmen und/oder eine „bessere“ Ehefrau sein müsste. Ob es mir dann nicht besser gehen könnte. Unglaublich, aber diese und ähnliche Fragen stellte ich mir, was mir aus heutiger Sicht völlig abstrus erscheint.

Später trat die Angst zunehmend und auch am Tag und bei den verschiedensten Gelegenheiten auf. Und ich ängstigte mich davor, dass die Angst wiederkommt; hatte nun auch noch Angst vor der Angst! Das schränkte meinen Bewegungsradius in den folgenden Jahren mehr und mehr ein, bis ich mir kaum noch etwas zutraute. Ich fuhr nur noch selten alleine mit dem Auto, was ich eigentlich gerne machte. Ich fuhr fast gar nicht mehr in Urlaub, wollte einfach nicht woanders übernachten, am liebsten zuhause bleiben. Aber auch dort tauchte die Angst auf, auch dort plagten mich die Herzattacken.

Dem Rat meines Hausarztes folgend machte ich keine Therapie, sondern nahm Beruhigungstabletten (Psychopharmaka), ohne die ich bald nicht mehr auskam. Die Tabletten wirkten erst einmal beruhigend, dämpften die Angst oder beseitigten sie – zumindest vordergründig und für eine gewisse Zeit. Doch nach fast jeder Tabletteneinnahme quälte ich mich damit, dass ich es wieder nicht ohne diese geschafft hatte.

Irgendwann erzählte mir eine Freundin, dass ihr Stiefopa sie als ganz kleines Kind „sexuell missbrauchte". Mir fiel mit Schreck ein: Das hab ich doch auch erlebt! Mit meinen zwei ältesten Brüdern! Doch ich sagte nichts, weder gegenüber der Freundin noch sprach ich in der Folgezeit mit Anderen darüber. Ich packte diese kurze Erinnerung wieder weg.

Weil es mir immer schlechter ging, entschied ich mich doch für eine Psychotherapie – rund fünf Jahre nach der geschilderten Nacht. Die Therapie brachte tatsächlich eine wesentliche Verbesserung in meinem Befinden. Ich konnte nach einer gewissen Zeit die Tabletten weglassen, benötigte bei einer Tachykardie, also sehr heftigem Herzrasen mit einem Puls von mehr als 180, keinen Notarzt mehr und überstand die Herzattacken ohne diese immense Angst zu sterben. Die Gründe, warum es mir schlecht ging und warum ich Herzprobleme hatte, blieben leider ungeklärt, weil die Therapeutin wegen ihres Umzuges in eine weit entfernte Stadt die Therapie nach nicht einmal zwei Jahren beendete.

Jetzt werde ich wohl nie mehr Gelegenheit haben, über „das" mit meinen Brüdern zu reden!, dachte ich damals. Denn genau das hatte ich mir auch von der Therapie erhofft, dass ich zum ersten Mal darüber würde reden können. Es traf mich zutiefst, dass das nun nicht mehr möglich war.

In den ersten Jahren nach dieser Therapie ging es mir aber insgesamt ganz gut, die Herzprobleme tauchten immer mal wieder auf, doch ich konnte recht gut damit umgehen.

Einige Jahre später schaute ich mir eine von Wildwasser konzipierte Ausstellung zum Thema „sexueller Missbrauch" an bzw. nur zum Teil an, denn irgendwann musste ich den Rundgang unter zunehmend heftigerem Weinen abbrechen. Dadurch ausgelöst sprach ich zum ersten Mal mit einer Freundin, mit meiner Schwester Lisa und auch mit meinem Mann darüber, dass mich meine zwei ältesten Brüder „sexuell missbraucht" hatten. Darüber zu reden war äußerst schwierig, weil ich mich schämte, aber auch weil ich mich schuldig oder zumindest mitschuldig fühlte. Durch das Erwähnen der sexuellen Gewalt, ohne dass ich Details schilderte, erinnerte ich mich diffus an einzelne Übergriffe. Ich redete jedoch nicht mehr weiter darüber.

Bald darauf ging es mir zunehmend schlechter. Tagelang spürte ich starke innere Unruhe und erhöhte Herzfrequenz bzw. unregelmäßiges Herzschlagen. Sogar die Ängste tauchten wieder auf. Und das blieb so. Ich konnte mir einfach nicht erklären, weshalb es mir wieder so erging. Erst bei einem Gespräch mit einer Freundin erkannte ich die Gründe: Endlich hatte ich zum ersten Mal über die sexuelle Gewalt geredet, was befreiend wirkte. Ich war anschließend jedoch davon ausgegangen, ich könne weiter so leben wie zuvor. Also das Thema wieder wegpacken. Aber das ging nun nicht mehr!

Deshalb entschied ich mich dafür, mir wieder eine Therapeutin zu suchen, was eine gewisse Zeit beanspruchte. Ich wollte wissen, was mit mir los ist. Ich wollte die Gründe für meine Herzprobleme und für die Angstattacken herausfinden. Und dies ganz besonders: Ich wollte nicht noch weitere Jahre leiden! (Seit der oben beschriebenen Nacht waren bis zum Beginn meiner zweiten Therapie mehr als 14 Jahre vergangen.) Ich war es so leid zu erleben, dass es mir schlecht und immer schlechter ging und noch nicht einmal die Ursachen dafür zu kennen!

Also begann ich eine weitere Psychotherapie bzw. Analyse, ahnend und wünschend, dass es sehr in die Tiefe gehen wird. Die Analyse hat mir über Vieles Klarheit verschafft – auch darüber, weshalb in der geschilderten Nacht die Schmerzen, das Gelähmtsein und die Todesangst auftraten. Außerdem wollte ich endlich über die Übergriffe durch meine beiden Brüder reden.

Und dann sagte ich gleich zu Beginn der Therapie, ohne zu wissen, wovon ich rede: Und da war noch etwas … –

Einleitung

Vor mehr als 25 Jahren begann ich die bereits erwähnte Therapie, um insbesondere, aber nicht nur, die in meiner Kindheit erlebte sexuelle Gewalt zu bearbeiten. Ich erinnere noch deutlich die ersten, sehr schweren Therapiejahre, mit all meiner Verzweiflung und der Angst, mir wird es niemals besser gehen. Und dem Gefühl, ich komme da nicht mehr raus. Für längere Zeit ging es mir sogar schlechter als zu Beginn der Therapie, und ich dachte daran, und das nicht nur einmal, meinem Leben ein Ende zu setzen, weil das Gefühl überwog, ich kann das alles nicht aushalten! Es ist zu hart und zu viel!

Heute steht für mich im Vordergrund, dass es mir gelungen ist, genau dies zu tun: Mir alles eingehend anzuschauen, alle Gefühle und Erinnerungen zuzulassen, zu be- und verarbeiten und aus all dem gestärkt hervorzugehen – mit dem Ergebnis, mein Leben mit Freude gestalten und genießen zu können. Genau davon möchte ich berichten.

Ich möchte mit diesem Buch andere von sexueller Gewalt Betroffene ermutigen, sich ihre Geschichte, ihre traumatischen Erfahrungen ebenfalls anzuschauen und zu bearbeiten.

Aber ich möchte auch von der Vielschichtigkeit, der Schönheit und dem Reichtum dieses Prozesses berichten, und davon, wie der Dialog und die gemeinsame Arbeit zwischen Klientin und Therapeutin vonstatten gehen. Ich lege dar, wie wichtig die stabile Präsenz der Therapeutin ist, die das alles aushält und die über ein sehr komplexes Wissen sowie entsprechende Erfahrung verfügen muss. Von der Empathie ganz zu schweigen.

Ich richte mich mit meinem Buch auch an Fachkräfte – von Fachberatungsstellen bis hin zu PädagogInnen, PsychologInnen und TherapeutInnen –, um nützliche Anregungen für ihre Arbeit zu geben.

Zudem möchte ich mit meinen Ausführungen anderen Interessierten eine Vorstellung vom Ausmaß solcher Traumata und deren Auswirkungen vermitteln und aufzeigen, wie viel Geduld, Zeit und Unterstützung es braucht, diese zu bearbeiten.

Ich schreibe von sexueller Gewalt, benutze also nicht den im Allgemeinen verwendeten Begriff „sexueller Missbrauch", weil er verharmlosend und falsch ist, da er suggeriert, es gäbe einen „richtigen Gebrauch". Außerdem fehlt der Aspekt der Gewalt. Es handelt sich um Übergriffe und um Machtausübung!

Eine sehr liebe Freundin schlug mir vor vielen Jahren vor, ich solle doch meine Geschichte aufschreiben und von der erfolgreichen Therapie berichten. Damit könnte ich anderen helfen. Meine erste Reaktion war ablehnend, mit der Begründung: Wer will das denn lesen? Das ist doch alles viel zu hart! – Zum damaligen Zeitpunkt befand ich mich noch in Therapie und hatte nicht den nötigen Abstand. Später gefiel mir diese Idee zunehmend besser, sodass ich dachte, ja, ich will und werde das alles aufschreiben. Das ist mir doch so wichtig, ich bin doch so glücklich über das Erreichte! Das mache ich für mich! Aber nicht nur durch die Ermunterung von FreundInnen und Anderen entschloss ich mich, meine Geschichte nun doch zu veröffentlichen.

Ich habe über 10 Jahre Therapie bzw. Analyse gemacht. Über 10 Jahre – das klingt unglaublich lange, doch das Bearbeiten der traumatischen Erlebnisse meiner Kindheit nahm lange Zeit in Anspruch. Ich brauchte dafür meine Zeit! Dazu kommt noch dies: Wenn Frau über 20 Jahre im Elternhaus verbracht und ihr dort so Vieles eingetrichtert wurde, die Gebote, Verbote und Maximen der Eltern bis in jede Zelle des Körpers eingegangen sind, dann braucht es viele Jahre, um zu sich und zu eigenen Maximen zu kommen. Und um gesund zu werden!

Ich berichte von meinen Erfahrungen auf diesem Weg, davon, wie viel Kraft und Ausdauer es kostete, aber auch wie viel Kraft und Energie und Lebensfreude ich dadurch gewonnen habe. Und: Wie mich die therapeutische Arbeit gerettet hat. Ich hätte nie gedacht, dass es mir einmal so gut gehen, ich meine zuvor nicht ausgesprochenen Ziele der Therapie auch nur annähernd erreichen würde. Aber das ist mir gelungen. Nicht nur deshalb will ich versuchen, mit diesem Bericht Mut zu machen, den Willen und die Zuversicht hervorrufen und stärken, dass Frau auch aus furchtbaren Verhältnissen, aus schlimmen körperlichen und seelischen Verfassungen hervorkommen und dabei sehr viel für sich gewinnen kann.

Ich möchte Andere durch meine Schilderungen ermuntern und bestärken, und zwar nicht nur zur Aufarbeitung traumatischer Erlebnisse wie sexueller Gewalt. Ich habe während der Jahre der Therapie so viel über mich erfahren, für mich gelernt und mir viel von dem zurückerobert, was ursprünglich in mir war, was zu leben ich mich nicht traute oder wovon ich gar nichts (mehr) wusste.

Mir ist bewusst, auch aus eigener Erfahrung, dass eine langjährige Therapie nicht nur eine Frage der Ausdauer oder des Willens ist. Es ist auch eine Frage der Zeit. Frau muss sich die Zeit für sich nehmen können und nehmen.

Es ist aber auch und besonders eine Frage des Geldes. Die gesetzlichen und die privaten Krankenversicherungen ermöglichen nur eine klar begrenzte, viel zu geringe Stundenanzahl, je nach Art der Therapie. Doch dieser Rahmen kann durch beharrliches Beantragen und Begründen durch TherapeutInnen erweitert und erhöht werden. Die TherapeutInnen dürfen gegenüber den Krankenkassen nicht lockerlassen. Dies ist mir durch meine Therapeutin, aber auch durch entsprechende Aussagen anderer Therapeutinnen, bekannt.

Ich verfügte zum damaligen Zeitpunkt nicht über ausreichend Geld, meine Therapie, in den Zeiten, in denen die Krankenkasse nichts mehr zahlte, komplett selbst zu finanzieren. Dennoch gelang es mir, jahrelang Therapie zu machen. Zum einen hatte mir meine Therapeutin die Bereitschaft signalisiert, mir zeitweise geldlich entgegenzukommen. Dies musste ich einige Zeit in Anspruch nehmen. Zum anderen sagten mir FreundInnen Hilfe zu, waren sie bereit, mich mit Darlehen zu unterstützen. Hinzu kommt, dass ich lieber auf vieles Andere als auf Therapiestunden verzichtete. Mir war meine Therapie und damit mein Leben, mein Wohlbefinden zu wichtig. Wichtiger als alles andere! Dagegen kam kein Buch oder keine Reise oder was auch immer an. Und das habe ich nie bereut!

Außerdem ist während dieser Zeit ein stützendes Umfeld sehr wichtig. Aus meiner Erfahrung kann ich sagen, dass es gut ist, mehrere UnterstützerInnen zu haben, auch damit nicht die eine oder andere überbelastet wird.

Ich werde von dem Therapieprozess und von meinen Kindheitserlebnissen berichten. In den Jahren der Therapie habe ich direkt nach den jeweiligen Stunden detaillierte Aufzeichnungen angefertigt. Außerdem führte ich ausführlich Tagebuch. In den hier publizierten Texten zitiere ich aus den Therapieaufzeichnungen, mehr oder weniger gekürzt, und den Tagebuchnotizen, füge aber auch Kommentare aus heutiger Sicht hinzu. Mir ist sehr bewusst, dass die Therapienotizen nicht immer den tatsächlichen Verlauf oder die tatsächlichen Äußerungen seitens der Therapeutin oder meinerseits wiedergeben. Während der Therapiephase stellte sich nicht selten heraus, dass ich die Therapeutin falsch verstanden hatte oder manches gar nicht mehr wusste, wieder verdrängt oder nur verändert erinnert hatte. Außerdem war ich oft so beteiligt, so in meinen Gefühlen oder Erinnerungen versunken, dass ein wirklicher Bericht von diesen Stunden nur unzureichend gelingen konnte.

Ich schreibe in diesem Buch fast ausschließlich zum Thema *sexuelle Gewalt,* den Erlebnissen von damals und dem Bearbeiten in der Therapie. Die Vorstellung, die traumatischen Erlebnisse nochmals zu durchleben, die Erinnerungen daran zuzulassen, erscheint zuerst unmöglich und viel zu hart! Aber nach jeder Erinnerung, nach jedem Einlassen stellte sich bei mir große Befreiung ein, nicht nur und nicht immer gleich, aber in erster Linie. Und erst recht auf lange Sicht.

Durch die Therapie habe ich meine Geschichte angenommen, konnte ich mir mein Leben (wieder) aneignen.

Das Buch ist wie folgt aufgebaut: Ich beginne mit der Vorstellung der im Zusammenhang mit meiner Kindheit, den Taten und dem Bearbeiten zentralen Personen; das dürfte das Lesen und Einordnen erleichtern. Unter dem Titel *Die Familie* stelle ich insbesondere meine Ursprungsfamilie vor und kommentiere, wie ich sie früher jeweils erlebte.

In den folgenden Kapiteln schreibe ich über meine beiden ältesten *Brüder,* über *Bernd* und *Adolf.* Im Kapitel *Der Vater* beschreibe ich ausführlich den langwierigen Therapieprozess, der u.a. verdeutlicht, wie viel Zeit und Geduld es braucht, um zu klaren Erinnerungen und einer eigenen Sicht auf die Geschehnisse zu kommen. Dabei verschweige ich nichts! Diese Kapitel müssen nicht gelesen werden, wenn das der Einen oder Anderen zu heftig ist oder zu werden droht.

Alle folgenden Kapitel, in denen die Übergriffe nicht noch einmal geschildert werden, vermitteln dennoch das Wesentliche und sehr viel von dem, worauf es (mir) ankommt. In diesen Kapiteln wiederholen sich einige Passagen, die auch in den Kapiteln über die Brüder und den Vater zu lesen sind, um auch denen, die diese Kapitel nicht lesen (können), die notwendigen Zusammenhänge und Einblicke zu geben.

Natürlich stellt sich die Frage, welche Rolle *Die Mutter* einnahm. Wusste sie von den Übergriffen? Durch das Schreiben über sie kam ich zu einigen neuen und überraschenden Erkenntnissen.

Das Kapitel *Nicht darüber reden* folgt im Anschluss. Zwar ist die Unmöglichkeit oder Schwierigkeit, zu erinnern und zu reden, Auswirkung der sexuellen Gewalt; das familiäre Schweigegebot ist aber mehr noch „Teil der Taten", die ohne die Drohungen und Ignoranz zumindest nicht häufiger hätten stattfinden können.

Wie reagierten Familie und FreundInnen, nachdem ich endlich über die Gewalterlebnisse reden konnte und geredet habe? Die Reaktionen reichten von „was willst du denn jetzt noch damit?" und „wieso hast du nicht schon früher was gesagt?" bis hin zur Bedrohung meines Lebens. Wie habe ich es dennoch geschafft, die Eltern und die Brüder zur Rede zu stellen? Dies alles ist nachzulesen im Kapitel *Reaktionen.*

Bei den *Auswirkungen der sexuellen Gewalt* unterscheide ich zwischen denen, die bereits *vor* der Therapie aufgetreten waren, und denen, die erst *während* der Therapie auftauchten oder bewusst wurden. Auch in diesen Texten behandele ich sehr wichtige Punkte der therapeutischen Arbeit. Es geht u. a. um die furchtbaren *Ängste,* um meine *Wut* und um die *Zweifel* an meinen Erinnerungen. Ich berichte von dem lang anhaltenden *Misstrauen* der Therapeutin gegenüber. Dies alles hat mir sehr zugesetzt und mich zeitweise außerordentlich belastet. Auch die Fragen „bin ich Schuld an dem Vorgefallenen?" oder „wer ist verantwortlich?" und andere wichtige Auswirkungen werden eingehend behandelt – bis hin zu den beruflichen und finanziellen Auswirkungen.

Darüber zu schreiben, *Was mir während der Therapiezeit guttat,* bereitete mir besonders große Freude. All die dort aufgeführten kleinen und großen Dinge haben mir außerordentlich geholfen, das Schlimme auszuhalten und zu überstehen.

In fast allen Kapiteln wird das Verhalten der Therapeutin und ihre Arbeitsweise sehr deutlich, aber ich kommentiere oder erwähne dabei nicht immer explizit, wie das jeweilige Verhalten auf mich wirkte. Außerdem sind in diesen Kapiteln viele relevante Aspekte nicht erwähnt. Auch deshalb gehe ich in dem Kapitel *Die Therapie und die Therapeutin* nicht nur kurz auf den *äußeren Rahmen der Therapie* ein, sondern ausführlicher darauf, *wie ich die Therapeutin erlebte.* Im Mittelpunkt der Betrachtung steht dabei ihr Umgang mit mir im Zusammenhang mit dem Auftauchen von Erinnerungen und Bildern, ohne diese jeweils noch einmal zu schildern.

Im abschließenden Kapitel berichte ich davon, was *Der Schreibprozess* bei mir bewirkte und welche Bedeutung das Schreiben dieses Buches für mich hatte. Vorweg kann ich sagen, dass das Schreiben die Entdeckung für mich ist!

Im *Nachwort* wird meine Therapeutin, Frau Prof. Dr. Krüger-Kirn, einige Aspekte des Therapieprozesses beleuchten.

Noch einige redaktionelle Anmerkungen sowie Hinweise zum besseren Verständnis:

Alle zitierten Träume, Notizen aus meinen Tagebüchern sowie die Aufzeichnungen der Therapiestunden sind durch Kursivschrift (Träume serifenlos kursiv), Kleinschreibung und Einrücken kenntlich gemacht.

Während der Therapie/Analyse habe ich die Therapeutin als „Frau K." oder „die K." in meinen Aufzeichnungen erwähnt. Bald fühlte ich für sie wie für eine Mama, was sich in meinen Texten entsprechend wiederfindet.

Ich zitiere die Therapeutin oft mit einem „Ehem". Hierbei handelt es sich um eine Äußerung des Mitfühlens. Sie ist auch als Bestärkung oder Bestätigung zu verstehen.

Im Lauf der Therapie veränderten wir ein paarmal unsere Sitzpositionen. Es wird auch ein Spiegel bzw. das Anschauen über den Spiegel erwähnt, der sich uns gegenüber befand.

1. Die Familie

Meine Eltern kamen 1912 bzw. 1917 auf die Welt. Meine Geschwister und ich wurden zwischen 1933 und Anfang der 50er Jahre geboren. Ich werde sie nach der Reihenfolge ihrer Geburt aufführen, kurz charakterisieren und beschreiben, wie ich zu ihnen stand und was sie mir bedeuteten. Auf meine Eltern gehe ich etwas ausführlicher ein.

Mutter

Sie verfügte über keine Berufsausbildung, konnte weder Fahrradfahren noch Schwimmen. Ich könnte jetzt böse hinzufügen, dafür aber putzen und herumkommandieren. Aber das war zum Glück nicht alles. Sie hat viel gelesen und mich, zu meiner Freude, in das Lesen von Literatur eingeführt. Besonders gerne erinnere ich mich daran, wie wir zusammen klassische Musik anhörten. Natürlich mussten jeweils zuvor alle Hausarbeiten ordentlich erledigt sein! Meine Mutter war überhaupt an kulturellen und geschichtlichen Dingen sehr interessiert.

In der Erziehung hatte sie sich an meinem Vater und mit ihm teilweise noch an der Naziideologie orientiert. Sie war sehr kommunikativ, eher freundlicher Natur, konnte aber auch sehr kalt und eklig sein.

Grundsätzlich liebte ich meine Mutter, hatte vor ihr keine Angst. War sie verärgert, blieb das nicht lange so. Sie konnte bald sagen, komm', nun ist es wieder gut. Das schätzte ich sehr an ihr. Andererseits war sie unerbittlich was den geforderten Einsatz für den Haushalt betraf.

Vater

Er hatte eine handwerkliche Ausbildung, wurde später aber verbeamtet. Mit autoritärem Charakter ausgestattet verhielt er sich wie ein Patriarch. Sein Frauenbild war entsprechend. Seine Eltern waren Italiener, sozusagen Gastarbeiter, noch bevor es diese Bezeichnung gab, da sie bereits um 1910 nach Deutschland kamen. Mein Vater wurde durch absolute Assimilierung zu einem „richtigen" Deutschen: Gehorsam, Ordnung, Sauberkeit und Fleiß waren die Maxime – zumindest nach außen hin.

Ich liebte meinen Vater sehr, besonders als ich noch ein kleines Mädchen war. Andererseits hatte ich vor ihm sehr viel Angst. Wenn er mich, eine Schwester oder einen Bruder bestrafte, weil er sich über uns geärgert hatte oder wir nicht brav genug waren, dann strafte er uns mit absoluter Nichtbeachtung. Die konnte sehr lange andauern. Für uns Kinder war das kaum auszuhalten, sodass wir alles gemacht haben, um wieder in seiner Gunst zu stehen. Einfach damit die Kälte des Schweigens, des Ausgeschlossenseins aufhört!

Unsere Erziehung und das Familienleben an sich sind auch vor dem Hintergrund zu sehen, dass beide Elternteile katholisch waren, wir katholisch erzogen wurden. Erschwerend kam hinzu, dass meine Geschwister und ich entweder während oder direkt nach der Nazizeit auf die Welt kamen. Diese Ideologie wirkte sich ebenfalls negativ auf unsere Erziehung aus. Wenn ich heute von meinen Kinder- und Jugendjahren erzähle, kommt es mir manchmal vor, als wäre ich im Mittelalter aufgewachsen!

Maria

Sie kam auf die Welt, als meine Mutter gerade mal 16 Jahre jung war! Maria war für mich und meine jüngeren Geschwister wie eine zweite Mama. Da wir so viele waren, musste sie als Älteste, wie das damals recht häufig der Fall war, nach dem Schulabschluss ohne jede Ausbildung zuhause bleiben. Sie war sehr engagiert und liebevoll für uns „Kleinen" da. Ebenso für die anstehenden Haushaltsarbeiten. Durch diese Rolle erlebte nicht nur ich sie mehr zu den Eltern gehörend als zu uns Geschwistern. Sie setzte sich auch öfter bei den Eltern für uns ein. Ich liebte sie sehr!

Als sie heiratete und von zuhause wegzog, war ich neun Jahre alt. Wir Jüngeren vermissten sie schmerzhaft. (Als meine anderen Geschwister heirateten oder aus anderen Gründen die Familie verließen, war ich zwischen 15 und 20 Jahren alt.)

Adolf

Er, der erste Sohn, war ein eher schwieriger Typ, nicht so beliebt, auch weil er sich oft großmäulig, irgendwie schmierig, jedenfalls unangenehm verhielt. Er war nicht so brav und angepasst wie der größere Teil von uns Geschwistern, besonders nicht gegenüber meinem Vater. Das fand ich manchmal gut. Ich mochte ihn nicht so sehr, auch weil er nicht nur auf sexuellem Gebiet übergriffig war. Adolf wohnte lange bei den Eltern; er zog erst nach mir aus.

Birgit

Auch sie verhielt sich eher unangepasst, was natürlich keine Freude bei den Eltern auslöste. Gegenüber der sehr anerkannten und geliebten Maria hatte sie bei den Eltern einen schwereren Stand. Maria wurde mir und meiner anderen Schwester seitens meiner Eltern als gutes bis leuchtendes Beispiel vorgehalten, Birgit eher als das Gegenteil. Das wirkte sich negativ auf meine Haltung ihr gegenüber aus, was ich damals noch nicht recht durchschaute. Ich hatte sie lieb wie ein Kind nun mal die Schwester liebt, aber sie war mir nicht so nahe, was sicher auch mit dem Altersabstand von rund 8 Jahren zu tun hatte.

Bernd

Vielleicht ist er so gut zu beschreiben: Den Eltern und wohl auch den meisten Geschwistern, wie auch mir, sehr sympathisch; ein eher ruhiger und angenehmer Typ, nicht auffällig, sondern angepasst wie wir anderen. Obwohl bei den Eltern sehr beliebt, konnte leider auch er, wie Maria, nichts gegen das System der Eltern, gegen ihre Ausbeutung und Unterdrückung ausrichten.

Lisa

Mir war Lisa eine sehr wichtige und geliebte Schwester; wir waren uns sehr nahe. Auch weil sie nur ein paar Jahre vor mir geboren wurde, aber nicht nur deswegen. Doch konnte sie auch unangenehm auf Abstand gehen, sich distanziert und verschlossen verhalten, was mit den Verhältnissen an sich zu tun hatte, weniger mit mir. Wir hielten auch mal zusammen, wenn uns unsere Mutter zu sehr mit Hausarbeiten traktierte oder uns mit Strenge oder ungerechtem Verhalten zusetzte. Lisa erscheint mir von heute aus betrachtet den Eltern gegenüber besonders ohnmächtig und ausgeliefert gewesen zu sein, hierbei ist an erster Stelle mein Vater zu nennen.

Ich vermisste Lisa nach ihrem Auszug von zuhause sehr; ich war 16 Jahre alt. Doch wir blieben bis vor ein paar Jahren in gutem Kontakt.

Heiner

Er war mir ebenfalls sehr nahe, aufgrund der altersmäßigen Nähe, aber auch weil er so ein lieber und solidarischer Bruder war. Ihn erinnere ich als einen hilfs- und schutzbedürftigen Jungen, der gegenüber meinem Vater einen sehr schweren Stand hatte. Vielleicht weil er sich weicher und nicht so jungen- oder mannhaft zeigte wie die anderen Brüder. Manchmal nahm mich Heiner gegenüber unserer Mutter in Schutz, z.B. wenn sie meinte, ich würde nicht schnell genug putzen. Natürlich bestand zu ihm eine andere Nähe als zu Lisa, weil er ein Bruder war, dennoch waren wir uns sehr nahe. Auch ihn vermisste ich nach seinem Auszug.

Ann

Zu mir schreibe ich auch ein paar Zeilen. Ich war, im Gegensatz zu Lisa, offener, kommunikativer, fröhlicher – warum auch immer. Gut, dass es so war. Auf Fotos von mir als kleinem Mädchen ist das gut zu erkennen. Aber das blieb leider nicht so. Auf späteren Fotos ist das fröhliche Lächeln verschwunden. Schon relativ früh blickte ich mit kritischen Augen auf unsere Familie, besonders auf die Rolle, die wir als Mädchen und Frauen ausfüllen mussten, traute mich das und Anderes aber nur in beschränktem Maße und hinter eher witzigen Bemerkungen versteckt zu äußern – zu meinem Schutz, um böse oder sanktionierende Reaktionen oder gar völligen Liebesentzug zu vermeiden. Mit 22 Jahren heiratete ich und zog in eine andere Stadt und das war sehr gut!

Leo

Der Jüngste von uns und als solcher entsprechend verwöhnt. Besonders meine Mutter nahm ihn oft in Schutz. Er durfte fast alles, erlebte aber nie oder kaum Sanktionen, egal was er sich leistete. Dies und dass er schon früher als ich abends weggehen durfte, obwohl einige Jahre jünger, ärgerte mich über die Maßen. Aber er war ja ein junger Mann, keine junge Frau! Als kleiner Junge erlebte ich ihn als ganz süßen Burschen. Später änderte sich das leider, denn dann war er verlogen, oberflächlich und bedrohlich.

Auch sie gehör(t)en zu meiner Familie und deshalb muss ich sie hier unbedingt erwähnen.

Philipp

Er war mein Ehemann bis zu den ersten Jahren der Therapie. Mehr möchte ich hier nicht von ihm oder uns berichten, das ist ein anderes Thema. Jedenfalls war er mir lange ein ganz besonders lieber und zuverlässiger Mensch, der mich in den schweren Jahren liebevoll und solidarisch unterstützte.

Isabelle und Ira

Zwei Töchter meiner ältesten Schwester, mit denen mich ein sehr nahes und liebes Verhältnis verbindet.

Lea

Eine der Töchter meines Bruders Bernd, die ich zu meiner Freude während der Zeit des Schreibens kennenlernte.

2. Die Brüder

2.1 Die Therapie beginnt

Bereits beim Erstgespräch hatte ich meiner späteren Therapeutin gegenüber erwähnt, dass ich von meinen beiden ältesten Brüdern als kleines Mädchen sexuelle Gewalt erlebte. Schon bald nach Beginn der Therapie kamen wir darauf zurück. Die Therapeutin sagte dazu unter anderem, durch die Therapie würde ich die Brüder nicht mehr schützen, weil ich darüber redete, damit quasi an die Öffentlichkeit ginge. Zur Unterstützung für diesen Schritt würde ich sie, die Therapeutin, benötigen. Ja, Unterstützung brauchte ich! Sehr sogar, wie sich später herausstellte.

Ich hatte sie darauf angesprochen, ob sie es als normal ansieht oder als jugendliche Neugier, was meine Brüder mit mir gemacht haben. – Ändert das etwas in Ihrem Empfinden, wenn ich so oder anders denken würde?, fragte sie und fügte hinzu: Was denken Sie denn? – Ich verstehe es nicht! Und empfinde es als eine Frechheit! Ich fühle mich verletzt ... – Manchmal würde ich mich allerdings auch fragen, war es wirklich Gewalt? Oder mir sagen, es war doch nicht so schlimm! Nach so langer Zeit könne ich das doch jetzt nicht mehr hervorkramen; ich müsse halt damit leben. Sie betonte, das wäre sicher der Teil in mir, der immer noch im Ohr habe, man müsse sich zusammenreißen. Ich erkannte aber auch: Dass es mir so schlecht ging und geht und dass ich Therapie machen muss, zeigt doch, wie schlimm es für mich war! –

Ob ich denn bei Anderen das Gefühl hätte, die könnten mich nicht verstehen? – Ja! Aber noch schlimmer ist, ich kann das Geschehene selbst nicht fassen! Ich kann's einfach nicht fassen! – Dabei musste ich bitterlich weinen. Als ich sagte, ich fasse es nicht, meinte ich nicht nur das, was ich damals erlebte, sondern auch die Beschädigungen, Auswirkungen, alles, was danach kam. – Ich werde das nie vergessen und denen nicht verzeihen! Nie! – Mit großer Traurigkeit dachte ich aber auch daran, dass ich mich damals nicht beschützen konnte.

Zu Anfang haderte ich eine ganze Weile damit, Therapie zu machen. Ich sprach davon, Therapie machen zu *müssen.* Ich empfand die Therapie sogar als Strafe. Es ärgerte mich maßlos, dass ich die machen muss, meine Brüder aber nichts machen müssen: Erst tun die mir das an, keiner hilft mir, und jetzt muss *ich* deswegen Therapie machen! – Ich müsse mit den Tatsachen und den Taten der Brüder leben, könne weinen, toben, in die Therapie gehen oder auch nicht, ich würde das Thema nicht mehr loswerden! Aber die könnten weiter ihr Leben leben!, stöhnte und schimpfte ich in der Therapie. Ich solle doch bedenken, dass ich das für mich tun würde, hielt die K. dem entgegen. Das wüsste ich doch, dennoch sei es zum Heulen und zum Verzweifeln.

Die Therapeutin vermutete, es fiele mir so schwer, mich mit dem Erlebten auseinanderzusetzen, weil ich als Opfer jetzt aktiv werden müsse. Sie gab zu bedenken, wenn ich bei dieser Haltung bliebe, eigentlich keine Therapie machen zu wollen, wenn ich mich nicht auf diese und somit auf meine Gefühle einließe, bliebe doch alles wie bisher. Meine jetzige Haltung und Einstellung würde den Fortgang der Therapie behindern. Das sah ich ein, aber ich konnte mir das Gefühl ja nicht einfach verbieten und dann ist es weg. Ich konnte mir nur immer wieder klarmachen, die Therapie ist eine Chance für mich. Dies wurde mir noch klarer, nachdem die K. mich gefragt hatte, was ich beispielsweise einer 60-jährigen Frau sagen würde, die sich entschließt, eine Therapie zu machen, um sich mit ihren schlimmen Kindheitserlebnissen auseinanderzusetzen. Lächelnd, weil ahnend, worauf sie hinauswill, hatte ich geantwortet, das fände ich gut, wenn sie das nochmal anpackt. Zu dieser Frau konnte ich das ganz frei sagen. So empfand ich es auch. Aber da ging es ja nicht um mich. Für mich oder bei mir legte ich andere Maßstäbe an. Auf weitere Bemerkungen der K. hin, dass die Frau die Therapie begann, um wenigstens in ihren letzten Jahren besser leben zu können und ähnliche Aspekte, hatte ich doch tatsächlich gesagt: Ich finde es toll, dass sie das noch anfängt! – Daraufhin konnte ich auch von mir sagen, ich fände es schon irgendwie gut, dass ich das angehen würde, um dann hinzuzufügen, mit fester Stimme, dass ich es toll finde, es endlich geschafft zu haben, mich diesem Thema zu stellen. Toll und mutig!

Nun klärte sich, warum ich die Therapie zuvor sogar als Strafe angesehen hatte. Die Therapeutin dazu: Wenn Sie die Therapie als Strafe ansehen und denken, Ihre Brüder sind schuld, dass Sie hier sitzen, dann sind Sie selbst nicht verantwortlich für die Therapie. ... Fällt es Ihnen damit leichter, die Therapie zu machen, dies nach außen zu vertreten? – Genau so war es! Ich konnte nämlich nicht gut dazu stehen, ich will eine Therapie machen, ich will wissen, was mit mir los ist! Ich will, dass es mir besser und sogar gut geht! Nun hatte ich mich zwar dafür entschieden, doch das Problem, dazu zu stehen, blieb mir noch länger erhalten, auch in Form von diesen Gedanken: Wie lange willst du eigentlich noch Therapie machen? Wie lange willst du dich noch in den Mittelpunkt stellen? Dazu nochmal die K.: Von wegen in den Mittelpunkt stellen, das wäre ja allerhand, dass ich das dächte. Jetzt hätte ich das rund 30 Jahre nur mal versteckt hochkommen lassen, so viele Jahre gelitten und jetzt, wo ich endlich aktiv daranginge, es aufzuarbeiten, wären 3 Jahre oder wie viel auch immer schon zu viel. Recht hatte sie!

Fragen zum Alter und zur Schuld

Bevor ich von einer der Situationen mit den Brüdern berichten konnte, beschäftigte mich sehr die Frage, wie alt ich damals eigentlich war und wie alt meine Brüder; eine Frage, mit der ich mich eine Zeit lang sehr abquälte. Die konnte ich zu Anfang nämlich nicht recht beantworten, weil ich mich nicht genau erinnerte. Auch bezüglich der Häufigkeit der Attacken von Adolf erinnerte ich erstmal nur diffus, dass die über Jahre hinweg stattfanden.

Ich glaube, das Thema Alter tauchte so bald auf, weil sich für mich damit die Frage nach der Verantwortung für die Taten stellte. Bin ich schuld an dem, was passierte, was die gemacht haben? Hätte ich das verhindern müssen, weil alt genug?

Dabei gingen mir solche und ähnliche Gedanken durch den Kopf: Wenn ich schon 10 Jahre alt war, dann hätte ich mich doch wehren müssen! Dabei hatte ich jedoch ein Mädchen im Alter von 14 und mehr Jahren im Kopf; das fiel mir aber erst später auf. Andererseits, wenn ich erst 6 Jahre alt war, so eine weitere Überlegung, dann war Bernd gerade mal 12, also doch noch ein Kind! Mal davon abgesehen, ob ein Junge in diesem Alter schon „sowas" macht, ist es doch interessant, dass ich ihm den Kindstatus zuerkannte, mir umgekehrt aber nicht. Er hätte keine oder nur geringe Verantwortung gehabt, ich aber mit zehn Jahren auf jeden Fall, denn ich hätte mich wehren müssen.

Hier offenbarte sich also ein Gefühl, das immer noch in mir vorherrschte, ein Gefühl aus all den vergangenen Jahren: Ich bin schuld! Auch dies wollte ich überprüfen. Die Therapeutin meinte dazu unter anderem: Die Übernahme der Schuld könne auch Schutz gewesen sein, denn sie befreie das Mädchen von der furchtbaren und kaum auszuhaltenden Hilflosigkeit, weil es dann selbst handelt bzw. sich als Handelnde phantasieren kann und sich nicht mehr nur ausgeliefert fühlen muss.

Ich konnte die Frage nach dem Alter noch nicht beantworten, mich vorläufig nicht erinnern, wie alt ich war, weil ich die Antwort mit der Frage nach der Schuld und der Verantwortung für das Geschehene verband. Sicher hat auch die jahrelange Verdrängung mit dazu geführt, dass das exaktere Erinnern Zeit brauchte.

Missverständnisse und beginnendes Misstrauen der Therapeutin gegenüber

Unbewusst könnte bei dem Voranstellen der Frage nach dem Alter mit hineingespielt haben, auszutesten, welche Haltung die Therapeutin hinsichtlich der Verantwortlichkeit der Brüder oder auch der meinigen einnahm. Denn ihre Aussagen dazu und zu den

Brüdern überhaupt lieferten oft die Ursache für Misstrauen ihr gegenüber und Streit und Stress mit ihr.

So sprach sie einmal von eventuellen Doktorspielen, und die Brüder wären noch Kinder bzw. Jugendliche gewesen. Da ging bei mir die Sicherung durch, so formulierte ich es nach der Therapiestunde. Ich hatte sie daraufhin sofort unterbrochen und dagegen protestiert. Sie meinte, wenn ich 6 Jahre alt gewesen wäre, dann wäre Bernd 12, also noch ein Kind, und Adolf 16, also in der Pubertät. Das käme bei mir so an, als wolle sie die Brüder entschuldigen, entgegnete ich erbost. Das Thema hätten wir nun schon öfter gehabt, aber könnten dem auch nicht aus dem Wege gehen, betonte sie. Bei mir käme das als Entschuldigung und Verniedlichung des Erlebten an. Nein, so sei das von ihr nicht gemeint, sie wolle die beiden auf keinen Fall entschuldigen. Ich verteidigte weiter meine Reaktion, die ihre Gründe haben müsse, auch wenn ich diese nicht kennen würde. In dieser Stunde fühlte ich mich zwischendurch dem Thema hilflos ausgeliefert. Wie ein schutzloses, kleines Kind. Das hatte ich der Therapeutin gegenüber auch geäußert.

Sie hatte mich gefragt, ob meine Brüder und ich nicht mal geschmust hätten. Daran könne ich mich überhaupt nicht erinnern. Gerade kleinere Schwestern würden doch gerne mit ihren größeren Brüdern spielen und schmusen, meinte sie. Das sei ganz natürlich. Vielleicht hätten die beiden Brüder das dann ausgenutzt.

Bei einem Gespräch mit meiner Schwester Lisa erinnerten wir uns beide, dass wir nie mit den Brüdern schmusten oder uns auch nur mal umarmten. Wir hatten auch nicht mit ihnen zusammen gespielt, mit den ältesten Brüdern gar nicht spielen dürfen, das unterbanden die Eltern gleich. Heute frage ich mich, warum durfte das nicht sein? Ahnten sie oder wussten sie, was passieren würde oder bereits geschah?

Als ich in einer anderen Stunde sagte, ich hätte Zuneigung gewollt und dann *das* bekommen, meinte die K., Bernd hätte auch Wärme und Zuneigung gewollt und bekommen. – Was? Ich hab' doch gar nichts gemacht! – Er habe doch meine Haut gespürt, das wäre für ihn Wärme und Zärtlichkeit gewesen. Ich reagierte sehr empört und ärgerlich: Der wollte doch keine Zärtlichkeit! Der wollte ausprobieren, ob er sich das holen kann, ob er das kriegt, was er will! Das war Ausprobieren von Macht! – Sie gab mir zu bedenken, in meiner Familie, die so körper- und sexualfeindlich gewesen wäre, hätte jede Berührung etwas Sexuelles gehabt. Bernd habe sicher auch unter der wenigen Liebe und der Körperfeindlichkeit gelitten. Ich hätte das aber doch auch nicht getan, auch wenn ich diesen Mangel erlebte! Und meine Schwestern bestimmt ebenfalls nicht! Sie brachte das in Zusammenhang mit unserer Erziehung; wir seien doch in ganz typischen Geschlechterrollen erzogen worden und aufgewachsen. Genau das war das Problem, denke ich heute. Damals vermisste ich das Verständnis der Therapeutin sehr, hätte ich mir mehr Anteilnahme gewünscht. Jedenfalls kamen einige ihrer Aussagen und Reaktionen bei mir nicht so an, wie ich es mir wünschte.

Sie sprach circa ein Jahr später davon, auch Bernd, die Brüder insgesamt, seien Opfer der patriarchalen Erziehung gewesen. Da war es immer noch schwer, dies nicht als Entschuldigung anzusehen.

Aber zum Glück gab es auch ganz andere Bemerkungen und andere Verhaltensweisen der Therapeutin, die mir das Vertrauen in sie nach und nach ermöglichten. Die folgenden und ähnliche Aussagen halfen mir sehr. Eigentlich wäre es doch egal, wie alt ich und die Brüder gewesen seien. Wichtig wäre doch, dass ich das nicht gewollt hätte, das käme klar bei ihr an, und dass es für mich schlimm war und ist. – Ja, egal wie alt und wie schuldfähig die beiden waren und egal, was Andere darüber denken, für mich war und ist es furchtbar und hatte schlimme Folgen! – Ja! Und das ist für Sie das Entscheidende!, stimmte sie mir zu. Sie hatte noch betont, bei mir kämen erklärende Aussagen so an: Wenn es für das Verhalten der Brüder Erklärungen oder Entschuldigungen gibt, haben meine Gefühle, meine Verletzungen keine Geltung mehr. – Ja, deshalb werde ich dann wütend! – Das sei doch gut, dass ich dann wütend werden und um die Geltung meiner Gefühle kämpfen würde. (Wenn es früher um die Brüder ging, galten wir doch auch nichts! Es drehte sich doch alles um die!, erinnerten Lisa und ich später beim Gespräch über diese Stunde.)

Anlässlich einer der erwähnten Auseinandersetzungen mit der Therapeutin hatte sie mich bereits gefragt, ob ich Angst hätte, meine Gefühle und die von den Brüdern angerichteten Schäden würden nicht mehr beachtet, sobald sie, die Therapeutin, oder Andere über die Täter, ihr Alter oder ähnliches redeten. Das konnte ich nur bejahen. Sie ergänzte, ich würde dann den Eindruck vermitteln, als wäre ich ganz klein, wie nicht mehr vorhanden. Dann würde ich völlig verschwinden. Ich erinnere noch heute ganz deutlich dieses Gefühl des Verschwindens.

Abschließend kann ich die Frage nach meinem Alter beantworten. Die Therapeutin hatte einmal die Vermutung geäußert, dass ich zwischen 8 und 10 Jahre alt war. Damit lag sie richtig. Meine Schwester Lisa erinnerte sich, dass ich in diesem Alter sehr ernst geworden wäre und sie nicht mehr richtig angucken konnte, nur so vor mich hinschaute, mit nach unten gerichtetem Blick. Später wurde ich immer sicherer, dass meine Brüder nach dem Auszug meiner ältesten Schwester mit ihren Taten begannen. Ich war damals 9 Jahre alt. Wir Jüngeren erlebten und liebten Maria wie eine Mama. Sie war sehr für uns da, aber dann leider nicht mehr. Sie hätte vielleicht etwas bemerkt. Ob sie hätte helfen und einschreiten können, ist zu bezweifeln.

Die Frage, die ich mir schon bald stellte, ob denn keine oder keiner damals etwas von den Übergriffen bemerkt hatte, blieb noch offen.

2.2 Bernd

„Ich brauchte Liebe und bekam *das!*"

Nach dem vorläufigen Abschluss des Themas Alter konnte ich in der Therapie zum ersten Mal zaghaft und mit viel Scham von dem Erlebten mit Bernd berichten: Ich würde mich nur an eine Situation mit ihm erinnern. Ich hätte auf meinem Bett gelegen, wohl im Dunkeln, und er hätte mich an meiner Vagina berührt. Ich hätte noch keinen Busen gehabt, das wisse ich ganz sicher. Bernd sei eher behutsam gewesen. – Ich habe irgendwann zu ihm gesagt, Adolf macht das auch ... Bernd reagierte ganz verdutzt und sagte nach einer kleinen Weile, es wäre nicht in Ordnung und käme auch nicht mehr vor. Und hörte gleich auf. – Ich hätte beim Erzählen am liebsten geweint, das ging aber nur ein bisschen.

Ich will hier kurz einfügen, was ich bald erkannte und was mir sehr wichtig war: Indem ich zu Bernd gesagt hatte, Adolf macht das auch, habe ich mich in dieser Situation, entgegen meiner ersten Annahme, doch gewehrt, wenn auch eher indirekt. Ich hatte Adolf erwähnt, weil ich ahnte oder wusste, Bernd will nicht mit ihm auf einer Stufe stehen; Adolf, der in unserer Familie nicht so geliebt und „angesehen" war. Und so war es dann tatsächlich, es funktionierte. Bernd ließ von mir ab und hat mich dann nie wieder so berührt!

Nachdem die K. ein paar Fragen gestellt hatte, gestand ich ihr, dass ich mich damals fragte, als er mich danach in Ruhe ließ, ob er mich jetzt trotzdem noch mag. Das fände ich zum Kotzen, aber so hätte ich damals gedacht. Sie gab mir zu bedenken, dass ich ihn doch gerne gehabt hätte, die Frage wäre also sehr verständlich. Mir ging durch den Kopf, was ich bald aussprach: Vielleicht habe ich Zuwendung gebraucht und gewollt und deshalb nicht gleich nein gesagt ... – Mir war nach Schreien zumute. Ja, entgegnete die K., vielleicht sei es so gewesen, weil ich ihn gerne gehabt hätte; ich hätte ihn doch auch toll gefunden, den großen Bruder. Bei einem fremden Mann hätte ich es vielleicht nicht zugelassen. Wieder kam ein Gefühl von Schuld in mir auf. Sie betonte nochmals, das sei doch verständlich, ich hätte ihn gern gehabt; das habe mit Schuld doch nichts zu tun. So oder ähnlich äußerte sie sich. – Ich finde die Vorstellung furchtbar, ich hatte ihn gern, wollte Zuwendung und er macht sowas! – Ich war zwischendrin so erschrocken über diese Vorstellung, dass ich auf einmal nicht mehr wusste, wieso ich mich so erschreckt hatte. Dann fühlte ich es wie früher als Kind, diesen Widerspruch, ich habe ihn lieb ... und der macht etwas, was ich nicht will, nicht verstehe, was mir weh tut! Ich begann zu weinen. Sagte bald: Ich fühle mich wie beschmutzt! – Und noch ein schlimmer Gedanke tauchte auf, aber ich konnte ihn nicht aussprechen. Das schaffte ich erst nach einer Weile, unter heftigem Weinen: Ich brauchte Liebe und bekam *das!* Ich hab' doch viel zu wenig Liebe bekommen! ... Und dann das! –

Ich hätte die Therapeutin, während ich sprach, gerne angesehen, aber ich schämte mich zu sehr.

Über diese Stunde sprach ich später mit einer Freundin; dabei haderte ich sehr mit mir, weil ich Bernds Übergriff anfangs zugelassen hatte. Ich sei doch noch ein Kind gewesen, meinte sie und fügte ganz ernst hinzu, sie habe sogar als erwachsene Frau einiges über sich ergehen lassen. – Als erwachsene Frau!, hatte sie nochmals betont. Sie begründete ihr Verhalten u. a. damit, dass sie damals ihren Partner nicht verlieren wollte. Diese Aussage stimmte mich mir gegenüber ein bisschen milder. Ja, ich war doch noch ein Kind! Und ich habe mir doch zu helfen gewusst mit dem Satz, Adolf macht das auch. Das weitere Geschehen, ob es weitere Attacken von Bernd gab oder nicht, hing zwar immer noch von ihm ab, aber ich erreichte, was ich wollte. Er ließ mich danach in Ruhe.

Was mich aber weiterhin sehr quälte war der Gedanke, ich wollte Liebe und bekam *das!* Das ging sogar soweit, dass ich sagte: Ich hasse mich für das, was ich von ihm wollte, und weil ich mich nicht mehr gewehrt habe. – Dass er „dabei" auch irgendwie lieb war, das mache alles noch viel schlimmer!

Wie schwer fiel es mir, der K. zu sagen, Bernd war auch lieb, und dass ich mich freute, weil er mich wahrnahm, sich mir zuwandte! Ich befürchtete, nein, ich hatte richtig Angst davor, von der K. könnten ähnliche Vorwürfe kommen, wie sie mir meine Eltern gemacht hätten: Wenn Du von ihm Zuneigung willst, bist Du selbst schuld. Dann musst Du Dich nicht wundern, dass er das macht! – Aber diese kamen ganz und gar nicht. Im Gegenteil, die K. reagierte mit sehr viel Verständnis.

Eine exemplarische Therapiestunde

Nach einem halben Jahr kam ich auf die Situation mit Bernd zurück. Die Therapeutin und ich hatten über eine andere Sitzposition diskutiert. Sie regte an, mich eventuell auf die vorhandenen Matratzen zu legen. Das wolle ich nicht. Für mich selbst überraschend, hatte ich hinzugefügt, dann könne ich mich nicht wehren. Sitzen sei besser.

Daraufhin wurde ich immer ernster, saß ganz angespannt dort und plötzlich tauchte das Bild von der Situation mit Bernd auf. Erschreckt versuchte ich, es wieder zu verdrängen. Weil dies nicht gelang, erzählte ich, was in mir vorging. Durch die Bestärkungen der Therapeutin ließ ich mich auf die Erinnerung ein und beschrieb sie ihr, mein Gesicht dabei mit meinen Händen verdeckend: Ich liege quer auf meinem Bett, meine

Beine hängen über den Bettrand, und er sitzt neben mir. – Ich fügte hinzu, dass meine Arme seitlich neben mir lagen. – Können Sie Ihr Gesicht sehen? Wie gucken Sie? – Ich glaube so, wie ich im Moment gucke. – Ich hatte das Gesicht verzogen wie eine, der etwas unangenehm und eklig ist und die das nicht will. Zwischendrin sprach ich innerlich mit der kleinen Ann. Währenddessen hatte ich meinen Oberschenkel mit der linken Hand ein bisschen gestreichelt, die Hand so hin und her bewegt. Nun fragte die K., was die Hand machen würde. Ich nahm meine Hand wieder vors Gesicht und sagte, ich hätte das Gefühl, ich müsse die kleine Ann beschützen und trösten, ihr sagen, sie sei nicht alleine. Sie also wie an die Hand nehmen.

Auf einmal spürte ich Gefühle in meiner Vagina! Ich erschrak fürchterlich, verdeckte mein Gesicht komplett mit meinen Händen. Ich hab' auch damals was gefühlt!, schoss es mir durch den Kopf. Das hatte ich zuvor nicht gewusst, einfach vergessen, verdrängt. Entsetzt sagte ich der K., ich könne etwas nicht aussprechen, es sei so schlimm. In leicht strengem Ton meinte sie, ich würde doch nicht alleine Therapie machen. Ich solle es doch sagen. Auf meinen Protest hin meinte sie, in milderem Ton, sie sei doch nicht streng, und ermunterte mich wieder, es zu versuchen. Das gelang mir nach einer Weile. Ich sagte, dass ich erotische Gefühle spüren würde. – Ich kann's nicht fassen! Das hab ich ja gar nicht gewusst! ... Ich find's entsetzlich! –

Ich wolle noch etwas Anderes sagen, aber es ginge nicht. – Es geht Ihnen doch besser, wenn Sie es aussprechen!, sagte sie ganz mild. Ich nahm mir Zeit, bis ich auch das in Worte fassen konnte: Ich will und kann da nie mehr was fühlen! ... Ich könnte mir das rausschneiden! – Ich bedeckte weiter mein Gesicht so, dass ich kaum Luft bekam. Ich wollte nichts sehen, mich ganz auf mich konzentrieren und streichelte ein bisschen mein Gesicht, um mich zu trösten. Ob ich mein Gesicht auch zuhalten würde, damit sie mich nicht sähe?, fragte sie. – Ja! Ich könnte in den Erdboden versinken!, so sehr schämte ich mich.

Ob ich mich nicht auf den Boden setzen wolle? – Nein! Warum sollte ich? – Dann kann ich mich hinter Sie setzen. – Nein, das will ich nicht! Wenn ich Sie in meiner Nähe haben will, wollen Sie nicht. – Das sagte ich in Anspielung auf eine Situation, in der sie sehr auf Abstand ging und blieb. – Jetzt will ich nicht!, fügte ich hinzu. – Das ist es doch nicht! Sie wollen sich selbst bestrafen! – Ich will niemand in meiner Nähe haben! – Sie sprach wieder von Selbstbestrafung und dass ich mich quälen würde, wenn ich so für mich bliebe.

Ich kam nur schwer aus dem trotzigen Gefühl heraus. Aber dann wurde ich so wütend: Diese Sau!, schimpfte ich und vielleicht auch noch mehr. Mit großen Hemmungen und mit Unterstützung und Ermutigung von Seiten der Therapeutin gelang es mir nach einer Weile, meine Wut auf Bernd auszutoben. Ich stand auf und schlug mit einer kleinen Peitsche auf eine Matratze. Diese Peitsche hatte ich mitgebracht, als es schon einmal um die Wut auf Bernd ging, ich danach etwas Adäquates, mir Entsprechendes suchte, um sie angemessen austoben zu können. Dabei war ich auf die Peitsche gekommen. Die K. ermunterte mich nun, noch fester zuzuschlagen, und das machte ich. Ich tobte meine Wut ganz ordentlich aus. Und das tat gut!

Drei Tage in die Wanne legen, so wäre jetzt mein Gefühl, das würde ich brauchen, weil ich mich so beschmutzt fühlte. Ich saß, völlig außer Atem, auf dem Boden, als die K. langsam auf mich zukam. Ich solle sagen, wenn es mir zu dicht werden würde. Es sei ok, sagte ich, als sie sich mir gegenüber, ungefähr eine Armlänge von mir entfernt, hinsetzte. Sie streckte eine Hand aus, flach, mit der Innenseite nach oben, und forderte mich auf, meine Hand auf die ihre zu legen. Ich zögerte, war ganz verdutzt, während sie mir ihre Absicht erläuterte. Unsere Hände nun ganz flach aufeinanderliegend, demonstrierte sie mir, dass ich selbst bei einer nicht aktiven Berührung etwas spüre, also immer etwas fühle, wenn mein Körper berührt wird, auch wenn ich nichts fühlen will. Ich zog meine Hand bald zurück mit dem Hintergedanken, wenn ich sie wegziehe, tut es weniger weh, als wenn sie ihre wegnimmt. Weil ich es schön fand, ihre Hand zu fühlen …

Meine zweite Hand hielt ich weiterhin so vor mein Gesicht, dass die Therapeutin meine Augen nicht sehen konnte. Währenddessen guckte ich nach unten, also vor mich hin. Dann sah ich die K. an, kurz, empfand ihre Augen als ganz starr und streng. Versuchte es nochmals. Wieder nahm ich sie so wahr. Auf einmal fiel es mir wie Schuppen von den Augen. Ich erzählte, wie ich ihren Blick aufgenommen hatte und an was er mich erinnerte: So war es damals mit meiner Mutter. Ich hab' damals gedacht, die muss mir das doch ansehen! Die muss doch was merken! – Da die Stunde zu Ende ging, forderte die K. mich auf, alles dort zu lassen. Sie wisse, das sei schwer. Ich wolle noch ein bisschen im Raum bleiben, ob das in Ordnung sei? – Ja! –

Ich konnte nicht weinen, saß nur fassungslos dort. Sie kam nach einiger Zeit noch einmal in unseren Raum, schaute nach mir, was mich sehr freute und mich tröstete. Ich würde mich so alleine fühlen. Ja, das könne sie mir nicht abnehmen, es wäre hart, oder ähnlich ihre lieben Worte. – Ich weiß, dass ich das aushalten muss, aber es ist schwer. Ich fühle mich einfach sehr alleine. Brauchte es, mich nicht so zu fühlen. – Sie entgegnete, und das fand ich sehr schön: Sie sind doch nicht alleine! Deshalb bin ich doch noch einmal gekommen. *Jetzt* sind Sie doch nicht mehr alleine! – Es tröstete mich wirklich sehr, dass sie nochmal nach mir sah.

Während dieser Stunde hatte ich zweimal gesagt, ich halte das nicht aus. Und sie: Doch, das halten Sie aus! – In der folgenden Stunde sprach sie davon, dass die letzte eine sehr schwere und mutige Stunde gewesen sei, die schwer zu verkraften wäre und viel Kraft koste. Auch das tat mir gut.

Bald darauf fragte ich mich doch tatsächlich, ob ich das, was Bernd tat, vielleicht von ihm gewollt hatte. Ein ungeheuerlicher, furchtbarer Gedanke, den ich in der Therapie aussprach. – Kann ich so etwas gedacht haben? ... Aber das geht doch gar nicht! Ich hatte doch überhaupt keine Vorstellung davon!, beantwortete ich mir meine Zweifel selbst. Aber gleich darauf stellte sich mir eine andere Frage. Ob ich mir lieber das noch einmal gewünscht hatte, das, besser als gar keine Zuwendung? Ich konnte diesen Gedanken kaum zulassen. Was für ein Notstand, welch armseligen Zustände hatten bei uns geherrscht, dass ich eventuell selbst „sowas" in Kauf genommen hätte. – Es ist so ungeheuerlich! Dazu gibt es wirklich nichts mehr zu sagen! – Mir war kotzübel, und das verstehe ich heute noch sehr gut!

Obwohl die Situation mit Bernd im Vergleich zu denen mit Adolf nicht so schlimm war, beschrieb ich sein Handeln später so: Was er gemacht hat ist so verletzend! Es ist wie ein Schnitt in meine Haut! –

Meine Haltung Bernd gegenüber nach dem Bearbeiten in der Therapie

Es gab eine Zeit, in der ich mich fragte, warum habe ich ihn heute eigentlich noch lieb? Und noch später konnte ich sogar das Gefühl zulassen, dass er mir fehlt. Vielleicht will ich mit ihm irgendwann einmal über die sexuelle Gewalt reden. Vielleicht brauche ich das, so meine Gedanken, um mich mit mir und meiner Vergangenheit auszusöhnen. Um diese Vergangenheit anzunehmen. Dazu benötige ich noch Zeit.

Ich habe mich tatsächlich ein paar Jahre später mit ihm getroffen und ihn mit meinen Erinnerungen konfrontiert. Mit großer Angst bereitete ich mich in der Therapie darauf vor. Wie wird er reagieren? Egal was er sagt, ich weiß, dass er das gemacht hat! Da war ich absolut klar. Dennoch hatte ich Angst vor einer Begegnung. Und dennoch traf ich mich mit ihm. Alleine. In meiner Wohnung. Und das war gut. Anfangs sehr aufgeregt, wurde ich immer ruhiger, konnte ihm alles sagen und seine Reaktion war gut, angenehm. Das war nicht wirklich überraschend, dachte ich später; ich hatte sie ihm irgendwie zugetraut. Aber sicher kann frau sich dabei ja nicht sein ... Im Kapitel „Reaktionen" beschreibe ich die Vorbereitung und das Treffen mit ihm ausführlich.

Ich bin mir heute noch dankbar, dass ich ihn zur Rede stellte, und froh über den Ausgang dieses Treffens.

2.3 Adolf

Träume mit Adolf

Schon bald träumte ich sehr oft von Adolf und den Situationen mit ihm. Dabei stand im Mittelpunkt, dass er mich mit seinen Übergriffen bzw. seinen Versuchen richtig verfolgte. Er ließ mir mit seinen Attacken einfach keine Ruhe, und das über Jahre hinweg!

Ich trage hier ein paar Träume zusammen, die die Realität sehr gut beschreiben und die vorwegnahmen, was ich dann in den Therapiestunden erinnerte und bearbeitete. Das begann schon bald nach Beginn meiner Therapie.

> *Traum 1: Adolf kommt ins zimmer, berührt mich am arm, touchiert meinen busen, und will von mir geld. Er fordert mich immer wieder dazu auf und packt mich dauernd am arm oder an meinen schultern. Lass mich, ich will nicht, dass du mich berührst, ich gebe dir nichts!, so immer wieder meine abwehr. Du sollst mich in ruhe lassen und mich nicht anfassen! Akzeptiere das endlich mal! Erinnere dich, du hast mir schon genug angetan. Er ließ nicht locker. Ich nahm einen stuhl, um ihn abzuwehren, und lief dann weinend aus dem raum und sah Maria [meine älteste Schwester] im garten arbeiten, lief weinend an ihr vorüber in der hoffnung, sie fragt nach, was los ist! Aber da kam nichts!*

Nachdem ich diesen Traum in der Therapie erzählt hatte, fügte ich hinzu: Mein Mann hat mal gehört, dass ich im Schlaf sagte, irgendwann erschieße ich ihn noch! Ich bin richtig wütend, dass ich mich nicht gewehrt habe! – Die Therapeutin stellt sofort klar: Aber Sie haben sich doch gewehrt! – Ich habe aber nichts ausgerichtet! Er hat meinen Willen nicht akzeptiert. – Ein Wehren ohne Erfolg ist für Sie anscheinend kein Wehren. Wenn auch nur begrenzte Möglichkeiten bestehen, sich zu wehren, ist es doch immer ein Sich-Wehren! – Da kann ich ihr nur recht geben. Damals brauchte ich noch eine ganze Weile, um zu dieser klaren Sicht zu kommen. Und es ging weiter mit dem Träumen:

> *Traum 2: Adolf und ich in einer wohnung, alleine ... Er hat geschenke für mich, will mich berühren, mich dazu überreden, es zuzulassen. Ich will nicht. Er wolle mich nur berühren, nicht mit mir schlafen. Ich weiche dauernd aus; versuche, ihn auf dem balkon auszusperren, aber er hat den schlüssel. Mir fällt nichts mehr ein, wie ich ihn abwehren könnte, denke dann, lass' es zu, ist ja auch ein angenehmes gefühl dabei ...*

Durch den Traum, so schrieb ich danach in mein Tagebuch, kann ich mich besser verstehen. Ich versuchte, so notierte ich weiter, mich zu wehren, soweit meine Möglichkeiten reichten und tröstete mich mit dem Gedanken von wegen der angenehmen Gefühle über meine Hilflosigkeit hinweg. Dennoch war das mit den Gefühlen im Nachhinein schwer zu ertragen.

Sicher auch deshalb hatte ich Schwierigkeiten, den Traum in der Therapie zu erzählen. Der Traum käme der Realität sehr nahe, auch wenn ich darin eine erwachsene Frau gewesen sei, was ich so interpretieren würde: Ich habe ja schon als kleines Mädchen gesagt bekommen, Männer können nichts dafür, das sind ihre Triebe. Das musst du wissen. Ich bin also verantwortlich. – Ja, ja, die können nichts dazu, ist ihre Natur! Das ist ja wirklich toll!, so oder ähnlich reagierte die K. laut und ärgerlich. Später stellte sie nochmal heraus, dass ich das alles doch nicht gewollt und mich sogar dagegen gewehrt hätte.

In der folgenden Stunde wäre es mir am liebsten gewesen, die Therapeutin könnte mich nicht sehen, so sehr schämte ich mich wegen der im Traum vorkommenden angenehmen Gefühle. Ich wäre am liebsten unsichtbar gewesen. Noch besser: In den Erdboden versunken! Nach sehr langem Schweigen sagte ich zu der vorigen Stunde, dass es mir danach nicht gut ging, ich sei sehr ernst und traurig gewesen, geradezu fassungslos. – Mir ist, als hätte mir jemand das erzählt, als hätte ich zuvor nicht gewusst, was der gemacht hat. So ging es mir auch, als ich überhaupt zum ersten Mal darüber redete. Als hätte ich zum ersten Mal davon erfahren. –

> *Traum 3: Wieder bedrängt mich Adolf auf seine eklige weise, ich wehre ihn dauernd ab. Höre meine mutter draußen, sage lauter: Der soll mich in ruhe lassen! … Nein, das kann ich ihr nicht sagen, die regt sich auf, das ist gefährlich.*

Anschließend träumte ich, dass ich meiner Schwester Lisa erzähle, dass mir ein Mann etwas angetan hätte. Er habe mich vergewaltigt. Ich will sowas nicht träumen!, fügte ich im Tagebuch hinzu. Im nächsten Traum ging es dann wieder um Adolf.

> *Traum 4: Er bedrängt mich, lässt mir keine ruhe. Er soll das lassen, ich will das nicht. Er, in seiner unangenehmen art, tut, als hätte ich nichts gesagt.*

Nach dem Erzählen in der Therapie sagte ich: Danach erwachte ich mit heftigem Herzklopfen. So habe ich das wirklich erlebt! – Und dann musste ich etwas sagen, was mir sehr, sehr unangenehm war. Dass ich eigentlich auch Berührungen wollte, weil die auch was Angenehmes hatten. Das sei doch furchtbar, einfach zum Kotzen! Daraufhin sagte die K. Schönes und Entlastendes: Sie wollten ja nicht von ihm, sondern überhaupt berührt werden. Danach sehnten Sie sich ja! … Eigentlich wollten Sie es doch so haben: Ich will bestimmen, von wem ich berührt werden will! – Ja! – Vielleicht habe ich das Folgende damals schon gedacht oder gegenüber der Therapeutin ausgesprochen, es gehört jedenfalls dazu: Und ich will bestimmen, wo und wie ich berührt werden will!

Erinnerungen mit Adolf

Die ersten Erinnerungen tauchten nach den folgenden Träumen auf:

> *Traum 5: Er wieder mit seinem ekligen lächeln …*
>
> *Traum 6: Ich lasse meinen ärger über ihn raus. Auf einmal ist meine mutter dabei, die sagt, ich solle mich nicht so anstellen! Sie guckt, als wüsste sie etwas. Ich werde so wütend! Sage, du weißt ja nicht, was der alles macht!*

Nach dem letzten Traum notierte ich in mein Tagebuch:

> *Ich hab angst, wieder von ihm zu träumen, starkes herzklopfen, lasse das licht an, rede beruhigend auf mich ein! Dann: Ich muss das endlich angehen, in der therapie bearbeiten und dann abschließen! … Ich habe dauernd ein bild im hinterkopf, von einer situation mit ihm. … Schreibe alles auf.*

Die Therapeutin war zu diesem Zeitpunkt in Urlaub; da war es schon mal entlastend, die erinnerte Situation mit Adolf im Tagebuch festgehalten zu haben. Aber darüber in der nächsten Stunde zu sprechen fiel mir außerordentlich schwer. Bevor ich die Träume mit ihm schildern konnte, spürte ich Angst, auch Angst davor, dass das noch mehr Erinnerungen und Bilder auslösen könnte. Ich erzählte die Träume dennoch. Es könne wirklich so gewesen sein, dass ich meiner Mutter gegenüber etwas angedeutet hätte. Am Tag nach dem Traum sei dann die Erinnerung an eine Situation mit ihm aufgetaucht. Ich zitiere von dieser Stunde aus meinen Therapienotizen:

> *(…) Ich konnte erstmal nicht weiterreden, ich wollte das aber doch sagen! Es sei das bild, dass ich zu beginn der therapie in erinnerung hatte. Mein weiches shirt an mich schmiegend, an mein gesicht, wie eine schmusetuch, sprach ich weiter, sagte erstmal spontan: Ich möchte aber nicht, dass sie denken, ich bin schlecht! … Ich möchte selbst nicht so von mir denken! – Nach und nach sprach ich von meinen erinnerungen, sah die situation sehr deutlich vor mir: Wir sind in der küche*

... Ich sehe die küche ... Es ist sonst niemand da. Ich habe meine schürze an. Adolf drückt mich mit meinem rücken an sich, so, dass ich ihn im rücken spüre ... – Ich konnte nicht sagen, dass ich meine, bin mir ja nicht sicher, seinen harten penis gespürt zu haben. Ich konnte das nicht aussprechen, aber sie wird es verstanden haben oder sich denken können. – Ich weiß nicht, ob seine hände was gemacht haben ... –

Dass ich auch noch eine schürze anhatte, das passt doch!, stöhnte ich verzweifelt. – Ehem!, kam bestätigend von der K. Dann sagte ich, und das fiel mir am schwersten: Und dann hat er mir geld in die schürzentasche gesteckt! – Ich sagte es so leise, weil's mir doch so schwerfiel, ich mich sehr schämte, dass die K. mich nicht verstand, ich das wiederholen musste. Auch das noch! Ich wiederholte es ... – Ich hab' jetzt solche rückenschmerzen! – Ich streckte mich, setzte mich aufrechter hin. –

Das mit dem geld, das ist doch unerträglich! Das ist das schlimmste! – Jetzt sagte die K. mal wieder etwas; sie hatte mir lange schweigend zugehört: Das sehe ich anders! Das ist quasi wie eine anerkennung. Gegenleistung. Er hat sich damit etwas erkauft. Nein, erschlichen stimmt eher. – Sie denke, ohne das geld wäre es noch schlimmer für mich gewesen. So oder ähnlich erklärte sie mir ihre sicht. – Aber sie fühlen das so, ich will ihnen das auch nicht ausreden. – Das ist doch pervers!, hatte ich in ihre Erklärung hinein gesagt. – Ja, eigentlich ist es pervers! Aber auch entlastend. –

Ich hatte anschließend die Vermutung geäußert, dass es diese Situationen mehrmals gab. Und dass auch seine Hände „was" gemacht haben. Ich wäre bestimmt schon 8 Jahre alt oder älter gewesen, also kein ganz kleines Mädchen mehr. Ich zitiere weiter aus meinen Aufzeichnungen von der Therapiestunde:

Ich weinte, kam mir so verletzt vor. Sagte leise vor mich hin: Wie dreck! – Die K. fragte nach. – Dass der das mit mir machen konnte! Er, der in der männerhierarchie bei uns an letzter stelle stand, das konnte er sich dennoch rausnehmen! Ich war wie dreck! Mich kann man einfach benutzen! War doch nichts wert! – Ich dachte an die arme, kleine Ann, und musste noch mehr weinen. Ich sah mal kurz zur K. hin. – Und der hat mich behandelt wie eine frau auf unterster stufe, wie eine dirne! – Ich schwieg. – Nachdem ich mich letzte woche an diese situation erinnerte, hab ich gedacht, meine brüder und mein vater haben mir mein ganzes leben versaut! Bis zum heutigen tag! – Ich wisse, dass ich tierische wut auf Adolf hätte, aber im moment sei davon nichts da. Spürte sie aber tief in mir drin. Das bild mit ihm sei immer noch da, es gehe einfach nicht weg! Sie forderte mich auf, ich solle versuchen, das bild dort zu lassen. Schwierig! Ich war traurig, als sie ging, konnte aber noch ein bisschen in unserem raum bleiben …

Ich schloss meine Aufzeichnungen von der Stunde mit diesem Satz: Ich fühlte und fühle mich erleichtert, nachdem ich der K. das heute alles sagen konnte! – So ging es mir damals des Öfteren. Erst plagten mich Scham und/oder Angst, darüber zu reden, manchmal brauchte ich sogar ein paar Anläufe, ein, zwei Stunden, bis ich heftige Träume, aber besonders die Erinnerungen wiedergeben konnte. Danach aber fühlte ich, wie nach dieser Stunde, oft große Erleichterung und Entlastung. Manchmal aber auch schlechtes Gewissen, weil ich nicht mehr schwieg.

Wunschphantasien und Angst davor, sie auszusprechen

Ich will hier noch ein wunderschönes Bild beschreiben, das ich nach den oben erwähnten Träumen, also noch während der Ferien der K., phantasierte. Ich finde, das muss einfach erwähnt und beschrieben werden: Wenn ich ein kleines Mädchen wäre, würde ich auf die K. zugehen, sie mit meinen kleinen Armen umfassen, reiche ihr nur bis zum Bauch, und würde mich an sie schmiegen, mich geborgen fühlen. Sah mich von hinten, die K. umklammern, weinte fast beim Phantasieren. … Meinen Kopf an ihrem warmen Bauch gelehnt, gekuschelt. Wie Schutz suchend. – Halt mich fest, Mama! Bitte, halt mich lieb und fest! – Ich wünschte mir, ihre Arme würden mich auch lieb umschlingen, auf meinem kleinen Rücken liegen. So eine schöne Vorstellung!

Und dieses Bild war immer mal wieder da. Es machte mich todtraurig, und gleichzeitig fand ich es wunderschön!

Als ich der Therapeutin in der folgenden Stunde von dem schönen Bild mit ihr erzählen wollte, tauchte wieder Angst auf, was ich ihr sagte. Ich beschrieb es ihr dennoch. Fing zwischendrin an zu weinen; bald weinte ich sehr und spürte weiter heftige Angst. Für mich bedeute das Bild Schutz und Geborgenheit. – Als ich gestern darüber nachdachte, ob ich Ihnen das erzählen kann, dachte ich, nein, das geht nicht. – Und warum nicht? – Ich hatte Angst, dass Sie mich abschmettern, weil das Bild bedeutet, ich will etwas von Ihnen. Jetzt habe ich wieder so starke Angst. Aber ich will keine Angst haben! Ich will mir was wünschen können, ohne Angst zu haben!, schimpfte ich. – Ehemmm! – Ich fand's schön, dass an dieser Stelle ihr bestätigendes „Ehem“ kam. Für sie würde das Bild auch Vertrauen bedeuten: Sie wissen, ich komme wieder. Und Sie wissen, dass Sie wieder hierher kommen und mich umarmen können. – Ich lächelte, freute mich sehr über ihre Deutung.

Aber etwas später beschwerte ich mich bei ihr, weil sie nichts zu meiner Angst gesagt hatte, mich mit meiner Angst alleine ließ. Ihre Entgegnung lautete ungefähr so: Ich denke, Sie suchen nach Antworten und wollen die von mir haben. Aber die müssen Sie selbst finden. Sie müssen Ihre Antworten finden! Sie wollen sie von mir hören, aber das müssen dann nicht Ihre sein. – Ich glaube, ich verstand ungefähr, was sie mir damit sagen wollte.

Zu ihrer Frage, wofür die Angst denn gut sei, konnte ich nichts sagen. Mir fiel auf, sobald ich die K. ansah, ging die Angst weg. Auf meine Frage an sie, warum das wohl so sei, vermutete sie, weil ich merken würde, dass sie meine Angst und auch meine Wünsche annehmen könne. Und dass das von mir Befürchtete nicht geschehen würde.

Nach dieser Stunde dachte ich, ich muss doch viel Vertrauen zur K. haben, dass ich ihr das alles sagen konnte. Trotz der Angst auch viel Vertrauen!

„Er hatte mich im Griff!"

Nun komme ich wieder auf Adolf zurück. Meiner Schwester Lisa war zu ihm eine Begebenheit von früher eingefallen, von der sie mir erzählte. Adolf habe zu mir gesagt: Gell, Du bist mein Mädchen! Wir verstehen uns. – Ich weinte laut, als Lisa das schilderte und sagte bald: Der hatte mich im Griff! – Lisa fuhr fort, sie habe sich damals gewundert, dass Adolf „so einen Kontakt" mit mir hatte, weil er sonst doch mit niemandem nahe gewesen wäre. Es hätte so „privat" gewirkt, so drückte sie sich aus. Sie habe ganz stark das Gefühl, das hätten alle gehört. Und die Eltern wussten das! Unsere Mutter habe im Sessel gesessen und so wissend geschaut. Ich weinte heftig, konnte mich kaum beruhigen. In einem späteren Traum drückt sein Gesicht genau das aus: Ich gehöre ihm!

Lisa erinnerte sich an mich zum damaligen Zeitpunkt, an meinen Gesichtsausdruck. Ich hätte geschaut wie auf dem Foto, das wir uns mal zusammen ansahen. (Dieses Foto hatte ich der K. ebenfalls gezeigt. Ich war ungefähr 10 Jahre alt, mein fröhliches Kinderlächeln war weg!) Als ich von all dem in der Therapie berichtet hatte, sagte ich: Ich war ganz alleine, isoliert von den Anderen. Ich fühle mich jetzt so leer wie auf dem Foto: Es ist alles vorbei! Macht mit mir, was ihr wollt. Stellt mich hin, fotografiert mich oder was auch immer. Ist mir egal! – Während ich das aussprach war ich wie weg. Dann spürte ich plötzlich Hunger. Hunger nach Liebe, dachte ich. Und dann schaute ich die K. lange schweigend an, und als sie draußen war, weinte ich … – Früher …, das war eine Katastrophe für mich! – Dies hatte ich noch zur Therapeutin gesagt. Ja, eine Katastrophe!

Und ich träumte wieder und weiter von und mit Adolf.

Traum 7: Ich bin im bad, Adolf kommt rein. ... Wenn er mich nicht in ruhe ließe, würde ich alles sagen! ... Er droht mir gewalt an!

In der folgenden Therapiestunde erzählte ich den Traum, mit ganz viel Wut und Hass. Heiner habe gemeint, berichtete ich weiter, ich solle Adolf lieber nicht auf seine Taten ansprechen, sonst müsse ich mit Gewalt von ihm rechnen, Adolf könnte mit irgendwas zum Schlagen bewaffnet vor meiner Tür auftauchen. – Ach, du Schande! Das gibt's doch nicht! – Doch! So ist der! Ich hatte früher Angst vor ihm, jetzt nicht mehr! Ich würde ihn erschlagen! ... Wenn ich an früher denke, komme ich mir wie permanent verfolgt vor! – Das mit dem „ich würde ihn erschlagen, habe keine Angst mehr vor ihm", stimmte nicht ganz. Nur in diesem Moment fühlte ich es so, aus lauter Wut und Hass.

Mir wurde zunehmend klarer, dass seine Übergriffe viel zahlreicher und umfassender waren, als ich das zu Anfang der Therapie in Erinnerung hatte. Ich näherte mich immer mehr dem an, was damals geschah; meine Erinnerungen bekamen klarere Konturen.

Die Wut zulassen

Endlich spürte ich auch meine Wut, starke Wut! Auch wenn es mir schwerfiel, dazu zu stehen oder gar sie auszutoben.

In diesem Zusammenhang erwähnte ich in einer Therapiestunde einen Traum, in dem ich versuchte, einen Mann mit einem Stock zu verprügeln, was mir aber eher wie streicheln vorgekommen sei. Das habe den überhaupt nicht berührt. So käme mir das hier jetzt auch vor, ich sei so wütend, aber mir fiele es schwer, meine Wut zuzulassen. Die K. versuchte mich dazu zu ermuntern mit ihrer Aussage, das mache eine doch maßlos wütend, wenn jemand die eigene Wut nicht ernst nehme, wie es in den Träumen der Fall gewesen sei und wie sich Adolf früher verhielt. Das hatte ich zuvor nämlich erwähnt: Der hat doch sogar über mich gelacht, als ich mich wehrte! Als ich sagte, er solle mich in Ruhe lassen! Der hat sich noch über mich lustig gemacht! – Das müsse mich doch wütend machen!, betonte sie nochmals. – Ja, schon. Aber ... – Ich war weiter wie gelähmt und spürte Schwindel im Kopf, wie schon öfters, wenn ich wütend war, die Wut aber nicht zulassen konnte oder wollte.

Immerhin schaffte ich es, mal auszusprechen, warum ich so wütend war und zwar ungefähr in diesem Wortlaut, ich zitiere aus meinen Therapieaufzeichnungen:

Weil ich unter all dem, was diese säue gemacht haben, heute noch leide!
Was die mit so einem kleinen mädchen gemacht haben!

Weil ich mich dann nicht mehr als frau fühlen wollte! Meinen körper nicht mehr wahrnahm, nichts mehr fühlen wollte!

Weil ich es jetzt so schwer habe, meine gefühle zuzulassen! Mich überhaupt zu fühlen! Und meiner wahrnehmung zu vertrauen!

Weil es heute noch wie eine bombe ist, wenn mich jemand verbal angreift! Ich mich dann gar nicht mehr spüre!

Weil ich dagegen ankämpfen muss zu denken, ich bin immer noch mit diesem thema befasst, muss mich damit zurückhalten.

Weil ich jetzt dauernd spüre, welche auswirkungen das heute noch auf mich hat!

Weil sich das so vermischt bei meinen gefühlen für sie! Und dass ich keine klaren gefühle haben kann! –

Zwischendrin weinte ich fast. Die K., als ich geendet hatte, ganz energisch: Ich denke, sie müssen die verklagen! So ist es doch nur in dem kleinen kreis ihrer familie! Es gehört in die öffentlichkeit! – Aber das geht ja nicht mehr, ist zu spät! Es ist verjährt! – Es hat mir aber sehr gut getan, das alles mal ausgesprochen und damit dazu gestanden zu haben.

Im folgenden Traum wehre ich mich, werde sogar handgreiflich und das finde ich heute noch gut:

> Traum 8: Adolf wie er leibt und lebt. Der hat mich richtig gequält. War wie ein verrückter! Lässt mich nicht in ruhe. Ich sage, ich will nicht! Er lacht wie ein blöder! Er hört nicht auf! Dann szenenwechsel, ich bitte andere, ihn festzuhalten, und dann schlage ich ihn auf seinen nackten arsch, kratze mit meinen fingernägeln, ziehe die so richtig über seinen rücken! Du arsch! Als schon fast das rohe fleisch zum vorschein kommt, sage ich: so hast du mich gequält! (…)
>
> *Beim wachwerden sehr aufgewühlt, er so nah, seine bescheuerte art. Es lohnt sich nicht zu leben! Was frau schon als kleines mädchen alles abkriegt, es lohnt sich wirklich nicht! … Beim aufschreiben tränen. Fühle mich gerade völlig überfordert …, auch alleine. Will zu meiner mama [zur Therapeutin], und die soll mich liebevoll in die decke packen und mir eine schöne geschichte erzählen, und ich lausche ihrer stimme!*

In der folgenden Stunde erzählte ich der Therapeutin, dass mir erst im Zusammenhang mit Selbstbefriedigung klar wurde, dass das, was die Brüder mit mir gemacht hatten, mit Sexualität zu tun hat. Ich hätte in den Situationen ja etwas gefühlt, aber das doch noch nicht einordnen können. Aufklärung und Sexualität war kein Thema bei uns zuhause. Die Gefühle bei der Selbstbefriedigung hätten mich an die Gefühle und Situationen von damals erinnert. Dann hätte ich verstanden, wofür die mich benutzt haben! Diese Erkenntnis sei wie ein Schock gewesen.

Ich erzählte ihr auch den Traum mit Adolf und betonte, dass der mich damals wirklich so gequält habe! Mein Gefühl nach dem Traum: Es lohnt sich nicht zu leben. – Bitterlich weinend hatte ich geklagt, mir sei gerade alles zu viel, das von früher und das von heute! Ist doch auch verständlich!, kann ich jetzt nur hinzufügen. Aber ich kämpfte mich weiter durch diesen Sumpf, wenn auch mit kleineren oder größeren Pausen.

Weitere Erinnerungen

Ich kam viele Monate später in der Therapie nochmals auf eine Situation mit Adolf zurück, erinnerte noch weitere Details. Ich könnte es auch so formulieren: Nun war ich in der Lage, mich noch mehr auf meine Erinnerungen einzulassen, noch mehr davon zuzulassen. Wieder hatte er mich mit meinem Rücken an seine Vorderseite gepresst. Ich spürte seinen harten Penis. Und dann wollte er, dass ich den mit meiner Hand berühre. Er nahm meine Hand und legte sie auf seinen Schwanz ..., ich zog meine Hand sofort zurück. – Nein! Nein, das will ich nicht!, so oder ähnlich meine Reaktion. Er versuchte mich zu überreden. Ich blieb dabei. Ob er dann onaniert habe?, hatte mich die Therapeutin gefragt. Ich konnte mich nur undeutlich erinnern, dass da was war und dass er mich danach mit so einem glasigen Blick angesehen hatte. Und mir wieder Geld zusteckte ...

Nach dieser Stunde hatte ich weiter das Bild von dieser Situation in mir, ich dabei ganz starr!

So hat er mich immer wieder traktiert bzw. zu traktieren versucht. Ich brauchte lange, bis ich akzeptieren konnte: Ich habe mich gewehrt, auch wenn ich ihm lange keinen Einhalt gebieten konnte. Das ging über Jahre hinweg, bis ich es schaffte, mein Nein durchzusetzen. Ich weiß noch, dass ich damals dachte, auch wenn ich dann kein Geld mehr von ihm bekomme. Wir waren arm, es gab kein Taschengeld, das Geld reichte kaum zum Essen. Es hätte auffallen müssen, dass ich Geld besaß, um Kaugummi oder Ähnliches zu kaufen. Das schließt eigentlich die Frage mit an, hat denn wirklich niemand etwas bemerkt? Darauf komme ich später zurück.

Noch einige Bemerkungen wegen des Geldes: Ich erinnerte immer deutlicher, dass Adolf mir nicht von Anfang an Geld zusteckte, sondern erst, nachdem ich noch stärker versuchte, ihn abzuwehren. Ganz sicher ist, dass er mir das Geld nicht nur gab, damit ich seine Übergriffe duldete, sondern damit ich nicht darüber redete.

Natürlich war es sehr schwierig und mit Schuldgefühlen behaftet, in der Therapie zu gestehen, ich bekam Geld von ihm. Ich schämte mich abgrundtief und machte mich immer wieder selbst dafür runter. Damit sei ich doch nicht besser als er! – Aber nein!, widersprach die Therapeutin. Sie legte mir eingehend dar, dass ich als Mädchen damit – für mich! – wie zur Mittäterin geworden sei. Das Geld zu nehmen habe der kleinen Ann geholfen, so ungefähr drückte sie sich aus, die eigentlich unerträgliche Situation des Ausgeliefertseins und der Hilflosigkeit besser zu verkraften. Durch das Annehmen des Geldes hätte auch ich gehandelt, und das sei das Entlastende. Ich brauchte einige Zeit, bis ich diese Sicht verstand und annehmen konnte.

Abschließende Betrachtungen

Ich komme kurz auf meine Angst vor meinen Brüdern während der Jahre der Therapie zu sprechen. Irgendwann wies mich die K. darauf hin, dass meine Brüder doch weiterhin potentielle Täter seien. Sie hätten Kinder, Töchter, könnten auch mit denen..., wenn ich nicht gegen die Brüder vorginge, sie und deren Ehefrauen nicht darauf ansprechen würde. Ja, daran hätte ich natürlich auch schon gedacht. Die K. wurde zum Teil richtig energisch. Ich würde doch nichts gegen die unternehmen! Mit fast kindlicher Stimme hatte ich damals geantwortet: Aber ich hab' doch immer noch Angst vor denen! – Das konnte sie verstehen.

Kürzlich äußerte ich gegenüber meinem Bruder Heiner die Überlegung, ob ich damals nicht doch meine Brüder *und* ihre Ehefrauen mit der sexuellen Gewalt hätte konfrontieren müssen, trotz aller Angst, insbesondere um eventuelle Übergriffe der Brüder auf deren Töchter zu unterbinden. Heiner meinte daraufhin: Wenn Du das getan hättest, was meinst Du, was dann los gewesen wären! Das wäre für Dich nicht auch noch zu verkraften gewesen! Ich verstehe Dich, aber Du musstest Dich schützen. –

Die Angst vor ihnen war damals so groß, dass ich meinen Eintrag aus dem Telefonbuch und bei der Telefonauskunft streichen ließ und von meinem Klingelschild nicht ablesbar war, dass ich alleine wohnte.

Und nun zurück zu der Frage, hat denn damals keine und keiner was bemerkt? Während des ersten Jahres der Therapie äußerte ich in einem Gespräch mit meiner Schwester Lisa einen furchtbaren Gedanken: Vielleicht habe unsere Mutter oder haben gar beide mitbekommen, was meine beiden Brüder mit mir gemacht haben, und dachten, na ja, dann wissen die, was bei einer Frau „hinten und vorne" ist – ein sehr beliebter Ausspruch meines Vaters, wie auch der folgende! – und ich weiß auch gleich, „wo's lang geht"! Lisa schwieg eine kleine Weile, dann sagte sie: Das habe ich schon vor einiger Zeit gedacht und gehofft, dass Du das nicht auch mal sagst. –

Daher könnte frau durchaus davon ausgehen, dass meine Eltern etwas bemerkten und es billigten – ein kaum auszuhaltender Gedanke.

Dazu passt die folgende Begebenheit: Ich war vielleicht 18 Jahre alt, als mir meine Mutter erzählte, Adolf sei angezeigt worden, weil er ein junges Mädchen sexuell belästigt habe. Ich erinnere mich sehr gut daran, dass ich damals dachte: Mutti, warum erzählst Du mir das? Ich könnte Dir was ganz Anderes erzählen! – Meine Mutter hatte noch hinzugefügt: Wie diese jungen Dinger aber auch heutzutage rumlaufen!, nach dem Motto, da müssen die sich nicht wundern! Die sind doch selbst schuld! Spätestens nach diesem Kommentar war klar, dass ich keine Gnade finden würde, wenn ich von den Übergriffen der Brüder reden würde. Also sagte ich nichts. Ich bin mir ziemlich sicher, dass ich spätestens danach das Thema sexuelle Gewalt völlig verdrängte. Was blieb mir auch Anderes übrig.

Die Angst vor Adolf hielt lange an. Heute habe ich keine mehr vor ihm. Es war mir bisher nicht wichtig, Adolf auf seinen Taten anzusprechen. Auch weil von ihm nicht zu erwarten war, er würde sich diesen stellen und sich damit auseinandersetzen.

Aber kurz vor Fertigstellung dieses Buches nutzte ich eine Gelegenheit und sprach Adolf tatsächlich an. Seine Reaktion darauf gestaltete sich ganz anders als von mir erwartet. Davon berichte ich im Kapitel Reaktionen.

Bis heute ist mir immer wieder deutlich geworden: Mädchen und Frauen wird ihr Recht auf ihre Körper abgesprochen. Sie dürfen kein Körperbewusstsein haben, schon gar nicht selbst über ihre Körper bestimmen und verfügen. Auch deshalb ist für mich sexueller „Missbrauch" Gewalt, sexuelle Gewalt und Ausübung von Macht, um genau dies zu zementieren!

3. Der Vater

Ich will und muss zu diesem Thema auch über meinen Vater schreiben. Doch das ist sehr komplex, da gibt es so viel „Material". Und es ist nicht so wie bei Adolf und Bernd, da erinnerte ich Vieles bald sehr gut. Bei meinem Vater ist es wie nicht zu fassen. Nicht zu fassen ..., das passt in jeder Hinsicht.

Noch eine kurze Anmerkung: Ich habe mich an einigen Stellen für eine ausführlichere Wiedergabe meiner Therapieaufzeichnungen entschieden, weil ich damit besser vermitteln kann, wie ich um Klarheit gerungen habe. Wie schwierig das war, aber auch wie erfolgreich.

3.1 Erste Hinweise – Träume und indirekte Erinnerungen

Zu den ersten Hinweisen darauf, dass vielleicht auch mein Vater zu den Tätern gehörte, sind zwei Erinnerungen sofort da: Ich ging in die Therapie mit dem Wissen um die Taten meiner zwei ältesten Brüder. Doch schon in der ersten Stunde hätte ich gesagt, so später die Therapeutin, dass ich befürchten würde, es sei noch mehr gewesen als das, an was ich mich erinnern könne. Als sie mich nach Monaten an diese Aussage erinnerte, war ich zunächst ganz verblüfft. Erst später fiel mir ein, ja, das habe ich gesagt. Damals hatte ich, nachdem das Erstgespräch mit ihr beendet war, gedacht: Was sag' ich denn da? Ich weiß doch von nichts!

Der Hinweis der Therapeutin auf meine Aussage kam im Zusammenhang mit einem Traum, der die zweite Erinnerung betrifft und den ich unter dem Stichwort „Blutschande" noch sehr gut erinnere; auch meine heftigen Reaktionen darauf.

„Blutschande"?! – Ein Traum

Diesen Traum schilderte ich in der Therapie:

> *Eulen und hähne fliegen in der luft. Vielleicht waren es auch tiere, die beide anteile hatten. Zwei fliegen dicht übereinander, bei dem unteren tier fließt in großen strömen blut heraus. Sie kommen immer näher. Es fließt dick blut, platscht richtig zur erde. Ich liege, glaube ich, neben philipp im bett, kann nicht aufstehen, nichts*

machen, wie gelähmt. Sie fliegen immer hin und wieder zurück und nähern sich unseren betten. Wir bekommen etwas blut ab, aber nicht viel.

Mit den Eulen und dem Blut konnte ich nichts anfangen. Die Therapeutin hingegen meinte, …

… sie müsse bei diesem traum an blutschande *denken! Das wort traf mich wie ein schlag! Ich käme mir vor, als habe mir jemand in den magen geschlagen, sagte ich und dachte an die sexuelle gewalt durch die brüder. Ihre aussage von wegen blutschande schockiere mich, obwohl der missbrauch doch nicht neu für mich sei. Sie fragte, was ich unter „blutschande" verstehen würde. Wieder reagierte ich geschockt, wurde ganz unruhig, bewegte mich auf meinem stuhl hin und her. Ich konnte die K. nicht anschauen. Mir ging durch den kopf: Vergewaltigung durch meinen vater oder einen der brüder! Ich konnte das aber nicht aussprechen! Sagte nur: Das ist ja furchtbar!, oder ähnlich. Und: Ich kann nichts sagen! Ich kann's nicht aussprechen! – Mir sei kotzübel. Nach einer kleinen weile: Aber ich kann mich doch an gar nichts in dieser art erinnern! – Ich spürte auf einmal angst; panik stieg in mir hoch! Ich sagte es. Und wieder: Aber ich kann mich doch an nichts erinnern! – (…)*

Nach dem ende der stunde blieb ich in unserem raum. Ich weinte bitterlich, stand wie unter schock. Was bedeutet meine reaktion? Ist da noch etwas, an das ich mich überhaupt nicht erinnern kann? Anders kann ich diese reaktion nicht verstehen.

In der nächsten Stunde kam ich darauf zurück, was mir sehr schwerfiel:

Ich erzählte erst einmal, dass ich in der letzten nacht wahnsinnige rückenschmerzen und angst vor der heutigen stunde hatte. Ich berichtete von dem schock, den ihr wort „blutschande" bei mir auslöste. Und von dem schock über meine reaktion auf dieses wort. Beschrieb ihr, was in mir vorgegangen war, nachdem sie fragte, was ich unter dem wort „blutschande" verstünde. Ich wüsste auch jetzt noch nicht, was ich von meiner reaktion halten solle, hätte wahnsinnige angst, es bedeute, es gab tatsächlich noch mehr in der vergangenheit, mehr als das, an was ich mich erinnere. (…)

Ich sagte ihr auch, dass ich wütend auf sie war. So ein wort kurz vor ende der stunde zu sagen und mich damit „unverdaut" von dannen zu schicken. Sie habe bei dem wort an etwas ganz anderes gedacht, entgegnete sie, nämlich daran, dass nicht nur die brüder schuld hätten, sondern dass das die ganze familie beträfe, hauptsächlich die eltern. Von denen sei ich nicht geschützt worden. Sie hätten durch die erziehung ihren beitrag zu den taten der brüder geleistet.

Die therapeutin betonte, sie könne meine wut auf sie verstehen, aber sie habe ja nicht ahnen können, welche reaktion das bei mir auslöst, über die sie dann ebenso überrascht gewesen wäre wie ich. – Ich hab' gedacht, sie wollten mich mit dem wort provozieren. Sie haben mich doch schon einmal gefragt, ob ich das gefühl hätte, es war noch mehr. – Sie sagte, ich wäre damals in die Therapie gekommen und hätte gesagt, dass ich angst hätte, dass mehr wäre als das, an was ich mich erinnern könne. Sie fügte jetzt hinzu: Ich würde sie doch damit nicht provozieren, selbst wenn ich solche vermutungen hätte. –

Ich hätte in den letzten tagen häufiger überlegt, sie anzurufen, um sie zu fragen, wie sie meine reaktion interpretiere. Die angst davor, sie bestätigt meine befürchtungen, habe mich davon abgehalten. Ich sprach von meiner überlegung, gar nicht mehr darüber reden zu wollen: Mir genügt vielleicht zu wissen, dass noch mehr war, aber ich muss (will?) gar nicht genau wissen was. – Sie meinte, das thema auszusparen ginge wohl nicht. Erst einmal zu akzeptieren, wenn dem so wäre, dass etwas war, wir aber nicht wüssten was, das findet sie ok. Sie denke, dass ich sicher nicht mit dieser unklarheit leben könne; keine könne das auf dauer.

In einer zusätzlichen Stunde erklärte die Therapeutin nochmals, wir müssten warten, bis sich das Unterbewusste offenbare. Ich war sehr unzufrieden damit, meinem Unterbewussten so ausgeliefert zu sein, wann es was rauslässt, und haderte wieder damit, womöglich noch Jahre Therapie machen „zu müssen". Die Therapeutin vermutete hinter meiner Reaktion u.a. Angst und schlechtes Gewissen, da die Therapie eine Distanzierung und am Ende vielleicht sogar die Trennung von der Familie bedeuten könne.

Dies passte auch zu meinen Schuldgefühlen, die in diesem Zusammenhang auftauchten. Ich hatte Worte meiner Eltern im Kopf, in denen sie mir die Schuld geben würden für das, was meine Brüder gemacht hatten. Über meinen Vater sagte ich kurz darauf in der Therapie:

Er hat gesagt, ein beispiel, worin sich alles widerspiegelt: Wenn frauen abends um 11 auf der straße vergewaltigt werden, sind sie selbst schuld. Um diese zeit hat eine anständige frau nichts mehr alleine auf der straße zu suchen!!! Ich hasse ihn!! – Die K.: Damit ist wirklich alles gesagt! – Oder: Vergewaltigung gibt es nicht für ihn. Die frauen würden zwar nein sagen, wollten das aber! –

Außerdem empfand ich Schuldgefühle, weil ich nicht schon früher über alles geredet hatte. Die Therapeutin meinte dazu, es gäbe doch Gründe, warum ich so lange nicht hätte reden können. Ich hätte doch nach und nach alles vorbereitet, um jetzt darüber zu sprechen. In mein Tagebuch schrieb ich in diesem Zusammenhang:

Ich denke, dass meine annäherung an frauenkreise und mein engagement für frauenthemen auch eine vorbereitung für die aufarbeitung war, da bei frauen das größte verständnis zu diesem thema besteht, die größte bereitschaft, sich damit auseinanderzusetzen. Also habe ich das schon nach und nach vorbereitet: Die erste therapie diente der stabilisierung, des lernens, mit den herzproblemen umzugehen. Die phase danach: Abschluss des studiums, anfang der kontakte zu frauenbewegten frauen und auseinandersetzung mit frauenthemen.

Weitere Träume mit dem Vater

Ich schildere nun weitere Träume mit meinem Vater einschließlich der sich jeweils daran anschließenden Bearbeitung in der Therapie:

Mein vater hatte seine arme um meinen hals gelegt, wohl um sich von mir zu verabschieden. Im zurückgehen ließ er seine hände von meinen schultern richtung busen gleiten. Ich ahnte das, ging einen schritt zurück. Dadurch fielen seine hände wohl von mir ab, ohne meinen busen zu berühren.

Es fiel mir schwer, den Traum zu erzählen, und ich wusste ihn erst einmal nicht zu deuten. Aber er erinnerte mich daran:

Dass mich mein vater, noch als ich schon 30 und mehr jahre alt war, beim begrüßen und verabschieden auf den mund küsste; dass mir das irgendwann nicht mehr passte, ich aber lange zeit nicht wusste, wie ich das verhindern sollte. Ich löste das dann, indem ich ihn auf italienische art umarmte, also meinen kopf beim umarmen jeweils links und rechts von seinem kopf platzierte.

Die K. findet das verhalten von vater übergriffig. Ich meinte noch, mein vater habe sich mit dem kuss auf den mund seine macht oder verfügungsmacht bestätigt. – Ich finde es so furchtbar! Das ist mir alles zu viel!, stöhnte ich. – Aber fühlen sie nicht auch erleichterung, nachdem sie in letzter zeit einiges losgeworden sind? – Ich hab' auch gerade gedacht, dass ich froh bin, das alles mal loszuwerden. Ich will mich davon freimachen, will nicht mehr zu denen gehören! –

Mir fiel daraufhin ein anderer Traum ein, den ich noch gar nicht erwähnte, obwohl ich ihn lange zuvor geträumt hatte, nämlich nach dem Erstgespräch, aber noch vor Beginn der Therapie bei der K.:

> *Ich liege in einem bett, jemand versucht, an meinem busen rumzufummeln. Ich beiße ihm in den finger, ganz fest. Versuche, um hilfe zu rufen, aber meine stimme ist leise, wie erstickt. Es gelingt mir dann doch, laut hilfe zu rufen. Da erscheint der kopf meines vaters. Ich schlage ihm auf seinen kopf und rufe: Du bist ein schwein! Du schwein! –*

Ich schilderte der therapeutin meine damalige reaktion auf den traum: Ich wurde wach und hätte am liebsten geweint. Es ging nicht, hatte wie einen kloß im hals. War nur geschockt. Dann dachte ich an mein beißen, und ich spürte förmlich meine zähne sich in die haut, das fleisch bohren. Das beißen war selbst für mich schmerzlich. Und dann weinte ich sehr. (...)

Ich verstehe nicht, wie ich so was träumen kann! Ich kann mich nicht erinnern, dass mein vater so etwas getan hat. – Wir wissen ja nicht (oder noch nicht?), was er getan hat, meinte sie. Ich erschrak heftig über diese aussage. Sie meine mit ihrer aussage nichts konkretes, nur, dass wir wissen, mein vater war übergriffig. – Sie haben diesen biss ja richtig sinnlich erlebt. Vielleicht haben sie ihm damals wirklich in den finger gebissen. – Hätte ich ihn doch abgebissen, dieser sau! – Bei ihr kam das so an, als wäre ich mit meiner gegenwehr wieder nicht zufrieden, wie das in anderen situationen schon der fall war. – Sie haben sich doch gewehrt! Und, ich muss es mal so hart sagen, für eine, die gewohnt war, nur zu buckeln, für die stillhalten die erzwungene normalität war, ist ein ausweichen, sich zurückziehen doch schon toll. Das sollten sie mal wahrnehmen! – Sie haben ja recht! Aber ich habe so eine große wut! Das mit dem fingerabbeißen reicht auch nicht. In den reißwolf müsste man ihn packen und dann ganz langsam ... – Vielleicht sollten wir uns mal ihren phantasien widmen, was sie gerne mit ihm machen würden ... – Leise sagte ich: Nein, im moment nicht. Ich habe ihn als kind doch so lieb gehabt, ihn so toll und schön aussehend gefunden. – Das solle ich mir bewahren, erwähnte sie nicht zum ersten mal.

Wir kamen auf meine mutter zu sprechen. Die K. meinte, wenn mein vater so mit uns umgegangen sei, hätte er sicher auch gegenüber meiner mutter sehr viel macht ausgeübt. Und diese habe uns deshalb mit unterdrückt. Ich äußerte meinen gedanken, ob mein vater meine mutter nicht vielleicht sogar vergewaltigt habe. – Die haben sich doch schon so früh kennengelernt, meine mutter war wohl

14 jahre, und mit 16 ½ hat sie schon Maria bekommen. – Was? Das habe ich ja noch gar nicht gewusst! Die war dann ja noch ein kind! – Das hat anscheinend tradition in unserer familie, das schon mit kindern zu machen, meinte ich sarkastisch.

Wir müssten zum schluss kommen, sagte sie nach einer kleinen weile des schweigens und fragte, ob ich noch etwas sagen wolle. – Nein. – Doch dann fragte ich: Können sie verstehen, dass mir das zu viel ist? – Ja! Ja, das kann ich verstehen. ... Wenn sie mich brauchen, rufen sie mich an! –

Erst nach ein paar Monaten kam ich wieder explizit auf meinen Vater zu sprechen. Ich hatte mich gefragt, wieso ich ihn eigentlich so sehr liebte. Die Therapeutin erklärte es so:

Ich sei doch als kleines mädchen offen und fröhlich gewesen, hätte gerne geschmust. Meine mutter wäre wohl eher abweisend oder gar nicht für mich da gewesen. Aber mein vater wäre doch sehr darauf eingegangen. – Aber irgendwann wollte er nicht mehr mit mir schmusen, mich nicht mehr auf seinem schoß haben. – Was denken sie denn, warum sie ihren vater so geliebt haben? – Ich finde ihre erklärung schon einleuchtend, aber vielleicht musste ich ihn so lieben, um was anderes zu verdecken... Ich weiß es ja nicht, würde es aber so gerne wissen. Ich habe so große angst, dass da noch was war! ... Als ich dieser tage daran dachte, dass ich auch im elternschlafzimmer geschlafen habe, fiel mir der traum mit dem biss ein. Wer weiß! Dem traue ich alles zu. –

Eine Situation im Praxisbüro

Nun komme ich zu einem Vorfall in der Praxis der Therapeutin, den ich noch sehr gut erinnere. Ich kam in die Praxis, die Therapeutin bat mich, im Büro der Praxis einen Moment zu warten, da alle Räume belegt seien. Das Büro hatte eine Tür zum Flur, durch die ging ich ins Büro, und eine Tür zu unserem Raum. Ich war kurz alleine im Büro, aber bald kam ihr Kollege dazu; er schloss die Türe. Somit waren beide Türen geschlossen. Mir wurde zunehmend seltsamer zumute; ich fühlte Enge, so als ob mir jemand Herz und Lunge zusammenpressen würde. Dann konnte ich in unserem Raum. Die K. entschuldigte sich wegen des Chaos. In der nun beginnenden Stunde spürte ich heftige Angst! Und ich weinte sehr.

In der darauf folgenden Stunde berichtete ich ihr, was mir im Nachhinein zu der Situation eingefallen war:

Ich fühlte mich an das elternschlafzimmer erinnert! Und mir fiel ein, ich war ja die ganze zeit mit dem typ alleine da drin! Und bei dieser erinnerung wurde mir wieder ganz eng zu mute, als würde ich keine luft bekommen, als würde mir jemand den brustkorb zusammendrücken. Jetzt weiß ich, warum ich so starke angst hatte! Ich war die ganze zeit alleine mit dem im Zimmer! Während der situation habe ich das nicht kapiert. Das stinkt mir! Ich hatte zwischendrin das gefühl, ich will hier raus, aber wohin hätte ich gehen sollen? Erst hab' ich gedacht, der wäre nur einen moment mit im büro gewesen, aber dann fiel mir ein, der war ja noch da, als sie sagten, ich könne jetzt in unseren raum reingehen. Ich hätte dann an ihm vorbeigehen müssen, das konnte ich aber nicht. Das ging nicht, das wäre zu eng geworden. Also ging ich durch die andere tür über den flur in unseren Raum. Jetzt ist mir klar, warum ich so starke angst hatte! Und immer, wenn ich mich daran erinnere, taucht dieses erstickungsgefühl wieder auf. – Das war ja wirklich heftig!, meinte sie mitfühlend und verständnisvoll.

Sie hatte hinzugefügt, es wäre aber gut, mal eine heutige Situation zu erleben, die wir nun bearbeiten könnten. Auch wenn sie wisse, dass es für mich hart war, könnten wir dadurch doch weiterkommen. Ich hatte noch gesagt: Wenn der was gemacht hätte, ich hätte ihn kaputtgeschlagen! – Das fand sie gut. Von ihr kam aber keine Entschuldigung dafür, dass sie mich in diese Situation brachte, was mich sehr ärgerte, und diese meine Wut äußerte ich. Ich hätte mich bewusst für eine Therapeutin entschieden und keinen Therapeuten! Und was passiert? Ich käme trotzdem in die Situation, mit einem Mann alleine in einem Raum zu sein. Ich könne mich dort nicht mehr beschützt fühlen! Dies alles sagte ich mit geballten Fäusten. Sie reagierte absolut verständnisvoll und betonte, sie wolle in Zukunft auf jeden Fall dafür sorgen, dass ich nicht mehr in eine solche Lage käme. Sie fügte später hinzu, dass ich so stark reagierte habe aber mit der realen Situation im Büro nichts zu tun.

Ich kam darauf zurück, dass ich in der vorhergehenden Stunde doch Angst hatte, sie anzusehen, anzusprechen:

Sie aus scham nicht anschauen zu können, das kenne ich ja! Aber aus angst!? Das kommt mir neu vor. – Vielleicht hatten sie, als sie hier reinkamen, das gefühl, ich sehe ihnen das an, ich merke, was los war, wie sie das von ihrer mutter immer dachten. – Sie fragte, ob ich in ihrem kollegen meinen vater gesehen hätte. – Ja! Als er in seinem aktenkoffer wühlte, nichts sagte, das erinnerte mich an meinen vater. Das wurde mir später bewusst. Er machte das oft, tat dann so geheimnisvoll. –

Schweigen. Sie fragte nach einer weile mit lieber, weicher stimme: Wo sind sie denn jetzt? – Wie gut das tut, so von ihr gefragt zu werden!, dachte ich. – Ich frage mich, was das zu bedeuten hat, dass mir das elternschlafzimmer einfällt und dass ich panik hatte, weil ich mit dem typ im büro alleine war. Ich habe angst davor, was das heißen soll. Ich kann mich doch an nichts erinnern! –

Ich finde es heute noch sehr gut, dass sie mich am Ende der Stunde eindringlich auf das Folgende hinwies:

Ich möchte ihnen noch etwas sagen, weil ich denke, dass ihnen das im moment nicht bewusst ist. Sie haben sich in der situation des rausgehens doch zu helfen gewusst! Sie haben einen ausweg gewusst. Sicher wird der ihrem anspruch, dass sie sich wehren wollen, nicht genügen, aber sie haben sich zu schützen gewusst. Das sollten sie nicht übersehen! –

Auf diese Büroszene kam die Therapeutin bald nochmals zu sprechen: Dass Sie so starke Angst bekamen, das will uns ja etwas sagen. Wir sollten das nicht so sehen, das war wie früher, Sie sind immer noch nicht weiter. Diese Angst steht ja für etwas, was wir bisher noch nicht geklärt haben. – Wir sahen die Angst aber auch im Zusammenhang mit dem Urlaub der Therapeutin. Vielleicht löse die Therapiepause etwas in mir aus, was ich von früher kennen würde, nämlich mich isoliert zu fühlen, mit allem ganz alleine zu sein, hatte ich vermutet. Sie betonte, das sei doch heute anders. Ich hätte meine Freundinnen, könne alles aufschreiben und ich könne meine Gefühle schon besser zeigen. Sie fügte hinzu:

Als sie älter waren, sagen wir mit 14 jahren, war es ja wie ein verbot für sie, darüber zu reden. Als kleines kind konnten sie das sowieso noch nicht in worte fassen. Vielleicht erinnern sie die ferien daran, und es macht ihnen jetzt angst, wieder damit alleine zu sein. Möglicherweise ist es die angst von früher; die angst, als sie bemerkten, sie sind damit alleine, sie können mit niemandem darüber reden. –

Träume mit den Eltern und mit Lisa

Die Therapeutin hatte mich damals gefragt, was denn meine Schwester Lisa zu meinen Befürchtungen bezüglich meines Vaters sagen würde. Lisa, die zu dieser Zeit ebenfalls in Therapie war, hasse meinen Vater und hätte die gleichen Befürchtungen wie ich, also dass unser Vater auch mit ihr... Die Reaktion der Therapeutin: Puuuh... – Sie fragte nach meinen anderen Schwestern. Da sei wohl nichts. Warum sie frage? Weil sie denke, wenn ein Vater das mit einer Tochter macht und hat noch andere Töchter, belässt er es sicher nicht nur bei der einen. Daraufhin machte ich: Puuuh... – Die K. weiter, es bringe aber nichts, darüber zu spekulieren. Wenn noch etwas in mir wäre, dann käme das schon zum Vorschein, vielleicht durch einen Traum oder anderes Erinnern. Das müssten wir abwarten, könnten wir nicht beschleunigen. Ich würde mich fragen, ob es gut und sinnvoll sei, wenn ich alles weiß. Das nahm ich dann selbst wieder zurück. Die K. fügte hinzu, sie würde mich auch so einschätzen, dass ich keine Ruhe gäbe, bis ich zu bestimmten Dingen ein klares Gefühl hätte. – Ja, ich will und muss das wissen. Will das nicht mehr verdrängen! –

Schon in den nächsten Nächten kamen weitere Träume hinzu. Ich träumte von unserem Mädchenschlafzimmer ...

> *Furchtbarer traum, stichwort mädchen-schlafzimmer, große, graue gestalten, große frau. Kälte. Schnee. Wir sollten das große fenster aufmachen...*

Den kommentierte ich in meinem Tagebuch und fügte gleich einen weiteren furchtbaren Traum mit der Bemerkung hinzu: Kein Kommentar!

> *Für mich steht das riesige fenster für den mangel an persönlichen bereichen für uns töchter, für permanente grenzverletzungen. Und unsere mutter bekommt alles mit, sieht alles, hört alles, kann in mich reinschauen – so mein damaliges gefühl, unser gefühl. Der schnee steht für das eisige klima, die absolute gefühlskälte meiner mutter. ... Will nicht mehr leben! Sehr deprimiert. Angst. Liebevolle unterstützung von andrea. Ich solle mal eine woche zu ihr kommen, sie würde mich verwöhnen. Nur wir zwei. Das tat mir so gut!*

> *Lisa und ich bei den eltern, wir wollen ihnen endlich unsere meinung sagen! Wir lagen wohl alle zusammen, mein Vater neben mir. Irgendwann ist es auf dieser seite feucht. ... Wir, lisa und ich, sagten ihnen, wie sie uns gehalten haben: wie dienstmägde.*

In einer zusätzlichen Stunde erzählte ich diese Träume und dass ich danach unter Schlafstörungen und starken Herzproblemen litt, mir aber sagen konnte, ich bin doch jetzt nicht wieder alleine mit allem. Könne jetzt auch den Anderen zeigen, wie es mir geht, müsse nichts mehr vorspielen. Die K. freute sich sichtbar mit mir und wiederholte auf meinen Wunsch hin, was sie in der letzten Stunde gesagt hatte: Ich müsse mich nicht mehr zusammenreißen, nichts überspielen, sondern solle die traurigen und alle anderen

Gefühle annehmen und mir so viel wie möglich Schönes holen und mir selbst geben.

Zu dem Traum mit den Eltern, Vater neben mir im Bett, auf dieser Seite wird es dann feucht, sagte ich, der Traum sei für mich schockierend, aber auch entlastend. – Es muss keine Vergewaltigung gewesen sein, aber für mich wird immer klarer, irgendwie ist mir mein Vater zu nahe gekommen. Irgendwas war da! Das ist für mich nach diesem Traum klar! – Was denn das Gute und Entlastende im Traum gewesen wäre? – Dass ich meinen Eltern mal sagen konnte, was ich denke, also dass sie uns wie Dienstmägde hielten. – Und das fügte ich noch hinzu: Ich habe vor kurzem gesagt, es macht keinen Unterschied, wenn auch was mit meinem Vater war. Das stimmt aber nicht! Das ist ein großer Unterschied! – Das konnte die K. gut verstehen. Ob sie mir meine Annahme nicht ausreden wolle? Nein, das könne schon so gewesen sein: Ihr Vater war ja übergriffig. – Sie betonte an dieser Stelle wieder, Träume würden sich oft nicht gleich erschließen; es könne sein, dass wir erst sehr viel später wüssten, was sie bedeuteten. Manchmal hätten sie auch nur Symbolcharakter.

Zum Traum mit unserem Mädchen-Schlafzimmer berichtete ich ihr von meiner Deutung und dass die für mich absolut stimmig sei. Ja, so sei auch ihr Gefühl. Ich wollte dann nichts mehr sagen, lieber die Therapeutin einfach nur ansehen und das Dortsein genießen.

Dazu möchte ich einen kleinen, wunderbaren Einschub folgen lassen: Ich konnte die Therapeutin ganz lange gar nicht oder immer nur ganz kurz ansehen. Und dann erlebte ich im dritten Therapiejahr eine wahnsinnig schöne Stunde, die ich nie vergessen werde! Ich habe sie, die Therapeutin, anschauen, also richtig in ihren Augen schauen können. Es gelang mir für einige Sekunden, nachdem sie mir dabei half – und das war so wunderbar! Danach konnte ich mir auf diese Weise ganz oft so viel Liebes und Stärkendes bei ihr holen. Nur durch das Anschauen! Wenn also in der Folge erwähnt wird, noch langes Anschauen o. ä., dann ist dieser Hintergrund mitzudenken. Und auch der: Da Berührungen, wie in den Arm nehmen oder ihre Hand auf meinem Rücken, für längere Zeit nicht mehr möglich waren – auch dazu an anderer Stelle mehr –, war das Anschauen, in ihre Augen schauen oder von ihr angeschaut zu werden wie von ihr gehalten und gestützt zu werden. Das hat mir sehr, sehr viel gegeben.

Doch nun weiter zu meinem Vater. In der Folgezeit gab es längere Passagen in der Therapie, während der keine neuen Erinnerungen oder Erkenntnisse im Zusammenhang mit ihm auftauchten. Das waren Erholungsphasen, so sehe ich es zumindest heute, die ich sicher brauchte. Aber ich erwähnte, dass mein Vater wohl tatsächlich was mit Lisa *gemacht* habe; ihre Arbeit in der Therapie deute darauf hin, so Lisas Aussage. Bei mir lösten Lisas Worte Resignation und Hoffnungslosigkeit aus. Mir kam alles so sinnlos vor. Lisa bestätigte einige Zeit später, sie sei sich sicher, da war etwas. Sie erinnerte seine Aussage, wage ja nicht, jemand was zu sagen! Mir und meiner Therapeutin war klar, das ist Lisas Geschichte. Ich will und muss nach mir schauen, gucken, was bei mir war. Nur das ist für mich relevant.

„Sonst herrschte Finsternis!" – Über den Vater

Ich möchte hier eine Passage über meinen Vater ganz allgemein einfügen, die einiges über ihn und unser Leben früher aussagt:

> *Für vater war ich doch nichts! Für den hatte ich keine gefühle, die interessierten ihn überhaupt nicht! Was ich dachte und fühlte, das war dem absolut egal! Für den war ich ein nichts! Eine frau, mehr nicht! Eine frau, die man(n) benutzen kann! Zum wäschewaschen und schuheputzen! Der konnte einen angucken, so von oben herab, da kam man sich soooo klein vor! Dreck, wie dreck … war ich für den! Dreck, die eine saubere schürze tragen musste und an der er sich die schuhe abputzte! Mehr nicht! … Es wäre besser, ich wäre nicht auf der welt … – Ich weinte, weinte jämmerlich! Mir fällt gar nicht mehr alles ein, was ich sagte. Aber alles kam so richtig von unten aus dem bauch raus. Ich dachte nicht, sondern ich erinnerte, fühlte und sprach es aus. – Ich wollte doch, als ich das gefühl hatte, sie gucken mich von oben herab an, auch einmal auf sie runtergucken. Das erinnerte mich an meinen vater! Der konnte so kalt gucken! Das ist nicht zu beschreiben! Dass man nur angst spürte. (…) – Irgendwann schwieg ich; sie fragte, was jetzt wäre. Ich freute mich so, ihre stimme zu hören. – Ich habe wut, bin groggy … und traurig! – Sie saß jetzt auf dem boden vor der heizung! (Dort saß sie auch danach immer dann, wenn es um meinen Vater und mich ging. Dadurch waren wir auf einer Höhe, da ich auf einer auf dem Boden liegenden Matratze saß. Normalerweise nahm sie in einem Sessel Platz.) –*
>
> *Haben sie ihrem vater denn auch mal von sich erzählt? – Wie meinen sie das? Wie es mir geht, was ich gemacht habe? – Ja! – Da muss ich wirklich lachen! Das hat den doch überhaupt nicht interessiert! … Haben sie die frage wirklich ernst gemeint, ob ich ihm von mir erzählte? – Aber ja! – Ich muss nochmal lachen. Das ist so absurd! Das hat den überhaupt nicht interessiert, was ich gemacht, gedacht oder gefühlt habe. Ich war gar nicht da für ihn! – Das war aber doch nicht immer so? Ab wann war das so? – Ich überlegte: Die ersten jahre nahm er mich schon wahr, war er lieb zu mir. Bis ich 4 oder 5 jahre alt war, da bin ich mir relativ sicher. Aber danach …, ich weiß nicht, wann sich das änderte.*

Vorher war ich wie ein spielzeug für ihn. Dann musste ich nur noch funktionieren, habe ich ihn nicht mehr interessiert. ... Ich mag es eigentlich gar nicht sagen (weil es mir oft mit der K. auch so geht), aber wenn er mich lieb anguckte, mich lieb anlächelte, das kam ja selten vor, doch dann war es wie von der sonne beschienen zu werden. Sonst herrschte finsternis! – Ein liebes, tiefes „Ehem" kam von ihr. – Was hab' ich immer auf ein lächeln von ihm gewartet. ... Sie können sich nicht vorstellen, wie oft ich mir sagen muss, wenn sie in der stunde nichts sagen, dass das nicht bedeutet, sie reden nicht mit mir. Das kommt so oft vor! Dann muss ich mir das richtig eintrichtern. –

Die stunde war bereits vorbei, sie sagte aber noch nichts. – Können sie sich vorstellen, wie das ist, wenn sie mit jemandem in einem raum sind, den sie sehr lieb haben, und der nimmt sie überhaupt nicht wahr? Fernseher, radio, das nimmt er wahr, aber sie nicht! – Ehem!, wieder weich und verständnisvoll.

In dem gleichen Frühjahr erzählte ich in der Therapie den folgenden Traum, den ich während der Therapieferien geträumt hatte:

Mein vater hat im treppenhaus die wände voll geschrieben, von oben bis unten, mit sportergebnissen ...

(...) Mir ging es freitag früh so schlecht, ich hatte solch eine angst! – Erst hätte ich mit dem traum nichts anfangen können, dann wären ganz furchtbare gedanken aufgetaucht: Ich dachte, das treppenhaus ist mein körper und mein vater hat von oben bis unten spuren bei mir hinterlassen! Ich weiß nicht, wie ich darauf komme, habe nicht darüber nachgedacht, das war einfach da. – Ehem! – Ihre lieben bestätigungen kamen heute ganz oft, und das tat so gut! Ich erzählte von meinem versuch, den traum und die aufgetauchten gedanken zu vergessen, und wie schlecht's mir nach dem traum ging. Zu verdrängen ging nicht, deshalb hätte ich das jetzt erzählt. Hätte aber noch ein anderes, ein aktuelles thema, aber das thema „mein Vater" tauche immer wieder auf. Ich sei jetzt ganz aufgeregt: Da ist doch was! Ich bilde mir das doch nicht ein! – Bevor ich wieder auf das andere Thema zurückkommen konnte, wollte sie noch was sagen:

Ihre brüder auf jeden fall, vielleicht auch ihr vater, haben sie missbraucht, ihre grenzen nicht geachtet. Die angst steht doch wahrscheinlich im zusammenhang mit dem traum. Sie hatten danach angst, in die welt zu treten, rauszugehen. Aber sie haben sich entzogen, diesem sumpf ihrer familie entzogen und sich auf sich besonnen. –

Mit den letzten Worten kam sie auf meine Aussage zurück, ich hätte mich von dem Traum nicht runterziehen lassen wollen, mich entschieden, weiter meine Dinge zu machen und dass mir das ganz gut gelungen wäre. Aber nach dem Traum hatte ich notiert: Wie soll ich das verkraften, wenn der wirklich was mit mir gemacht hat? Wie überleben, das aushalten? Traurig und niedergeschlagen, so meine Stimmung in dieser Zeit, und immer wieder das Gefühl, es war was mit meinem Vater! Ich empfand die gleiche Trauer wie damals, als ich zum ersten Mal über die sexuelle Gewalt meiner Brüder sprach.

Mit dem Abstand von heute kann ich sagen, ich näherte mich immer mehr diesem Thema an, in kleinen Schritten, denn das musste ich ja verkraften können, was an Erinnerungen und Gefühlen so nach und nach auftauchte. Ich finde es spannend, dass das Unterbewusste das so steuern kann. Damals fand ich das natürlich überhaupt nicht spannend, im Gegenteil!

3.2 Bilder, physische und psychische Erinnerungen tauchen auf

Ich sprach in vielen Therapiestunden unbewusst, aber auch ganz bewusst von Anderem, um ja nicht am Thema „Vater“ weiterzuarbeiten. Einmal bemerkte ich in einer Therapiestunde so nebenbei, da wäre ja noch etwas aus meiner Vergangenheit... Die K. daraufhin: Haben Sie Angst, dass ich das Ablenkungsmanöver nicht durchschaue? – Das fand ich gut. Sie ist aufmerksam.

In einer weiteren Therapiestunde erinnerte ich plötzlich, dass ich als kleines Mädchen öfter auf dem Schoß meines Vaters saß. Während ich das schilderte, war ich ganz aufgeregt, bewegte mich unruhig hin und her und spürte leichte Angst:

Er flüsterte mir worte wie scheiße und andere, die man eigentlich nicht sagen durfte, ins ohr, und ich lachte. – Ich wollte eine vermutung äußern, das ging aber nicht! Zitternde knie, gefühle im unterleib, was ich ihr sagte. Ich könne meinen vater doch nicht verdächtigen ... – Aber sicher können sie das! – Sie half mir mit ihrer frage: Ihr vater hat sie angefasst? – Ja! Und meine vermutung ist, diese geflüsterten worte sollten kaschieren, dass er seine hand zwischen meine beine legte ... –

Die K. sagte dann einiges, auch dazu, dass ich meinte, irgendwann durfte ich nicht mehr auf seinen schoß, habe er sich von mir abgewandt. Das wäre aus heutiger sicht ja gut! Damals wäre es sicher furchtbar für mich gewesen, weil er mich gar nicht mehr beachtete. Aber damit habe er meine grenzen doch wieder respektiert.

Bald darauf fragte ich die K., ob es möglich sein könnte, dass etwas mit meinem Vater war, ich das aber völlig vergessen habe? Das wollte sie natürlich nicht beantworten. Später erinnerte sich mich jedoch daran, ich hätte ja auch Anderes verdrängt und irgendwann die Herzprobleme bekommen. – Man sagt ja, der Verstand vergisst, der Körper nicht! – Sie betonte, das würde mich sicher so lange beschäftigen, bis ich für mich ein klares Gefühl dazu hätte.

Sie erinnerte mich auch daran, dass ich nicht nur wegen der sexuellen Gewalt in die Therapie gekommen sei, sondern doch auch, weil ich mir meine Geschichte insgesamt vornehmen wollte. Und genau das wollte ich: Mich genau anschauen, meine Kindheit, mein damaliges Leben, alles auf den Prüfstein stellen. Irgendwann später hatte ich das Gefühl, durch die Analyse bleibt kein Stein mehr auf dem anderen! Der Kommentar der K. dazu: Dann kann frau aber auch etwas Neues aufbauen. – Recht hatte sie.

Kontakt zur kleinen Ann

Jetzt komme ich zu etwas sehr Schönem und Hilfreichem, nämlich zu dem Kontakt zu der „kleinen Ann". Durch die Arbeit mit der K. gelang es mir immer mehr, Zugang zu der kleinen Ann zu bekommen! Ich „sprach" mit der kleinen Ann, redete ihr auch mal Mut zu oder versprach, sie zu beschützen. Durch den liebevollen Umgang der Therapeutin mit mir konnte auch ich immer liebevoller, fürsorglicher und besser mit mir umgehen. Darüber kam ich auf die folgende Idee:

> *Mir würde es so schwerfallen, das folgende zu sagen, würde mich kaum trauen. (Besonders aus scham!!) Seit den sommerferien sei ich in „kontakt" mit der kleinen Ann, hätte sie also während der therapieferien beruhigen können: Ich stellte mir vor, ich schmiege sie an mich, streichele ihr den rücken, sehe ihr strahlen... Und dann strahlte ich auch. – Ich berichtete weiter recht ausführlich, und dann wurde es noch schwerer. Ich sprach von meiner idee, auf diesem weg herauszubekommen, ob etwas mit meinem vater war. – Ich fragte die kleine Ann, ob ihr papa was mit ihr gemacht habe? Hatte dann ein bild von ihr, wie sie die augen niederschlägt ... und ganz komische gefühle im unterleib. Das erschreckte mich sehr. Ich wollte nicht weitermachen aus angst. Schickte die bilder wieder weg ... – Ich erwähnte noch, dass ich bei der vorstellung, das zu berichten, angst vor ihr, der therapeutin, bekam. (...) – Ich würde gerne versuchen,* hier *den kontakt mit der kleinen Ann aufzunehmen... – Ja, sie sei auch dafür, das nicht alleine zu tun, meinte die K.*

In dieser Stunde erwähnte ich Traumfetzen, bei denen hätte ich an Vater und seine großen Hände denken müssen, und beim Erwachen ganz tiefe Angst gespürt und das Gefühl, ich verliere den Boden unter den Füßen.

Als ich mich wieder einmal so oder ähnlich fühlte, dachte ich: Es ist doch alles sinnlos, wofür leben? Es lohnt sich nicht! Ich will nicht mehr leben! Weinte. Aber dann: Doch, es lohnt sich doch! Für meine schönen Gefühle! Ich hab' doch so schöne Gefühle, und dafür lohnt es sich doch! Ich war selbst überrascht von dieser Wendung. Ich sah es später so, dass nach den körperlichen Übergriffen, der sexuellen Gewalt, etwas in mir gestorben war, nämlich meine Gefühle, besonders meine lieben Gefühle. Die sind erst wieder aufgetaucht, als ich anfing, liebe Gefühle für die Therapeutin zu empfinden. Und dann auch für mich! Jetzt kann ich wieder lieben!, so meine wunderbare Entdeckung. Ich erinnere das mit so viel Freude, weil es so etwas Wichtiges, Schönes und Stärkendes für mich war.

„Ich will nicht in mein Bett!"

Das brauchte ich auch, denn die Bilder häuften sich und wurden konkreter. Damals erwachte ich oft und dachte, ist die Nacht denn immer noch nicht vorbei? Nachdem ich das in der Therapiestunde erwähnt hatte, brach aus mir heraus: Ich will mich an nichts erinnern!, und dass die Therapeutin sicher denke, ich spinne. Im weiteren Verlauf der Stunde hatte ich das Bild, ich liege im Bett, und da ist ein Kopf. ... Sofort waren wieder Gefühle in meiner Vagina aufgetaucht, näher habe ich sie damals nicht beschrieben, und ich erinnerte mich an den Traum mit dem Biss in den Finger. Äußerte Zweifel, sagte, ich wolle „sowas" nicht von ihm sagen, könne ihm doch nichts anhängen. Aber das bliebe doch erst einmal bei uns, hatte die K. betont. Dennoch sollten wir meine Bedenken ernst nehmen. Ob denn Bilder auf einmal so da sein könnten?, hatte ich sie gefragt. Ja, es gäbe Erinnerungen, die körperlich und seelisch da sind, die man aber vom Kopf her nicht mehr weiß. Vom Kopf her ginge es mir so, ich weiß nichts. Aber ich würde es so fühlen und mich so verhalten, als ob ich etwas wüsste. Sie sagte zum Abschluss dieser Stunde, es wäre ein Zeichen von Kraft und Stärke, dass ich all dies jetzt zulassen könne. Ja, das hätte ich auch schon gedacht, sogar gedacht, ich bin stark genug, das auszuhalten.

Eines Abends, im 4. Therapiejahr, tauchten beim Fernsehen Erinnerungen auf. Ich fühlte Beklemmungen, hatte ein Gefühl, als würde ich ersticken. Ich weinte bitterlich. Versuchte mich zu beruhigen, streichelte mich selbst, sprach tröstend mit der kleinen Ann, dass ich sie doch beschütze. – Ich hab' so eine Angst!, so meine Worte damals.

Und: Ich will nicht in mein Bett! Ich gehe nicht in mein Bett! ... Ich will nicht, dass es dunkel wird. Hab' so eine Angst! – Philipp, mein Mann, hielt mich. Ich wollte aus diesem Gefühl raus und hatte die gute Idee, mich schön zu pflegen und mich sehr auf mich zu konzentrieren. Guckte mich selbst lieb im Spiegel an. Gut, dass ich dazu endlich in der Lage war, nämlich liebevoll mit mir umzugehen. Ich schlief dann ganz gut! Jammerte aber nachts beim Wachwerden ein paarmal „Mama, Mama", wie eine, die Schmerzen hat. Ich berichtete davon in der nächsten Therapiestunde:

> *Nach dem starken weinen samstag spät abends und meiner aussage, ich wolle nicht ins bett, wolle nicht, dass es dunkel ist, hätten Philipp und ich an ein erlebnis aus der vergangenheit gedacht. Mir wurde jetzt ganz eng, ich konnte nicht weiterreden. Das sagt schon wieder alles!, dachte ich. Ich schaffte es dann doch, erzählte von der nacht vor vielen jahren, Philipp und ich im campingbus (März 1977) in P. und dass meine eltern nebenan in ihrem campingbus schliefen. Während meiner schilderung pustete die K. hörbar luft aus. –*
>
> *Philipp und ich sehen diese nacht als den anfang meines absturzes an. – Ehem! – (...) Ich hatte (nach diesem märz!) in unseren urlauben nachts oft große schwierigkeiten. Ich erinnere mich an nächte, in denen ich am ganzen körper zitterte vor angst. Ich konnte mich kaum beruhigen. Ich wusste nicht zu sagen, warum oder wovor ich angst hatte. Ich hatte nur ganz furchtbare, tiefsitzende angst. Damals habe ich Philipp aber geweckt. – Ehem. –*

Die Hände der Therapeutin, die Hände des Vaters

Nun komme ich zu etwas ganz Unangenehmem. Manchmal tauchten plötzlich Phantasien oder Bilder mit der Therapeutin auf. So hatte ich eines Tages die Phantasie, sie steckt mir etwas in meine Vagina; ich spürte sofort Gefühle dort. Für mich war das ein totaler Schock. Das wolle ich doch gar nicht!, äußerte ich mich entrüstet, nachdem ich ihr das nur unter größten Mühen und mit sehr, sehr viel Scham sagen konnte. Sie fragte, was sie denn dazu benutzt hätte? – Ihre Hände. – War das schwer auszusprechen! Wieder dachte ich an meinen Vater. Ich spürte auch erotische Gefühle und Brennen. Und gleich noch mehr Scham und schlechtes Gewissen. Ich erinnere noch heute ihre lieben und ermutigenden Aufforderungen, ich solle meine Gefühle doch zulassen, aber auch, wie schwer mir das fiel.

Dieses Brennen und andere Gefühle spürte ich in der Folge immer wieder bei Erinnerungen und Bildern im Zusammenhang mit meinem Vater. Und dabei „sah" ich seine Hände. Als ich in einer der nächsten Stunden über meine Eltern redete und sagte, ich könne ihnen im Moment nicht gegenübertreten, fügte ich hinzu: Ich würde meinem Vater noch nicht mal mehr die Hand geben wollen …

> *Hand!!! Großer schreck bei dem wort „hand"! Und dann das gefühl, als wäre eine große hand zwischen meinen beinen. Brennen in mir, in meiner vagina! Ich kuschelte mich in die decke. Der K. das sagen? Bedenken, sie glaubt mir das nicht. Aber ich fühle das doch! Will das ernst nehmen. Sagte es ihr dann doch, aber auch, wahrscheinlich würde ich mir das nur einbilden. Dann: Ich bilde mir das nicht ein! Ich fühle das so! – Wieder zweifel, fragte, ob sie mir glaube? Sie antwortete nicht. Ich erkannte, ich wollte mal wieder, dass sie mir meine gefühle bestätigt. Aber das tut sie ja nicht, und das ist auch gut so! Ich sagte:* Ich habe diese Gefühle! *Ich bilde mir das nicht ein! Egal was sie bedeuten! Ich habe die hand gespürt und das brennen, und das ist immer noch da! – Kuschelte mich wieder richtig in die decke. Sah die K. an, angst stieg in mir hoch. Sagte es ihr und dass die nicht wegginge. Die K. dazu: Die angst ist doch oft da, wenn sie etwas nicht aussprechen, was sie eigentlich sagen wollten! – Ja? – Ja! – Darauf habe sie mich doch schon öfter aufmerksam gemacht. Ich konnte mich nicht daran erinnern.*

Ich berichtete in der folgenden Stunde, wie es mir am Abend nach dieser Stunde ergangen war, und sagte weiter:

> *Von wegen der hand meines vaters …, dann war auf einmal da: Es war etwas mit meinem vater! Ich weiß, dass was war! Nein, ich weiß es nicht, aber ich fühle es, dass was war! Ich hätte das am liebsten in meinem zimmer an die große weiße wand geschrieben, damit ich es nicht vergesse und nicht mehr daran zweifle! Und dann musste ich ganz viel und lange weinen! … – Ehem, ehem! –*
>
> *Ich schwieg. Wurde immer trauriger. Mama! Mama, lass mich nicht alleine!, so mein gefühl. Die K. fragte, was bei mir sei. Ich sei traurig. Bald darauf sagte ich, ich würde mich gerne hinlegen, aber ich hätte beim gedanken daran gleich gefühle gespürt … – Aber das ist doch gut! Das hilft uns doch! Die gefühle sind doch sowieso da! Da ist es doch besser, sie zuzulassen, sonst geistern sie woanders rum!, so ungefähr ihre aufmunternden worte. Ich versuchte es mit dem hinlegen. Nach einer ganzen weile beschrieb ich, wie es mir geht: Ich kann ihnen nur sagen, ich liege hier wie bei dem „bild" mit meinem vater und spüre ein brennen ganz tief in mir drin … [in meiner Vagina]. – Ich lag zeitweise auf dem rücken mit weit ausgebreiteten beinen. Es kamen keine weiteren bilder hoch, ich wollte mir aber auch keinen druck machen! –*

Ein teil von mir will wissen, was war, und ein anderer nicht! – Mich nerve das mit dem wissen und nicht wissen, das lähme mich. – Das braucht doch zeit! Das kann man nicht forcieren! – Ich solle mich keinesfalls unter druck setzen, das sei ja dann auch manipulation, wie beim missbrauch! Solle es auf mich zukommen lassen, es komme ganz von selbst.

Anonyme Anrufe – Angst von früher

Seltsamerweise kam es zu dieser Zeit einige Male vor, dass ich anonyme Anrufe erhielt, zum Teil mit obszönen Bemerkungen. Davon berichtete ich in der Therapie. Zuvor kam ich auf die vorige Stunde zu sprechen:

Nachdem ich das gestern hier mit meinem vater ausgesprochen hatte, fühlte ich mich irgendwie ein bisschen ... größer, stärker ... und erwachsener! – Ehem! –

Ich erzählte von einem anruf vor einiger zeit und von dem letzte nacht. – Und heute morgen gleich der nächste schock. Ich hatte überlegt, ob ich Philipp von dem anruf erzähle; entschied mich dagegen und legte mich wieder in mein bett. Auf einmal spürte ich, wie aus dem bauch hochkriechend, angst! Sie zog wie nerven durch meinen ganzen körper, bereitete sich aus bis zu den händen. Es war fürchterlich, grausam. ... Das kann man nicht lange aushalten, da wird man verrückt! – Wärmflasche auf den bauch, weil von dort die angst kam, reden mit der kleinen Ann. – Es war wie ein schock! –

Die K. sagte, sie fühle es so, dass das angst und unsicherheit von früher gewesen wären: Damals wussten sie auch nicht, ob und wann irgend etwas von ihrem vater oder ihren brüdern kam. Und jetzt haben sie angst, dass auf einmal etwas kommt, was sie nicht erwartet haben. Eine unwägbarkeit, hart ausgedrückt: wieder ein missbrauch! – Ich schlug erschreckt die hände vor mein gesicht! ... Sie sprach noch davon, ich sei heute doch sehr eingebunden, hätte feste beziehungen, sei regelmäßig mit ihr in kontakt, also mit mir in kontakt! Also nicht zu erwarten, dass irgendwelche überraschungen ... – Aber sie haben von früher herrührend jetzt angst, dass doch was kommen könnte, was sie nicht abschätzen können. – Ich konnte sie kaum verstehen, konnte kaum denken!

Angst vor der Angst und liebe Gefühle, die retten

So ging es mir immer wieder einmal. Ich konnte nicht denken oder verstand nicht oder kaum, was die Therapeutin gesagt hatte. Dann seien wir nahe an dem, worum es gehe, oder dann stimme das, was sie vermutet habe; so erläuterte sie irgendwann diese Phänomene. Nun weiter von der Therapiestunde:

Ich erzählte ihr von meinem gedanken heute früh: Ich würde ihnen alles geld geben, was ich habe, wenn sie etwas machen, dass ich nie mehr diese überwältigende angst spüren muss! – Sie lächelte mich lieb an. – Das kann man wirklich nur dem aller ärgsten feind wünschen! –

Sie sagte das noch: Sie hatten ja lange zeit angst, dass ich nicht wiederkomme. Im ersten und zweiten therapiejahr war das so. Das ist ganz normal, da fehlt noch das vertrauen. Wenn sie einer einjährigen sagen, die mama kommt wieder, versteht sie das nicht. Das ist auch bei einer vierjährigen so, und so ist es in der therapie dann auch, je nach dem stand, wo sie sich gerade befinden. – Aber jetzt weiß ich doch, dass sie wiederkommen! – Ja! Das sehe ich auch so! Also hat diese angst jetzt eine andere qualität. –

Nun hatte ich wieder, wie vor vielen Jahren, Angst vor der starken Angst, dass die wieder zurückkommt. Konnte mir jetzt aber sagen, ich weiß doch, dass es etwas Altes ist! Das beruhigte mich einigermaßen. Die starke Angst erinnerte mich aber auch an die Situation, als ich alleine mit dem Kollegen der K. im Büro der Praxis war. In der sich daran anschließenden Therapiestunde spürte ich doch auch ganz tiefe Angst. Aber die Angst jetzt, das war wohl das Schlimmste, Härteste was ich bis dahin erlebt hatte. Danach stand ich wie unter Schock. Ein Gefühl der Leere in mir, fühlte mich ohne jeden Antrieb.

Gleich zu Beginn des neuen Jahres konnte ich wegen meines schlechten Zustandes während der Therapieferien der K. eine zusätzliche Stunde in Anspruch nehmen.

Aber die K. kam und kam nicht. Hat sie unseren termin vergessen? Ist sie doch schon weggefahren? Mir wurde ganz anders. ... Nach längerer zeit kam sie, ich wollte gerade gehen. Missverständnis mit der Uhrzeit. Ich sprach von meiner phantasie, sie hat mich vergessen! Schweigen. Sitze bei ihr, bei meiner mama. Eine ganze weile saßen wir ganz still und ruhig zusammen, sie wieder vor der heizung, also auf dem boden.

Ich erzählte ihr ziemlich ausführlich, für mich selbst überraschend, von dem, was eine freundin mir letzten dienstag berichtet hatte: Eine junge lesbe habe sich an weihnachten aufgehängt. Zuvor krisensitzung bei ihrer therapeutin. Die freundin sei jetzt auf die therapeutin sauer. Meine erste reaktion, man trage für sich selbst verantwortung, es sei ihre autonome entscheidung. Dann sei mir das alles doch sehr nahe gegangen: Ich fühle mich doch selbst so unsicher! ... Ich kann das nicht verstehen, frau k., dass mein vater so was gemacht hat! Ich kann's nicht fassen! Ich hab ihn doch mal so lieb gehabt! ... Alles schöne hat keinen wert mehr für mich! – Ich berichtete, hätte das thema ruhen lassen wollen, aber es ginge mir nicht aus dem kopf. Starke kopfschmerzen, rückenschmerzen, in die badewanne und ganz viel geweint und immer wieder mama, mama in mir.

Hätte auch an sie gedacht, meine lieben gefühle gespürt und hinterher gedacht: Die lieben gefühle retten mich! Wenn die nicht wären, gäbe es keinen sinn, weiterleben zu wollen! –

Mich ängstigte selbst, was ich erzählte und dabei fühlte. Sie schaute liebevoll nach mir und sagte bald, sie denke, ich würde mich wertlos fühlen! Ich hätte wohl in mir, wenn mein vater das mit mir gemacht hat, bin ich nichts wert! Vielleicht würde ich mich auch wertlos fühlen, weil meine mutter mich nicht beschützte, nicht für mich da war. Vielleicht habe die gar nicht gedacht, dass er so was macht, ihm vertraut. – Aber er hat das nicht gemacht, weil sie nichts wert sind!! Weil er sie nicht mochte! – Dann würde sie, die therapeutin, auch noch in urlaub gehen, gäbe mir damit auch das gefühl, ich sei ihr nichts wert, sei wertlos! All das würde sich vermischen. – Für mich ist es eher so, was ist alles noch wert, wenn ein vater so etwas macht? Das ist doch wie krieg, wenn man noch nicht mal vor seinem vater ruhe hat! Für mich ist das wie krieg! –

Sie kam auf die geschichte mit der jungen lesbe zurück. Die junge frau habe ja sehr viel wut auf ihre therapeutin gehabt, aber ... – Denken sie, ich habe wut auf sie? – Ich spürte keine. – Es wäre mir jedenfalls lieber, sie hätten wut auf mich, als wenn sie die wut gegen sich richten. – Sie denke, es gehe bei mir eher um schuldgefühle. Schuldgefühle, weil wir schon zwei sonderstunden während ihrer ferien machen mussten. Und dann bestärkte sie mich ungefähr so: Sie wissen, was sie brauchen und übernehmen die verantwortung für sich. – Sie strahlte zuversicht und sicherheit aus, und das war gut für mich. Ich fragte nach ihrer telefonnummer im urlaub, die sie mir eigentlich geben wollte. Aber dann äußerte sie bedenken, und ich entschied mich dagegen, auch nachdem sie mich darauf hinwies, diese furchtbare angst hätte ich doch auch ausgehalten, und die sei, im gegensatz zu den attacken früher, doch nicht lange anhaltend gewesen. Außerdem würde sie, die K., keine wunder bewirken. Ich *würde doch alles machen, schaffen! Ich sagte ihr das noch: Als ich letzte woche hier wegging, hatte ich den impuls, sie zu umarmen. Aus dankbarkeit! Und aus freude, dass sie so lieb für mich da sind. Gerade jetzt! ... Wenn ich es schon nicht tue, will ich es doch wenigstens sagen! – Ehem!, lächelnd.*

Ja, die lieben Gefühle für die Therapeutin haben mich (mit) gerettet, so empfinde ich das heute noch! Mit dem Abstand von heute ist mir das noch klarer. Sie gaben mir ein Gefühl von Geborgenheit und Aufgehobensein, es war wie lieb gehalten werden. Diese Liebe gegen meine Verletzungen. Und über die Liebe zu der Therapeutin kam ich, wie bereits kurz erwähnt, auch zu lieben Gefühlen für mich. Ich lächle beim Schreiben, weil das wirklich eine wundervolle Erinnerung ist. Auch wenn ich jahrelang mit der Liebe für die Therapeutin sehr große Probleme hatte, ich immer dachte, das ist doch nicht normal, ich bin nicht normal. Heute weiß ich diese Gefühle einzuordnen und bin dankbar, mir dankbar, dass ich sie zulassen konnte, zugelassen habe.

„Als hätte mir Vater die Flügel gebrochen …"
Bald darauf träumte ich wieder von Vater und sagte nach dem Wachwerden, weinend: Mama, rette mich! – Ich träumte, …

> *… dass ich mit meinem vater im campingbus nach B. [Wohnort meiner Eltern] fahre. Er versucht, mit seiner hand in meine bluse zu kommen. Ich herrsche ihn an, schlage seine hand weg. (…) Meine mutter erwartet uns, ich gehe zu ihr hin, sage: Ich fahre nicht mit ihm zurück, nehme ein taxi!, sage es ganz fest. Ich drücke sie und sage was liebes zu ihr. Sie: Es wird alles gut! – Sie guckt so wissend. Am bahnhof, kein taxi, vater kommt hinter mir her, kommandiert zwei riesige fahrzeuge, die auf mich zufahren. Ich versuche zu entkommen. … Weitere bedrohung durch meinen vater, die ich aber nicht mehr genau erinnere.*

Mir ist, schrieb ich im Anschluss an den Traum in mein Tagebuch, als hätte ich Vater im Traum so erlebt wie er wirklich ist. Ich fügte nach dem Erzählen in der Therapie hinzu: Mein Gefühl ist, als hätte mir mein Vater den Arm gebrochen. Nein, als hätte er mir meine Flügel gebrochen! – Ich schwieg. … Und sagte dann: Es wäre besser gewesen, nicht auf die Welt gekommen zu sein. – Die K. hielt dagegen:

> *Sie hätten dann aber viel schönes nicht erlebt! … Und sie hätten nicht erlebt, dass sie sich für sich einsetzen! – Mich zog das alles runter; ich sprach von meinem gefühl, mein vater wollte mich umbringen und jetzt will ich nicht mehr … – Ich hab so eine angst! – Mir war ganz schwindlig im kopf. – Haben sie angst, damit an die öffentlichkeit zu gehen?, fragte sie, weil ich im traum zweimal zu vater gesagt hatte, ich würde darüber reden, wenn er mich nicht in ruhe lässt oder so ähnlich. Ob ich angst hätte, er könne mich umbringen wollen, wenn ich darüber rede?, fragte sie auch. Ich hätte am liebsten geweint, konnte nichts sagen, nur dass ich traurig bin und angst habe und das gefühl, mama, beschütze mich, rette mich! (…)*

Misstrauen und Stillstand

Ich wollte mir nicht weiter angucken, was mit meinem Vater war, ich hatte einfach unermessliche Angst davor. Und wenn doch etwas hochkam, hatte ich wohl schon öfter der K. gegenüber geäußert, sie würde mir das sowieso nicht glauben. Darauf ging sie irgendwann einmal fast energisch ein:

> *Das geht doch schon einige jahre so, dass sie, wenn sie vermutungen oder ihre gefühle äußern, mir unterstellen, ich würde sagen, das stimmt doch nicht! Bilden sie sich nur ein und ähnlich. Wofür ist das gut? Was wollen sie damit abwehren? ... Irgendwie ist stillstand. Es bewegt sich nichts. – Ja, ich habe auch das gefühl, ich bin wie gelähmt. Kann nichts machen! ... Aber ich habe doch so eine riesige angst davor! – Wovor? – Wenn sich rausstellt, es war was mit vater, wie ich das verkraften soll! – Wenn sie sich dem stellen, wissen, ob und was war, werden sie sich freier fühlen, befreit sein. Und sie haben die autonomie, sich frei zu entscheiden, will ich kontakt oder keinen, hab ich wut oder nicht, und und und. – Ich frage mich, ob mein vater mich vergewaltigt hat und wie ich damit leben soll. Dann will ich nicht mehr leben! Das überleb' ich nicht! – Das haben sie aber schon mal überlebt! Und leben bis heute damit! ... Sie lassen nichts an sich rankommen! Es ist wie watte um sie herum! –*
>
> *Ich sagte bald darauf: Vielleicht geht es nicht mit reden, sollte ich doch malen oder mich irgendwie anders ausdrücken. – Ja, das habe sie auch gerade gedacht. Sie machte vorschläge; ich entschied mich für fingerfarben. Zum schluss der stunde musste ich ihr das noch sagen: Ich will, dass wir meine ängste ernst nehmen! – Damit ich mir keinen druck mache, dachte ich später.*

3.3 Die Erinnerungen werden konkreter

Schon vor langer Zeit hatte die Therapeutin den Vorschlag gemacht, zu malen oder mich mit anderen Mitteln auszudrücken. Sie schlug auch mal vor, als ich mich wie in einer Ecke fühlte, aus der ich nicht rauskam, mich wirklich in eine Ecke unseres Raumes zu setzen. Das erschien mir damals alles völlig albern; das wolle ich nicht, würde ich nicht machen, ich sei doch kein Kind! Aber das veränderte sich, ich veränderte mich, meine Einstellung dazu; ich konnte mich in der Folgezeit darauf und auf Anderes einlassen. Nun schlug ich selbst vor zu malen.

Gefühle mit Fingerfarben malen

Über die Notizen zur Stunde mit den Fingerfarben schrieb ich: Habe meine Gefühle mit Fingerfarben gemalt. War gut bei mir und konnte gut in mich rein fühlen. Ich notierte weiter:

> *Schön gekleidet und mit dem gefühl, ich will meine wünsche und gefühle zulassen, kam ich dort an. Nach ein paar worten von mir sagte sie bald, sich zu mir hinbeugend: Ich habe gestern noch fingerfarben besorgt! – Sie lächelte vielsagend und ich über sie. Ja, ich wolle das versuchen. – Ich möchte ihnen vorher noch etwas sagen, aber das ist nicht so leicht. – Dann, fest, bestimmt: Ich will das mit meinem vater angehen! Ich will wissen, ob und was mit ihm war! Ich will da raus! – Ja! –*
>
> *Dann holte sie die farben, papier und andere materialien. Mir kamen die tränen; sie fragte nach. Ich sagte ihr, dass dauernd in mir sei: Mama, ich hab doch so eine angst! – Ehemmmm! – Sie habe leider kein rot bekommen. Das würde ich aber unbedingt brauchen, sagte ich, weil schon wieder so ein brennen in meiner vagina. Sie holte rote wachsmalkreide. Da war feuriges rot dabei! – Jetzt geht es erst einmal nicht darum, ob da ein missbrauch war oder nicht, sondern um ihre gefühle. Ihr befinden. Das andere hat noch zeit, das kommt dann schon! – Ja, sagte ich leise.*
>
> *Vor dem brett mit papier und den anderen Utensilien sitzend, ganz bei mir, in mich rein fühlend... malte ich mit den fingerfarben, mit schwarz, einen breiten kreis. Kam eher zufällig mit meinen schwarzen fingern über den kreis ... und musste weinen. Dann fuhr ich von außen nach innen ganz langsam mit den fingern von allen seiten in die weiße fläche, bis kaum noch weiß übrig blieb. Sie fragte, ob ich etwas dazu sagen wolle, was ich male, also was es bedeute. – Ich musste schwarz nehmen. Das war gleich klar. Es bedeutet angst! Und dunkelheit! – Und warum ist der weiße fleck dann auch schwarz geworden? – Der kreis ist mein bauch. Alles voller angst! Und von außen kommt noch angst hinzu! Nimmt jemand besitz von mir.... Und da, wo noch weiß ist, müsste noch alles rot sein! Und da unten auch. Weil es da [im Bauch] und da [Vagina] brennt. – Dann malen sie das doch noch!, forderte sie mich mit lieber stimme auf. Ich nahm den roten stift: Alle weißen flecken müssen rot sein ... und da auch! – Malte rotes dreieck unterhalb des kreises ... Das war einfach so da, ohne zu denken. Hörte sofort auf zu malen. – Warum hören sie so abrupt auf? – Weil ich mich erschreckte beim malen des dreiecks. – Ich sagte auf ihre frage noch, dass es im bauch und „dort" (konnte kein wort „dafür", für meine vagina, finden) brenne. – Aber*

warum haben sie den stift so fallen lassen? – Das hab ich gar nicht bemerkt! ... Ich spüre jetzt nur noch das brennen und hab das gefühl, als ziehe sich alles zusammen. ... Ist alles ganz verspannt. – Ich beendete das malen.

Ob ich das bild behalten wolle oder ob sie das mitnehmen solle? – Es ist besser, wenn sie es mitnehmen. – Ja, das glaube ich auch!, und sie fügte hinzu: Da ist ganz viel leben drin! Ganz viel!, und schaute mich so lieb an! Als sie draußen war, musste ich erst einmal weinen. Mama, mama. Ich blieb noch fast eine stunde in meinem sonnendurchfluteten kinderzimmer! Hätte ihr am liebsten diese worte hinterlassen: Danke, mama! Mama, ich hab' dich so lieb!, weil sie mich doch so unterstützt!

Kinderzimmer – so nannte ich zeitweise unseren Therapieraum, weil ich dort noch einmal Kind sein durfte und die Therapeutin sich mir gegenüber wie eine liebe Mama verhielt. Zur nächsten Stunde hatte ich weitere Fingerfarben mitgebracht.

Die K. fragte lächelnd: Also kann es weitergehen mit dem malen? – Ja! – Ich hatte auch blau mitgebracht, brauchte blau! – Mir ging es in den letzten beiden tagen richtig gut, aber heute bin ich so unruhig, habe angst! – Ich wolle ihr aber zuvor von schönem berichten: Ich habe zum ersten mal mädchentraining gegeben. [Nachdem ich selber in einer Frauensportgruppe Selbstbehauptung und Selbstverteidigung gelernt hatte, begann ich, nach entsprechender Ausbildung, Mädchengruppen zu trainieren.] Das hat mir so viel freude gemacht, und ich kam gut zurecht! – Die lächelnde Therapeutin freute sich sichtlich mit mir. Bevor ich versuchte zu malen, sprach ich wieder von meiner angst: Ich weiß nicht, ob ich das noch mal schaffe. Würde am liebsten mama schreien! ... Und mir ist so kalt! ... Es ist so kalt! – Musste von den malsachen weg, musste zurückweichen, als drohe mir von dort schlimmes. Sagte jetzt hörbar: Mama! Mama!, und weinte. Endlich. –

Mir ist so kalt, als ob jemand die tür aufmacht und kalter wind hereinkommt. Eisig kalt! Meine beine ganz kalt! – Ich deutete an, auch am unterleib kalt. Und dann, dass ich „dort" gefühle spüren würde. – Da ist was!, deutete zwischen meine beinen. Sie fragte irgendwann, wie alt ich sei, wie alt ich mich fühlen würde. Erst konnte ich gar nicht darauf eingehen, dann: Noch klein, keine 4, keine 6 jahre, kleiner. Etwas später: Es ist was ganz schweres auf mir ... Der will mich umbringen! Ich hab so eine angst! Mir ist richtig schwindlig im kopf. Und ganz schlecht! ... Mein gefühl, ich kann nicht weg! Der bringt mich um. – Ich musste aufstehen, mich anders setzen. Dann legte ich mich auf den rücken, arme, beine ausgestreckt. Mein gefühl, ich gebe auf! ...

Bald darauf ende der stunde. Ich könne noch in unserem raum bleiben, sie würde später noch einmal nach mir sehen. Sie forderte mich auf, am wochenende mit dem, was ich heute gefühlt und erinnert habe, gut umzugehen!

Und ich nahm mir ganz fest vor, genau das zu tun! Und meinen gefühlten Erinnerungen zu glauben. *Allen* gefühlten Erinnerungen! An diesem Abend war ich mit meiner Schwester Lisa zum Essen verabredet. Dabei berichtete sie von einer Situation von früher, die ich nach unserem Treffen in meinem Tagebuch festhielt:

Sie war ungefähr 13 jahre alt, mit vater im wohnzimmer, auch birgit und ich dabei. Nachdem ich rausgegangen war, sagte vater ungefähr: Das mache gar nichts, wenn man als kind „dahin" (Vagina) gepackt bekäme, dann sei das später nicht so neu und man würde sich dann als erwachsene frau nicht so anstellen!

Einige Zeit danach fiel Lisa noch ein, dass auch meine Brüder Adolf und Bernd bei den Worten meines Vaters zugegen waren. Nachdem ich diese in der Therapie wiedergab, fügte ich hinzu:

Ich kann das nicht verstehen! Dass ein vater so etwas machen kann. Das mein vater so etwas gemacht hat! – Ich sehe es so, dass ihr vater vielleicht wirklich dachte, das steht ihm zu. Kein unrechtsbewusstsein besaß ... – Das sehe ich anders! – Ich hielt meine augen geschlossen, war ganz konzentriert: Mein vater wusste genau, dass das nicht in ordnung war! Der hat sich das nur zurechtgelegt, um sein tun vor sich zu rechtfertigen. Das passt zu ihm! So war er! – Ich sah ihn jetzt so klar! Sagte: Innen verbrecher und drecksau, nach außen der feine herr! –

Mut zur Konfrontation – geträumt und real

In dieser Therapiestunde erzählte ich der Therapeutin aber auch noch etwas ganz Anderes, nämlich dass ich beim Frauensport zwei Bretter mit dem Fuß durchgetreten hatte. Ich erläuterte ihr, das gehe nur, wenn frau sich vorstellt, sie will sich von den Brettern nicht aufhalten lassen. Ich hätte dabei an jemanden gedacht. Die Therapeutin fragte nach: Und an wen? – An meinen Vater! Und schon waren sie durch! – Ich denke, sie wird daraufhin geschmunzelt habe, die liebe Frau K. Das finde ich heute noch beglückend, das geschafft zu haben. Damals hat mich das sehr gestärkt und mir Auftrieb gegeben. In dieser Stunde erkannte ich: Ich will doch mein Leben leben! Und dafür ist es notwendig, das mit meinem Vater zu klären und mich dann von ihm frei zu machen. Sonst schaffe ich das nicht! –

Im folgenden Traum nahm ich vorweg, was ich nach der letzten Stunde überlegte, meinem Vater am Telefon sagen zu wollen:

Bernd war auch dabei, vielleicht auch Adolf. Frage vater, ob er sich erinnern könne, was er, als ich klein war, mit mir gemacht habe? Ich könne mich daran erinnern! Hätte es nicht vergessen! An reaktionen habe ich nur dies in erinnerung: Dass Bernd rausgeht, was ich so empfinde, er weiß, was ich meine, und er weiß auch, warum ich mich heute nicht mehr bei ihm melde. Von meinem vater keine erinnerte reaktion ...

Ist auch egal, denke ich nach dem erwachen. Dass ich es aussprach, das finde ich gut!

Von meinem Impuls, meinen Vater anzurufen, sprach ich in der Therapie, aber auch von meiner Wut auf ihn, die ich bald darauf zuließ. Vorher spürte ich heftige Angst. Das würde ich doch kennen. Ich hätte doch selbst dann Angst, wenn ich nur malen wolle!, erinnerte mich die Therapeutin. Ich hatte eine kleine Weile gebraucht, auch um rauszufinden, was zu meiner Wut passt; aber dann tobte ich und das nicht zu knapp! Ich schlug auf ein Kissen, malträtierte es, auch mit meinen Fäusten, biss vor Wut hinein und feuerte es zum Schluss gegen die Tür. – Ja, am besten auch noch die Treppe runter!, bestätigte ich ihre Vermutung. Nach dem Toben stellte ich mich ans Fenster, ließ die wärmende Sonne auf mich scheinen und sagte: Ich fühle mich gut, aber ich bin auch traurig. – Ja! Ehem! – Aber das war noch nicht alles!, betonte ich. Danach konnte ich weinen; weiter mit viel Wut und Hass in mir. – Gut, dass ich das geschafft habe. Auch dass ich selbst spürte, was ich machen will, wie ich meine Wut austoben will. – Ja! –

Nach der Stunde weinte ich ununterbrochen; hatte die Therapeutin zuvor gar nicht gehen lassen wollen. Ich notierte: Die kleine Ann ist so traurig und hat Angst. Ich hab' sie lieb in den Arm genommen, ihr den zarten Rücken gestreichelt und wir haben zusammen geweint. Ich passe auf dich auf! Ich bin deine Mama! Und wir haben zusammen eine liebe Mama! – Damit meinte ich die Therapeutin.

Ein paar Tage später rief ich tatsächlich meinen Vater an. Dazu notierte ich in meinem Tagebuch:

Unerträgliche kopfschmerzen, rücken völlig verspannt. Immer wieder den impuls, die alten anzurufen. Ich will das loswerden! Aber das bringt doch nichts und ähnliche gedanken! Um 12 uhr rief ich an, meine mutter war dran, liebe stimme, TV im hintergrund. Ich wolle vater sprechen. Er begrüßte mich freund-

lich. (...) Kannst du dich daran erinnern, was du mit mir gemacht hast, als ich ein kleines mädchen war? ICH kann mich noch erinnern! Habe es nicht vergessen! – Was soll ich denn mit dir gemacht haben? – Du hast mit mir gemacht, was man „sexuellen missbrauch" nennt! – Nein! So was habe ich nicht gemacht, da bin ich mir ganz sicher! – Dann kann ich nicht weiter mit dir reden! – Ich brach das telefonat zitternd ab.

Mein vater sprach ganz nett und lieb, auch nachdem ich mit meinen fragen deutlicher geworden war. Oh gott, was hab' ich jetzt gemacht? Vielleicht bilde ich mir das doch alles ein! Ich rief Lisa an, die war sehr geschockt. Tat mir gut, mit ihr zu reden, bestärkte mich wieder, meinen gefühlten erinnerungen zu glauben.

Nun die Aufzeichnungen der darauffolgenden Therapiestunde:

Es schoss aus mir raus: Ich habe gestern meinen vater angerufen! – Sie strahlte mich richtig an, überrascht. Und ganz gespannt forderte sie mich auf: Erzählen sie! – Ich berichtete, aufgewühlt, auch dass ich danach unsicher war durch seine liebe art zu reden. Ich müsse jetzt erst einmal einen moment schweigen. Musste mich wieder spüren, fühlen, was ich sagen will und dass ich dort bin. Ich berichtete dann vom gespräch mit Lisa: Wir denken beide, gut, dass das endlich mal auf den tisch kommt! Ich musste das loswerden! Das war wie ein zwang! Ich will, dass diese verlogenheit unserer familie endlich mal ein ende hat! – Ehem! –

Birgit sei, während ich anrief, bei den eltern gewesen, laut aussage von Maria. – Gut, dann weiß die auch gleich bescheid!, verkündete die K. ganz locker. Ich sprach von meiner angst vor reaktionen der familie. Aber auch das sei da: Ich musste meine gefühle und erinnerungen mal mit der wirklichkeit konfrontieren, also mit meinem vater, um zu sehen, ob sie dem standhalten. ... Ich bin froh, dass es raus ist! – Die K. schlug vor, wegen meiner angst vor der familie, ich könne doch meine mutter anrufen, fragen, ob sie das vom vater weiß. Das sei unmöglich! Könne mir nicht vorstellen, mutter anzurufen.

Ich sprach noch über mögliche reaktionen seitens der familie. (...) – Ich denke, egal wer sich meldet, es ist wichtig, dass ich dann ganz bei mir bleibe. Nicht darüber rede, wie es meinem vater geht oder so, sondern nur über mich! – Ja! – Mir war dann ganz schwindlig im kopf. Ich schwieg. – Hab' ich die verantwortung, wie es den anderen damit geht, meinem vater oder wem auch immer? – Nein!! Es ist nicht ihre aufgabe, auf deren befinden zu achten. –

Die K. meinte noch zur reaktion meines vaters am telefon, er habe das vielleicht verdrängt. Vielleicht sei es nach dem gespräch wieder da, vielleicht bleibe es verdrängt aus schlechtem gewissen. – Es ist ganz wichtig, dass sie bei sich bleiben! – Ja ... –

Ich erzählte damals den Sportfrauen von meinem Anruf bei meinem Vater. Sie waren begeistert. Wir tranken später Sekt. Sie gaben mir noch gute Tipps, wie ich mit den eventuellen Reaktionen aus der Familie umgehen könnte. Ich hätte nicht die Verantwortung, sondern die Täter. Solange ich nichts sage, würde ich die weiterhin schützen. Die Angst vor den Folgen verhindere oft das Aussprechen und schütze die Täter. Wie gut, dass ich damals diese Sportgruppe, diese Frauen hatte! Am Abend rief meine Schwester Birgit an. Ja, sie wolle wissen, was ich zu unserem Vater gesagt hatte. Ich sagte es ihr. Danach in mir: Ich hab's gesagt!, zittrig, aber sehr froh.

Die Zeichen häufen sich

Ich bin jetzt mit meinen Schilderungen im vierten Jahr der Therapie. In diesem Jahr war ich nun wirklich an dem Thema mein Vater. Es verging kaum eine Stunde ohne Erinnerungen oder Bezug auf etwas, was ich in den Tagen dazwischen erlebte, erinnerte oder was auch immer.

Ich sprach von meiner angst, mein vater könne vor unserer wohnung stehen: Der hat noch immer macht über mich! Immer noch! – (...) Ich fragte die K.: Finden sie es eigentlich gut und richtig, dass ich meinen vater angerufen habe? – Sie lächelte: Das möchte ich jetzt nicht beantworten. – Ich würde mich in einem prozess befinden, in den sie nicht eingreifen wolle. Außerdem hätte ich doch ihre reaktion erlebt. Müsse nur die augen aufmachen! (...)

Ich betonte meine freude nach dem anruf: Ich habe es gesagt! – Und noch etwas schönes berichtete ich ihr: Ich habe mit der kleinen Ann gesprochen, was sie davon hält, dass ich es vater sagte. Sie hat über das ganze gesicht gestrahlt! Die kleine Ann ist unglaublich froh darüber! – Ja ..., ehem ... –

Wieder ernster erzählte ich einen traum von der letzten nacht:

Es klingelt an unserer wohnungstüre. Mehrere menschen. Ich will nicht öffnen. Sie schlagen die scheibe ein und strecken ihre hände durch die zerbrochene scheibe. Angst! Sie brechen dann die türe auf. (...) Ich kann nicht laut um hilfe rufen, versuche es, bringe kaum einen laut raus ... Ein mann betatscht mich, meinen ganzen körper begrapscht er. Und ich denke: Es ist ja nicht so schlimm! Es ist ja nicht so schlimm! Wieder versuche ich, um hilfe zu schreien. Angst, herzklopfen, wach!

Zu dem „es ist doch nicht so schlimm" sagte ich: Ich denke, dass ich mir das damals gesagt habe, um es auszuhalten... –

Mir war wieder und weiter so kalt. Hüllte mich jetzt ganz in die decke ein. Meinen dicken, weichen pulli ans gesicht oder an meine brust schmiegend, sagte ich, was ich fühlte: Mama, mama! Ich muss dir noch was sagen! – Hm?, weich von ihr. Sah noch einmal zu ihr hin, dann schloss ich meine augen. Gestern wäre ich zur massage gewesen, dabei hätte ich wohl an sie gedacht. ... Dann in mir: Mama, ich will bei dir bleiben! Schick mich nicht weg! – Ich fing an zu weinen. Wann hat mich meine mutter denn weggeschickt?, hätte ich mich gefragt: Und mir fiel ein, wenn wir besuch bekamen, musste ich bei ihr schlafen, weil der besuch in meinem bett schlief. Und ihr war das oft zu warm mit mir. Ich fand's schön. Schmiegte mich gerne an sie, weil sie dick und dadurch so weich war. Aber weil es ihr zu warm wurde, schickte sie mich dann zu meinem vater in sein bett. ... Ich erschrak bei diesem gedanken. Dachte, meine mutter war doch dabei! ... Erinnere mich, hätte mich am liebsten an sie geklammert, damit ich nicht weg muss! – Ehem! – Ich erinnerte die K. an meinen traum vor längerer zeit, biss in den finger meines vaters. Der sei mir gerade eingefallen. Jetzt wieder brennen in meiner vagina. Mir sei so kalt, würde die zähne fest zusammenbeißen. – Irgendwas war im bett meines vaters! Das weiß ich! – Ich weinte: Ich will nicht weggeschickt werden! Will bei meiner mama bleiben! – Schweigen.

Ich war dann sehr erschöpft. Wäre doch auch viel gewesen heute, meinte die K. mit liebevoller stimme: Jetzt müssen sie der mutigen Ann aber noch ein paar liebe worte sagen! – Ich schaute sie an. Das tut der kleinen Ann doch so gut! Sah in die augen der therapeutin und die tränen liefen mir übers gesicht.

Später, zuhause, erinnerte ich mich ganz genau an das Gefühl, ich will nicht in Vaters Bett! Ich schrieb in mein Tagebuch:

Spürte den Wunsch nach umfassender geborgenheit und versorgtsein ... und erkannte: Das muss ich mir selbst geben! Das ist es! Sich selbst so wertvoll empfinden, so beschützenswert, und sich selbst so ernst nehmen und dann entsprechend liebevoll und fürsorglich mit sich umgehen. Also sich selbst mama sein!

Welch wunderbare Gedanken im Anschluss an diese wirklich nicht leichte Stunde! Um die umzusetzen benötigte ich allerdings viel Zeit und langes daran Arbeiten. Damals konnte ich das immerhin schon ein bisschen.

In einer der folgenden Stunden berichtete ich von einem Telefonat mit Lisa, von bescheuerten Aussagen ihres Mannes bezüglich unseres Vaters und wie schlecht es mir damit ergangen sei. Ich hätte es kaum ertragen können.

Die K. sagte nichts. Ich hätte mich danach ins bett gepackt und deutlich gefühlt, wie damals: Ich will nicht zu meinem vater ins bett! – Schweigen. Mir sei in den letzten tagen noch anderes eingefallen, auch wie ich mit ihrem kollegen alleine im büro war. Da hätte ich doch an das elternschlafzimmer gedacht. Auch weil ich hier ja nie ins büro gehen würde, habe es mich daran erinnert, so wie wir früher nicht im elternschlafzimmer sein durften. ... Die K. schwieg weiter. Ich unsicher. Warum sagt sie nichts? Mir fiel es so schwer, bei mir zu bleiben! Forderte mich öfter dazu auf!

Berichtete weiter, im bett liegend hätte ich irgendwann an hier gedacht: Und auf einmal hatte ich ganz deutlich das gefühl, etwas in mir drin zu spüren. Ganz tief in mir drin (in meiner vagina). – Ich schwieg. Warum kommt überhaupt nix von der K.? Glaubt sie mir nicht? Aber Ann, die kann doch nicht eingreifen! Du *musst dir glauben! Ich sagte es ihr: Ich muss die ganze zeit gegen das gefühl ankämpfen, gleich kommt von ihnen, was das ganze soll oder so. Aber ich hab' auch selbst zweifel! Stimmt das, was ich sage? Bilde ich mir das nicht nur ein? ... Aber ich hab' das so gefühlt! Und das ist mir alles dazu eingefallen! – Mir wurde klar, ich will was abwehren. Fror wieder an den beinen. Kuschelte mich in meine decke.*

Ich schwieg. Plötzlich hörte ich ihre kollegin in die praxis kommen, was mich zu dieser aussage veranlasste: Ich muss ihnen sagen, lassen sie niemanden hier rein! – Was verbinden sie damit? An was denken sie? – Endlich sagt sie was!! Sie ist doch da! Und nimmt meine aussagen ernst! Was freute ich mich! – Ich hörte, dass draußen jemand ist; weiß ja, dass niemand hier reinkommt. Dennoch hatte ich ganz stark das bedürfnis, ihnen sagen zu müssen, sie sollen das verhindern! – Ich dachte: Es ist zweideutig, wenn ich sage, sie soll dafür sorgen, dass niemand „reinkommt". Schweigen.

Dann brennen, ganz starkes brennen. Sagte es ihr. – Können sie sagen, wofür das brennen steht? Was es bedeutet? Oder wodurch es weggehen könnte? – Sie hilft mir! – Ich hab das gefühl, als wäre was in mir drin. Steckt was ganz in mir drin! Und ich habe gefühle dabei, leichte gefühle, aber auch ganz starkes brennen. Das tut richtig weh! Würde es am liebsten aus mir rausreißen! Und ganz weit weglaufen! – Weinen. Angucken. – Mama, mama!, sagte ich leise. Rückenschmerzen. (...) Ich legte meine hand auf meine vagina, wollte sie spüren, liebhalten, beruhigen. Sagte weich, leise: Das ist doch mir! – Ja!, die K. mit lieber stimme. – Das gehört doch mir! Da soll niemand hinfassen! – Schweigen. – (...)

Zum schluss dann: Ich will's nicht wissen, will nicht wissen, was war! – Und wer will das nicht wissen? Die große oder die kleine Ann?, fragte sie, und ich genoss für einen moment ihre stimme. – Ich glaube, die große Ann! – Und warum nicht? – Ich hab doch angst! Angst davor, wie ich das verkraften soll! – Ich schmiegte das kissen und die decke an mich. – Aber es ist auch anderes da! ... Will's auch wissen! Es geht hin und her. – Da fragte sie mit lieber stimme: Ist da nicht sowieso ein ES, das es weiß? – Ja..., leise, mit kindlicher stimme und sehr berührt von ihrer frage, ihrer feststellung. (...) –

Ich hab das gefühl, ich muss mir das angucken! Das bin ich der kleinen Ann schuldig! – Ja!!, kam wieder lieb, aber auch bestimmt von ihr. Ich sah die K. an: Aber ich würde eigentlich lieber weggucken. – Wie ihre mutter?, trocken. Ich erschrak! Ich wolle doch wissen was war, hätte aber angst davor, sagte ich wieder. Lehnte meinen kopf an die wandmatratze, sah in die augen der K. ...

Eine Notiz aus meinem Tagebuch, die ich im Anschluss an die zitierte Stunde anfertigte:

Mein lieber, wundervoller Freund K. rief an; ich erzählte ihm die neuigkeiten. [Anruf bei Vater, Gespräch mit Birgit] Er: Ann, du musst dich jetzt sehr gut schützen! Vor möglichen angriffen deines vaters, deiner familie! ... Aber ich sehe auch licht! Blauen himmel! Sonne! – Es wühlte mich zwar sehr auf, darüber zu reden, aber es tat auch so gut, ihn zu hören, seine besorgnis zu spüren und seine liebevollen hinweise, wie ich lieb mit mir umgehen könne.

Wieder träumte ich von Vater. Er verhält sich so, als hätte ich ihm nichts vorgeworfen. Er nimmt mich nicht ernst. Wie im richtigen Leben.

Ich wollte der Therapeutin in der folgenden Stunde etwas mit meinen Händen „sagen", weil es mir mit Worten nicht möglich schien. Aber das fiel mir so schwer. Sie hatte mich aufgefordert, ich solle doch meine Hände „sprechen", sich erinnern lassen. Mit großem Schamgefühl ließ ich mich darauf ein. Stellte mir vor, lege meine Hände „dorthin". Und spürte, erinnerte, dass Vater *meine* Finger in mich reinsteckt! Gefühle! Schreck! Zappelte herum. Erinnerungen an Versuche, ihn wegzustoßen. Konnte das später doch noch in Worte fassen. Schwieg danach, schmiegte das dicke, weiche Kissen an mich und weinte: Wie soll ich mich da jemals wieder selbst berühren, meinen Körper spüren wollen? –

Die Therapeutin hatte mich in einer der vorigen Stunden schon einmal gefragt: Wissen Sie denn, was war und können es nur nicht sagen? – Es war was im Bett meines Vaters. Und er hat etwas mit seinen Händen gemacht! – Ich konnte nicht sagen, er war mit seinen Händen, seinen Fingern in meiner Vagina. Schon gar nicht sagen, er hat mich damit vergewaltigt. Ich konnte es einfach nicht aussprechen!

Zu den Händen meines Vaters fällt mir noch ein, dass ich lange Zeit immer wieder die Hände der K. ansehen musste; das erwähnte ich dort auch. Erst als mir klar wurde, was mein Vater mit seinen Händen gemacht hatte, veränderte es sich.

Trotz der Härte konnte ich zwischendrin wieder vom Mädchentraining berichten, wie viel Freude und gute Laune mir das brachte. Ich sei danach richtig glücklich gewesen, wenn auch weiterhin sehr traurig. Sei so froh über das, was ich im Moment alles mache.

Ein Urlaub in Italien

Ich komme nun zu einem von meiner damaligen Philosophinnengruppe geplanten Italienurlaub. Wir waren an politischer Theorie interessierte Feministinnen und nannten uns bald die PhiloSisters. Ich wollte gerne mitfahren, konnte es mir andererseits nicht vorstellen. Das Wegfahren und gar das woanders Übernachten waren schon in der Vorstellung purer Horror für mich. Einfach unvorstellbar! Aus Angst vor auftauchender Angst, vor Panik- und Herzattacken.

Obwohl ich mich dagegen entschieden hatte, sprach ich das Vorhaben in der Therapie an. Das deutete schon daraufhin, es gibt eine Seite in mir, die will mitfahren. Das erlebte ich öfter: wenn ich dort solche Themen anspreche, dann will ich etwas für mich erreichen und verändern.

Aber in der Stunde sagte ich, ich wolle nicht wegfahren, niemand so nahe haben, wolle keinen Menschen sehen und nicht gesehen werden. – Ich fahre nicht weg! Ich will hierbleiben! – Die Therapeutin dazu:

> *Wenn sie nicht fahren, dann deshalb, weil sie sich wertlos fühlen! Wertlos, weil sie die sexuelle Gewalt erlebt haben und weil ihre mutter nicht für sie da war. Sie gehen damit der wut auf ihre mutter aus dem weg! Und sie haben schuldgefühle und bestrafen sich selber, weil sie ein kind waren, das sich nicht helfen konnte, dass das alles nicht verhindern konnte. ... Ich habe die befürchtung, dass sie auf sich und mich wut haben werden, wenn sie nicht fahren. –*
>
> *Ich war völlig verwirrt, fiel mir sehr schwer, über all das nachzudenken. Ich käme mir wieder vor, wie weggeschickt zu werden, sagte ich, wisse aber, dass das nicht stimme. Mir fiel nichts mehr ein. Nur: Ich will nicht weg! – Sie sagte lächelnd, das habe ja was trotziges! – Ja! Ich lasse mich nicht wegschicken! Ich bleibe hier!! – Nach der stunde wut auf sie und dann weinte ich bitterlich!*

In der nächsten Stunde berichtete ich von meiner Idee, Kontakt mit der kleinen Ann aufzunehmen und sie zu fragen, was sie denn wolle? – Die kleine Ann will zusammen mit den Frauen wegfahren!, sagte ich mit kindlicher Stimme. Die K. sah mich freudig lächelnd an. – Das war für mich das stärkste Argument, doch mitzufahren! ... Die Freude der kleinen Ann war die Freude eines kleinen Mädchens. So ganz natürlich! – Ehemmmm. –

In der folgenden Stunde kam ich auf meine Angst vor Panik und Herzproblemen während der Reise zurück:

> *Was mache ich denn dann? – Ich sprach eine Weile darüber. Bald bot mir die therapeutin an, sie aus italien anrufen zu können, wenn ich das brauchen würde. Und sie unterbreitete mir einen weiteren wunderbaren Vorschlag: Wir können auch einen festen termin für eine stunde vereinbaren, in der wir auf entfernung miteinander verbunden sind. Dann sagen sie das, was bei ihnen ist und was sie mir sonst in der stunde sagen würden. Und stellen sich meine antworten vor. ... Wir können auch beides vereinbaren. – Ich lächelte. Das mit der stunde fände ich gut.*
>
> *Sie holte ihren kalender. Ich staunte: Wie meinen sie das? Wollen sie sich wirklich hier hinsetzen und an mich denken und so? – Ja! Damit sie, während sie dort sitzen, das gefühl haben, ich bin bei ihnen! Als würde ich ihnen zuhören! – Ich sprach von meinem problem, das anzunehmen, wegen mir so einen aufstand zu machen. Sie aber sagte: Jetzt geht es doch um das innere kind! Darum, dass es dem gut geht! ... Das mache ich doch mit meinen kindern auch so, wenn sie alleine wegfahren. Die brauchen den kontakt mit zuhause. Die können mich dann jederzeit anrufen. Sie können mich ja auch anrufen! Es ist doch wichtig, dass diese sicherheit da ist! – Ich konnte es kaum fassen!*

Ich finde es heute noch wunderbar, wenn ich daran denke, ich fragte die kleine Ann, und sie strahlte, ja, sie will fahren. Nur so konnte ich das für mich herausfinden und mich dafür entscheiden. Manchmal ziehe ich die kleine Ann auch heute noch zu rate! Das kommt aber nur ganz selten vor, weil ich heute meist bald und sicherer weiß, was ich will oder nicht will oder was auch immer gerade ansteht. Das war mir früher fast gar nicht möglich!

So fuhr ich also mit den PhiloSisters nach Italien, wissend, ich kann die K. anrufen und am vereinbarten Termin auf die Ferne eine Stunde bei ihr sein. Und nun folgt das, weswegen ich davon ausführlich berichte: Die erste Nacht in Italien und meine dramatische Angst, die ich in den folgenden Stunden und Wochen in entsprechenden Zusammenhang bringen konnte.

Eine woche in Italia mit den PhiloSisters! Ich hatte zum glück ein zimmer für mich alleine. Musste nachts erst mal weinen. Schlief ein paar stunden, wach, wurde immer unruhiger! Angst beschlich mich immer mehr! Angst = wut?, fragte ich mich. Immer stärkere angst, panik, zittern, meine beine schlotterten. Wie früher, wie vor jahren! Es war furchtbar!!!! Ich kam nicht an mich ran. Es wurde immer dramatischer! So eine panik! Dachte ganz viel schlimmes.

Ich ging zu vera und karin ins zimmer: Vera, ich habe so eine angst! Ich will nach hause! – Ich fahre mit dir! Kein problem! – (So hatten wir es vorher ausgemacht, sonst wäre ich nicht mitgefahren!) Ich kam weiter nicht an mich ran. Wir redeten dann ganz viel. Beim reden ging es mir besser, schwiegen wir, so schlich sich die angst wieder an. (...)

Kopfschmerzen, immer stärkere wut auf die alten! Und langsam spürte ich: Ich will doch hierbleiben! Wir werden frühstücken und dann rufe ich die K. an. Und werde erst einmal hierbleiben! ... Ja, ich müsste eigentlich beschließen, ich bleibe hier! Dann klappt es mit der selbstbestrafung nicht! Verpufft es, dass es mir schlecht geht!, so meine gedanken.

Abends rief ich tatsächlich die K. an, erzählte ihr von der letzten nacht. Sie ermutigte mich bald zu bleiben! Sie denke auch, es ist selbstbestrafung. – Machen sie das, was sie sich vorgenommen haben! – Ich sprach von der furchtbaren angst und davon, ich krieg die angst nicht in den griff. Die angst so heftig, das war wie ein schock! Ja, sicher sei ich geschockt, auf einmal so starke, ganz alte(!) angst zu spüren! Aber sie glaube nicht, dass das noch einmal so heftig werden würde. Und sie ermunterte mich wieder: Bleiben sie! Beschließen sie: Ich bleibe! – Ich brauchte ihre bestärkung, dass sie mir mut machen!, sagte ich weich. – Ja! Ehem! Ist doch auch ok! –

Kuschelte mich danach an vera und weinte bitterlich! Mama, mama, mama! War so schön, ihre stimme zu hören! Ich spürte wirklich, ich will bleiben!

Die anschließenden Urlaubstage erlebte und notierte ich so:

Zusammen kochen, sonnen auf dem balkon, abends spiele und viel, viel lachen. Auch mal in einem restaurant, mal am meer, mal in einer größeren stadt in der markthalle alle zusammen einkaufen, frischen fisch, gemüse, salat, das war wirklich toll. Aber auch immer wieder mal ängste, nicht so gutes schlafen. Sagte mir bald: Angst, du kannst ruhig kommen, ich bleibe hier! Das half! (...)

Immer wieder druck- und engegefühl. Aber auch lachen, schönes machen, lesen, dortsein. In einer nacht wurde es dann wieder schlechter, nachdem zuvor auch mal richtig gutes schlafen angesagt war. Zitterte wieder, war sehr unruhig, hellwach nach nur wenig schlaf. Wieder kam ich nicht an mich ran. Las dann in meinem tagebuch, wie ich sonntag früh mit der angst umgegangen war und über das telefongespräch mit der K. Das half mir tatsächlich! Ich wurde ruhiger! Spürte mich langsam wieder und schlief dann auch bald ein. Beim erwachen am morgen musste ich sehr und viel weinen.

Am Mittwoch dann die fern-therapiestunde. Vera hatte ein tolles plätzchen in der natur für mich gefunden, schöner frühlingstag. Ich ging mit herzklopfen los, schon auf dem weg liefen mir die tränen übers gesicht: Mama! Mama! Mama! Es ist so schwer! – Ich fand den schönen platz, an einem bach, viele veilchen, und einen stein, auf den ich mich setzte. Die vögel zwitscherten. Ich stellte mir vor, dass meine mama in unserem raum auf dem boden vor der heizung sitzt und an uns, an die kleine und die große Ann denkt. – Mama! Mama!, weinend. Hielt den grünen glitzerstein, den sie mir eines tages von sich gegeben hatte, in meiner hand. Ich sprach noch mal über diese schlimme erste nacht. Ich weinte und weinte. – Es ist so schwer, mama! – Aber auch, es ist so schön, dass sie jetzt an mich denkt, es tut so gut! Stellte mir manchmal ihre bestätigungen vor, dass sie mich verständnisvoll anlächelt. Sagte ihr unter starkem weinen: Am schlimmsten ist die dunkelheit! Ich warte immer sehr darauf, dass es endlich wieder hell wird! – Stellte mir auch vor, dass die K. sagen würde: War doch klar, dass es auch schwierige phasen geben wird, aber sie haben die gemeistert! Und sie haben immer wieder viel spaß mit ihren freundinnen! –

Dann stellte ich mir vor, dass sie gehen muss, die zeit rum ist. Sie hat an mich gedacht! Das ist so eine wunderschöne vorstellung! Ich fühlte es wie ein kleines Mädchen, dessen mama an sie denkt! Ich blieb dort noch sitzen, in meinem natur-kinderzimmer; weinte noch und ging dann ein bisschen spazieren. Das alles hat mir sehr gut getan!

Es ist viel trauer in mir, könnte dauernd weinen. Wie schlimm war alles für mich, dass es mir so geht?!

Wir haben hier so viel gelacht, es war zum teil so köstlich! Und andererseits war es die hölle! Diese angst und panik! Ich werde nie vergessen, wie schwer das für mich war! Und nur durch das schöne war es überhaupt auszuhalten! Wie nah diese angst ist! Furchtbar!

Finde es gut und mutig, dass ich gefahren bin, auch wenn es knochenhart, einfach zu hart war!

Nach dem Urlaub der K. berichtete ich in der Therapie von meiner Italienreise:

> *Ich war sehr aufgeregt, als ich in die praxis kam. – Ich bin nicht früher nach hause gefahren! – Sie lächelte mich so lieb an. Ich wurde ernst: Es war die hölle! Das mache ich nicht noch einmal! Ich fahre nie mehr weg aus angst, das nochmal zu erleben! (...) Die nächte waren das schlimmste! Aber: Es war auch ganz toll! Wir hatten auch sehr viel spass! ... Es hat mir so gut getan, sie anzurufen! Danach hab ich ganz viel geweint. ... Es war so ein wunderschönes gefühl zu wissen, ich kann sie anrufen, sie sind da! Ich fühlte mich wie ein kleines mädchen, das seine mama anrufen kann, und die mama ist da! ... Das kenne ich von früher nicht! Das habe ich zum ersten mal so erlebt, und das hat mir sehr geholfen! ... – Sie schaute mich weiter lieb und intensiv an. – Über die ganz schlimme angst samstag nacht möchte ich jetzt nicht sprechen ... – Weil jetzt nur diese stunde, dann ist sie noch mal eine woche nicht da! – Die dunkelheit war das schlimmste! Obwohl ich ein nachttischlämpchen anließ. –*
>
> *Und dann berichtete ich ausführlich von der fern-therapiestunde. (...) – Das war wirklich eine geniale idee! Das hat mir so gut getan! Ich hab' ihnen alles erzählt. Hab' die ganzen 50 minuten gebraucht. Ich hab' mir vorgestellt, dass sie hier sitzen. Das hat mir so gut getan! –*
>
> *Mir ging es jetzt richtig schlecht, mein herz polterte sehr. Sagte ihr, dass es mir schwerfiel, ihr das alles zu sagen. Ich hatte gleich eine idee dazu: Weil ich angst habe, mich ausgeliefert zu fühlen. Dass sie mich dann ausnutzen können. Weil ich sie brauche, ich ihnen gesagt habe, dass mir das gutgetan hat, können sie alles von mir verlangen! Und sie können mit mir machen, was sie wollen!, sagte ich bitter und traurig. An was ich dabei denken würde. – Ich will nichts denken, nichts sagen! Es ist wie eine wand, will nicht weiterdenken! – Ich sah sie an, dachte: Ann, guck sie an, sie macht doch nichts! Die nützt mich doch nicht aus! Schlug auf die matratze: Ich will mir das schöne gefühl nicht nehmen lassen! Das war so toll zu spüren, sie sind für mich da! –*
>
> *Ich konnte ihr dann noch von dem schönen erzählen. (...) – Das hat es mir möglich gemacht, das andere auszuhalten. – Ehem! –*

Bevor ich weiter von damals berichte, will ich das Folgende, am liebsten dick und fett gedruckt, hier einfügen: Heute fahre ich weg, heute kann ich das! Es ist nicht immer leicht, aber ich habe nachts meist keine Ängste mehr. Wenn ich mich in der ersten Nacht nicht so wohl fühle, weil ich in einer neuen, ungewohnten Umgebung bin, dann kann ich gut damit umgehen. Ich weiß, woher diese Unsicherheit noch herrührt. Dafür brauchte ich aber Jahre! Es wurde erst nach und nach besser; ich musste immer wieder an mir arbeiten und öfter wegfahren. Nur so konnte sich das verändern, konnte ich Vertrauen entwicklen – zu meiner großen Freude!

Ich hatte damals um eine weitere, zusätzliche Stunde gebeten, bevor die Therapeutin nochmal für eine Woche nicht da war.

Nach der gestrigen stunde hätte ich mit herzklopfen und schwindel reagiert und mich wie krank gefühlt. Durch das erzählen sei alles wieder hochgekommen. Nach vielem weinen sei es mir dann besser gegangen.

Ich wollte ihr was sagen, angst stieg in mir hoch, aber ich sagte es dennoch: Ich wusste ja schon vor dem italienurlaub, dass sie für mich da sind. Aber ich habe das noch nie so gefühlt wie dort! Und ich habe richtig vertrauen gespürt! Und als ich wieder hier war, war das immer wieder da. Fühlte das vertrauen in mir, und das machte mich innerlich ganz ruhig. – Froh und gutes gefühl, ihr das zu sagen. – Das war jetzt ganz schön mutig von mir, sagte ich froh und kühn. Kuschelte das dicke kissen an mich. Dann wurde ich traurig. – Die angst in den ferien, das war wie ein schock! ... Ich muss darüber reden, wenn sie wieder da sind! – Jetzt weinte ich sehr, konnte mich kaum beruhigen. Schaute sie an, die tränen liefen mir übers gesicht. Sie guckte mich lieb und vielleicht bedauernd oder mitfühlend an. – Ich fühle mich wie krank. –

Während dieser Therapiepause rief ich meine älteste Schwester und meine Mutter an. Ich sagte beiden, dass auch was mit meinem Vater war. Ich wollte wissen, ob sie, besonders meine Mutter, nichts bemerkt hatten. Und warum sie mich nicht beschützte(n)! – Ich musste das meiner Mutter unbedingt sagen! Das war ich der kleinen Ann schuldig!, so meine Aussage dazu gegenüber der Therapeutin, als ich später von dem Gespräch mit meiner Mutter berichtete. Und dass ich es toll fände, das geschafft zu haben! Meine Mutter hätte mir nicht geglaubt, aber das hätte ich auch nicht erwartet. – Ich hab sie auch aus Wut angerufen!, fuhr ich fort, ihr das aus Wut gesagt! Ich musste das meiner Mutter mal zurückgeben! Und ich bin froh, dass ich es getan habe. –

In den folgenden Therapiestunden kam ich auf die Angst in Italien zurück, auch weil ich wieder diese starke Angst spürte, nun aber in der Therapie!

Plötzlich angst in mir! Und diese tauchte im zusammenhang mit schönen, lieben gefühlen für die Therapeutin auf. Aber warum in diesem zusammenhang? Die jetzige angst erinnere mich an die in italien. Es wurde immer schlimmer: Die angst nimmt total besitz von mir. ... Es geht mit mir wie eine spirale nach unten. – Haben sie keine macht mehr über sich? Wie hilflos? – Ja! – Ich sah die K. nur noch mit großen, ängstlichen augen an. Ihre frage, ob die angst von innen oder außen komme, konnte ich nicht recht beantworten. Zum schluss der stunde meinte sie: Ich bin froh, dass die angst da ist! Also ich meine, dass sie mal hier *auftaucht! Ich denke, die angst hat was mit den (lieben) gefühlen (für sie, die K.) zu tun! – Sie fügte hinzu, es wäre gut, wenn ich noch eine halbe stunde dort bliebe. Das machte ich. Sobald sie draußen war weinte ich sehr. Spürte sehr viel vertrauen zur K. Bin bei ihr mit meiner angst gut aufgehoben.*

Zu Beginn der folgenden Stunde war ich sehr aufgeregt.

Sagte dann, ganz in mich gekehrt, würde an meinen vater und an die angst in italien denken: Ich glaube nicht, dass die nur mit selbstbestrafung zu tun hat! – Ehem! – Ich fühle es so: Die angst ist eigentlich immer da! Die schleppe ich immer mit mir rum! ... Ich bin selbst überrascht, dass ich das sage, aber so fühle ich es. –

Ich wolle jetzt von der nacht in italien berichten. Und schon wieder tauchte die angst auf. Wolle die zulassen, wolle wissen, warum ich diese angst habe! Erzählte vom erst ruhigen schlafen, dann gegen sechs wach, noch dunkel, angst, immer schlimmer werdend. Panik, krieg das nicht in den griff.

Im flur der praxis auf einmal krach; die angst wurde stärker. Die K. forderte mich auf, die angst zu beschreiben. – Die kommt so aus dem bauch raus. ... Zieht mich runter, ich hab keine macht mehr über mich. Keine kontrolle mehr! – Und dann immer stärkeres brennen, in meiner vagina. Angst und brennen immer stärker. Beschrieb ihr das alles. Mein gefühl: Mama, rette mich! Ich fühle mich wie ausgeliefert! – Fühlen sie doch mal, was die mama in ihnen jetzt gegen die angst tun kann, was sie brauchen! – Erstmal die augen aufmachen, damit ich licht sehe!, sagte ich spontan. Schmiegte das dicke kissen an mich, meinen bauch. Ich war so erschöpft. Die K. guckte mich sehr lieb an, lieb und mitfühlend. – Die angst war jetzt so stark, dass mir fast schwarz vor augen wurde, und mir ist ganz schwindlig in meinem kopf. Und dann der krach draußen! Ich fühle mich wie verletzt dadurch! – Ehem! – Anschauen. Sah ihr ganz lange in ihre augen. Die schloss sie dann zärtlich, weich, öffnete sie nach einer kleinen weile und sah mich wieder und weiter so lieb an! Das ist wie umarmt und gehalten werden.

Als wir einige Monate später auf den Italienurlaub und diese furchtbare Angst zurückkamen, erklärte mir die Therapeutin diese mit den Worten: Zu diesem Zeitpunkt konnte sich der Missbrauch nur so zeigen! –

Mit den Händen „sprechen“

Bald darauf fuhr ich wieder weg, dieses Mal nur nach KL zu einer Freundin, und dennoch wurde ich wieder von Ängsten geplagt.

Berichtete ausführlich vom wochenende in KL, von der angst vor der angst. Die angst wie eine schlange am hals, als würde sie darauf warten, rauskommen zu können. In den ersten stunden in KL starke unsicherheit, wie früher. Dann war es gut, viel schönes gemacht. Immer wieder in mir, angst vor der rückfahrt. Zweite nacht, angst im bett. Während des redens hatte ich das gefühl, als würde mir jemand den hals zudrücken; sagte es der K. Konnte erst nicht weitersprechen. Sie angucken, auftanken. Das dicke kissen an mich geschmiegt. Erzählte weiter, von den verschiedenen gefühlen, der angst, dem brennen und den erotischen gefühlen. Auch dort wieder das brennen. – Als ich nachts in KL diese angst hatte, fühlte ich: Ich muss ihnen wieder was mit meinen händen sagen, muss was mit meinen händen formen! – Ein ganz weiches, zärtliches „Ehem“ kam von unserer mama. – Ich hab auch eine vorstellung davon, was ich formen will, aber ich kann es nicht sagen, nicht aussprechen! – Sie fragte, was ich für das formen denn brauchen würde: Knete oder tücher? – ... Auf einmal starke angst! Sagte es. – Weil's konkret wird? – Ja! – (...)

Erwähnte später, nach einem traum heute morgen wieder gefühle in mir drin, als wäre was in meiner Vagina. Was ich schon einmal gedacht hätte: Ich würde das am liebsten in einem symbolischen akt aus mir rausreißen! – Ganz viel angucken, schweigen, wollte nichts mehr sagen.

Aber sie sagte dann: Ich sehe es so, dass sie jetzt in der situation sind, dass sie den missbrauch wieder erleben! Egal welchen. Dass sie alle gefühle von damals wieder erleben und dass ihr körper das ausdrückt. Und es ist gut, *dass ihr körper das endlich einmal ausdrücken darf! – So ungefähr ihre worte. Ich war sehr berührt von dem, was sie sagte, und ihr dankbar, und mir, weil ich so mutig bin!*

Meinem Impuls, ihr etwas mit meinen Händen „sagen“ zu wollen, ging ich gleich in der nächsten Stunde nach:

Wieder brennen und erotische gefühle. – Ich kann nichts sagen, kann nicht sprechen! – Und was geht statt dessen? – Vielleicht was mit den händen machen ... – Ich würde gerne vorher wissen, wie's ihnen gestern nach der stunde ging. – Das ist jetzt soweit weg... – Ok! Ich hole ihnen etwas, und sie gucken dann, was sie brauchen können! – Sie ging ins büro, kam mit einigen utensilien wieder. Ich kam auf ihre frage zurück: Ich kann mich nur daran erinnern, dass ich gestern dachte, ich bin froh, dass ich jetzt da bin, wo ich bin! – Sie breitete vor mir aus: ein kaleidoskop, eine papierrolle, fingerfarben, eine decke und anderes. Kaleidoskop ... Ich rückte wieder in meine ecke. Angst! Brennen! ... – Ich bin klein! Ich bin ganz klein! .... – Wieder zweifel in mir. – Es geht nicht! – Aber sie haben eine vorstellung von dem „es" von „es geht nicht"? – Ja! Aber dann denke ich, das kann doch nicht sein! – Kurzes schweigen. –

Ich würde gerne etwas machen, aber ohne dass sie zugucken! – Ehem!, kam sofort von ihr. Nicht wegdrehen oder weggehen, sie möge nur ihre augen schließen, bat ich sie. – Ist schon passiert! – Ich spürte angst und wieder und weiter brennen. Nahm nach einer weile das kaleidoskop, mit geschlossenen augen, dann packte ich es in die kleine decke. Hielt es. Kein gefühl dazu. Legte es schnell wieder weg, schmiegte mich wieder an die wandmatratze. Sie könne ihre augen wieder öffnen. – Ich hab so eine angst! – Sagte dann traurig mehrmals leise und lauter: Mama! Mama! ... Mama! – Das war angenehm, erleichternd, beruhigend, es dort jetzt zu sagen.

Angucken. Sie wandte sich mehr zu mir hin und schaute mich dauernd an. Ich in ihre augen! – Ich kann nichts mehr sagen, hab' nur das gefühl, als würde ich mit weit aufgerissenen augen hier sitzen. ... Und ich habe sooo mama-bedürfnisse!, sagte es mit kindlicher, trauriger stimme. – Ehem! – Sie fragte, ob ich nicht doch noch etwas sagen wolle. Sie wolle mit der Frage verhindern, dass ich weggehe, ohne etwas zu sagen, und mich das hinterher belaste. Solle alles bis donnerstag ruhen lassen, und wenn es nicht gehe, sie anrufen, sagte sie mit eindringlichem blick. Ich wolle noch knete besorgen, weil ich die brauchen würde. – Aber nicht für sich alleine damit etwas machen?, hatte sie gleich gefragt. – Nein, hier! – Ja, gut! –

In der nächsten Stunde ließ ich meine Hände weiter „sprechen":

Ich hatte knete mitgebracht. Berichtete zuerst, wie es mir nach der letzten stunde ergangen war. Schon während der stunde schlecht, danach noch schlimmer! Wäre doch nicht gut gewesen, dass ich das für mich behielt, was ich „formte". Nach der stunde mein gefühl, ich sterbe! Sehr, sehr viel geweint. – Wenn das stimmt, will ich nicht mehr leben! Ich war völlig weg! Angst, ich werde verrückt! – Weinte auch jetzt. Dass Vera auf meine bitte hin kam. Ich hätte mir später was leckeres gekocht, kerzen und musik dabei. Die K. fragte nach, was ich mir kochte. Ich berichtete. Und obwohl es mir weiter so schlecht gegangen sei, wäre mein wunsch gewesen, meine dinge zu tun. Das hätte ich auch geschafft. – Ich wollte meine dinge dagegen stellen! Das war gut! ... Und ich habe gemerkt, dass es mir sehr hilft, wenn ich meine gefühle spüre! Meine mamagefühle spüre! – Ehem, ehem! – Angucken, auftanken. Erwähnte, dass ich das manchmal zwischendurch einfach brauche, diese gefühle zu spüren. Sie dann: Sie wollten sich doch was von der seele reden! – Das hätte ich heute schon zweimal gesagt. Ich konnte mich gar nicht erinnern. Ich wüßte nicht, ob ich das aussprechen oder lieber mit meinen händen ...? – Mit den händen! Das ist näher! Sind sie näher bei sich!, meinte sie gleich.

Ich würde mich fragen, ob es nicht besser sei, weil so viel, so hart, eine pause zu machen. Sie glaube, dass es jetzt stimmig sei mit dem weitermachen. ... Dann würde ich auch das kaleidoskop und die decke brauchen, die sie gleich holen ging. Damit hätte ich dienstag nämlich was gemacht, als sie ihre augen geschlossen hielt.

Jetzt schloss ich meine augen, nahm die knete, knetete einfach so drauf los. Es entstand ein stab. Das kann nicht sein! Machte den weg, knetete wieder und es entstand das gleiche! Sagte ihr, ich würde nichts spüren, keine gefühle, sei wie leer. – Es ist alles weg! Fühle mich wie tot! – ... – Vielleicht ist sich tot fühlen, nichts zu fühlen, manchmal besser!, sagte sie hier oder später. Sie wolle sich mehr in meine nähe setzen, ob das gehe? – Ja! – Weiter mit geschlossenen augen, den gekneteten stab in der hand. – Darf ich die decke dazunehmen? – Ja, klar! – Ich packte die decke um den stab, hielt ihn dann durch die decke eingehüllt fest. – Was bedeutet das, dass die decke darüber ist? Haben sie das nicht gesehen? – Nein! – Aber sie haben gefühlt, dass es da ist? – Ja! – Und was macht das ding? Was ist noch damit? – Weiß nicht! ... Als ich die idee hatte, will ihnen was mit meinem händen zeigen, hatte ich diese vorstellung. – Ehem! – Und das hab ich dienstag auch mit dem kaleidoskop und der decke gemacht. Und hab' dann so 'ne angst gekriegt, angst, was das bedeutet, wenn ich das mache! Fragte mich, ob mich mein vater vergewaltigt hat! Aber das kann doch nicht sein! Ich war doch noch klein! Dann müsste doch alles kaputt sein! ... Das ging alles durch meinen kopf, sagte ich verzweifelt. – Und danach dachte ich, dann will ich nicht mehr leben! ... Ich weiß es ja nicht, ob das so war. ... Ich weiß aber, dass was in mir war! – Sie fragte nach meinem alter. Noch klein. Vielleicht 4 jahre, vielleicht auch kleiner. Hätte kein bild von mir, alles weg! – Ich weiß nur, dass es mein vater war! Und dass ich noch ganz klein war! – Schweigen.

Die K. sagte nun ungefähr folgendes: Es sei verständlich, dass ich jetzt denken würde, dann will ich nicht mehr leben. Aber vielleicht sei das auch ein gedanke, ein gefühl von damals! – Sie sagten ja, sie fühlten sich wie tot. ... Lieber tot, als das zu fühlen! Das kann ja sein. Aber vielleicht heißt es auch: Dann soll er lieber tot sein!? – Schweigen. Ich hielt weiter meine augen geschlossen. Hörte ihre stimme so nah, angenehm! Ich sagte ihr, würde mich immer noch wie weg fühlen. – Bin gar nicht hier, nicht da! – So ging es mir noch, nachdem ich meine augen geöffnet hatte. Tiefes durchatmen.

Sie setzte sich dann frontal zu mir hin, hielt mir ihre hände ausgestreckt entgegen und sagte: Ich möchte ihre hände halten ... und dass sie mich ansehen. ... Damit sie merken, wir sind wieder hier! Ganz hier! – Ich war so überrascht! Und glücklich! Drehte mich zu ihr hin und legte meine hände in ihre. Sie lächelte lieb, und ich war so froh! Und dann ganz viel ansehen, in ihre augen, ihr gesicht. ... – Meinen sie, dass ich das überlebe? – Sie guckte mich fest an: Ja! ... Sie haben das ja schon einmal überlebt! Und das überleben sie jetzt auch! – Ich will doch leben! – Ja! Auch wenn sie mal nichts fühlen, ihre gefühle nicht spüren ... Wir gucken uns das alles in ruhe an!, so ungefähr ihre worte. Ich dachte, wir müssen gut auf mich achten. Sagte noch, manchmal hätte ich angst, dass ich etwas sage, was sich nachher gar nicht bewahrheite, nicht stimme. – Das macht doch nix! Ist doch nur hier! Das klärt sich schon! – Diese befürchtungen hätte ich doch schon öfter gehabt. Das beruhigte mich. Ich dachte, aber das hat sich immer bewahrheitet. ... Weiter händehalten, ihre hände spüren, sie angucken. Das alles hat mir so geholfen, mir so gut getan! Ich war nach der stunde sehr, sehr groggy, erschüttert und auch erleichtert! Fühlte mich so viel besser als nach der Dienstagsstunde!

Nach dieser Stunde wurde mir klar, wenn ich ausspreche, was in mir ist oder es auf andere Weise verdeutlichen kann, geht es mir viel besser, als wenn ich es für mich behalte. Trotzdem zweifelte ich wieder, kann das denn alles stimmen? Vom Kopf her weiß ich es doch nicht! Aber dann dachte ich, an meinen Gefühlen, meinem Befinden wird es doch deutlich, dass es so war! An dem Schlimmen, was ich fühle, der Verzweiflung, der Trauer, den Ängsten und an der Erleichterung danach ist doch abzulesen, dass es stimmen muss!

Zu dem mit meinen Händen „Gesagten" möchte ich kurz zeitlich etwas vorgreifen und von Besuchen bei meiner Zahnärztin berichten. Ich war bei einer Behandlung auf einmal traurig geworden, sehr traurig. Ganz komische Gefühle waren aufgetaucht und trotz größter Mühe, das abzuwenden, liefen mir Tränen übers Gesicht. Danach todtraurig!, notierte ich in meinem Tagebuch. Davon berichtete ich allerdings erst ein paar Monate später der Therapeutin, weil es mir am Vortag wieder so ergangen war:

Gestern bei der zahnärztin erging es mir wie schon mal vor monaten, tränen und gedanken an missbrauch... Mir liefen einfach die tränen, obwohl ich mir sagte, das ist doch jetzt, heute! – Ich weinte sehr! – Und es war dann noch was furchtbares! ... Weil ein bohrer undicht war, wasser raus spritze, wickelte die zahnärztin ihn in ein handtuch ... – Ich weinte so bitterlich! – ... und mir fiel das „bild" ein, was ich hier mal gemacht habe, mit meinen händen. ... Jetzt, wo ich ihnen das gesagt habe, wird's mir noch deutlicher, warum es mir bei der zahnärztin so erging! – Ehemmmm! – Das ist doch wie beim missbrauch, man liegt da und jemand macht was mit einem... – Leise, ganz leise sagte ich paarmal mama ..., mama, mama! Setzte meine brille wieder auf, musste kontakt zur K. aufnehmen. Sah sie an, sie nahm mich mit ihren augen in den arm. Und ich nahm sie in mich auf.

Nachdem ich meine Hände hatte „sprechen" lassen, und die uns viel sagten, war zum einen weiter der Impuls vorhanden, noch mehr mit meinen Händen „sagen" zu wollen. Andererseits fragte ich mich, ob es nicht besser ist, nicht weiter zu gucken, was mein Vater alles gemacht hat. Ich weiß, dass was war, das reicht doch. Ist doch so schon schlimm genug und schwer auszuhalten. Dies äußerte ich in der Therapie und fügte hinzu, mein Gefühl sei, ich müsse mich schützen. Ja, entgegnete die K., sich schützen zu wollen sei doch gut. – Es ist nur die Frage, ob sie *sich* schützen wollen oder ihren Vater? – Das saß!

Ich ließ das auf mich wirken, schweigend. Ich erzählte der K. von meinen Gedanken vom Abend zuvor. Vielleicht habe mein Vater mich auch „richtig" vergewaltigt, nicht nur mit seinen Händen. – Wenn das so wäre, das halte ich nicht aus! Das verkrafte ich nicht! ... Das ist so ein furchtbarer Gedanke! – Was ist daran anders gegenüber dem, was Sie jetzt schon wissen, dass er etwas mit Ihren Händen gemacht hat und mit seinen? Versuchen Sie doch einmal, das zu beschreiben? – Es fiel mir sehr, sehr schwer, das in Worte zu fassen. Ich antwortete ungefähr so: Ich finde es noch erniedrigender! Das bedeutet noch mehr Macht! ... Ich war doch noch ein Kind! Ich hab' das Gefühl, das halte ich nicht aus! ... – Ehem! Ehem!, kam immer wieder von ihr. – Das, was Bernd gemacht hat, das ist ja schon so verletzend! Das ist wie ein Schnitt in meine Haut. – Ehem! – Mit den Händen in mich eindringen, das ist noch schlimmer. Die haben doch überhaupt nicht geguckt, was das für mich bedeutet, wie das für mich ist! Für ein kleines Mädchen! ... Das ist für mich schon vergewaltigend! Aber dann DAS noch, das ist doch furchtbar! ... Wenn ich mir vorstelle, dass mein Vater mit seinem Schwanz in mir war ... Das ist ja noch erniedrigender! Zeigt noch mehr seine Macht! – Und ist noch gewalttätiger!, schrieb ich dazu in mein Tagebuch. Später hatte ich dazu noch gedacht, vielleicht ist es gar nicht schlimmer. Ist doch so schon Vergewaltigung, was der gemacht hat! Dann aber: Nein, das ist noch schlimmer!

Allumfassende Angst

In dieser Stunde kam ich noch auf meine Ängste vor dem und beim Wegfahren und woanders Übernachten zu sprechen:

> *Die angst in KL und die in italien, das war doch jeweils totale angst, das war doch ganz schlimm! Diese fürchterliche, überwältigende angst taucht immer dann auf, wenn ich von zuhause wegfahre! ... Dazu hatte ich dann folgende gedanken: Früher schlief ich bei meinen schwestern im zimmer, war geborgen, beschützt. Auch erst mal, wenn ich bei meiner mutter im bett lag. ... Aber dann nicht mehr! Dann war ich bedroht! ... Vielleicht ist das jetzt ähnlich, sobald ich von zuhause wegfahre. Dann weiß ich nicht, was auf mich zukommt ... – Ehem! –*

Nachdem ich der Therapeutin den folgenden heftigen Traum unter Tränen erzählt hatte, sprach ich von meiner großen Angst, dem Gefühl des Alleineseins und der Angst, ich drehe durch:

> Liege im rechten bett, meine mutter links. ... Auf einmal draußen geräusche; hab angst, mein vater kommt ins zimmer. Krieche zu meiner mutter, wecke sie: Der kommt! Ich hab angst! – Der kommt nicht! – Doch die türe geht auf, ich hab panische angst, angst, der könnte reinkommen, mich umbringen! Das war mein gefühl! Ich war wie von sinnen! ...

> *Die angst kam nicht wie sonst so aus dem bauch raus. Die kam von außen! Wie eine bedrohung! – Kurzes schweigen. – Ich hab' das gefühl, das bin gar nicht ich, die ihnen das alles erzählt, nicht die Ann von heute. – Sie fragte, was ich mit „ich drehe durch" meinen würde. Ich versuchte es zu beschreiben, aber das fiel mir sehr schwer. Dann sagte ich: Ich hab keine kontrolle mehr! Hab' noch weniger sicherheit als sonst! Es ist, wie wenn man im wasser ist und kann nicht schwimmen! – Ehem! Ehem ... –*

Wir sprachen noch darüber, dass und warum ich mich so alleine fühlte, was ich dazu unbewusst beigetragen hatte, um das noch einmal zu erleben, was ich von früher kenne. Dazu sagte die K. sehr Spannendes und Erhellendes:

> *Man macht das nochmals so, wenn man etwas früher nicht verarbeitet hat und das jetzt nachholen will. – Dann hab ich mich so gefühlt, als ich ganz klein war! DAS hab ich dann als kleines mädchen ausgehalten?! – Ich konnte es nicht fassen.*

Die K. betonte, ich solle versuchen, die angst anzunehmen: Das ist doch eine ganz tiefe, allumfassende angst! – Ich solle sie annehmen als einen teil von mir, und sie mir dann angucken. Ich versuchte, die angst zuzulassen, das dicke kissen an mich geschmiegt, lutschte auf einmal am daumen und mir wurde wieder sehr kalt an den beinen. Da die zeit bald vorbei war, fragte sie, ob ich mit der heutigen stunde auskommen würde oder ob eine zusätzliche stunde nötig sei. Ich dachte darüber nach. – Ich will alleine damit fertig werden! Hab' heute nacht schon gedacht, ich bin doch für dich da, kleine Ann! –

In meinem Tagebuch fügte ich hinzu:

Nach dem traum und auch noch in der stunde: Angst, unsicherheit und das gefühl, keinen boden unter den füßen zu haben. Das war wie vor jahren, in meinen schlimmsten zeiten! Sagte mir, die angst hat doch nichts mit jetzt zu tun, liebe Ann! Wenn sie auftaucht, lass' sie zu, guck sie dir an! Und sorge selbst sehr gut für dich!

Ich erinnere mich noch gut daran, dass die Therapeutin in der folgende Stunde von ihrer Überlegung sprach, ob sie mich nicht in eine Klinik überweisen solle, weil ich in letzter Zeit sehr oft gesagt hätte, ich wolle nicht mehr leben. So auch in dieser Stunde, in der ich auch die folgenden Träume erzählte, die ich am Tag zuvor nicht erwähnt hatte:

Ein baby krabbelt in unserem raum bis vor den spiegel. Setzt sich. Die K. versorgt es irgendwie, aber es ist wohl nicht ihr kind! Sie muss es zu einer freundin bringen oder so; jedenfalls muss sie kurz weg. (...) Ich fahre dann auch los, um etwas zu erledigen. Wir treffen uns auf der straße. Sie fährt mit ihrem auto rückwärts, mir, ich auf dem rad, vor die füße. Ich protestiere, lächelnd.

Im zweiten traum ist die K. dann wieder da, sagt so etwas wie, ob ich mein herz schon abgehört bekommen hätte, sie könne das machen. Ich freute mich, fühlte mich gut aufgehoben!

Sie fragte mich mit fester stimme: Warum nehmen sie die träume eigentlich nicht ernst? Die sprechen doch eindeutig dagegen, sich das leben zu nehmen, nicht mehr leben zu wollen! Und ich sehe es auch so, die kleine und die große Ann wollen etwas von mir. Aber ich sehe die träume auch lebensbejahend! Wir hatten spass miteinander in der situation ich im auto, sie auf dem rad. Der traum spricht auch für vertrauen. Dass sie mit allem hierherkommen können! Sie wol-

len mit allem gesehen und angenommen werden. Und sie wollen sich *auch komplett sehen, mit allem annehmen. –*

Bald darauf sagte sie, sie habe mich in den letzten tagen mit sorge beobachtet: Bisher habe ich immer gespürt, wenn es bei jemandem eng wurde. Aber ich kann mich auch täuschen. Deshalb will ich von ihnen wissen, ob sie mir sagen wollen, ich soll sie davor bewahren, sich das leben zu nehmen. Bisher hatte ich nicht das gefühl, aber jetzt will ich es von ihnen wissen! – Sie habe daran gedacht, mich in eine klinik zu schicken. Schreck! Sie beschrieb die klinik, betonte, es gäbe eine besondere station für „missbrauchte" frauen. Dort wäre ich gut versorgt, müsse nicht, wie hier, immer wieder gehen. ... Hat sie solche angst um mich? Kann sie mir nicht mehr helfen? Ich sagte ihr, dass mich das erschrecke und mir angst mache. – Ich gehe aber nicht in die klinik! – Das sollen sie ja auch nicht! Das wäre ja ihre *entscheidung! – Nach kurzem schweigen erinnerte ich sie an meine worte nach der letzten stunde: Ich will das alleine schaffen! – Ja. –*

Ich kam dann auf den folgenden Traum zurück, den ich in der vorherigen Stunde erzählt hatte, und schilderte der K. meine Gedanken und Gefühle dazu:

Bei einem haus, in dem ich wohne, wird von außen das ganze kellergeschoss aufgerissen! Ich sehe einen VWBus mit handwerkern in das kellergeschoss (wie garageneinfahrt) reinfahren. Frage meine mutter, wieso sie das machen lässt? Sie reagiert gleichgültig. Sie kommt mir vor wie eine fremde frau.

Ich hab' danach gedacht, das haus ist mein körper. Meine mutter hat mich nicht beschützt! Und hatte heute morgen ein bild dazu: Meine mutter im bett, abgewandt, mit dem rücken zu meinem Vater und mir! Und gefühle in mir drin, dass was in mir ist ... (in meiner vagina). Sie hat ja selbst immer gesagt, wenn sie schläft, bekäme sie nix mit. ... Ich denke, sie wollte nichts mitkriegen! Und dass sie mich zu ihm ins bett geschickt hat, damit sie ruhe vor ihm hatte! ... Das denke ich schon länger! – Betroffenes, betretenes schweigen.

Ich sprach noch von der angst, dass wieder diese furchtbare angst auftaucht wie nach dem traum, nachdem ich fühlte, mein vater will mich umbringen.

Das möchte ich noch einfügen: Mir wurde nach und nach klarer, was die K. damit meinte, ich solle vielleicht lieber in eine Klinik gehen: Sie ist für mich da, während der Stunden und auch mal dazwischen oder telefonisch. Außerhalb dessen muss ich selbst für mich sorgen und gut nach mir schauen. Sie kann das nicht rund um die Uhr leisten.

Ich schrieb in mein Tagebuch: Ich will mir selbst immer mehr und besser Mama sein, weil ich merke, dass mir das so sehr hilft. –

Noch später erkannte ich, dass mich die Therapeutin vor mir selbst schützen wollte, damit ich die Wut nicht selbstzerstörerisch gegen mich richte.

Die Menschenfresser-Geschichte

Ich las immer sehr viel, schon in meiner Jugend. Während der ersten Jahre der Therapie konnte ich jedoch für lange Zeit keine Literatur lesen. Ich konnte und wollte keine Geschichten von Anderen in meinem Kopf haben. Nun veränderte sich das wieder – zu meiner großen Freude. Manche Bücher bekamen eine ganz besondere Bedeutung für mich. So las ich Erzählungen der Italienerin Susanna Tamaro aus dem Buch „Love". Davon berichtete ich in der Therapie:

> *Ich gab, völlig ernst, kurz den inhalt wieder. Als ich den schluss zitierte: ... es heißt, dass es keine menschenfresser mehr gäbe. Dabei gibt es noch menschenfresser. Mein papa ist tagsüber rechtsanwalt und nachts menschenfresser", atmete die K. laut aus: Puuuh! – Und dann kam ein liebevolles, weiches: Ehem! – Ich konnte kaum weiterreden, mir blieb fast die luft weg! Schwindelgefühle! – In dieser erzählung ist wirklich messerscharf beschrieben,* wie *die mutter wegguckt! ... Diese geschichte möchte ich meinen eltern schicken! – Ich war völlig fertig und voller wut und hass und gleichzeitig ganz leer.*

Die Therapeutin bat mich um eine Kopie dieser Erzählung, die ich ihr beim nächsten Mal mitbrachte. Ich fertigte eine zweite Kopie an und sandte sie meinen Eltern mit der Bemerkung: Ich möchte, dass ihr beide diese Geschichte lest! Ann.

Als ich den Umschlag in den Briefkasten warf, war ich sehr aufgeregt und dachte, ich will mich meiner Wut stellen, denen was davon abgeben, da gehört die doch hin! Meine Eltern reagierten nicht – noch nicht! Kurze Zeit später sagte ich zu meiner Therapeutin:

> *Ich kriege sein lächeln und das, was er noch gemacht hat, einfach nicht zusammen! Der konnte mich so lieb anlächeln! – Das kann man auch nicht zusammenkriegen!, sagte sie ganz bestimmt und fuhr fort: Dass ihr vater lieb zu ihnen war und auf der anderen seite sie so missbrauchte, sie ihm so ausgeliefert waren, das ist nicht zusammen zu bringen! Das ist aber eine spaltung in* ihm! *Das hat nichts mit ihnen zu tun! – Ich verstand!*

Wut und Erinnerungen

Nun ging es um Wut auf meine Mutter, aber nicht nur. Es war mir sehr schwer gefallen, mich darauf einzulassen. Ich hatte gedacht, ich könne auf sie gar keine Wut spüren. Nachdem ich die Situation erinnerte, meine Mutter in ihrem Bett, sie dreht sich weg, konnte ich sehr viel Wut auf sie zulassen und austoben.

Gleich im Anschluss tauchten wieder Bilder und erinnerte Gefühle auf. Während dieser Erinnerungen sprach die K. mich mit Du an, sprach sie mit der kleinen Ann! Ich wusste danach noch weniger als sonst oft, was sie alles gesagt hatte, weil ich so „weg" und so beteiligt war. Jetzt möchte ich beschreiben, wie es nach den Erinnerungen weiterging:

Ich legte mich auf den rücken, die hände auf meiner vagina, wie zum schutz. Dann die hände seitlich neben mir. – Warum liegst du jetzt so da? – Ich glaube, ich sagte: Ich fühle mich wie tot! Fühle nichts mehr! Bin wie weg! Bin nicht da! ... Ich will weg! – Wohin willst du? – Raus aus dem zimmer (elternschlafzimmer). – Mit so lieber stimme fragte sie: Willst du zu Lisa? – Nein! – Oder willst du zu Maria? – Ich stand auf, ging zu meiner matratze, sagte: Ich will das nie mehr fühlen, frau K.! – Ehemmm! – Kuschelte mich auf meine matratze. Vielleicht sagte ich da schon: Ich bin ganz alleine! – Setzte mich mit dem rücken gegen die wandmatratze, deckte mich ein bisschen zu. Ich glaube, dann hockte sich die K. neben meine matratze. – Ich will zum schluss aber noch zeit haben! – Ehem! – Ich kringelte mich ganz zusammen, die decke noch mehr über mich ausbreitend. Mir war so kalt. Dann sagte ich leise: Ich will zu meiner Maria! – Ehem! – Und was macht Maria? – Ich will mich an ihren bauch schmiegen. – Ehem! ... Ist doch auch schön!?, mit leiser, weicher stimme. Ich spürte leichte, erotische gefühle. – Ja, sagte ich. Und dann: Ich gehe weg von Maria! – Und warum? – Die versteht mich nicht! Die versteht nicht, was los ist! Die merkt auch nichts! ... Gehe zu meiner mutter, in ihre arme. ... Aber auch das nützt nichts! Ich spüre nichts! Ist nichts da! ... Ich bin ganz alleine! – Ehem! –

Ich sehe mich, dass ich ganz alleine bin! Alle sind da, aber ich bin ganz alleine! ... Wir hatten so einen ganz großen flur, so groß wie das zimmer hier. Von der küche ins wohnzimmer musste man über diesen großen flur. Der war meist völlig dunkel. Und ich hatte oft große angst, durch den flur zu gehen. – Ehem! – Ich sehe jetzt den flur, sehe alle um mich herum, sie gehen hin und her, aber ich bin ganz alleine, obwohl alle da sind. Die sehen mich an, aber sehen mich nicht richtig. – Weil sie nichts wissen? – Die wissen nicht, was los ist! Keiner weiß es, nur mein vater! – Sieht er sie an? – Ich spüre auf einmal wahnsinnige angst. Ihre frage ..., das war jetzt wie ein schlag in den bauch! Ich sehe meinen vater nicht, aber er ist da! Er ist überall. Ich hab solche angst! ... Ich sehe Lisa! ... Die *sieht mich richtig an! Sehe sie als mädchen vor mir (mit zöpfen, 10–12 J alt?), und sie guckt mich richtig an, auf mich runter! – Ich versuche, die kleine Ann zu sehen, mir wird komisch! Spüre wut aufsteigen, sage: Mir ist jetzt ganz schlecht! – Ehem! – Mir ist so schlecht! Und ich spüre wieder wut! – Wusste nicht auf wen. – Ich will jetzt nicht mehr! Kann nicht mehr! – Ja, ehem! – Es war so viel heute! Will schluss machen! – Ja! Das war auch sehr viel!, bestätigte die K.*

Ich schaute mir den raum an, war noch nicht wieder richtig da. – Was fühlen sie jetzt? Wie fühlen sie sich? – Irgendwie leer! ... Aber auch größer! Erwachsener! Und ich bin groggy! – Spürte auch leichte angst, wie ich das wohl verkrafte.

Dann sagte ich: Ich würde gerne mal kurz ihre hand haben und ihnen was sagen. – Keine reaktion, doch, ein vielsagendes lächeln. Dann die K.: Sagen sie es so! – Ich wurde sauer. Blöde kuh! Lass es, Ann, oder willst du dir die stunde versauen? Ich sah sie an: Gut! Geht auch so: Danke! – Sie lächelte: Sagen sie erstmal zu sich *danke!, forderte sie mich fröhlich auf. Hatte ich auch schon gedacht, dass* ich *das doch geschafft habe. Lächelnd entgegnete ich: Habe ich vielleicht schon gemacht?! – Sie machte ein verschmitztes gesicht; sagte noch: Ich möchte ihnen meine hand nicht geben, damit die kraft bei ihnen bleibt! – Das ist eine! – Ich hatte gleich den impuls, ihre hand kurz zu drücken und ihnen danke zu sagen! – Haben sie sich denn schon die hand gedrückt?, ließ sie nicht locker und lächelte mich lieb an. – Ja! Vorhin schon! – Ich zeigte ihr meine hände, dass ich meine linke faust mit der rechten hand umschloss. Sie lächelte. Und dann sah ich ihr in die augen! Sah sie komplett an, ihre hände und wieder in die augen. – Ich finde toll, dass ich das geschafft habe. – Sie freute sich sichtlich mit mir. (...) – Lassen sie es hier!, kam zum schluss lieb von ihr. Ich konnte nur leicht lächeln.*

Ein Kommentar in meinem Tagebuch aus diesen Tagen:

Ohne meine lieben gefühle (für meine mama, die K.) könnte ich die härten der therapie nicht aushalten! Ohne diese schönen gefühle wäre das sicher nicht zu verkraften. Das wurde mir so klar!

Das hatte ich nicht zum ersten Mal so gefühlt und erlebt! Aber auch nicht zum letzten Mal.

Das Nein von damals

In der zuletzt beschriebenen Therapiestunde hatte die Therapeutin etwas leise zu mir gesagt, auf das ich in der folgenden Stunde zurückkommen wollte.

Ich hab jetzt angst wie mittwoch, als ich dort lag, beim bild im bett meiner eltern! – Ehem! – Da ist eine erinnerung aus der mittwochstunde, kann es nicht aussprechen. ... Hab' so angst! – Ich ermunterte mich. Will es doch sagen! Ich erinnerte die K. an die situation, als ich mir vorstellte, mich an Maria zu schmiegen. Dazu habe sie wohl gefragt, ist das schön? – Und das war, als würden sie mir das ins ohr flüstern. Ich spürte einen schauder in meinem körper und erotische gefühle. ... Und dann war da, dass mein vater mir so was „dabei" ins ohr flüsterte. Das war mir sofort klar! Ich bin mir ganz sicher! – Können sie die stimme ihres vaters hören? – Nein! Nicht schon wieder! Ich will das nicht alles wieder erleben!, u. ä. meine abwehr. Das verstand sie, hielt aber dagegen: Wenn sie das zulassen, kommt doch auch ihr NEIN hervor! Das haben sie damals doch gespürt, dass sie das nicht wollen! –

Bei mir sofort genau das gefühl, ich wollte nicht, dass er das macht! Wieder engegefühl, ängste. Die K. forderte mich auf, in den bauch einzuatmen. Plötzlich spontan von mir: Aber man darf doch nicht nein sagen! – Das war aber trotzdem da! In mir! – Fühlen sie mal, was in ihnen ist, kam mit lieber stimme. ... – Der soll mich in ruhe lassen! Ich will das nicht! – Ja!, fest und bestätigend von der K. Ich rutsche unruhig hin und her: Lass mich in ruhe! ... Ich will nicht, dass du mich anfasst! – Noch lauter, fast brüllend: Hau ab! – Hielt mir die ohren zu! Wieder auf dem bauch liegend: Ich will das nicht hören! Sei still! – Trampelte, sagte wieder, ich will dich nicht hören und ähnliches. Meine hände auf einmal zu fäusten geballt, damit schlug ich auf die matratzen, trampelte mit den füßen auf die matratze. Ganz fest. – Ja! Das ist gut!, ermunternd von der K. – Ich will nicht, dass du das machst! – Ich strampelte und schlug und trat wie ein zorniges kleines mädchen. Dann weinte ich. –

Ich will zu meiner mama! Mama! Mama! – Ja ..., die K. weich und verständnisvoll. – Mama, lass mich doch zu dir! – Kuschelte mein gesicht in die decke, ganz dicht an der wand. Jetzt setzte sich die K. zu mir, neben mich; irgendwann legte sie eine hand auf meinen rücken. – Meinen sie, dass ihre mama sie hört? – Nein! Ich kann nicht rufen. Krieg keine luft! Angst. Hab das gefühl, ich ersticke. – Wenn man das gefühl hat zu ersticken, kann man auch nicht um hilfe rufen. Warum kriegen sie keine luft? Haben sie ein bild? – Es ist was auf meiner brust. Ganz schwer! – Und wo ist dein vater? Was macht dein vater? –

Vielleicht hatte sie schon früher wieder „du" gesagt ... – Schwarz ... Dunkelheit ... In meinem kopf ist alles schwarz. Ich hab so angst! Ich sehe ihn nicht, aber er ist überall! ... Mein vater ist überall um mich rum, überall – Ehem! – Ich spürte ihre hand, ihren arm auf meinem rücken. Ihre hand berührte leicht meinen hals. Gut! Fühlte mich geborgen mit meinem schmerz. Aber der blieb. – Die haben mich umgebracht! – Wer? – Meine eltern! Die haben mich umgebracht. Ich bin tot! –

Die K: Ja, tot sein ist besser, als das zu fühlen! – Ich bin tot, fühle mich wie tot ..., fühle nichts mehr. Es ist alles weg. – Ich denke, wir machen jetzt für heute schluss. Ist genug! – Ja! Ich will wieder hier sein! – Ja!- Sie nahm ihre zweite hand dazu, streichelte mir sanft meinen rücken!! Wie soll ich das beschreiben? Es ist so schön, dass sie das tut! –

Ich bin völlig platt, krieg kaum luft! Und mir ist schwindlig im kopf! – Ja! Das ist die wut! – Schweigen. Angucken. Dann irrten meine augen wieder im raum herum. Schaute wieder die K. an. – Mir ist so, als hätte ich das wirklich alles meiner mutter gesagt. ... Das kommt, wie aus einer anderen welt. Ich will wieder ganz hier sein! – Ehem! ... Wie war das wochenende? Wie ist es ihnen ergangen? – Tagsüber ganz gut, aber nachts z. t. harte träume. Ein traum mit meinem vater, danach mit schreck erwacht; ein traum mit brennendem flugzeug, das über unser haus fliegt, in der nähe abstürzt. Es ist krieg ... – Ich berichtete noch mehr. – Fühlen sie eigentlich, dass es ihnen guttut, wenn das alles rauskommt? – Ja! Manchmal nicht so, aber schon, ja ... Ich will, dass das alles von meiner seele runterkommt! – Ja! –

Verstehen und Benennen

Ängste von damals, Ängste von heute

Die Therapeutin sagte danach Spannendes zu meinen Ängsten. Und ich konnte ihr noch etwas Schönes von mir berichten.

Zu meinen diversen Ängsten oder angstsituationen meinte sie: Da hat sich doch jetzt schon einiges geklärt. Sie wissen zumindest ungefähr, womit sie in bezug stehen, was es damit auf sich hat! ... – Ja, das hat mir in letzter zeit auch geholfen. Bin ja auch meist alleine damit zurecht gekommen. –

Sie fuhr fort: Die angst wie gestern mit dem druck auf der brust, ich denke, dass die im zusammenhang mit einer konkreten situation steht. Ich nehme an, mit dem „missbrauch" des vaters! ... Die andere angst ist ja eine alles umfassende angst, eine existentielle angst! Ich denke, dass sie die hatten, weil sie ja nicht wuss-

ten, wann wieder so eine situation auf sie zukommt. Sie mussten ja immer damit rechnen, dass ihr vater wieder was macht. Und dadurch entstand bei ihnen das gefühl des absoluten ausgeliefertseins, der absoluten hilflosigkeit! Sie hatten ja keinen einfluss!, so ungefähr ihre worte. – Ihre existentiellen Ängste haben sicher mit dem „missbrauch" zu tun, aber vielleicht auch noch mit anderen erlebnissen, in denen sie sich ausgeliefert fühlten. Das müssen wir noch klären. –

Sie fragte noch, ob sich meine eltern auf die zusendung der erzählung hin gemeldet hätten. – Nein! Für meine eltern spinne ich sicher! – Ich spürte dann so einen hass auf meinen vater; sagte es, wisse aber immer noch nicht, was bei ihm passen würde, also wie die wut, den hass austoben. (...)

Trotz druck in der brust und beklemmungen wolle ich ihr noch was schönes sagen: Ich spürte gestern nachmittag ganz deutlich, dass es mir viel besser geht, wenn ich die erinnerungen und gefühle und alles zulasse und rauslassen kann! Selbst wenn die stunde so schwer ist wie gestern, geht es mir trotzdem besser. Anders z. b. als ich den hass auf meine mutter nicht zulassen konnte. Das war ja sehr bedrohlich, da ging's mir doch bedrohlich schlecht! Gestern nach der stunde fühlte ich mich erwachsener, größer! ... Und gleichzeitig spürte ich die kleine Ann in mir, spürte sie in der großen, gewachsenen Ann! Und das hat mir sehr gut getan! – Ehem! Ehem! –

In dieser Zeit hatte ich wieder heftige Träume und anschließend starke Angst. Die K. bemerkte dazu:

Es ist wichtig, und sie sollten daran arbeiten, dass sie die angst nach den träumen, die mit diesen im zusammenhang steht, von dem heute trennen! Die angst gehört zu den träumen! Sie brauchen ja heute keine angst mehr zu haben. ... Es ist doch gut, dass sie die träume haben. Sie sind wertvoll für uns. Und die angst hat mit den träumen zu tun!, betont sie nochmals. – Das sollten sie trennen! Damit die nächte nicht so angstbesetzt für sie sind. Sie nicht immer denken, ich will das nicht und ähnliches, denn dadurch wird es nur schlimmer! –

Das Geschehene in Worte fassen

In einer der nächsten Stunden fragte mich die K., was mein Vater eigentlich genau gemacht habe?

Ich dachte, ich höre nicht richtig! Was soll das denn jetzt? Sie fuhr fort, sie habe das gefühl, dass ich mehr wisse als das, was ich bisher gesagt oder gezeigt hätte; dass wir beide also nicht auf dem gleichen stand wären. Sie wisse auch nicht, ob mich der vater aus dem mädchenschlafzimmer „dafür" herausholte oder ob ich mich dort hätte sicher fühlen können. Oder ob er das „nur" im elternschlafzimmer gemacht habe. Warum will die das alles wissen? Die hat das doch mitbekommen, was ich alles erinnerte!, fragte ich mich völlig verwirrt. Glaubt sie mir nicht? Ich versuchte, mich zu besänftigen und antwortete ihr: Ich bin mir ziemlich sicher, dass mein vater mit seinen und mit meinen fingern in mir drin war … (in meiner Vagina!) – Was fiel es mir schwer, das auszusprechen! – Und das war immer im elternschlafzimmer, da bin ich ganz sicher! Und meine mutter war dabei! … Wie oft das geschah, weiß ich nicht, aber es war öfter. Ich weiß auch nicht, wie alt ich war. – Ich dachte ungefähr vier, vielleicht auch sechs Jahre. Ich fragte, ob das auch das sei, was sie „wisse". Ja, aber bei ihr sei noch mehr da, sagte sie zu meiner überraschung.

Ich ergänzte, dass ich annehme, er hat mich auch „richtig" vergewaltigt (also mit seinem Schwanz!). Erinnerte an ein nächtliches bild mit ihm, er liegt auf mir, und meine anschließende angst. Deshalb hätte ich doch an dem thema nicht weiterarbeiten wollen.

Erst mit großem zeitlichen Abstand wurde mir klar, dass die K. wollte, dass ich einmal alles konkret ausspreche, also die Erinnerungen in Worte fasse. Damals hatte ich gedacht, sie glaubt mir nicht. Heute finde ich es gut, dass sie mich das über Träume, Gefühle, gefühlte Erinnerungen und mit meinen Händen „Ausgesagte" einmal klar benennen und aussprechen ließ.

Eine Tagebuchnotiz aus dieser Zeit:

Wenn ich mich dafür entscheide, nicht mehr zu leiden, sondern mein leben zu leben, entscheide ich mich für mich, gegen ihn, gehe ich von ihm weg! Hört die sexuelle gewalt auf, noch auswirkungen auf mich zu haben!

Das war und ist eine gute und wichtige Einsicht! Auch deshalb füge ich den damaligen Kommentar der Therapeutin dazu an:

Das hieße ja auch, dass man selbst entscheidet, ob man traurig oder ausgelassen ist, als beispiel. Also dass man die freiheit hat, sich jeden augenblick selbst zu entscheiden! – Ja, aber wenn ich z. b. angst habe, dann geht das nicht so einfach! … – Weiter die K.: Wenn sie in so einen strudel mit ihrem vater reingeraten, angst auftaucht, bedeutet die angst doch auch, dass sie nicht das machen können, was sie wollen! Nicht von ihm weggehen können, klein bleiben! Die angst steht doch für die verbote ihres vaters! Sie wollen ihre dinge tun, dann taucht die angst auf, damit *sie das unterlassen! Wenn sie aber trotzdem ihre dinge machen, ihr leben leben, gehen sie von ihrem vater weg! Hören nicht mehr auf ihn, sondern hören auf SICH! Jetzt geht es nach mir! So! Das bedeutetet ja, NEIN zu sagen! – Ich konnte ihr kaum folgen, war ganz benommen. (...)*

Erst nach der Stunde und noch mehr nach einem größeren zeitlichen Abstand verstand ich ihre Aussagen.

3.4 Die Erinnerungen vervollständigen sich

In den folgenden Monaten wurden die Erinnerungen und Bilder und dabei auftauchenden Gefühle noch deutlicher. Und dann konnte ich endlich weinend sagen, was ich erinnerte: Mama, der ist in mir! Hilfe, Mama! – Zuvor hatte ich gesagt, dass ich mich so sehr schämen würde. Sie fragte nach dem Grund des Schämens. – Wegen dem, was ist! – Und bald darauf: Ich will zu meiner Mama ins Bett. Die soll mich lieb halten. Mich beschützen, dass das nie wieder passiert! ... Und die soll den wegjagen, ganz weit weg! Und verprügeln. – Ich weinte und weinte.

Irgendwann zu dieser Zeit sagte ich, ich würde Zeit benötigen, wolle und müsse mir Zeit nehmen mit dem Zulassen der Erinnerungen. – Sie bestimmen das Tempo!, so die klare, feste und wohltuende Aussage der Therapeutin dazu. Ich erzählte von einem Anruf bei meiner Mutter, musste sie mal wieder anrufen:

(...) Das hat mir richtig gutgetan! Ich wollte meine wut mal dort abladen, wo sie hingehört! – Ehem! – Aber ich habe meine mutter früher nicht erreicht und heute auch nicht. Die ist wie beton! ... Aber es geht nicht um die, sondern um mich! – Ja! Ich kann sie nur darin unterstützen, dass es nicht um die, sondern um SIE geht! Darum, dass sie sich glauben! Ich will ihnen nicht die illusion rauben, aber die werden das weiter verdrängen. Aber es geht ja um sie! Sie sehen selbst, wie schwer es ist, zu sich zu stehen! Wie viel kraft das kostet! ... Sie haben sich jetzt geoutet!, meinte sie lachend. Ich sagte bald, dass ich jetzt erst richtig wut hätte! Herzklopfen vor wut! – Ja, das ist doch ok, das ist doch gut so! Das muss doch mal raus! –

Die Therapeutin betonte anschließend, es sei ja auch ein politischer Akt, wenn Frauen ihre Geschichte angehen würden. Sich gegen diese „Normalität“ stellten, dass Töchter und Frauen den Männern gehörten. Das verbinde mit der Vergangenheit anderer Frauen, auch weltweit – wie eine Art Schwesternschaft. Ich schmunzelte, da ich das schon lange so sah.

„Ich bin gar nicht da!" – Ein Lied und eine Skizze

In der folgenden Stunde berichtete ich davon, dass mir irgendwann und dann öfter das Lied durch den Kopf ging: „Ganz Paris träumt von der Liebe". Dies wurde gesungen von Catarina Valente, und meine Mutter und meine älteste Schwester Maria liebten es sehr. Maria sang oder pfiff es gern vor sich hin, als ich klein war. Und nun tauchte das in meinen Erinnerungen auf. Vielleicht nicht zufällig?, fragte ich mich und bemühte mich herauszufinden, wann es erschienen war. Dazu wieder meine Therapieaufzeichnungen:

> *Ich erwähnte das lied von der Valente, dabei kurzatmig, aufgeregt. – Es muss eine bedeutung haben! War wohl um 1954/55, da war ich 6 jahre alt ... – Schweigen. Gedanken. Ich weinte sehr. – Ich sehe jetzt andauernd unsere alte wohnung in B. Maria singt das lied. ... Ich sehe niemanden, nur die leere wohnung. Aber nur die küche, das bad, das mädchenschlafzimmer und das zimmer meiner eltern. ... Es ist niemand da! Ich sehe mich nicht, bin aber da! – Das kam alles so nach und nach aus mir raus, zwischendrin ihre bestätigungen oder fragen. Ich hielt dauernd die luft an. Irgendwann auf einmal das: Ich sag nix! Ich darf nix sagen! ... Ich bin gar nicht da! Dann kann ich auch nix sagen! Und nix fühlen ... Ich* darf *nix sagen, hab's versprochen. – Bewegte irgendwie meine arme, sagte dann, es ist, als wären die gar nicht mir! – Das ist doch klar, wenn sie nicht in ihrem körper sind, sind es auch nicht ihre arme! – Ich bin gar nicht da! – Ich lächelte bisschen: Ich schwebe einfach davon! Denke mir was schönes aus, schöne bilder ... – Gerade lächelte ich noch und dann fing ich so sehr an zu weinen, ganz, ganz heftig! – Ich hab so einen hunger! Mama! Mama! Hilf mir!, weinte bitterlich. Nahm mir das dicke kissen, schmiegte es auf meinen bauch.*
>
> *Setzte mich aufrecht hin, will da raus, will nicht mehr! – Wie geht's ihnen jetzt? – Mir ist ganz benommen, wie nach einem schlag auf den kopf. Am besten kalte dusche, sonst kippe ich gleich wieder weg! – Ich öffnete meine augen, schaute mir das zimmer genau an. Will wieder ganz hier sein! – Auch wenn ich es versprochen habe, nichts zu sagen. Ich will, dass das raus ist! – Ja! Außerdem ist es dann ja nur hier! – Ich sprach von meiner angst gestern, dass wieder die ganz starke angst kommt. – Wenn die wieder auftaucht, fahre ich nach B. und erschieße ihn! – Dann kommen sie aber vorher zu mir!, meinte die K. (...) Sie bot mir einen termin für den nächsten tag an, ich könne es mir überlegen, solle sie anrufen.*

Mein Tagebuch weist nach dieser Stunde folgende Notiz auf:

> *Ich muss mir selbst sagen, lass es dort! Muss es abschütteln, wieder im jetzt leben! Will mich nicht davon runterziehen lassen! Lenkte mich ab. Wenn ich doch*

daran dachte, sofort wieder benommen und die gefahr, wegzudriften. Versuch, zur kleinen Ann in kontakt zu gehen. Sie will es loswerden, will es unserer lieben mama sagen! Morgen!

In dieser zusätzlichen Stunde konnte ich wieder nur mit meinen Händen, nämlich malend, mein Erinnertes wiedergeben.

Ich weinte wohl, sagte, meine mama solle mich nicht alleine lassen. Schwieg ganz, ganz lange, ganz starker druck auf meiner brust, kaum luft. Wieder nach einer pause: Es ist alles ganz schwarz! Ganz, ganz schwarz! – Schweigen, gefühle in mir, komische gefühle im hals, schleim. Musste mich immer wieder auffordern, lass es zu. Von der k. kam gar nichts. Mein gefühl wäre, sie glaube mir das doch nicht. … Legte beide hände um meinen hals. Überraschend kam ihre frage nach dem warum. Wusste nichts zu sagen, weinte ganz heftig los. – Meine mama soll mir das glauben! – Viel weinen. Husten. Ich könnte kotzen! Schleim im hals! – Ich hab ein gefühl, von dem was war, aber ich kann es nicht sagen, nicht aussprechen! – Und warum nicht? – Das sagt man nicht! … Ich schäme mich!, so ungefähr meine antwort. Ich „malte" dann auf einmal mit der linken hand auf die matratze, stellte mir vor, „das" zu malen. Äußerte das, sie holte block und malkreide. Sie hockte sich in meine nähe. – Und wenn es nicht stimmt? – Das macht doch nix! Dann wissen sie es irgendwann, dass es nicht stimmt! – Das folgende musste ich sagen: Sie müssen mir aber versprechen, es nicht weiterzusagen! – Ich hörte ihre stimme ganz nahe, hatte weiter die augen geschlossen: Ich sage nie etwas weiter! –

Ich nahm die malutensilien, sie setzte sich wieder vor die heizung. Eine hand seitlich vor meinen augen, damit sie mein gesicht nicht sieht, malte ich mit rot meine lippen, mit braun einen pimmel, ganz einfach, ziemlich schnell. Setzte mich wieder zurück, meinen kopf, mein gesicht ins kissen gedrückt: Ich schäme mich so! – Sie: Ich muss sie was fragen? Sind das ihre lippen oder ist das ihre scheide? – Meine lippen! … Aber es kann da auch noch gewesen sein … Ich weiß es nicht! – Versuch, wieder zu mir und ins jetzt zu kommen. Sah sie an: Ich muss das noch mal sagen, dass das niemand wissen darf! – Es geht doch darum, dass SIE es wissen! – Ich brauchte eine weile, den satz zu verstehen. – Ja, ich will wissen was war! … Vielleicht stimmt es ja nicht! – Jetzt ist es erstmal draußen, sind sie es losgeworden! Das klärt sich schon noch! – Schweigen. – Sie müssen meine zeichnung verstecken! – Ich nehme sie an mich. –

Sie deutete das ende der stunde an. Ich bat um eine weitere minute. War ok. Angucken. Sie schaute mal weg, atmete tief durch, als wäre es ihr zu viel! Kann ich ja verstehen.

Ich fühle mich von oben bis unten wie eine offene Wunde!, notierte ich in mein Tagebuch. Und: Ich will aus diesem Sumpf raus! – Ich war noch nicht soweit, aber auf dem Weg dorthin…

Heute morgen, als ich dachte, sie dürfen jetzt nicht weggehen, hatte ich wieder das gefühl, dass was in mir ist, ganz weit drin, bis in den bauch… Und dachte, jetzt bin ich endlich da, wo ich immer hinwollte. – Immer wieder kurzes schweigen, dann sprach ich weiter: Und fühlte, mama, geh jetzt nicht weg! – Vielleicht sei meine mutter auch mal nicht dabei gewesen. Ich weinte jetzt ganz viel. Also dass sie mich mit vater alleine ließ. (…) Hätte jetzt das gefühl, als könne ich vaters gesicht gleich sehen: Der ist jetzt irgendwie mehr „da" als sonst, ist nicht nur alles schwarz. Und ich spüre seine nähe. … –

Etwas später sagte ich, dass ich Maria jetzt das lied von der Valente pfeifen hören würde. Fühlte mich dabei mit meinem vater ganz alleine, Maria irgendwo in der wohnung. Ich atmete mal schneller, hielt mir dann mit beiden händen den mund zu, strampelte mit den beinen, brach ab. Mein gefühl sei, der habe mir den mund zugehalten. – Ich will wieder hier sein! – Ehem! – Ich sah sie an. In ihre augen. – Ich brauche sehr viel zeit dafür! … Es ist so schwer, sich zu glauben! Wenn nur eine winzige erinnerung auftaucht, kommt sofort ein riesiger schwall von zweifeln hinterher! – Vielleicht stehe das im zusammenhang damit, vermutete sie, dass ich doch davon sprach, ich brauchte viel zeit. – Vielleicht schützen sie sich damit … – (…) Ich blieb noch über eine stunde in unserem raum. Viel vertrauen zu der K. Und zu mir!

Umgang mit widersprüchlichen Gefühlen

Für mich war es sehr schwer zu verkraften, dass ich bei Bildern von Situationen mit meinem Vater auch schöne Gefühle erinnerte; die Therapeutin half mir, auch das einordnen und annehmen zu können:

Erzählte von gestern früh, den lieben gefühlen für sie und dabei wohlige, erotische gefühle. Und dann das sichere gefühl, dass bei den manipulationen und übergriffen meines Vaters auch solche gefühle dabei waren!

Zwischendrin berichtete ich vom mädchentraining, wie sehr die kleinen mädels klar haben, dass sie beispielsweise keinen arm um die schulter gelegt haben wollen. Die K. meinte, heute sei das thema einfach präsenter, bei mir damals ja nicht. Also damals dieses wissen, das will ich und das nicht, fast unmöglich; aber sie glaube, dass kinder trotzdem schon eine ahnung haben, was ihnen guttut und was nicht! (...)

Nun kam ich wieder auf die wohligen gefühle zu sprechen: Das ist doch furchtbar, einfach nicht zu verkraften, dass dabei schöne gefühle auftauchen oder vielleicht nur schöne! – Sie, ganz engagiert, dass sei ja gerade das schlimme! Ohnmacht, hilflosigkeit, das wäre ja schon schlimm, da sei wut möglich. Bei den schönen gefühle entwickele man die jedoch nicht! Und deshalb sei das noch schlimmer für ein kleines mädchen. – Für all das würde ich meinen vater am liebsten auspeitschen! –

Wieder „hörte" ich das Lied der Valente, von Maria gesungen, gepfiffen, bei der Erinnerung, dass mein Vater auf mir liegt und ich versuche, ihn wegzustoßen.

(...) Ich will damit nicht leben! – Das hat aber die kleine Ann gesagt! Das ist doch ein gefühl von damals! – Ich will wieder hiersein! – Ehem! – (...)

Die K. streichelte meinen arm: Wollen wir die kleine Ann mal in den arm nehmen? – Ich sagte doch tatsächlich: Nein! – Ist das zu viel?, fragte sie liebevoll. Ich weinte: Ich bin ja doch alleine! Das nützt doch alles nichts! – Sie nahm meine hand, drückte sie. – Es ist wie ein kühlschrank in mir drin. – Aber wenn man die tür aufmacht, wird es warm!, sagte sie weich und nahm noch meine zweite hand. Ich schaute in ihre augen. Ich spürte, ich liege im schutz meiner mama! Schwindel im kopf, als würde ich gleich ohnmächtig. Fragte, ob das wut sein könne? – Ja, vielleicht auch! Das sind die starken gefühle, das ist zu viel für ihren körper! – Sie streichelte mir über die schläfe. Angucken. – Glauben sie, dass ich das alles aus mir rauskriege? Dass ich mal ein ganz klares gefühl habe, was war? – Ja ..., wenn sie die kühlschranktüre immer weiter aufmachen! – Ich musste über sie lächeln. Nach dem aufstehen von den matratzen gingen wir zusammen zur balkontüre, frische luft einatmen. Das war notwendig und gut!

Nach einer der Stunden, in denen es sehr schwer war, notierte ich:

Das ist einfach alles zu viel! Es ist unmenschlich! Guter, wohltuender gedanke: An vaters campingbus auf beiden seiten – da ist ja viel platz – in großbuchstaben mit einem schraubenzieher reinzuritzen: Ich habe meine töchter sexuell missbraucht! Ich bin ein vergewaltiger!

Das war eine wunderbare Phantasie! Die Idee hatte ich einem Buch entnommen. Der Täter, ein Lehrer, fuhr damit durch die Stadt, da er das von einer seiner Schülerinnen an der Beifahrerseite Eingeritzte nicht bemerkt hatte. Ich sprach darüber mit Freundinnen. Meike sagte, kurz und knapp: Ich fahre mit dir und stehe Schmiere! – So etwas muss hier doch auch erwähnt werden! Ich habe dieses Vorhaben nicht realisiert, habe mich nicht getraut. Schade eigentlich. Aber schon die Phantasie, das zu tun, stärkte mich und munterte mich sehr auf. Und dann noch die zugesagte Unterstützung!

In zwei weiteren Therapiestunden ging es um meine Angst, auch sie, die K., lässt mich nicht gehen. Wieder fiel es mir schwer, offen zu sagen, was ist.

Ich könne es einfach nicht sagen. Mit ihrer lieben stimme fragte sie: Was brauchen sie? Was könnte ihnen helfen? – Mit kindlicher stimme: Ich hab so angst, wenn ich es sage, dass ich dann überhaupt nichts mehr von ihnen bekomme. – Mich schüttelte es am ganzen körper und mama, mama in mir. Ich hörte sie sich bewegen. Sie kam in meine nähe und sagte ganz fest: Ich möchte, dass sie mich angucken. – Ich öffnete meine augen, sah sie an, in ihre augen; sie sah mich ganz eindringlich an: Das wird aber nicht so sein! – *Mit weicherer stimme: Ist das klar? – Ja ..., weich, dankbar.*

Ich schaute vor mich hin: Ich möchte es ihnen doch so gerne sagen ... – Ehem!, weich. Versuchte, meine zuvor gefühlten erinnerungen in worte zu fassen: Ich hatte auf einmal das gefühl, sie lassen mich nicht los, wollen nicht, dass ich von ihnen als mama weggehe. ... Ich bekam angst! Erinnerung an meinen vater, dass er mich von meiner mutter wegbrachte, mich auf seine seite zog. Mich nicht von sich wegließ! – Sie fragte nach, wie er das gemacht habe. Ich antwortete ungefähr so: Die hat doch keine zeit für dich, aber ich! Und ich hab dich doch lieb! ... Und du hast doch schöne gefühle. Das ist doch schön für dich! – Ich verzog mein gesicht: Mir ist jetzt richtig schlecht. Das war überhaupt nicht schön, nur eklig! ... Der hat mich richtig von meiner mutter weggelockt. Wie im märchen, wenn die kinder zu irgendwas in den wald gelockt werden. – Schweigen. –

Ich hatte solche angst, wenn ich ihnen das sage, dass sie mich nicht gehen lassen wollen, wenn ich ihnen das unterstelle, dass sie mich dann alleine lassen. – Die K. mit der überraschenden frage: Ist das so wie mit dem kühlen blick ihres vaters? Dass er sie auf einmal nicht mehr wahrnahm? Haben sie da auch etwas zu ihm gesagt? Dass sie nicht mehr wollen. Oder was war da? – Mein herz flatterte, ich legte meine hand dorthin. Wieder schüttelte es mich am ganzen körper, meine beine zitterten. – Da hat sich was verändert, da ist irgendwas passiert, aber ich weiß nicht mehr was. – Ich kam da nicht dran. Dann das gefühl, ich will nicht mehr leben, das hält man doch nicht aus! Aber ich will doch leben! Sagte ihr, es wäre jetzt genug für heute! – Ehem!, weich. Schweigen, setzte mich aufrechter. – Sind sie jetzt wieder hier? – Ja, so ziemlich! – Ich bin auch da!! *Auch wenn ich gleich weg muss!, sagte sie lächelnd.*

Lisas Erinnerungen nach meinen Schilderungen der letzten Therapiestunde:

Dass meine mutter öfter zu mir sagte, dass du immer so krank bist! Irgendwie seltsam, wie mutter das gesagt habe. Wir erinnerten uns beide, wenn frau krank war, bekam sie wenigstens mal aufmerksamkeit! Lisa habe sich damals gefragt, wieso kümmert sich unsere mutter so um mich? Hatte sie ein schlechtes gewissen? Weitere erinnung von Lisa, dass unser vater uns mit bonbons gelockt hat, sie aus seiner tasche holte, guck mal, was ich hier habe. Die anderen müssen davon aber nichts mitkriegen! – Mir fiel mein traum von vor monaten ein, die aktentasche meines vaters, voll mit bonbons. Ich solle still sein, die tasche zumachen!, seine bemerkung dazu.

Erinnerungen an nicht-sexuelle physische Gewalt

In der folgenden Stunde erinnerte ich dann nicht nur sexuelle, sondern auch physische Gewalt durch meinen Vater. Kann das wirklich sein?, fragte ich mich zweifelnd. Erinnerte, er zieht mich an meinen Haaren, damit an meinem Kopf.

Am wochenende hätte ich mich so getrennt von ihr gefühlt, sie so weit weg und jetzt sei dauernd da: Mama, ich will dir was sagen! ... Nein, ich will nichts sagen, will nicht reden. – Ihre nachfrage, ob das schweigen sei wie, will nur hier sein, nicht reden, oder eher, könne es nicht sagen, weil ...? Ich fühlte mich sehr verstanden, sagte: Ich kann nichts sagen! Ist besser, ich sage nichts! – Schweigen. Bei mir das gefühl, als zöge was meine stirn hoch, zöge jemand an meinen haaren. Was ist denn das jetzt? Rieb meine stirn, zog die brille ab, hände vors gesicht. Bild: Der zieht an meinen haaren, ganz fest! Eher erinnertes gefühl als bild. ... Und steckt mir seinen pimmel in den mund. Verzog mein gesicht, sagte vor mich hin: Das kann doch nicht sein! Das glaub' ich nicht! ... Hab jetzt so ein bild, ein gefühl, aber das kann nicht sein! ... Mama, mama, sagte ich leise vor mich hin.

Sagte ihr, mein gefühl sei, dass jemand an meinen haaren zieht, so richtig fest! Meine stirn mit hoch zieht. – Und dann ist da noch was ..., aber das kann ich nicht sagen!, leise von mir. – Doch! Sagen sie es! – Und was ist dann? – Dann weiß ich es auch, und sie sind es losgeworden! ... Ist die kühlschranktür wieder ein bisschen weiter auf, wird es wärmer!, sagte sie mit lieber stimme. Ich konnte nix sagen. Sie fragte nach, wer an meinen haaren gezogen habe? – Mein Vater! Ich sehe ihn nicht ..., hab die augen zu. Aber ich rieche ihn! Fühle, dass er es ist! Ich will ihn nicht sehen! ... Er hat mich an den haaren gezogen und gegen was geknallt ... – Aber ich konnte es selbst nicht glauben! Enge am hals, druck auf der brust. Wadenkrämpfe. – Ich hab mich ganz klein gemacht ..., ganz klein! Aber das hat auch nichts genützt! – Ich saß ganz zusammengekauert dort. Sie fragte nach; ich sagte: Mein gefühl ist, mama, ich bin tot. Seitdem bin ich tot! – Ehem! – Wieder dachte ich, das kann doch nicht sein und so ähnlich. Schaffte nach einer weile auch das noch zu sagen: Ich hatte gleich zu anfang das gefühl, dass er an meinen haaren zieht, damit an meinem kopf zieht ... und dann das bild, was ich hier mal gemalt habe ... – Ehem, kam wieder sanft von meiner mama. – Zieht er sie von seinem schwanz weg? – Nein! Er hält meinen kopf damit fest, sagte ich ganz aufgewühlt, ... und dann haut er meinen kopf irgendwo gegen. – Ich schluchzte, konnte nicht richtig weinen. Mit kindlicher stimme: Ich habe keine mama und keinen papa mehr! – Ja ... Ja, so kann man das heute sehen! Verstehe ich! –

Ich saß weiter ganz zusammengekauert dort, waden verkrampft. Irgendwann: Vielleicht will ich ja nur aufmerksamkeit... – Also nein! Das glaube ich wirklich nicht! – War eher so ein gefühl von früher, dass meine mutter das gesagt hat. –

Die stunde fast zu ende, ich streckte und reckte mich. – Glauben sie, dass ich das aushalte? Dieses bild, all das. Ob ich das ertrage? – Wenn sie mich brauchen, ICH bin da! ... Haben sie keine wut? Ich spüre ganz viel wut! – Nein, nur so ein komisches gefühl im kopf ... und angst! Aber vielleicht steckt dahinter ja die wut? – Sie lächelte, stand auf, kam auf mich zu, verabschiedete sich nicht, sondern hockte sich direkt vor mich hin, hielt mir ihre beiden hände geöffnet hin. Und ich legte meine hände in ihre. Sie drückte meine hände, hielt sie. Ich sah in ihre augen, ihr gesicht. Leise: Sie gehen doch nicht weg? Nochmal halte ich das nicht aus! – Nein! Ich gehe nicht weg! ... Und wenn sie wieder das trennende zwischen uns spüren, dass sie sich getrennt von mir fühlen, dann denken sie daran: Das ist die wut! – Ja! – Angucken. Sie drückte wieder meine hände. Das weggehen war heute schwer!

Ich spüre, dass meine Mama (die Therapeutin) für mich da ist und dass ich auch für mich da bin!, so eine Notiz in meinem Tagebuch. Wie schön!

Konfrontationen

Während dieser Phase der Therapie rief ich meine Mutter und meinen Vater an. Konfrontierte beide mit meinen Erinnerungen, seinen Taten, ohne Einzelheiten zu nennen.

Nachdem ich klar und deutlich meinem Vater sagte, ich erinnere mich genau, du hast das gemacht, kam tatsächlich von ihm: Und was ist, wenn ich sage, ja, ich habe das gemacht? – Das schaue ich dann. Werden wir sehen. – Hat er es mit seiner Frage schon zugegeben?, fragte ich mich damals. Meine Therapeutin sah es so.

Lange Zeit war mein Wunsch, wenn ich ihn darauf anspreche, wenn ich das überhaupt schaffe, dann hätte ich gerne, dass er es zugibt. Ich erkannte aber, ich brauche das, weil ich mir nicht glaube – oder mir noch nicht glauben will! Aber das veränderte sich nach und nach. Und das kam hinzu: Ich will die alle drei nicht mehr schützen!

Nach dem deutlichen Telefonat mit meinem Vater kam bald ein Schreiben seines Rechtsanwaltes mit der Androhung einer Unterlassungsklage, ich dürfe diese Behauptung nicht wiederholen, sowie der Aufforderung, einen seriösen Facharzt (also Psychiater) aufzusuchen. Eine Frechheit! Die werden mich noch kennenlernen!, war eine meiner ersten wütenden Reaktionen. Ich holte mir Rat bei einer Rechtsanwältin und kam zu dem Schluss: Das ist doch ein einziger Einschüchterungsversuch! Dennoch blieb dieses Schreiben leider nicht ohne Wirkung, besonders deshalb, weil mein jüngster Bruder, Leo, in der Folgezeit mit verstellter Stimme nachts auf meinem Anrufbeantworter Todesdrohungen und Beschimpfungen der übelsten Art hinterließ. Bei mir löste der Telefonterror natürlich große Angst aus.

Nach einer langen Phase des Überlegens und entsprechender Beratung mit der Rechtsanwältin und meiner Therapeutin reagierte ich überhaupt nicht auf das Schreiben des Rechtsanwaltes. Ich hielt mich zwar für eine Weile zurück, nahm aber nichts zurück! Die Wahrheit war ausgesprochen und in der Welt, das Schreiben des Rechtsanwaltes meines Vaters konnte daran nichts mehr ändern.

Ich berichtete in der Therapie, dass ich gegenüber meiner Nichte mit klaren Worten von den Taten meines Vaters sprach, mich also nicht zurückhielt, trotz des Schreibens seines Rechtsanwaltes.

> *Endlich habe ich zu mir gestanden! – Ja! – Und es war ganz selbstverständlich! Ich ließ keinen zweifel daran, dass mein vater das gemacht hat! Ich hab' ganz klar zu Ira gesagt, es interessiert mich nicht mehr, wie es den eltern geht! (Weil die immer leiden, es ihnen schlecht geht usw., besonders nachdem sie kritisiert oder angegriffen wurden!) Und: Ich wolle nicht mehr leiden! Sondern leben! … Das hab ICH gesagt! – Ja!, von der strahlenden K. –*
>
> *Was mir noch klar geworden ist: Auch wenn meine mutter so etwas sagte wie stell dich nicht so an oder so…, mich alleine ließ. Auch wenn sie mich zu ihm ins bett geschickt hat.* ER hat das gemacht! *ER ist der täter! ER trägt die verantwortung! –*

Wichtige Erkenntnisse

Im Laufe dieses Therapiejahres konnte ich weitere wichtige Erkenntnisse aus den zum Teil sehr harten Stunden ziehen. So sprach ich in einer Stunde davon, dass ich jede Nacht ganz spät ins Bett ginge, könne einfach nicht früher zu Bett gehen. Danach wieder Erinnerungen mit meinem Vater. Er liegt auf mir und ganz viel schwarz breitet sich über mich und um mich herum aus. Der Therapeutin fiel meine Angst vor der Nacht ein und meine Freude darüber, wenn es wieder hell ist, von der ich mal gesprochen hätte. – Haben Sie also auch daran gedacht? Dieses Dunkle, Schwarze vorhin, als das auftauchte, dachte ich sofort daran! –

Ich hatte dort wohl schon öfter erwähnt, dass ich immer sehr spät ins Bett gehe. Ich erinnerte mich nun an mein heftiges Weinen vor circa einem Jahr, als ich immer wieder spürte: Ich will nicht in mein Bett. Ich gehe nicht in mein Bett! – Jetzt also die noch deutlichere Erkenntnis, *deshalb* ist das so schwierig für mich! Wegen der Angst vor der Nacht. Und die hat mit den Attacken meines Vaters zu tun!

In einem Buch hatte ich gelesen, dass eine Frau jahrelang nachts zwischen drei und fünf erwachte. Erst als sie ihre Geschichte bearbeitete, erkannte sie, warum. In dieser Zeit war immer ihr Vater in ihr Zimmer gekommen. Davon sprach ich ein paar Monate später in der Therapie und fügte hinzu: Ich gehe doch schon seit Jahren nicht mehr vor zwei Uhr ins Bett. Das ist für mich wie eine magische Zahl! – Die K. verstand.

Wir sprachen auch über meine Angst vor dem geplanten Urlaub mit den PhiloSisters. In diesem Jahr wollten wir nach Frankreich fahren. Ich wolle einfach nicht wieder solche Angst und solche Nächte wie in Italien erleben. Die K. beruhigte mich, auch mit diesen Worten: In Italien hatten sie das mit dem „Missbrauch“ ihres Vaters noch nicht klar. Da konnte er sich nur so zeigen! Jetzt sind sie doch weiter, wird das nicht mehr so sein! – War es dann auch nicht.

Außer vom dem Schweren und Schrecklichen sowie den Erkenntnissen kann ich auch aus dieser Zeit von Schönem berichten. Nach der oben erwähnten Stunde phantasierte ich, dieses Bild zu malen:

> *Eine kleine, ganz bunte fläche, die kleine Ann, und darüber weit ausgebreitet alles schwarz! Und weitere bilder, auf denen das schwarze immer weniger wird, das bunte immer mehr zum vorschein kommt, bis das schwarze in weite ferne rückt! So wünsche ich es mir!*

Ich schrieb sogar auf: Ich bin stolz auf mich! Das ist etwas, was ich zuvor wohl noch nie dachte. Ich bin stolz darauf, was ich in letzter Zeit alles geschafft habe, und stolz, weil ich zu mir stehe!

Das Mädchentraining machte mir zunehmend Freude, ich wurde immer selbstbewusster dabei. Und beim Training mit den anderen Frauen trat ich mehrmals Bretter durch!

Zurück zu den Erinnerungen. Wir sind im 6. Jahr der Therapie. Nachdem ich aus der Erzählung von Susanna Tamaro zitiert hatte, wie sie den Vater beschrieb, am Tag Rechtsanwalt und in der Nacht Menschenfresser, hatte ich mich noch so geäußert, dass ich bei meinem Vater diese beiden Seiten einfach nicht zusammen bekäme. In der folgenden Stunde veränderte sich meine Sicht; dazu ein kleiner Ausschnitt:

> *Jetzt war ich ganz weit weg, richtig weg. Fiel mir wieder so schwer, das, was bei mir war, ernst zu nehmen: Der hält mich fest, lässt mich nicht gehen! ... Der sagt, das war doch schön! Ich hätte das doch auch gewollt. Aber das stimmt nicht! Das weiß ich ganz genau!, kam so nach und nach aus mir raus. – Das ist doch gut, dass du das weißt! Hast du ein bild, wie alt du bist? – Nein! Ich glaube, ich will mich nicht sehen! – Schweigen, sammeln, wieder im jetzt sein.*
>
> *Ich sagte dann zu dem gefühl, das kann doch nicht wahr sein, dass der so war, mit fester stimme: Doch, wenn ich richtig die augen aufmache, passt das auch zu ihm, wie er am tag war! Wenn ich ihn richtig sehe, nicht so, wie ich ihn gerne haben wollte. – Ehem! –*
>
> *Und weiter, mit fester stimme: Ich hab' das so gefühlt, dass er mich festhält und sagt, ich soll nichts sagen! Und ich hätte das gewollt. – Sie sah mich direkt an: Jetzt haben sie das* gefühlt, *was sie ja schon lange wussten, worüber sie schon öfter geredet haben! – (...)*

In der folgenden Zeit kam ich nur noch selten auf meinen Vater und seine Taten zu sprechen. Mein berufliches Weiterkommen und andere wichtige Veränderungen standen nun im Vordergrund – und das war auch gut so!

Aber es gab, mit größeren Abständen, doch noch sehr wichtige Stunden, die das bisher Erinnerte bestätigten, festigten und sogar noch erweiterten. In mein Tagebuch schrieb ich:

> *Ich bin stark genug, mir das anzuschauen, die Wahrheit zu (er-)tragen. Mein körper, meine seele, mein herz, die wissen es ja sowieso! Die haben nichts vergessen!*

Das Wehren und Verstehen nachholen

So konnte ich mich in der Folgezeit beim Erinnern zunehmend selbst „sehen", wollte das auch, auch ihn. Aber das war sehr schwer, ging nur ganz zaghaft. In dieser Stunde ging es erst einmal darum, mich nachträglich gegen meinen Vater zu wehren:

Ich beschrieb, wie meine mutter sich wegdreht: Wir sind im elternschlafzimmer, liegen in deren betten. – Wo ist ihr vater? – Hier und auf mir und überall. Ich will uns nicht sehen! – Strampelte mit den beinen. – Stoßen sie ihn doch weg! – Ich stützte meine hände gegen die Wandmatratze, drückte dagegen. – Keine chance, der ist so stark! – Schubsen sie ihn weg! Sagen sie NEIN! *Dass sie das nicht wollen! – Ich möchte ihn an den haaren ziehen! – Ja! – Und mit den fäusten auf den kopf schlagen! – Ja! Machen sie das! Spüren sie die kraft in ihren fäusten. – Ja! Auf seinen kopf!, und ich schlug mit meinen fäusten auf die matratze. – Und auf seine augen! Damit er nicht mehr so eklig gucken kann!, weinend. Ich fügte noch hinzu, dass seine haut so eklig wäre. – Ich möchte ihm das gesicht zerkratzen und seine augen ausstechen. – Ja! – Mit meinen fingern kratzte ich auf der matratze, aber wie!*

Dann schweigen. – Rache! Rache! – Ja! – Mir ist so schlecht, richtig schlecht!, sagte ich paarmal. Und etwas später: Da ist die kleine Ann im bett ihrer eltern, ganz alleine! – Aber jetzt ist sie doch nicht mehr alleine! – Sie fragte noch, was ich denn nach der stunde mit der kleinen Ann machen wolle …

In der folgenden Stunde sagte ich dann, ich würde meinen Vater und mich gerne mal „dabei“ sehen, hätte aber riesengroße Angst davor. Und …

… mir fiele jetzt der traum vom anfang der therapie ein, beiße ihm in den finger. Das hätte ich damals ja sehr deutlich gespürt, das beißen. Immer, wenn sie, die K., mich bei erinnerungen gefragt habe, ob ich mich sehe, wie alt ich sei, wäre ja sofort alles weg gewesen… – Ich hab solch eine angst, mich zu sehen! – Ja. … Damals war es sicher so, dass sie das sofort in ihrem alltag vergessen haben, vergessen wollten. Das ist verständlich und auch wichtig! Ich denke, dass sie dann auch daran zweifelten, ist das wirklich passiert? Ihr vater wird das unterstützt haben, wird das auch verdrängt haben wollen. Ich könnte mir vorstellen, dass er sie dann einige zeit nicht beachtete, um sich daran zu hindern, es wieder zu tun. Das könnte ihre zweifel, ist das wirklich geschehen, bestärkt haben. Und dass es ihnen dadurch jetzt noch schwerer fällt, es zu glauben! Andererseits ist es auch eine frage, wie sehr man das zulassen kann! Das muss man auch verkraften! – (…)

Gestern nach der stunde hätte ich das alles sehr von mir weggehalten! – Ja! Das ist doch gut! Das muss ich jetzt auch mal sagen! Die rache und die wut sind ja sowieso gut. Das muss sein! Aber das andere muss man auch mal wieder wegschieben. Das haben sie gestern hier ja auch gemacht, ihn mit den armen weggedrückt, weil sie ihn sonst nicht loswerden konnten. Und das war doch gut! Das kostet doch alles sehr viel kraft. Aber das muss ich auch noch sagen, das ist ja nicht alles in ihrem leben! Das war zwar ganz furchtbar, aber sie haben auch anderes erlebt! –

Ich weinte ein bisschen. Sprach davon, sie sei gestern so hell gekleidet gewesen, ich wäre mir ganz dunkel vorgekommen. Schwarz. Wie tot! Wüsste nicht, wie ich es besser beschreiben solle. – Sie beschreiben das doch ganz genau! Das ist doch verständlich! So muss sich die kleine Ann doch gefühlt haben. Tot! Dann kriegt man nichts mehr mit! Die kleine Ann muss doch gedacht haben, das geht jetzt das ganze leben so weiter! Dann lieber tot sein! – Gestern war nach der stunde auch mal wieder in mir, ich will nicht mehr leben. Und jetzt wieder! – Ehem! Ehem! –

Ich wollte da raus, wollte mich selbst aus dieser stimmung herausholen. – Jetzt muss ich ihnen noch was schönes erzählen! – Lächelnde ma: Ja! – Ich hätte ein prospekt vom frauenbildungshaus altenbücken entdeckt. Im sommer könne frau dort urlaub machen, sich verwöhnen lassen! – Ich würde gerne einmal alleine wegfahren, dort hinfahren! – Wie lieb sie lächelte! – Aber ich traue mich doch nicht! Würde das wirklich gerne machen, ganz alleine wegfahren! [Das hatte ich noch nie in meinem Leben gemacht!!] Und das dachte ich gestern nach DER stunde! – Die K. reagierte richtig begeistert. – Es wäre ein guter rahmen für mich, nur frauen, kann ich mich geschützt fühlen. – Ja… –

Bald darauf sprach ich in der Therapie davon, dass ich endlich das Thema mein Vater abschließen wolle. Daraufhin die Therapeutin:

Solange noch etwas auftaucht, egal was, solange ist das thema noch nicht abgeschlossen! – Das sah ich ja nicht anders! (…) Und dann fiel mir ein traum ein. Wieder schwer zu erzählen, aber ich erkannte rasch, wenn es mir schwerfällt, zeigt das doch, da ist was! Das wußte ich inzwischen. Ich erzählte den traum:

Dicke, lange teile, wie korken, weiß gar nicht, wo die alle in mir drin waren, im mund, den ohren, überall in meinem kopf.

Danach fühlte sich mein gesicht wie gelähmt an. Das war das schlimmste! Furchtbar! Beim aufschreiben dachte ich an den schwanz meines vaters … und daran, was ich hier doch schon mal erinnert hatte, dass er ihn in meinen mund gesteckt hat. (Mich so vergewaltigte!) Aber ich frage mich auch, konstruiere ich da was? Dachte, du spinnst! – Die therapeutin ganz fest: Gerade wenn sie das sagen, sie dachten an den penis ihres vaters, sagen: Du spinnst! Gerade das zeigt, dass es was damit zu tun hat! – Ich staunte. – Sicher kann man den traum auch anders sehen. Man hat ihnen alle sinne zugestopft! Aber ist das besser?, fragte sie mich mit eindringlichem blick. – Nein!, sagte ich leise und erschreckt. – Ich habe das gefühl, es geht im moment darum, dass SIE ihren gefühlen glauben, dem, was bei ihnen ist! Und welche bedeutung SIE dem beimessen! Welche bedeutung sie darin sehen! … Und dass sie SICH GLAUBEN! – Mama, mama!, ging's dauernd in mir. War ihr so dankbar. Nach einer kleinen weile ihre frage, ob es mir jetzt besser gehe, nachdem ich es doch gesagt hätte. – Oder nehmen sie noch was mit? – Nein, nehme nichts mit. Und ja, es geht mir besser, weil ich es gesagt habe. Und wegen dem, was sie sagten … –

Ich hatte also davon gesprochen, ich wolle das Thema „mein Vater" endlich abschließen. Dagegen erhob die Therapeutin berechtigte Einwände. Dennoch sagte ich bald darauf, ich wolle dazu nichts mehr sagen, kein Wort! Wäre deshalb am liebsten gar nicht in die Therapiestunde gegangen. Ich hatte nochmals betont, ich wolle dieses Thema nicht mehr! Darauf reagierte die Therapeutin so:

Das ist nicht gedacht. Das ist ihr gefühl! Ein starkes gefühl kommt da raus, wenn sie sagen, ich will dieses thema nicht mehr! Nie mehr! Und das heißt ja, wir sind nahe dran! … Wir beide können übereinkommen, dass sie darüber im moment nicht reden, damit jetzt nicht weitermachen können. Dass da was ist, was es ihnen unmöglich macht. Aber wir beide wissen davon. … Aber darüber nicht zu reden wäre verrat!, sagte sie ganz fest und eindringlich. – Mama, mama, sagte ich leise vor mich hin. – Das thema zu vergessen geht ja nicht!, betonte sie. Und von wegen verrat: Das wäre verrat an der seite, die doch wissen will was war! Die das klären will! Diese seite würden sie verraten und damit einen teil von sich. – Ich war betroffen … und ihr so dankbar!

Ich möchte nicht unerwähnt lassen, dass ich mich auch gefragt hatte, warum ich eigentlich als kleines Mädchen dennoch wieder zu meinem Vater ging, immer noch Liebes von ihm wollte, trotz seines furchtbaren Verhaltens. Darauf kam ich in der folgenden Stunde zu sprechen.

Ich fühle es so: bei dem gefühl der leere mit meiner mutter (die emotional einfach nicht für mich da war; wegen der vielen kinder sicher überfordert), dass da eine verbindung zu meinem vater besteht, dass das auch mit ihm zu tun hat. – Ehem!, weich, verständnisvoll. – Ich hab' nämlich wieder gefühle in mir drin … – Pause. Schweigen. – Dass ich vor der leere mit meiner mutter zu meinem vater geflüchtet bin. – Ehem! – Hände auf meinen unterleib, meine vagina, dann ins gesicht, an meinen mund. Ganz komische gedanken. Mag sie nicht sagen, aber ich tue es dann doch. Mein gefühl sei: (...) Dass er im wahrsten sinne des wortes das loch, die leere in mir, zugestopft hat! – Ehmmmm! –

Ich habe das bild von früher, alle waren da, aber ich war ganz alleine! Wie im weltall. Und danach das gefühl der angst! Seitdem!!! – Ehem!, immer wieder von ihr. – Seitdem hab ich das gefühl, ich bin ganz alleine auf der welt! – Schweigen.

Dann: Ich bin vor der leere mit meiner mutter zu meinem vater geflüchtet, und der hat das gemacht. Und ich hab' jetzt diese gefühle ..., schöne gefühle. ... Einerseits die gewalt und andererseits diese gefühle. Das ist doch pervers! – Ja, das ist pervers! ... Aber es ist auch ganz normal. Da war berührung, nähe! Das war ja auch da! – Ja, ich weiß! Dennoch... Ich fühle es jetzt so: Mein wunsch nach nähe, berührung... und dann kam die gewalt! Und deshalb meine angst! SEITDEM habe ich diese angst! – Ehem! ... Sie haben nähe und wärme gesucht. Da hätte ihr vater sie auf seinen schoß nehmen sollen, wie ein lieber vater, in seine arme! Es ist ganz normal, kinder machen das so. Sie brauchen wärme, berührung und den kontakt, und wenn sie das nicht bekommen, nicht von der mutter, dann suchen sie woanders danach. – Ob es so war, lieber das, als gar nichts? – Aber nein!, kam weich und dennoch ganz, ganz fest und klar von meiner mama! Ich hatte weiter meine augen geschlossen, hörte ihre angenehme stimme. Aber heute war ich doch eher weg, bei mir oder weit weg.-

Aber wieso bin ich dann überhaupt noch zu jemandem, also zu ihm hingegangen? Das hätte ich nie mehr tun sollen!, traurig, bitter. – Doch!, kam mit so weicher, liebevoller mamastimme. Ich weinte. – Ich hatte doch so einen hunger! – Ja!, weich. – So einen großen hunger! – Ich weinte sehr. Hatte mir vorher schon die gelbe decke genommen, an meinen bauch und mein gesicht geschmiegt. Nahm sie jetzt vor den mund, weinend: Ich hab doch so einen hunger, so einen großen hunger! – Eheeeemmmm! – Lutsche unter der decke an meinen fingern. Kopf, hinterkopf ganz verspannt. Gefühl, ich gucke wie früher als kleines mädchen. Sagte es der K. und dass ich jetzt aufhören wolle. Wolle da raus! – Ja!!, fest, aufmunternd. Augen auf, setzte mich in ihre richtung. Angucken. – Gell, ich bin mutig? – Wie sie strahlte: Ja!! – Und dann fragte die K., ob sie die decke mal mitnehmen dürfe, zum waschen? – Das ist ja ihre decke! – Ja? – Ja! Ganz alleine ihre decke! Und manchmal darf man die doch dann nicht waschen, muss die so bleiben ..., kam so liebevoll von ihr. So wie eine liebevolle mama mit ihrem kleinen mädchen umgeht! Ich lächelte, kinderlächeln: Das ist meine decke! Das ist schön! – Decke an mein gesicht, gesicht in meine decke. Angucken. Leider immer noch angst in mir. – Wenn ich jetzt gehe, sie sind doch dann noch da!?, fragte ich mit kindlicher stimme, als würde sie wirklich verschwinden, wenn ich sie nicht mehr sehe. – Ich? Ja!, lächelnd, fest. – Ja, SIE ja! – Angucken …

Dies schrieb ich nach der Stunde in mein Tagebuch:

Ich ging raus, tiefes durchatmen. Sonne! Gefühl der befreiung. War es so etwas? Sah mir die welt mit etwas anderen augen an. Stolz auf mich! … Mir geht es besser, wenn ich weiß! Wenn ich alles raus- und zulasse, und sei es noch so knüppelhart. Ingesamt ist es besser! Ich fühle mich stärker dadurch.

Die kleine Ann „sehen"

Nach Monaten kam ich noch einmal zurück zu meinem Vater und zu der kleinen Ann. Und ich „sah" die kleine Ann, konnte auch das noch zulassen.

Gleich zu beginn der stunde wurde ich ganz müde. Gefühl der leere. Dann: Ist da was? – Ich ließ mich darauf ein; beschrieb alle meine gefühle und das, was auftauchte.

Ich spürte meine füße, meine zehen. Meine beine ganz weit auseinander. Sagte, dass er da wäre. – Ich verkrampfe meine beine, meine füße. Liege ganz steif hier, angespannt. Und er ist da! Er liegt hier!, deutete auf meine mitte. Ich nahm mir zeit. Zu meiner aussage, mein herz, mein körper würden etwas sagen, aber der kopf käme dazwischen, meinte sie, ich solle doch hingucken, damit ich endlich der kleinen Ann glauben würde. – Dann verstehst du die kleine Ann! Kannst du zusammen mit der kleinen Ann weinen! Und dann kannst du dir endlich selbst glauben! – Alles sehr, sehr ungefähr. –

Sagt er was zu dir? Oder ist es still? Sagt er, dass es doch schön ist? – Es wäre absolute stille. – Stille und dunkelheit … Und ich bin ganz alleine, so alleine!, sehr weinend, manchmal jammernd, wimmernd. Mir wäre ganz schlecht, auch im mund ganz schlecht. – Ich will ihn nicht sehen! Ich will nicht sehen, dass er da ist! – Ehem. – Ich bin so alleine! – Ja! Aber jetzt *bist du nicht mehr alleine damit! JETZT nicht mehr! – Ich sagte immer mal mama zu ihr oder einfach so: Mama, es tut doch so weh! – Ja … – Ich kann oder will ihn nicht sehen, aber ich weiß, dass er da ist. … Ich rieche ihn. … Mein papa war doch so groß! So schwer! … Ich sehe seinen rücken, seine haut, seinen nackten rücken und rieche ihn. … Und im mund ist auch was, aber ich weiß nicht was. Er ist überall an meinem körper! Ich spüre ihn überall. – Irgendwann sagte ich: Ich weiß, dass er es war und dass er das gemacht hat! Ich hab's nicht vergessen. – Ich glaube, sie fragte, ob die kleine Ann etwas sage. – Nein! Der sieht mich so an. ... Wenn der einen ansieht, ist man still! ... Seine augen gehen durch mich durch wie pfeile, bis zur erde. Da* kann *man nichts mehr sagen! – Ehemmm, ehemmm!, weich von der mama.*

Und dann spürte ich, fühlte es: Meine mutter ist auch da. Sie hat sich rumgedreht, weggedreht. Guckt nicht zu. – Ja, sie guckt weg!, böse von der K. – Ich will es nicht glauben, aber ich weiß es ganz genau, der hat das gemacht! – Ehem! – Und es fühlt sich so an, als wären da haare dran... –

Und dann war ich bei der kleinen Ann, ließ ich mich zum ersten mal darauf ein, die kleine Ann zu sehen, wollte *sie sehen! Langsam kamen bilder hoch; ganz viel weinen dabei. – Sie hat ganz zarte haare. Und ich sehe ihre augen! Sie hat ganz große augen. Reißt ihre augen weit auf. – Ehem! – Ich sehe ihre braunen augen und ganz viel weiß drumherum ... Meine süße! Meine kleine! Ach, meine süße, kleine Ann! – Und ich weinte sehr. – Ich sehe ihre stirn ... und mir fällt ein foto ein, wo ich in die schule komme ..., da bin ich 6 jahre alt. War ich damals 6 jahre? Ich weiß es gar nicht! Ich war doch noch so klein! – Ja ... – Aber kein baby mehr, schon ein mädchen. – Ehem! – ... Ich sehe jetzt ihre schultern, ihre ganz schmalen schultern. – Ehem!, wieder so weich von meiner mama. – Sie ist doch so schmal! Kleine mädchen haben doch einen ganz schmalen körper. – Ja ... – Ich spüre sie, als ob ich meine hände auf ihre schultern lege; ihre zartheit, ihre knochen. ... Nur das sehe ich: Ihre großen augen, ihre stirn und ihre haare. Und ihre schmalen schultern. – Ehemmm! – Ach, meine süße! ... Die kleine Ann! ... DAS BIN JA ICH!, mit heftigem weinen. – Das bin ja ich! Ich sehe ja mich! – Vielleicht sagte meine mama hier was, aber ich weiß es nicht mehr. –*

Wir nehmen sie mit zu uns!, sagte ich plötzlich. – Wir nehmen sie mit hierher! – Ja, das ist doch eine gute idee! – Wir lassen sie nicht mehr dort! Nehmen sie mit zu uns! – Ja! – Dann ist sie nicht mehr alleine! – Ja! – Ich hatte jetzt diese phantasie ... Und fühle es so: Jetzt ist sie hier bei uns! – Ja! – Weinend: Ich hab mich *gesehen ... –*

Ende der stunde. Es ist genug! – Ja!, von meiner mama. Sie atmete tief aus. – Was machen wir denn jetzt mit der kleinen Ann?, fragte ich weinend. – Jetzt packe ich sie erstmal in die decke! Jetzt decken wir sie erstmal zu. Ja? – Ja! – Sie kam zu mir, packte die decke ganz leicht und weich um mich rum, um meine schultern, meine beine. Hockte sich zu meinen füßen. Ich öffnete meine augen: Die sonne scheint! – Ja!, die mama lächelte. Ich war noch so verwirrt. Sie fragte: Sie wollen doch heute nicht hierbleiben? Ich möchte sie jetzt nicht gerne hier las-

sen!, so liebevoll und ganz intensiv sah sie mich an. – Ja, das stimmt, das ist nicht gut! ... Sie sind doch morgen wieder da? – Ja! – Angucken, in ihre augen, ihre haare. Mama, meine mama! ... Wir gingen dann gemeinsam aus der praxis, die treppe runter, bis zur straße hin. Das war so schön, nebeneinander herzugehen. Sie fasste mich an der schulter: Und, was macht die Ann heute noch schönes? – Sport und dann zusammen mit den frauen essengehen und trinken ... – Gut! – Ich ging aufrecht nach hause, fühlte mich eigentlich ganz gut.

Am nächsten Tag die nächste Stunde:

Ich fing gleich an zu weinen. Sah die K. zuvor an und dann weinte ich, zum Teil sehr heftig. Und ganz viel mama in mir. – Ich hab es ja schon lange gewusst, dass mein vater mich vergewaltigt hat. Und gestern habe ich es mit aller deutlichkeit gespürt, dass ich es weiß! Dass ich es ganz genau weiß! ... Und trotzdem ist es, nach der stunde gestern, noch mal anders. – Ja... – Ich weinte sehr und sagte: Danach kann man eigentlich gar nichts mehr sagen! – Ja... Ehem! –

Nach einer kleinen pause sagte sie noch etwas zur gestrigen stunde, was ganz stark bei ihr da war bezüglich meines vaters. Sie erinnerte an seine äußerung, es ist gut, wenn kleine mädchen ..., dann wissen sie, was auf sie zukommt. Sie gehe davon aus, dass er das wie einen initiationsritus ansah, sich in einer tradition sah. Sie meine das auf keinen fall im sinne von ihn entlasten! Nein, das habe er sich so zurechtgelegt! Patriarch, er darf das! Die K. weiter: Und ich spürte, sah ganz deutlich das bild, dass er das mit all ihren schwestern gemacht hat! Missbrauch! Vergewaltigung! – Ich hörte ihr schweigend zu. Sie fühle oder sähe es so, dass das die erste szene mit ihm war. Sie hätte mich gefragt, ob er was sagte, weil sie so spürte, der sagte, stell' dich nicht so an! Ist doch nicht schlimm! Das, was ich mal erinnerte, du hast es ja selbst gewollt, habe er wohl nicht beim ersten mal und nicht gleich, sondern erst später gesagt.

Weil ich so sehr weinen musste, meinte sie, ganz weich und liebevoll, es wäre doch sehr gut, dass die traurigkeit jetzt da wäre, dass die endlich rauskäme! Ja, so fühlte ich es auch.

Wir sprachen über meine ausgestreckten, starren beine, die verkrampften zehen während der gestrigen stunde; so war es auch schon in anderen stunden. Die K. sagte, dadurch hätte ich den anderen schmerz nicht so gespürt, aber mehr noch: Das sollte den vorhandenen impuls der beine verhindern! Die wollten doch treten! Sich wehren! – Sie hatte gestern schon gesagt, das wäre doch gut, das mit den füßen, den beinen. ... Ach, geliebte mama! ... Die reaktion, sich quasi tot zu stellen, sich ganz steif zu machen, sei typisch bei traumatischen erlebnissen. –

Und das ist eine absolut gute reaktion! Die psyche und der körper schützen eine damit!, ich glaube, so drückte sie es aus.

Ich erzählte ihr, gestern nach der stunde hätte ich mich erstmal „dort" waschen müssen. Und am abend beim sport, hätte mich sehr auf die frauen gefreut, wäre ich rumgetobt und hätte auch spaß gehabt. Danach nach hause, wollte alleine sein. Dann geduscht, ganz heiß und ganz lange. Weinte wieder sehr. – Und dann hab ich die kleine Ann und mich mit weichen handtüchern sanft abgetrocknet und uns ganz liebevoll eingecremt! – Ehem!, weich von meiner mama. – Und dann haben wir uns noch schön gemacht, ganz schön angezogen, obwohl es schon spät war! Ganz weiche, schöne sachen zum anziehen rausgesucht. – Ich weinte und weinte. – Warum ich denn jetzt so weinen müsse?, fragte sie liebevoll. – Ich frage mich gerade: Warum? Warum hat er das gemacht? Ich will es einfach nicht glauben! Er ist doch mein vater! Ich will nicht, dass es wahr ist! Mein kopf begreift es, aber mein herz, meine seele, die verstehen das nicht! – So ungefähr fühlte ich.

Die K. sagte noch, ich wäre ja zuvor schon in kontakt mit der kleinen Ann gewesen, aber der habe sich durch die stunde gestern noch einmal verändert: Was sie vorhin sagten, wie sie gestern duschten: Dann habe ich „uns" schön eingecremt! „Uns" schön gemacht! Das ist jetzt noch intensiver! – Sie sagte es so liebevoll! – Ja! Und die kleine Ann ist jetzt hier bei uns! Nicht mehr dort! Ich hab sie gestern mit hierher genommen ..., sagte ich weich, während ich die K. ansah. – Ja ... –

In der darauf folgenden Stunde betonte ich, dass die vorigen Stunden wirklich sehr hart gewesen seien und dass ich danach sehr viel weinte:

Aber ich spürte auch klar, das macht mich stark! – Ehem! – Wenn ich genau weiß, was war, das macht mich stark! Damit nehme ich dem die macht über mich! – Ehem! – Ich fügte hinzu, bisher wären ja nur gefühlte erinnerungen aufgetaucht, aber jetzt hätte ich ihn und mich auch deutlich gesehen, also bilder dazu! – Sie sah mich direkt und ernst an; mit fester stimme sagte sie: Das bedeutet ja, dass sie den bildern mehr trauen als ihrem körper und ihren gefühlen! – Huch! – Ja ..., stimmt!, sagte ich nachdenklich.

Ein Traum, den ich einige Monate später träumte, spiegelt gut meine Entwicklung wieder:

Traum mit meinem vater. Er will mich auf meine linke brust küssen. Nein! Ich stoße ihn zurück. Er ist völlig verwundert. Später will er mich auf den mund küssen! Ich beiße ihm fest in die oberlippe. Trotz großem widerwillen beiße ich fest zu. Er, völlig erschreckt, stöhnt, hält sich die lippe. Ich denke, jetzt kann er sehen, wie er die blutende lippe erklärt! Ich stehe hoch erhobenen hauptes vor ihm. Er hält sich die lippe, stöhnt! Klasse!

Und dies gehört ebenso dazu – eine Notiz aus meinem Tagebuch:

> *„Erklärt Pereira" gelesen, sehr beeindruckt. Wie schön, dass ich wieder solche lust am lesen habe! Und trotz aller traurigkeit liebe ich mein leben, so wie es gerade ist! Und obwohl ich traurig bin, spüre ich auch all die anderen gefühle: Meine freude, meine liebe für alle meine lieben, meine freude an meinem jetzigen leben und, und, und. Das ist ja relativ neu für mich, dass ich das so wahrnehme, dass alle gefühle gleichzeitig da sind. ... Und das spüre ich jetzt ganz klar: Mir ist es nicht mehr wichtig, ob mein vater seine taten zugibt! Dass ich es glaube ist nicht mehr davon abhängig! Ich glaube mir!*

Ich stellte meinen Vater einige Jahre später nochmals in aller Ruhe persönlich zur Rede, blieb beharrlich, stand gut zu mir. Das war für mich ein wichtiger und passender Abschluss dieser harten Zeit.

3.5 Zum Abschluss

Am Ende dieses Kapitels möchte ich noch einige wichtige Aspekte erwähnen, die auch mit dem Schreiben des Kapitels in Verbindung stehen.

Ich habe mich dafür entschieden, den Therapieprozess in zeitlicher Folge (und nicht nach den verschiedenen Formen des Erinnerns gegliedert) zu schildern. Das Bearbeiten in der Therapie, der Analyse, stellte sich als ein langer Prozess dar. Diesen näher zu beschreiben macht deutlich, wie langwierig das Erinnern sein kann und wie viel Zeit und Geduld es dafür braucht; wie schwierig es war, all die Erinnerungen zuzulassen; welche Härten das Bearbeiten mit sich brachte, aber auch wie befreiend die einzelnen Schritte und der Prozess insgesamt waren.

Erst beim Schreiben bzw. weiteren Überarbeiten des Kapitels wurde mir bewusst, dass sich in meinem Therapieprozess verschiedene Formen der Erinnerungen zeigten:

Körperliche und psychische Erinnerungen, Träume und indirekte Erinnerungen; außerdem ließ ich meine Hände „sprechen", meine Erinnerungen auf diese Weise ausdrücken.

Als *körperliche und psychische Erinnerungen* bezeichne ich die Herzsymptome, die Angst- und Panikattacken sowie Schlafstörungen. Gerade die Herzsymptome, die verschiedenen Ausdrucksformen meines Herzens, haben mich überhaupt erst darauf aufmerksam gemacht, dass „mit mir etwas nicht stimmt“! Mein Herz hat mir gesagt: Guck’ nach dir! – Sie waren, zusammen mit den Angst- und Panikattacken, der Auslöser, der Antrieb, mir anzuschauen, was mit mir los ist.

Die *Träume* unterstützten das Erinnern, halfen mir, wie schon im Zusammenhang mit meinem Bruder Adolf, mir nach und nach die Taten meines Vaters anzuschauen und zu bearbeiten. Der Traum, den ich unter dem Stichwort „Blutschande“ beschrieben habe, war sogar der Ausgangspunkt, der Auslöser dafür!

An *„indirekten“ Erinnerungen* gab es einige. Ich meine damit unter anderem die folgenden Situationen: Als ich mit dem Kollegen der Therapeutin alleine im Praxisbüro war; mein „ich will nicht in mein Bett“, „will nicht, dass es dunkel wird“, meine Angst vor der Nacht – siehe die erste Nacht im erwähnten Italienurlaub. Außerdem meine Phantasien bezüglich der Hand der Therapeutin; auch die Situation bei der Zahnärztin. Und ich denke an das Lied der Valente „Ganz Paris träumt von der Liebe“, das ich „hörte“ bei Erinnerungen an Situationen mit dem Vater.

Eine weitere Form oder Station des Erinnerns sehe ich darin, *meine Hände „sprechen“ zu lassen,* wenn ich meine Gefühle und Erinnerungen nicht anders ausdrücken konnte.

All diese Formen der Erinnerung lösten wiederum Bilder und direkte Erinnerungen an Situationen der sexuellen Gewalt und der Bedrohung durch meinen Vater aus – meist begleitet von (körperlichen) Reaktionen der verschiedensten Art, die durch die Übergriffe entstanden waren:

Erstickungsgefühle, Enge im Brustkorb; Brennen in meiner Vagina, aber auch schöne und erotische Gefühle; Kälte; alles schwarz, will lieber tot sein; es liegt was Schweres auf mir …; und Wut, Traurigkeit, Gefühle des Ausgeliefertseins, der absoluten Hilflosigkeit und nicht zuletzt auch Schmerzen.

Das theoretische Betrachten des Erinnerungsprozesses, das ich gerade schilderte, hat mir geholfen, noch mehr Abstand zu dem Geschehenen zu gewinnen.

Zuletzt möchte ich noch einen weiteren wichtigen Aspekt erwähnen, der mir beim Abfassen dieses Kapitels gewisse Probleme bereitet hatte, der mir dann jedoch zu großer Klarheit verhalf.

Mein erster Impuls war, nicht genau zu beschreiben, was mein Vater im Einzelnen mit mir gemacht hatte. Entsprechend sah die erste Fassung dieses Kapitels aus. Ich schilderte meine Erinnerungen, Bilder, Gefühle, also das, was in der Therapie jeweils auftauchte, beschrieb aber nicht, welche Bedeutung ich den verschiedenen Bildern und all dem Anderen beimaß. Ich entschied mich also für „Leerstellen“, so nannte ich es, wenn ich es bei Andeutungen beließ. Ich wollte die Lesenden nicht erschrecken, nicht ab-schrecken, wollte vielleicht nicht nur sie schonen, sondern auch mich, indem ich nicht genau in Worte fasste, was geschehen war.

Nach einem Gespräch mit meiner damaligen Therapeutin kam ich jedoch zu einer anderen Ansicht. Ich entschloss mich, die Taten ganz genau zu benennen, zu „versprachlichen". Dieser Entschluss hatte etwas sehr Befreiendes! Und er löste einen Traum aus, auf den ich im Kapitel zum Prozess des Schreibens noch näher eingehen will; hier würde das zu weit führen.

Hatte ich noch in der ersten Fassung, dies als Beispiel, davon geschrieben, ich habe „was" gemalt oder mit Knete einen Stab geformt, hatte ich es den LeserInnen überlassen, sich das Offensichtliche zu denken. Nun war und ist es mir wichtig, genau zu benennen, was das Offensichtliche war! Also in Worte zu fassen, zu versprachlichen, dass ich, um bei dem Beispiel zu bleiben, den Penis des Vaters abbildete. Ich habe mich für diese Form entschieden, weil ich genau dokumentieren will, was war! Was mein Vater mit mir gemacht hat! Ich will mich nicht und schon gar nicht ihn schonen! Es ist mir wichtig, meine Erinnerungen zu versprachlichen, auch damit weder ich noch Andere, weil wir ja gerne Schlimmes und Ähnliches verdrängen, genau dies nicht tun können!

Von dem, was das Versprachlichen wiederum bei mir auslöste, will ich zum Abschluss kurz berichten:

Während ich das Kapitel für die Veröffentlichung überarbeitete, um die Taten genau zu benennen, bemerkte ich, dass ich den Text fast wie eine unbeteiligte Dritte las. Und auf einmal dachte ich: Wieso habe ich so lange zweifeln können an dem, was ich erinnerte? Es ist doch ganz klar, tritt doch deutlich hervor, was mein Vater gemacht hat! Dass er mich mit seinen Fingern und mit seinem Penis vergewaltigt hat, oral und vaginal. Wie konnte ich daran so lange zweifeln? Besonders zu meinem Schutz, das ist mir schon klar. Aber nun gibt es gar nichts mehr zu bezweifeln!

Zuerst war ich wirklich nochmal erschüttert. Aber dann: Befreit! Es ist befreiend, weil so klar!

4. Die Mutter

Zu Beginn dieses Kapitels muss ich davon berichten, wie ich mich dem Schreiben näherte bzw. wie es mir dabei erging, denn das gehört mit zu dem, was es zu meiner Mutter zu berichten gibt.

Beim Durchsehen meiner Aufzeichnungen der ersten Therapiejahre erschien mir meine Mutter konturlos. In den Notizen zu den folgenden Jahren veränderte sich das. Ihr Verhalten, ihr Anteil an dem Geschehenen und an den Folgen treten mit den immer deutlicheren Erinnerungen an die Taten meines Vaters entsprechend mehr in den Vordergrund. Das überraschte mich nicht. Aber dass es so schwierig werden würde, mit dem Arbeiten an diesem Kapitel zu beginnen, das hatte ich nicht erwartet! Nach dem sehr schweren und harten Kapitel zu meinem Vater nahm ich an, das zur Mutter würde mir, wenn auch nicht leicht, so doch wesentlich leichter fallen. Aber so war es nicht!

Die Vorarbeiten gestalteten sich schon nicht leicht. Zudem bekam ich plötzlich Herzschmerzen! All die verschiedenen Arten von Herzsymptomen sind mir vertraut, in meinem Buch auch beschrieben, aber Herz*schmerzen* gehörten bisher eher nicht dazu. Meine Mutter hatte früher oft Herzschmerzen, hielt uns damit, wie mir später klar wurde, im Zaum. Wir verhielten uns dann still aus Angst davor, sie könnte sterben. Das war ihr Druckmittel, damit meine Geschwister und ich brav und angepasst waren und blieben. Und nun tauchten bei *mir* Herzschmerzen auf! Nach einer Stunde musste ich aufhören zu schreiben, war es zu unangenehm. Was soll das denn jetzt?, fragte ich mich und nahm mir Zeit, genauer hinzuschauen.

Ich stellte bald erstaunt fest, dass ich eigentlich gar nicht so recht wusste, was ich zu meiner Mutter schreiben will, was mir zu ihr wichtig ist. In meinen Aufzeichnungen gibt es einiges an „Material“, keine Frage! Aber was ist in meiner Erinnerung zentral, was gibt es mit dem großen Abstand von heute zu ihr zu berichten?

In einem Gespräch mit meiner ehemaligen Therapeutin wurde mir der Hintergrund meiner Schwierigkeiten klarer: Bei meinem Vater und den Brüdern ging es sehr um meine Wut, mehr um meine Wut. Bei meiner Mutter steht jedoch der Schmerz im Vordergrund! Bei ihr geht es darum, wie schmerzvoll es für die kleine Ann war, dass ihre Mutti nicht für sie da war. Dass ihre Mutti die Taten nicht verhinderte!

Und nun weinte ich, weinte so sehr, und das wiederholte sich. – Mama, Mama! – Ganz oft sagte ich es, leise oder lauter, bitterlich weinend. Diese Gefühlsausbrüche überraschten mich wirklich. Ich habe das doch alles gründlich bearbeitet! Das ist doch für mich abgeschlossen! Doch nun musste ich, nicht zum ersten Mal, feststellen, auch wenn alles gut bearbeitet ist, kann so ein Schmerz immer wieder spürbar sein.

Ich versuchte es erneut mit dem Kapitel. Aber dann: Ich will zu meiner Mutter nichts schreiben! Es ist zu hart! Ich unterbrach schon wieder. Es ging einfach nicht. Es war zum Verzweifeln! Und ich war ratlos. Nach einer Weile des Zauderns kam ich zu dem Schluss, mir wieder Zeit zu nehmen, das Schreiben ruhen zu lassen. Aber die Gedanken an ihre Verantwortung ließen mich nicht los.

Während der Therapie und auch danach vertrat ich, wie an anderer Stelle bereits erwähnt, deutlich die Position, dass mein Vater die Verantwortung für seine Taten und die Folgen trägt. Das sehe ich auch heute noch so.

Aber nun tritt die Verantwortung meiner Mutter stärker in den Vordergrund. Jetzt sehe ich das deutlicher als zur Zeit der Therapie. Vielleicht musste ich damals meine Mutter in gewissem Maße schonen. Oder eher mich selbst, um mir meine Mutter noch etwas zu bewahren! Vielleicht wäre es ansonsten zu heftig für mich geworden, hätte ich das nicht auch noch verkraftet.

Andererseits kann ich gar nicht sagen, wie oft ich mich während der Jahre der Therapie gefragt habe: Warum hat meine Mutter mir nicht geholfen? Warum hat sie mich nicht beschützt? Das ist doch einfach unbegreiflich, dass ein kleines Mädchen von seiner Mama nicht beschützt und geschützt wird!

Doch nun treten ihr Verrat und mein Schmerz über ihren Verrat deutlich hervor. Sie hat mich und meine Liebe für sie verraten! Und dafür und für die Folgen trägt SIE die Verantwortung!

Diese Erkenntnis versetzte mich in die Lage, mit dem Schreiben des Kapitels zumindest beginnen zu können.

Bevor ich mit den Erinnerungen im Hinblick auf das Verhalten meiner Mutter beginne, will ich von weiteren Schwierigkeiten beim Schreiben und den Überlegungen dazu berichten. Später erging es mir nämlich wieder so: Ich will und kann zu meiner Mutter nichts schreiben! Was mach' ich denn jetzt? Das ist doch wichtig! Ja, aber es tut mir zu weh!

Ich entwickelte dann diese Idee: Ich schreibe nichts zur Mutter, sondern über die Therapeutin, die wie eine liebe, wundervolle Mama für mich da war. Ich beschreibe, wie sie mit all dem Schlimmen umging, wie liebevoll sie die kleine Ann auffing, begleitete und stärkte. Und wie sehr sich die kleine Ann bei ihr geborgen fühlte. An der Schilderung all dessen wäre dann zu ersehen, was ich früher mit und von meiner Mutter nicht hatte und was ich als kleines Mädchen in den bedrohlichen Situationen so dringend gebraucht hätte. Was ich deshalb in der Therapie und bei der Therapeutin gesucht, von ihr ersehnt und bekommen habe: das Aufgehoben sein, das liebevolle Verständnis, die klare Haltung, sie ist auf meiner Seite, und besonders die Geborgenheit, die ich noch heute in mir spüre. Und: Dass *sie* mich nicht verraten hat!

All das habe ich mit und bei der Therapeutin erlebt. Aber nicht mit meiner Mutter! Das wäre dann eine indirekte Aussage zu dem Verhalten meiner Mutter, zu ihrer Verantwortung und ihrem Versagen.

Als ich ein weiteres Mal mit der K. über meine Schreibschwierigkeiten sprach, sagte ich: Tief in mir drin gibt es immer noch die kleine Ann, meine verletzte Seele. Und damit die sich nicht alleine fühlt, ist es gut, dass ich heute noch Mama fühle, die Mamaliebe in mir habe. – Ich würde das so fühlen, als würden die Mamagefühle meine verletzte Seele umhüllen, ihr Geborgenheit geben. Genau darum gehe es, bestätigte die K.: Es ist alles bearbeitet, aber die Seele ist ja trotzdem noch verletzt! – Und die Mamagefühle seien dafür existentiell.

4.1 Erinnerungen an das Verhalten der Mutter

Ohne nochmals auf die Taten des Vaters einzugehen, komme ich nun zu meinen Erinnerungen bezüglich des Verhaltens meiner Mutter in diesem Zusammenhang. Sie tauchte dabei insbesondere so auf: Ich liege bei ihr im Ehebett, mein Vater in seinem. Ich bin ein kleines Mädchen, ungefähr vier bis sechs Jahre alt. Meine Mutter sagt, ihr sei es mit mir zu warm, ich solle mich zu Vater ins Bett legen. Was dort geschah, wollte sie wohl nicht wissen. Jedenfalls gibt es die öfter wiederholte Aussage von ihr, wenn sie einmal eingeschlafen sei, würde sie nichts mehr mitbekommen – was ich jedoch sehr bezweifle, auch weil sie das immer wieder betonte. Es gab vielleicht auch andere Gründe, warum ich zu meinem Vater ins Bett wechseln sollte, nicht nur, weil es ihr mit mir zu warm war. In einer Therapiestunde fühlte oder erinnerte ich es so, dass sie uns, meine Schwestern und mich, opferte, wenn sie nicht mehr konnte. In dem folgenden Ausschnitt aus dieser Therapiestunde wird das ganze Drama deutlich:

> *Mir fiel ein traum ein, in dem ich einem mädchen sagte: Nein, er kommt nicht runter. Ich gehe hoch und sorge dafür! Nach dem traum stellte sich mir die frage: War meine mutter nicht manchmal doch auf unserer seite? ... Jetzt fühle ich, das war so! – Ich schwieg, ließ mich darauf ein. Sah dann ganz deutlich unsere damalige wohnung vor mir. Dachte wieder an den traum und erinnerte: Manchmal hat sie sich geopfert ... und manchmal uns! – Ehem! – Wenn sie nicht mehr konnte, opferte sie uns! ... Das ist doch furchtbar! Dass es keine andere möglichkeit gab! Keinen anderen ausweg als das! Das ist doch furchtbar! – Ja! –*

Ich hab das bild in mir wie in dem traum, ich gehe hoch, besänftige ihn wie ein tier, eine bestie, die besänftigt werden muss, dann macht sie nix. Ja, so fühle ich es, dass meine mutter das so gemacht hat. Und er wie ein tier. (...) Ich glaube wirklich, dass meine mutter auch manchmal auf unserer seite stand! –

Aber leider nur manchmal!

In einer anderen Therapiestunde phantasierte ich, meine Mutter an die Hand zu nehmen, ihr zu sagen, sie soll dafür sorgen, dass das nicht wieder passiert – zuvor hatte ich Erinnerungen mit meinem Vater geschildert. Aber meine Hand blieb leer: Es ist niemand da! Ich bin ganz alleine und mir ist so kalt! Da ist *niemand!* Warum hilft sie mir denn nicht? – Und ich erinnerte dieses Bild:

Meine mutter liegt in ihrem bett, dreht sich um. ... Dreht sich einfach um! Ich sehe ihren rücken. ... Die hat sich einfach weggedreht und lässt mich allein. – Ehem! – Weinen, weinen. – Warum hilft sie mir denn nicht? Ich sehe deutlich ihren großen, breiten rücken! Ich bin doch noch so klein! – Das bild mit ihr so deutlich vor augen! – Warum hilft sie mir denn nicht? ... Ich bin ganz alleine! Sie lässt mich alleine! – (...) – Ich hab' das gefühl, dann hab' ich nicht mehr gelebt! (...) Ein teil von mir hat dann nicht mehr gelebt. –

Am Ende dieser Therapiestunde wollte ich gar nicht gehen: Ich möchte bei Ihnen bleiben! Ich will nicht von Ihnen weggehen. Ich würde gerne bei der Mama bleiben!, hatte ich leise hinzugefügt; bei der Mama, der Therapeutin. Ich stellte noch diese Fragen: Aber WIR vergessen die kleine Ann doch nicht? Lassen sie doch nicht alleine? – Nein, antwortete die K. mit weicher, lieber Stimme.

Und gerade nach dieser Stunde, nach diesen Fragen erreichte mich die Absage der nächsten Therapiestunde, eventuell auch mehrerer Stunden, über eine Kollegin der K., worauf ich zuerst mit großer Angst reagierte. Mit Verlassenheitsängsten. Und ich weinte sehr. Aber dann konnte ich mir selbst sagen: Das sind doch Gefühle von früher! Das würde auch die K. sagen. Ich will es gut packen ohne sie! Ich habe doch einen starken Willen. – Dies alles und noch mehr sagte und notierte ich mir. Und diese Mama kam wirklich wieder, war wirklich wieder für mich da! Wegen eines nahen Todesfalles hatte sie absagen müssen, konnte sie mich nicht persönlich davon informieren.

In der folgenden Stunde weinte ich wie ein kleines Mädchen, das ohne Mama so tapfer war, so notierte ich es in mein Tagebuch. Ich berichtete ausführlich, sprach auch von meiner mich an früher erinnernden Panik. Ich hätte es aber geschafft, mir zu sagen: *Sie* kommen wieder! – Andererseits hätte ich nicht glauben können, dass sie tatsächlich wiederkommt. Auf meine Äußerung, ich müsste doch eigentlich weiter sein, reagierte die K. so: Sie müssten Ihre Mamagefühle annehmen! ... Und nicht weiter sein! – Nach unserer letzten Stunde, gerade dieser Stunde, wäre meine Reaktion darauf, dass sie so unerwartet nicht für mich da war, doch völlig verständlich. Mir blieb dazu in Erinnerung: Diese Mama lässt mich nicht alleine! Die nicht!

In einer anderen Stunde erinnerte ich ein weiteres Mal sehr deutlich, wie meine Mutter sich wegdrehte. Und die kleine Ann hatte doch so auf ihre Hilfe gehofft:

> *Ich will zu meiner mutti ins bett! Die soll mich lieb halten! Mich beschützen, dass das nie wieder passiert! ... Und die soll den wegjagen. Ganz weit weg! Und verprügeln! –*

Aber meine Mutter hat mich nicht beschützt, sondern ...

> *... sie hat mich an meinen vater ausgeliefert! Mein gefühl ist, als hätte sie gesagt, stell dich nicht so an! Also nicht, dass sie das sagte, eher mir das so vermittelte. – Ehem! – Schweigen. – Ich hab' jetzt ein bild, meine mutter nimmt mich an die hand, bringt mich zu meinem vater, liefert mich aus. – Ja? – Nicht, dass es wirklich so war. Das ist ein bild für mein gefühl! – Ehem! –*

Oder hat sie mich sogar an ihn verraten, wie ich es erinnerte?

> *Dass ich es meiner mutter gesagt habe, und sie irgendwie furchtbar mit mir umging, mich damit alleine ließ. Oder mich sogar an ihn verraten hat! ... Da ist was ganz tief in mir drin! Und ich habe angst davor! Meine mutter hat mir im zusammenhang mit meinem vater ganz furchtbar weggetan! ... Es zerreißt mir das herz! –*

Ich erinnerte mich auch daran, dass sie uns Kindern, und damit auch mir, mit dem Vater drohte, der Vater als letztes Disziplinierungsmittel, was bei mir furchtbare Assoziationen auslöste. Ich fühlte es so, als habe sie mich in der Hand, könne mich zum Schweigen zwingen:

> *Dass meine mutter mir mit meinem vater gedroht hat! Dass der „das" macht! ... Oh, nein! – Ich weinte sehr. So sehr. – Nein, ich glaub das nicht! – Bitterliches weinen. – Aber ich fühle es wirklich so! – Ehem! – Entweder hat sie mir damit gedroht, dass er „das" macht, sie hat ja öfter mit ihm gedroht, wenn man nicht spurte, oder ich hab' gedacht, dass er „das" dann wieder macht! – Wieder ihr ganz weiches: Ehem! –*
>
> *Glauben sie, dass ein kind sich so was phantasieren kann, dass der „das" macht? – Ja! Ihre mutter hatte sie ja alleine schon damit in der hand, dass sie mit*

ihrem vater drohte! Und es kann schon sein, dass sie dann dachten, er macht „das" wieder! Dass das bei ihnen dann immer da war, bei der drohung ihrer mutter. –

Bald darauf sagte ich, dass ich wohl nur annahm, er macht „das" wieder, meine Mutter das aber nicht so gemeint habe.

Ein Körper sei doch so zerbrechlich, der sei es doch wert, beschützt zu werden, hatte ich einmal zur K. gesagt. Dem stimmte sie zu und meinte, er brauche eine gute Mutter, von außen, aber auch eine gute Mutter von innen. Ich fasste das damals so auf, dass ich selbst wie eine gute Mutter für mich sorgen solle.

Dazu fällt mir ein „Märchen" ein, das ich mir ausgedacht und wovon ich der Therapeutin berichtet hatte:

Ein märchen mit einer löwin: Die kleine löwin will, dass die löwen-mama den alten löwen-vater bestraft. Verdrischt! Zerreißt! Aber die löwen-mama verhält sich so wie die menschen. Die macht das auch nicht!, so formulierte ich, schön verpackt, meine an sie, die therapeutin, gerichteten wünsche. Die K, fest: Das bild ist ja auch falsch! Was sie beschreiben war ja die wirklichkeit. Das haben sie so erlebt, dass ihre mutter das nicht getan hat. Aber jetzt, heute, müssen sie SELBST die löwen-mama sein! – Und das kam später noch von ihr: Selbst wenn die mama den löwen-papa verdrischt, das nimmt dem kind nicht seine wut, nicht seine trauer, seine schmerzen! Die gehen davon nicht weg! –

Von Beginn der Therapie an ging es immer wieder um die Frage, ob meine Mutter denn wirklich nichts bemerkte. Auch im Zusammenhang mit den Übergriffen der Brüder sprach ich darüber. Ich hätte damals gedacht, die muss mir das doch anmerken! Sie bekam doch sonst alles mit! Aber sie war nicht für mich da …

4.2 Meine Mutter war nicht für mich da

Wo war meine Mutter? Sie schien mir in der Erinnerung wie abwesend. Ich schrieb bereits zu Anfang, dass mir meine Mutter bei den Vorarbeiten zu diesem Kapitel konturlos vorkam. Das erinnere ich jetzt, und mir fällt weiter dazu ein, dass ich meine Mutter in einer Therapiestunde einmal „darstellte". Ich hatte dafür Zeitungen und ein Betttuch mitgebracht.

Warum zeitungen?, fragte die K. – Die entschwinden, weichen aus, geben nach. – Passe vielleicht zu meiner mutter. Ich wolle versuchen, etwas damit zu ma-

chen. Aber das war schwierig. – Überlassen sie sich ihren händen! Schließen sie ihre augen … und gucken sie mal, was dann entsteht. … Lassen sie die kraft in ihre hände fließen. –

Es fiel mir sehr schwer, mich darauf einzulassen, aber dann zerknüllte ich die zeitungen. Als ich ganz viele zeitungen zusammengeknüllt und zu einem haufen gepackt hatte, legte ich das bettuch darüber. Ich konnte aber nichts damit anfangen. Ich solle ein stück davon wegrücken, fühlen, was ich gemacht habe und es beschreiben. … Das sei der rücken meiner mutter, so, wie ich es schon ein paarmal beschrieben hätte, nachdem sie sich wegdrehte!

Dieses „luftige“ Gebilde sollte also meine Mutter darstellen. Meine Mutter war irgendwie nicht zu fassen. Konturlos. Aber das trifft es auch nicht ganz.

Damals ließ ich sehr viel Wut auf meine Mutter raus. Dabei sagte ich unter anderem, weinend: Mutti, warum hast Du nichts gemerkt? Warum hast Du mich nicht beschützt? – Und: Die kriegt nix mit! Es ist zum Verzweifeln. Das kommt einfach nicht an! –

Als ich im ersten Jahr der Therapie davon sprach, dass für die Taten der Brüder mein Vater mit verantwortlich sei, wies die K. mich darauf hin, dass ich in diesem Zusammenhang immer nur von ihm reden würde. Meine Mutter wäre doch auch nicht für mich da gewesen. Sie hätte das kleine Mädchen nicht beschützt, ihr nicht genug Liebe gegeben und nicht bemerkt, was mit der kleinen Ann los ist!

Nach dieser Stunde saß ich eines Abends im Wohnzimmer und begann zu weinen. Ich zitiere aus meinem Tagebuch:

Das weinen wurde immer heftiger. – Warum war damals keine da, die mich beschützt und die mir geholfen hat? Ich habe keine mutti gehabt, die mir half. Ich habe keine mutter mehr! – Ich konnte mich nicht beruhigen, es wurde immer schlimmer. Ich bekam angst. Versuchte nach gut einer stunde, die K. zu erreichen. Sie war nicht da! Es war entsetzlich.

Mein Mann und meine Schwester Lisa standen mir an diesem Abend bei und dennoch spürte ich Panik, dass ich mich nicht mehr beruhigen und wieder, wie damals, mit allem alleine sein könnte. Am nächsten Tag war ich dann so froh, dass ich das durchgestanden hatte, aber weiter sehr, sehr traurig.

Ich nahm mir vor, die Therapeutin in der folgenden Stunde zu fragen: Können Sie verstehen, wie schwer und schlimm das für mich war, dass Sie an dem Freitag Abend nicht da waren, wo mein Thema gerade ist, dass damals niemand für mich da war? – Natürlich verstand sie das.

Später meinte sie, meine Mutter sei damals für mich seelisch abwesend gewesen, dass diese Rolle eher meine Schwestern erfüllt hätten, besonders Maria. In einer anderen Stunde sagte die K.: Wenn Ihre Mutter damals richtig für Sie da gewesen wäre, hätte sie Sie beschützen können! –

Und wieder ging es mir so, dass ich zuhause bitterlich weinte:

> *Dabei wahnsinnige herzprobleme, mehrmals fast eine tachykardie. Philipp hielt mich, seine hand zart auf meinem rücken, streichelte meinen kopf, mein gesicht, das war so gut! Ich hab' bestimmt hundertmal „mama, mama" gesagt, gerufen, gejammert. Mama, warum bist du nicht für mich da? Warum lässt du mich alleine? Warum guckst du nicht nach mir? Ich hatte keine Mama! Ich hätte so gerne eine Mama! Ich bin so alleine! Und weiter jammern, weinen, herzprobleme bis zu herzschmerzen, dass ich schon ängstlich wurde. – Mir zerspringt das herz!, sagte ich, und das passte wirklich. Sagte mir dann selbst: Ich hab' dich so lieb, kleine Ann! – Und fühlte es auch so! ... Und dann wieder das gefühl, ich bin rettungslos verloren.*

Als mich die K. fragte, was meine Mutter denn so gemacht habe, wenn sie keine Zeit hatte, sagte ich nach einer kleinen Weile:

> *Meine mutter war für meinen vater da! – Ich ließ ihre frage auf mich wirken. – Ich sehe sie gar nicht. ... Sehe unsere große küche, aber meine mutter ist nicht da. Das ist ja komisch! Ich sehe meine mutter gar nicht. Weiß doch, sie hat gekocht, wäsche gewaschen und so. Sehe sie aber nicht, nur mal kurz, dann verschwindet sie wieder. Sie ist nicht zu packen! ... Und sie ist im schlafzimmer mit meinem vater. ... Meine mutter gehört meinem vater! ... Aber sonst ... Ich sehe sie nicht! –*

In der darauf folgenden Stunde spürte ich einen solchen Mamahunger, und den ließ ich zu.

> *Dauernd „mama, mama" in mir. Tränen. – Ich hab jetzt so einen hunger nach mama! – Ein so weiches, liebesvolles „Ehem" von der K. – Schon seit der letzten stunde geht es mir so! Ich habe so einen hunger nach mama! Wie wenn sie im urlaub sind! ... Hunger nach ihrer körperlichen anwesenheit. ... Das ist eine ungewöhnliche formulierung, aber so fühle ich es. –*

Relativ am Anfang meiner Therapie hatte ich gesagt: Ich habe schon ganz lange so eine Sehnsucht in mir. ... Ich kann Ihnen gar nicht sagen, wonach. – Später wusste ich es: Sehnsucht nach einer Mama! Sehnsucht nach einer lieben Mama, die für mich da ist! Auch Sehnsucht nach Mutterliebe.

Jetzt geht es weiter um die reale Mutter, meine Mutter von damals und um ihr Desinteresse, aber nicht nur darum. Ich war in einer Therapiestunde an ein Gefühl von früher erinnert worden:

> *Da ist ein ganz altes gefühl tief in mir drin... – Ehem!, ganz, ganz weich von meiner mama. – In mir geht's jetzt dauernd „mama, mama", aber es ist nicht an sie gerichtet. ... Das ist ganz lange her, liegt ganz weit zurück... – (...)*
>
> *Und dann: Ich will das alles aus mir raushaben! – Ich weinte so sehr. Dann Angst, Schreck! – Hatte gerade das gefühl, dann hätte mich meine mutter nach der geburt auch gleich an die wand klatschen können! (Puuuh!) – Weinte so sehr. Herzbeklemmungen, angst, bekam kaum luft. Sagte ihr das. Sie fragte, wie ich jetzt darauf käme, also von wegen an die wand klatschen? – Da war etwas ganz tief in mir drin, was lange her ist. ... Vielleicht ist es nur ein bild für etwas. – Ein bild für ihre hilflosigkeit, einsamkeit oder für die ablehnung, das desinteresse? – Ja, dazu passt mein gefühl: Desinteresse! – Ich weinte.*

Auch meine Schwester Lisa konnte sich daran erinnern, dass Mutter sagte, wenn sie von uns genervt war: Ich könnte dich an die Wand klatschen! – (Kleine Kätzchen, wenn es zu viele waren, wurden auf einem Bauernhof früher so „entsorgt".) Mutter hatte auch andere Drohungen parat:

> *Die hat doch gesagt, ich soll still sein. Sie will nichts mehr hören! ... Aber ich will nicht still sein! Will das nicht hinnehmen! – Weinte ganz heftig: Ich hab' jetzt auf einmal in mir, meine mutter sagte, wenn ich nicht still bin, nicht vernünftig, dann schickt sie mich weg! Komme ich in ein heim oder so! – Zweifel, kann das sein, dass die das gesagt hat? Die K. vehement: Das haben die doch zu dieser zeit alle gesagt! Du bringst mich noch ins grab! Oder: Wenn du nicht brav bist, kommst du ins kinderheim! –*
>
> *Sie forderte mich auf, ich solle mich darauf einlassen von wegen ins heim, gucken, welche gefühle in mir auftauchten. Hände vors gesicht, fest. Ganz dunkel. Bat die K., die lampe über mir auszumachen: Jetzt ist es nämlich ganz dunkel! Schwarz! ... Ich bin ganz alleine! ... Und ich könnte in die hose pinkeln. – Schweigen. Mehr ging nicht. Sie mit ihrer lieben mamastimme: Wollen sie lieber wieder hier sein? Sollen wir eine zäsur machen? – Ja, weinend. Unter vielem weinen das noch von mir: Ich hab das gefühl, das bricht mir das herz! Das tut mir in der seele weh! – Ehem!, ganz weich. Ich solle meine füße aufstellen, den boden spüren. Bald konnte ich mich beruhigen. (...)*

Die Drohung stand ja im Zusammenhang damit, ich war nicht still, ich ließ nicht locker. Es ging während der Therapie öfter darum, habe ich meiner Mutter von den Übergriffen erzählt? Als ich noch klein war, ganz sicher nicht. Das hatte die K. einmal betont, dass so kleine Mädchen dafür noch keine Worte haben. Aber später, als Adolf mich attackierte, mir keine Ruhe ließ, vielleicht habe ich dann der Mutter gesagt, dass der „was" mit mir gemacht hat. Das habe ich ihr zumindest auf kindliche Art zu verstehen gegeben.

Aber es gab nicht nur Drohungen, sondern auch Aussagen wie: Stell Dich nicht so an! Was Du immer hast! – Und: So ist das Leben! – Und diese Aussagen erinnerte ich in der Therapie und zwar so: Die K. und ich hatten einen kleinen Disput. Ich hatte, als ich zur Therapie kam, ein zweites Mal geklingelt, weil sie nach dem ersten Klingeln nicht öffnete. Dadurch fühlte sie sich genervt und machte beim Diskutieren darüber eine Bemerkung dieser Art: So ist das Leben! – Ich reagierte sehr verärgert. Das müsse man mir nicht sagen! Ich wisse sehr wohl, wie das Leben sei!

Sie blieb gelassen und wies mich daraufhin, dass das doch jetzt kein drama gewesen sei. – Sie meinen, die wut passt nicht dazu, gehört nicht dahin? – Ja. – Ich könne nicht mehr denken, wüsste nicht, was jetzt mit mir los sei. Ich schwieg. ... Dann: Mir fällt ein, dass ich heute morgen fühlte, ich möchte getröstet werden... – Ehem!, ganz weich. Ich sah meinen verletzten daumen an und sagte, sicher mit eher kindlicher stimme: Ich hatte den wunsch, dass sie sagen, ich möchte mir ihren daumen jetzt mal ansehen. Dass sie nach meinem daumen gucken! – [Ich hatte mich am Morgen geschnitten.] Daraufhin sagte meine geliebte mama so etwas liebes, verständnisvolles: Ich denke, dass die kleine Ann das kennt. Den zusammenhang kennt von dem satz: Stell dich nicht so an! Was du immer hast! Und dem wunsch nach trost. Dann soll die mama wenigstens nach dem daumen gucken. (...) Wenigstens das will die kleine Ann sagen! Wenn sie das andere schon nicht sagen darf. – Ich konnte es kaum fassen, dass sie das sagte, das spürte. Mein kopf sank noch tiefer, ich war ganz weg! Sagte leise: Ja ... –

Ich möchte die kleine Ann in die decke packen, richtig einpacken und halten! – Aber zuvor erzählte ich ihr von einer beobachtung meiner freundin: Wenn ich glücklich sei, hätte ich kleine augen. Aber wenn es mir nicht gut gehe, große! ... Die haben doch früher immer gesagt: Du mit deinen großen augen! Besonders die mutter. Und: Das hält man ja nicht aus, wie du einen anguckst! – Und ich erinnerte, meine mutter offen angesehen zu haben und dass sie „dazu" sagte: So ist das leben! So ist es nun mal! – Ich sprach mit ganz ernster stimme. Meine mama atmete hörbar aus. Einen moment stille, dann sagte sie: Wollen wir jetzt die kleine Ann in die decke packen? – Ja... –

In einer der darauf folgenden Stunden kam ich auf den Ausspruch meiner Mutter, so sei nun mal das Leben, zurück:

Dass es mir in den letzten ein, zwei wochen oft so geht, dass ich mich bedrückt fühle und herzprobleme habe, das steht für mich im zusammenhang mit den sätzen meiner mutter: Stell dich nicht so an! So ist nun mal das leben! – Ich war dann wieder weg, weit weg. Schilderte der K., was ich erinnerte: Ich bin damals zu meiner mutter gegangen. Sie soll mich retten! Aber sie lässt diese sätze los! Lässt mich alleine! – Ich nahm ernst, was weiter in mir geschah, was ich erinnerte: Ich hab' das gefühl ich falle, falle..., hab' keinen boden mehr unter den füßen. ... Meine mutter lässt mich fallen. – Ich fühlte es wie ein körper, der fällt und fällt. – Mein gesicht hat sich jetzt verändert. Ich gucke jetzt so wie früher als kleines mädchen. – Ehem!, kam ganz weich. Endlich kam was von ihr. Sie glaubt mir! – Seitdem ist dieser hass in mir! – Schweigen. Wollte wieder dort sein. Strampelte mit den beinen. Sie fragte nach. Angst, ich stürze wieder ab. – Ich könnte meine mutter totschlagen! Ich möchte sie treten! In den bauch und überall hin! ... Und ich will wieder hiersein! –

Die K. sagte: Ich hab jetzt so ein bild, dass sie als kleines mädchen dann zu ihren schwestern gingen. War das so? – Konnte erst nichts fühlen, nichts sagen. Dann: Vielleicht zu maria, weil sie mich trösten sollte. ... – Ich wollte auf ihre frage nicht so recht eingehen, sie verstand auch gleich warum. – Ich kam auf die frage, weil kinder so reagieren. Die geben nicht gleich auf. Und es bedeutet auch kraft, nicht gleich aufzugeben! Und: Das heißt nicht, dass ihre mutter sie nicht *alleine ließ! – Ich hab das gefühl, ich muss das verteidigen. – Ja, das ist ja auch ok! Ihre wut ist ja wut auf ihre mutter, die ihnen ihre gefühle ausreden wollte. Aber auch auf mich. Aber ich tue das doch nicht. Ich weiß doch, dass es so war. ... Auch wenn sie dann zu ihren schwestern gingen, das ändert nichts an dem verhalten ihrer mutter! ... Und wegen ihrer angst, wieder so abzudriften: Früher waren sie ein hilfloses kleines mädchen. Aber das sind sie doch jetzt nicht mehr! – Ich dachte, heute ließe ich nicht mehr so mit mir umgehen! Spürte kraft, widerstand! Wut! Sie fuhr fort: Früher hatten sie sogar halt durch ihre schwestern. Und jetzt haben sie doch auch halt! – Ich gab nicht auf: Ja, aber vor jahren ging es mir doch wieder so, diese ängste! Keinen boden unter den füßen und das alles! Also ich kenne das auch als erwachsene! – Ja, das stimmt! Aber das war doch so, weil sie noch nicht wussten, was mit ihnen ist! Weil sie noch niemand hatten, der ihnen sagen konnte, was mit ihnen ist!, sie fast vehement.*

Ich notierte nach dieser Stunde in mein Tagebuch:

Jetzt weiß ich, woher dieses gefühl kommt, ich stürze ins nichts, trifte ab, hab' keinen halt mehr. Weil mich meine mutter damals alleine gelassen hat! Also ist es ein gefühl von früher. Und deshalb brauche ich davor heute keine angst mehr zu haben!

4.3 Wut und Schuldgefühle

In diesen Stunden kam auch meine Wut auf die Mutter zum Ausdruck, die ich verschiedentlich austobte. Als ich mich wieder einmal mit meiner Wut auf die Mutter abquälte, hatte ich gesagt:

Ich weiß gar nicht, was bei ihr passt! ... Da ist nur leere und kälte. Gefrierschrank! – Dann schlug ich aber doch wütend mit dem schläger auf das kissen, das mutter darstellen sollte. Dann, weinend: Mama! Mama! ... Warum? Warum, mama? ... Bei ihr geht es einfach nicht so richtig mit dem austoben! – Es ist ja auch nicht zu verstehen, warum sich ihre mutter so verhalten hat. Das heißt es doch, ihr „warum, mama?"! Das kann man auch nicht verstehen. Noch nicht! Weil wir auch noch nicht wissen, ob sie nur weggeguckt hat oder ob sie sogar mitgemacht hat, das noch unterstützt hat. Deshalb *ist es bei ihr so schwer!* –

Ich hatte nicht nur Wut auf meine Mutter, sondern auch Schuldgefühle ihr gegenüber. Schuldgefühle gerade wegen der Wut auf sie, die sich besonders nach dem Austoben meiner Wut zeigten. Aber auch meine Arbeit in der Therapie, dass ich mich kritisch mit ihr und ihrem Verhalten auseinandersetzte, verursachte schlechtes Gewissen und Schuldgefühle. All das seien Schritte von ihr weg, erläuterte die Therapeutin.

Schuldgefühle waren wohl schon da, als ich noch ein kleines Mädchen war. Dazu ein Auszug aus einer weiteren Stunde:

Angst tauchte auf, als ich der K. das sagen wollte: Es ist mir schon länger aufgefallen, wenn ich an sie denke oder auch einfach so, dass ich auf einmal fühle: Mama, ich hab' dich doch lieb! ... Das ist immer wieder so da! – Die K. äußerte daraufhin die vermutung, dass ich vielleicht meiner mutter gegenüber schuldgefühle empfunden hätte, weil bei den berührungen ja auch angenehme gefühle da waren, was ganz natürlich sei. Vielleicht hätte ich mir die ablehnung der mutter auch inszeniert, um mich zu bestrafen oder um wieder zu meinem vater gehen zu können. Durch ihre ablehnung hätte ich dann eine entschuldigung dafür ge-

habt. Ich sagte, ich bekäme jetzt ein schlechtes gewissen deswegen. Die K.: Aber das ist doch ganz normal! Ein kind braucht liebe und zuwendung! Wenn es das von der mama nicht bekommt, sucht es woanders danach! –

Meine schuldgefühle blieben; mir wäre so, als hätte ich etwas gemacht oder „das" gewollt! – Nein!, kam ganz fest von der K. Mit ganz lieber stimme fuhr sie fort: Stellen sie sich doch mal die kleine Ann vor! Die geht zu ihrer mama, will von ihr beschützt werden, in den arm genommen werden, will, dass die mama merkt, dass was nicht stimmt. ... Ihre mutter hätte doch aufmerksam sein müssen! ... Aber die merkt nichts! Hilft der kleinen Ann nicht! – Sie sprach noch von schuldgefühlen meiner mutter gegenüber, weil vater sich mir zuwandte, und auch schuldgefühlen meinen geschwistern gegenüber. – Ja! Deshalb spreche ich hier nie über eifersucht auf meine geschwister, weil ich ihnen gegenüber deshalb ein schlechtes gewissen habe! – [Also schlechtes Gewissen, weil mein Vater mich aus der Masse der Geschwister ausgewählt hatte. Verrückt, aber so war's!]

Zum schluss der stunde sprach ich noch die vermutung aus, dass ich damals vielleicht dachte, meine mutter erwarte, dass ich das zulasse, was mein vater gemacht hat. Weil ich doch alles für vater und mutter tun musste...

4.4 Ich stelle sie zur Rede – Eingeständnisse meiner Mutter

Nun komme ich auf einige Konfrontationen mit Mutter während der Therapiephase zu sprechen. Meinen ersten Anruf bei Mutter, im 4. Jahr der Therapie, habe ich in meinem Tagebuch festgehalten:

Ich fragte sie: Weißt du, was ich mit vater gesprochen habe? – Ja, aber das könne sie nicht glauben. Und gleich ging's los. Ich hätte doch immer mit ihm geschmust, warum ich das weiterhin getan hätte? – Ich weiß, sagte ich, dass er und deine zwei ältesten söhne mir gegenüber übergriffig waren als ich klein war. Ich habe es nicht vergessen! Und ich will wissen, ob du das mitbekommen hast, weil ich denke, als mutter kriegt man doch viel mit! – Nein! – Und wieder, das könne sie nicht glauben. Sie wollte näheres wissen, ich beschrieb nur vage, was vater gemacht hatte. Ich erinnerte sie an adolfs geschichte (der hatte sich, als ich noch zuhause lebte, gegenüber einem fremden Mädchen mit geöffneter Hose gezeigt!), dass sie damals sagte, wie die jungen dinger aber auch rumlaufen. Da hätte ich ihr von mir doch gar nichts zu erzählen brauchen. Ich sagte, dass vater doch auch mit Lisa „was" gemacht habe! – Ich frage mich, wo war meine mutter damals? Du hast doch sonst immer alles mitgekriegt! – Ich war bei meiner familie!, kam hart und vorwurfsvoll, sonst hätte ich das unterbunden. – Außerdem glaube sie das mit vater nicht. – Und von wegen deiner brüder ... Ich habe mit jemand kompetenten darüber gesprochen, der sagte, das käme vom engen zusammenleben. – Ich protestierte sofort energisch dagegen. Sie kam wieder damit, warum ich damals nichts gesagt hätte. – Du glaubst es mir doch auch jetzt nicht. Wie hätte ich das damals verkraften sollen? Wie soll man das aushalten, wenn einem die eigene mutter nicht glaubt? Mutter, ich weiß es! – Ich erwähnte meine schon mit circa 17 jahren aufgetretenen herzprobleme, obwohl mein herz völlig gesund sei. Die habe man nicht einfach so! Vater habe ihr versichert, er habe nichts gemacht. Ich sprach noch davon, dass die drei mir mein leben versaut hätten, das müsse ich ihr auch mal klar sagen. Sie wäre ja meine mutter.

Danach weinte ich sehr und hatte wahnsinnige kopfschmerzen und verspannungen. Philipp hatte ein tonband mitlaufen lassen. Als ich das abhörte, war ich dann doch sehr zufrieden mit mir. Ich bin stolz auf mich. Das war ich der kleinen Ann schuldig!

Eine freundin meinte dazu, dass ich meinen eltern damit sagte, dass ich nicht mehr zu der angeblich tollen, heilen familie gehören wolle!

Meine Schwester Lisa hatte unsere Mutter ebenfalls mit den Taten des Vaters konfrontiert, da auch sie als kleines Mädchen von ihm sexuell traktiert worden war. Bei diesem Gespräch begründete Mutter die Übergriffe von Adolf mir gegenüber unter anderem damit, dass ich früher im Unterrock über den Flur gegangen sei. Da müsse ich mich nicht wundern! Zuvor hatte sie noch gemeint, es wäre infam, ihr zu unterstellen, sie habe von den Taten des Vaters oder der Brüder gewusst. Ich rief sofort meine Mutter an: Selbst wenn ich im Unterrock über den Flur ging, hatte der mich nicht zu begrapschen! – Ich war so ärgerlich, und das ließ ich zu. Danach tranken Lisa und ich erst einmal einen Sekt!

Ein knappes Jahr später äußerte sich meine Mutter so: Sie wolle nicht, dass Adolf davon erfahre, was ich unserem Vater vorwerfen würde. Sie befürchte, dass Adolf dann dem Vater „an den Kragen ginge"! Damals reagierte die Therapeutin, für mich völlig überraschend, mit dieser Annahme: Das ist doch ein Eingeständnis ihrer Mutter, dass sie es weiß! Sie befürchtet, der große Bruder verprügelt den Vater, um seine kleine Schwester zu rächen. – Vielleicht wünsche sie es sich sogar, dass Adolf den Vater verprügelt. Ich konnte das erst einmal nur hören, nichts dazu sagen.

Nach rund zwei Jahren kam ich in einer Therapiestunde darauf zurück, was meine Mutter damals am Telefon gesagt hatte:

Dass ich mich noch als junge frau zu meinem vater ins bett gelegt hätte. – Was? Das hat ihre mutter gesagt? Das wusste ich ja gar nicht! – Ja! Und sie hat auch zu Lisa bezüglich Adolf gemeint, ich wäre immer ohne BH rumgelaufen. Lisas reaktion: Das wäre doch quatsch! Als Adolf das machte, hätte ich doch noch gar keinen BH gebraucht! (Da war ich doch noch ein kleines Mädchen!) Also hat mutter damit gesagt, ich bin ja selbst schuld. – Ehem! – (...) Die K. sagte fest: Das bedeutet doch, damit hat sie den „missbrauch" zugegeben! – Ja? – Ja! Wenn sie das sagt, du hast dich ja als erwachsene..., heißt das doch, du bist doch selbst schuld! Und das heißt doch, ja, da war was. Damit hat sie den „missbrauch" zugegeben. –

Die annahme der K. bewegte mich sehr. – Meinen sie, dass das das wirklich heißt? – Ja!, ganz fest von ihr.

Damals schwieg ich, betreten, betroffen, hatte nichts weiter dazu sagen können. Die K. überließ es mir, die Aussage meiner Mutter letztlich selbst zu beurteilen.

Von wegen noch zu meinem Vater ins Bett gelegt: Ich erinnere mich gut daran, dass ich als junge Frau manchmal im Schlafzimmer meiner Eltern war, wenn wir etwas zu besprechen hatten und dass ich mich dabei auch mal auf den Bettrand zu Vater setzte. Es war auf keinen Fall so, dass ich mich zu meinem Vater ins Bett gelegt hätte. Und selbst wenn ich mich dazu gelegt hätte! Das erklärt oder entschuldigt doch nicht das Verhalten meines Vaters mir gegenüber als ich ein kleines Mädchen war! Außerdem hatte ich die Taten meines Vaters doch völlig verdrängt. Unglaublich, was Mütter so konstruieren können, um sich selbst zu entlasten!

Einige Jahre später sprach ich in der Therapie davon, dass ich meine Mutter gerne treffen und ihr das sagen wolle:

Das, was ich von meinem vater und meinen brüdern sagte, stimmt! Das hat immer noch bestand! Und dass sie das weiß und damals mitbekommen hat. Auch wenn sie jetzt nichts mehr davon wissen will! ... Und dass es ganz, ganz hart war, mir das alles anzugucken, was war!... Aber dass es absolut gut für mich ist! – Das sagte ich alles sehr fest. Mit geballten fäusten!

Ich phantasierte weiter, jetzt sehr weinend, mutter dies zu sagen: Warum hast du in den letzten jahren nicht mal nach mir geguckt, mutti? Niemals nach mir gefragt? ... Nie mal angerufen? Ich bin doch deine tochter! – Ich sagte noch mehr und dann: Ich hab' jetzt das bild, wenn ich mich dann von ihr verabschiede, lasse ich sie ja ganz alleine mit dem! *– Ich weinte heftig. – Und sie guckt wie ein kind! Das darf ich doch nicht machen, sie mit dem alleine lassen! – Schweigen. – Früher, als jugendliche, junge erwachsene und später noch, habe* *ich sie auch oft gegen ihn verteidigt, beschützt. Der war dann manchmal so von oben herab zu ihr. Richtig eklig!* –

Hier füge ich einen kleinen Ausschnitt aus einer anderen Therapiestunde ein, weil er zu den letzten Sätzen passt. Ich hatte davon gesprochen, dass sie mir ihre Probleme, ihr Leid mit meinem Vater noch überstülpte. – Und das war mir einfach zu viel! – Die K. meinte, sie glaube nicht, dass mir das zu viel gewesen sei, sondern: Ich glaube, dass Sie dann Ihr Leid nicht mehr spüren konnten! Dass Ihre Mutter das damit wegmachte! –

Ich komme zurück auf meinen Wunsch nach einer Begegnung mit meiner Mutter. Einige Monate später lag meine Mutter im Krankenhaus. Ich nutzte die Gelegenheit, sie alleine antreffen und offen mit ihr reden zu können. Beim dritten Besuch im Krankenhaus schaffte ich es, was mir vorher absolut unmöglich erschien, ihr nochmals ausführlich zu sagen, dass sie mich nicht beschützt hatte, nicht für mich da war, wie furchtbar das für mich war und was mir dazu alles auf dem Herzen lag. Und sie bejahte zu meinem großen Erstaunen und meiner großen Freude jeweils meine Aussagen. Sie hatte sogar gegenüber Lisa gesagt, weil diese Vater nicht mehr sehen wollte, ja, das könne sie verstehen! Meine Mutter war wirklich auch für positive Überraschungen gut!

4.5 Ich erlebte auch Schönes mit meiner Mutter

Dies sei hier zuerst erwähnt: Ich liebte meine Mutter sehr. Liebte auch ihren Humor und ihre körperliche Weichheit. Als kleines Mädchen kuschelte ich mich gerne in ihre Arme, weil die so weich und warm waren. Aber das war ihr leider oft zu viel.

Zu dem Schönen mit meiner Mutter gehört ganz besonders, dass sie mir, vielleicht ab meinem vierzehnten Lebensjahr, aus ihren Büchern diejenigen auswählte, die ich, wie sie meinte, lesen und verstehen könne. So führte sie mich nach und nach in die Literatur ein. Ich las sogar anspruchsvolle russische Literatur und das mit großer Begeisterung – auch wenn ich nicht immer alles verstand. Natürlich war das Lesen wiederum mit der Schwierigkeit verbunden, überhaupt Zeit dafür zu haben. Erst musste alles getan sein. Vom teuren Stromverbrauch beim abendlichen oder nächtlichen Lesen ganz zu schweigen! Denn aus diesem Grund wurde mir das Lesen des Öfteren untersagt. Manchmal konnten Mutter und ich über die jeweiligen Bücher reden. Leider war das zu selten der Fall aus Mangel an Zeit.

Und das Zweite, was mir einfällt, ist das gemeinsame Hören von klassischer Musik. Ich erinnere mich an den Gefangenenchor aus Nabucco von Verdi, auch und besonders gerne an Suiten von Edvard Grieg, Musik von Jean Sibelius. Natürlich musste auch hierbei zuerst die Hausarbeit erledigt sein.

Diese, wenn auch geringe, gemeinsame Zeit, in der wir zusammen Musik hörten oder über Bücher sprachen, habe ich trotz all dem Furchtbaren wie einen kleinen Schatz in mir bewahrt. Auch heute noch liebe ich es, (nicht nur) klassische Musik zu hören und Literatur zu lesen.

An dieser Stelle will ich unbedingt noch erwähnen, dass ich in der Therapie nicht selten von gerade gelesenen Büchern berichtete, die mich bewegt und beschäftigt hatten. Das Spannende daran war, dass die Therapeutin und ich dann schauten, in welchem Zusammenhang die Bücher, beziehungsweise das mich Bewegende, zu meinen jeweiligen aktuellen Themen in der Therapie standen. Oder wir gingen der Frage nach, warum ich davon erzählte, weil uns das wiederum zu weiteren Erkenntnissen oder bestimmten Themen führte.

Das alles kommentierte die Therapeutin, mit Bezug auf meine Mutter und mich und die Bücher, einmal so:

Jetzt sehe ich da auch einen zusammenhang zwischen uns und den büchern! Sehe das jetzt in einem ganz neuen licht! Damit haben sie es doch geschafft, hier zwischen uns beiden diese gute muttersituation herzustellen. – Sie sprach fast begeistert. – Haben sie zwischen uns auch so eine verbindung geschaffen. –

Beim Schreiben lächle ich, berührt, weil ich damit an positive Erlebnisse mit meiner Mutter anknüpfen konnte.

Schon während der Therapiezeit kamen die bereits geschilderten und andere positive Erinnerungen zum Vorschein. So erzählte ich von einem Traum mit meiner Mutter und fügte hinzu, beim Erwachen das liebe Lächeln von ihr erinnert zu haben. Es sei schön, auch eine solche Erinnerung an sie zu haben. Das quittierte die Therapeutin mit einem ganz besonderen Lächeln, wohl weil sie sich freute, dass ich das zulassen konnte.

Bei einer anderen Gelegenheit hatte ich in der Therapie erzählt, dass ich in einem Café saß als Belohnung für etwas, was ich geschafft hatte.

Das sei einfach etwas, was mir sehr entspräche: Im café, alleine und ganz für mich … – Tränen. Und dann erinnerung an meine mutter, dass sie mir dieses bedürfnis austrieb. Dass das aber etwas war, was wir beide liebten, also auch sie. Wie ein urbedürfnis, das uns verband. Und dass sie mir das austrieb, wie sie es sich selbst ausgetrieben hatte. Erinnerungen an gemeinsames musikhören, klassische musik: Aber sobald der alte zu erscheinen drohte, war schluss damit! Strammstehen angesagt! –

Also schöne Erinnerungen, aber auch verbunden mit dem Schrecken! In einer weiteren Stunde hatte ich, wohl zum ersten Mal, erwähnt, dass meine Mutter nach einem Streit schnell wieder auf uns Kinder, also auch auf mich, zugehen konnte – im Gegensatz zu meinem Vater. Hätte ich mir also irgendetwas geleistet, wäre böse oder nicht folgsam gewesen, dann hätte sie zwar ärgerlich reagiert, was aber nicht lange anhielte. Dann hätte sie meist gesagt, „komm, jetzt ist es doch wieder gut" oder so ähnlich; wäre sie nicht mehr böse, manchmal sogar sehr lieb gewesen. Die K. machte mich damals darauf aufmerksam, dass bei meiner Schilderung meine Augen richtig gestrahlt hätten.

Es fiel mir sehr schwer, während der ersten harten Jahre der Therapie mir das Schöne zu bewahren; es gegenüber all der Enttäuschung und ihrem Verrat noch bestehen zu lassen gelang zeitweise gar nicht.

Eine Freundin sagte vor kurzem zu mir: Du musst auch Schönes zuhause erlebt haben, sonst wärest Du nicht so, wie Du bist, und sonst hättest Du das alles nicht geschafft. – Ähnlich hatte sich die K. schon ganz früh geäußert, aber da wollte ich das gar nicht hören! Später veränderte es sich. Und darüber bin ich heute sehr froh, auch weil mich das innerlich ruhiger sein lässt.

4.6 Mögliche Gründe für das Verhalten meiner Mutter

Warum hat sich meine Mutter so verhalten? Warum hat sie mich nicht beschützt, das alles nicht verhindert? Warum stand sie nicht zu uns Töchtern? Gibt es nachvollziehbare Gründe für ihr Verhalten?

Diesen Fragen versuche ich nachzugehen, wohl wissend, dass ich sie nur unvollständig beantworten kann, nur Vermutungen dazu habe, auch weil darüber leider kein Gespräch mit meiner Mutter möglich war.

Die Therapeutin wies mich mehrmals darauf hin, dass sie es für wichtig erachte, mir die Wurzeln der Eltern anzuschauen. Mir meine Mutter anzuschauen, um ihr Handeln (oder eher ihr Nichthandeln) zu verstehen. Meine Reaktion: Ich wolle mir die Eltern nicht anschauen, nicht über sie in diesem Zusammenhang reden. Es solle nur um mich gehen! Aber es gab noch andere Gründe, warum ich mir die Mutter nicht anschauen wollte. Ich befürchtete, wenn es etwas zu verstehen gäbe an ihrem Verhalten, dürften meine Verletzungen, meine Gefühle nicht mehr da sein, würden die nicht mehr gelten. Dagegen betonte die Therapeutin genau das: Auch wenn ich wüsste, warum sich meine Mutter so verhielt, hätte ich ein Recht auf meine Gefühle.

Im Zusammenhang damit, dass ich mir meine Mutter doch anschauen solle, sprach die K. einmal von „selbst gemachten Gefühlen". Meine Mutter sei ja nicht für mich da gewesen, was ich, nach Ansicht der Therapeutin, so interpretieren würde: Also hat sie mich nicht lieb! – Es könne sich jedoch herausstellen, dass meine Mutter mich liebte, aber *trotzdem* nicht hingucken und mich nicht beschützen konnte. Auch deshalb sei es wichtig, genauer hinzuschauen.

Nachdem ich lange einen großen Bogen, wie die K. es nannte, um dieses Thema gemacht hatte, stellte ich mich dann doch dieser Herausforderung.

Ich beginne mit einigen Überlegungen zum Verhältnis der Mutter zu meinem Vater. Sie selbst betonte immer wieder, für sie stehe Vater auf einem Podest. Er kam für sie an allererster Stelle. Aber warum? Und warum nicht ihre Kinder, gerade weil sie unter anderem (schutz-)bedürftig sind?

Weiter oben erwähnte ich meine Erinnerung: Meine Mutter gehörte dem Vater! – Das sagt ja schon einiges! Aber warum bildeten die beiden so eine Einheit? Hatten vielleicht beide in ihrer Kindheit ebenfalls sexuelle Gewalt erfahren? Das zieht sich ja oft über Generationen hinweg durch die Familien. Hat sich meine Mutter deshalb nicht mit uns Töchtern solidarisiert, weil sie nicht genau hingucken *konnte?* Denn wenn sie selbst sexuelle Gewalt erfahren hatte, war das Thema für sie ein Tabu! Das könnte ein Grund für ihr Verhalten sein.

Ich suche nach weiteren Gründen und komme dabei zu meiner Mutter in Bezug auf mich. Ich wurde als siebtes Kind geboren. Siebtes Kind! Das muss frau sich einmal vorstellen! Da waren schon sechs andere da! Und nach ein paar Jahren kam noch ein achtes Kind hinzu! Wie kann eine Mutter dann für ein kleines Mädchen da sein?

Ich bezweifele sogar, dass sie mich überhaupt haben wollte. Dass ich ihr zu viel war, ist mir jedenfalls klar! So war ja auch mein Gefühl: Ich bin ohne Mutter aufgewachsen. –

Ich will an dieser Stelle eine spannende Aussage aus dem Buch I.M. von Connie Palmen, das ich damals las, einfügen: Wenn wir von den Eltern zu früh im Stich gelassen werden, bevor wir uns *selbst* von ihnen gelöst haben und *selbst* von ihnen haben weggehen können, dann bleiben wir das ganze Leben an sie gebunden. So ungefähr hatte es auch die K. ausgedrückt bezüglich meiner Mutter und auf Parallelen zu mir hingewiesen. Das „zu früh" hieße aber nicht als Jugendliche, sondern viel früher, nämlich bevor man selbst fühle: ICH. Ich will mal probieren! ... Mir mal selbst das Brot schmieren!, hatte sie lachend hinzugefügt. Also so früh sei das gemeint. Und das hätte ich genauso erlebt.

Eine Mutter war zum Zeitpunkt meiner Kindheit die zentrale Bezugsperson für die Kinder, war mehr oder weniger alleine für diese zuständig – mehr noch als heute. Mütter waren, wie auch meine Mutter, meist nicht berufstätig, die Väter verdienten das Geld. Die Mütter trugen in hohem Maße die Verantwortung für das Aufziehen der Kinder und deren Erziehung, hatten also auch die Aufgabe zu schauen, was mit den Kindern los ist.

War die Sozialisation von Frauen oder speziell die meiner Mutter so, dass sie davon ausging, eine Frau hat das zu akzeptieren oder zumindest hinzunehmen, wenn der Ehemann die gemeinsamen Kinder sexuell traktiert? Die K. äußerte eine ähnliche Vermutung:

> *Bei ihnen ging es ja sehr frauenfeindlich zu. Darunter hat auch ihre mutter gelitten. Sie war auch darin eingebunden. Wenn sie selbst unterdrückt und missbraucht wurde, musste sie ja weggucken. Dadurch baut man viele tabus auf. Aber trotzdem können sie wut auf sie haben! Die ist trotzdem berechtigt! Von einer mutter kann man verlangen und erwarten, dass sie ihre tochter beschützt! Ich denke, dass ihre mutter davon ausging, eine frau muss ihrem mann dienen! ... Wenn diese hypothese stimmt, dann hat sie den umgang ihres vaters mit ihnen allen als normal angesehen. –*

Vielleicht war es für meine Mutter auch deshalb keine Option, von meinem Vater wegzugehen. Wohin hätte sie sich auch wenden sollen mit den vielen Kindern? Damals gab es sicher für solche Fälle überhaupt keine soziale Absicherung. Meine Mutter hatte keinen Beruf gelernt. Wovon hätten wir leben sollen ohne entsprechende Hilfe durch den Staat? Ich denke, zur damaligen Zeit war die Vorstellung, als Frau mit Kindern ohne Mann zu leben, völlig undenkbar, nicht nur sozial und finanziell.

Meine Mutter wollte sicher verhindern, dass unsere Familie auseinander bricht, dass sie ihren Mann verliert. Von der Scham, dass so etwas in unserer Familie geschah, gar nicht zu reden!

Auch dieser Aspekt ist zu berücksichtigen: In der Zeit meiner Kindheit bestand vorherrschend die Ansicht, die Kinder gehören den Eltern. Die können mit den Kindern machen was sie wollen – besonders die Väter!

Wie bereits erwähnt wurde meine Mutter schon mit fünfzehn Jahren zum ersten Mal schwanger, bekam ihr erstes Kind mit sechzehneinhalb Jahren!

In diesem Zusammenhang ist eine Erinnerung meines Bruders Bernd von großer Wichtigkeit. Er habe aus dem Schlafzimmer der Eltern die Worte der Mutter vernommen, mehrmals wiederholt, sie wolle keine Kinder mehr! Neun Monate später sei unser jüngster Bruder geboren worden.

Sie wurde nicht gefragt, ob sie will oder nicht! Und ihre Ablehnung nicht akzeptiert! Vielleicht auch schon in meinem Fall nicht.

Das alles sind für mich mögliche Gründe für das Verhalten meiner Mutter. Ich versuche sie und ihr Verhalten zu verstehen, was für mich aber nicht bedeutet, sie zu entschuldigen. Aber mit dem Versuch, sie zu verstehen, mit den erwähnten möglichen Begründungen lebt es sich besser. Ich bin mir durchaus bewusst, dass es noch mehr an Wissen und Hintergründen zur Mutter, und auch zum Vater, bedarf, um ihr jeweiliges Verhalten wenigstens einigermaßen zu verstehen oder einordnen zu können. In diesem Zusammenhang ist es mir nur wichtig, dass ich mich den erwähnten Fragen stellte, wie es mir damit erging und zu welchen Schlüssen ich gelangte.

Abschließend sei noch diese Aussage meiner Therapeutin erwähnt: Meine Mutter habe mir nicht mehr geben können, warum auch immer. Das habe aber nichts mit mir zu tun, sondern nur mit der Mutter! Es sei ganz wichtig, mir das klarzumachen. Ich verstand.

4.7 Das Kapitel (mit der) Mutter ist für mich abgeschlossen

Einige Monate nach den Begegnungen mit meiner Mutter und unserer Aussprache sagte ich in der Therapie, das Thema Mutter sei für mich irgendwie abgeschlossen. Ich hätte nicht vergessen, was sie gemacht habe, wie sie sich früher verhielt. Aber ich könne mit ihr zu tun haben. Mit meiner Mutter hätte es etwas von annähernd versöhnt zu sein, was mit dem Vater unvorstellbar sei. Versöhnt war aber nicht gemeint im Sinne von verzeihen! Ich hatte damals hinzugefügt, wie gut es sei, dass ich diese Begegnungen mit ihr haben und sie nochmals so nahe erleben konnte.

In den folgenden Monaten und Jahren fühlte ich mit ihr, tat sie mir leid, wenn sie wieder in die Klinik musste, es ihr schlecht ging – trotz all der Enttäuschungen, die davon unberührt blieben. Ich besuchte sie ein paarmal im Krankenhaus. Aber es kam auch vor, dass ich sie vor lauter Wut nicht besuchen konnte. Die K. hatte dazu erklärt, es wäre doch klar, dass Wut und Hass immer wieder hochkommen könnten. Das sei ganz natürlich. Aber ich wisse doch jetzt, dass auch noch andere Gefühle für die Mutter da seien. Das bejahte ich.

In einer anderen Therapiestunde berichtete ich von einem meiner neuerlichen Besuche im Krankenhaus:

> *Ich schilderte ausführlich, wie sehr meine mutter abgebaut habe, was mich sehr traf. Vater würde die gemeinsame wohnung auflösen. Dazu habe mutter gemeint, ich solle mir doch auch etwas holen, bücher, aber nicht mit den anderen streiten, einfach tauschen! Die K. sagte lächelnd, sie wundere sich, wie offen meine mutter sein könne. – Als ich mutter sah, dachte ich sogar, wenn man dich hier so sieht, könnte man dir alles verzeihen! – Ehem! – Aber nach dem besuch waren auch die anderen, die negativen gefühle für meine mutter wieder da!, stellte ich klar.*

Durch meine sich daran anschließende Frage, wieso ich es nicht schon früher schaffte, als Mutter noch gesünder und kräftiger war, mit ihr in einen anderen Kontakt zu kommen, nahm die Stunde folgenden spannenden Verlauf:

Ich fuhr ungefähr so fort: Die hatte ja noch haare auf den zähnen! Mit der ziege, so wie sie damals war, konnte das ja nicht gehen. Leider ist sie erst jetzt weicher. – Die K. ganz fest: Aber lag das nicht an ihnen? – Weil ich noch so viel wut auf mutter hatte? – Nein! Weil sie noch nicht soweit waren! Als sie ihrer mutter mit ihrer wut begegneten, da konnte die ja nicht anders reagieren, als absolut dicht zu machen und abzuwehren. ... Und dann haben SIE SICH ja geändert! Das haben SIE doch ganz alleine geschafft, dass sie sich vor zwei jahren in der klinik begegnen konnten! ... Also für mich haben ganz alleine sie das möglich gemacht. Sie mussten erst soweit sein! ... Und erst in der klinik war ein gespräch alleine *mit ihrer mutter möglich. Das muss man bedenken! Da war sie nicht mehr unter dem bann ihres vaters. – Diese worte taten mir so gut! –*

Ich verstehe, dass sie traurig sind. Dahinter stehen der wunsch und die enttäuschung, dass es nicht früher möglich war! ... Außerdem war ihre mutter damals, als sie zu ihr in die klinik gingen, noch fit! So ist es doch nicht! – Ja, das stimmt. – Also, das hätte auch ganz anders ablaufen können. Aber weil sie sich verändert haben, ging es so gut! Das war ganz alleine ihr verdienst! – Ich bin ja auch sehr froh, dass ich das geschafft habe! –

Die K. meinte noch, dass meine mutter diese offenheit anscheinend nur haben konnte, weil sie alleine, also vater nicht dabei war. Es sei wirklich erstaunlich und toll, dass meine mutter sich so darauf einließ. – Als sie das hier erzählten, musste ich mit ihnen weinen, weil es auch mich sehr berührte. – Ich sah sie verwundert an. Das hatte ich damals gar nicht bemerkt, weil ich so sehr bei mir war. (...)

Weitere erinnerungen an meine mutter: Einerseits spürte ich dabei liebe für sie. Sie konnte mich so lieb anschauen! Zum anderen erinnerte ich, wie kalt und herrisch sie sein konnte, schrecklich! Dieser kalte, harte blick! – Ich hätte doch schon einmal erwähnt, dass ich als kleines mädchen nicht verstand, wie das so wechseln konnte.

Die K. sieht mutter so, dass sie ihr fähnchen nach dem wind hing, sich gut anpassen konnte. Bezogen auf den besuch von mir in der klinik, bei dem ich sie mit ihrem verhalten konfrontiert hatte, meinte die K., dass mutter da offen war, ihre gefühle zuließ, sich auf mich einließ. Ich würde mich fragen, ob das ehrlich war. – Ja! Ehrlich und authentisch! Das waren keine lügen! – Außerdem hätte ich das doch sicher meiner mutter angemerkt; davon sei sie überzeugt. – Ja, an ihren augen kann man das gut ablesen, ob sie ehrlich ist oder nicht, bestätigte ich. –

Von wegen fähnchen nach dem wind hängen, dahinter stehe ja angst, sehr große angst und unsicherheit. Warum auch immer, aber so wäre es. Die K. fuhr fort: Aber was ich toll finde, dass ihre mutter sich dennoch einen rest an zugang zu sich selbst, zu ihren gefühlen bewahrt hat. Das finde ich wirklich erstaunlich und toll! Dass sie nichts wegen ihres vaters zurücknimmt, sondern sich auf diese schöne situation mit ihnen einlassen kann, genau wie sie. – Ach, wie sehr freute ich mich über die worte der K., die noch meinte, meine mutter sei doch, wie ich, eine genießerin. – Ja... – Ich lächelte vor mich hin. Ganz froh und glücklich. Liebe für meine mutter... und für die ma. Dankbar für ihre worte!

Bald darauf träumte ich, die Türe zu meiner Mutter wäre geschlossen. Dazu meinte die K., vielleicht wolle ich gar nicht mehr zu ihr gehen. Darüber reflektierte ich einige Zeit später:

Ich hätte mich gefragt, warum ich nicht öfter zu meiner mutter fahre. Verschiedenes dazu in mir. Vielleicht um nicht zu spüren, dass ich sie noch lieb habe. ... Daraufhin erinnerungen an früher: Ob ich mich nicht schämen würde, jetzt noch, wo ich schon so groß sei (aber noch ein Kind, füge ich hinzu!), etwas liebes, so wie schmusen oder so von ihnen zu wollen? Das passe doch nicht mehr! Schon gar nicht, bei meinem vater noch auf den schoss zu wollen. Ich solle mich schämen! ... Ich weinte. ... Also am besten die beiden gar nicht sehen, damit ich nicht spüre, dass ich noch liebe in mir habe, sogar für meinen vater! Vor jahren hätte ich mich dafür noch sehr verurteilt, aber jetzt sähe ich das anders, also so, dass es ein kind trotzdem immer wieder versucht und immer noch hofft, doch was liebes zu bekommen. Aber ich hätte auch tierische wut auf die! Die K dazu: Ich denke, sie fahren nicht öfter zu ihrer mutter, weil sie angst haben, sie spüren nur *die liebe, und die wut ist weg! Dass sie also ihre wut nicht mehr spüren. Und das würden sie als verrat empfinden! – Ja, die wut soll nicht weggehen! – Die ma lachte auf ihre schöne art, nein, die solle auch nicht weggehen! Es ginge doch darum, dass* beide *gefühle da sind und das auch zusammenpasse! – Vielleicht haben sie angst davor, dass sie spüren könnten,* beide *gefühle sind da? Aber besonders die lieben gefühle. Dass trotz all der berechtigten wut, und die ist ja berechtigt, auch die lieben gefühle da sind! –*

Vielleicht hätte ich auch angst, vermutete die K. später noch, dass ich beide gefühle zulassen könnte, weil das nicht meinem bild von mir entspräche. Ich meinte, es sei auch anderes dazu in mir: Es ist gut für mich, tut mir gut, beide gefühle zuzulassen. Macht mich innerlich ganz ruhig. – Das hätte ich zumindest schon mal im kopf. Das wäre ein langer prozess gewesen bis dahin, meinte sie. Und das jetzt auf die eltern zu übertragen, darum ginge es, das wäre schwierig. In unserer familie habe es ja bisher nur entweder kontakt oder null kontakt gegeben.

Als ich einen weiteren Besuch im Altenheim plante, mein Vater war inzwischen ebenfalls dort, hatte ich darüber nachgedacht, wie es gut gehen könnte, den beiden zu begegnen; dabei auch meine Bedenken geäußert. Aber dann spürte ich Leichtigkeit in mir, lächelte, sagte, vielleicht sei es gar nicht so schwer, weil ich doch mit all meinen Gefühlen dort sein könne. Ich phantasierte, ich gehe zu ihnen und habe ein Netz voller Päckchen dabei:

In dem einen ist die trauer, im anderen die wut. In einem sind die schönen dinge, die ich mit ihnen erlebte, die lieben gefühle. ... Also dass ich alles dabei habe mit dem gefühl, ich trage alle päckchen, habe alle gefühle präsent, und es ist ok! Leicht! – (...)

Die K.: Das ist doch ein schönes bild! – Aber das verschwand auch gleich wieder.

So ambivalent und wackelig war es und ist es manchmal noch heute. Aber vielleicht passt der Ausdruck „ambivalent" gar nicht. Sind doch das Schöne und das Schlimme, der Verrat, der Schmerz und alles Andere weiter vorhanden, kann ich das doch alles nebeneinander stehen lassen.

Zum Abschluss dieses Kapitels noch eine aktuelle Begebenheit: Vor ein paar Tagen fuhr ich am Friedhof, auf dem die Eltern begraben liegen, vorbei und ging, einem spontanen Impuls folgend, zu ihrem Grab. Muss mal dahin!, so mein Gefühl. Ich stand nur kurz dort, dachte an meine Mutter, drehte mich gleich wieder um und wollte gehen. Es schien mir nicht zu passen, dort zu sein. Während ich mich wegdrehte, noch an meine Mutter denkend, fühlte ich: Ich hab' die immer noch lieb! ... Das gibt's doch nicht! ... Ich hab' die ja immer noch so lieb! – Völlig überrascht wandte ich mich wieder dem Grab zu und wieder: Ich hab' die einfach lieb! Trotz all dem Schlimmen! – Nach einer kleinen Weile ging ich zurück zu meinem Auto, sehr berührt von den lieben Gefühlen. Und diese blieben noch einige Tage ganz nahe, waren wie etwas ganz Weiches weiterhin sehr spürbar. Es fühlte sich sehr schön an.

Und nun kann ich sagen, ich habe meine Mutter noch immer sehr lieb – auch wenn mir ihr Verhalten und ihre Verantwortung so viel deutlicher geworden sind und der Schmerz darüber noch da ist.

5. Nicht darüber reden

Es gab und gibt ein nicht darüber Reden*können* und ein nicht darüber Reden*dürfen*. Und das gilt leider meist bis weit ins Erwachsenenalter!

Ich will mich dem Thema unter einem zeitlichen Aspekt annähern: Es geht zum einen jeweils um die Zeit, als die Übergriffe meines Vaters und die meiner Brüder geschahen. Und es geht um die Zeit, als ich Jugendliche, junge Erwachsene war bis hin zu der Zeit, in der ich über die Taten zum ersten Mal sprach, somit nicht mehr schwieg, hier besonders die Zeit der Therapie, also bei mir bis zum Alter von Mitte 40 und darüber hinaus.

Zu Anfang der Therapie hatte ich keine Erinnerung, ob ich als Kind überlegte, mich an meine Mutter zu wenden. Die Therapeutin fragte mich bald, welche Antwort ich erwartet hätte? – Dass mir meine Mutter die Schuld gibt, sagt, ich hätte vielleicht geträumt oder ich hätte nicht so leicht bekleidet rumlaufen dürfen. – (Also zum Beispiel im Unterrock über den Flur ins Bad gehen! Mehr bzw. weniger an Kleidung war da nicht! Außerdem: Auch leichte Bekleidung rechtfertigt nicht sexuelle Gewalt!) Ich fuhr fort: Ich weiß es eigentlich nicht. Ich kann mir nicht vorstellen, etwas gesagt zu haben. Ich glaube, die sexuelle Gewalt war für mich so ungeheuerlich, dass ein darüber Reden gar nicht in Betracht kam. –

Ich erinnere dazu eine Situation in der Therapie, die ich im Folgenden zitiere. Ich hatte einige Monate zuvor über den Übergriff durch Bernd gesprochen.

> *Ich sagte weinend: Ich würde gerne meiner mutter sagen, was passiert ist. – Die therapeutin schob einen stuhl seitlich vor mich hin, setzte ein dickes kissen darauf und meinte, ich solle mir vorstellen, das sei meine mutter, der ich jetzt das sagen könne, was ich ihr sagen wolle. Ich bekam einen riesigen schreck, hielt mir die augen zu, weinte. Daraufhin stellte die K. sich neben mich und fragte, wie alt denn die kleine Ann wäre, die das ihrer mutti sagen wolle, 10 oder 8 oder 6 jahre? – 6 jahre. – Ich fühlte mich sehr klein, sagte, meine mutter solle sich hinsetzen und mir zuhören. Dann konnte ich nichts mehr sagen. Ich fand keine Worte. Mir war richtig übel, ich bekam rückenschmerzen. – Ich kann nichts sagen. – Will die kleine Ann jemanden zu hilfe rufen? Vater, geschwister? – Nein! Meinen vater? Nein!! ... Die Lisa soll mir helfen. – Dabei musste ich wieder weinen. – Was soll die Lisa machen? – Sie soll der mutti sagen, sie soll mal sitzen bleiben und mir zuhören. Ich müsse ihr was furchtbares erzählen. – Ich fand aber keine formulierung, keine worte! – Soll die Lisa es sagen? – Ja! –*

> *Dann: Nein! Ich will es selber sagen! – Aber es ging nicht, mir war weiter übel, weiter rückenschmerzen.*
>
> *Bald darauf sagte die K., sie müsse jetzt der* großen *Ann etwas sagen: Mädchen in dem alter können dieses schlimme nicht in worten ausdrücken. Sie haben keine worte dafür. Sie können das höchstens in ihrem verhalten zeigen. Oder in bildern, die sie malen. –*

Ich begann zu verstehen, warum ich weder meiner Mutter noch jemand anderem etwas davon hatte sagen können. Das traf auf jeden Fall für die Zeit zu, als ich noch ein kleines Mädchen war, also zu Zeiten der Übergriffe meines Vaters.

Nach einigen Jahren Therapie, nachdem ich auch die Übergriffe meines Vaters bearbeitet hatte, erinnerte ich in einer Therapiestunde, dass ich als kleines Mädchen zu meinem Vater sagte, er solle mich gehen und in Ruhe lassen: Ich versprach ihm, ich sage auch nix! … Aber er sagte, ich hätte das doch selbst gewollt! – Auch damit hat er mich dazu bringen wollen, nichts zu sagen. Von seiner Androhung von Gewalt ganz zu schweigen!

Ich komme zurück zu der Frage, habe ich als Kind darüber geredet bzw. warum habe ich nicht darüber geredet. Als ich 9 und mehr Jahre alt war, als meine Brüder mit ihren Übergriffen begannen, wie war es da? Auch zu diesem Zeitpunkt war damit zu rechnen, dass meine Mutter, meine Eltern mir die Schuld geben würden. Außerdem hätte ich mich doch selbst schuldig gefühlt, sagte ich in der Therapie, wie hätte ich dann etwas davon sagen können? Vom Schamgefühl, das auszusprechen, ganz zu schweigen! – Ich dachte damals doch, mit mir stimmt was nicht, dass die das mit mir machen! Das muss mit mir zu tun haben. Anscheinend bin ich schlecht! Warum sollte ich dann darüber reden? – Ja, das stimme, so die Therapeutin, die Schuldproblematik, aber auch die katholische Erziehung dürften wir nicht außer Acht lassen. Die Religion habe ja verhindert, dass überhaupt über Körper, Sexualität und Intimität geredet wurde. Da hätte ich erst recht nicht über Verbotenes reden können.

Und dennoch beschäftigte mich immer mal wieder die Frage, habe ich mich nicht doch an meine Mutter gewandt? Als es eines Tages darum ging, dass eine Freundin nicht wollte, dass ich weiter über die sexuelle Gewalt redete, tauchte in einer Therapiestunde dieses Bild in mir auf: Ich habe meiner Mutter was gesagt. Und ihre Reaktion war: Du hast es mir jetzt gesagt, jetzt ist es gut! Ich will nichts mehr davon hören! – Meine vier Jahre ältere Schwester Lisa konnte sich tatsächlich daran erinnern, dass ich neben meiner Mutter stand, die sowas sagte wie, ich solle jetzt still sein! Dass es Versuche von mir gab, meiner Mutter gegenüber etwas anzudeuten; sie habe jedoch abweisend reagiert mit den Worten, ich solle sie in Ruhe lassen, sie habe keine Zeit oder so ähnlich. Diese Situationen erinnerte Lisa mehrfach. Nach solch vergeblichen Versuchen sei ich immer krank geworden. Sie, Lisa, hätte auch gerne etwas zu unserer Mutter gesagt, aber sie habe nur wie betäubt daneben stehen können.

Nach und nach wurde mir nun klarer, erinnerte ich, wie es dazu kam, dass ich schwieg. Wichtig ist, dabei mitzubedenken, dass die Abhängigkeit von der Familie für ein Kind, für eine Jugendliche immens groß ist! Ich brauchte meine Familie. Ich brauchte meine Mutter! Ich konnte doch nirgendwo anders hingehen! Also lieber schweigen.

So kamen wir zwangsläufig in der Therapie auf meine Angst um meine Familie und um meine Mutter zu sprechen und auf meine Angst, wenn ich darüber rede, zerstöre ich die Familie. Ich erinnerte mich, dass meine Mutter mir zu verstehen gab, ich müsse ihr helfen, sie unterstützen. Sie brauche mich. Sonst würde die Familie auseinander brechen, stehe sie mit uns Kindern alleine da. Sonst sei alles aus. – Ja, ich habe versucht, die Verantwortung zu übernehmen, damit unsere Familie nicht auseinander bricht. –

Ich konnte damals aber erstmal nicht still sein, wollte das nicht hinnehmen. Aber meine Mutter drohte mir damit, dass ich in ein Kinderheim komme oder in die Psychiatrie. Deshalb sagte ich bei einem Telefongespräch mit meinem Vater, auf seine Frage, wieso ich nicht schon vor 40 Jahren was gesagt hätte: Dann hättet ihr mich doch in die Klapse gesteckt! Aber jetzt bin ich zu stark! –

Irgendwann habe ich dann nichts mehr gesagt, auch für meine Mutter geschwiegen, um unsere damaligen Lebensverhältnisse aufrecht zu erhalten und um mich zu schützen. Aber auch aus Angst um meine Mutter. Ich hatte früher, wie alle meine Geschwister, sehr viel Angst um sie wegen ihrer Herzprobleme. Also durfte ich sie nicht aufregen, musste brav und still sein. Still und angepasst. So hat das funktioniert. Das will ich hier noch anfügen, brav und still auch deswegen: Muss mich ruhig verhalten, dann passiert mir nichts. Wenn ich brav bin, nicht auffalle, dann lässt Mann mich in Ruhe. Das hat aber nicht funktioniert!

Dieses Schweigen hat sich dann fortgesetzt bis ins jugendliche Alter; auch als junge erwachsene Frau sprach ich nicht darüber. Zum einen wegen der zu erwartenden Reaktionen meiner Eltern, zum anderen wegen der Angst, aus dem Familienverbund ausgestoßen zu werden. Wie hätte ich solche Reaktionen als so junger Mensch verkraften sollen? Ich habe mich mit meinem Schweigen also weiterhin geschützt. Zudem war die Scham viel zu groß und die Schuldgefühle waren keinesfalls weniger geworden.

Es kam noch ein weiterer, neuer Aspekt hinzu. Ich hatte die Erlebnisse irgendwann völlig verdrängt; die Erinnerung daran vergraben. Auch die Erinnerung an die Übergriffe meiner Brüder. Ab wann das so war, weiß ich nicht mehr, aber ich vermute, dass ich spätestens nach dem folgenden Ereignis alles endgültig verdrängte:

Meine Mutter erzählte mir, als ich ungefähr 18 Jahre alt war, dass Adolf sich gegenüber einem fremden Mädchen übergriffig verhalten habe. Das Mädchen hatte ihn angezeigt. Ich erinnerte noch sehr genau, dass ich dachte: Mutti, warum erzählst Du mir das? Ich könnte Dir noch was ganz Anderes erzählen! – Meine Mutter habe seinerzeit hinzugefügt, so berichtete ich weiter in der Therapie: Wie diese jungen Dinger aber auch heute rumlaufen! – Also nach dem Motto, die sind selbst schuld! Spätestens nach diesem Kommentar sei klar gewesen, dass ich bei ihr keine Gnade finden würde.

Wie im Prolog bereits erwähnt, ließ sich das Erlebte nicht mehr verdrängen, nachdem ich einige Jahre später eine Ausstellung von Wildwasser zum Thema sexuelle Gewalt sah. In meiner ersten Therapie wollte ich dann endlich darüber reden, aber die Therapeutin beendete unsere Arbeit, bevor ich das tun konnte, weil sie wegzog. Jetzt werde ich wohl nie mehr Gelegenheit haben, mit jemand darüber zu reden!, dachte ich verzweifelt, niedergeschlagen und hoffnungslos.

Und dann schaffte ich es doch noch – in meiner zweiten Therapie! Wie lange und wie nachhaltig jedoch ein Verbot bzw. die Unmöglichkeit, darüber zu reden, wirken kann, zeigte sich in dieser Analyse. Ich werde nur auf einige Beispiele eingehen.

Vorweg sei noch einmal erwähnt, dass ich mir die Therapeutin eine Zeit lang als Mama phantasierte, dass ich in ihr eine Mama sah; die Psychoanalyse spricht dabei von Übertragung.

In einer Therapiestunde sagte ich zu der Therapeutin, ich hätte Angst, wenn ich das alles mit meinem Vater rauslasse, alles sage, dass sie dann weggehe, mich auch damit alleine lasse. Dass ich mir aber sagen würde, *die* Mama lässt „uns“ (die große und die kleine Ann) doch nicht alleine, auch wenn „wir“ *darüber* reden. Das wüsste ich doch ganz sicher! Aber die Angst wäre dennoch da. Die Therapeutin erklärte mir das u.a. so: Sie *sollen* ja nicht darüber reden! Und wenn Sie es dennoch tun, tun wollen, haben Sie ein schlechtes Gewissen, weil es wie ein Weggehen von Ihrer Mutter ist. Die hat ja auch verlangt, dass Sie schweigen. Und hinter der Angst verbirgt sich auch die Wut auf Ihre Mutter, weil sie Sie nicht beschützte. – Meine jetzigen Schritte seien auch ein Stück Weggehen von der realen Mutter, also meiner Mutter. Auch deshalb hätte ich Schuldgefühle.

Es gab auch eine Seite in mir, die es nicht oder nicht mehr hören wollte, so wie es früher niemand hören wollte. Die Therapeutin machte mich darauf aufmerksam; aber auch darauf, wenn ich nichts sagen, es nicht öffentlich machen würde und die Brüder das vielleicht auch mit ihren Töchtern machten, dass es dann immer weiter ginge! Ja, daran hätte ich auch schon gedacht, aber ich hätte einfach Angst vor den Reaktionen meiner Brüder. Das könne sie sehr gut verstehen, aber das sei doch genau das, womit wir dazu gebracht werden würden, weiter zu schweigen und die Täter damit zu decken und zu schützen. Ich dachte genauso wie sie, aber die Angst war da! Darüber ärgerte ich mich selbst maßlos.

Zu einem etwas späteren Zeitpunkt sagte ich, obwohl ich ein immer klareres Gefühl bezüglich der Taten meines Vaters hatte: Ich will nix sagen! Ich will nicht! ... Ich werde das nie in meinem Leben sagen! – Das hatte ich sehr deutlich genauso in mir. Die Therapeutin ging verständnisvoll damit um, aber ich blieb dabei: Ich kann nichts sagen. Kann es nicht sagen. – Und wen schützen Sie damit? Das überlegen Sie sich mal! ... Wen Sie damit schützen! –

Mit der Schilderung von sehr wichtigen Stationen auf dem Weg, doch darüber zu reden, will ich diesen Punkt abschließen. Ein kleiner, aber hilfreicher Schritt war das Erkennen, ich sage es doch erstmal nur der Therapeutin. In der Therapiestunde, in der mir das klar wurde, hatte ich zuvor noch gesagt: Ich werde nix sagen! Ich hab's versprochen! Ich darf nix sagen! – Doch dann veränderte sich etwas bei mir, denn nun fügte ich hinzu: Auch wenn ich es versprochen habe, nichts zu sagen, ich will, dass es raus ist! Dass das endlich mal aus mir raus ist! – Das zu sagen hatte schon etwas Befreiendes. Und daraufhin folgte der entlastende Satz der Therapeutin: Ja! Außerdem ist es dann ja erstmal nur hier! –

In einer anderen Stunde tauchte ein Bild mit meiner Mutter auf, die gesagt hatte, das sei ja unerhört, was ich da behaupten würde. Also was ich von meinem Vater gesagt hatte. Ich wollte das der Therapeutin nicht sagen, sagte nur, ich hätte mir doch fest vorgenommen, es nie wieder zu sagen. Niemandem! Und daraufhin die Therapeutin: Doch! ... Sagen Sie es mir! ... Und sagen Sie es sich selbst! ... Sagen Sie es für uns beide! – Ich notierte anschließend in mein Tagebuch: Das sagte sie mit ihrer lieben Mamastimme. Fand es so schön, dass sie sagte, sagen Sie es für uns! Sagen Sie es uns beiden! – Im Anschluss an diese liebe Aufforderung redete ich, sprach von meinem Vater und wie das für mich war. Zum Ende dieser Therapiestunde betonte die Therapeutin nochmals, mich dabei eindringlich anschauend: Es ist mir wichtig, dass Sie sich klarmachen, Sie haben es mir und *sich selbst* gesagt! Dass sie es auch *sich* gesagt haben! Damit Sie das für sich annehmen können. –

Auch aus heutiger Sicht muss ich unbedingt festhalten, wie schön es für mich war, dass die Therapeutin in die Rolle der Mama geschlüpft ist und mir damit eine Mama gegeben hat, die es hören wollte. Die nachfragte und die da war.

Und zum Schluss füge ich noch diese Erkenntnis an: *Die* haben das gemacht! Ich sage es doch nur! Wenn ich nicht darüber rede, ist es doch wieder wie früher! Aber diese Klarheit hatte ich lange nicht. Leider!

Wie die Reaktionen insbesondere der Familie ausfallen, wenn frau als inzwischen Erwachsene dann doch darüber redet, sei bereits kurz angedeutet, bevor ich darauf im folgenden Kapitel ausführlich eingehe: Zum einen reduziert das späte darüber Reden ihre Glaubwürdigkeit. Andererseits, was noch schlimmer ist, sind nicht die Taten und die Täter das Problem, sondern diejenige, die das Geschehene ausspricht!

Anlässlich einer Tagung der Evangelischen Akademie Tutzing zum Thema Kindheitsverletzungen wurde mir noch klarer, wie sehr das Schweigen des Umfeldes, besonders der Familie, zur weiteren Traumatisierung einer Betroffenen beiträgt! Das familiäre Schweigegebot ist als Teil der Taten anzusehen, die ohne die Drohungen und die Ignoranz zumindest nicht häufiger hätten stattfinden können.

6. Reaktionen

In den Texten zu meinen Brüdern und meinem Vater kamen schon einige Reaktionen auf mein Öffentlichmachen der sexuellen Übergriffe zur Sprache. Ich möchte nun ausführlicher auf die Reaktionen, besonders seitens meiner Eltern und Geschwister, eingehen. Die wahrscheinlichen oder befürchteten Reaktionen sind ein wichtiges Kapitel beim Offenlegen von sexueller Gewalt. Die Ängste und die Gefahren, die damit verbunden sind oder befürchtet werden, erschweren oder verhindern genau dies. Auch deshalb will ich davon berichten.

Damit komme ich zu einer zu Beginn der Therapie und des Redens über die sexuelle Gewalt oft gehörten Frage: Was willst Du jetzt noch damit? Das ändert doch nichts mehr! Und bald folgten Bemerkungen wie: Das müsse man doch mal ruhen lassen und nach vorne schauen. Das Erlebte sei sicher schlimm, könne ich jedoch nicht ungeschehen machen. Ich könne es mir doch gut gehen lassen, hätte einen netten Mann, keine Kinder, ausreichend Geld; warum ich das nicht einfach genießen würde? Und auch das: Es ginge mir doch immer schlechter, also warum diese Therapie? Die würde doch nichts ändern. Ob es nicht ausreiche, mit Freundinnen darüber zu reden?

Abschließend zwei weitere Beispiele aus der Reihe „einfühlsamer" oder besser gesagt abwehrender Äußerungen: Wenn Du beruflich mehr ausgelastet wärest, hättest Du nicht so viel Zeit zum Nachdenken. Und eine andere Aussage lautete: Komisch, dass die „das alles" mit Dir gemacht haben! Die Betonung lag auf *dir*. Es tat sehr weh, das zu hören. Aufkommende Schuldgefühle stoppte ich mit meiner Antwort: Wieso? Ich weiß doch nicht, ob mein Vater „das" nicht auch mit den ältesten Schwestern gemacht hat, wenn ich an deren Reaktionen denke. – Außerdem habe doch meine Schwester Lisa ebenfalls sexuelle Gewalt durch ihn erlebt. Daraufhin nachdenkliches Schweigen ...

Es gab aber auch von Anfang an sehr viel Positives und Unterstützendes – ohne das wäre das Aufarbeiten für mich nicht zu bewältigen gewesen. Meine FreundInnen haben mich zum Teil jahrelang sehr unterstützt, waren solidarisch, hilfsbereit, so sehr für mich da – jede auf ihre Weise. Aber es war ihnen auch mal zu viel, dann wollten sie nichts von diesem furchtbaren Thema hören und mitbekommen. Es fiel mir nicht immer leicht, das zu akzeptieren, auch wenn ich es verstehen konnte. Mir war es oft selbst zu viel, viel zu viel. Eine Freundin brach sogar den Kontakt zu mir ab, nachdem sie mich wirklich wunderbar, liebevoll und mit so großem Verständnis unterstützt hatte. Sie wollte nichts mehr von diesem Thema hören, und ich wollte andererseits nicht wieder dazu schweigen. Dieser Abbruch war für uns beide sehr schmerzhaft, aber Jahre später konnten wir wieder aufeinander zugehen. Unsere wunderbare Freundschaft besteht auch weiterhin – zu meiner großen Freude! Aber damals traf mich das wirklich sehr, steigerte es meine sowieso vorhandene Angst, dass ich noch alle verliere, weil ich das Thema aufgegriffen hatte, weil ich darüber redete und nicht mehr schweigen wollte.

6.1 Reaktionen aus der Familie

Ich konnte zuerst nur über die Taten meiner beiden ältesten Brüder reden, denn die Erinnerungen daran, was mein Vater mir angetan hatte, waren ja nicht sofort da, geschweige denn klar. Lisa war eine der Ersten, der ich davon erzählte, weil wir uns sehr nahestanden. Sie reagierte sehr, sehr liebevoll, aber auch geschockt. Sie selbst hatte keine solchen Erlebnisse mit den Brüdern.

Birgit, meiner zweitältesten Schwester, sagte ich es erst während des ersten Therapiejahres in einem Telefongespräch. Ich weinte und sie auch. Sie würde mich umarmen, wenn ich in der Nähe wäre. Nach dem Gespräch bekam ich Angst, was ich damit auslöse. Beim nächsten Telefonat erzählte mir Birgit, dass sie nach dem ersten Gespräch einen Nervenzusammenbruch erlitten habe. Das traf mich sehr. Nach einer Weile wurde mir klar, das kann nichts damit zu tun haben, dass ich das erzählte. Das alleine kann es doch nicht sein! Birgit meinte, sie sei dennoch froh, dass ich mit ihr darüber gesprochen hätte, dass sie es jetzt wisse. Wenn ich das früher als Kind oder Jugendliche ausgesprochen hätte, so Birgit weiter, hätten die Eltern das abgestritten, das gab es nicht, du musst was falsch verstanden haben oder ähnlich. Mir wurde dadurch klarer, warum ich das auch als Jugendliche nicht ansprechen konnte. Ich war so froh für diese Erkenntnis, nachdem ich zeitweise sehr mit mir gehadert hatte, an den Auswirkungen zum Teil selbst schuld zu sein, weil ich so lange schwieg.

Bevor ich auch Maria, meine älteste Schwester anrief, fragte ich mich in der Therapie, warum ich das den beiden unbedingt erzählen will. Ein Geheimnis dauernd zu bewahren würde sehr viel Kraft und Energie kosten, vielleicht wolle ich die nicht mehr aufwenden, vermutete die Therapeutin.

Ich rief sehr aufgewühlt bei Maria an. Ich fragte sie, ob sie sich an Veränderungen bei mir als kleinem Mädchen erinnern könne, was sie verneinte. – Du Arme!, so Marias erste Reaktion und gleich anschließend die Frage: Wer, Vater oder Brüder? – Brüder. Adolf und Bernd. – Auch Bernd?, sie ganz erschüttert. – Ja! – Sie reagierte liebevoll. Dafür gäbe es keine Entschuldigungen, egal wie alt die waren. Die Brüder und die Eltern seien verantwortlich. Sie wäre froh, dass ich ihr das alles sagte, aber es wäre unfassbar. Sie mache sich Vorwürfe, weil sie damals nichts bemerkt habe, und fragte, ob ich nie überlegt hätte, ihr davon zu erzählen. Nein. Ich hätte als Kind dafür noch keine Worte gehabt, erläuterte ich ihr. Ich erwähnte meine Schuldgefühle und die Scham, selbst mit ihr darüber zu reden.

Maria weinte immer wieder, sprach von ihrer maßlosen Erschütterung und was das für mich die ganzen Jahre bedeutet habe. Sie spüre Erleichterung, weil ich in Therapie sei und auch mit Anderen darüber redete. Ich sagte, ich würde die Solidarität meiner Schwestern dringend benötigen. Nach dem Gespräch weinte ich sehr, war gleichzeitig froh und erleichtert und fix und fertig. Am Abend reagierte ich mit heftigen Panik- und Herzattacken. Ich verdammte mich dafür, dass ich es Maria sagte, befürchtete schlimme Reaktionen aus der Familie. Erst spät konnte ich mich einigermaßen beruhigen.

Bei einem weiteren Gespräch mit Maria sagte ich, irgendwann wolle ich die Taten der Brüder und all das Verlogene bei uns zuhause den Eltern auf den Kopf zusagen. Maria befürchtete, die könnten das nicht verkraften; dann müsste man gleich einen Arzt oder Psychiater dazu holen. Wahrscheinlich wollte sie mir damit sagen: Das kannst Du nicht machen! Du bist schuld, wenn denen etwas passiert!

Und nun komme ich zu den Reaktionen auf das Ansprechen der Taten meines Vaters.

Nach Lisa waren es wieder meine ältesten Schwestern, mit denen ich darüber reden wollte und konnte. Zuvor sprach ich in der Therapie über meine Angst, wie die beiden darauf reagieren, besonders Maria, weil sie sich einmal so äußerte, sie habe „das mit den Brüdern“ noch immer nicht verkraftet. Die Therapeutin energisch, was mir sehr gut tat: Sie werden doch nicht aus Angst vor der Reaktion von Maria *nicht* zu sich stehen? –

Zuerst sprach ich wiederum mit Birgit. Ich versicherte mich: Willst Du es wirklich wissen? – Ja! Etwa das gleiche wie mit Adolf und Bernd?, fragte sie mit ganz normaler Stimme. – Ja! Und Lisa hat das auch erlebt. – Birgit reagierte gar nicht geschockt. Nach dem Telefonat zitterte ich richtig, war aber so froh: Ich hab's gesagt! Ich hab's gesagt! – Und sehr erleichtert über die ruhige Reaktion von Birgit.

Dabei blieb es aber leider nicht. Ein gutes halbes Jahr später äußerte sie sich so: Sie könne das nicht glauben. Traue das unserem Vater nicht zu. Es sei für sie schwierig, mit den Eltern *und* mit mir in Kontakt zu sein. Es gehe ihr schlecht. Ob ich nicht verstehen könne, wie schlimm das für sie sei? Ich würde schon verstehen, dass es auch für sie, Birgit, schwierig sei, das zu verkraften. Ich beschwerte mich bei ihr, dass sie nicht für mich da sei, noch nicht mal frage, wie es mir gehe. Keine Unterstützung anbiete. Und ich hätte das schon früher sagen sollen? Wie ich das damals hätte verkraften sollen, wenn auch damals ohne Unterstützung? Dann hätte ich mir sicher das Leben genommen. Selbst heute hätte ich Last, das nicht zu tun. Stille. Birgit hatte aufgelegt, wie mir bei einem Rückruf ihr Sohn bestätigte. Sie wolle nicht weiter mit mir reden.

Zuerst machte ich mir Vorwürfe, aber nach einem Gespräch mit Lisa kam ich zu der Erkenntnis: Wenn Maria und Birgit die Wahrheit nicht verkraften können, ist das nicht meine Verantwortung! Ich kann nicht auf sie zählen. Sie sind auf der Seite der Eltern. Birgit brach den Kontakt zu mir ab.

Natürlich sprach ich in der Therapie über das Telefonat. Ich erwähnte meine Angst, Birgit könne sich was antun. Die Therapeutin stellte klar: Wenn Birgit sich etwas antut, dann deshalb, weil Ihr Vater das gemacht hat. Und nicht, weil Sie es ihr sagten! – War es ein Fehler, sie anzurufen? – Auf keinen Fall! Schweigen und nicht Handeln ist auch mittun!, so die Therapeutin. Maria und Birgit seien doch um die 60 Jahre alt, hätten ihr Leben gelebt. – Aber sie reagieren wie 10-jährige, denen man die Eltern wegnehmen will! – So wären Lisa und ich früher auch drauf gewesen. Wir hätten nur durch und für die Eltern eine Existenzberechtigung gehabt. – Ja! Und wenn Sie Maria und Birgit die Eltern jetzt wegnehmen, haben sie keine Existenzberechtigung mehr. –

Nachdem ich auch Maria von den Übergriffen unseres Vaters erzählt hatte, fragte sie nach meinem Alter, auch danach, was er gemacht habe. Sie weinte. Ich erwähnte, Lisa habe mit ihm Ähnliches erlebt, und er habe uns gedroht. Maria sprach von ihrer Angst, ich könne nichts mehr von ihr wissen wollen, weil sie den Kontakt zu den Eltern nicht abbrechen könne. Sie mache sich auch Sorgen um mich. Nach dem Gespräch spürte ich Angst um Maria. Aber dann sagte ich mir: Denke nur an den (Italien-)Urlaub, Ann, die Angst vor der Dunkelheit usw., das reicht doch, um deine falsche Rücksichtnahme aufzugeben! Am nächsten Tag fiel mir ein, dass Maria von Selbstmord sprach. Wieder ängstigte ich mich um sie. Doch bald sagte ich mir selbst: Du hast nur gesagt, was man(n) dir angetan hat! So funktioniert der Schutz der Täter, wenn du aus Angst nichts sagst. So hat das schon jahrzehntelang funktioniert. Jetzt ist es aus damit! Das bist du der kleinen Ann schuldig! –

Bei einem weiteren Telefonat mit Maria sagte ich ihr, dass ich unsere letzten Gespräche überhaupt nicht gut fand. Dass es auch für sie ein Schock war, sei mir klar. Es sei aber nur noch um sie gegangen, aber *ich* hätte das doch erlebt! Ich müsse sie aber permanent schonen. Sie sagte, sie fühle sich verantwortlich. Das sei sie aber nicht, entgegnete ich. Eine Freundin habe sie gefragt, wieso ich ihr das denn hätte sagen müssen? Ich reagierte ärgerlich: Die Frage müsste doch lauten, warum hat der das gemacht? Außerdem, jetzt, wo alle erwachsen sind und in gefestigten Verhältnissen leben, selbst jetzt soll ich das noch nicht mal meinen Geschwistern sagen können! –

In den darauf folgenden Tagen tauchten bei mir immer wieder Ängste davor auf, was ich mit all dem auslöse. Ich dachte auch daran, dass meine Eltern doch jetzt schon alt sind.

Ungefähr einen Monat später entschloss ich mich, Birgit und auch Maria anzurufen. Beim Gespräch mit Maria weinte ich und sagte, ich würde mich alleine fühlen. Ich könne nicht verstehen, warum sie nicht mal anruft, nach mir fragt. Sie wollte wieder wissen, was er genau gemacht habe und ihn darauf ansprechen. Es könne ja ein Versehen gewesen sein, was ein Kind nicht auseinanderhalten könne. Ich reagierte sehr ärgerlich: Der hat mir gedroht, mich umzubringen! Und: Was der gemacht hat, das war kein Versehen, Maria! –

Einige Zeit später hörte ich von meiner Nichte Isabelle, dass ihre Mutter, also Maria, mit unserem Vater gesprochen habe. Ich rief Maria dazu an. Vater habe gesagt, er könne sich an „sowas" nicht erinnern. Ich müsse was falsch verstanden haben. Meiner Mutter gegenüber habe er mehrfach versichert, da war nichts. Also war meine Mutter misstrauisch ihm gegenüber, wenn sie mehrmals nachfragte? Weil ich nichts mehr von Maria hörte, rief ich sie nach rund zwei Monaten an und fragte, wieso sie keinen Kontakt mehr zu mir haben könne. Es wäre so schlimm für sie, sie könne nicht mehr richtig schlafen, würde es am liebsten gar nicht wissen. – Und ich hätte es lieber nicht erlebt! – Ich fragte sie, wie es für sie sei, wenn sie ihn jetzt sieht, er sie beim Begrüßen berührt. Ob sie dabei mal daran denke, was er mit ihrer kleinen Schwester …? Ob ich Kontaktabbruch zu den Eltern von ihr verlange? Daran hätte ich überhaupt nicht gedacht. Ginge für sie auch nicht, auch wenn ich dann den Kontakt zu ihr abbrechen würde. Sie müsse sich schützen. Ob sie nicht eher unseren Vater damit schütze?, fragte ich sie. Und das sagte ich noch: Das geht jetzt erst richtig los, Maria. Der wird meine Rache zu spüren bekommen! Ich werde die zwei nicht mehr schonen. Vielleicht an die Öffentlichkeit gehen, auch wegen Bernd und Adolf. Es könnte doch sein, dass sie auch bei ihren Töchtern übergriffig sind. – Das könne ich doch nicht machen! Damit könne ich die Verbindung zwischen ihnen zerstören. Ich entgegnete: Wir schützen doch mit unserem Schweigen *die Täter!* – Ich war sehr fest, bestimmt, aber zwischendrin auch lieb und verständnisvoll.

Wiederum zwei Monate später schrieb mir Maria eine Karte. Sie könne keinen Kontakt mit mir haben, weil sie weiter im Kontakt mit den Eltern sei und den nicht abbrechen könne. Sie brauche Zeit. Und sie habe mich lieb. Ich antwortete ungefähr so auf ihre Karte: Dass sie mit unserem Vater, der mir so viel Leid zugefügt habe, noch Kontakt haben kann und will. Mit meinem Peiniger! Aber nicht mit mir! Sie stehe also auf seiner Seite! Schütze ihn und verrate mich – und damit sich selbst!

Isabelle erzählte mir einige Zeit später, Maria habe ihr meine Karte zu lesen gegeben und gesagt, sie fände meine Worte sehr hart. Isabelle entgegnete ihrer Mutter, aber so sei es doch! Maria habe Angst, wenn sie sich von den Eltern zurückzöge, würden die sterben. So dachte ich früher auch. Nun bemerkte ich, wie absurd das ist! Wenn ich autonom bin, nicht mehr so, wie die Eltern mich haben wollen, sterben sie!

Unser Kontakt blieb lange unterbrochen. Nach mehr als zwei Jahren schrieb mir Maria einen Brief, in dem sie mich über den Krankenhausaufenthalt unserer Mutter informierte, die verwirrt sei und verwirrt von mir geredet habe. Ich rief Maria an und bedankte mich. Ich bin ihr heute noch sehr dankbar, dass sie mich informierte, denn so hatte ich die Gelegenheit, die ich auch wahrnahm, meine Mutter ohne meinen Vater zu treffen.

Einige Monate später rief Maria mich an. Sie sprach mit aufgeregter, fast kindlicher, aber auch lieber Stimme. Sie habe sich schon längst einmal melden wollen. Ich begann zu weinen, was ich ihr sagte. Ich sei so berührt von ihrem Anruf. Und: Ich habe dich vermisst. – Sie habe mich auch vermisst, entgegnete sie. Sie hätte das alles bis heute nicht verkraftet, wüsste nicht, wie sie das hinbekommen solle. Ich sagte ihr, ich hätte Zeit und auch Verständnis für sie, sei aber auch wütend und enttäuscht. Ich spürte während des Telefonats sehr viel Gelassenheit und eine gewisse Sicherheit. Bin ganz klar, bin auf dem richtigen Weg!

Aber es gab auch wieder andere Telefonate. Eines dieser Gespräche und den Bericht der darauf folgenden Therapiestunde möchte ich hier wiedergeben, weil beides so deutlich zeigt, welche Facetten die Reaktionen haben können. Bemerkenswert finde ich heute noch, dass meine Schwester davon sprach, *ich* hätte den Kontakt zu ihr abgebrochen, nicht umgekehrt.

Sie könne es nicht glauben, dass unser Vater das gemacht habe. Birgit ginge es ebenso. – Ich weiß es aber, Maria! Es ist für mich nicht von Bedeutung, ob er es zugibt, auch nicht, ob Ihr mir glaubt. Ich weiß es. – Ich äußerte Verständnis für sie, sagte aber deutlich, dass sie und Birgit nicht für mich da seien. Ich wisse noch nicht, wie ich mit ihren Reaktionen umgehe wolle. Danach spürte ich so eine Wut auf sie und Birgit, rief Maria deshalb nochmal an. Sie denke, so Maria, weil Adolf und Bernd das machten, unser Vater mich nicht beschützte, würde ich das auf ihn übertragen. – So ist es aber nicht, Maria, sondern so, wie ich es Dir gesagt habe! – Ich wäre es leid, dass sie sage, sie könne das nicht glauben! Das sei eine unglaubliche Reaktion von ihr! Ich hätte auch keine Lust mehr auf Verständnis für sie und Birgit. – Ich habe schon Verständnis dafür, dass es hart ist für Dich, für Euch. Aber *ich habe das erlebt!* Ich spreche es nur aus. Und was kriege ich ab? Wo seid Ihr? Hat eine von Euch mich auch nur einmal gefragt in den letzten Jahren: Wie geht es Dir jetzt damit, Ann? Mal nach mir geguckt? Wo ward Ihr? Sieben Jahre Therapie, jahrelang ging es mir sehr schlecht. Keine von Euch beiden war für mich da! Maria, ich habe es erlebt, und ich soll Rücksicht nehmen! Und, Maria, es ist doch so, Dir geht es jetzt seit fünf Jahren schlecht, obwohl wir uns nicht sahen. Du kannst in die Antarktis gehen, dass Thema ist trotzdem da! Und ich schweige auch nicht mehr. Damals musste ich still sein. Jetzt bin ich nicht wieder still! Unsere Brüder haben das auch gemacht, machen das vielleicht jetzt mit ihren Töchtern. Das geht doch immer weiter, wenn frau schweigt! Mir ging es jahrelang so schlecht, ich musste damit fertig werden, dass Vater das gemacht hat. Aber jetzt ist es raus! Und jetzt geht es mir so gut wie noch nie in meinem Leben, weil es raus ist, ich mir das angeguckt habe! – Ich fuhr fort, ich würde sie doch gut kennen. Wenn es um eine andere Frau ginge, hätte sie sich nie so verhalten. Ja, das hätten Isabelle und Ira auch gesagt. Ich betonte zum Schluss, ich wolle Kontakt mit ihr, aber nur unter der Bedingung der Offenheit.

Was war ich danach froh. Endlich! Endlich habe ich die Stärke, die es einfach braucht, und die Sicherheit in mir, dass ich mir selbst glaube und das alles offensiv vertreten kann!

Davon berichtete ich ausführlich in der folgenden Therapiestunde. Die Therapeutin betonte, dass es gut war, einerseits zu mir zu stehen, andererseits aber auch Marias Befinden im Auge zu behalten: Wenn Maria sagt, sie würde sich am liebsten das Leben nehmen, dann ist da ja was bei ihr, dass es zu viel für sie ist! – Es wäre wichtig, mir weiterhin darüber im Klaren zu sein, dass es zum einen meine Position als erwachsene Person gäbe, die zu sich stehen wolle, aber auch im Blick zu haben, dass Maria nur bedingt belastbar sei. – Und es ist wichtig, dass Sie sich nicht von ihr abhängig machen. – Ob sie mir das glaubt oder nicht, meinen Sie das? – Ja! – Also es ist ja wirklich spannend und auch phantasievoll, wie Maria das für sich zurechtlegt. Es waren die Brüder. Nur Übertragung auf den Vater. Damit ist der gerettet. Und sie kann dann auch Mitleid mit Ihnen zulassen, weil Sie es erlebt haben. – Für Maria sei es existentiell, die Eltern für sich zu behalten. Die K. sah mich eindringlich an: Die können Sie Maria nicht wegnehmen! Weil sie dann nichts mehr hat, weil sie keinen Ersatz für sie hat. Das kann sie nicht aushalten. –

Einige Monate später schrieb mir Maria, sie brauche Zeit, ob ich auf sie warten könne? Das bejahte ich. Nach weiteren vier Monaten telefonierte ich länger mit ihr. Sie meldete sich danach nur selten. Wir kamen in der Therapie zu der Erkenntnis, dass ich für Maria eine Bedrohung bin und bleibe. Die Therapeutin formulierte es ungefähr so: Weil Sie nicht mehr die Augen verschließen. Weil Sie die Familie mit Abstand betrachten. Weil Sie die Dinge ansprechen und nicht mehr schweigen. –

Heute stehen Maria und ich in gutem Kontakt, erfreuen uns daran, wieder zueinander gefunden zu haben. Dabei ist unser Verhältnis ein ganz anderes, viel erwachseneres als zu der Zeit, bevor ich mich diesem Thema stellte. Und das ist gut so!

Natürlich habe ich mir Gedanken darüber gemacht, warum meine beiden ältesten Schwestern so reagierten. Meine Schwester Lisa, ich, die Töchter von Maria, auch meine Therapeutin mutmaßen fast zwangsläufig, dass Maria und Birgit ebenfalls sexuelle Gewalt mit unserem Vater erlebten. Dass sie deshalb alles abblocken mussten. Ihr Weggucken und Abtauchen habe aber vielleicht auch (noch) andere Gründe, so damals meine Therapeutin, nämlich: Sonst müssten sie sich damit auseinandersetzen, bei ihren Kindern, ihren Töchtern vielleicht selbst nicht richtig hingeschaut zu haben ... –

Und nun komme ich zu meiner lieben und unterstützenden Schwester Lisa. Ich erwähnte bereits, dass sie ebenfalls sexuelle Gewalt durch unseren Vater erlebte und dies zeitgleich mit mir in einer langjährigen Therapie bearbeitete. Wir unterstützten uns gegenseitig. Sie konnte sich oft an Begebenheiten o.ä. genauer erinnern, insbesondere was mich anbetraf, weil sie vier Jahre älter ist. Ihre Erinnerungen halfen mir sehr. Als ich sie einmal fragte, ob es ihr nicht zu viel sei, wenn ich mit all meinen Erinnerungen oder Schilderungen der Therapiestunden zu ihr komme, sagte sie zu meiner großen Freude, im Gegenteil! Sie könne gar nicht genug darüber reden. Endlich würden wir über all das reden.

Aber es gab auch Enttäuschungen. So wäre ich gerne zusammen mit ihr zu unserem Vater gefahren, hätte ihn gerne gemeinsam mit ihr zur Rede gestellt. Wir beiden Schwestern, die sich nichts mehr gefallen lassen, die sich endlich getrauen, gegen unseren vermeintlich übermächtigen Vater vorzugehen! Das wäre doch was gewesen! Aber das wollte sie nicht. Es fiel mir sehr schwer, das zu akzeptieren, doch ich erkannte, dass jede ihren Weg geht und dass diese Wege nicht die gleichen sein müssen.

Dennoch bleibt für mich ihre wunderbare Unterstützung, dass sie mir oft mit viel Liebe und Verständnis half, all das Schlimme zu ertragen, dass wir viel gemeinsam gekämpft haben und uns dann auch gemeinsam freuten und das Erreichte feierten.

Mit meinem Bruder Heiner habe ich während der Jahre der Therapie nicht oft über dieses Thema gesprochen. Lisa hatte ihm schon recht früh von den Taten unserer Brüder erzählt, mit meinem Einverständnis. Sie bat ihn auf meinen Wunsch hin, er solle nicht darüber reden wegen meiner Angst vor Reaktionen; das versprach und hielt er. Er fragte nach meinem Alter, überlegte, wie alt die Brüder waren und meinte zu Lisa: Die wussten, was sie taten. Das waren keine kindlichen Spiele. – Er versuchte aber auch, die Brüder zu verstehen von wegen „bei dem Vater"! Die hätten versucht, ihrer männlichen Rolle zu entsprechen. Auch unsere engen, kleinbürgerlichen Verhältnisse hätten dazu beigetragen, unser mangelnder Kontakt nach außen. Bei mir kam das erstmal so an, er hat mehr Verständnis für die Brüder als Solidarität mit mir und Mitgefühl für mich. Die Therapeutin vermutete, dass er sich damit schütze, um nicht alles einzureißen, was die Familie anbeträfe.

Als ich ihn nach ein paar Jahren fragte, wieso er sich nie nach mir erkundigt habe, nie nach mir fragte, antwortete er, er habe so lange gebraucht, das zu verdauen. Es sei doch auch sein Vater! Er habe, wenn auch nicht selbst betroffen, damit fertigwerden müssen, dass unser Vater mich und unsere Schwester Lisa sexuell traktiert hatte. In einem weiteren Gespräch sagte er, für ihn sei es völlig unverständlich, wie ein Mann, zudem der Vater, „das" mit Kindern, mit seinen Töchtern machen könne. Er frage sich aber auch, wie die Frauen es dann später schafften, überhaupt etwas mit Männern zu tun zu haben. Tja ...

Irgendwann erzählte mir Heiner, dass er mit unserem Vater über Lisa und mich geredet habe. Er habe unserem Vater gesagt, dass er ihm das nie verzeihen könne. Natürlich habe unser Vater alles abgestritten. Nachdem ich davon in der Therapie berichtete, meinte die Therapeutin: Jetzt bin ich ganz berührt. Jetzt haben Sie doch das erlebt, was Sie sich immer von Heiner wünschten. Dass er gegenüber Ihrem Vater für Sie eintritt! – Ja!, antwortete ich freudig lächelnd.

Die nächtlichen Anrufe meines jüngsten Bruders Leo und seine Androhungen von Gewalt erwähnte ich bereits kurz im Zusammenhang mit den Erinnerungen an die Taten meines Vaters. Bald nach dem Eingang des Rechtsanwaltsbriefes meines Vaters begannen diese Drohanrufe. Sie lösten große Angst bei mir aus. Ich löschte daraufhin meinen Eintrag im Telefonbuch und bat meine „guten" Geschwister, Leo nicht meine Telefonnummer zu geben. Er hatte jeweils mit verstellter Stimme gesprochen, war aber nicht nur von mir, sondern auch von meinem Mann sowie von Lisa und ihrem Mann klar erkannt worden. Beim ersten Anruf war ich gegen Mitternacht nach Hause gekommen, alleine in der Wohnung, hörte den Anrufbeantworter ab und ihn mit verstellter Stimme: Du Nutte! Ich mach dich kalt! – Schreck und Schock und Herzklopfen bis zum Hals. Zu meinem Glück erreichte ich eine Freundin, die mir half, mich einigermaßen zu beruhigen, und die mir anbot, zu mir zu kommen und bei mir zu schlafen. Ich weinte erstmal sehr. Fühlte mich sehr bedroht! Das war kaum zum Aushalten. Ich will vor denen aber keine Angst haben! Aber ich muss und werde mich gut schützen. Lisa meinte am nächsten Tag, nachdem ich ihr die Aufzeichnung auf dem Anrufbeantworter vorgespielt hatte, der sei zu allem fähig. Wir sollten lieber nichts mehr unternehmen.

Meine Freundinnen reagierten schockiert und absolut unterstützend. Eine Freundin meinte, ich könne immer, auch nachts, zu ihr kommen. Und bestärkte mich: Ann, Du darfst jetzt nicht aufgeben! – Ich könne auch, solange Philipp nicht da sei, bei ihr einziehen. Eine andere Freundin holte mich zu einer Feier ab und brachte mich wieder nach Hause. Ich bin nicht alleine damit! Das werde (und habe!) ich nie vergessen! Auch meine Therapeutin unterstützte mich. Ich dürfe jetzt nicht aufgeben! Die müssten ja viel Angst haben, erst der Rechtsanwaltsbrief, jetzt dieser Anruf. Aber es ginge jetzt auch darum, dass ich mich gut und sicher fühlte. Ich wolle auch nicht aufgeben. Wolle aber genau schauen, jeden Schritt sehr gut überdenken. Müsse mich doch beschützen! – Ja! –

Zwei Wochen später war Leo nachts wieder auf dem Anrufbeantworter, wieder mit unflätigen Worten und Todesdrohungen. Und wieder erschreckte mich das sehr! Dennoch reagierte ich insgesamt viel ruhiger. Ich sprach kurz darauf mit meiner Rechtsanwältin darüber, was jetzt am besten zu tun sei. Ich kam zu der Ansicht, erst einmal nichts gegen ihn zu unternehmen. Meine Therapeutin erzählte mir damals von einer Klientin, die ihren Onkel auf seine Taten angesprochen hatte, der sie jetzt ebenfalls bedrohte. Und eine Klientin ihres Kollegen habe ihren Peiniger angesprochen, der sie jetzt verklagt habe, weil sie angeblich falsche Behauptungen aufstelle. Sie, die Klientin, wurde daraufhin erkennungsdienstlich behandelt. – Das muss man sich mal vorstellen! Sie wird so behandelt!, entrüstete sich zu Recht meine Therapeutin.

Zum Jahresende, an einem späten Abend, erreichte mich wieder ein Anruf, wieder ein Mann mit verstellter Stimme. Dieses Mal hatte ich den Hörer abgenommen, weil ich zuhause war. Und wieder erkannte ich Leos Stimme, sagte deshalb bald: Du Arschloch!, und legte wütend auf. Mein Mann war dabei. Er gab keine Ruhe: Lass' uns Leo anrufen! – Und das machte er. Leo stritt ab, angerufen zu haben. Ich nahm den Hörer, sagte, dass man genau seine Stimme erkennen könne. Ich würde ihn anzeigen. Solle ich doch tun, entgegnete er. – Du kannst mir nichts vormachen! Ich kenne Dich!, sagte ich lachend und legte auf. Das war gut. Aber ich spürte eine unglaubliche Wut! Danach gab es keine weiteren Anrufe mehr, was nochmal bestätigte, es war Leo!

Als Heiner von den Drohanrufen erfuhr, reagierte er sehr geschockt. Auch Maria, der ich Jahre später von Leos Drohanrufen erzählte. Die Gründe für Leos Anrufe kann ich nur ahnen oder vermuten: Entweder hat er selbst auch Übergriffe erlebt und will nicht daran erinnert werden oder er war selbst übergriffig.

Bei einer Jahre später stattfindenden Geburtstagsfeier meines Vaters im Altenheim reagierte Leo sofort wieder unflätig. Mich hatte niemand zu dieser Feier eingeladen; ich erfuhr davon durch Lisa. Ich wolle nicht, dass das wieder zum Tabuthema wird, sagte ich in der Therapie. Wegen mir selbst nicht, aber auch wegen meiner Nichten, sollten die wirklich Ähnliches erlebt haben. Dann solle ich das auch ernst nehmen, bestärkte mich die Therapeutin. Ich wolle keine Randale machen, aber dort präsent sein. – Wieso haben die mich nicht eingeladen, z. b. Maria und Birgit? Das passt doch zu meinem Gefühl, dass das Thema wieder zum Tabu wird. Nicht eingeladen zu werden ist wie nicht da sein, als gäbe es mich nicht! Aber wenn ich mir vorstelle, Vater, Adolf und Leo sind dort ... Das ist hart! Das ist zu viel! – Das verstehe sie, aber ich hätte doch jetzt schon Übung darin, würde sicher auch etwas sagen, meinte die K. – Ich würde gerne ruhig, gelassen, groß und aufrecht dort erscheinen. Mich gibt es noch! Ich will mir Euch alle mal ansehen. Ich bin da! Und damit ist auch das Thema noch da. – Ja, das wäre doch gut. –

Nach dieser Therapiestunde notierte ich in meinem Tagebuch: Nur wenn ich daran denke, bin ich sofort aufgeregt. Wenn ich dahin gehe, muss ich etwas sagen, von wegen nicht eingeladen, aber mich gibt es! Vielleicht auch Leo ansprechen, seine Drohungen etc. Der Entschluss ist gefasst, ich fahre! Nicht um dort herum zu wüten, das würde für mich nicht stimmen. Außerdem solidarisieren die sich dann und wenden sich alle gegen mich. Und dann notierte ich dies in mein Tagebuch:

Ich habe es geschafft, gut *geschafft!*

Und so berichtete ich in der Therapie davon:

Ich war gestern tatsächlich in B.! – Oh, oh!, sie lächelte. Ich erzählte ausführlich, zuerst von der entscheidungsfindung, dabei immer wieder angst und nee, das schaffe ich nicht. Bald klar, besser ohne begleitung durch eine freundin dort hinzufahren: So war es eigentlich von anfang an in mir, ich will und muss das alleine machen! – Ehem! – Das war auch gut so! ... Ich hatte solche angst, als ich auf die tür zuging, die aufmachen wollte. – Ja! Ja, das glaube ich!, die K. fast stöhnend.

Leo: Wer hat denn die eingeladen? – Ich nicht!, Adolf. – Nee, wusste ich nicht!, Maria. Ich ging zu vater und mutter, begrüßte beide. – Ich sah zur K. hin, sie war „ganz dabei"! Als ich von Leos wüster beschimpfung sprach von wegen mich rauswerfen, mir „aufs maul" hauen zu wollen, weil ich sagte, ich komme auch ohne einladung, da staunte sie. Erzählte auch von Marias unterstützenden worten nach Leos frechheiten. Nun hätte ich es doch geschafft zu kommen, so Maria, solle jetzt nicht zurückweichen, weil ich fast gegangen wäre. Die K. staunte sehr. – Ja, das hätte ich Maria auch nicht zugetraut! – Sie habe mir damit geholfen, denn ich wäre mir kurz sehr alleine vorgekommen; hätte es dann aber geschafft, wieder gut bei mir zu sein.

Ich hätte später leo zur rede gestellt. Er wieder so aggressiv. Wovor er angst habe? Er habe keine angst! Habe mir auch nicht auf den Anrufbeantworter gesprochen. Hätten ihn aber einige erkannt. Ich dabei sehr gelassen.

Mein vater habe sich bei der verabschiedung für meinen besuch bedankt und gesagt, mehr müsse er doch nicht sagen? – Nee, aber ich ja wohl auch nicht! – Die K. lächelte. – Danach war ich völlig platt. – Ja, das ist doch auch verständlich. – Jetzt hätte ich wieder so ein wattegefühl im kopf wie damals, wenn ich von den übergriffen redete. – Ehem! – Und ich fühle mama und hab' mamabedürfnisse ..., sagte ich leise und nahm mir das kleine rote kissen und schmiegte es an mich.

Aber es wäre auch in mir, ich ballte die faust: Ich hab's geschafft! Total klasse! Und es war genau richtig, dass ich dort war und zu mir stand! – Ehem! – Ich

> *will, dass das thema* nicht *wieder verschwindet! – Ja! – Aber jetzt bin ich doch platt. – Ja, das glaube ich! Das war doch auch so anstrengend! – Sie pustete die luft aus, schüttelte den kopf: Mensch, mensch! Das war ja was!, und meinte, das werde viel bewegen, bei den anderen, aber auch bei mir. Und sagte dann, lächelnd: Toll! Toll! –*
>
> *Wir schwiegen. Ich schloss meine augen, sie hatte ihre auch geschlossen. Bin bei der mama, so mein gefühl. Am Ende der stunde sagte sie, ich könne noch bleiben, sie würde in den anderen raum gehen. Ich musste erst einmal heftig weinen.*

Mir war es lange Zeit sehr wichtig, dass das Thema nicht wieder „unter den Teppich" gekehrt wurde. Ich musste nicht mehr darüber reden, wollte aber auch nicht, dass es in Vergessenheit gerät, wieder verdrängt wird. Der Wunsch seitens der Familie, das Thema wieder „zu vergessen", war und ist z. T. weiterhin da.

Bevor ich nach den Reaktionen meiner Geschwister zu denen meiner mir wichtigen Nichten komme, will ich auch kurz meinen damaligen Mann Philipp und sein Verhalten erwähnen. Er reagierte sehr geschockt, als ich ihm von den Taten meiner Brüder und später von denen meines Vaters erzählte. Er weinte mit mir und unterstützte mich liebevoll und über Jahre hinweg. Natürlich reagierte er auch wütend. Er wäre meinen Vater und die beiden Brüder am liebsten angegangen, auch körperlich. Aber ihm war auch klar, dass das keine wirkliche Option war. Auf die Auswirkungen auf unser Zusammenleben will ich hier nicht weiter eingehen.

Meine Nichten Isabelle und Ira, die Töchter meiner ältesten Schwester Maria, sind mir seit ihrer Geburt sehr nahe. Insbesondere ab ihrem Alter von rund 10 Jahren habe ich sie oft gesehen, weil ich zu diesem Zeitpunkt in ihre Nähe zog und die beiden an manchen Wochenenden bei mir waren, wir öfter was gemeinsam unternahmen. Unser Kontakt ist seit diesen Jahren sehr liebevoll und von gegenseitiger Achtung geprägt. Natürlich habe ich auch ihnen von der sexuellen Gewalt meiner Brüder und meines Vaters nach und nach erzählt. Beide glaubten mir ohne jeden Zweifel. Isabelle drückte es einmal so aus: Wer Dich gut kennt, der weiß, dass es stimmt, wenn Du das sagst. – Für sie seien Männer, die so etwas mit Mädchen machten, Verbrecher. Sie wolle mit denen nichts zu tun haben. Also könne sie auch mit meinem Vater nichts mehr zu tun haben. Ira äußerte sich ähnlich. Aber für beide war es auch furchtbar, da es ja um ihre Onkel und besonders um ihren Opa ging. Meine offenen Worte bzw. die geschilderten Übergriffe beeinträchtigten die beiden sehr. Sie brauchten Zeit, dies alles zu verkraften. Ich überließ es ihnen, wie sie sich zu allem verhalten, ob sie meinen Vater treffen wollten oder nicht. Ich machte ihnen keinerlei Druck, das war mir selbst sehr wichtig. Insgesamt waren meine Nichten sehr, sehr solidarisch und unterstützend.

6.2 Ich konfrontiere Bernd mit seinen Taten

Erst im achten Jahr meiner Therapie konnte ich diesen Schritt unternehmen, den ich in der Therapie gut vorbereitete. – Ich muss Bernd ansprechen, ich will ihn konfrontieren. Ich muss das für mich tun! – Ja! – Aber ich habe so eine Angst davor! ... Vielleicht nehme ich jemanden mit, zur Unterstützung. – Ich solle nachspüren, wer dafür in Frage käme. Das konnte ich erst einmal nicht beantworten. Ich stellte Überlegungen darüber an, was ich mir von diesem Gespräch und von ihm erwartete oder wünschte. Weinend hatte ich gesagt: Am liebsten wäre mir, er würde es zugeben und sagen, es tue ihm leid. – Ja... Ehem!, weich von der K.

Bald darauf war ich ganz klar: Ich muss das alleine machen! Und ich muss ihn bald anrufen. Und egal, wie er reagiert, ich weiß, was er gemacht hat. – Ich wolle ihn nicht dämonisieren, damit ich ein Recht auf meine Verletzungen hätte. Ich wolle ihm aber sagen, wie furchtbar das für mich war. Von all den Jahren der Angst und Panik reden, was das für das Leben eines kleinen Mädchens und dann einer Frau bedeutet. Von all den Auswirkungen wolle ich reden. – Ja! – Aber ich hab' auch Angst, dass er blöd reagiert, dass ich dann nicht weiter weiß. – Vielleicht sollten Sie sich jetzt schon vornehmen, dass Sie dann in einen anderen Raum gehen und schauen, ob noch alles gut für Sie ist, ob es für Sie noch stimmt. Und wenn nicht, das Gespräch beenden. Sie können ja sagen, es geht nicht mehr, lass' uns ein anderes Mal weiterreden. – Ja. –

Ich hatte, entgegen meiner ersten Vermutung, doch kein Problem damit, dass wir uns bei mir zuhause treffen und ich dann mit ihm alleine bin. Ich spürte weiterhin große Angst und doch war klar, ich will das, muss das für mich machen! Das ist für mich wie Wiedergutmachung!

Und Bernd kam. Ich hatte angekündigt, ich wolle mit ihm darüber reden, warum wir seit Jahren, nach zuvor gutem Kontakt, nun keinen mehr hatten. Aufgeregt und dennoch ruhig, nach anfänglichen Schwierigkeiten, wie damit beginnen, sprach ich von meiner Erinnerung mit ihm, seinem sexuellen Übergriff. Daran könne er sich nicht erinnern. Ich betonte, dass ich damals zu ihm sagte, Adolf mache das auch. Bernd wiederholte, er habe keine Erinnerung, wirklich nicht, sonst würde er sich sofort entschuldigen. – Ich hab Dir das damals mit Adolf gesagt, weil ich hoffte, Du willst nicht so sein wie er. Und entsprechend reagiertest Du, sagtest, es sei nicht ok und käme nicht mehr vor. Und daran hast Du Dich gehalten. – Wieder sein Erstaunen: Nein, ich kann mich wirklich nicht daran erinnern. Muss ich wohl verdrängt haben, sonst hätte ich Dir spätestens nach ein paar Jahren gesagt, tut mir leid, dass ich das gemacht habe. –

Du weißt Bernd, wenn ich das heute zu Dir sage, dann stimmt es! – Ja, das weiß ich. – Er fragte, was Adolf denn gemacht habe. Ich sagte nur, dass er mich jahrelang nicht in

Ruhe ließ. Aber auch unser Vater ... – Wie, der auch? – Ja! – Auch so wie ich? – Nein, viel schlimmer! Vergewaltigt. – Nachdem er noch etwas zu unserer Mutter sagte, wiederholte er: Auch wenn ich mich wirklich nicht daran erinnere, es tut mir leid. Das tut mir wirklich leid! – Ich schwieg, berührt, froh. Bald darauf sagte ich, ich wolle das Gespräch jetzt gerne beenden, es wäre genug für heute. Ihm ging es sicher nicht anders. Ich war insgesamt ganz ruhig und ganz klar. Beim Verabschieden gaben wir uns die Hand. Er fragte, ob er mich umarmen dürfe. Nach kurzem Zögern: Ja. – Leichte Umarmung. Anlächeln. Er: Schwesterchen! – Ich, lächelnd zurück: Ja, Brüderchen. –

Ich war sehr berührt. Konnte jedoch nicht fühlen, ob es gut war. Erst am nächsten Tag und noch mehr beim Berichten in der Therapie wurde mir bewusst, was ich geschafft hatte. Ich erwähnte auch, dass Bernd mich immer ansah, mir in die Augen habe schauen können. Irgendwann sagte ich sehr weinend: Er hat es also doch gesagt! Er hat gesagt: Es tut mir leid! – Ja. – Ich hätte ihm alles sagen können, auch wie schlecht es mir jahrelang ging; hätte von der Therapie erzählt. Zum Abschluss, wieder sehr weinend, sagte ich: Was wir dadurch alles verloren haben, er und ich! – Ja! Ehem! – Die Therapeutin reagierte berührt und äußerte sich erfreut über den Verlauf des Treffens und darüber, wie gut ich das Gespräch geführt hatte. Auch dass Bernd so toll reagieren konnte. Und ich spürte immer mehr meine Zufriedenheit, fühlte mich zunehmend glücklicher.

6.3 Ich konfrontiere Adolf mit seinen Taten

Kurz vor Fertigstellung dieses Buches traf ich Adolf anlässlich einer Geburtstagsfeier von Maria. Fast alle Geschwister nahmen daran teil. Mir war zuvor nicht wohl zumute und während der Feier spürte ich leichte Traurigkeit. Und bald war klar: Ich muss Adolf ansprechen!

Schon am nächsten Tag erbat ich mir bei Heiner die Telefonnummer von Adolf und rief ihn mit starkem Herzklopfen an.

Ich wolle ihm sagen, warum ich auf Abstand zu ihm gegangen sei. Ob er sich erinnere, was er früher, als ich ein kleines Mädchen war, mit mir gemacht habe? Dass er mich begrapscht habe! Ich beschrieb kurz, was er gemacht und von mir gewollt hatte. – Nein, das habe ich nicht gemacht! – Doch, Adolf, das hast Du! – Nein! Auf keinen Fall! Ich kann mich jedenfalls nicht erinnern. ... Das tut mir leid. Das tut mir wirklich leid! Dann entschuldige ich mich. – Ich war völlig perplex über seine ruhige und angenehme Reaktion. Ich konnte es kaum glauben. Er wiederholte seine Entschuldigung. Ich erwähnte, dass er mir Geld zusteckte. Dass auch unser Vater, als ich noch kleiner war... Adolf wieder, es tue ihm leid. Ich beendete bald darauf das Gespräch, immer noch überrascht und aufgewühlt. Gerade von ihm hatte ich doch etwas ganz Anderes erwartet.

Nach einigen Minuten rief Adolf mich an, betonte, wie sehr es ihm leid tue. Er fragte, etwas verschreckt, ob ich ihn jetzt anzeigen wolle? Nein, das sei nicht mehr möglich, da verjährt. Ob ich das seiner Frau und seinen Kindern gesagt hätte? Nein. Daraufhin fragte ich ihn, ob er das auch mit seiner Tochter gemacht habe? Er reagierte sehr erschüttert. Nein, auf keinen Fall, so etwas würde er nicht machen. Ich sagte ihm noch einmal, was unser Vater mit mir und Lisa gemacht hatte. Schilderte ihm, wie schlecht es mir jahrzehntelang ging, sprach von meiner Therapie. Adolf dann völlig aufgelöst und hilflos.

Nach vielen Entschuldigungen seinerseits sagte ich ihm, dass ich diese nun annehmen könne, was ihn etwas beruhigte. Ich war so, so froh, auch das geschafft zu haben.

Später rief mich Heiner an, den Adolf zuvor angerufen hatte. Adolf habe gesagt: Dann bin ich wohl ein Schwein?! – Ja, das sei er, hatte Heiner geantwortet. Er erläuterte Adolf, warum dieser sich nicht an seine Taten erinnern könne. Dass es sich um Verdrängung handele, wie die funktioniere und dass Adolf dadurch mit sich selbst habe weiterleben können.

Heiner sagte ihm auch, er fände Adolfs Verhalten sehr gut, hätte, wie ich, eine ganz andere, nämlich schlimme Reaktion erwartet. Adolf hätte sich viel besser verhalten als unser Vater, der sogar einen Rechtsanwalt eingeschaltet habe. Und auch im Gegensatz zu Leo, der mich mit dem Leben bedrohte. Heiner berichtete Adolf von mir, von all den schlimmen Auswirkungen, den schweren Jahren, der Therapie und wie lange ich gelitten hätte. Er forderte Adolf auf, mich erst einmal in Ruhe zu lassen, weil das auch für mich heftig sei.

Ich war außerordentlich berührt davon, wie toll mich Heiner vertrat. Das sagte ich ihm natürlich. Es sei ihm eine Herzensangelegenheit gewesen, entgegnete er.

Ich dachte im Nachhinein, wie schwer es war, Adolf anzusprechen, und erinnerte mich daran, wie völlig unvorstellbar es damals war, meinen Vater konfrontieren zu wollen.

Heiner und ich sorgten uns, dass Adolf sich etwas antun könnte. Heiner hatte ihm zugesichert, er sei immer für ihn da; sie telefonierten deshalb mehrmals miteinander. Heiner ermahnte ihn, er solle die Verantwortung für sein Tun übernehmen, sich jetzt nicht selbst zum Opfer machen. Er müsse das aushalten, sich wenn nötig professionelle Hilfe holen.

Bei einem Gespräch mit meiner ehemaligen Therapeutin sagte sie, es könne sein, dass Adolf schon bei bzw. nach seinen Übergriffen diese sofort abgespalten, also verdrängt habe.

Monate später rief mich Adolf anlässlich meines Geburtstages an, er wiederholte, wie sehr es ihm leid tue und dass er mich gerne treffen wolle. Dies sagte ich ihm zu. Wegen eines Krankenhausaufenthaltes von Adolf konnte das Treffen noch nicht stattfinden.

Nachdem ich Adolf konfrontiert hatte, schon nach dem ersten Gespräch, bemerkte ich, dass ich ihn nicht mehr verteufeln musste wie während der Therapiezeit. Dass ich ihn kompletter, über seine Taten hinaus sehen kann. Das zuzulassen fühlt sich sehr gut an!

6.4 Ich konfrontiere meine Eltern – ihre Reaktionen

In den Kapiteln zu den Eltern habe ich bereits von ein paar Telefonaten mit ihnen berichtet. Nun gehe ich ausführlich auf die Reaktionen meiner Eltern ein. Mir ist wichtig zu zeigen, dass sich die Konfrontationen mit ihnen über Jahre hinzogen und welche Entwicklung sie genommen haben. Ich staune heute noch, wie hartnäckig ich dabei war – trotz aller Ängste und Bedrohungen. Ich gab nie auf, auch wenn ich zwischendrin lange Zeit nichts unternahm.

Zuerst sprach ich meinen Vater, innerlich sehr aufgewühlt, telefonisch an, nachdem ich mir darüber, was er mir angetan hatte, immer klarer geworden war. Wie nicht anders zu erwarten, sagte er, er habe das nicht getan. Anfangs war ich noch fest in der Stimme, später änderte sich das, leider. Deshalb brach ich das Gespräch ab. Davon berichtete ich ausführlich in der nächsten Therapiesitzung. Ich sprach über meine Furcht vor Reaktionen der Familie, dass es mir aber wichtig sei, dass ich gut bei mir bliebe. – Ja!, bestätigte die Therapeutin und sie betonte, ich trüge keine Verantwortung dafür, wie es den Anderen jetzt damit gehe.

Auf diesen Anruf kamen keine Reaktionen seitens meiner Familie. Einige Monate später schaffte ich es, auch meine Mutter anzurufen, was mir vorher kaum möglich erschien. Wieder war ich sehr aufgeregt. Vater hörte wohl zu. Ich sagte zu meiner Mutter: Du hast das doch mitgekriegt! Hast Dich rumgedreht, es nicht sehen wollen. Ich finde Euch so zum Kotzen! Und Vater ist für mich ein Verbrecher! – Sie glaube mir nicht, wolle aber nicht sagen, dass ich lüge. Ich sprach von meinen Herzproblemen, seit ich Jugendliche bin. Die hätte ich doch nicht von ungefähr! Sie wolle davon nichts mehr hören. Es ging hin und her, teilweise sehr scharf. Ich betonte, dass es in erster Linie um Vater ginge, nicht um sie. Er sei der Täter. Was ich mir davon verspräche? – Dass endlich mal die Wahrheit über unsere Familie ausgesprochen wird! Und das, was ich erlebt habe, was mich kaputt gemacht und jahrelang fast in den Selbstmord getrieben hat! – Ich sei enttäuscht von ihr als Mutter. Ich sprach auch von Adolf, dass er mir jahrelang keine Ruhe ließ mit seinen Übergriffen. Warum ich nicht früher was gesagt hätte? Sie hätte das dann unterbunden. – Weil ein Kind das nicht aussprechen kann. Aber ich habe es Dir zu verstehen gegeben. Du fragst mich, warum heute erst? Dazu muss frau sehr stark sein. Du siehst ja, was heute abgeht, wo ich das sage! Du fragst ja jetzt auch nicht nach mir, wie das für mich war, welche Auswirkungen das hatte usw. Warum kannst Du das nicht mal fragen, Mutter? – Ich weiß es selbst nicht. Kann es Dir nicht sagen. – Es ist wie früher. Damals war ich damit alleine und jetzt auch. Bis auf Lisa. Im Grunde will es keiner wissen, wird weiter verdrängt, sagte ich traurig. Und fügte noch hinzu: Sage Vater, ich weiß, dass er das nicht vergessen hat. – Ich war dann irgendwann so sauer und ihr Geschwätz so leid, dass ich einfach auflegte.

In der darauf folgenden Therapiestunde berichtete ich aufgewühlt und wütend von diesem Telefonat. Ich hätte meine Wut mal da abladen wollen, wo sie hingehört. Das habe mir sehr gut getan, aber: Ich habe meine Mutter früher nicht erreicht und heute auch nicht. Die ist wie Beton. Aber es geht nicht um die! Es geht um mich! – Ja! – Die Therapeutin weiter dazu:

> *Ich kann sie nur darin unterstützen, dass es nicht um die, sondern um* sie *geht! Darum, dass sie sich glauben. Ich will ihnen nicht die illusion rauben, aber die werden das weiter verdrängen. Aber es geht ja um sie! Sie sehen selbst, wie schwer es ist, zu sich zu stehen. Wie viel kraft das kostet. – Ich sagte, dass ich jetzt erst richtig wütend sei! Herzklopfen vor wut! – Ja, das ist doch gut so! Die muss doch mal raus! –*

Einige Tage später rief meine Mutter an. Mehrmals sprach sie auf meinen Anrufbeantworter; sie hätten sich überlegt, eine Aussprache wäre gut. Ich fühlte mich nicht in der Lage zu einer Aussprache. Aber das muss ich doch, muss doch zu allem stehen! Nein, ich bestimme, wann ich eine Aussprache will. In der Therapie sprach ich von meiner Angst, den Eltern zu begegnen, Angst, mich wieder so klein zu fühlen wie früher. Bei mir habe sich doch viel verändert, so die Therapeutin, also verändere sich jetzt auch was, wenn ich denen gegenüberträte. Es sei wichtig zu gucken, was ich jetzt wolle. Ich solle das nur angehen, wenn es für mich stimme.

Kaum war ich nach dieser Stunde zuhause, rief meine Mutter wieder an; sie sprach auf den Anrufbeantworter. Ich zitiere aus meinem Tagebuch:

> *Ich hörte einen moment zu. Sie hätte ihn immer schützen wollen, das wäre aber nicht gut, besser, wenn er selbst mit mir spräche, das wolle er jetzt tun. Ich schaltete mich ein. Das wäre ja ganz was neues. Wieso sie dann anrufe? Ich war hart und unerbittlich und sie zahm, wenn es ihr auch merkbar schwerfiel. – Wenn er mit mir reden will und mir sagen will, er hat nichts gemacht, dann brauchen*

wir nicht miteinander zu reden. Wenn vater endlich zu dem steht, was er getan hat, dann soll er *mir das sagen! Ich rede mit ihm also nur, wenn er dazu steht. – Meine mutter musste sich sehr bemühen, sich zurückzuhalten, weiter ruhig zu bleiben.*

Zu meiner großen Überraschung äußerte sie sich dann so, dass sie damit rechneten, ich würde meinen Vater anzeigen – was wegen Verjährung leider nicht möglich war. Sie würde auch weiterhin zu ihm stehen, auch den letzten Weg mit ihm gehen. Ich antwortete, äußerlich gelassen, dass mich das nicht mehr interessiere. Auch nicht mehr, wie es Maria und Birgit ginge. – Ihr fragt auch nicht, wie es mir geht! – Sie betonte noch einmal, „Vati" würde mich anrufen.

Auch davon berichtete ich in der Therapie: Ich finde es so klasse, dass ich das geschafft habe, sagte ich strahlend, das hatte ich mir nicht zugetraut. Ich habe zu mir gestanden! Und zur kleinen Ann! Und dass die mal richtig Schiss haben, das ist so gut! – Die Therapeutin freute sich mit mir. Aber ich sei auch sehr aufgeregt nur beim Gedanken, der ruft an. – Wieso? Sie können das doch! Ging doch richtig gut mit Ihrer Mutter. – Ja, aber mit ihm ist das doch anders. – Zum Schluss der Stunde sagte die K. noch etwas zu einem möglichen Telefonat mit meinem Vater: Stellen Sie sich vor einen Spiegel, und gucken Sie sich bei dem Gespräch an. Legen Sie ein Bild von sich daneben, auf dem Sie sich stark fühlen. – Beides solle dafür gut sein, wenn ich mich während des Gesprächs nicht mehr so gut fühlen würde, mich zu stärken. Wenn das Gespräch aber nicht mehr so laufe, wie es gut für mich sei, solle ich es beenden.

Mir war bald klar, welches Foto ich nehmen wollte: Ein Foto von meinem 40. Geburtstag, auf dem ich mir gerade eine Zigarette anzünde und dabei fast ein bisschen hochnäsig schaue; jedenfalls fühlte ich mich damals sehr gut und stark. Ich notierte in mein Tagebuch: Mein Ziel ist nicht, dass Vater es zugibt. Ich weiß, er hat das gemacht. Ich bin nicht von seiner Bestätigung abhängig. Ich will die drei nicht mehr schützen! Das ist mein Ziel!

Paar Tage später versuchte ich, meinen Vater anzurufen, weil er sich nicht meldete, aber meine Mutter ging ans Telefon. Er solle mich zurückrufen, noch heute Abend. Ja, das mache er. Ich zitiere aus meinem Tagebuch:

Er meldete sich mit seiner scheiß harten paschastimme. Ich war sehr aufgeregt, sprach aber von anfang an hart, massiv, kompromisslos. Ließ ihn kaum zu wort kommen. Während der gesamten zeit stand ich vorm spiegel, sah mich an, das bild von mir seitlich hingeklebt. Hatte einen zettel mit stichworten in der hand.

Er sagte bald, er habe nichts gemacht. Hätte morgen termin beim rechtsanwalt. – Ich weiß aber, dass du was gemacht hast und du weißt es auch! – Dann würde er es zugeben, meinte er. – Das glaubst du doch selbst nicht! ... Mindestens zwei deiner töchter hast du missbraucht! Vater, du bist die mieseste sau, die ich kenne! Ein verbrecher! Ich schweige nicht mehr! – Wieso ich nicht schon vor 40 jahren ...? Weiß er doch genau, wann das war? – Dann hättet ihr mich in die klapse gesteckt. Aber jetzt bin ich zu stark! Und von wegen rechtsanwalt, das ist völlig egal, was der sagt, ich gehe an die öffentlichkeit. Du wirst meine rache zu spüren bekommen, das verspreche ich dir. – Er auf einmal, leise, mit brüchiger, unsicherer stimme: Hm ... Hm ... Und was ist, wenn ich sage, ja, ich habe das gemacht? – Ich war perplex. Überlegte. – Dann steht das im raum, nehme ich es zur kenntnis. Und dann gucken wir, was dann ist. Ich hab nichts mehr zu sagen. – Hm! Hm!, er fast aufgeregt. Ich beendete das gespräch.

Meine knie zitterten, ich war völlig fertig. Aber ich sprang vor freude in der wohnung herum. Und trank einen grappa, ex! Bisschen weinen. Angst, war das richtig? Mit lisa telefoniert. Sie: Herzlichen glückwunsch! – Aber die kleine Ann konnte sich nicht freuen. Aus angst? Oder weil sie ihn auch lieb hat? Ich beschütze dich doch! Später bei anrufen erschreckt, angst und herzklopfen, auch in der nacht. Dann besser. Ein foto von ihm mit strengem blick auf die erde gelegt, drauf rum getrampelt und mit einem nagel in seine augen und in seinen kopf gestochen und dann in kleinste fetzen zerrissen.

Von der darauf folgende Therapiestunde zitiere ich ebenfalls:

Ich berichtete der K., wie's mir vor dem anruf erging, heftiges herzflattern, ganz schlimm! Wut und hass auf den vater. Da die herzprobleme nicht weggingen, hätte ich den entschuss gefasst, ich rufe ihn an. Ich konnte kaum reden. Die K. forderte mich lieb auf, ich solle eine pause machen, in ruhe atmen. Als ich seine frage erwähnte, und was ist, wenn ich sage, ja, ich habe das gemacht, da stieß die K. einen laut des erstaunens aus. Ich berichtete weiter, auch wie es mir danach erging. – Ich finde es wirklich klasse, dass ich das alles sagen konnte, dass ich das geschafft habe! Es ist das mindeste, was der verdient hat! – Ich begann zu weinen: Aber am liebsten wäre mir doch, das wäre alles nicht geschehen. Würde es am liebsten aus mir rausreißen. –

Nach einer kleinen pause fragte ich die K.: Aber warum geht es mir jetzt so schlecht? – Sie deutete das so: Indem sie ihm das sagten, zweifeln sie ja nicht mehr an dem, was er gemacht hat. Jetzt ist es offiziell! Jetzt sind sie damit in die öffentlichkeit gegangen, indem sie es ihm sagten, dazu gestanden haben. Und ihr vater hat es ja zugegeben! Bestätigt! – *Haben sie das so verstanden? – Ja, das verstehe ich so. Und das müssen sie jetzt erst einmal verkraften. –*

Sie hatte noch ganz fröhlich gesagt: Ich hab gestern schon gespürt, dass das ansteht, sie ihn anrufen werden. Es war fällig! – Wir schwiegen. Ich begann zu weinen. Sie fragte nach. Ich hätte weiter herzprobleme. Sie: Ich hab vorhin schon gedacht, man müsste ihr kleines herz beschützen. Beschützend in beide hände nehmen … – Sie guckte so lieb. Was ich denn jetzt fühlen würde? – Mama … – Leise sagte ich es, beschämt, und nochmal: Mama … – Was könnte ihr herz denn jetzt gebrauchen? Was würde ihnen gut tun? – Die mama soll es in den arm nehmen, mich in den arm nehmen. – Ja … Ja! Das machen wir! Das ist gut! – Sie kam lächelnd auf mich zu, streckte mir ihre arme entgegen, kniete sich mir gegenüber auf die matratze, nahm meine beiden hände und drückte sie und ich ihre. Wir hielten uns so, sahen uns in die augen. Und sie lächelte und schaute mich so lieb an. (…) Wollte sie später gar nicht loslassen, sagte es auch. Sie lächelte nur und hielt weiter meine hände. Nachdem sie unseren raum verlassen hatte, weinte ich sehr.

Nach der Stunde rief ich eine Freundin an; auch sie verstand die Aussage meines Vaters als Eingeständnis. Ich war unsicher. Vier Tage später erreicht mich der bereits erwähnte Brief des Rechtsanwaltes. Und noch einige Tage später erfolgte der erste Drohanruf meines Bruders Leo, von dem ich berichtete.

In den darauf folgenden Monaten ließ ich den Brief auf mich wirken; von der ersten heftigen Wut, ich lasse mich nicht einschüchtern, blieb zeitweilig nicht viel übrig. Nicht nur der Brief des Rechtsanwaltes, auch die Bedrohungen durch meinen Bruder Leo hatten mich zeitweise in große Angst versetzt. Ich entschied mich, erst einmal Ruhe zu bewahren, arbeitete in der Therapie weiter an mir und an diesem, aber auch an vielen anderen Themen. Einige Monate später beriet ich mit meiner Rechtsanwältin einen von mir verfassten Briefentwurf an meine Eltern. Der sei zu scharf formuliert, es bestehe die Gefahr einer Klage seitens meiner Eltern, so die Rechtsanwältin. Vor Gericht hätte ich keine Chancen. Ich reagierte wütend und frustriert. Wieder ließ ich alles auf mich wirken und notierte dann in mein Tagebuch: Das, was ich wollte, habe ich doch erreicht. Ich habe ihn und meine Mutter konfrontiert, und ich habe es so gesagt, wie ich es fühlte, also offen und wütend. Ich habe mir und meinen Erinnerungen geglaubt. Ob er es zugibt oder nicht ist unerheblich.

Bei dieser Haltung blieb ich, und es bestätigte sich: Das Thema war nun auf dem Tisch. Ich schwieg nicht bei Gesprächen mit dem Teil der Familie, mit dem ich zu diesem Zeitpunkt in Kontakt war. Das Tabu war gebrochen. Daran konnte auch der Brief des Rechtsanwaltes nichts mehr ändern.

6.5 Kontaktwünsche bzw. Konfrontationen mit meiner Mutter und ihre Reaktionen

Was Konfrontationen mit meinen Eltern anbetrifft unternahm ich ein paar Jahre nichts mehr. Aber dann rief mich meine Mutter an meinem 50. Geburtstag zu meiner großen Überraschung an. Sie sprach auf meinen Anrufbeantworter, sie habe sich schon lange einmal melden wollen. Leider rief sie danach nicht noch einmal an. Nach langem Hin und Her schrieb ich ihr eine Karte. Ich würde sie gerne treffen. Ja, bei mir hatte sich was verändert. Nun musste ich nicht mehr wüten, nun wollte ich einfach wieder in Kontakt mit meiner Mutter kommen und schauen, wie es mir und uns damit geht. Leider reagierte sie nicht auf meine Post, was mich immer verzweifelter und wütender machte. Seit Längerem tauchten sogar wieder solche Gefühle auf: Es nützt doch alles nichts. Am besten tot sein. ... Aber ich will doch leben! Die Therapeutin wies mich darauf hin, dass es doch eigentlich darum gehe, dass ich mit meinen Eltern nicht weiterkäme. Besonders nicht mit meiner Mutter. Dass die sich nicht bei mir melde. Das würde doch hinter meiner Verzweiflung stehen. – Ja! Bin ich ihr das nicht wert? Werte ich mich ab, weil ich ihr das nicht wert bin, dass sie sich mal meldet? –

Es fiel mir schwer, nach all den Jahren voller Traurigkeit, Wut und Hass nun dazu zu stehen, meine Mutter treffen zu wollen, aber für mich war das noch nicht abgeschlossen. Meine Therapeutin bestärkte mich, meine Gefühle und Bedürfnisse ernst zu nehmen, egal, wie meine Mutter reagieren würde.

Darüber sprachen wir eingehend in einer Therapiestunde, rund zwei Monate nach meinem Geburtstag. Nach dieser Stunde fand ich zuhause den Brief meiner Schwester Maria vor, in dem sie mich wissen ließ, dass Mutter in der Klinik läge und von mir als kleinem Mädchen gesprochen habe. Ich rief sofort bei der Therapeutin an und las ihr den Brief vor. – Ach, das rührt mich aber jetzt! Das rührt mich wirklich!, so ihre Reaktion. Wir sprachen darüber, dass ich meine Mutter in der Klinik aufsuchen wollte. Nun konnte ich sie doch endlich einmal ohne meinen Vater treffen. – Ich muss da jetzt hingehen! – Ja! – Ich notierte in mein Tagebuch: Ich freue mich, meine Mutter wiederzusehen!

Am nächsten Morgen hörte ich eine von mir sehr geliebte Callas-Arie zweimal hintereinander und fuhr dann ins Krankenhaus zu meiner Mutter. Ich war erstaunlich ruhig, aber je näher ich dem Krankenhaus kam, dem Zimmer meiner Mutter, um so aufgeregter wurde ich. Ich hatte sie 8 oder 9 Jahre nicht mehr gesehen. Ich fasse das Treffen kurz zusammen: Meine Mutter reagierte ganz aufgeschlossen auf meinen Besuch. Nach meiner Karte an sie habe sie zuerst gedacht, nein, ein für alle Mal sei es aus zwischen uns! Aber sie habe auch sehr geweint. Sie habe eigentlich nichts mehr mit mir zu tun haben wollen, andererseits würde uns doch etwas verbinden. Wir hätten uns doch immer gut verstanden. Sie fragte, wie es mir gehe, nach meiner Arbeit usw. So redeten wir eine Weile miteinander. Zum Schluss fragte ich sie, ob sie möchte, dass ich wiederkomme? Dazu müsse sie doch nichts sagen. Doch! Ja, sie wolle, dass ich wiederkomme. Ich war hinterher so froh, so glücklich, endlich wieder mit ihr in Kontakt gekommen zu sein. Sofort der Wunsch in mir, sie bald wieder aufzusuchen, da sie noch eine Weile im Krankenhaus bleiben musste.

Ich war so froh, berichtete entsprechend freudig davon in der Therapie. Gestand der Therapeutin auch, dass ich schon vor der letzten Therapiestunde fühlte, ich habe alle, meine ganze Familie, doch irgendwie noch lieb. Selbst für die beiden Brüder und den Vater würde ich noch liebe Gefühle empfinden. Und ich betonte, wie froh ich war, dass ich meine Mutter aufsuchte, dass ich das geschafft hatte. Es berühre sie sehr, so die Therapeutin: Auch, dass Ihre Mutter so reagierte, sich so darauf einließ. Sie müssen nochmal zu ihr! – Ja, das habe ich vor! –

Auf den Rat meiner Therapeutin hin kündigte ich am nächsten Tag meinen Besuch für den Abend bei Mutter an und bat sie, dafür zu sorgen, dass wir alleine sind. – Du brauchst keine Bedenken zu haben, komm nur! Wir sind alleine. –

Ich war fast zwei Stunden bei meiner Mutter – und mit ihr alleine. Sie hatte schon auf mich gewartet, sich schön gekleidet, frisiert. Ich war weicher gestimmt als bei unserem ersten Treffen, sie ebenfalls. Mir kamen immer wieder Tränen. Es war auf eine schöne Art nah zwischen uns. Ich sprach von meinen Gedanken, weil sie sich nicht meldete, ich sei ihr wohl nichts wert. Nein, das sei nicht so, das wisse ich doch. Ich sagte ihr auch, dass ich sie in den letzten Jahren sehr vermisst hätte. Wie es ihr denn damit gegangen wäre, dass wir keinen Kontakt hatten? Das habe sie einfach verdrängt. Beim Verabschieden sagte ich: Die zwei Stunden werden wir doch nie vergessen? – Nein, die vergessen wir nicht! – Was war ich mir danach so dankbar, dass und wie ich das geschafft hatte! Es kam mir vor, als habe ich mir damit selbst ein schönes Geschenk gemacht.

Die K. mutmaßte, dass meine Mutter alles verdrängt habe. Das hätte sie wohl gemusst, denn wie hätte sie das sonst aushalten können. Sonst hätte sie sich von meinem Vater trennen müssen. – Aber es ist ja nicht alles weg. Ihre Mutter weiß ja schon irgendwie, was los ist. Sie sagte doch, Du kannst kommen, wir sind alleine, brauchst keine Bedenken zu haben. Da hat sie ja an Ihren Vater gedacht. Aber ich bin überrascht, wie Ihre Mutter jetzt in der Lage war, mit Ihnen in Kontakt zu kommen. Das finde ich erstaunlich und schön. – Sie habe eher damit gerechnet, dass meine Mutter mich abschmettert.

Bei mir kamen nach der Freude starke Bedenken auf, berechtigte, wie ich noch heute finde, nämlich diese: Habe ich mich mit diesen Besuchen nicht selbst verraten? Was ist damit, dass sie früher so hart war, dass sie nicht für mich da war, mich nicht beschützte? Was ist mit meiner Enttäuschung? Mir wurde immer klarer, ich muss noch einmal offene Worte an sie richten. Damit stehe ich zu mir!, erkannte ich.

Ich ging am nächsten Tag, gleich am Morgen, wieder zu meiner Mutter ins Krankenhaus. Ich hatte das Gefühl, ich muss zu ihr! Tue es für mich! Und wenn sie blöd reagiert, bleibe ich gut und ruhig bei mir. Sie war überrascht, dass ich kam. Sie würde heute entlassen. Ihr ginge es nicht so gut. Oh je, kann ich denn dann ernst mit ihr reden? Verkraftet sie das? Ein Arzt sagte aber bald, ihr Zustand habe nur mit einer Tabletten-Umstellung zu tun.

Es war schwierig zu beginnen. Ich gebe meine Tagebuchnotizen von diesem Besuch gekürzt wieder:

Vorher hatte sie schon gesagt, wie schön es wäre, dass wir uns jetzt einige Male sahen, wieder kontakt haben. Das würden wir nicht vergessen. Ich solle sie doch zuhause anrufen. – Bevor wir deine tasche packen, will ich dir noch was schönes und was ernstes sagen, mutti! Und ich möchte deine hände halten. – Ja. –

Das schöne sei, dass wir diese stunden zusammen erlebt haben. – Und das schwere?, fragte sie sofort. Schwierig zu beginnen, aber ich schaffte es: Dass ich in den letzten jahren keinen kontakt zu euch haben konnte, das hatte auch mit dir zu tun. In erster linie mit dem, was ich dir gesagt habe, so ungefähr begann ich. Sie gleich dazwischen: Ja, aber man muss doch auch mal einen schlussstrich ziehen! Jetzt sind wir ja wieder zusammen. – Moment, mutter, höre bitte erstmal zu! – Ja. – Ich bin hierhergekommen, aber das, was ich dir gesagt hatte, was vater, adolf und bernd gemacht haben, das gilt immer noch! Und das war der hauptgrund dafür, dass ich mich zurückzog. Aber es hat auch mit dir zu tun. – Wir sahen uns dauernd an, in die augen. – Es war früher für mich ganz schlimm, dass du mir nicht geholfen hast. – Ja. – Dass du mich nicht beschützt, nicht geschützt hast! – Ja, wieder ernst von ihr. – Dass du nicht genug für mich da warst. – Ja. ... Aber wir waren doch so viele. – Moment, mutter! – Ja. – Ich weiß, dass es schwer für dich war, aber ich hätte deine hilfe gebraucht! – Ja. – Und dass du in den letzten jahren wieder auf der seite von vater warst. Nicht auf meiner! Auch deshalb hat es auch mit dir zu tun. – Ja. ... Man kann es ja nicht vergessen, aber wir können doch trotzdem jetzt einen schlussstrich ziehen. Jetzt sind wir doch wieder zusammen. – Ja, mutti! Ich bin ja auch hier! (...) Aber es war sehr schwer für mich, auch in den letzten jahren. –

Ich sagte noch, ich hätte die erinnerung, dass ich sie schon als kind gestützt hätte, sehr für sie da war. Ob sie das auch so erinnere? – Ja! – Wir sprachen über ein anderes Familiengeheimnis, das mit ihrer Kindheit zu tun hatte. Wir könn-

ten das aber doch anders machen, deshalb müssten wir darüber reden. – Es ist doch in deinem herzen und in meinem. Es ist doch sowieso da! Und jetzt haben wir es angesprochen, ausgesprochen. Das ist doch besser. Es geht doch anders auch nicht weg! – Ja. – Wenn man jemanden lieb hat, kann man auch das schlimme sagen. Das schöne bleibt doch da. – Alles so oder ähnlich von mir. – Jetzt haben wir uns ja wieder. Und das, was jetzt hier war, das vergessen wir nicht!, betonte sie wieder. – Nein! – Ich hielt weiter ihre hände, schaute sie an. Ach, war das wundervoll! Ich sagte ihr, dass ich in den letzten jahren sehr gelitten hätte; hätte es doch jahrelang am herz gehabt, psychische ursachen; viele jahre therapie, dadurch ginge es mir jetzt aber viel besser. – Ja, du wirkst auch ganz ruhig und ausgeglichen. – (...)

Wir waren dann beide so froh! Saßen noch eine weile zusammen, hielten uns an den händen, redeten und verabschiedeten uns dann. Ich war unbeschreiblich glücklich. Zufrieden mit mir. Lief durch den tiefen schnee. Bus fahren? Nein, ich muss mich bewegen! Ich strahlte, lief los, ging und hüpfte durch den schnee und strahlte! Ich habe es geschafft, meiner mutter alles offen zu sagen!

In der anschließenden Therapiestunde berichtete ich natürlich ausführlich. Auch dass mir zuerst danach war, am liebsten keine offenen Worte zu sagen, aber ich hätte es doch nicht weiter mit mir rumtragen wollen: Ich musste ihr das sagen! Das war so stark! – Ja. – Zum Ende meines Erzählens meinte ich zu der K.: Das ist doch auch für uns beide schön, oder?, also für die K. und mich. – Ja ..., weich, lächelnd.

Nach diesen Treffen spürte ich ein paarmal richtig Sehnsucht nach meiner Mutter, nach der schönen Nähe. Ich telefonierte mit ihr, besuchte sie während eines weiteren Klinikaufenthaltes und auch im Pflegeheim, in das sie später kam. Zu diesem Zeitpunkt war sie leider schon sehr verwirrt.

Ich möchte noch ein Gespräch mit Maria wiedergeben, in dem sie ihre Verwunderung darüber äußerte, dass ich mit unserer Mutter zu so einem Kontakt fähig war, dass ich so auf sie zugehen konnte. Gerade mit dieser Veränderung meiner Gefühle und Wünsche hatte ich ja selbst sehr gehadert, mich doch selbst gefragt, verrate ich mich nicht? Meine Antwort gegenüber Maria lautete damals ungefähr so: Ja, ich hatte jahrelang große Wut, sogar Hass und war unendlich traurig und maßlos enttäuscht von unserer Mutter. – Dass es sehr gut war, all diese Gefühle zuzulassen. Nun hätte ich diesen Teil meiner Vergangenheit angenommen, aber nichts von dem, was ich unserer Mutter vorwarf, wäre vergessen. Deshalb könne ich jetzt auch ihre guten und lieben Seiten wieder sehen.

Diese Problematik hatte die Therapeutin in den Jahren zuvor immer mal wieder angeschnitten und betont, wie wichtig es sei und wie sehr sie mir wünsche, dass ich zu einer solchen Haltung kommen könne.

Bis heute geht es mir so mit den Zusammentreffen mit meiner Mutter wie schon damals beschrieben, ich habe mir ein Geschenk gemacht, ihr und mir. Wie gut, dass ich es auch noch schaffte, offen mit ihr zu reden, dass ich ihr nicht nur am Telefon meine Vorwürfe gemacht, sondern sie im persönlichen Gespräch nochmal mit all dem konfrontiert und so gut zu mir gestanden habe. Das alles hinterlässt bis heute in mir ein ganz ruhiges und schönes Gefühl, wenn auch mal, durch irgendeine Erinnerung angestoßen, wieder Wut, Traurigkeit und Enttäuschung da sind. Ich habe allerdings lange arbeiten müssen, die lieben und schönen Gefühle und die Wut anzunehmen, beides nebeneinander stehen zu lassen. Und das möchte ich noch hinzufügen: Als meine Mutter wieder in der Klinik lag, tat sie mir sehr leid, weil es ihr so schlecht ging. Und das konnte ich gut annehmen, und nicht nur das. Ich sagte damals in der Therapie:

Ich finde es gut, dass ich das so fühlen kann. Und geradezu spannend, dass auch weiterhin alle meine enttäuschungen, verletzungen, wie sehr ich mich in all den jahren über sie geärgert habe, dass das alles auch noch da ist. Es ist schon verwirrend, dass sie mir jetzt so leidtut. Aber ich finde es gut, dass ich beides fühlen kann. Wäre doch furchtbar, wenn ich jetzt, wo es ihr so schlecht geht, alles andere wegmachen würde! – Ja. – Das fühlt sich richtig gut an. Das macht mich ganz ruhig, dass beides da ist! –

6.6 Weitere Konfrontationen mit meinem Vater und seine Reaktionen

Ich hatte gegenüber meinem Vater erst einmal Ruhe gegeben. Hatte in der Therapie weiter an mir und an aktuellen Entwicklungen bei mir gearbeitet, mich mit meiner Mutter auseinandergesetzt und einen positiven Abschluss mit ihr gefunden. Aber das Thema „mein Vater“ war für mich noch nicht abgeschlossen. Mir wurde zunehmend klarer, ich muss ihn persönlich konfrontieren! Natürlich tauchten vorher Zweifel und Bedenken auf, ob und wie ich das schaffe. Hätte es am liebsten gehabt, so äußerte ich mich auch der Therapeutin gegenüber, dass sie mit mir zu ihm fährt. Aber es wäre auch in mir, ich muss das alleine bewältigen. Sie meinte, ich könne doch daran denken, dass sie gedanklich bei mir sei. Und wir würden das doch alles vorher besprechen. Klar war: Ich muss das jetzt machen! Das steht jetzt an! Ich will ihn mit mir konfrontieren und mich mit ihm! –

Da meine Mutter wieder im Krankenhaus lag, mein Vater somit alleine in deren Wohnung, würde ich ungestört mit ihm reden können. In meinem Tagebuch notierte ich auf fast zwei Seiten, was ich meinem Vater sagen wollte. Hier ein paar Punkte daraus: Kannst Du dir überhaupt vorstellen, was das mit mir gemacht hat? Dass Du mich fast zerstört hast! Mein Leben nahezu zerstört hast! All die Jahre mit Ängsten, Schmerzen. Meine Verzweiflung! Ich wollte nicht mehr leben! Nicht mit dem Wissen leben, mein Vater hat das mit mir gemacht! Ich hab Dich doch so geliebt! Du bist doch mein Vater! ... Wie gerne hätte ich einen anderen Vater gehabt, der mich liebt und achtet, meinen Körper achtet und mich beschützt und nicht das Gegenteil tut!

Wie ich diesen Schritt in der Therapie vorbereitete und wie ich das Zusammentreffen mit meinem Vater gestaltete, das möchte ich hier durch meine Notizen von damals wiedergeben. Erst einmal zu der Therapiestunde:

Ein zusammentreffen mit meinem vater sei schon ein anderes vorhaben als das treffen mit meiner mutter. Zu ihr hätte ich doch fast spontan hingehen können. Die K. erinnerte mich aber daran, dass ich mich schon lange zuvor damit auseinandergesetzt hatte, dass ich meine mutter sehen wollte; das sei hier doch jahrelang thema gewesen. Ja, das stimme. Ein treffen mit meinem vater dagegen noch nie. – Ich fühle es glasklar, ich will ihn treffen! – Wenn das so klar sei, dass ich das wolle, dann würde ich das auch schaffen, meinte die K.

Ich ging mit ihr verschiedene punkte durch. Was ist, wenn er mich nicht in die wohnung lässt, schimpft oder mich sogar hinauswirft? Mir sei wichtig, immer wieder zu überprüfen, stimmt es noch für mich oder will ich gehen. Und: Es geht mir nicht darum, dass er was zugibt! Ich will mich nicht davon abhängig machen, wie er reagiert! – Ich sagte aber auch: Wieso muss ich immer kämpfen? – Darum gehe es doch nicht. Das habe was depressives, wenn ich das so sehen würde. – Es geht doch darum, sich für sich einzusetzen! – Ich jammerte dann noch rum, weshalb sie bald sagte, wenn ich so jammern würde, könne ich gar nicht meine eigene stärke spüren. Eigentlich wäre es doch so, ich wäre sauer auf sie, dass sie mir das nicht abnimmt! Sie machte mich mit jammernder stimme nach. Das müsse ich jetzt aushalten, so ihre reaktion auf meinen protest. Ich wolle bedauert werden: Du hast es ja so schwer, u. ä. von ihr. Sie könne jetzt auch 3 stunden mit mir so jammern und leiden und erst danach kämen wir zu dem treffen mit meinem vater. – Und dann würden sie denken, nur weil ich das alles mitgemacht habe, deshalb haben sie das dann geschafft! Und darum geht es mir: Dann würden sie nicht ihre eigene kraft spüren! Nicht spüren, dass SIE das alles geschafft haben! ... Haben sie mich verstanden? – Nein, ich habe sie überhaupt nicht verstanden! – Sie lachte. Sie könne doch mal zugeben, dass das ein schwe-

rer schritt sei. Das habe sie mir doch schon 100, 150mal klar zu verstehen gegeben, dass sie wisse, wie hart das sei. Sie sah mich direkt an: Können sie sich nicht vorstellen, dass wir uns hinterher gemeinsam freuen? – Blick in ihre augen. Berührt. – Oder auch gemeinsam traurig sind? – Ich fühlte Mama … Ach, kam das gut.

Ich wollte nun ihre frage beantworten, welches bild ich dazu habe, meinem vater zu begegnen. Erst war gar nichts da. Aber dann: Ich habe angst zu spüren, dass ich ihn noch lieb habe. – Pause. – Und dass ich vielleicht mitleid mit ihm habe wegen seines alters. – Pause. Dann richtete ich mich auf, setzte mich ganz aufrecht auf meine matratze, und sagte mit geschlossenen augen langsam und ruhig:

Ich will mich mit meinem vater konfrontieren. Mit dem, was ich jetzt weiß, vor ihn hintreten, ihn angucken, in seine augen schauen. Mit dem, was ich erinnerte, mit meinen gefühlen, meinem körper erinnerte vor ihn hintreten und vor ihm damit bestehen können. – Ehemmm, weich von ihr. – Ja, das ist mir ganz wichtig! Ich will gucken, wie es mir damit geht. – Ich öffnete meine augen. Als ich mir das alles vorstellte, spürte ich totale aufregung. – Ich will ihm sagen, welchen schaden er damit bei mir angerichtet hat. Wie schlimm das für mich war! – Ehemmm! – Das ist ganz wichtig für mich, dass ich auch das zu ihm sage! Dass ich das kaum überlebte! –

Dann stellte sich mir die frage, wie komme ich in die wohnung, wenn er das nicht will? – Ach, ich komme da schon rein! … Aber was ist, wenn er das abstreitet? Und noch sagt, er habe mich doch lieb gehabt oder so etwas? – Wir redeten eine weile darüber. Die K. sagte u. a., das sei ja abspalten, verdrängen! So wäre das dann auch für ihn, hat er nicht getan! So würden es selbst triebtäter machen, die ihre opfer umbringen; die wüssten das dann wirklich nicht mehr! Sie erinnerte mich daran, dass ich das ja auch verdrängt hätte: Wenn vor jahren jemand zu ihnen gesagt hätte, dein vater hat dich missbraucht, da hätten sie geantwortet, nein, auf keinen fall! – Ich verstand. Sie schlug vor, auf seine eventuelle verneinung zu sagen: Ich weiß aber, dass du das gemacht hast! Und ich weiß, dass du auch eine andere seite hattest … –

Sie fasste es zusammen mit den worten, es gehe mir ja darum, ihn zu treffen, gehe um die begegnung mit meinem vater. – Ja! Und ihm zu sagen, was er damit angerichtet hat! – Ja, weich von ihr.

Ich fuhr nach der Stunde nach Hause, sehr aufgewühlt, hatte Kopfschmerzen. Bald war klar: Ich will noch heute zu ihm fahren! Erinnerte die Worte der K., es ginge darum, dass ich es alleine schaffe! Dass ich spüre, ich habe diese Kraft! Gut, dass sie so stur ist! Ich zitiere wieder aus meinem Tagebuch:

Ich, die Ann, ich habe es geschafft! Ich war in B. bei meinem vater!

Aber der reihe nach: Ich fuhr zuerst zu [meiner Freundin] Andrea und ihrer kleinen tochter. Komm, setz dich zu uns. Eis essen. Spielte mit Linda. Das war so schön. Ihre lachenden, strahlenden augen. Und wenn sie mama sagt, wundervoll! Dachte an meine geliebte mama, die therapeutin! Um 18 uhr, aufgeregt: Ich fahre jetzt. – Andrea drückte mich lieb: Wenn du willst, komm doch hinterher wieder her! – Ja! –

Ich fuhr los, sehr aufgeregt! Ann, denk an Italien! Nur an diese eine furchtbare nacht in Italien! Das sagte ich mir zwei-, dreimal, das genügte. Aufgeregt, aber auch irgendwie ruhig, wahrscheinlich weil ich so klar war.

Mein Vater ließ mich ohne Probleme in die Wohnung. Ich fasse das wesentliche zusammen: Er wünschte, dass ich mich aufs sofa setze. Ich zöge den sessel vor, damit ich ihn ansehen könne. Fragte ihn, ob er mir nichts zu sagen habe? Erinnerte ihn an das, was er mit mir gemacht hat. Er sofort, fast aggressiv, hätte er nicht, wenn, dann aus versehen. Stellte zusammenhang her, nur Leo erbt, wir alle nicht, deshalb käme ich jetzt mit dem vorwurf. – So ein unsinn! Das geht mir am arsch vorbei! ... Ich weiß, dass du das gemacht hast! – Ich blieb meist ganz ruhig, sah ihn immer an. Er guckte oft weg, aber wir sahen uns auch mal richtig in die augen. Wieder, er hätte das nicht gemacht. Ich betonte, er hätte das auch mit lisa gemacht! Er sagte, er habe, als ich klein war, nicht gewollt, dass ich zu ihm ins bett komme. Ich stellte ganz ruhig klar, es sei doch ganz natürlich, dass ein kleines mädchen zu seinem papa ins bett, sich an ihn kuscheln wolle. Es käme aber auf den papa an, was der dann mache!

Ich sagte wieder: Ich weiß, dass du das gemacht hast! – Sprach auch ausführlich davon, wie schlimm das für mich war und noch ist. Meine herzprobleme, die vielen nächte mit angst- und panikattacken! Zum schluss: Ich weiß, dass du das gemacht hast. Und ich habe es dir jetzt gesagt. Ich wollte dich mal sehen, mir meinen vater mit diesem wissen angucken. Du bist ein alter mann, hättest jetzt die möglichkeit, mit deinem leben ins reine zu kommen. – Das lehnte er aggressiv ab! Ich nahm meine lederjacke und ging, aufrecht, ruhiger.

Ich fuhr zurück, weinte fast, aber spürte bald große Erleichterung. – Mama, ich hab's geschafft! – Danach wieder zu Andrea. Gemeinsames Abendessen, nachdem ich ihr kurz berichtete. Spielte noch mit Linda und brachte sie in ihr bettchen. Das war sehr schön! Fuhr dann nach hause, erschöpft, aber auch unbeschreiblich froh: Ich hab's geschafft!

Am nächsten tag das gesicht meines vaters vor mir, eklig! Bedürfnis, meine wohnungstüre abzuschließen, Angst vor Leo. ... War gut, dass die K. sagte, besser alleine zu meinem vater zu gehen, damit ich meine kraft spüre, erlebe, ich habe das ganz alleine geschafft! In ihren gedanken hat sie mich ja begleitet, und ich hab sie ja auch in mir! ... Heute wieder das gefühl, das war noch nicht das letzte gespräch mit ihm.

In der darauf folgenden Therapiestunde berichtete ich ausführlich:

Ich sah die K. über den spiegel an, sie lächelte, und ich sagte, deutlich, langsam: Ich war bei meinem vater! – Sie lächelte. – Ich hab's doch tatsächlich geschafft! – Und dann redete ich gleich los, oft mit geschlossenen augen; sah sie zwischendrin aber auch an und sie mich, lächelnd, froh. Immer wieder ihre bestätigungen, ihr nicken. Schön! Ich saß ganz aufrecht und berichtete. Sie fragte zwischendrin, wie ich denn gekleidet gewesen wäre? – Schwarze jeans, das hellblaue kaputzenshirt und die kleine schwarze lederjacke so seitlich über der schulter getragen. – Gut! – Ich berichtete weiter. Er sei aggressiv und arrogant gewesen. Seine aussage von wegen vergewaltigung, vielleicht aus versehen. Meine entrüstung: Ein kleines mädchen aus versehen vergewaltigen, das gibt's ja wohl nicht! – Nein, habe er nicht gemacht. Ich hätte ihn eklig gefunden, sein gesicht, seinen kopf, seine augen! Hätte ihm nicht die hand geben können, so eklig!
Ich hätte alles sagen können, was ich mir vorgenommen hatte. – Und ich bin ganz alleine hingefahren, und das war passend! – Ihr blick, wundervoll!

Danach wieder zu Andrea, das habe sie mir angeboten. – Hab mich bei Andrea sehr geborgen gefühlt. – Ehem! –

Schon bald sei die idee entstanden, noch mal mit ihm reden zu wollen. Mein gefühl, mit meiner mutter ..., das wäre abgeschlossen. – Habe nicht vergessen, was sie gemacht hat, wie sie sich verhielt. Aber ich kann mit ihr zu tun haben. Für mich fühlt sich das rund an. Da bin ich ganz ruhig. – Ehem! – Mit meinem vater ist versöhnung unvorstellbar! – Ja!, fest von der K. Ich solle doch mal nachspüren, welcher teil von mir sowas wie versöhnung gerne hätte. – Vielleicht die kleine Ann, die sich mit einem teil von ihm versöhnen will ... Aber es ist auch wut da bei der kleinen Ann! Sie ist traurig, aber auch unermesslich wütend auf ihn! – Ehem! – Ich hatte das bild, ihn mit meinen kinderfäusten auf seine brust zu schlagen, zu trommeln. Die große Ann wolle sich nicht versöhnen, sagte ich noch. Nein, das passe auch nicht. Die große Ann, die erwachsene, sei ja die, die an gerechtigkeit, recht und gesetz glaube und selbst danach lebe. Der vater müsse eigentlich bestraft werden.

Also mein gefühl, das thema mein vater ist noch nicht abgeschlossen. – Jetzt seien sie doch erstmal stolz auf sich! Und freuen sie sich, dass sie das geschafft haben. Das andere kommt dann noch. –
Ich kann ihnen gar nicht sagen, wie froh ich bin, dass ich das geschafft habe! – Ja. – Wir sprachen noch davon, dass er das verdrängt haben mag. Ich würde das aber nicht glauben. –
Unglaublich, dass ich ihm, meinem vater, diesem patriarchen, so gegenüber trat! – Ja! – Dass ich ihn anguckte, immer in seine augen sah! Das alleine ist schon der hammer!, lachend. – Wo er doch früher der herr über unseren blick war. Guck mich an! Guck unter dich! Also nach dem motto: Wie kannst du es wagen, mich anzusehen! Dich erdreisten! ... So ging's doch zu bei uns. Und jetzt guckte ich ihn dauernd an! – Sie lachte: Alleine schon die formulierung, guck unter dich! – Ja, aber das ist mir so vertraut. Wie ein geflügeltes wort! – Ja, ja, verstehe ich schon, aber was für eine formulierung! – Ich sprach wieder davon, dass er nur aggressiv und arrogant war. Und dann die therapeutin: Das muss ich jetzt noch sagen: A... Arrogant! Aggressiv! ... Arschloch! – Wunderbar! Fand ich klasse, sagte: Ja, arrogantes arschloch! Das ist er! – Sie lachte: Das war vorhin schon in mir, als sie sagten, er war aggressiv, arrogant. Da dachte ich schon dauernd, das dritte A fehlt, Arschloch! – Diese tolle mama!
Wir strahlten uns an. Kurz ganz bei mir, dann: So ist gerade mein gefühl: Mama, ich hab's geschafft! – Sah sie bisschen verlegen an. Sie lächelte mich so wundervoll an ...

Nach diesem Gespräch mit meinem Vater war das Thema für mich tatsächlich noch längst nicht abgeschlossen, aber einen ersten großen Schritt dazu hatte ich getan. Das stärkte mich sehr. Im Herbst des gleichen Jahres kam ich auf diese Thematik zurück. Es ging erst einmal um die mangelnde Solidarität von Maria, Birgit und Heiner, dass sie mich nicht unterstützten. Die Therapeutin deutete diesen Wunsch auch dahingehend, dass ich gerne weiter zur Familie gehören wolle. Jeder wolle in seiner Familie bleiben, auch wenn es da noch so heftig zuginge. Aber das müsse sie auch sagen: Die Auseinandersetzung mit Ihrem Vater können nur Sie selbst leisten. Und Sie müssen für sich einen Weg finden, wie Sie das mit Ihrem Vater weiterverfolgen und abschließen wollen. –

Aber das brauchte noch Zeit, *ich* brauchte noch Zeit.

Im März des darauf folgenden Jahres ging es wieder darum, wie will ich das mit meinem Vater abschließen? Dabei kam ich in der Therapie an Gefühle, die mir lieb und nicht so lieb waren:

Ich hätte gerne was versöhnlicheres. Dass was von ihm käme! – Die K. forderte mich auf zu phantasieren, was ich mir von ihm wünschen würde. – Ich sag mal ..., im idealfall ... – Ich weinte sehr und sprach weiter: Dass er sagt, es tut mir so leid! Ich verstehe auch nicht, warum ich das gemacht habe. – Die therapeutin meinte, dahinter verberge sich ja, dass ich weiß, es gab auch schönes zwischen ihm und mir. – Ja, das weiß ich schon noch. Ich wollte es jahrelang nicht wahrhaben, dass ich ihn auch noch irgendwie lieb hab'. Ich verstehe es ja selbst nicht! – Sie sagte noch einiges dazu, u. a., dass er mich doch auch lieb hatte. Ich, weinend: Aber das macht es ja noch schlimmer, dass er dann so was gemacht hat! Das ist doch erst recht nicht zu verstehen! – Nein, das ist auch einfach nicht zu verstehen. –

Die K. sagte zu meinem wunsch, dass er sich entschuldigen möge: Ich kann das so gut verstehen, dass sie sich das von ihm wünschen. Aber damit machen sie sich sehr abhängig von ihm! – Sie sah mich dabei lieb und eindringlich an. – Ja, das stimmt. – Auf Bernd wäre ich ja zugegangen, bei meinem vater müsse ebenfalls ich das tun, wenn ich da was für mich wolle: Und damit machen sie sich auch unabhängig von ihm. Sie *können das, auf die zugehen. Die nicht! Die haben so eine abwehr aufgebaut! –*

Es fiel mir ziemlich schwer, diese Seite anzunehmen. Die Therapeutin erinnerte mich wieder daran, er sei nun mal mein Vater. Aber sie könne es wirklich gut verstehen, dass es sehr schwer für mich sei. Einerseits seine Taten, die Verletzungen und dann diese Wünsche. Das zu integrieren, diese gegensätzlichen Gefühle, das wäre einfach kaum zu leisten.

Sie vermutete sogar, dass ich mir nicht nur was Liebes von ihm wünschte, sondern ich ihm auch was Liebes geben wolle. Das sei noch schwerer zu akzeptieren, so meine erste Reaktion. Wieder half sie mir, das zu verstehen und anzunehmen, indem sie sagte: Wenn jemand einen so verletzt hat, ist es fast nicht möglich, so einen Wunsch zu akzeptieren. Das ist ja geradezu verrückt! Aber es ist so! Und daran merken Sie, dass Sie noch Gefühle für Ihren Vater haben, an diesen Wünschen. Und es ist sehr gut, dass es so ist! –

Ich stellte mir in der Therapie vor, ihm ein weiteres Mal zu begegnen und dabei gut zu mir zu stehen. Ich würde es so fühlen, dass uns das trenne, meinen Vater und mich. Ja, das sei ein ganz wichtiger Punkt, mich getrennt von ihm zu fühlen, bestätigte die K. Und ich erkannte: Würde ich nicht nochmal vor ihn hintreten, das wäre für mich, als würde ich alles wieder zurücknehmen. – Ich müsse und wolle mich ein weiteres Mal mit ihm auseinandersetzen, auch wenn von ihm nur Abwehr käme, womit die K. rechnete.

Ich eigentlich auch. – Es geht mir darum, dass ich mir glaube und ihm gegenüber zu mir stehe. Damit übernehme ich für mich die Verantwortung. So fühle ich es: Er hat die Verantwortung für seine Taten. Und ich dafür, wie ich damit zurechtkomme. –

In einer weiteren Stunde kam ich darauf zurück, was ich mir von ihm wünsche: Ich hätte ihn gerne mal leise. Dass er von seinem Thron runterkommt und hinguckt!, so o.ä. beschrieb ich es. Ich hätte mir doch immer gewünscht, ihm groß und erwachsen und fest entgegentreten zu können. Jetzt, bei der Phantasie, er gibt nichts zu, käme er mir kleiner vor, ich eher stärker ... – Ehem!, betont. Und ich fühlte ganz deutlich, wenn er sich auf meine Worte einlassen, sich auch entschuldigen könnte, dann würde ich ihn nicht klein und schwach erleben. Aber wenn er so hart und unerbittlich bleibt wie bisher, dann empfinde ich ihn schwach.

Ich kam zu dem Schluss: Annehmen, dass er das gemacht hat, ja, aber verzeihen, nein! Verzeihen, das geht nicht! – Die Therapeutin daraufhin, ganz vehement: Nein, verzeihen, das ginge zu weit! ... Vielleicht sollte es heißen: Verstehen! Ja, verstehen, warum er das gemacht hat, warum er so geworden ist. Das hatten wir ja schon öfter, angucken, woher der Vater kommt, seine Geschichte anschauen ... –

Noch heute erinnere ich mich daran, wie der Entschluss immer mehr in mir reifte, zu meinem Vater zu fahren. Ein paar Tage nach der gerade erwähnten Therapiestunde war es dann so weit. Ich fuhr zu ihm, er war alleine. Davon berichtete ich am nächsten Tag in der Therapie. Die Tagebuchaufzeichnungen dazu gebe ich gekürzt wieder:

Bevor ich mich auf die couch legte, sagte ich: Ich war gestern bei meinem vater! – Oh! Neuigkeiten!, sie lachte, holte ihren fußschemel herbei und sah mich gespannt und lächelnd an. – Ja, neuigkeiten! – Ich legte mich auf die Couch und begann zu berichten:

Nach der letzten stunde war klar, ich muss das bald machen! Aber auch, ich würde es am liebsten rausschieben. ... Ich wollte ihn nicht aggressiv angehen, damit er nicht nur abwehren kann und muss. – Ehem! – Wollte in ruhe mit ihm reden. So habe ich das gespräch dann auch begonnen. –

Zuerst meine frage, wie es ihm mit dem tod meiner mutter gehe? Hätte mein gespräch mit ihr von vor zwei jahren erwähnt, unser reden über die zeit, als ich kind war, was für mich schwer war damals usw. Dass wir gut miteinander geredet hätten, so dass ich, als sie starb, ganz ruhig war mit dem gefühl, wir haben einen guten abschluss hinbekommen. Ich wäre heute bei ihm, weil ich nach ihm sehen und auch mit ihm über meine vorwürfe von vor zwei jahren reden wolle. Er hätte vergessen, worum es dabei ging. – Sexueller „Missbrauch“! – Er hätte sofort angefangen zu schwitzen, sich ganz oft den schweiß abgewischt. Will er nicht drüber reden, hat er nicht gemacht, gleich massivst abblockend. Und dabei blieb er. Irgendwann dann bei mir der punkt, jetzt kann ich eigentlich nur noch gehen. Aber dann wäre ich doch sehr unzufrieden! Ich berichtete weiter, er habe zwischendrin von leuten aus b., meiner heimatstadt, erzählt. Ich sei aber immer wieder auf den grund meines dortseins zurückgekommen. Mein entschluss: Nein, ich gehe jetzt nicht! Ich bleibe! Hielt auch das zeitweilige schweigen aus. Ich halte das aus, dass wir hier zusammensitzen, und er muss es auch aushalten!

Hätte ihm von Bernd erzählt, der sich auch nicht erinnerte, weil er seine Taten verdrängt habe; der sich dann aber entschuldigte. Mutter habe es auch geschafft! Ich hätte ihm alles erzählt, auch von ihren Jas berichtet. Könne er nicht verstehen. Ich blieb bei meiner klarheit. Auch wenn er sich nicht erinnere, ich wisse es!

Wieso er, wenn er das nicht gemacht habe, nicht wenigstens mal frage, was denn Adolf und Bernd gemacht hätten, seine söhne, mit mir, seiner tochter? Er wolle über dieses thema nicht reden. Sei für ihn ein tabu-thema. – Finde ich erstaunlich, dass er das sagte, tabu! – Ehem, ehem!, immer wieder zwischendrin von der K. – Also ich blieb. Und ich blieb beharrlich. Fragte ihn, ob er mal so ins grab gehen wolle, wir so verbleiben sollten? Ich hätte ihm aber auch, fast weinend, gesagt, wie schlecht es mir jahrelang ging, wie oft in mir war, vom balkon zu springen. Seine stimmung sei dann irgendwann gekippt. Er habe sich schon so oft und lange damit beschäftigt, aber das habe er nicht gemacht!

Ich berichtete weiter, dass ich bewegt zu ihm sagte, ich hätte ihn doch mal so lieb gehabt! Ob er sich daran nicht erinnere? Klar würde er sich daran erinnern. Und dass er dann nicht mehr laut, sondern berührt reagierte, doch auf das gespräch eingehend. Und dann, er, ganz emotional: Ja, dann entschuldige ich mich! Tut es mir leid! ... Die hände vor der brust gegeneinander. – Die K. ließ ihr berührtsein vernehmen. Vater weiter sehr emotional: Aber das nütze doch nichts oder nichts mehr. Und er habe das doch nicht gemacht! –

Ich sagte, dass mir hinterher klar wurde, wie gut es war, nicht gegangen zu sein. Auszuharren, das auszuhalten. Und dann habe sich ja auch was verändert. – Ja, sonst wären sie hinterher sehr frustriert gewesen! – Ja! ... Ich habe mir doch gewünscht, dass er mal leiser, sich mal berührt zeigt. So ungefähr war es dann. Er nicht mehr abblockend, sondern sichtlich aufgewühlt. – Ehem. –

Er und meine mutter hätten, so die aussage meines vaters, mein jahrelanges nicht mehr melden als schlechtes gewissen meinerseits gedeutet. Die K. lachte: Ja, so machen die das! Die fanden doch schon immer entschuldigungen oder deutungen, so wie es für sie passte. – Ja! – Ich habe gegenüber meinen eltern kein schlechtes gewissen! Überhaupt keins! – Ja! –

Ich erwähnte noch seine überraschende aufforderung, mir von dem schmuck meiner mutter etwas auszusuchen. Ich hätte mich für einen schönen ring entschieden.

Zum schluss des gesprächs habe mein vater sich in die hose geschissen, was für mich ein zeichen dafür war, dass er nicht mehr so kontrolliert war. – Klar, ist alter mann; passiere ihm auch sonst, wie er sagte. Aber es passte! – Ehem! –

Mein gefühl, besonders jetzt beim erzählen: Gut! Das war gut! – Ja! – Gut, dass ich nicht ging, sondern blieb, beharrlich war! – Ja! – Für mich habe sich mit dem gespräch wirklich was verändert! – Ja! – Dass ich mich unabhängig von ihm fühlte, klar war und dass ich zu mir stand und nicht wegging. Das hat einfach viel ausgemacht. – Ja! Gestern konnten sie bleiben. Zuvor war es ja anders gewesen, dass sie quasi gehen mussten. – Es habe sich auch deshalb was verändert, weil er sich doch irgendwann berührt zeigte, sich damit auf ein gespräch einließ. – Wenn er jetzt sterben würde, wäre es für mich anders als zuvor. – Ja. – Vor ihm zu mir zu stehen wäre dann doch nicht so schwer gewesen wie gedacht!

Er sei zum schluss sehr aufgewühlt gewesen, so dass ich fast ein schlechtes gewissen bekommen hätte, weil es ihm sichtlich nicht gut ging. Ich hätte aber gedacht, das muss er jetzt mal aushalten, ich habe auch so viel ausgehalten!

Am Ende der stunde setzte ich mich, sah die K. an. – Schön!'... Gut!, sagte sie, lächelte mich an und ich sie. – Jetzt sind sie aber stolz auf sich! Sie strahlen mich so sehr an! – Sie strahlte und lächelte auch sehr! – Ja! Ja, das stimmt, von wegen stolz! Aber sie haben es doch auch miterlebt, nur die letzten wochen jetzt hier. Wie schwer das war! – Ja! – Sie schaute mich sichtlich berührt an. Mir kamen fast die tränen ...

In der folgenden Stunde sprach ich darüber, wie froh ich über das Erreichte war: Das kommt besonders dadurch, dass ich zu mir stand, mal ganz abgesehen davon, wie mein Vater reagierte. – Ja! Weil Sie standhielten! – Dass ich meine gefühlten Erinnerungen, erinnerten Bilder gegen meinen Vater vertreten konnte, Wahnsinn!, so schrieb ich ein paar Tage später in mein Tagebuch, und dass es einfach unbeschreiblich ist, was ich damit für mich erreichte. Ich war mir selbst dankbar. Das war ich mir selbst schuldig, mir und der kleinen Ann. So sehe ich das noch heute!

7. Auswirkungen sexueller Gewalt

Ich weiß gar nicht, wie ich dieses Thema zu Papier bringen kann, womit ich anfangen soll. Vielleicht wieder so: Was fällt mir eigentlich zuerst dazu ein? Da sprudelt es geradezu aus mir heraus: Auswirkungen auf die Beziehung zur Therapeutin – keine Berührungen möglich, Vermischung der Gefühle, liebe und kindliche Gefühle und erotische Gefühle, großes und immer wieder auftretendes Misstrauen ihr gegenüber. Ganz furchtbar die Ängste und Panikattacken, die katastrophale Angst in der Nacht und vor den Nächten. Auf keinen Fall mehr Kind sein oder gar kindliche Bedürfnisse haben. Kein Mädchen, keine Frau sein wollen. Ich kann mich nur auf meinen Verstand verlassen, nicht auf meine Gefühle, deshalb meine Herzprobleme! Wut und Zorn und Hass und Rachephantasien! Das Verschwinden meines Körpers! Nur meinen Kopf nahm ich wahr, der Rest blieb mehr oder weniger außen vor. Und oft dachte ich während der Jahre in der Therapie: Ich will nicht mehr leben! Ich kann das alles nicht ertragen, nicht aushalten! Ich will einfach nicht mehr! Mit entsprechenden Phantasien, wie ich meinem Leben ein Ende setzen könnte. Aber zum Glück letztendlich: Aber ich will doch leben!

Schon im ersten Jahr der Therapie sprach ich davon, wie sehr ich überrascht sei von den schlimmen Auswirkungen; ich nannte damals die Herzprobleme, aber auch Trauer, Schmerz und Wut. Viele andere Auswirkungen waren mir zu diesem Zeitpunkt nicht bewusst, schon gar nicht das Ausmaß der erlebten Gewalt.

7.1 Auswirkungen, die schon *vor* der Therapie auftraten

Körperliche Auswirkungen

Eine der heftigsten und noch heute anhaltenden Auswirkungen sind die Herzattacken, die ich bereits zuvor erwähnte.

Im Alter von 15 bis 17 Jahren erlebte ich zum ersten Mal eine Tachykardie. Hierbei handelt es sich um Herzrasen mit 180 und mehr Pulsschlägen in der Minute. Später kamen noch andere Herzsymptome hinzu, wie schnelleres kontinuierliches Herzschlagen oder sogenannte Extrasystolen, d. h. normales Herzschlagen wird unterbrochen von unregelmäßigem und geht wieder über in den normalen Rhythmus und das wechselt sich ab. Mit zunehmendem Alter traten die Proteste meines Herzens heftiger und häufiger auf. Nachdem anlässlich einer Tachykardie die hinzugerufene Ärztin meinte, und das mir gegenüber auch aussprach, ich hätte einen Herzinfarkt – ich war ungefähr 27 Jahre alt –, dachte ich, jetzt muss ich sterben. So kam in der Folge zu den Herzattacken noch Todesangst hinzu. Ab diesem Zeitpunkt wurden die Herzattacken von zunehmender Angst begleitet; manchmal tauchten zudem heftige Panikattacken auf. Vom Hausarzt verschriebene Beruhigungstabletten sollten diese beseitigen. Das war auch mehr oder weniger der Fall. Aber damit begann eine gewisse Tablettenabhängigkeit. Ich ging nie mehr ohne Tabletten aus dem Haus. Ein Kreislauf von Angst vor der Angst vor einem Herzanfall begann.

Herzattacken habe ich noch heute. Durch die Therapie weiß ich aber zum einen, dass und aus welchem Grund ich diese überhaupt habe. Zum anderen kann ich mit den Herzattacken meist gut umgehen, erkenne fast immer rasch, warum mein Herz „sich meldet" und sorge dann entsprechend für mich. Außerdem kommen sie seltener vor – und zwar besonders dann, wenn ich nicht zu meinen Gefühlen und zu mir stehe.

Aber nicht erst mein Herz zeigte, dass mit mir oder bei mir etwas nicht stimmte. Schon als Kind quälten mich sehr oft starke Migräneanfälle. Sie stehen für mich für Wut und Hass, die ich ganz sicher als Kind auch schon hatte, auch wenn ich sie als solche nicht wahrnahm. Irgendwann wurde die Migräne abgelöst von Rückenschmerzen, die nicht nur körperliche Ursachen hatten. Meine Wut tobte sich nämlich an und mit meinem Rücken aus. Das erkannte ich während der Therapie. Sobald ich meine Wut und meinen Zorn dort ausagierte, mit Schlagen, Schreien u.ä., sobald ich also eine Form fand, dies alles zuzulassen, verschwanden die Rückenschmerzen. Manchmal brauchte ich noch eine Spritze, damit sich die Muskulatur wieder leichter entspannen konnte, aber oft reichten meine Wutanfälle und das Schreien aus.

Psychische Auswirkungen

Neben der Angst, der Panik und Todesangst, die die Herzattacken begleiteten, hatte ich Angst, die ohne für mich erkennbaren Anlass auftauchte. Eine Situation schildere ich im Prolog. Eine weitere möchte ich nun beschreiben.

Ich war mit meinem Mann bei Maria, meiner ältesten Schwester. Wir sitzen alle bei Maria, reden, lachen, trinken und essen. Und ich habe diese mit nichts begründbare, maßlose Angst – und sage nichts davon. Ich kann nichts sagen! Eine Angst, als würde ich mich auflösen; nichts gilt mehr! Ich habe keine Sicherheit mehr. Das Grundvertrauen und die Grundsicherheit sind weg! Ich konnte einfach nichts sagen, nicht darüber reden! Ich habe dann bestimmt eine Tablette genommen (Psychopharmaka). Damals konnte ich mir überhaupt nicht erklären, warum ich so starke Angst habe; schon gar nicht sagen, wovor! Ich spürte einfach tiefe, maßlose Angst! Nicht zu wissen, warum ich diese Angstzustände bekam, machte es noch schlimmer und noch schwerer, sie zu ertragen. Das war sicher mit ein Grund, warum ich nichts von meinen Ängsten sagte; ich konnte sie nicht erklären!

7.2 Auswirkungen, die erst *während* der Therapie auftauchten oder bewusst wurden

Angst

Angst, ausgeliefert zu sein

Die Gründe für meine Ängste deuteten sich zu Beginn des zweiten Jahres der Therapie an. Ich spürte zum ersten Mal während der Therapiestunde Angst. Es war mir in dieser Stunde sehr schwer gefallen, aber ich hatte dann doch über meine Bedürftigkeit reden können, über meine „Sehnsucht nach Mutterliebe" und dass ich es am liebsten hätte, wenn sie, die Therapeutin, wie eine Mutter für mich (da) wäre, und ich nochmal Kind sein könnte. Sie hatte gefragt, was ich mir denn wünschen würde, nachdem sie betonte, es sei wichtig, meine Wünsche und Bedürfnisse auszusprechen. Ich sagte u.a., ich wolle geliebt werden, wie ich bin, und auch mal in den Arm genommen werden. Ob ich mir das nicht bei meinem Mann oder meinen Freundinnen holen könne. Sie könne verstehen, dass ich das brauche, aber sie benötige für ihre Arbeit einen gewissen Abstand; ungefähr so äußerte sie ihre Haltung dazu.

Ich reagierte mit Schweigen, war in mich gekehrt und plötzlich spürte ich Angst. Angst davor, von der Therapeutin ausgenutzt zu werden. Das sagte ich ihr und fuhr fort: Weil ich Ihnen sagte, was ich gerne hätte; weil ich mich damit schwach gezeigt habe ... Ich fühle mich jetzt ausgeliefert. – Mir war klar, das hat mit meinen Erfahrungen von früher zu tun, aber ich sprach weiter: Ich habe ganz stark das Gefühl, Sie können mich jetzt ausnutzen ... Nicht die Scham über meine Wünsche machte es mir so schwer, Ihnen diese zu sagen. Viel schlimmer war die Angst, das wird mir jetzt ganz klar! – Ich hatte beide Hände zu Fäusten geballt, zeigte sie ihr; so würde ich jetzt hier sitzen. Sie glaube, dass ich mich beschützen könne. Es wäre doch sehr mutig, dass ich das alles trotz meiner Ängste geäußert hätte. Ich überdachte ihre Äußerungen; fand es dann auch gut.

Und dann sagte ich, für mich selbst überraschend und nach der Stunde erst richtig gewahr werdend, was ich gesagt hatte: Und wenn Sie was machen ..., ich bringe Sie um! – Ganz ernst, fest und ruhig sagte ich das. Sie antwortete, ebenso ruhig: Und was ist, wenn ich Ihnen sage, dass ich auch auf Sie aufpassen werde? – Ich brauchte eine Weile, bis mir dieser liebe Satz, nach meinem drohenden, einging. – Ich würde es mir wünschen! –

Nach der Stunde fiel mir meine Aussage wieder ein. Erst zweifelte ich, ob ich das tatsächlich gesagt hatte. Aber so war es, stellte ich erschüttert und entsetzt fest. Ihre Reaktion fand ich sehr lieb. Sie tat mir so gut, und ich fühlte auf einmal großes Vertrauen und Verständnis zwischen uns. In der nächsten Stunde war ich jedoch wieder vorsichtig; sagte mir, ich muss aufpassen, dass sie mich nicht packt und denkt, sie hätte mich jetzt rumgekriegt.

Ich erwähnte einmal, ich würde mich eigentlich bei niemandem trauen, mich ganz schwach zu zeigen, aus Angst, dass ich mich damit ausliefere.

Angst vor der Angst

Bald sprach ich davon, dass ich Angst vor der Angst hätte, die ich nicht beschreiben könne, die aus dem Bauch kommend nach oben steige und von mir Besitz ergreife. Ich hätte dann keinen Zugang mehr zu mir, bekäme Panik. Dies war im Zusammenhang mit ihren anstehenden Therapieferien. Ich hätte Angst, dass ich in Situationen käme, die ich nicht aushalte. Und ganz besonders Angst vor dieser tiefen Angst.

Nach einem Traum mit meinem Vater tauchte genau die auf. Er hatte in dem Traum das ganze Treppenhaus vollgeschrieben. Für mich ein Bild dafür: Das Treppenhaus ist mein Körper, und mein Vater hat von oben bis unten Spuren bei mir hinterlassen. – Inzwischen deutete sich immer mehr an, dass mich, als ich Kind war, nicht nur meine Brüder sexuell traktiert hatten, sondern auch mein Vater. Die Therapeutin sah die Angst ebenfalls im Zusammenhang mit diesem Traum. Ich hätte danach Angst gehabt, in die Welt zu treten, rauszugehen. Davon hatte ich ausführlich berichtet. – Es ging ihnen schlecht, wie lange nicht mehr, aber sie haben es alleine geschafft! Sich auf sich bezogen! Und sind da rausgekommen. – Ich würde das vielleicht deshalb noch nicht wahrnehmen, dass ich das schaffte, weil ich noch nicht wisse, was hinter der Angst stünde. Ihr sei es wichtig, mich darauf hinzuweisen, dass ich diese Kraft und dieses Potential in mir habe.

Angst vor Berührung

Während einer sehr intensiven Stunde wollte die Therapeutin ihre Hand auf meinen Rücken legen, um mich so symbolisch zu stützen. Auf ihre Frage, ob das für mich in Ordnung sei, antwortete ich sofort: Nein! ... Ich hab so eine Angst, dass die Hand meinen Körper berührt! – Ich fragte sie, ob sie denke, dass ich mir diese Berührungen wünschen würde. – Nein! Ich denke, dass Sie die nicht wollen, aber Angst haben, ich könnte das tun. – Was war ich froh, dass sie das sagte. – Ja! Genau das war mein Gefühl! –

Angst vor Arztbesuchen und Krankenhausaufenthalten

Auch dabei können Auswirkungen der sexuellen Gewalt sichtbar werden! Bei einem Termin bei meiner Frauenärztin nahm sie eine Ultraschalluntersuchung mit einem „Stab" vor, den sie in meine Vagina einführte. Ich kannte diese Art der Untersuchung noch nicht. Die Ärztin hatte nicht angekündigt, wie das heute zum Glück meist der Fall ist, welche Art von Untersuchung sie vorzunehmen gedachte. Ich konnte nichts sagen, mich nicht dagegen wehren, ich lag nur einfach da. Irgendwann hinterher musste ich weinen. Damit fertig zu werden, dass ich das erlebte und es nicht verhindern konnte, fiel mir wirklich schwer. Ich rief, nachdem ich in der Therapie darüber gesprochen hatte, die Ärztin an und erzählte ihr alles. Sie reagierte absolut positiv. Sie entschuldigte sich und sagte, sie habe bisher nicht daran gedacht, dass die Untersuchung diese Problematik berühren könne; das wolle sie zukünftig beachten. Sie bedankte sich für den Hinweis, Ich musste sie anrufen, das musste ich für mich tun und war hinterher sehr froh darüber.

Als ich einige Jahre später wegen einer Operation an der Nase ins Krankenhaus musste, kam ich wieder mit meinen Erlebnissen in Berührung. Ich hatte unbeschreibliche Angst davor und dachte tatsächlich, ich schaffe es nicht, dorthin zu gehen. Mich erinnerte das an ein weiteres, traumatisches Ereignis, das ich als 15jährige erlebte. Damals lag ich in der Hautklinik. Eine Schwester nahm mir Urin ab. Sie nahm dazu einen Katheter o. ä., auch sie erklärte nichts. Ich weiß noch, dass ich dachte: Ich muss das aushalten! Ich muss das aushalten! Ist doch nicht so schlimm! So erging es mir auch bei der Frauenärztin: Ich muss das jetzt aushalten.

Meine Therapeutin sagte zu dem Geschehen in der Hautklinik, das alleine sei schon traumatisch gewesen, aber noch mal mehr vor dem Hintergrund der erlebten sexuellen Gewalt. Allerdings würde ich es mir mit dem jetzt geplanten Klinikaufenthalt schwerer machen, als es sei; aber bei dem, was ich erlebt hätte, wäre das sehr verständlich!

Meine Eltern waren damals, als ich in die Hautklinik musste, trotzdem in Urlaub gefahren, nur meine Schwester Lisa kümmerte sich um mich, und das sehr liebevoll. Ich lag über 5 Wochen im Krankenhaus, kleinere operative Eingriffe am Auge und der Nase, es wurde Krebs vermutet, wurden vorgenommen. Ich war dabei immer alleine, davor, danach. Und dazu noch diese Katheter-Untersuchung. Die Therapeutin konnte es kaum fassen, dass meine Eltern mich in der Klinik alleine ließen, mit gerade mal 15 Jahren! Ich weiß noch, dass ich das jahrelang für ganz normal hielt.

Ich ging dann doch in die Klinik und ließ mich erfolgreich operieren. Zuvor sprach ich offen mit der Anästhesistin über meine Angst, dass ich mich bei einer Vollnarkose, meiner ersten, doch nicht beschützen könne und vor welchem Hintergrund dies alles zu sehen sei. Sie sagte, sie würde währenddessen auf mich aufpassen. Was war ich froh, das Gespräch mit ihr geführt zu haben und über ihre bemerkenswerte und tolle Reaktion!

Die Angst während der sexuellen Gewalt wird erfahrbar

In den Therapiestunden tauchten auch Ängste auf, wenn ich z.B. nicht aussprechen wollte oder es noch nicht konnte, was ich erinnerte oder was in mir vorging. Auf diese Zusammenhänge hatte mich die Therapeutin erst aufmerksam machen müssen; mir war das gar nicht aufgefallen.

Nachdem mich nachts ein Typ mit einem Anruf weckte und ekliges Zeug mit sexuellem Hintergrund von sich gab, reagierte ich geschockt und maßlos wütend. Ich fühlte mich verletzt und weinte. Ich hatte meinem Mann am folgenden Morgen erst einmal nichts davon erzählen wollen. Als dann die Angst so langsam aus meinem Bauch herauf kroch und immer stärker wurde, erzählte ich ihm doch von dem Anruf. – Die Angst zog wie Nerven durch meinen ganzen Körper, breitete sich bis zu den Händen aus! Es war fürchterlich! Grausam! Als ich mich wieder hinlegte, kam die Angst wieder hoch. Das kann man nicht lange aushalten, da wird man verrückt!, so schilderte ich es in der Therapie. Ich erzählte aber auch, wie ich mir durch das Reden mit der kleinen Ann helfen und mich beruhigen konnte.

Die K. meinte, dass es sich sicher um Angst und Unsicherheit von früher handelte. Damals, als Kind, hätte ich ja auch nicht gewusst, ob und wann irgendeine Attacke von meinem Vater oder meinen Brüdern kommt. – Und jetzt haben Sie auch Angst, dass auf einmal etwas Unvorhergesehenes geschieht, eine Unwägbarkeit. Hart ausgedrückt: Wieder ein Missbrauch! – Ich schlug meine Hände vor mein Gesicht, verwirrt, erschreckt.

Nun erläuterte die Therapeutin, dass wir das doch schon öfter hatten, wenn ich etwas nicht sage, dass dann Angst auftauche. Und heute Morgen sei das auch so gewesen, ich hätte meinem Mann zuerst nichts sagen wollen. Das sei aber sicher nicht der Grund für die tiefe Angst gewesen. Sie fügte einiges Beruhigendes hinzu, wie: heute sei ich doch sehr eingebunden, regelmäßig mit ihr im Kontakt, also mit mir, hätte feste Beziehungen usw. Es seien somit nicht irgendwelche Überraschungen zu erwarten. Also würde diese Angst von früher herrühren. – Ich würde Ihnen alles Geld geben, was ich habe, wenn Sie was machen, dass ich nie wieder so eine Angst spüren muss! – Sie hatte mich daraufhin lieb angelächelt, und ich fuhr fort: Das kann man wirklich nur dem aller ärgsten Feind wünschen! – (Da fallen mir heute noch welche ein …)

Diese Angst vom Vormittag habe eine andere Qualität als die z.B. vor den Therapieferien, betonte sie. Ich sprach später von der Angst als der schlimmsten, der härtesten, die ich bis dahin erlebt hätte. Die Folge dieser starken Angst war wiederum die Angst genau davor!

Bald darauf malte ich in einer Therapiestunde meine Angst mit Fingerfarben. Es war zuvor um die sexuellen Übergriffe meines Vaters gegangen. Mir fiel es schwer, Erinnerungen und Bilder davon zuzulassen. Deshalb kamen wir auf das Malen, und es mussten Fingerfarben sein. Als ich beginnen wollte, betonte die Therapeutin, jetzt gehe es erst einmal um meine Gefühle, mein Befinden, das solle oder könne ich damit ausdrücken. Ich malte mit schwarz einen Kreis. Dann malte ich von außen nach innen, von allen Seiten, in die weiße Fläche hinein schwarz, bis kaum noch weiß übrig blieb. Schwarz bedeute Angst, sagte ich. Und Dunkelheit. Das sei mein Bauch, und da sei alles voller Angst. Und von außen käme noch mehr Angst hinzu, nähme Besitz von mir.

Und Schwarz und Dunkelheit und Angst erlebte ich bald darauf besonders schlimm in einer Nacht in Italien. Ich war mit Freundinnen für eine Woche dort. Erst nach großem Zögern, ob ich mich das überhaupt traue vor lauter Furcht vor Angst und Panik, war ich doch mitgefahren. Im Text zu meinem Vater schildere ich diese furchtbare Nacht, den Urlaub und die Erkenntnisse, die wir daraus gewannen, ausführlich. Hier möchte ich kürzer davon berichten. Ich war gleich in der ersten Nacht nach dem Einschlafen wieder erwacht, wurde immer unruhiger und so langsam beschlich mich Angst, die immer stärker wurde. Panik. Zittern am ganzen Körper. Meine Beine schlotterten. Es wurde immer dramatischer ... Irgendwann ging ich zu Vera, mit der ich abgesprochen hatte, dass ich zu ihr kommen könne. Sie wäre sogar mit mir frühzeitig, auch schon am nächsten Tag, nach Hause gefahren. Ich überstand die Nacht, die furchtbaren Ängste und die Panik.

Diese Nacht und die Ängste bearbeiteten wir eingehend in meinen Therapiestunden. Bei diesem Bearbeiten sprach ich in einer Stunde von meinen lieben Gefühlen für die K. Kurz darauf nahm wieder Angst, wie ich es ausdrückte, Besitz von mir, Angst so heftig wie damals in Italien. Es gehe mit mir wie in einer Spirale nach unten. Ob ich keine Macht mehr über mich spüren, mich hilflos fühlen würde?, fragte sie. – Ja! – Zum Schluss dieser Stunde meinte die Therapeutin, es wäre gut, dass die Angst da sei, also dass sie jetzt in der Stunde aufgetaucht sei, dann könnten wir damit arbeiten. Auch in der nächsten Stunde, als ich endlich ausführlich von der Nacht in Italien erzählen wollte, spürte ich die Angst aus dem Bauch aufsteigen. Und am darauffolgenden Wochenende hatte ich wieder Angst vor dieser Angst. Es fühlte sich an, als würde die Angst darauf lauern, auftauchen zu können.

Zu all dem sagte die Therapeutin zum Schluss: Ich sehe es so, dass Sie jetzt in der Situation sind, dass Sie den „Missbrauch" innerlich wieder erleben. Egal welchen. Dass Sie alle Gefühle von damals wieder erleben und dass Ihr Körper das ausdrückt. Und es ist gut, dass Ihr Körper das jetzt endlich alles ausdrücken darf! –

Angst annehmen

Sie hatte mich schon bald und immer wieder im Laufe der Therapie aufgefordert zu versuchen, die Angst anzunehmen. Ich solle sie annehmen als einen Teil von mir, und sie mir dann zusammen mit ihr anschauen. Das sagte ich mir dann auch selbst. Es ist doch was von früher! Angst von früher! Das half mir tatsächlich in konkreten Angstsituationen. Aber für das Annehmen der Angst brauchte ich Zeit und die Erkenntnis über den Zusammenhang mit der Wut und dem Hass, aber noch mehr über die Gründe meiner Angst. Erst dann gelang mir das besser bis gut bzw. dann verschwanden die Ängste!

Die K. erläuterte mir meine verschiedenen Ängste, wie sie diese zu dem damaligen Zeitpunkt sah, was mir sehr beim Annehmen half: Da habe sich doch schon ein bisschen was geklärt. Ich wisse zumindest ungefähr, womit sie in Bezug stünden, was es damit auf sich habe. Sie unterschied zum einen die Angst, die im Zusammenhang mit einer konkreten Situation mit meinem Vater stehe. Die andere Angst sei eine alles umfassende Angst, eine existentielle Angst. Sie denke, dass ich die früher hatte, weil ich nicht wusste, wann wieder eine solche Situation auf mich zukommen würde. Ich hätte ja immer damit rechnen müssen, dass mein Vater wieder was mit mir macht. Und dadurch sei bei mir das Gefühl des absoluten Ausgeliefertseins, der absoluten Hilflosigkeit entstanden.

Durch die erinnerten Bilder und Gefühle von Situationen der sexuellen Gewalt mit meinem Vater konnte ich meinen Ängsten auf den Grund gehen. Nun verstand ich, warum ich besonders vor den Nächten Angst hatte, selbst wenn ich zuhause schlief, in meinem eigenen Bett. Noch schlimmer war es, wenn ich wegfahren wollte, woanders übernachten musste. Nach dem Italienurlaub war mir das immer klarer geworden. Sobald ich woanders schlafen wollte oder musste, erinnerte mich das unbewusst an die Situationen, als mich meine Mutter ins Bett meines Vaters schickte und an seine Übergriffe, an die sexuelle Gewalt.

Angst bindet an den Vater

Einen weiteren spannenden und für mich ganz wichtigen Aspekt der Angst habe ich in der Therapie besprochen: Wenn es mir so schlecht geht, ich mich runterziehen lasse, nicht meine Sachen mache, das heißt doch, ich komme nicht von meinem Vater los! Wenn ich aber alles das mache, was „Meins“ ist, dann gehe ich von ihm weg, entscheide ich mich für mich! – Die Therapeutin formulierte es kurz darauf so: Wenn Sie in so einen Strudel mit Ihrem Vater geraten, Angst auftaucht, dann hat die Angst doch auch zur Folge, dass Sie nicht das machen können, was Sie wollen! Dass Sie nicht von ihm weggehen können und klein bleiben! Die Angst steht doch für die Verbote des Vaters! Sie wollen Ihre Dinge tun, dann taucht die Angst auf, *damit* Sie das unterlassen! – Wenn ich aber trotzdem meine Dinge machen, mein Leben leben würde, ginge ich damit von meinem Vater weg! Würde nicht mehr auf ihn hören, sondern auf mich! Nach dem Motto: Jetzt geht es nach mir! Und das bedeute doch, Nein zu sagen!

Heute erlebe ich nur noch selten Ängste, die denen von früher nahekommen. Wenn das doch der Fall ist, eigentlich nur in der Nacht, kann ich gut damit umgehen, auch wenn es sehr unangenehm ist. Ich überlege, was wohl der Grund sein mag, dass die Angst jetzt auftaucht. Manchmal kann ich es mir erklären, manchmal nicht; dann habe ich vielleicht etwas geträumt und weiß es nicht mehr. Jedenfalls bleibe ich meist ruhig.

Misstrauen – Vertrauen

Misstrauen gegenüber der Therapeutin

Mein Misstrauen der Therapeutin gegenüber hat mich leider sehr lange außerordentlich gequält. Vielleicht habe ich auch die Therapeutin manchmal damit genervt, ich weiß es nicht. Für mich jedenfalls war es eine Qual! Ich selbst bemerkte mein Misstrauen anfangs gar nicht. Aber die Therapeutin machte mich schon bald darauf aufmerksam; sie sagte, ich würde sie austesten und sei ihr gegenüber misstrauisch. Darauf reagierte ich verständnislos, verblüfft, verneinte das und wurde ärgerlich, als sie wieder davon sprach. Ich nahm mich nicht misstrauisch wahr. Wusste gar nicht, was sie von mir wollte. Ihr „ich würde sie austesten“ kam bei mir fast wie eine Abwertung an. Nein, das meine sie nicht wertend, sie wolle mich damit auch nicht verärgern, mich nur darauf hinweisen.

Sie verdeutlichte mir, was sie unter „austesten“ verstand, als sie mich einmal auf meine Weigerung hinwies, ein bestimmtes Thema weiterzuverfolgen, obwohl sie meinen gegenteiligen Impuls wahrnahm. Nun wolle ich testen, so die K., hakt sie nach, nimmt sie das ernst, ist sie aufmerksam oder ist es ihr egal. An anderer Stelle sprach sie wieder von austesten und dass ich misstrauisch wäre. Das sei kein Vorwurf, stellte sie klar. Es wäre wichtig zu schauen, warum ich misstrauisch bin. Sie fügte hinzu, mein Vertrauen wäre ganz schlimm gebrochen worden, deshalb bestünden doch Gründe für mein Misstrauen. Und: Ihr Misstrauen hat doch mit dem Thema zu tun! – Ich glaube, damals verstand ich diese erläuternden Worte nicht ganz oder wollte sie noch nicht an mich herankommen lassen.

An anderer Stelle wies sie mich darauf hin, dass ich schon mit sehr viel Misstrauen in die Therapie gekommen sei. Sie habe mir z.B. zu Beginn der Therapie einen Fragebogen vorgelegt, den ich nicht hätte ausfüllen wollen mit der Begründung, ich wisse ja nicht, ob ich dort bleiben würde. Das habe sie akzeptiert. Ich entgegnete, ich könne das heute noch verstehen, fände es auch ok.

Tatsächlich fragte ich mich in der ersten Zeit der Therapie des Öfteren, bin ich bei ihr richtig? Ist sie die Richtige, besonders beim Thema sexuelle Gewalt? Ob sie mir wirklich helfen kann? Mir geht's doch immer schlechter. Das nützt doch alles nichts! Die Therapie hilft mir auch nicht! Sie hat wahrscheinlich selbst keine Ahnung davon, was mit mir los ist! Das Zweifeln an ihr und ihrer Kompetenz setzte mir sehr zu. Obwohl ich doch beim ersten Gespräch bald das Gefühl hatte, ja, hier bin ich richtig, ich will bei ihr Therapie machen, zweifelte ich in der Folgezeit oft daran.

Darüber zu reden war absolut schwierig. Aber die K. bat mich mehrmals eindringlich darum, nicht nur bei diesem Thema, ich solle sagen, was in mir vorgehe, auch wenn es Kritik an ihr sei. Das bringe uns weiter, helfe uns bei unserer Arbeit. Wenn ich mich dann getraute, sie zu kritisieren o.ä., fragte ich sie ganz oft, ob ich nun gehen müsse, ob ich nicht wiederkommen dürfe, weil ich sie doch kritisiert hatte. Immer wieder ihr klares: Aber nein! –

Durch meine Zweifel an ihr bemerkte ich, dass ich mir selbst gegenüber ebenfalls misstrauisch war. Ich bezweifelte nämlich, ob ich überhaupt einschätzen kann, ob sie gut und richtig für mich ist oder nicht. Es war wirklich so, dass ich mir damals nicht zutraute, das entscheiden zu können. Ich war mir meiner Gefühle einfach nicht sicher! Um mich nicht weiter damit zu quälen, kam ich nach einiger Zeit zu dem Schluss: Ich mache jetzt erstmal bei ihr weiter, irgendwann werde ich Klarheit haben. Ich weiß doch nicht, wie es bei einer anderen Therapeutin wäre. Also bleibe ich weiter bei ihr in Therapie. Darüber sprach ich offen mit ihr. Sie beeinflusste mich nicht, überließ es mir und das war gut.

Mein Misstrauen kam deutlich zur Sprache, nachdem sie eine Stunde absagte, die sie eigentlich in ihren Ferien hatte halten wollen. Das war nach gut einem halben Jahr Therapie. Ich äußerte meinen Ärger darüber und sagte, ich würde gerne noch viel mehr von mir sagen, aber aus Misstrauen und vor Scham könne ich das nicht. Nun hätte ich ein bisschen Vertrauen gespürt, das sei aber jetzt durch die Absage angeknackst. Ich sei bei der Ankündigung der Stunde gleich skeptisch gewesen. Mein Misstrauen habe also zu recht bestanden. Ich fügte hinzu: Wie soll ich eigentlich überhaupt noch jemandem vertrauen, wo ich noch nicht mal meinen Eltern vertrauen konnte! – Fast weinend fuhr ich fort: Wie soll ich dann Ihnen, einer Fremden, vertrauen? –

Die K. reagierte, trotz meines geäußerten Ärgers, mit großem Verständnis. Mein Misstrauen habe ganz viel mit meiner Kindheit, meinen Eltern und dem Erleben der sexuellen Gewalt zu tun. Das müssten wir erstmal so hinnehmen, es wäre doch ok. Sie würde mich hier so erleben, wie sie glaube, dass ich mich als Kind gefühlt habe, nämlich: Ich kann niemand vertrauen! – Sie glaube, ich würde an mich die Forderung stellen, ich dürfe ihr gegenüber nicht wütend und nicht misstrauisch sein, da ich befürchte, dass sie dann gehe, sie mich alleine lasse. Ich sagte es wohl nicht, aber in mir war ganz sicher: Ja! –

Es ging immer wieder hin und her, mal spürte ich ein bisschen Vertrauen, dann war ich wieder misstrauisch. Auch darüber sprach ich offen: Ich kann wohl nicht vertrauen und nicht glauben, dass bei diesem Thema Vertrauen gerechtfertigt sein könnte. Ich frage mich nur, wie ich diesen Wechsel von Vertrauen und Misstrauen aushalten soll. – Das wäre nämlich überhaupt nicht gut zu ertragen. Ich könne versuchen, so die K., an die Momente zu denken, in denen ich Vertrauen spürte, mich an Situationen zu erinnern, in denen ich mich verstanden und aufgehoben fühlte. Ja, so würde ich zum Teil auch schon damit umgehen.

Ein anderes Mal bemerkte ich, dass ich ihr schon sehr viel von mir gesagt hatte und fragte mich, ob ich deswegen jetzt Angst haben müsse. Verneinte es, aber leichte Zweifel blieben: Ich hoffe, sie enttäuscht mich nicht!, schrieb ich in mein Tagebuch. Bald darauf staunte ich, dass ich schon auf Adolf und seine Taten und dass er mir Geld gab zu sprechen kommen konnte. – Weil ich jetzt Vertrauen in die K. habe!, notierte ich dazu.

Ich fragte mich damals sogar, ob sie es ernst meint, wenn sie nett oder gar liebevoll mit mir umgeht. Ob das einmalig ist oder ob sie wieder so sein wird. Diese Zweifel tauchten ganz oft auf, obwohl ich es immer wieder erlebte, wie verständnisvoll und zugewandt sie war.

Es kam auch vor, dass ich bezweifelte, ob sie in der nächsten Woche wieder für mich da ist. Und ganz besonders quälendes Misstrauen befiel mich, was ich dann auch als solches erkannte, wenn es um ihren Urlaub ging. Ich konnte mir ganz lange nicht vorstellen, dass sie danach wiederkommt. Besonders in der Zeit, als ich mich wie ein kleines Mädchen fühlte, regredierte, waren ihre Ferien für mich sehr bedrohlich. Aber das hatte nicht nur mit Misstrauen zu tun, so erklärte es mir die K. später. Kinder könnten sich in diesem Alter einfach nicht vorstellen, dass die Mama, oder der Papa, wiederkomme. Das wäre ganz natürlich.

Bei einem Gespräch mit der mich in den ersten Jahren an die K. delegierenden Ärztin erwähnte ich mein Misstrauen. Auch sie meinte, bei dem Thema sexuelle Gewalt sei das ganz „normal" und sehr verständlich. Im Allgemeinen seien drei Jahre dafür notwendig, wieder vertrauen zu können. So lange!, dachte ich. Andererseits beruhigten mich ihre Worte sehr. Sie hatte noch betont, dass meine Therapeutin das doch wisse und mein Misstrauen einordnen könne. Auch das entlastete mich. Nun verstand ich mich und mein Verhalten besser. Es war mir oft eine große Hilfe, wenn ich nachvollziehen konnte, warum es mir so oder so ging oder warum ich wie reagierte.

Der kalte Blick oder: Wie Übertragung funktioniert

Als es in der Therapie mehr und mehr um meine Erinnerungen an die Situationen mit meinem Vater ging, kamen weitere Gründe für Misstrauen und Zweifel an der Therapeutin hinzu. So erlebte ich sie im Umgang mit mir kühl oder sogar kalt, oder ich kam mir von oben herab behandelt vor und fand, dass sie auf mich herabschaut. Wenn sie wenig sagte, mehr zuhörte, wenig kommentierte, war das für mich ganz furchtbar. Das kam bei mir an, als würde sie nicht mehr mit mir reden und ich dachte dann, wahrscheinlich bin ich ihr egal! Sie verhält sich, als gäbe es mich gar nicht, als wäre ich nicht da! Als ich es geschafft hatte, ihr das offen zu sagen, fügte ich hinzu: Sie können sich gar nicht vorstellen, wie oft ich mir sagen muss, wenn Sie nichts sagen, dass das nicht bedeutet, Sie reden nicht mit mir! Das kommt ganz oft vor! Dann muss ich mir das richtig eintrichtern. – Das alles setze mir heftig zu und erinnere mich sehr an meinen Vater: Der konnte so kalt gucken! Das ist nicht zu beschreiben. So kalt, dass man nur noch Angst spürte! – Ich erwähnte weiter, dass ich mir sogar wegen ihres so kalten Blickes ein paarmal vor Angst fast in die Hose gemacht hätte.

Die Therapeutin sprach von Übertragungen. Und ich erkannte nach und nach, sobald wir auf das Thema „mein Vater“ kommen, sehe ich sie so, nehme ich sie so wahr wie ihn. Das ärgerte mich maßlos! Ich musste mir ganz oft sagen: Du weißt es doch, das hat nicht wirklich was mit ihr zu tun! Da geht's doch um den Vater (oder auch mal um meine Mutter), aber nicht um die K.! Ihr kannst du doch vertrauen! – Als das mit dem Vertrauen noch wackelig war, meinte sie, ich würde das doch nur einem Teil von ihr unterstellen, aber vom anderen Teil wüsste ich, dass er oder sie nicht so ist. Das half mir, zumindest zeitweise, meine widersprüchlichen Gefühle besser zu ertragen und auseinanderzuhalten.

Nicht nur im Text zu meinem Vater erwähnte ich bereits meine Zweifel an mir, an meinen gefühlten Erinnerungen und Bildern. Ich hatte mich doch immer wieder gefragt: Kann das wirklich sein? Woher kommen denn diese Erinnerungen auf einmal? Das hätte ich doch wissen müssen! Bilde ich mir das nicht nur ein? Ich konnte bzw. wollte diesen Erinnerungen zuerst nicht glauben! Entsprechend unterstellte ich der Therapeutin, dass sie mir nicht glaubt, was ich an Gefühlen und Bildern erinnerte. Sobald nur eine Nachfrage von ihr kam, dachte oder sagte ich ihr, sie glaubt mir nicht. Auf diese Unterstellungen reagierte sie manchmal harsch, doch letztendlich immer mit Verständnis. Es dauerte eine ganze Weile bis ich verstand, *ich* war es doch, die mir nicht glaubte! Und diese Zweifel übertrug ich auf sie. Eigentlich wolle ich, so die Therapeutin, dass sie mir meine Gefühle bestätigt. Aber gerade das wolle sie nicht tun! Sie wolle nicht auf mich einwirken, sonst hänge es von ihr ab, was ich glaube und was nicht. Einerseits wurde ich sauer auf sie, weil sie so stur reagierte, andererseits war unterschwellig da, gut, dass sie sich so verhält.

Mir wurde zunehmend klarer, dass es doch nicht verwunderlich ist, dass ich gerade der Therapeutin nur schwer Vertrauen entgegenbringen kann – ansonsten erlebte ich mich nämlich nicht misstrauisch. Mit ihr ließ ich mich auf meine Gefühle und die Erinnerungen an meinen Vater und meine Brüder ein. Ließ Erinnerungen zu, wie ich deren Taten als kleines Mädchen erlebte und dass meine Mutter mir nicht half, mich nicht beschützte. Die von mir geliebten Menschen, denen ich doch vertraut hatte, verrieten mich. Ja, so fühlte ich es, die haben mich verraten, meine Liebe zu ihnen verraten! Wie solle ich da noch jemandem vertrauen? Und gerade um diese Themen ging es im Besonderen in den ersten Jahren der Therapie. In der Übertragung sah ich in der K. meinen Vater, meine Mutter. Dann gehört auch das Misstrauen in die Therapie! Dies hatte die K. des Öfteren betont.

Meinen Gefühlen vertrauen

Als die Therapeutin bald darauf meinte, ich wisse jetzt, dass ich mich auf sie verlassen könne, kommentierte ich das nicht gleich, sondern ließ es auf mich wirken. In der folgenden Stunde wiederholte ich diese Aussage von ihr und überlegte laut: Wenn das so ist, dann müsste *ich* ja vertrauen können!? – Da lachte sie: Ja! – ... Das hieße doch, auch *meinen Gefühlen* vertrauen können!? – Jaaaa ... – Sie betonte, ich hätte doch auch andere Erfahrungen gemacht, schon früher, aber auch in meinem jetzigen Leben.

Damit deutete sich für mich zum ersten Mal an, dass, wenn ich Anderen vertraue, es auch bedeutet, ich vertraue meinen Gefühlen. Noch jetzt spüre ich die große Verwunderung bei diesen Gedanken und möchte ausrufen, wie ich es damals vielleicht empfand: Huch, was ist das denn für eine neue Wendung bei diesem unsäglichen Thema? ... Das bedeutet Vertrauen? Mir selbst vertrauen zu können, meinen Gefühlen und meiner Wahrnehmung!? Unglaublich! Daran hab ich ja noch gar nicht gedacht! –

Aber nun war mir das zumindest schon mal begegnet. Im weiteren Verlauf der Therapie kamen wir darauf zurück. Ich spürte tatsächlich, zu meiner großen Freude und Entlastung, zunehmend Vertrauen zu der Therapeutin. Dazu schilderte ich ihr einmal dieses Bild: Sie gehen auf der anderen Straßenseite, aber ganz weit weg. Dennoch fühle ich, dass mich etwas mit Ihnen verbindet. Und ich dachte, das ist mein Vertrauen zu Ihnen! – In einer anderen Therapiestunde sagte ich ihr, obwohl dabei wieder Angst in mir hochstieg: Ich wusste ja schon vor dem Italienurlaub, dass Sie für mich da sind. Aber ich habe das noch nie so gefühlt wie dort. Ich habe richtig Vertrauen gespürt! Und als ich wieder hier war, war das immer da. Fühlte ich das Vertrauen in mir. Und es machte mich ganz ruhig. – Und ich erkannte: Ich kann meine Angst bei ihr zulassen, ich bin gut bei ihr aufgehoben. Und wenn zwischen uns etwas auftaucht, sind es doch Übertragungen! –

Ich finde es so wunderbar, entdeckt zu haben: Ich vertraue! Ich kann doch vertrauen! Auch und gerade meinen Gefühlen und meiner Wahrnehmung, also mir vertrauen. Dass da ein Zusammenhang besteht. Und damit fahre ich gleich fort, denn darum ging es zunehmend. Ich hatte davon gesprochen, die Therapie als Sicherheit im Hintergrund – da könnte ich doch noch mehr riskieren, also mich noch viel mehr trauen. Wieder tauchte Angst auf, als ich darüber reden wollte. Dennoch sprach ich darüber. Die Therapeutin wiederholte, ja, ich könne mich mehr trauen. – Und das hat mit Vertrauen zu tun. Aber nicht, ob Sie mir vertrauen können, sondern ob Sie sich selbst vertrauen können! Ob Sie Ihren Gefühlen vertrauen können; ob das, was Sie tun, wie sich hier auf die Therapie einlassen, gut für Sie ist! –

Nun hatte ich bei all dem Arbeiten an mir und dem Thema Klarheit gewonnen: Es geht nicht nur darum, der Therapeutin (oder Anderen) vertrauen zu können, sondern auch darum, ob ich meinen Gefühlen vertrauen kann. Wenn ich fühle, sie ist zuverlässig oder was auch immer, kann ich diesem Gefühl und damit mir trauen? Darum geht es, gerade *darum* geht es! Das ist doch eine spannende Entdeckung!

Zweifel

Im Text zu meinem Vater kamen meine Zweifel an meinen gefühlten Erinnerungen, an den Bildern, die im Zusammenhang mit ihm auftauchten und deren immer deutlicher werdende Bedeutung bereits zum Ausdruck. Ich komme nun ein weiteres Mal und explizit auf diese Zweifel zu sprechen, weil ich diese Thematik für besonders wichtig erachte. Ich möchte aufzeigen, wie sehr mich diese Zweifel quälten, wie ich mit ihnen umging, welche „Vorteile“ sie hatten und wie ich vom „bilde ich mir das nicht bloß ein?“ hin zu der klaren Haltung gelangte: Ich weiß es! – Zum Schluss sogar zu dieser Überzeugung: Ich brauche nicht, dass er es zugibt! Ich glaube mir! –

Von Anbeginn an, schon bei den ersten vagen Vermutungen, es könnte auch etwas mit meinem Vater gewesen sein, zweifelte ich daran. Ich konnte nicht glauben, dass mein Vater „das“ gemacht hat.

Ich wolle, so sagte ich in einer Therapiestunde, lieber viel reden, damit keine Bilder auftauchten.

Aber sie tauchten auf! Bilder, Gefühle und gefühlte Erinnerungen. Und wieder meine Zweifel. Ich versuchte, wenn es mir so erging, mir gut zuzureden, alles ernst zu nehmen: Nichts ist hier zufällig, alles ist wichtig! Was du fühlst, denkst und was du erinnerst und erzählen willst! – Obwohl ich noch unsicher war, konfrontierte ich meinen Vater und meine Mutter mit dem Thema. Ich wollte meine Erinnerungen mit der Realität konfrontieren, schauen, ob sie Bestand haben. Nachdem ich von meinen offenen Worten meiner Mutter gegenüber, die natürlich alles abstritt, in der Therapie berichtet hatte, sagte die Therapeutin, es gehe nicht um die Eltern, sondern um mich! – Es geht darum, dass Sie sich glauben! Ich will Ihnen nicht die Illusion rauben, aber die werden das weiter verdrängen. Doch es geht ja um Sie! Darum, dass Sie sich glauben!, betonte sie nochmals. Bevor ich bald darauf Erinnertes malen wollte, weil ich es nicht aussprechen konnte, fragte ich sie: Und wenn es nicht stimmt? – Sie beruhigte mich: Das macht doch nichts! Dann wissen Sie irgendwann, dass es nicht stimmt! – Ich musste noch hinzufügen: Aber Sie müssen versprechen, es nicht weiterzusagen! – Ich sage nie etwas weiter! – So handhabten wir das auch in der Folgezeit, was mir sehr half. Ich ließ meine Erinnerungen in den Therapiestunden zu, und das blieb erst einmal unter uns.

Warum ist es nur so schwer, mir zu glauben? – Diese Frage beantwortete ich mir einmal so: Das, was mein Vater nachts gemacht hat, und sein Verhalten am Tag ist einfach nicht zusammenzubringen. Am Tag war er meist ein lieber Vater und (soweit ich weiß) ein überaus korrekter Beamter. Und nachts Kinderschänder und Vergewaltiger! Das macht es so schwer, mir zu glauben! Weil das doch gar nicht sein kann! –

Ich schilderte der Therapeutin meine Überlegungen, wie ich besser mit meinen Zweifeln umgehen könnte. Damit mich mein Zweifeln nicht mehr so quält, wolle ich mir in Zukunft sagen, ich *vermute*, dass mein Vater mich vergewaltigt hat. Es ist eine Vermutung! Es wird sich klären. Außerdem müsse ich doch auf niemanden Rücksicht nehmen. Es gehe doch um mich! – Ja! Es geht um Sie! – Ich sprach auch davon, dass ich sehr viel Zeit brauchen würde, um mir das alles anzuschauen. – Es ist so schwer, sich zu glauben! Wenn nur eine winzige Erinnerung auftaucht, kommt sofort ein riesiger Schwall von Zweifeln hinterher! – Vielleicht schützen Sie sich damit, vermutete die Therapeutin. Zu dieser Sicht kam ich selbst immer mehr. Wenn ich zweifle, gewinne ich Zeit; Zeit, die ich brauche, um das zunehmend sichere Wissen, um das Schlimme zu verkraften. Und dennoch war das Zweifeln weiterhin quälend, besonders dieser Wechsel von zweifeln und, nein, ich zweifle nicht mehr, ich weiß es.

Manchmal rief ich mir meine Ängste und Panikattacken in Erinnerung – dann verschwanden die Zweifel, zumindest für eine gewisse Zeit.

Als es wieder einmal darum ging, dass ich immer noch an diesem Thema bin, meinte die Therapeutin dazu und zu meinem Zweifeln ganz vehement: Dass Sie das, was von Anderen kommt, so wichtig nehmen! Dass das für Sie mehr Bedeutung hat als das, was Sie fühlen! Die Anderen wollen nicht weiterdenken, nicht gucken und fühlen. Und Sie lassen sich davon beeindrucken! Stellen das, was Andere sagen, höher als das, was Sie fühlen! –

Sie mutmaßte bald darauf, dass ich damals sehr wahrscheinlich in meinem Alltag sofort vergessen hätte, was geschehen war; dass ich das vergessen wollte! Das sei verständlich und wichtig. – Ich denke, dass Sie dann auch daran zweifelten, ob das wirklich passiert war. Ihr Vater wird das unterstützt haben, wird das auch verdrängt haben wollen. Ich könnte mir denken, dass er Sie einige Zeit nicht beachtete, um sich daran zu hindern, das wieder zu tun. Das wiederum könnte Ihre Zweifel, ist das wirklich geschehen, bestärkt haben. Und dass es Ihnen dadurch jetzt noch schwerer fällt, das zu glauben! –

Als ich einige Zeit später darüber sprach, dass meine Mutter mir früher mit meinem Vater gedroht habe und ich das so interpretierte, der macht „das" dann wieder, sagte die K.: Das beweist noch mal, dass Ihr Vater Sie sexuell missbraucht hat! Sie zweifeln ja manchmal noch daran. Aber dass Sie Angst davor hatten bei der Drohung Ihrer Mutter, ganz abgesehen davon, ob sie Ihnen das wirklich androhte oder ob Sie bei der Drohung mit dem Vater daran dachten, dass er „das" dann wieder macht. ... Dass Sie *davor* Angst hatten ist ein weiteres Indiz dafür, was er gemacht hat! – Ich zweifele ja gar nicht mehr daran! Ich würde lieber zweifeln, aber es gibt nichts mehr zu bezweifeln. Ich habe es so gefühlt. Es stimmt!, fügte ich damals traurig und niedergeschlagen hinzu.

Im Anschluss an diese Stunde ging ich wie benommen nach Hause. Im Nachhinein fand ich es gut, dass ich meine Gefühle wieder zugelassen hatte, sie so ernst nahm! Gegen alle Zweifel.

Im Laufe der Therapie konnte ich weitere und deutlichere Erinnerungen und Bilder mit meinem Vater zulassen. Wie ich bereits im Text zu meinem Vater schilderte, war das oft mit großer Angst und Abwehr verbunden. Als ich wieder hochkommende Bilder und Gefühle abwehren wollte, forderte mich die Therapeutin mit lieber Stimme auf, ich solle doch hingucken, damit ich endlich der kleinen Ann glauben würde. Und dann sprach sie mich mit *Du* an, was mich sehr rührte: Dann verstehst Du die kleine Ann! Kannst Du zusammen mit der kleinen Ann weinen! ... Und dann kannst Du Dir endlich selbst glauben! –

Ich ließ im Verlauf dieser Stunde noch deutlichere Erinnerungen zu, „sah" sogar zum ersten Mal die kleine Ann in einer Situation mit ihrem Papa. Dazu äußerte ich mich in der folgenden Stunde am nächsten Tag: Ich weiß es doch eigentlich schon lange! Aber seit gestern ist es nochmals anders! –

Ich erkannte: Wenn ich genau weiß, was war, das macht mich stark! Damit nehme ich dem die Macht über mich! – Bisher wären es nur Gefühle und gefühlte Erinnerungen gewesen, aber gestern hätte ich mich und ihn „gesehen", jetzt also auch Bilder dazu!

Daraufhin meinte die Therapeutin, sie sah mich ernst dabei an, sprach mit fester Stimme: Das bedeutet ja, dass Sie den Bildern mehr trauen als Ihrem Körper und Ihren Gefühlen! – Huch! ... Ja, stimmt! –

Ein weiterer wichtiger Aspekt, warum ich an meinen gefühlten Erinnerungen lange zweifelte, ist mir inzwischen klar: Im Gegensatz zu anderem Vergessenen stellte sich bei den Situationen mit meinem Vater die Erinnerung nicht einfach wieder so ein. Da war etwas, das habe ich vergessen, und nun ist es wieder da. So war es leider nicht. Aber in meinen Gefühlen, meinem Körper, da war und ist alles noch sehr deutlich vorhanden!

Ich hätte es bei den Konfrontationen mit meinem Vater zu Anfang gerne gehabt, er hätte gesagt, ja, das habe ich gemacht! Aber das sagte er nicht, nicht in der Deutlichkeit. Doch das war irgendwann kein Thema mehr für mich, denn ich spürte ganz klar: Mir ist es nicht mehr wichtig, ob mein Vater es zugibt! Dass ich mir glaube ist nicht mehr davon abhängig! Ich glaube mir! –

Wut – Hass – Zorn

Wie kann ich dieses umfassende Thema angehen? Mir fielen ziemlich rasch einige wichtige Aspekte dazu ein, die ich meinen eigentlichen Ausführungen voranstellen will.

Abwehr der Wut! Wie und mit welchem Zweck? Mein „ich will nicht mehr leben" gehörte auch dazu – ich richtete die Wut lieber gegen mich als gegen die Anderen. Oder Wut, dann spüre ich meine Traurigkeit nicht mehr!

Wie habe ich die Wut ausgetobt, wie entwickelte sich das Austoben? Vorher z.B. sehr viel Scham! Negative Folgen (Schuldgefühle, Selbstbestrafung) und positive (befreit, stolz auf mich, Wiedergutmachung) der Wut und des Austobens der Wut. Meine Rachephantasien gehören auch hier hin!

Gegen wen richtete sich meine Wut, und wie habe ich sie jeweils speziell zum Ausdruck gebracht?

Werde ich noch geliebt, wenn ich wütend bin? Ich konnte mir zuerst nur Wut *oder* Liebe vorstellen!

Verantwortung für meine Wut übernehmen, dazu stehen! Es ist allein *meine* Wut! Was kommt nach der Wut?

Und eine ganz wichtige Entdeckung: Wut ... und meine Angst verschwindet! Früher, vorher, meine furchtbaren Ängste. ... Waren die dafür „gut", um meine Wut, meinen Hass nicht zu spüren, nicht zuzulassen?

Und ganz besonders lang und ausdauernd: Wut gegen die Therapeutin – um der Wut auf die Eltern aus dem Weg zu gehen!

Wut auf die Therapeutin

Die lange Wutphase auf die Therapeutin begann zu Anfang ihrer Sommerferien im dritten Jahr der Therapie, also zu einem Zeitpunkt, zu dem das Thema sexuelle Gewalt durch meinen Vater bereits länger präsent war. (Doch für mich wurde erst ab dem folgenden Jahr ganz klar, oder ich konnte es ab dann immer mehr annehmen, dass auch mein Vater Täter war.) Ich spürte sehr viel Wut auf die Therapeutin, weil sie meine Regression nicht abstoppte, und es mir dadurch mit den Ferien nicht leichter machte.

Und nun kam sie wieder. Ich sprach gleich in der ersten Stunde nach den Ferien meine starke Wut auf sie an. Ich sagte sogar: Ich hätte Sie verprügeln können! – Sie nahm das alles gelassen auf und schilderte mir ihre Sicht. Zum einen könne man eine Regression nicht einfach so stoppen, zum anderen habe sie mich so wahrgenommen, dass ich Kraft habe, diese Zeit zu überstehen, auch wenn es nicht leicht werden würde.

In der folgenden Stunde beschäftigte mich erst einmal meine Angst, wie sie auf meine Wut reagiert. Wird sie mir irgendwann meine Wut heimzahlen, so wie meine Mutter? Sie wird mir bestimmt nicht mehr entgegenkommen und mir keine zusätzliche Stunde anbieten, wenn sie Urlaub hat. Das gibt es sicher nicht mehr. Nein, sie sei mir nicht böse, und nein, sie komme mir auch wieder entgegen, so die vorübergehend beruhigenden Worte der Therapeutin.

Nach einem „Bild“ mit meinem Vater, also plötzlich aufgetauchten Erinnerungen an sexuelle Übergriffe durch ihn, und sehr starkem Herzklopfen nahm ich telefonischen Kontakt zur K. auf. Wir telefonierten ein weiteres Mal und vereinbarten eine Sonderstunde am Wochenende, weil es mir so schlecht ging. Ich erlebte die K. während dieser Stunde wieder sehr kühl und sprach davon, sie könne mir wohl auch nicht helfen. Sie entgegnete trocken, dann könnten wir die Stunde ja beenden. Aber sie erinnerte mich auch daran, dass ich nach den Ferien doch oft mit Misstrauen ihr gegenüber reagiert hätte. Dann wären doch wieder ruhigere Zeiten gekommen. Es ging hin und her und mir nach der Stunde doch besser. Und dennoch bekam ich so eine starke Wut auf die Therapeutin, und die schrieb ich mir „kraftvoll von der Seele“, wie ich es im Tagebuch beschrieb. Ich würde ihr auch gerne mal so wehtun, wie sie mir wehgetan habe durch den letzten Urlaub; die letzten Stunden davor habe sie nicht mit mir geredet, mir nicht geholfen. Hätte mich doch auch mal in den Arm nehmen können! Sie sei so kalt, einfach eklig! Schimpfte auf sie, nannte sie eine schlechte Therapeutin. Haben sie überhaupt kein Herz?, auch das war in mir, und dass ich sie hassen würde für ihre Kälte, für ihre abweisende Art, ihr überhebliches Getue. Dass es mich maßlos ärgere, dass sie bestimmt, ob wir lieb, ärgerlich, nett oder kalt miteinander umgehen. Ich hätte darauf keinen Einfluss! Ich notierte weiter: Wissen Sie, warum es mir wahrscheinlich so schlecht geht? Ich will Ihnen zeigen, dass Sie mir nicht helfen können! Will Sie von Ihrem hohen Ross runterholen! Das ist meine Art der Bestrafung, auch wenn ich selbst dabei so sehr leide, dass mir alles sinnlos erscheint und ich nicht mehr leben will! –

In der folgenden Stunde berichtete ich von dem wieder sehr starken Herzklopfen und von meinen Wut- und Hasstiraden auf sie. Auf ihre Aufforderung, meine Gefühle doch zuzulassen, entgegnete ich, meine Hasstiraden könne ich nicht wiedergeben, aus Angst, dass sie sagt, ihr reiche es. Diese Angst sei schlimmer als die Peinlichkeit. Als Therapeutin sähe sie doch den Zusammenhang, auch dass die Wut nicht nur mit ihr zu tun habe. Ich sagte, dass ich Wut, aber auch weiterhin liebe Gefühle für sie empfingen würde. Liebe und Hass gehörten doch zusammen, so die K.

In der nächsten Stunde behauptete die K., ich hätte Wut auf sie, weil sie mir meine Schuldgefühle wegen meiner Wut, dem Schreien usw., nicht nähme, nicht die Verantwortung dafür übernähme und mir nicht helfe, die Schuldgefühle loszuwerden. Da bekam ich wirklich Wut auf sie, vorher hatte ich keine gespürt. Das würde doch gar nicht stimmen! – Ich denke, jetzt lassen Sie mich auch allein! – Sie blieb zu meiner Überraschung gelassen, sie sei doch da, sei ganz wach, für das, was ich ihr sage und das, was sie dahinter sehe oder fühle. Sie fragte mich überraschend: Haben Sie Angst, dass nichts Schönes mehr übrig bleibt? – Ja, das denke ich schon lange! – Dann war es irgendwie zäh. Zum Schluss der Stunde, beim Verabschieden, fragte ich sie traurig und zu meiner eigenen Überraschung, ob sie Mittwoch wieder für mich da sei. – Ist Ihre Frage ernstgemeint? – Ja! – Ich fing fast an zu weinen. Sie hielt weiter meine Hand: Aber natürlich bin ich wieder für Sie da! – Ich ging und fing an zu weinen.

Es folgten drei bittere, schwere Stunden mit viel Wut auf die Therapeutin. In einer dieser Stunden reagierte sie sehr erbost. Ich hatte gesagt, ich hätte Wut auf sie, weil sie mich nicht von meinen Schuldgefühlen befreie, sie mir nicht genug helfe usw. Dann auf einmal die K.: Das ärgert mich jetzt wirklich! Ich kann machen, was ich will, das kommt bei Ihnen anscheinend nicht an! Entweder wollen Sie mich nicht verstehen oder Sie verstehen mich wirklich nicht. Ich bin mit meinem Latein am Ende. Ich kann Ihnen nicht mehr helfen. – Ich reagierte geschockt, bekam Angst. Aber ich fand es auch gut, dass sie mal wütend wurde, und das sagte ich ihr. Sie erläuterte mir dann, dass sie heute mal eine ruhigere Stunde hätten brauchen können und was ihr (ganz privater) Hintergrund dafür war. Außerdem sei ich seit den Ferien, also schon drei Wochen, wütend. Die Wut wegen der Ferien könne sie verstehen, aber heute hätte sie meine Wut mal nicht mehr annehmen und aushalten können. Ich sei doch selbst am Ende, hätte auch keine Lust mehr auf Wut, klagte ich. Es wurde dann ruhiger zwischen uns. Sie würde mich auch nicht rausschmeißen, versicherte sie mir.

In der nächsten Stunde sprach ich davon, die Wut gehöre wohl eher nach B. (Wohnort meiner Eltern). Die K. reagierte energisch: Die Wut ist aber hier und nicht in B. aufgetaucht, deshalb muss sie hier bearbeitet werden! – Es sei sehr wichtig, fuhr sie fort, dass ich meine Wut annehmen könne. Ob ich schon einmal auf jemanden eine solch starke Wut gehabt hätte?, fragte sie. – Nein, ich glaube nicht! Nein! Ich bin so ärgerlich, ich kenne mich gar nicht wieder. Ich stehe wie unter Strom. – Sie wiederholte, ich solle meine Wut annehmen, mich selbst mit meiner Wut annehmen.

In der dritten Stunde sagte ich, ich wolle nochmal alles rauslassen, was mich auf sie wütend gemacht habe; mir schiene, das hätte ich bisher noch nicht getan, als wäre das bei ihr noch gar nicht angekommen. Dem widersprach sie, forderte mich aber dennoch auf zu beginnen. Und ich legte los, angefangen mit den beiden Stunden vor den Ferien, wo sie nicht mit mir gesprochen habe, bis zum heutigen Tag. Ich schaute sie zum Schluss an, wollte zu allem stehen. Stille. Ob sie nichts dazu sagen wolle? Nein! Sie habe sich wirklich nicht so erlebt wie ich sie, betonte sie nochmals. Sie wolle mir meine Gefühle nicht ausreden. Es könne sich um eine Projektion handeln, es könne aber auch etwas Persönliches zwischen uns sein. Ich wolle jetzt gehen, vor dem Ende der Stunde, ich sei fix und fertig. Was das hieße?, fragte sie, sichtlich überrascht von meinem Wunsch zu gehen. – Ich kann Ihnen gar nicht sagen, was mich das gekostet hat, Ihnen das alles zu sagen! Bevor ich anfange zu weinen, will ich lieber gehen. – Und das machte ich dann auch.

Nach der zuletzt erwähnten Stunde quälten mich wieder starke Schuldgefühle. In der darauf folgenden Therapiesitzung machte ich mich selbst runter. Die Therapeutin meinte, ich hätte doch nichts Schlimmes gesagt, nur meine Gefühle geäußert. Sie verstehe gar nicht, dass ich so starke Schuldgefühle habe. Ich konnte das kaum glauben. Erleichterung und Verwunderung wechselten sich bei mir ab. Ich sprach darüber, wie sie meiner Wahrnehmung nach meine Beschimpfung aufgenommen hatte. Sie habe ernst und verärgert geschaut. Auch heute würde sie so gucken nach dem Motto, mir reicht's! Ich hätte ihren Blick wieder als cool und auch mal so von oben herab wahrgenommen. Außerdem sei sie gestern und heute zu spät gekommen, das sei für mich ein Zeichen, ich bin ihr lästig, mich kann man nicht ertragen. Die K. konnte es kaum fassen, war fast sprachlos und meinte, sie wisse ja, woher das komme, aber es wäre kaum zu glauben. Aber ich hätte früher ja nicht wütend sein, nie meine Meinung sagen dürfen usw., fügte sie hinzu. Und: Die Reaktionen Ihrer Eltern waren ja so, dass Sie das irgendwann ganz gelassen haben und das war sicher die richtige Reaktion, wenn man sonst damit rechnen muss, fertiggemacht, quasi vernichtet zu werden. – Diese Sätze erleichterten mich.

Die Therapeutin fragte mich: Glauben Sie eigentlich, dass ich Ihre Wut aushalten kann? – Ich habe einerseits sehr viel Angst, andererseits denke ich, ich muss auch viel Vertrauen zu Ihnen haben, sonst würde ich das Risiko nicht eingehen. – Ich freue mich, dass Sie das so sehen. Aber Sie haben nicht nur Vertrauen in mich, sondern auch in sich selbst, dass Sie die Kraft haben, das anzugehen. Was Sie gestern äußerten, war ja Kritik an mir, und ich verstehe, dass es schwierig ist, mich zu kritisieren, mich anzugreifen, Wut gegen mich zu äußern. Aber Sie sind doch nicht hier, um mich zu loben, mich zu beglänzen, um nur lieb zu sein! … Verstehen Sie das? – Ja! Aber ich kann es eigentlich nicht glauben, dass Sie das so hinnehmen. – Danach dachte ich tatsächlich, so entnehme ich es meinem Tagebuch, irgendwann wird ihre Rache kommen!

Jetzt schon 4 Wochen Wut und Schuldgefühle!, notierte ich zur nächsten Stunde, in der ich von meiner starken Angst in der Nacht berichtet hatte. Ich hätte so eine Wut auf meine Eltern, würde denen am liebsten eine Bombe schicken! – Ich halte das nicht mehr aus! Ich bin am Ende meiner Kraft! Und ich halte die Distanz zu Ihnen nicht mehr aus! Ich habe das Gefühl, ich verhungere! – Am liebsten hätte ich gleich losgelegt mit toben! Ich hätte einen Gürtel an, aber noch lieber wäre mir, die auszupeitschen. Die K. forderte mich auf, doch erst einmal nachzuspüren, auf wen ich Wut hätte. Schon verflog meine Wut. Ich sprach nochmal darüber, ich würde doch Nähe brauchen, die Distanz zu ihr nicht mehr aushalten! Ob ich mir nicht vorstellen könne, nachdem ich die Wut auf sie doch ausgedrückt hätte, wieder auf sie zuzugehen. Nein, das sei für mich gleichbedeutend mit alles wieder zurückzunehmen. – Können sie sich nicht vorstellen, wieder Nähe zu mir zuzulassen, *ohne* etwas von Ihrer Kritik usw. zurückzunehmen? Sie müssen doch deshalb nichts zurücknehmen! – Sie fragte, wie lange unsere Eltern denn früher nicht mit uns sprachen, wenn eine nicht brav oder nicht angepasst war? Nur mein Vater habe sich so verhalten, stellte ich klar. Ob denn die Geschwister die Lücke hätten füllen können, wenn ich mich damals so allein fühlte? Nein, die wären dann eher sauer gewesen, wenn ich, oder eine Andere, die Stimmung versaut hätte. Dann hätten sich eigentlich alle abgewandt. – Es herrschte dann absolute Kälte! Man war dann ganz alleine. – Gell, das ist wie verhungern?, sagte sie. – Ja! – Verhungern, das traf es.

Bald darauf wollte ich endlich meine Wut auf die Eltern rauslassen, das gelang aber nicht so recht. Die K. hatte gefragt: Ist es Wut von früher oder von heute? Und auf wen haben Sie Wut? – Und schon spürte ich wieder Wut auf sie! Das machte mich völlig hilflos. Sie wies mich darauf hin, durch das Toben würde ich meine Schuldgefühle nicht loswerden. Außerdem bliebe ich mit Schuldgefühlen abhängig, und diese Abhängigkeit hielte ich aufrecht aus Angst davor, was kommt, wenn frau sich trennt, also wenn ich mich von den Eltern frei machen würde; so verstand ich sie zumindest.

Sie erklärte noch einmal, wie sie die letzten Wochen sah, sagte u.a., sie habe mich ja darauf hingewiesen, dass es *meine* Wut, *meine* Verantwortung und *meine* Schuldgefühle seien. Auch das mache ärgerlich, schaffe Distanz; und auch darüber sei ich wütend geworden. Außerdem bekäme ich meine lieben Gefühlen für sie und andererseits meine Wut auf sie irgendwie nicht zusammen, würde von früher her nur entweder oder kennen. Ich fragte tatsächlich erstaunt, ob das normal sei, in Ordnung wäre für sie. Aber ja, so sei das doch bei allen. Zum Schluss wies sie mich noch darauf hin, was ich in den letzten Wochen alles geschafft und gearbeitet hätte. Wieder sprach ich von meinen Wünschen, nochmal Kind sein zu dürfen, aber damit sei es ja jetzt vorbei. – Fühlen Sie es also wieder so, Sie können nicht bestimmen, haben keinen Einfluss? – Ja! Genau so! Stimmt das nicht? – Das stimmt ganz und gar nicht! –

Dass ein Teil von mir dennoch wusste, dass ich mitbestimmen kann und Einfluss habe, zeigt ein Traum aus dieser Zeit, der mich heute noch zum Schmunzeln bringt:

> *Die K. sitzt in einem breiten sessel, von oben bis unten in ein tuch eingehüllt, und sagt, sie müsse sich schonen, ich solle sie also nicht so belasten. Ich antworte eher ruhig: Ich überweise ihnen ja auch nicht 85 mark, sondern 100! – Das scheint sie zu überzeugen, sie nimmt das tuch weg, ihr kopf damit wieder frei, wir können arbeiten.*

Da muss ich nicht nur schmunzeln, sondern lachen! Der ist doch wirklich köstlich! Sie schmunzelte damals ebenfalls, nachdem ich den Traum erzählte.

In einer der nächsten Stunden kam ich mit einer wichtigen Erkenntnis: Mir ist klargeworden, *ich* mache das, was hier abläuft! Also ob ich Sie böse oder wie auch immer sehe, das hängt von mir ab! Wenn das so ist, dann müsste ich das ja auch ändern können... – Ja... – Diese Einsicht habe mir gutgetan.

Aber ich spürte auch wieder Wut auf sie. In der folgenden Stunde ebenso, auch wieder Herzklopfen. Ich fragte sie, ob sie wirklich glaube, dass ich das hier die ganze Zeit so mache, also auf vieles Schöne verzichte, um mich für meine Wut zu bestrafen? So hatte sie sich geäußert. Ich könne das gar nicht fassen. – Ja, das glaube ich! Sie waren ja eine sehr lange Zeit wütend auf mich und dafür müssen Sie sich bestrafen. Aber es spielen auch andere Aspekt hier mit rein. – Auf die ging sie dann noch näher ein. Sie fragte mich, wie ich die letzten Wochen erlebt hätte. – Ich fand es furchtbar! Einfach fürchterlich! Das alles so lange auszuhalten! Aber ich dachte auch, ich muss schon Kraft haben, sonst hätte ich das nicht so geschafft! Aber manchmal war es hart an der Grenze! – War? Das ist doch noch so, oder? – Ja!, sagte ich betreten und berührt, dass sie das auch so sieht. – Gibt es keinen leichteren Weg? – Nein! Es geht nicht anders!, sagte sie lieb und mitfühlend. Ich wäre lange wütend gewesen, wütend, misstrauisch, zweifelnd. Sie finde es gut, dass ich selbst bemerkte, ich müsse Kraft haben. – Kraft haben Sie! Und zäh sind Sie, ganz schön zäh! – Wir lächelten beide.

Nach einer kleinen Pause fügte sie hinzu: *Sie* müssen da durch! *Ich* muss da durch! *Wir* müssen da durch! – Sie sagte noch anderes Hilfreiches, Versöhnliches, machte mich auch darauf aufmerksam, dass ich mich mit meiner Wut nur sehr schwer annehmen könne. Das sei ein großes Problem für mich. Ich wolle mich von der gewohnten Struktur nicht verabschieden, denn das hieße, sich in diesem Punkt von den Eltern zu verabschieden, also von ihnen loszusagen: Ich lebe nicht mehr nach eurem Muster, habe mein eigenes! – Das bedeute ja Trennung. Und Trennung löse bei mir Angst aus, so wie jetzt ihr Urlaub. Eine alte Struktur aufzugeben sei sehr schwierig, dauere lange und brauche viel Kraft und Ausdauer. Ja, kann ich dazu heute nur laut und vernehmlich sagen!

Diese Worte griff ich in den Therapieferien für mich auf: Ich will diese alte Struktur nicht mehr! Ich will wütend sein können und mich nicht dafür bestrafen! Will mir das holen, was ich brauche! Ich will es unbedingt schaffen, anders mit mir umzugehen, nicht mehr wie diese kalten, lieblosen Eltern! –

Bald darauf kam ich (endlich) in die Therapie mit dieser Aussage: Ich möchte mal den ganzen Frust der letzten Monate hier rausschreien! – Das schaffte ich tatsächlich! Bevor ich anfing zu toben, bat ich sie, dabei besser hinauszugehen. Sie blieb an der Tür stehen. Ja, das sei gut, bestätigte ich. Ich solle erstmal im Raum herumlaufen, munterte sie mich zu Anfang auf. Dann schlug ich auf ein dickes Kissen, brüllte: Ich will diese Scheiße nicht mehr! – Oder schrie einfach nur. Zwischendrin setzte ich mich auf den Boden. Das solle ich nicht tun, sondern weiter im Raum herumgehen. Das machte ich und brüllte auch wieder los. Danach war mir schlecht und schwindlig. Jedes Mal, wenn ich mich hingesetzt hätte, sei ich in mich zusammengesunken, beschrieb mich die Therapeutin. Ich solle nach der Stunde losgehen, bis ich erschöpft sei! Davon würden das Herzklopfen und der Schwindel weggehen.

In der nächsten Sitzung wollte ich gleich weitermachen mit dem Schlagen, Toben und Brüllen, aber ich schaffte es nicht. Irgendwann sagte die K. zu meiner Überraschung: Sie können wohl dem kleinen Mädchen immer noch nicht verzeihen, dass es sich damals nicht mehr wehren konnte, dass es damals keine anderen Möglichkeiten hatte? – Nein! – Weil es mit dem Toben nicht ging, sagte ich, vielleicht ginge es besser im Wald. – Ja, gut, gehen wir in den Wald! – Aas!, sagte ich doch tatsächlich. – Sie müssen ja nicht, war nur ein Vorschlag!, entgegnete sie trocken. Ich solle nach der Stunde rausgehen, joggen, egal, wie kalt es sei; könne mich ja warm anziehen! – Ich jogge auch mit Ihnen! Dabei kann man gut schreien. – Es ist nicht zu fassen, zu was sie alles bereit ist! Ob sie spüre, ich will da raus?, fragte ich. Ja, aber der Wunsch zu schreien sei nicht zielgerichtet. Da fehle noch etwas.

Was noch gefehlt hatte, war in der nächsten Stunde da! Ich notierte in großen Buchstaben in mein Tagebuch: Ich hab's geschafft! Ich hab' so geschrien! Ich habe getobt und geschlagen! Endlich spürte ich Wut und Hass auf meine Eltern! Endlich! –

Erst sagte ich jedoch, das seien doch meine Eltern, das gehe doch nicht! Aber die Wut und der Hass seien doch gerechtfertigt, meinte die K. Ich gebe die verkürzten Notizen dieser Stunde wieder:

Ich stand auf. Nein, sie soll nicht an die bürotüre gehen, erst mal hierbleiben. Ich ging im raum umher. Bleibe bei dir! Achte nicht auf sie!, sagte ich mir selbst. Ich versuchte, gerade und aufrecht zu gehen. – Wollen sie, dass ich ihnen helfe? – Ja! – Sie stellte sich in meine nähe. Ich solle in den bauch rein atmen und beim ausatmen leise töne von mir geben, dann lauter. Ich brachte keinen ton heraus, obwohl ich mich innerlich aufforderte. Sie half, atmete ein und mit leisem anhaltendem ton aus. Ich machte mit. Meine hände hielt ich am rücken, zu fäusten geballt; sie nahm sie nach vorne. Ich solle weiter fäuste machen! Wir atmeten weiter ein und hörbar und immer hörbarer aus. Dann, wir waren schon ganz schön laut, hörte sie auf. Ich machte alleine weiter. Noch lauter werdend. Hörte auch mal auf. Sie half mir weiter und hörte wieder auf. Ich dann alleine, sehr laut. Hörte sie irgendwas machen, sie holte die matratzen, hielt sie. Ich solle draufschlagen. – Das geht nicht, wenn sie in meiner nähe sind. – Sie stellte die matratzen gleich neben mich an die wand. Ich zögerte, griff den ton wieder auf, schrie laut, ohne worte, und schlug zu!

Ich schrie aus dem bauch heraus und schlug so fest ich konnte mit den fäusten. Schrie keine worte, nur so. Machte zwischendrin eine pause. Ermunterte mich selbst und weiter ging's. Ich schrie: Ich hasse euch! So eine scheiße! Ich will da raus! Ihr schweine! Ich verzeihe euch nie! – Aber meist schrie ich einfach nur meinen hass und meine wut und meinen frust über die letzten wochen und monate raus und schlug weiter mit meinen fäusten auf die matratzen. Dann war ich erschöpft. Nein, das war noch nicht genug!, erkannte ich und machte tatsächlich weiter, bis es im hals kratzte. Ich bat um ein glas wasser. Sie meinte, ich solle auf meine kraft achten, es müsse nicht alles auf einmal sein, o.ä. Ich setze mich auf den stuhl. Ich war fix und fertig. Sagte ihr, zu anfang hätte ich zweimal fast eine tachykardie bekommen, wohl wegen meines schlechten gewissens. Ich schwieg, schaute immer mal zu ihr hin, sie lächelte mich an. Und ich wurde dann immer froher. Ich hab's geschafft!

Sie ging, ich blieb noch, weinte ein bisschen und dann freute ich mich so sehr!

Ich berichtete ihr in der nächsten Therapiesitzung, wie ich mich nach dieser Stunde fühlte: Ich war sehr froh! Und wurde immer froher! Wie befreit fühlte ich mich. Größer. Ging aufrechter. Auch beim Selbstverteidigungstraining schlug und brüllte ich nochmal rum. – Die K. strahlte mich an. So nahm ich sie wahr! – Wie schwer es ist einzusehen, dass man selbst etwas tun muss, damit man da rauskommt! Und es dann auch zu tun! Wenn es einem schlecht geht, merkt man gar nicht, dass man zumindest mit verantwortlich dafür ist, wie es einem geht. Und dass man selbst etwas ändern kann! – Ja, aber das sei normal, meinte sie tröstend.

Von wegen Wut auf die Therapeutin, dazu möchte ich noch diese Worte von ihr zitieren: Durch Ihre Wut auf mich sind wir doch immer vorangekommen. Die hat uns doch sehr geholfen bei unserer Arbeit! – Damals war mir das noch nicht so klar. Mit dem Abstand von heute kann ich es nur bejahen.

Es ist mir wichtig, hier nochmals zu erwähnen, dass ich auch Wut auf die Therapeutin hatte, wenn sie z.B. nach meiner Ansicht meine Brüder verteidigte, etwa sagte, die waren doch selbst noch jung, noch Jugendliche. Diese Wut habe ich selbstverständlich nicht durch Toben oder Brüllen rausgelassen. Mit der Wut auf die K. musste ich mich mit ihr auseinandersetzen, alles bearbeiten und klären. Eine gute Übung für den Alltag, kann ich dazu nur sagen.

Damit schließe ich diese Monate der Wut auch hier ab. So lange war ich der Wut und dem Hass auf meine Eltern aus dem Weg gegangen, hatte ich mich selbst gequält und die Therapeutin sicher belastet und genervt. Unglaublich, aber so war es. Das Thema Wut war damit natürlich noch lange nicht abgeschlossen. Auch nicht meine Versuche, dieser aus dem Weg zu gehen. Ich versuchte zeitweise wirklich alles, mal den Hass auf meine Mutter, mal den auf meinen Vater abzuwehren.

Wut auf die Mutter

Bevor ich auf die Wut auf meine Mutter zu sprechen komme, gehe ich nochmal an den Anfang der Therapie zurück. Ich brauchte nämlich einige Zeit, um überhaupt meine Wut wahrzunehmen. Die K. hatte zuvor das eine oder andere Mal gesagt, sie spüre Wut. Ob ich denn keine Wut spüren würde? Das verneinte ich jedes Mal und staunte über ihre Frage. Aber das veränderte sich, wie aus den Schilderungen zu meiner Wut auf die Therapeutin zu entnehmen ist.

Auch mit dem Austoben der Wut tat ich mich anfangs sehr schwer, da es mit viel Scham besetzt war. Es fiel mir schwer, mich „so gehen zu lassen“, wie ich das damals nannte. Und wenn ich dann vielleicht auch noch schreien würde und nebenan wären Andere, das sei doch peinlich! Aber Scham könne ich doch mal aushalten, hatte die Therapeutin dem entgegengehalten. Ich hätte aber auch Angst, was das dann bei mir auslöse, bekäme vielleicht ein schlechtes Gewissen, wenn ich meine Familie auspeitschen würde. – Aber Wut und Hass entwickeln sich doch! Die haben doch ihre Berechtigung! Die sind doch nicht von Geburt an da! – Als ich die ersten Versuche unternommen hatte, meinte ich, das sei doch alles nichts, das sei doch primitiv. Es fiel mir weiter schwer, meine Wut anzunehmen und etwas damit zu machen. Durch die Ermunterungen der Therapeutin und weil dann so viel Zorn und Wut in mir war, schaffte ich es doch!

Ich möchte eine Begebenheit schildern, bei der es um Wut auf meine Mutter ging. Ich wollte dieser Wut unbedingt ausweichen, aber dieses Mal erkannte ich das selbst! Wir kamen darauf, nachdem die Therapeutin meinte, ich würde etwas von ihr wollen wie von einer Mama. Was ich denn von ihr wollen könne?, fragte ich sie. – Zum Beispiel in den Arm genommen werden. – Ja, das wäre doch verständlich, sagte ich etwas trotzig. Sie begründete ihre Vermutung mit diversem Verhalten und Äußerungen meinerseits.

In der folgenden Stunde erwähnte ich eine Vermutung, die ich nach der gestrigen Stunde hatte: Ich glaube, wenn Sie mir das geben würden, was ich mir von Ihnen wünsche, brauchte ich meine Wut auf meine Mutter nicht zuzulassen! Könnte ich damit die Wut zudecken! – Die K. begeistert: Das ist ja wirklich klasse, dass Sie das herausgefunden haben! Was Sie da gestern geschafft haben! – Sehen Sie das auch so? – Ja! Absolut! – Was war ich erleichtert!

Die Therapeutin betonte später, es gehe aber nicht nur um die Wut auf die Mutter, sondern eventuell auch darum, dass ich mir nicht eingestehen wolle, sie auch geliebt zu haben und dass sie auch für mich da war. Außerdem würde ich nur von meiner Mutter reden, eigentlich ginge es doch um meinen Vater!

Dennoch ging es bald darauf wieder um meine Mutter und um meine Mamawünsche an die K., die ich nach ihrer Meinung stellvertretend an sie richten würde. Wieder sprach ich davon, ich wolle nicht mehr leben! Andererseits wolle ich das doch nicht denken! Irgendwann sagte ich: Ich hasse Sie, weil Sie mir das nicht geben, und mich, weil ich das von Ihnen will! – Daraufhin die K. vehement: Das ist doch der Hass auf Ihre Mutter, Ihr „Ich halte das nicht aus! Ich will nicht mehr leben!“, wenn Sie an dem Thema sind, was Sie von mir wollen! Sie halten den Hass nicht mehr aus! Sie spüren wahnsinnigen Hass auf Ihre Mutter, und diesen können Sie nicht aushalten! – Als ich einige Stunden später sagte, ich hätte so einen Hass in mir, sagte die K.: Endlich! – Gleich darauf spürte ich Angst, fragte die K., ob man auch Angst vor der Wut, dem Hass haben könne? – Oh ja! Meist ist die Angst gerade Angst vor der Wut, dem Hass! Ich denke, dass Sie Ihre Angst in einem anderen Licht sehen werden, wenn Sie sich Ihrem Hass gestellt, ihn zugelassen haben! – Doch so weit war ich noch nicht.

Es sei wichtig, so die Therapeutin, anzunehmen und zu akzeptieren, dass ich hasse und wütend sei! Und ich sagte: Mir ist klar, wenn ich den Hass nicht rauslasse, kann ich nicht leben! Nicht ich sein, nicht glücklich leben! –

In der folgenden Sitzung konnte ich mich dem Rauslassen schon etwas annähern. Ich hatte, um meine Mutter darzustellen, Zeitungen und ein Betttuch mitgebracht. Es war schwierig zu beginnen. Ich ließ mir Zeit. Dann zerknüllte ich nach und nach die Zeitungen, packte diese zu einem Haufen zusammen und legte das Tuch darüber. Das sei der Rücken meiner Mutter, wie bei dem Bild, sie liegt in ihrem Bett, hat sich weggedreht, und ich liege bei Vater im Bett. Da ich Zeit brauchte, vereinbarten wir, die nächste Stunde zu verlängern. Ich solle alles liegen lassen, sie würde sich darum kümmern.

Ich konnte am nächsten Tag an diese Stunde anknüpfen, brauchte allerdings wieder eine Weile. Sie holte das von mir Geformte, breitete es so aus wie am Vortag von mir hinterlassen. Ich packte noch ein dickes Kissen unter das Tuch, damit ich draufschlagen konnte. Und das gelang mir dann auch, aber wie! Ich brüllte und schrie, saute meine Mutter verbal runter. Auf einmal bekam ich ganz starkes Herzklopfen; sagte es. Dann: Ich lasse mich davon nicht abhalten! ... Ich höre dich nicht!, weil die Sätze der Alten auftauchten. – Halt die Fresse! Es interessiert mich nicht, was du denkst! – Ich brüllte wie ein Tier. Manchmal nur Töne, aber wie! Wollte dann meinen Gürtel zum Draufschlagen nehmen, die K. intervenierte. Ein Kind nähme die Fäuste oder die Füße! Ich schlug dann mit meinen Fäusten auf das Gebilde, schrie: Warum hast du nicht nach mir geguckt? – Trat mit den Füßen nach ihr, auf ihr herum. Pause. Aber es war noch nicht genug, also machte ich weiter mit Schreien und Schlagen und Treten. Dann war es erst einmal genug. Mein Hals schmerzte vom lauten Schreien und Brüllen.

Ich setzte mich etwas entfernt hin und begann zu weinen: Mutti, warum hast du nichts gemerkt? Warum hast du mich nicht beschützt? – Die K. kam dazu, setzte sich neben mich, legte ihrem Arm um meine Schultern, hielt mich. Ich weinte sehr. – Die kriegt nichts mit! Es ist zum Verzweifeln! Das kommt einfach nicht an. – Meinen Sie nicht, dass sie das gehört hat?, fragte die K. – Nein, die ist wie Beton! – Dann hat sie es auch verdient, geprügelt und getreten zu werden! – Das kam gut. Die K. hielt mich weiter. ... Ja, das war geschafft, und das war gut so!

Das war nicht das erste, aber auch nicht das letzte Mal, dass ich meine Wut auf meine Mutter zugelassen und ausgetobt und rausgeschrien habe. Aber es war ein weiterer wichtiger Schritt und von einer Intensität, die notwendig und angemessen war.

Wut auf den Vater

Kurz darauf ging es um meinen Vater, nicht zum ersten Mal! Ich hatte ein Bild, was ich am liebsten mit ihm machen würde, als Strafe dafür, wie er mit uns Frauen und Mädchen umging, für all seine blöden und ekligen Sprüche und sein Verhalten ganz allgemein. Ich begann in der Therapie dann so, ich wolle ihm mal alles sagen, hätte aber Angst. Sie holte einen Stuhl und packte das dicke, feste Kissen darauf. Mir schlotterten die Knie, als ich mit der Peitsche in der Hand davorstand. Erst konnte ich nur heftig auf das Kissen schlagen, brachte jedoch keinen Ton hervor. Nach einer Weile schrie ich, trat gegen den Stuhl, schlug zu, mehrmals, aber wie! Und dann ging die Peitsche kaputt und mir fuhr es schrecklich in den Rücken mit ganz starken Schmerzen dort. Die K. reagierte sehr besorgt. Schlagen ging nicht mehr, aber nach ihrer Frage: Wollen Sie sich dadurch davon abhalten lassen, ihm zu sagen, was sie denken?, stellte ich mich vor den Stuhl und sagte ihm mit ruhiger Stimme ganz viel – wovon ich in meinem Tagebuch interessanterweise keine Notizen hinterließ. In der folgenden Stunde kamen wir darauf zurück. Mir sei eingefallen, dass ich dabei nicht das Wort „Missbrauch“ hätte aussprechen können, wüsste sowieso nicht mehr, was ich ihm eigentlich alles gesagt hätte. Sie erinnerte, ich hätte unter anderem gesagt, er sei dafür verantwortlich, dass die Brüder mich „angefasst“ hätten.

Ich komme auf diese Stunde zurück. Nach dem Schlagen und den an meinen Vater gerichteten Worten packte ich die Decke um mich herum. Die K. setzte sich zu mir. Ich sprach von Gewissensbissen. – Aber wird es nicht Zeit, dass Sie die Loyalität mit Ihren Eltern aufkündigen und mit *sich* loyal sind? – Sie fuhr mich nach Hause, weil ich vor Schmerzen nicht mehr gehen konnte. Beim Arzt stellte sich später heraus, dass ich mir einen Wirbel ausgerenkt hatte. Das Einrenken war sehr, sehr schmerzhaft. Die Verletzung sei Ausdruck meines Hin- und Hergerissenseins, der Zwiespältigkeit; zum einen die Wut und der Hass, zum anderen meine Bedenken, die zuzulassen und auszutoben, so die K. Aber es war ein Anfang.

Ich schaffte es bald, auch ohne die Ermunterung und Hilfestellung der Therapeutin, meine Wut auf meinen Vater rauszulassen. Dabei bekam ich ob der Heftigkeit fast Angst vor mir selbst. Mir war es unheimlich, dass ich so schreien und hassen konnte. Aber ich machte weiter. Und das passte, auch weil ich die sexuellen Übergriffe meines Vaters immer deutlicher erinnerte.

Die Therapeutin hatte mich darauf aufmerksam gemacht, als ich von meinem Hunger nach einer Mama sprach, dass es doch nicht nur Mama, sondern auch Papa heißen müsse: Ich denke, Sie rufen nach einem Papa! Und Sie haben ein Recht auf einen Vater, der Sie lieb hat, sie achtet. Und der gut mit Ihnen umgeht. Und deshalb müssten Sie doch wahnsinnige Wut auf Ihren Vater haben! Vielleicht taucht der Wunsch nach einer Mama auf, damit sie Sie vor Ihrer Wut beschützt... – Ich hatte einfach Angst vor meiner eigenen Wut. Dennoch ließ ich sie in einer der nächsten Stunden zu. Ich schrie und brüllte. Ohne Worte! Schlug mit meinen Fäusten auf das dicke, feste Kissen und biss hinein. Das kam besonders gut! Dann trampelte ich auf dem Kissen herum und war anschließend sogar stolz auf mich. Aber auch sehr traurig. Und stellte fest: Das war aber noch nicht alles! – Nach der Stunde weinte ich sehr; spürte weiter heftige Wut, war aber zufrieden mit mir.

Ich erwähnte bereits, dass ich mehrere Male Wut auf die Therapeutin hatte, weil ich die Wut auf meine Mutter noch nicht zulassen konnte. Bei der Wut auf meinen Vater war das nicht anders. Mehrmals ging es darum, ich gehe meiner Wut auf ihn aus dem Weg. Aber dann gelang es mir, sie zuzulassen und heftig auszutoben. Zu diesem Zeitpunkt war es sogar so, im Gegensatz zu sonst, dass es mir leichter fiel, meinen Hass auf ihn zuzulassen als den auf meine Mutter. Damals meinte die K.: Vielleicht ist das so, weil wir noch nicht wissen, ob Ihre Mutter „nur" wegschaute oder gar mitgemacht hat… –

Jetzt geht es aber weiter um meine Wut auf meinen Vater. Ich war durch ein Gespräch an den Anfang meiner Leidensjahre erinnert worden, an die „totale, katastrophale, nicht zu beschreibende bodenlose und jahrelange Angst", so notierte ich es. Das griff ich in der Therapie auf und meinte: Und dann fahre ich zu meinem Vater (meine erste persönliche Konfrontation) und spreche so ruhig mit ihm. Hätte ich nicht viel eher wütend sein müssen? Ihm das alles sagen müssen? Schonungslos! – Ich würde jetzt nicht denken, es war nicht gut dort oder dass es falsch war, nein! Aber da müsse noch was Anderes kommen. Ich müsse meine Wut auf ihn auch noch rauslassen.

Die K. vermutete wohl, ich hätte Angst vor meiner Wut. Und sie fragte mich, ob das nicht Wut auf mich selbst sei, weil ich mich 20 Jahre lang von dieser Geschichte, von meiner Vergangenheit abhängig gemacht hätte. Dass ich dafür selbst die Verantwortung tragen würde. Na toll, und der Alte ist dann fein raus!, dachte ich wütend.

Bald ging es wieder einmal darum, dass ich etwas von ihr wollte und dass ich mich damit wieder sehr abhängig von ihr machen würde. Das führte sie näher aus. Hätte ich mich doch nie so auf meine Gefühle eingelassen!, haderte ich mit mir. In mir stieg eine Wut auf sie auf wie selten zuvor! Tolle Therapeutin, knallt mir das alles an den Kopf und dann beendet sie die Stunde und geht! Das gebe ich ihr nicht mehr, dass ich was von ihr will! Mich von ihr abhängig mache und sie ist die Starke! Nee, nie wieder! – Zornig auf diese grandiose Therapeutin!, notierte ich im Tagebuch. Aber auch: Flüchte ich damit nicht vor der berechtigten Wut auf meinen Vater, auf diese alte Sau? – Es ging hin und her in mir.

In der folgenden Sitzung löste es sich schon mal ein bisschen auf, warum ich so besonders wütend auf sie reagiert hatte.

Die K. meinte, ich sei nicht wütend wegen meiner wünsche an sie, sondern weil sie mir die nicht erfülle. – Das ist doch wie eine wiederholung! Ihre wut auf mich, ihr „ich will nichts mehr von ihnen usw." Das hätten sie doch zu ihrem vater sagen können: Obwohl du das gemacht hast, habe ich dich immer noch lieb gehabt! Bin immer noch zu dir gekommen! Wollte trotzdem was von dir! Aber jetzt will ich nichts mehr von dir! Danach nicht mehr! … Sie machen sich ja jetzt

> *runter, weil sie das wieder erleben, also von wegen „die gefühle aus mir rausreißen, wie wegschneiden!" – Ja! – Hier geht es doch um wiedergutmachung!, ganz fest und sicher von ihr. (...) – Wiedergutmachung für das, was sie mit ihrem vater erlebten! Und das haben sie ja hier mit mir reinszeniert, um das wieder gut zu machen. Aber nicht nur deshalb, sondern auch, um das zu bearbeiten. Und ihre wut ist ja angst, weil sie sich nicht vorstellen können, dass und wie sie diese gefühle integrieren können! – Ich versuchte zu verdauen, was sie alles gesagt hatte. Dankbar.*

Ich quälte mich aber erstmal weiter mit dem Hass auf die K. herum. Schon einige Sitzungen zuvor hatte ich erkannt, ich muss meine Wut schneller bzw. gleich zu- und rauslassen. Ich schleppe die viel zu lange mit mir herum! Immer noch! – Aber diese Einsicht brachte (noch!) keine Veränderung.

Zwischendrin schien mir, als sei auch die Therapeutin ratlos. Aber dann stellte sie fest: Das alles hat doch mit dem Treffen mit Ihrem Vater angefangen! Ich glaube, dass Sie maßlose Wut auf ihn haben, weil er nichts zugibt! Ihnen nicht die restlichen Zweifel nimmt! – Ich bekam sofort so einen Hass auf ihn. Ich bringe ihn um! Doch erstmal richtete ich weinend die Bitte an sie, sie solle mir das alles abnehmen, mich davon erlösen! Ja, Erlösung sei das passende Wort! Meine Mutter habe ja was Erlösendes für mich getan, indem sie es zugab und sich auf mich einließ. Aber bei meinem Vater sei das ganz und gar nicht der Fall. Sie kam auf eine Erinnerung von mir von ganz früher zurück, als ich ein Kind war und gebetet hatte, lieber solle Gott mich sterben lassen als meine Eltern. Ich hatte dazu in der Therapie vermutet, vielleicht sei damals in mir der Wunsch entstanden, die Eltern umbringen zu wollen. Hatte noch von einer Spur von einer Ahnung geredet, dass ich die gehasst habe, weil ich damals schon wusste, wie schlimm es war. Nun meinte die K., es wäre notwendig zu akzeptieren, dass ich jemand so hassen könne, dass sogar Mordphantasien auftauchten.

Nach dieser Stunde war nur noch Wut, Hass, Tränen und Verzweiflung in mir und ein Gefühl, mich selbst vernichten zu wollen, weil ich das alles nicht mehr ertragen konnte und wollte! Aber dann erkannte ich: Der Hass gehört doch nicht zur K., sondern zu meinem Vater! Hör auf mit dem Hass auf sie! Der gehört doch zu ihm! Nur das führt dich aus dieser beschissenen Situation heraus! – So war es dann auch. Diese Einsicht half!

Verantwortung für die Wut übernehmen

Es ging bisher viel um Abwehr und Zulassen meiner Wut. Es gibt einen weiteren wichtigen Aspekt: Selbst die Verantwortung dafür zu übernehmen. Anfangs unterstützte mich die Therapeutin sehr, damit ich meine Wut überhaupt zulassen und etwas damit machen konnte. Dennoch brauchte ich oft sehr viel Zeit vor lauter Scham und Hemmungen und Angst. Davon berichtete ich bereits. Nun hatte ich in einer Stunde davon gesprochen, welche Bilder mir zu meiner Wut eingefallen waren. Hätte mich wie ein verletztes Tier, wie eine verletzte Wildkatze gesehen!, war eine meiner Aussagen. Ich kam aber nicht weiter voran, konnte mit meiner Wut nichts anfangen.

Die K. sah dazu mehrere Ebenen bei mir, die sie wie folgt beschrieb: Zum einen soll ich Ihnen sagen, lassen Sie die Wut doch zu! Zum anderen soll ich Ihnen bestätigen, dass es doch in Ordnung ist, dass Sie Wut haben! Soll Ihnen also Ihre Gefühle bestätigen. Wenn ich ja zu Ihrer Wut sage, dann können Sie sie annehmen. Und ich soll die Verantwortung für Ihre Wut übernehmen, weil Sie sie nicht tragen wollen. – Sie fügte verständnisvoll hinzu, dass ich mich nach dem Austoben alleine fühlen würde und dann aufgehoben sein wolle. Das konnte ich nur bestätigen. Sie ergänzte, ich würde das doch gar nicht mehr brauchen, dass sie mich ermuntere oder bestärke. – Außerdem ist es doch egal, was Andere zu Ihrer Wut denken! Es ist doch *Ihre* Wut! ... Wenn es davon abhängt, dass ich sage, ja, machen Sie das doch hier, würde ich die Verantwortung für Ihre Wut übernehmen. Und Ihnen damit einen Teil Ihrer Wut wegnehmen. Aber es ist doch Ihre Wut! Zur Wut gehört ein *Ich!* ... Muss man *ich* sagen. Und man ist alleine! Man kann darüber mit Anderen reden, die können die Wut mitfühlen, aber es bleibt Ihre Wut. Und weil es ganz alleine Ihre Wut ist, grenzen Sie sich gegen die Anderen ab. ... Zur Wut gehört auch Mut! – Tja...

Rachephantasien

Es brauchte noch ein Weilchen, aber das gefiel mir gleich sehr: *Selbst* die Verantwortung für meine Wut zu übernehmen. Und dazu passen für mich meine Rachephantasien und das Thema Wiedergutmachung. Ja, ich hatte Rachephantasien! Köstliche Rachephantasien! Und beides tat mir sehr gut, die Wut und die Vorstellung, Rache zu üben.

So phantasierte ich, meinen Vater zu quälen, richtig zu quälen. Ich dachte an Bücher von Karl May, die ich als 12- bis 14-Jährige gelesen hatte. Darin wurden Gegner gequält, anschließend ließ man sie sich erholen, um sie dann weiter zu quälen. Oder ich stellte mir vor, meinen Vater zusammen mit mehreren Frauen zu verprügeln. Oder die Autoreifen zu zerstechen – sein Auto war etwas ganz Besonderes für meinen Vater. Als ich davon in der Therapie sprach, lachte die K. und meinte, das sei doch eine schöne Phantasie! Besonders gut fand ich die Vorstellung, die Idee hatte ich einem Buch entnommen, an der Beifahrerseite in den Lack zu kratzen: Ich habe meine Töchter sexuell missbraucht! Ich bin ein Vergewaltiger! – Eine Freundin wollte sogar Schmiere stehen, damit ich das nachts verwirklichen könnte.

Diese Rachephantasien, auch wenn nicht in die Realität umgesetzt, hatten etwas Befreiendes und etwas von Wiedergutmachung. Ebenso wie die Vorstellung, Geld für die Therapie von Vater und den Brüdern zu verlangen. Das habe ich wirklich ernsthaft überlegt, weil ich einige Jahre die Therapie selbst bezahlen musste.

Wiedergutmachung sei sehr wichtig!, so äußerte sich die Therapeutin. Jeder habe ein Gefühl für Recht und Unrecht. Genau darum gehe es, dass es wieder zurechtgerückt wird, wieder gut gemacht wird. Auch wenn ich meine Wut rausließe, wäre das Wiedergutmachung. Damals hätte ich die Taten nicht verhindern können, aber jetzt könne ich mit meiner Wut etwas wieder gutmachen. Und so empfand ich das auch.

Was verbirgt sich hinter der Wut?

Abwehr, Zulassen und die Verantwortung für meine Wut übernehmen – das waren wichtige Stationen. Ich komme zum Schluss noch zu einer sehr schmerzlichen und zu einer anderen, spannenden Erkenntnis. Dem einen Aspekt näherte ich mich auf diese Weise: Ich will nur noch hassen und Wut haben!, so meine Aussage, nachdem wir über das Thema Männer geredet hatten. – Und was ist hinter der Wut?, fragte die Therapeutin. – Wut! Und dahinter wieder Wut! Nur Wut! Und das ist gut so! Ich fühle es so, ich habe mich mal entschieden, nur noch Wut zu haben, dann muss ich dabei bleiben! – Dann bleibt die Wut aber bei Ihnen! Und Sie lassen nur einen Teil Ihrer Gefühle zu, den anderen spalten Sie ab. Dann frisst Sie die Wut auf! – Sie sah mich sehr eindringlich an. Ich reagierte erst nach einer Weile auf ihre Worte, zitiere nun direkt aus meinen Aufzeichnungen:

> *Ich hab schon das gefühl, dass dahinter noch was ist, aber das soll niemand wissen. – Wer soll die gefühle hinter der wut nicht wissen?, fragte sie mit ihrer liebevollen mamastimme. Ich spürte in mich rein. – Ich will es nicht wissen!, fest, überzeugt. Sie, ganz ruhig, ganz klar: Also ich fühle es so: Ihr vater soll es nicht wissen! Und ihre mutter soll es nicht wissen! – Ganz langsam, mit pausen, redete sie weiter, ich lauschte ihren worten, ihrer stimme: Und ich soll es nicht wissen! Aber auch doch. Bei mir ist es ambivalent für sie. Aber SIE wollen es wissen! SIE wollen diese gefühle spüren. Und es sind ja IHRE gefühle! Und sie bestimmen darüber, wem sie das sagen! SIE ganz alleine, ob sie darüber reden wollen oder nicht und wem sie das sagen wollen!, alles so ungefähr. Ich war so berührt davon, dankbar! Tränen. Weinte. Schwieg. Stille. Aber auch, sei doch still, mama, du machst es mir ja noch schwerer! –*
>
> *Wenn man das für sich behält, das ist besser! Das würde sonst so unendlich wehtun! Wären unendliche schmerzen!, sage ich weinend. Schweigen. Es kam wohl noch mehr von ihr, ich weiß es nicht mehr. Wieder, sei doch still, … noch*

schwerer ... Ich sagte dann: Das macht einen aber auch stark! Fühlt sich gut an die wut! Power! – Vor wem ich denn so stark sein müsse? – Für mich! Oder für alle? Ich weiß es nicht! – Immer wieder ging mir durch den kopf, was wohl hinter der wut ist. Liebe?

Ich kam in der folgenden Stunde auf ihre Frage zurück, sprach davon, bei mir sei sofort da gewesen: Liebe! ... Ich hab' die doch mal so lieb gehabt! Jetzt ist nichts mehr davon da. Nur noch Hass und Wut und alles schwarz! Tod! ... Hinter der Wut Liebe, aber ...

... da ist auch noch was anderes, ich weiß aber nicht was. Ich möchte es gerne wissen, habe aber angst davor. ... Ich will aber nicht an der wut festhalten. – Ich war jetzt ganz bei mir. Sie fragte: Und wie fühlt sich das an, das andere? – Nach einer weile: Wie eine leere. Wie eine ganz große leere! Wie ein großes loch. – Ehem! – Leere, alles schwarz. ... Nichts da! – Ich sprach fast wie von selbst, so, als würden meine gefühle sprechen. Alles war klar in mir. – Fühle das jetzt deutlich. Hinter dem hass ist noch was. ... Leere. Fühle jetzt nur leere. ... Fühle mich klein. – Fühlte mich wie ein kleines mädchen. So klare gefühle! Fühlte mich weich und war mir meiner gefühle so sicher! – Eine ganz große leere ist da. Nur das! Was ist, wenn ich mich darauf einlasse? ... Dann spüre ich schmerzen. ... – Schweigen. – Und diese schmerzen spüre ich jetzt an meinem ganzen körper. Meinem rücken. – Ehem!, liebevoll, wohltuend. Ich brach dann ab, sah sie an. Ich wusste, die Stunde ist gleich zu ende. – Das gefühl war jetzt ganz klar da! Die leere ... und ein riesengroßes loch. Ich kann's gar nicht richtig beschreiben. – Ehem! –

Und dann meine mama, die therapeutin, mit eindringlichen blick: Dafür gibt es ja vielleicht auch keine worte! – Wofür? – Für die schmerzen. Die leere. Die verletzungen. Dafür sind vielleicht keine worte angemessen. – Ja, berührt, dankbar.

Wir sahen uns an, ich sah in ihre augen, die augen meiner mama. – Jetzt ist wieder da, ich könnte nur mama rufen! Mama, Mama! Schreien! Nur das fühle ich jetzt. – Mit liebevollem blick, weich: Ja... Das passt doch auch. – Angucken. Setzte mich in ihre richtung. Wir sahen uns ganz, ganz lange an. ... Sie sagte dann: Sie sind aber auch sehr mutig! ... Wenn sie mich freitag brauchen, rufen sie mich an! Oder überhaupt an den tagen. Ich bin da! – In ihre augen... – Ja, sagte ich weich, dankbar, froh. Sie, lächelnd: Oder auch, wenn es was zum freuen gibt. – Ja! Gute idee!, ich lächelte auch. Jetzt war ich wieder richtig dort, im jetzt! Fühlte mich erwachsener. Gut.

Wut und Angst

Und nun komme ich zu einer weiteren spannenden Erkenntnis. Ich erwähnte bereits die Aussage der K., dass Angst oft Angst sei vor Wut und Hass. Und dass ich meine Angst in einem anderen Licht sehen würde, wenn ich mich meinem Hass gestellt, ihn zugelassen hätte. Zur Angst gehöre auch Wut; dahinter sei immer auch Wut. – Sie hatten ja als Kind Angst, dass Ihr Vater kommen und was mit Ihnen machen könnte. In diesen Situationen hatten Sie sicher nicht nur Angst, sondern dachten doch auch, der soll wegbleiben, Sie in Ruhe lassen! Hatten sicher starke Wut auf ihn. Also die war damals sicher schon dabei. Vielleicht half Ihnen die Wut, die Angst auszuhalten... –

Ich näherte mich also der Erkenntnis an, dass ein Zusammenhang besteht zwischen Angst und Wut. Zum einen ist Angst oft Angst vor Wut und Hass. Zum anderen sind hinter der Angst oft die Wut und der Hass versteckt. Letzteres fand ich selbst heraus und zwar so: Als ich nachts wieder große Ängste hatte, fielen mir auf einmal die Taten meines Vaters ein. Die Folge: noch mehr Angst! Und Panik! Die K. war in Urlaub. ... Ich kämpfte lange gegen die Angst und die Panik an. Und hatte dann diese wunderbare Idee: Richte die Wut gegen den! Lass die Angst nicht zu! – Es wurde besser mit der Angst. Wut, Hass, die hängen doch mit der Angst zusammen! ... Mit der Angst hat der immer noch Macht über mich! ... Ich malte mir meine Rache aus. Telefonterror, Kinderschänder auf sein Auto zu schreiben. Phantasierte, ihm mit der flachen Hand eine zu knallen: Ich hab's nicht vergessen! – Ihn solange zu prügeln und anzuschreien, so weiter meine Phantasie, bis er vor mir zittert! ... Plötzlich war nur noch die Wut da! ... *Und meine Angst verschwand!* Unglaublich, aber die Angst verschwand! Ließ ich die Wut abflauen, tauchte die Angst wieder auf! Ich holte meine Wut und meinen Hass wieder hervor ... und die Angst verschwand! Nach den Ferien berichtete ich davon in der Therapie:

> *Ich bekam so eine wut und einen hass auf meinen vater! ... Und dann ging die angst weg! Die ging wirklich weg! – Die K. strahlte. – Das kann doch nicht sein! Ich konnte es nicht glauben. – Lächeln. Freude. – Und wenn die angst wieder auftauchte, ließ ich wieder meine wut zu ... und die angst verschwand wieder! – Wir freuten uns beide. – Das hieße dann doch, dass die angst da ist, damit ich meine wut nicht zulasse, die nicht rauslasse! – Ja! – Und mit der angst kehre ich die wut ja gegen mich. Das hat doch was selbstzerstörerisches! Die hat mich in der vergangenheit doch fast umgebracht! Anstatt die wut rauszulassen, nach außen zu richten, richtete ich sie mit der angst gegen mich! –*

Bald darauf notierte ich in mein Tagebuch, als wieder Angst auftauchte: Ich habe doch eine gute Waffe, meinen Hass! – Ich bin jetzt noch überwältigt von dieser Erkenntnis! Ich ging danach immer wieder auf diese Weise mit aufkommender Angst um, und das bezeichnete ich so: Ich setze meine Wut gegen die Angst! – Wenn heute tatsächlich nochmal Ängste hervorkommen, hilft die Wut auch weiterhin. Nicht immer, aber meistens. Wie erleichternd und befreiend war damals diese Entdeckung! Ich bin nicht mehr der Angst ausgeliefert, ich bin wieder Handelnde!

Schuldgefühle

Habe ich mich gewehrt?

Diese Frage hat mich lange sehr beschäftigt und mir nicht wenige Schuldgefühle bereitet, denn, so dachte ich damals, wenn ich mich nicht gewehrt habe, bin ich mit verantwortlich. Auch deshalb musste ich dieser Frage nachgehen.

Schon gleich zu Anfang der Therapie fragte ich mich, ob ich mich vielleicht doch gewehrt habe, das aber nicht mehr weiß, oder ob ich es nicht als ein sich Wehren ansehen kann, weil meine Abwehr lange erfolglos blieb. Ich dachte dabei insbesondere an meinen Bruder Adolf.

In einer der ersten Stunden meiner Therapie kam ich darauf zu sprechen. Ich hatte einen Traum mit Adolf erzählt, indem er zudringlich wurde und ich ihm sagte, er solle mich in Ruhe lassen, ich wolle nicht, dass er mich berührt. Immer wieder wehrte ich ihn so ab. Das solle er endlich mal akzeptieren. Aber er ließ nicht locker. Ich nahm einen Stuhl, um ihn abzuwehren, und lief dann weinend aus dem Raum. Nach der Wiedergabe des Traumes sagte ich, ich sei wütend darüber, dass ich mich nicht habe wehren können. Die Therapeutin wies mich zu Recht darauf hin, dass ich mich doch gewehrt hätte. Ein Wehren ohne Erfolg sei dennoch ein sich Wehren!

Nach circa einem Jahr ging es in der Therapie darum, eine andere Sitzposition zu finden. Ich gehe an dieser Stelle nicht weiter darauf ein, wie ich zuvor saß und warum die Therapeutin vorschlug, mich auf Matratzen zu legen. In der folgenden Stunde hatte ich dazu gesagt, ich wolle mich nicht hinlegen. Für mich selbst überraschend fügte ich hinzu, dann könne ich mich nicht wehren! Sitzen sei besser. Ich wurde dabei ganz ernst, angespannt. Und erinnerte plötzlich den Übergriff von Bernd.

Einige Stunden später meinte die Therapeutin, sie habe ein Bild von mir, dass ich versuchte, die Brüder abzuwehren, mir irgendwie zu helfen, damit so wenig wie möglich geschieht. Ich musste an die Situation mit Bernd denken, als ich zu ihm sagte, Adolf mache das auch, mit dem Hintergedanken, dass Bernd sicher nicht auf einer Stufe mit Adolf stehen wollte. Ich drückte es so aus: Dass er merkt, er ist so eine Sau wie Adolf. – Bernd habe daraufhin gesagt, das sei auch nicht in Ordnung und käme nicht mehr vor. Und habe mich danach in Ruhe gelassen. Also hätte ich mich doch gewehrt. Ja, meine Bemerkung habe etwas bewirkt, bestätigte die K., aber ob sie was bewirkte, hätte von Bernd abgehangen, also ob er dann aufhört oder nicht. Sie fügte hinzu, das will ich hier ebenfalls erwähnen, damals wäre das gut und in Ordnung gewesen, aber heute solle es von *mir* abhängen. Ich würde mich auch heute noch so verhalten, dass die Wirkung meiner Aussagen u.ä. von dem Verhalten der Anderen abhänge. Das war mir damals gar nicht bewusst; mir war jedoch schnell klar, das will ich ändern!

Bald darauf erzählte ich in einer Therapiestunde einen Traum mit meinem Vater, in dem ich ihn damit abwehrte, dass ich ihm in die Fingerkuppe biss. Weil ich diesen Biss im Traum sehr real spürte, meinte die K., es könne ja sein, dass ich ihm als Kind wirklich mal in den Finger gebissen hätte. Eine Äußerung von mir kam dann bei ihr so an, als wäre ich mit dieser Gegenwehr wieder nicht zufrieden. – Sie haben sich doch gewehrt! Und, ich muss es mal so hart sagen, für eine, die gewohnt war, nur zu buckeln, für die Stillhalten die erzwungene Normalität war, ist ein Ausweichen, sich zurückziehen doch schon toll!, meinte sie, auch in Anspielung auf die zuvor beschriebenen Situationen. Das solle ich doch wahrnehmen! Ich gab ihr recht. Aber meine Wut sei so groß, dass mir der Biss in den Finger nicht ausreiche!

An anderer Stelle wies mich die Therapeutin darauf hin, dass eine Frau, die keine Autonomie habe, sich auch nicht wehren könne; ganz abgesehen von den Übergriffen wäre das so.

Nicht nur wegen der eventuellen Veränderung der Sitzposition, die ich bereits erwähnte, sondern auch wegen anderer Begebenheiten war das sich Wehren (können) auch ein aktuelles Thema in unseren Stunden. So hatte die Therapeutin eines Tages angemerkt, ihr wäre in der letzten Zeit aufgefallen, dass ich entweder blaue oder schwarze Jeans tragen würde und dafür gäbe es sicher Gründe. Sie wolle auch unbewusste Signale aufnehmen, und deshalb spräche sie das an. Das war mir einerseits gar nicht recht. War mir fast zu nahe. Andererseits fand ich es auch ziemlich gut, was sie alles bemerkte, ansprach und wie wichtig ihr das war. Nun entgegnete ich, ich müsse diese Jeans tragen: Und am besten noch die Lederjacke! – Ja, und die Stiefeletten heute passen auch dazu!, ergänzte sie. Ich müsse mich so kleiden, denn da wäre so ein Gefühl in mir von: Ich muss mich wehren können! – Ich ballte bei diesen Worten die Fäuste vor meinem Oberkörper. Ich käme aus diesem Gefühl im Moment nicht heraus! Wieder war ich überrascht von meiner Reaktion, konnte sie jedoch sehr gut verstehen. Das Gefühl war so stark: Ich muss mich wehren können! – Ich trug, das will ich hier auch noch erwähnen, zeitweise nur schwarze Kleidung, schwarze Jeans, Shirts und meine schwarze Lederjacke! Das gab mir ein Gefühl von Stärke. Und: Schwarz schafft Abstand!

Als es wieder um meine Brüder ging, kamen wir auf ein Gebot, so will ich es mal nennen, unserer Familie zu sprechen: Für diejenigen, die frau liebt, muss sie alles tun. Für die Familie sowieso! – Das sei gerade das Fatale, meinte die Therapeutin. Ich erkannte kurz darauf selbst: Es ist doch klar, dass mir das Nein schwerfiel, wo ich doch auch sonst *alles* für die tun musste! Vom Schuhe putzen angefangen bis hin zum Hintern abputzen bei meinem jüngsten Bruder, und das noch, als er schon 5 Jahre alt war!

In der Folgezeit kam ich zunehmend zu einer klareren Haltung. Gegenüber Bernd hatte ich mich erfolgreich gewehrt, hatte ich es clever angestellt, ihn mit Adolf auf eine Stufe zu stellen – und ich hatte mit dieser Strategie Erfolg – auch wenn der Erfolg von Bernds Reaktion abhing.

Bei Adolf haben meine Abwehrversuche lange nichts ausgerichtet, was schwer zu verkraften war. Aber ich schaffte es dann doch, dass er mich in Ruhe ließ, auch wenn er es erst einmal weiter versuchte. Ich hatte mein Nein erfolgreich durchsetzen können!

Bei meinem Vater ist es anders. Zum einen war ich noch ein kleines Mädchen, zum anderen spielten dabei noch viele andere Aspekte eine Rolle, die ich hier nicht ausführlich darlegen will, auch weil ich zu der Schuldfrage und an anderen Stellen näher darauf eingehe.

Ich möchte noch erwähnen, dass die Therapeutin bei einem aktuellen Anlass zu mir sagte: Sie sind im Moment an einem Punkt in ihrer Entwicklung, an dem es für sie wichtig ist, nein zu sagen und bei ihrem Nein zu bleiben! – Damit fängt ein sich Wehren an!, denke ich jetzt.

Und wie geht es mir heute mit dem Thema sich wehren können? Ich merke sehr schnell, wenn mir jemand zu nahe kommt, wenn es mir zu eng, zu nahe wird. Zum Beispiel beim Einkauf, an der Kasse stehend. Egal ob Männer oder Frauen, manche haben überhaupt kein Distanzgefühl, keine Grenzen und verhalten sich entsprechend. Ohne zu überlegen trete ich dann einen Schritt zurück oder spreche es an. Manchmal für die Anderen verblüffend, für mich selbst auch mal überraschend, wie klar ich dann bin, wie deutlich ich spüre, das ist mir zu nahe und wie energisch ich mich dann und wann dagegen verwahre. So war es einmal an einem Marktstand, ich sagte zu einem Mann, er möge doch bitte etwas Abstand halten. Er meinte daraufhin tatsächlich, ihm sei es nicht zu eng. Mir aber!, entgegnete ich, fest und deutlich.

Bin ich schuld?

Noch eine Frage, die mich sehr quälte und länger beschäftigte! Wenn ich die Frage stellen würde, wer war verantwortlich, wäre schneller klar, wer das ist. Aber ich habe mich in der Therapie und schon vorher dies gerade nicht gefragt, sondern: Ist es meine Schuld, dass die das gemacht haben? – Wenn eine Freundin dies alles in der Kindheit erlebt und mich gefragt hätte, ob sie schuld sei, hätte ich sofort geantwortet: Nein! Auf keinen Fall! Das ist die Verantwortung der Täter, deines Vaters und der beiden Brüder, aber auch deiner Mutter, die das nicht verhindert hat. Ein Kind trägt dafür keine Schuld! – Die K. hatte mich einmal aufgefordert, mir zu überlegen, was ich bei Anderen dazu denken würde. Das war ziemlich hilfreich!

Aber bei mir selbst konnte ich das erstmal gar nicht so sehen und sagen. Dass ich mich schuldig fühlte, hatte sicher auch damit zu tun, dass mir mein Bruder Adolf Geld gab. Da muss frau sich doch zwangsläufig schuldig fühlen! Aber es gab noch ganze andere Gründe.

Ich meine mich zu erinnern, dass die Therapeutin irgendwann sagte, durch die Übernahme der Schuld hätte ich auf meine Brüder nicht wütend zu werden brauchen. Als hätte die Schuldübernahme mich vor meiner Wut bewahrt bzw., da ich keine Wut zulassen durfte oder konnte (ein Mädchen darf nicht wütend sein!), musste ich die Schuld übernehmen, um all meine Gefühle zu bewältigen.

Was ich ganz sicher erinnere und was mich sehr entlastete waren die folgenden Aussagen der Therapeutin: Die Schuld zu übernehmen könne ein Schutz sein, um die eigene Hilflosigkeit auszuhalten. Es wäre mit das Schlimmste, besonders für Kinder, wenn sie spürten, sie sind absolut hilflos. Das könne kaum jemand aushalten. Das Übernehmen von Schuld verdränge das Gefühl der Hilflosigkeit. Aber nicht nur das, sondern damit hätte ich mit Verantwortung übernommen und das führe dazu, sich nicht mehr so ausgeliefert zu fühlen. Verantwortung zu übernehmen bedeute ja, man handele auch selbst und das entlaste. Also habe ich mich mit schuldig gefühlt, auch um meine Hilflosigkeit und das Ausgeliefertsein nicht spüren zu müssen, diese furchtbaren Gefühle zumindest abzumildern.

An anderer Stelle hatte sie mich darauf hingewiesen, dass die meisten kleinen Mädchen das Gefühl hätten, es müsse irgendetwas (Schlechtes) an ihnen sein, weil die meist männlichen Täter „das" mit ihnen machten. Dieses Gefühl wäre sehr, sehr stark. Dennoch würden die Mädchen wahrnehmen, dass diese Männer etwas Verbotenes tun.

In Bezug auf meine Brüder wurde ich mir immer klarer. Die haben das mit mir machen wollen und gemacht! Die haben mir das angetan! Ich hab das nicht gewollt! Und so konnte ich dann auch sagen: Ich denke nicht mehr, dass ich schuld bin! – Da hatte sich also etwas verändert.

Was meinen Vater anbetraf blieb es jedoch schwierig, zu einer klaren Meinung zu kommen. Meine Mutter würde sagen, so drückte ich es einmal in einer Stunde aus, dass ich selbst schuld wäre, weil ich immer wieder zu ihm hingegangen sei. Aber ich hätte ihn doch lieb gehabt und nur was Liebes von ihm gewollt, betonte ich in der Therapie. Dennoch würde sie das sagen. (Und genau das kam dann auch von ihr!) Darauf die K.: Ja, das sind die Sätze der Mutter. Aber Sie waren doch noch ein Kind! Ihr Vater war der Erwachsene! *Der* hat Ihre Grenzen überschritten! *Der* hat zu wissen, wo Ihre Grenzen sind und die hätte er wahren müssen! – Ich sei auch zu meinem Vater hingegangen, weil meine Mutter nicht für mich da gewesen sei. – Das ist doch ganz normal! Ein Kind braucht Liebe und Zuwendung. Wenn es die von der Mama nicht bekommt, sucht es woanders danach! –

Eine weitere Frage, die ich mir in der Therapie stellte: Habe ich das nicht selbst gewollt? – Aus heutiger Sicht eine völlig abwegige Frage, aber damals tauchte sie auf. Dazu die Therapeutin: Dass Sie sich selbst gegenüber misstrauisch sind, dahinter steht doch die Aussage Ihres Vaters: Du hast es ja selbst gewollt! – Ja, das hatte er zu mir gesagt! Sie erzählte mir von einem Gerichtsprozess, in dem der Stiefvater und der Onkel aussagten, das damals sechsjährige Mädchen habe das ja selbst gewollt. Sie hätten keine Gewalt ausüben müssen. Die K. betonte nochmals, dass meine Verwirrung mit der Aussage meines Vaters „du hast es ja selbst gewollt" zu tun habe: Deshalb haben Sie das Gefühl, ich kann mir ja selbst nicht trauen! –

In der darauf folgenden Therapiestunde berichtete ich, dass mir schon während der letzten Stunde und am Morgen immer klarer wurde: Ich habe das nicht gewollt! Ich weiß es ganz genau! Das wollte ich nicht! – Und es war gut, das genau so und so klar zu fühlen und zu wissen. Die Therapeutin ergänzte: Das Schlimme sei, dass mein Vater mir mit dieser Aussage ein schlechtes Gewissen gemacht und mir eine Mitverantwortung übertragen habe. Das wäre geradezu fatal! (...) – Aber warum bin ich dann trotzdem wieder zu meinem Vater hingegangen? ... Vielleicht wollte ich es einfach mal anders erleben!, beantwortete ich mir meine Frage selbst. Ja, und das sei doch auch verständlich!, so die K.

Es gab noch einen weiteren, mir gar nicht angenehmen Punkt. Die K. hatte gemeint, es sei ganz natürlich, dass ein kleines Mädchen dem Vater gefallen wolle, auch dass es mit den Schwestern und der Mutter konkurriere. Das wäre notwendig. Da hätte der Vater aber sagen müssen, ja, ich hab dich lieb als meine Tochter. Aber die Mutter ist meine Liebste! – *Er* hätte Grenzen ziehen müssen, er, der Erwachsene! – Ich fühlte mich schlecht beim Gedanken, ich habe konkurriert. Meine Mutter hätte doch gesagt, ich sei selbst schuld, ich hätte ihn in Ruhe lassen, nicht mehr zu ihm hingehen sollen. – Egal, ob Sie kokettiert oder konkurriert haben oder was auch immer, es bleibt dabei, die Verantwortung trägt Ihr Vater! *Er* ganz alleine! – In Bezug auf mich könne man nicht von Verantwortung reden. Bei mir ginge es um Gefühle. Ich hätte das Recht, mich ihm gegenüber so zu verhalten, wie ich mich verhielt, auch zu kokettieren. Er hätte nein sagen müssen! Wenn ich mich schuldig fühlen würde, stehe dahinter der Gedanke, hätte ich mich anders verhalten, wäre das nicht passiert. – Als hätten Sie es verhindern können! Das stimmt aber einfach nicht! Da bin ich mir absolut sicher! Das hätte Sie nicht geschützt! – Für solch klare Worte war ich ihr oft sehr, sehr dankbar, wie auch über diese!

Zur Untermauerung erzählte sie von ihren Kindern und deren Verhalten ab einem gewissen Alter gegenüber ihr und ihrem Mann. Das sei eine ganz natürliche Entwicklung, ein ganz natürliches Verhalten. Aber nicht das meines Vaters! Mir wurde dadurch Vieles klarer. Ich liebte meinen Vater doch sehr, freute mich über seine Beachtung. Aber das wollte ich doch nicht! Außerdem war ich doch noch ein kleines Mädchen, das auf keinen Fall „so etwas" von ihm wollte!

Und damit bin ich bei der Antwort auf die Frage, bin ich schuld? Nein, ich war nicht schuld! Die Verantwortung trägt mein Vater! Die K. erläuterte es so: Wir wären zuhause alle in diesem Klima aufgewachsen. Meine Brüder seien Täter und Opfer. Klar hätten sie Verantwortung für ihr Tun, aber sie seien auch diesem inzestuösen Klima ausgesetzt gewesen. Mein Vater habe dieses Klima geschaffen. Aber auch meine Mutter. Sie habe es zumindest geduldet. Es sei die Verantwortung der Erwachsenen, der Eltern!

Als ich mich darauf vorbereitete, mich mit meinen Eltern zu konfrontieren, ging es auch darum. Und das formulierte ich in der Therapie so: Ich würde meinen Eltern gerne sagen, dass allein *sie* verantwortlich sind! Alle beide. Aber besonders mein Vater! – Und das habe ich dann auch gemacht!

Schuldgefühle und Selbstbestrafung

Die auftauchenden Schuldgefühle während der Therapiephase kann ich so einteilen: In Schuldgefühle, die mit dem in der Kindheit Erlebten im Zusammenhang standen, und in solche, die mit dem Bearbeiten des Erlebten und meinen Reaktionen darauf zu tun hatten. Aus schlechtem Gewissen und Schuldgefühlen folgte manchmal, dass ich mich selbst bestrafte. Selbstbestrafung war sowieso bei mir leider oft der Fall, was mich heute noch, wenn ich an manches denke oder davon lese, in fast ungläubiges Erstaunen versetzt. Erst in der Therapie bin ich überhaupt darauf aufmerksam geworden, dass ich mich selbst bestrafte, z.B. wenn ich einen weiteren Schritt, eine weitere Entwicklung gemacht oder irgendetwas erreicht hatte. Unglaublich, aber so war es!

Auf die Schuldgefühle, die mit dem Erlebten in Zusammenhang standen, werde ich zuerst eingehen.

Dazu wieder ein kurzer Ausschnitt aus einer Therapiestunde. Ich musste mich entscheiden, ob ich mit Freundinnen verreise. In dieser Therapiephase ging es besonders um meinen Vater und das, was er mit mir gemacht hatte. Zu dem möglichen Verreisen sagte ich, ich wolle nicht wegfahren, wolle hierbleiben. Ich wolle niemand sehen und niemand so nahe haben. Die Therapeutin reagierte u. a. so darauf: Wenn Sie jetzt nicht wegfahren, bestrafen Sie sich damit selbst! Wenn Sie nicht fahren, dann deshalb, weil Sie sich wertlos fühlen. Wertlos, weil Sie das erlebt haben. Und wertlos, weil Ihre Mutter nicht für Sie da war. ... Und Sie haben Schuldgefühle und bestrafen sich selbst, weil Sie ein Kind waren, das sich nicht helfen konnte. Dass das alles nicht verhindern konnte. –

Von anderen Ursachen für Schuldgefühle schrieb ich bereits. Besonders die Äußerungen „du willst das doch, du hast das doch selbst gewollt, das gefällt dir doch auch" wirkten nachhaltig und lange. Außerdem waren mir schon von Kind auf durch die katholische Kirche, durch den katholischen Glauben, Schuldgefühle sehr vertraut, wurden mir wie eingeimpft. Gerade die katholische Kirche „arbeitet" ausgiebig damit.

Wie Schuldgefühle außerdem entstehen können, zeigt der folgende Ausschnitt. Wir kamen in einer Therapiestunde auf die Wärme und den Körperkontakt bei den Übergriffen zu sprechen. Dass also auch Angenehmes dabei zu spüren war, etwas, was ich nur schwer eingestehen und zulassen konnte. Die Therapeutin betonte: Das heißt aber nicht, dass das Mädchen diese Situation will! – Und weiter: Das ist ja gerade das, was Eine verrückt machen kann! Diese schlimme Situation und dabei auch Angenehmes! Und gerade das löst Schuldgefühle aus! –

Und nun komme ich zu den Schuldgefühlen, die durch das Bearbeiten in der Therapie und das Öffentlichmachen ausgelöst wurden.

Ich erzählte der Therapeutin von den guten Gesprächen mit meinen ältesten Schwestern über die Taten meiner Brüder. Ich war guter Laune wegen des weiteren Schrittes, den ich damit für mich getan hatte. Eine Bemerkung der K. und ein Lachen von ihr im weiteren Verlauf der Stunde fand ich überhaupt nicht witzig. Und schon wurde ich ganz ernst und meine gute Stimmung war dahin. Die K. stellte klar, wie sie ihre Bemerkung gemeint hatte. Ich versuchte, aus diesem Loch wieder rauszukommen, konnte aber nur mit knappen Sätzen weiter von den Gesprächen berichten. Die K. fragte mich, was denn gut sei an dem Zerstören meiner guten Stimmung? Ich ahnte, dass ich mir selbst meine Freude über das Erreichte genommen hatte. Die Therapeutin formulierte es so: Ich hätte, was das Thema sexuelle Gewalt anbetrifft, mein Leben in die Hand genommen, wäre aktiv geworden, würde die Familie, die Brüder nicht mehr schützen. Ich hätte gesagt, was los ist. Die (ersten) Reaktionen meiner Schwestern seien sehr positiv ausgefallen, was zuvor nicht zu erwarten gewesen wäre. Hätten die unangenehm oder blöd reagiert, hätte ich mich nicht selbst bestrafen müssen. Dann wäre ich enttäuscht und traurig gewesen. Nun hätte ich eigentlich glücklich sein können. Aber das ließen mein schlechtes Gewissen und die Schuldgefühle meiner Familie gegenüber nicht zu. Ich hätte Schuldgefühle, weil ich die nicht mehr schütze und mich dabei auch noch wohl und groß und stark fühlte. Mein schlechtes Gewissen habe bei mir öfter die Auswirkung, dass ich mir das Erreichte wieder wegnähme, mich selbst bestrafen würde.

Das empfinde ich heute noch als richtig bitter; das hat mich damals schon sehr genervt und traurig gemacht und nicht wenige Therapiestunden habe ich mir auf diese Art und Weise „zerstört"!

Schuldgefühle zeigten sich auch, wenn ich in der Therapie Wut und Hass auf meine Eltern äußerte oder gar austobte, also schlug und trat und schrie, was mir eigentlich sehr gut tat. Zumindest bei den ersten Wutattacken reagierte ich jedoch mit Schuldgefühlen und daraus resultierender Selbstbestrafung. Diese Selbstbestrafung sah dann wieder so aus, dass ich eine Haltung oder Aussage der Therapeutin gegen mich gerichtet wahrnahm, an ihr herumzerrte, wie sie es nannte, und schon verschlechterte sich die Stimmung zwischen uns. Manchmal musste ich dann mit dieser für mich nicht gut auszuhaltenden Stimmung aus der Stunde gehen, die nächsten Tage überstehen, bis wir das in der Folge- oder den Folgestunden auflösen konnten.

Also Selbstbestrafung aus schlechtem Gewissen und Schuldgefühlen konnte ich gut! Traurig, aber wahr!

Und dann gab es noch dieses Gefühl, diese Aussage von mir: Ich will nicht mehr leben! – Das, was dieser Aussage vorweg ging, will ich hier nicht wiedergeben, aber die Reaktion der Therapeutin auf meine Verzweiflung ist mir wichtig: Sie wollen endlich mal Ihre Schuldgefühle loswerden! Sich nicht mehr verantwortlich für das Geschehene fühlen! – Auch deshalb würden diese Gefühle auftauchen.

Einschränkung und Verwirrung der Gefühle

Liebe und Angst

In einer Therapiestunde sprach ich über meine lieben Gefühle. Wenn ich für jemanden Liebe empfinden würde, hätte ich Angst davor, mich dann ausgeliefert zu fühlen; also dieser Person gegenüber ausgeliefert zu sein. Während ich davon redete, spürte ich diese Angst ganz heftig. Und ich erinnerte mich, dass ich als Kind richtig Liebe empfunden habe. Mit richtig meinte ich, ich wusste nicht nur, ich habe jemand lieb, sondern ich spürte diese Liebe fast körperlich. Dann sei das nicht mehr so gewesen. – Und was ist dann passiert?, fragte die K. Darauf ging ich nicht ein, bekräftigte nur, dass ich das ganz genau erinnere. – Aber als Erwachsene kenne ich das nicht! Spüre ich die Liebe nicht mehr so, nicht mehr so klar, so deutlich. – Als Kind hätte ich für Maria und für eine Tante meine Liebe gespürt, die hätte ich sehr geliebt. Irgendwann hätte ich mir gesagt, ich will nie mehr jemanden lieb haben! Nie mehr! – Das habe ich jetzt zum ersten Mal mit Ihnen erlebt, sagte ich, mit geschlossenen Augen, da habe ich zum ersten Mal wieder meine lieben Gefühle gespürt! So richtig Liebe gespürt! Ja, da habe ich mich endlich wieder gespürt. Meine Gefühle! – Erst in der nächsten Stunde konnte ich auf ihre Frage eingehen: Sie fragten gestern, als Kind hätte ich doch noch liebe Gefühle gespürt und dann nicht mehr. Was dann passiert sei? ... Heute früh dachte ich, dann ist die Katastrophe passiert! … Und ich musste sehr viel weinen… – Ja!, mit liebem Blick. Sie verstand.

Bedürftigsein

Mein Vater hat mir mit seinen Taten aber nicht nur meine lieben Gefühle ausgetrieben, sondern auch meine Bedürftigkeit. Und nicht nur die Bedürftigkeit damals als kleines Mädchen, sondern auch die Bedürftigkeit schlechthin. Für mich ging überhaupt nicht zusammen, dass ich eine erwachsene Frau und gleichzeitig bedürftig bin. Jahrelang haben wir dieses Thema in der Therapie bearbeiten müssen. Für mich galt nur, ich muss funktionieren, nur noch Kopf sein! Und wenn ich funktioniere, kann ich nicht bedürftig sein. Beides ging für mich einfach nicht zusammen! Es hat mich sehr belastet, wie schwer es mir fiel, ein verändertes Bewusstsein darüber zu erreichen. Heute empfinde ich es als sehr schön und wesentlich, es geschafft zu haben, meine bedürftige Seite gut zulassen zu können. Vom Lieben ganz abgesehen!

Und da ich nun schon die lieben Gefühle und die Liebe zu meiner Therapeutin erwähnte, passt es sehr gut, dass ich zum folgenden schwierigen Thema komme, das jedoch mit meiner Liebe zu ihr eher weniger zu tun hatte.

Verwirrung und Vermischung der Gefühle

Ich fühlte noch gar keine Liebe für sie, zumindest nicht bewusst, da tauchten zum ersten Mal ganz andere Gefühle auf, und das geschah so: Ich hatte meine Wut auf meine Brüder ausgetobt. Saß dann auf dem Boden, die Therapeutin kam in meine Nähe, legte ihre Hand auf meinen oberen Rücken. Das war schön, aber mich schauderte auch irgendwie. Ich notierte damals: Ein komisches Gefühl ging durch meinen ganzen Körper. Das kommt mir unnormal vor. Ich kann ihr das doch nicht sagen! – Ich sprach nur davon, dass mir ihre Berührung gutgetan hätte, was ich aber gar nicht verstehen würde. Sie ging darauf ein, erläuterte u. a., es gäbe Situationen, da könne frau mit Worten nicht so stark ausdrücken, dass sie mitfühlt. Da würden Worte nicht passen. Deshalb ihre Berührung. Sie betonte klar und deutlich, wann immer ich das Gefühl hätte, ich wolle eine Berührung nicht, ob nach einer Sekunde oder nach längerer Zeit, müsse ich das sofort sagen! Es sei wichtig, dass ich merke, hier geschieht nur das, was ich wolle. Ob ich das verstanden hätte?, fragte sie eindringlich. – Ja! –

Nun befürchtete ich, es würde sich etwas zwischen ihr und mir verändern; ich traute mich nicht, offen darüber zu reden. Sie versuchte mir klarzumachen, dass ich oft als kleine Ann dort säße und die brauche Geborgenheit, Zuneigung usw. Später schaffte ich es dann doch, ihr zu sagen, ihre Berührungen seien mir durch und durch gegangen. Ich sprach über meine Sehnsucht nach Mutterliebe und sogar davon, dass ich manchmal glaubte, ich könne auch Frauen lieben. Ich verfluchte mich gleich danach, dass ich das auch noch offenbarte. Dachte, nun wird sie mich ganz bestimmt nicht mehr berühren. Als Beweis, dass sich nichts geändert habe, bat ich sie um eine Umarmung. Sie reagierte darauf so:

> *Ich solle sie mal ansehen. Es hätte sich von ihrer seite nichts geändert. Ich konnte sie nicht lange anschauen. – Gucken sie mich doch mal an, frau salandre! – Dies sagte sie ganz eindringlich. Ich versuchte es erneut, irgendwie fand ich ihren blick so hart, starr, gar nicht so lieb. Vielleicht weil ich was anderes wollte? Sie wolle nichts tun, keine umarmung oder ähnliches, damit nichts zugedeckt werden würde. Das wäre zwar hart für mich, das auszuhalten, aber ich solle gucken, was bei mir ist, was das bei mir auslöst. Die K. weiter: Diese gefühle und diese situation jetzt haben mit dem sexuellen missbrauch zu tun. Sie gehen etwas an, nehmen das in die hand, indem sie das jetzt angesprochen haben. –*

Auf meine Angst, von der Therapeutin nun abgelehnt zu werden, kam ich nach einiger Zeit zurück. Ich warf ihr vor, es habe sich doch etwas verändert! Mir käme es so vor, als rücke sie immer weiter von mir weg, nachdem mir ihre Berührungen so guttaten, ich das auch sagte; ich auch über die Angst vor Abhängigkeit und davor, dass etwas Furchtbares passiere, gesprochen hatte und von meiner Angst vor Strafe. Von ihr aus habe sich nichts verändert, betonte sie wieder. Vielleicht habe sich bei mir etwas verändert, vermutete ich. Ich würde mich fragen, wieso ich Angst hätte, dass etwas Furchtbares geschehen könne? Weil ich es als angenehm empfand, von ihr, einer Fremden, Zuwendung zu bekommen? Ich beschrieb die Gefühle als „angenehm", weil ich mir noch nicht eingestehen konnte, dass es erotische Gefühle waren. Unbewusst war das schon da, denn ich hatte hinzugefügt, das sei doch unmoralisch. – Weil ich eine Frau bin?, fragte sie. – Nein, das wäre bei einem Mann nicht anders! – Das wollte sie nicht recht glauben. Sie erinnerte mich daran, ich hätte doch von homoerotischen Gefühlen gesprochen, also davon, vielleicht auch Frauen lieben zu können. Mir wurde immer peinlicher zumute, aber ich gab zu, auch schon daran gedacht zu haben. Ich würde mich ärgern, dass ich das überhaupt aussprach. Ich hätte sie am liebsten gefragt, ob sie sich deshalb jetzt zurückhalte. Sie meinte noch, auch wenn ich nichts gesagt hätte, sie würde auch so einiges wahrnehmen, mir ansehen usw. Oh, wie war mir das alles peinlich! Das spüre ich noch heute. Ich weiß aber auch, dass ich es schätzte, dass sie aufmerksam war.

Bei einer anderen Situation, viele Monate später, sie hatte mich in eine Decke gepackt, saß in meiner Nähe und legte irgendwann ihre Hände auf meine Füße. Wieder schauderte mir so wohlig. Unbeschreiblich!, notierte ich dazu. Sinnliche Gefühle, die mir durch und durch gingen. Sie legte eine Hand auf meinen Rücken, aber das war mir zu nahe, ich sagte: Bitte nicht! – Sie nahm ihre Hand sofort weg. Ich zog auch meine Füße weg und sagte, ich hätte mich doch eigentlich nicht mehr von ihr berühren lassen wollen. Wir lächelten beide. Sie hatte mich in der Stunde zuvor an diese meine Aussage erinnert, auch an das folgende Misstrauen ihr gegenüber. Jetzt ließ ich meine Gefühle wieder zu, streckte meine Füße wieder aus und genoss es, dort zu liegen. Ich hatte mir von ihr gewünscht, etwas vorgelesen zu bekommen; vielleicht aus einem Märchenbuch. Ich war damals in einer Phase, in der ich mich sehr klein fühlte. Sie ging liebevoll auf meine kindlichen Wünsche ein. Tatsächlich hatte ich es geschafft, diesen Wunsch zu äußern! Und in der nächsten Stunde brachte ich ein Kinderbuch von Janosch mit – Traumstunde für Siebenschläfer. Und lauschte ihrer angenehmen Stimme, als sie mir dieses vorlas. Zuvor hatte sie sich wieder zu meinen Füßen gesetzt, diese in die Decke gepackt und ihre Hand zeitweise darauf gelegt. Zum Schluss bat ich sie darum, dass doch nochmal zu tun. Ja, das könne sie machen, kam in einem irgendwie seltsamen Tonfall, glaubte ich.

Bis hierhin war erst einmal alles sehr schön. Darauf folgten zwei furchtbare und peinliche Stunden. In der ersten sagte die K., es gäbe etwas, worüber wir sprechen müssten:

> *Mir ist nicht so klar, ob ihre wünsche wirklich nur kindliche wünsche sind oder ob das nicht auch homoerotische wünsche sind. – Zong! Ich war wie vor den kopf geschlagen! Kaum noch fähig zu denken! – Wenn es ganz klar ist, dass es kindliche wünsche sind, bin ich klar die mutter, übernehme ich die mutterrolle. Wenn das aber nicht klar ist, dann wird es schwierig. Wenn sich die bedürfnisse vermischen, müssen wir aufpassen!, so o.ä. Ich war sehr verwirrt, sagte, ich hätte das gefühl, dass es eher kindliche wünsche seien, die ich an sie habe. – Ich habe mich dabei eigentlich als kind gefühlt. – Nachdem ich meine wünsche wohl aufgezählt hatte, meinte ich, vielleicht würde es dadurch klarer. Sie verneinte das und sagte: Es sind nur zwei kleine worte, aber ich denke, das sagt uns schon etwas. Sie sprachen davon, dass es* eher *bzw.* eigentlich *kindliche wünsche seien. – Mir war das alles so peinlich, ich fühlte mich zurückgestoßen. Hätte ich doch gar nichts von meinen wünschen gesagt! Fühlte mich erwischt. Sie versuchte, mir klarzumachen, dass sie nicht dafür da sei, meine bedürfnisse zu befriedigen, sie wäre keine partnerin für mich. Diese bedürfnisse, wohl die homoerotischen, wie sie das ausdrückte, gehörten nicht in unsere beziehung, in die Therapie. Außerdem müsse das thema „missbrauch“ mit bedacht werden, bei der konstellation ich das kind, sie die mutter; besonders dann, wenn ich nicht nur kindliche wünsche an sie hätte.*

Zwischendrin sagte ich, ich fände das alles sehr schwierig, zum kotzen, verstünde gar nichts mehr, würde mich hilflos fühlen. Das könne sie verstehen, aber es wäre wichtig, das anzusprechen, zu klären. Und dass wir uns bei anderer gelegenheit meine homoerotischen wünsche noch näher ansehen sollten, z.b. warum die gerade jetzt aufgetaucht seien.

Erst fühlte ich mich sehr vor den kopf gestossen, verwirrt usw., dann hatte ich das gefühl, ich verstehe so ungefähr, worum es ihr geht. (...) Ich fragte noch, ob sie denke, dass das trotzdem noch mal möglich sei, dass ich mich dort hinlege. – Ja, theoretisch ist das sogar gleich möglich, aber die stunde ist jetzt zu ende. – Ich drückte noch einmal meine verwirrung aus, meine hilflosigkeit und was weiß ich noch alles.

Darauf folgte die zweite furchtbare Stunde, so bezeichnete ich sie damals in meinen Aufzeichnungen. In dieser betonte die Therapeutin, sie müsse beim ersten Anzeichen reagieren, wenn meine Wünsche bzw. Gefühle *nicht* kindlicher Natur seien. Es wäre besser, solange die Ebenen noch unklar seien, dass sie mich nicht berühre. Ich sagte: Strafe! – Sie: Wenn Sie das als Strafe ansehen, deutet es für mich darauf hin, dass Ihre Wünsche doch eher auf der Beziehungsebene liegen. – Ich fragte sie zum Schluss der Stunde tatsächlich, ob ich wiederkommen dürfe, was sie klar mit Ja beantwortete. Mein Gefühl, als ich ging: Nur raus hier. Hier bekommst du keine Hilfe, schon gar nicht das, was du brauchst! –

Die darauf folgende Stunde gestaltete sich sehr viel versöhnlicher. Ich möchte meine Aufzeichnungen davon mehr oder weniger komplett zitieren, weil sie das Komplizierte dieses Themas gut wiedergeben.

Mir war, wie schon vor der stunde, nicht danach, irgendetwas zu sagen aus angst, es geht wieder so dramatisch ab. Schwieg sehr lange, sagte ihr dann, warum ich schwieg, auch um nicht verwirrter zu sein als vorher. Schweigen. Dann sagte ich, mein gefühl sei, ich darf nicht mehr kommen. – Dazu muss ich aber etwas sagen, auch wenn sie nicht viel reden wollen. Wie kommen sie darauf, dass sie nicht mehr kommen dürfen? – Wegen meiner gefühle für sie! Ich habe sie so verstanden, die gehören nicht hierher. Dann muss ich wohl gehen. – Sie beugte sich vor, sagte ganz ernst: Das ist aber falsch! ... Haben sie das verstanden? – Ja, sagte ich leise, weich und sehr froh. – Diese gefühle dürfen sie doch haben! Es ist doch sogar gut, sich zu verlieben, egal ob in einen mann oder in eine frau! Man verliebt sich doch in jemanden, der etwas hat, was man selbst gerne hätte, der aber auch was hat, was man an sich selbst sehr mag. Das ist doch sogar gut! Diese gefühle dürfen sie doch haben! – Sie sagte noch mehr, ich bekam's nicht richtig mit.

– Ich hab das gefühl, ich muss weglaufen, muss gehen. – Können sie sich auch vorstellen, dass sie hierbleiben und mit mir zusammen versuchen zu schauen, was dahintersteckt? Dass wir das zusammen rausarbeiten? – Ich konnte es kaum glauben, dass sie auf einmal wieder so lieb mit mir umging. – Ich bin doch hier, bin doch wiedergekommen! Ich will doch selbst wissen, was mit mir los ist! Ich möchte doch hierbleiben! –

Etwas später: Mir ist, als wäre mir das thema schon lange bekannt, ohne dass ich weiß, was das problem ist. Das ging los, als ich merkte, ich mag sie. Auch bei meiner ersten Therapeutin ging es mir so, und ich dachte, irgendetwas stimmt nicht mit mir. ... Irgendwie kommt mir das lange bekannt vor. – Sie fand das sehr interessant. Durch ihre gute stimmung ermutigt, fragte ich sie, was sie mit ihrer aussage meinte, sie habe angst vor meinen homoerotischen gefühlen? Sie fragte zurück, was ich dazu gedacht hätte? – Ich weiß nicht, vielleicht, übertrieben ausgedrückt, dass sie denken, ich überfalle sie. Oder dass sie ekel davor empfinden. – Nein, das wäre es nicht. Angst wäre auch nicht der richtige ausdruck. Sie habe sorge, dass ich annähme, eine solche beziehung bestünde zwischen uns. – Ich will und werde diese gefühle nicht erwidern! Und mir ist wichtig, dass es immer klar ist, auf welcher ebene wir uns befinden. – Sie wolle mich davor bewahren, dass auch nur der anschein bestünde, wir hätten eine solche beziehung. ... Ich hatte wieder einmal schwierigkeiten, sie zu verstehen, hatte immer noch das gefühl, ich müsse gehen, was ich ihr sagte. Schwindel im kopf.

Sie denke, dass hier ganz stark meine erfahrungen mit sexueller gewalt mit reinspielen würden! Dadurch bei mir die verwirrung der gefühle, dass ich nicht feststellen kann, sind es kindliche oder erwachsene gefühle. Ich war sehr erleichtert über ihre erklärungen, dann kann ich die verwirrung besser ertragen. Sie betonte nochmals, dass wir im auge behalten sollten, dass mir dieses thema bekannt vorkomme. Und dann sagte sie etwas sehr schönes; sie legte eine hand unterhalb ihres halses und meinte: Ich habe ein ganz warmes gefühl dabei, dass dieses thema bei uns aufgetaucht ist! Dass wir beide in diesem thema stecken! Nicht ihre erste therapeutin, auch niemand anderes, sondern sie und ich. Und ich denke, dass wir das auch zusammen lösen werden. – Wie froh bin ich, dass sie das sagte!

Ich notierte danach: Ich freue mich so, dass ich meine Gefühle für sie nicht wegmachen, nicht unterdrücken muss! Wenn ich das tun müsste, wäre es so, als dürfte ich mich nicht selbst lieben. So kommt mir das vor. – Wie sehr ich sie lieb habe, gestand ich ihr dann ganz deutlich ein paar Stunden später. Doch jetzt geht es weiter um das Thema Berührungen, dass die nicht mehr möglich waren, was das bei mir auslöste und wie wir damit umgingen. Zwei Monate später kamen wir darauf zurück und zwar so:

Ich wollte mir gerne von ihr vorlesen lassen, mich dazu hinlegen, und das sagte ich ihr und fügte dann hinzu, es gehe aber nicht. Auf ihre nachfrage antwortete ich: Ich fühle mich von ihnen bestraft und habe deshalb wut auf sie! Darum geht es nicht. Bestraft, weil sie damals im zusammenhang mit dem thema berührungen sagten, wegen der sexuellen gewalt könne ich die kindlichen und die erwachsenen gefühle bzw. wünsche nicht auseinanderhalten. Deshalb sei es besser, wenn sie mich nicht mehr berührten. Weil meine brüder das mit mir gemacht haben, werde ich jetzt von ihnen bestraft! Das ärgert mich![Die Übergriffe meines Vaters begannen zu diesem Zeitpunkt erst nach und nach Thema zu werden.] – Ich habe nicht gesagt, dass es keine *berührungen mehr geben darf, sondern dass wir behutsam damit umgehen müssen. – Die klarstellung freute mich. Sie fragte, ob ich sie auf distanz halten wolle mit der annahme, sie bestrafe mich, vielleicht weil jetzt so viel nähe zwischen uns da war. Das konnte ich nicht beantworten. Meine wut war dann weg. – Ich dachte gestern, wenn sie mich doch berühren würden, knalle ich ihnen eine und spürte richtig hass! – Sie sagten doch gerade, sie haben angst, dass ich sie nicht berühre. – Sie forderte mich auf zu phantasieren, was ich ihr gegenüber fühle oder machen könnte o.ä. Bei mir tauchten dann nur liebe gefühle für sie auf. Sie machte mich darauf aufmerksam, dass die lieben gefühle weiter da sind, obwohl ich auch hassgefühle habe. Die gingen deshalb nicht weg! Ich freute mich, dass sie mich darauf aufmerksam machte. Aber es sähe für sie danach aus, dass ich mich für meine wünsche nach berührungen und für meinen hass selbst bestrafen wolle. Ich schimpfte wieder: Die haben das mit mir gemacht, und ich bekomme die strafe! – Sind das vielleicht gefühle, die sie früher ihren brüdern gegenüber hatten, wut und hass, weil sie etwas von ihnen wollten, die aber ganz was anderes taten? – Ich war wie geschockt und wieder wütend. Sagte dann, ich wolle nicht daran weitermachen, wolle was anderes.*

Sie fragte nach. Ich wolle mich hinlegen und was vorgelesen bekommen, hätte ein buch mitgebracht. „Frederik" von Leo Lioni (ein so schönes Buch!)

Das vorlesen setzen wir dann um. Ich legte die matratzen hin, sie holte die decke und kissen. Ich gab ihr das buch, nahm mein lieblingshalstuch, schmiegte später mein gesicht hinein. Sie kannte das buch, lächelte; es gäbe auch eine kassette dazu mit schöner musik, die müsse ich mir auch noch besorgen. – Ich hab's geschafft, ich hab's geschafft, sagte ich weich, als ich lag. Fing vor freude fast an zu weinen. Sie setzte sich ans ende der matratze, zu meiner freude. Ich legte mei-

nen kopf auf die zwei kissen, so konnte ich sie während des vorlesens ansehen. Sah ihr liebes gesicht. Ab und zu schaute sie zu mir hin und lächelte. Und sie las mir das buch vor. Ich genoss es sehr! Danach schweigen, genießen.

Ich fragte sie irgendwann wieder, ob sie mich nicht mehr berühre, weil ich vielleicht auch was von Frauen wolle? Nein, das sei nicht so, das könne sie klar ausschließen. In einer Stunde, in der ich von starken Herzproblemen sprach, von totaler Unsicherheit und Hilflosigkeit, fragte sie, was mein Herz denn brauchen könne? So kamen wir darauf, dass ich von ihr gehalten werden oder ihre Hand in meiner spüren wollte. Wieder tauchte Angst in mir auf. Ob sie mit ihrer Hand kommen solle? Ich antwortete sofort, fest: Nein! – Und fügte hinzu: Ich hab so eine Angst, dass die Hand meinen Körper berührt! – Ich fragte sie, nachdem ich weit von ihr weggerückt war, ob sie denke, dass ich mir diese Art der Berührungen wünsche? – Nein! Ich denke, dass Sie die nicht wollen, aber Angst haben, ich könnte das tun! – Was war ich froh, dass sie das sagte. – Ja! Genau das ist mein Gefühl! – Sie gab mir dann nicht ihre Hand, sondern etwas von sich, das ich mitnehmen durfte. Eine unbeschreiblich liebe Geste, für mich damals so wertvoll.

Wieder ging es um meine Gefühle für sie. Wieder war ich ganz mutig. Ich sprach davon, dass ich manchmal an sie denke und dann erotische Gefühle spüren würde. Oder dass ich an sie denke, damit ich die spüre. Ob sie sich dann missbraucht fühle? Darauf kam ein klares Nein von ihr. Ich fragte sie, was denn dann der Unterschied dazu sei, wenn sie mich berühre und ich erotische Gefühle habe? Sie, ganz ernst: Ich will sie nicht berühren, *wenn* sie erotische Gefühle haben! Und ich will sie nicht berühren, *damit* sie erotische Gefühle bekommen! – Das habe ich auch nicht gemeint!, erwiderte ich, nicht minder ernst und fest. Wir redeten eine Weile darüber. Sie nannte mir noch ein Beispiel: Ich würde mütterliche Unterstützung brauchen, dass sie mich hält usw., sie mich also berührt. Das sei in Ordnung. Aber wenn ich diese Berührung haben wolle, um erotische Gefühle zu spüren, das gehe nicht!

Auf die Vermischung meiner Gefühle, der lieben und der erotischen, zurückkommend, sagte sie zum Schluss dieser Stunde: Ich denke, dass sich das alles für sie klärt, wenn sie wissen und es annehmen können, dass das *ihre* Gefühle sind, dass die nichts mit mir zu tun haben! Es geht um *ihre* Gefühle, *ihre* Erotik und *ihre* Sexualität. –

Ich würde bei den Wünschen an sie, wenn sie die Mutterrolle übernehmen soll, so beschrieb es die Therapeutin später einmal, versuchen, die „erotische Komponente" herauszuhalten. Ich wolle das jetzt anders erleben, den Missbrauch damit ungeschehen machen. Ich wolle auch diese Art der Zuwendung wert sein und wert sein, dass es klar ist.

Einige Monate später fragte mich die Therapeutin, ob ich damit einverstanden sei, wenn sie bei einem Therapiekongress zum Thema „Körpertherapie und sexuelle Gewalt" über unsere Arbeit hier berichten würde. Ich könne es mir überlegen. Anwesend seien ausschließlich KollegInnen von ihr. Meine spontane Antwort war: Ja! – In der folgenden Stunde kam ich darauf zurück:

> *Ich weinte und sagte: Als sie mich gestern wegen des therapie-kongresses fragten, ... – Ich sah sie an und sagte fest: Ich finde es schön, dass sie mich vorher gefragt haben, ob ich damit einverstanden bin. – Wieder ganz bei mir, weinend, aufgewühlt: Nachdem sie mich das gefragt haben, hab' ich gedacht, dann können sie ihren kollegInnen sagen, was mit das schlimmste war! ... – Hm? – Das ist mir nämlich eingefallen. Das mit den berührungen! Dass das dann erstmal nicht mehr ging!, weinend und zornig! – Und dass ich damals das gefühl hatte, ich bin es nicht wert, von ihnen berührt zu werden! – Ihr liebevolles: Ehem! – Und damals dachte ich, schon alleine dafür, dass das deshalb nicht möglich war, schon alleine dafür müsste ich die umbringen! – Ehem! – (...)*
>
> *Am ende der stunde sagte sie, lächelnd, locker: Also dass da bei ihnen auch erotische gefühle waren, das war ja schon länger so. Das habe ich schon gespürt, nicht erst in dieser situation. Aber dann musste es angesprochen werden. – Ich hätte keine lust mehr auf dieses thema. Das verstehe sie. Aber dafür müssten wir das erst einmal abgeschlossen haben!*
>
> *Mein gefühl, bei aller verwirrung, es hat sich auch was geklärt.*

Sie berichtete später von diesem Kongress. In dem Workshop „Körpertherapie und sexueller Missbrauch" seien Berührungen ein zentrales Thema gewesen. Dadurch kamen wir nochmal auf die Situation zurück, als ich dort lag, sie mir vorlas. Die K. dazu: Da waren Sie ja im Gefühl des Kindes, das was von der Mutter wollte. Und wollten auch nur als das berührt werden!, kam ganz fest von ihr. – Aber dann tauchten die erotischen Gefühle auf. Und das sehe ich im Zusammenhang mit Ihren Erlebnissen als Kind. –

Später meinte sie, wenn wieder erotische Gefühle auftauchten, würden wir das nicht beenden, sondern schauen, was dann passiert, was das auslöst. Ich fragte, sehr erstaunt, warum sie das jetzt so anders sähe als damals? Sie, fest: Ich denke, dass Sie jetzt soweit sind, dass Sie sich darauf einlassen können. Da hat sich was verändert! Sie sind jetzt klarer! –

Zuvor ging es sogar wieder mit Berührungen, tauchten keine erotischen Gefühle auf. Aber es war auch wieder anders, was mich sehr ärgerte. Und irgendwann war es ganz klar, das Thema bearbeitet und abgeschlossen. Dennoch war es ein sehr schwieriger Prozess, das erinnere ich noch genau, aber auch meine Dankbarkeit und Freude, dass die Therapeutin so aufmerksam und sorgsam war. Trotz aller Wut und Härte war auch diese Einsicht damals schon da.

Berufliche und finanzielle Konsequenzen

Diese Folgen der traumatischen Erlebnisse sind mir erst in den letzten Jahren im Zusammenhang mit meiner künftigen Rente nochmal richtig deutlich geworden. Ich konnte aufgrund meiner seelischen und körperlichen Verfassung einige Jahre überhaupt nicht bzw. nur in geringem Umfang arbeiten. Als es mir im Laufe der Analyse besser ging, arbeitete ich zwar nach und nach mehr Stunden; bis ich jedoch wieder voll arbeiten konnte, benötigte ich weitere Jahre. Die Folgen: Behinderung des beruflichen Weiterkommens, auch das hat finanzielle Folgen, sowie zeitweise geringer bzw. kein Verdienst und dadurch verminderte Rentenansprüche. Ich glaube, an diese Folgen denken die wenigsten. Sie werden auch in Diskussionen oder Abhandlungen zum Thema sexuelle Gewalt grundsätzlich nicht erwähnt, obwohl sie doch auf Jahre hinaus die Betroffenen beeinträchtigen – wie erwähnt bis in die Rentenzeit.

Außerdem entstanden mir durch die Therapie hohe Kosten. Diese führe ich hier an, auch wenn ich sofort denke, aber es hat sich doch gelohnt! Ja, es hat sich gelohnt. Dennoch finde ich es wichtig, diesen Aspekt zu erwähnen.

Es gab und gibt für Frauen, die sexuelle Gewalt in ihrer Familie erlebt haben, keine Möglichkeit, Schadenersatz für finanzielle Einbußen zu bekommen. Schon gar nicht als Wiedergutmachung. Einige Betroffene aus kirchlichen Einrichtungen haben in den letzten Jahren finanzielle Hilfe oder Entschädigungen erhalten, was ich außerordentlich gut finde. Interessant dabei ist, dass es sich dabei meistens um Jungen bzw. Männer handelt. Doch für alle Anderen kommen die finanziellen Folgen zu den übrigen noch hinzu!

7.3 Ein Ende der Auswirkungen?

Nach einigen Jahren Therapie kam ich zu dieser tollen Einsicht: Die Auswirkungen werden dann aufhören, wenn ich mich dafür entscheide, ich will nicht mehr leiden! Ich will mein Leben leben! Damit entscheide ich mich gegen meinen Vater, gehe ich von ihm weg!

Später sprach ich davon, mich nicht mehr als Opfer fühlen zu wollen, mich nicht mehr als Opfer zu sehen. Und das gelang mir. Das war ein sehr, sehr wichtiger Schritt!

Diese Veränderungen im Denken, im Sein sind so hilfreich und wichtig, auch damit die Auswirkungen an Intensität verlieren, nur noch selten oder auch gar nicht mehr auftreten.

Zu dem, was ich heute noch an kleineren Schwierigkeiten oder Besonderheiten erlebe, gehört z. B., dass es für mich ganz wichtig ist, eine gute Schlafmöglichkeit zu haben. Wenn ich FreundInnen besuche, dort oder in einem Hotel oder einer Pension übernachte, schaue ich, dass ich mich mit meinem Schlafplatz wohl fühle. Dazu gehört, nicht im absolut Dunklen schlafen zu müssen. Auch dafür sorge ich.

Leider gibt es aber auch noch stärkere Beeinträchtigungen wie die bereits erwähnten Herzattacken. Diese können sich zeigen, wenn das Thema direkt berührt wird, ich an etwas von früher erinnert werde – durch Filme, Bücher oder irgendetwas Anderes. Ich hatte gehofft, wenn ich nur lange genug das Thema bearbeite, bin ich davon befreit, bin ich das alles los. Leider ist das nicht der Fall, wenn sich auch so viel zum Guten gewendet hat, ich heute so viel freier, unbeschwerter und sogar glücklich leben kann. Überhaupt gut leben und mein Leben selbst gut gestalten kann. Das habe ich immerhin geschafft.

Und ich erkannte: Es ist illusorisch, ein falscher Anspruch, gar nicht mehr von meiner Geschichte berührt und beeinträchtigt zu werden! Oder mit den Worten von Didier Eribon (Rückkehr nach Reims; Berlin 2016, S. 218): Die Vergangenheit ist noch immer unsere Gegenwart.

8. Was mir half, mir gut tat während der Zeit der Therapie

Ich schreibe an dieser Stelle über das, was mir half, all das Schwere und Schwierige und Schlimme auszuhalten, zu verkraften und auch mal wieder zu verdrängen. Ich schreibe einen gesonderten Text dazu, weil ich gut erinnere, wie wichtig bis hin zu überlebenswichtig das eine oder andere für mich war. Mir fiel, besonders während der Zeit, als ich mich immer mehr auf meine Gefühle und meine Erinnerungen einließ, und es mir besonders schlecht ging, manchmal gar nichts ein, was mir guttun würde oder womit ich mich ablenken könnte. Da war von den im Folgenden beschriebenen Möglichkeiten nicht viel bis gar nichts da! Damals kam meine Freundin Brigitte auf die äußerst hilfreiche Idee, ich solle, wenn es mir gut gehe, all das aufschreiben, was ich gerne mache und was mir in schweren Zeiten helfen könnte. Das setzte ich um. Ich fertigte eine Liste an, aus der ich, wenn ich es brauchte, auswählte, was für den jeweiligen Moment, die jeweilige Situation, in der ich mich befand, gut passte.

Manche der hier erwähnten Hilfen und hilfreichen Möglichkeiten standen mir von Anfang an „zur Verfügung", wie das Tagebuchschreiben und die Unterstützung durch meine Lieben. Anderes kam nach und nach hinzu bzw. lernte ich erst durch die Therapie.

Vieles, was ich im Folgenden beschreibe, wende ich heute noch an. Damit bin ich gleich bei meinem ersten Punkt, dem ...

8.1 Tagebuch schreiben

Ich begann mit dem Tagebuchschreiben schon vor dieser Therapie, aber nicht lange davor und nur mit großen zeitlichen Abständen. Mit Beginn meiner Therapie veränderte sich das, machte ich mir bald jeden Tag Notizen.

Tagebuchschreiben bedeutet für mich, all das, was ich bin, was mich beschäftigt, was mir Sorgen, Mühe oder Freude oder was auch immer bereitet, ernst zu nehmen. *Mich* ernst zu nehmen. Das war und ist für mich von besonderem Wert. Aber es gab, besonders damals, noch ganz andere und wichtigere Faktoren. Ich konnte meine Nöte, meinen Kummer, meine Ängste, meine Wut, einfach alles niederschreiben, im Schreiben analysieren, im Schreiben versuchen, nach Lösungen oder Auswegen zu suchen und mir durch das Aufschreiben Entlastung verschaffen. Besonders wenn ich Angst- und Panikattacken oder Herzprobleme hatte oder nachts nicht schlafen konnte, dann schrieb ich und schrieb, schrieb mir alles von der Seele, soweit das möglich war.

Über die einzelnen Therapiesitzungen fertigte ich jeweils nach den Stunden Kurznotizen an, die ich, meist am späten Abend, in mein Tagebuch übertrug; auch das zunehmend ausführlicher. Ich schrieb mir sogar Aussagen der Therapeutin wortwörtlich auf, soweit bzw. wie ich mich daran erinnerte. Dass das nicht immer mit den tatsächlichen Worten von ihr übereinstimmte, war mir damals schon klar.

Das Notieren der Therapiestunden kam einer Erweiterung der Therapie gleich. So sehe ich das heute. Ich habe es gebraucht, alle Einzelheiten, die ich erinnerte, selbst Gesichtsausdrücke oder die Kleidung der Therapeutin oder wie es in unserem Raum aussah und vieles Andere mehr, festzuhalten. Die Therapienotizen später nachzulesen, mir die Stunden damit wieder in Erinnerung zu rufen, besonders während der Therapieferien, war von immenser Bedeutung für mich. Und jetzt, beim Schreiben, konnte ich darauf zurückgreifen. Es berührt und freut mich, dass ich alles so ernst nahm und diese Aufzeichnungen anfertigte.

Ich möchte noch kurz auf einen besonderen Aspekt des Tagebuchschreibens eingehen. In einer Therapiestunde zitierte ich folgende Aussage meines Homöopathen: Ich hätte meinen Körper bisher nicht beachtet, damit die im Zusammenhang mit den Übergriffen erlebten Schmerzen nicht wieder auftauchten. Weinend hatte ich hinzugefügt, ich hätte diese Aussage nicht in mein Tagebuch aufnehmen können. Das interpretierte die Therapeutin so: Damit sie es nicht selbst übernehmen. Als Eigenes übernehmen. –

Ich führe auch heute noch Tagebuch, im Umfang viel geringer als damals. Im Schreiben denken und erkennen, aber nicht nur das, auch bewahren! Ja, auch darum ging und geht es mir.

8.2 Meine lieben und treuen UnterstützerInnen

Sie kommen bereits in den übrigen Texten vor. Dennoch will ich sie gerade hier besonders erwähnen, weil sie von Anfang an mit dabei und für mich da waren.

Dies waren, außer der Therapeutin, meine lieben FreundInnen Andrea, Brigitte, Meike und P.I. und P. und Vera sowie Elfriede und Klaus, die leider beide nicht mehr leben; außerdem meine Schwester Lisa und mein damaliger Mann Philipp. Und nicht zu vergessen meine lieben Nichten Isabelle und Ira. Sie alle und noch einige mehr haben mich, jede auf ihre Weise, wundervoll unterstützt, mich gehalten, beschützt und nicht allein gelassen. Sie haben all das Schwere hören und wissen wollen, mal mehr, mal weniger, auch mal gar nicht (mehr). Besonders wichtig war mir damals das miteinander

Reden und dass sie Tag und Nacht für mich da waren, wenn ich sie brauchte. Sie konnte ich anrufen, wenn ich es denn vermochte, denn Hilfe holen fiel mir nicht leicht, das musste ich erst lernen. Auch dass wir gemeinsam Schönes unternahmen war eine große Stütze.

Ohne diesen Kreis von UnterstützerInnen hätte ich es wohl nicht geschafft, mich all dem Beschriebenen zu stellen. Meine „Wahlfamilie" war für mich existentiell. Sie ist mir heute noch sehr, sehr wichtig, und ich bin ihnen allen noch immer von Herzen dankbar.

8.3 Körperliche Betätigung

Mir fällt keine angemessenere Bezeichnung ein. Ich hätte Sport schreiben können, aber das wäre zu hoch gegriffen. Es ging tatsächlich um körperliche Betätigung, darum, mich auszutoben, meinem „Gefühlshaushalt" einen Ausgleich zu verschaffen. Nur einige Kilometer durch den Wald zu joggen, meist mit meiner Freundin Andrea, war zum Austoben sehr gut. Besonders dann, wenn ich Wut hatte, die ich nicht oder noch nicht anderweitig rauslassen konnte, stampfte ich geradezu durch den Wald. Auch Radfahren diente diesem Zweck. Ich fühlte mich oft sehr bedrückt, wie in einem Korsett. Wenn ich mich dann auf die gerade beschriebenen Weisen betätigte, fühlte ich mich viel besser, war die Bedrückung weg.

Später lernte ich, an Felsen oder in einer Kletterhalle zu klettern. Zuerst an einem kleinen Felsen, dann ging es immer höher hinauf. Das war eine wunderbare Ablenkung! War ich am Klettern, vergaß ich fast alles Andere. Beim Joggen oder Radfahren war das nicht unbedingt der Fall, aber beim Klettern. Ich musste mich sehr konzentrieren, damit ich Halt fand, keinen Fehler machte, auch wenn ich angeseilt war.

Ein weiterer spannender Aspekt des Kletterns: Ich kletterte einen Felsen hoch und hatte Angst, was fast natürlich ist. Besonders Angst vor der Höhe, selbst anfangs, als es nur ein paar Meter waren. Hierbei handelte es sich um Angst vor etwas Realem, also vorm Abstürzen oder mir wehzutun. Die Angst, die mir während der ersten Jahre der Therapie begegnete und aus den Jahren davor vertraut war, stellte meist keine Angst vor etwas Realem dar. Die war wie nicht zu fassen. Ich kann es gar nicht recht beschreiben, warum es mir guttat, nun mit konkreter Angst umzugehen, aber so erlebte ich es.

Und das Klettern stärkte mich! Wenn ich mich dem einen oder anderen Kletterziel annäherte, spürte ich meinen starken Willen. Spürte ich Kraft und Energie. Manchmal dachte ich beim Klettern, besonders wenn es schwierig war, voranzukommen: Ich will da rauf! ... Und ich will raus aus diesem Familiensumpf! – Das Klettern war wie ein

Symbol dafür! Besonders wenn ich es schaffte, bis nach ganz oben zu kommen und auf dem Felsen zu stehen. Dann konnte es vorkommen, dass ich meiner Freude lauthals Ausdruck verlieh!

8.4 Selbstverteidigung

Auch dies war eine Idee meiner Freundin Brigitte. Ein Kreis von Frauen, die zum Teil selbst Trainerinnen waren, traf sich zum Sport und gegenseitigen und gemeinsamen Trainieren. Dazu gehörten auch Intuitionsübungen, Rollenspiele zu Themen wie abgrenzen und nein sagen können u. v. m. Wir machten Ringkämpfe oder boxten und traten, auch Stimmübungen bis hin zum lauten Schreien – gerade Letzteres konnte ich anfangs überhaupt nicht. Mir war es außerordentlich peinlich, meine Stimme zu erheben, nur Töne zuzulassen oder gar zu brüllen. Da kam erst einmal gar nichts aus mir heraus. Aber das entwickelte sich. Dabei entdeckte ich, wie bei entsprechenden Gelegenheiten in der Therapie, wie toll es sich anfühlt, richtig laut zu sein und meine Gefühle herauszubrüllen. Oder laut „Nein!" oder „Stopp!" zu sagen. Wie gut es tut, besonders für mich als Frau, die genau dies nicht tun durfte, die Stimme zu erheben! Und Wut zuzulassen! Ich nutzte jede Gelegenheit, sobald ich dazu in der Lage war.

Diese Trainingseinheiten waren anstrengend und aufregend und anfangs verstörend, von wegen brüllende Frauen u. ä. Aber ich spürte zunehmend meine Kraft und Energie. Das Abgrenzen und Nein sagen, so, dass ich eine Wirkung beim Gegenüber erzielte, fiel mir zuerst ziemlich schwer. Die Rollenspiele, die wir zum Üben einsetzten, bei denen ich mal in der Rolle war, Nein zu sagen, und mal in der Rolle, das Nein zu hören (und zu erleben, kommt es als Nein an), halfen sehr. Besonders beglückend fand ich, Bretter mit der Hand (Seitenkante) durchzuschlagen oder durchzutreten. Letzteres fühlte sich mit bloßen Füßen besonders gut an. Es war mir sogar möglich, drei Bretter auf einmal durchzutreten! Ich musste nur die richtigen Gedanken dazu im Kopf haben, vielleicht an meinen Vater denken ... Beim Schreiben überkommt mich an dieser Stelle ein gewisses Grinsen.

Wir machten auch schöne und leichte Übungen, wie Phantasiereisen, Meditationen, also Einheiten, die nicht mit Kämpfen, sich wehren und Kraft zu tun hatten. Manche Übungen vermittelten das Gefühl, gehalten oder getragen zu werden.

Einige der Trainerinnen gaben Kurse in Selbstverteidigung für Mädchen ab ungefähr 6 Jahren. Nachdem ich lange mit den Frauen trainiert hatte, ließ ich mich im Training für die Mädchen ausbilden und gab, zusammen mit einer weiteren Trainerin, ebenfalls solche Kurse. Das war ein außerordentliches Erlebnis für mich! Ich staunte, wie klein Mädchen von 6 oder 8 Jahren sind; dachte dabei natürlich an mich früher. Zum anderen überraschte und freute mich über die Maßen, dass selbst diese Kleinen Bretter durchhauen können. Und sie selbst staunten nicht minder! Auch für sie war das ein prägendes Erlebnis.

Ich denke, es ist gut zu erkennen, dass das Thema Selbstverteidigung in den hier geschilderten verschiedenen Facetten frau sehr viel bringen kann. Hinzu kam ein weiterer Aspekt: Ich bezog mich immer mehr auf Frauen. Auch das stärkte mich.

8.5 Mir selbst Gutes tun und Zeit für mich alleine

Ich fasse hierunter viele kleine oder größere Aktivitäten, die mir zusätzlich halfen, schwierige Zeiten zu überbrücken und zu überstehen. Bis auf eine Ausnahme haben alle die Voraussetzung, allein zu sein, mir Zeit für mich zu nehmen und mich mir selbst liebevoll zuzuwenden.

Ausblenden und Verdrängen ging, zumindest zeitweise, besonders gut, wenn ich an meinem *Kalender* saß, den ich über Jahre hinweg jeweils für das Folgejahr *selbst gestaltete.* Dazu verwendete ich ein linienfreies Buch in DIN A5 oder kleiner, schwarze Tusche und Federn in verschiedener Breite sowie Bilder oder Fotografien in schwarz-weiß, und ich malte selbst grafische Elemente hinein. Daran saß ich am späteren Abend, oft bis in die Nacht, malte, kreierte, schnitt Bilder aus und klebte und bastelte – und war dann vollkommen bei mir und meinem Tun!

Ähnlich erging es mir beim *Sonnen* und *Musikhören.* Ich legte mich gerne auf meinen Balkon und genoss die Wärme und die Musik. Das konnten Callas-Arien sein, die etwas Erhebendes haben und so viel ausdrücken. Ich wählte jeweils das aus, was am ehesten meinen Gefühlen und meiner Stimmung entsprach – von Popmusik bis zu Klavierkonzerten.

Ich schrieb damals in mein Tagebuch: So viel wie in der letzten Zeit habe ich wohl noch nie in meinem Leben geträumt, gefaulenzt und Zeit für mich gehabt und mir auch genommen und das genossen! – Dann war zwar immer wieder das Schlimme da, aber es ließ sich besser ertragen, weil ich durch das Schöne Kraft geschöpft hatte.

Auch *tanzen* lenkte mich gut ab. Ich ging auf Frauenfeten, weil ich mich dort geschützt fühlte, was zu diesem Zeitpunkt sehr wichtig für mich war. Ich überließ mich der Musik und meinem Körper, genoss die Bewegungen, den Rhythmus und tobte mich aus. Ein weiterer willkommener und wunderbarer Ausgleich.

Oder *spazierengehen* in der Natur bzw. überhaupt die *Natur genießen* – ich konnte mich nicht sattsehen an den Farben, genoss die Sonne, die Stille oder das Vogelgezwitscher und die Bewegung an der frischen Luft, selbst wenn ich dabei weinen musste. Natürlich war das Genießen der Natur nicht immer möglich. Manches Mal ging es mir so schlecht, dass ich ihr gar nichts abgewinnen konnte. Doch das Bewegen an der frischen Luft hatte meist beruhigenden Einfluss.

Meinem größten Interesse, dem *Lesen* von Literatur, konnte ich leider lange Zeit überhaupt nicht nachgehen. Ich konnte keine Geschichten von anderen Menschen an mich herankommen lassen, schon gar nicht Geschichten von anderen Betroffenen. Sachbücher, meist zu Frauen- bzw. feministischen Themen, waren das einzige, was mir in dieser Zeitspanne zu lesen möglich war. Dazu gehörte ein Sachbuch, das sich mit dem Thema sexuelle Gewalt auseinandersetzte, in dem beispielsweise erklärt wurde, warum es so lange dauert, bis Betroffene darüber reden können, in dem aber keine Geschichten von Betroffenen wiedergegeben wurden („Trotz allem" von Ellen Bass/Laura Davis). Dieses Buch half mir, mich selbst zu verstehen und anzunehmen.

Erst nach Jahren war es mir wieder möglich, mich der Literatur zuzuwenden – zu meiner großen Freude!

Leckeres Essen, auch das half mir. Das ging schon mit dem *Essen kochen* bzw. zubereiten los, weil ich mich darauf völlig konzentrierte und das Nachdenken und Grübeln damit unterbrach. Wenn es mir nicht gut ging, das gilt auch heute noch, dann brauche ich nicht viel, aber gutes, am besten sehr herzhaftes Essen. Es konnte auch ein bekömmliches und ausgiebiges Frühstück sein. Ich deckte mir den Tisch meist sehr sorgfältig, zündete Kerzen an und hörte Musik.

In meinen Tagebüchern finden sich des Öfteren Einträge wie: Jetzt koche ich mir erstmal was Gutes! – Oder: Ich hab mir was gekocht. Danach ging es mir besser! – Oder ich schrieb, ich wolle mich mit gutem Essen verwöhnen, wolle was Leckeres in meinem Mund(!) haben.

Und ich erkannte, aber auch dafür brauchte ich eine Weile, wie sehr es meiner Seele guttut, wenn ich *meinen Körper* ganz besonders und bewusst *pflege* und überhaupt *verwöhne.* Nach sehr heftigen Therapiestunden bzw. Erinnerungen kam es vor, dass ich selbst am späteren Abend noch duschte, mir dabei viel Zeit nahm, meinen Körper anschließend mit weichen Handtüchern ausgiebig und liebevoll abtrocknete und ihn anschließend ebenso fürsorglich eincremte.

Ich entdeckte ganz allgemein, ich will und muss genau schauen, was ich jeweils brauche, was mir guttut, und dann dafür Sorge tragen. *Mich mir selbst liebevoll zuwenden,* darum ging es.

Und noch etwas Anderes entdeckte ich: Ich brauche *Zeit für mich* – im wahrsten Sinne des Wortes! Wenn ich Zeit für mich habe, dann geht es mir gut! Zeit, in der ich *alleine* bin, egal wie ich sie dann fülle. Vor der Therapie konnte ich nicht gut allein sein, hatte ich oft Angst davor. Nun war ich zunehmend gern allein, weil ich dann genauer nach mir schauen und besser für mich sorgen konnte, aber auch um meine Gedanken ordnen und aufschreiben zu können.

Auch für *meine lieben Gefühle* für die Therapeutin benötigte ich Zeit. Ich musste und wollte oft allein sein, um diese zu spüren, mich ihnen ausgiebig hinzugeben, wie ich das damals nannte. Und das Besondere daran war: Wenn ich die lieben Gefühle für sie zuließ, konnte ich mich selbst besser spüren. Ich brauchte ungestörtes Alleinsein, um die innere Verbindung zur Therapeutin leichter halten zu können, was besonders in den sehr schwierigen Phasen hilfreich war. Dadurch fühlte ich mich gestärkt und nicht so isoliert.

Und das lernte ich im Laufe dieser Jahre: *mir selbst eine liebevolle Mama zu sein.* Das ist es! Sich selbst als so wertvoll empfinden, so beschützenswert und sich selbst so ernst nehmen und entsprechend liebevoll und fürsorglich mit sich umgehen. Ich notierte: Ich will mir selbst immer mehr und besser Mama sein, weil ich merke, dass mir das wahnsinnig hilft; vielleicht ist es die einzige, wirkliche Hilfe. – Darunter verstehe ich, genau nach mir zu schauen, zu schauen, was ich brauche, was ich will und nicht will und jede Idee oder jeden Wunsch, alles was auftaucht, ernst zu nehmen und zu versuchen, mein Leben entsprechend zu gestalten.

Daran anknüpfend komme ich zu einem weiteren wichtigen Punkt, nämlich zu dem liebevollen und fürsorglichen *Umgang mit der kleinen Ann.* Den Umgang mit der kleinen Ann musste ich allerdings erst mühevoll in der Therapie erlernen. Aber die Mühe lohnte sich! Indem die K. oft sehr liebevoll mit mir umging, manchmal explizit die kleine Ann in mir ansprach oder mit der kleinen Ann redete, eröffnete sie mir diesen Zugang.

Zur kleinen Ann und meinem Umgang mit ihr äußere ich mich besonders gerne, weil mir die Erinnerung daran noch immer sehr nahe ist. Ich las der kleinen Ann aus Kinderbüchern vor. Das waren Bücher von Janosch, ein anderes Buch hieß „Rosa Riedl Schutzgespenst“ und ein drittes, an das ich mich noch sehr gut erinnere „Gefühle sind wie Farben“. Ich las der kleinen Ann vor und redete ganz lieb mit ihr. Oder phantasierte, ich streichle sie liebevoll oder creme sie zart ein. Ein anderes Mal erzählte ich ihr von meiner Phantasie, dass wir über eine Wiese mit bunt blühenden Blümchen gehen, uns auf die Wiese legen und ich erzähle ihr Geschichten. Wir hörten den Vögelchen zu und dachten gemeinsam lieb an unsere Mama, die Therapeutin. Oder ich ging mit ihr in ein Café. Auf dem Weg dorthin hielt ich sie an der Hand, sprach innerlich mit ihr und dann tranken wir Kakao, aßen Eis oder Kuchen.

Anfangs brachte ich das nicht mit mir als erwachsener Frau zusammen, aber das änderte sich.

Dazu passte es vortrefflich, dass meine liebe Freundin Andrea eine Tochter bekam. Das war zeitlich etwas später, ich hatte bereits die schlimmsten Jahre hinter mir, aber nun konnte ich mit meinem *Patenkind Linda* Vieles von dem unternehmen, was ich gerade von der kleinen Ann beschrieb. Von Anfang an bzw. als Linda einigermaßen groß genug war, gingen wir ins Café, sie spielte und ich las oder wir malten gemeinsam; natürlich gab es Kakao dazu! Oder wir kauften vorher ein Kinderbuch, und ich las ihr vor. Ich machte außerdem mit großer Freude Linda-Sitting, wenn ihre Eltern nicht da waren, brachte sie zu Bett, las ihr vor und kraulte ihr den Rücken bis sie einschlief. Eines Abends forderte sie mich auf, wir hatten das Kinderbuch „Mama Muh“ zu Ende gelesen, ich solle mir dazu weitere Geschichten einfallen lassen. Das kann ich doch gar nicht!, so mein erster Gedanke. Und ob ich das konnte! Zu meiner eigenen großen Überraschung und sehr zu meiner Freude. Linda gab mir Stichworte, und ich schmückte sie zu Geschichten aus. Wir waren beide beglückt. Es war wunderbar zu erleben, wie so ein kleines Mädchen aufwächst, wie sie alles um sich herum entdeckte, welche tollen Ideen sie entwickelte, wie sie auf alles Neue reagierte und: Wie viel Freude ich mir selbst damit schenkte. Und wie wichtig und leicht es ist, ihre Grenzen zu wahren!

8.6 Was noch hinzu kam

Eine wesentliche Erkenntnis führe ich bewusst noch einmal an: Wenn ich meine Erinnerungen zulasse, wenn ich alles in mir Verborgene und Verdrängte rauslasse, geht es mir besser. Wenn ich weiß, was war, wenn ich nicht wieder von mir weggehe und zu mir stehe, das stärkt mich!

Diese Erkenntnis half mir außerordentlich, auch wenn sie mir zweitweise wieder abhanden kam.

Und ich traf, wie bereits erwähnt, eine wesentliche Entscheidung. Ich hatte die Wahl, mich weiter wie ein Opfer zu fühlen oder dem ein Ende zu setzen. Ich entschied mich für Letzteres. Dies war eine ganz zentrale und sehr wirksame Entscheidung, die zu treffen absolut notwendig ist und die mir durch die Therapie möglich wurde.

Genauso wie die Tatsache, dass es mir immer besser gelang, mein Leben so zu gestalten, wie es für mich gut war. Wahrzunehmen, wie und was ich leben will und das mehr und mehr umzusetzen. Das war mir vor der Therapie nicht in ausreichendem Maße möglich. Mir war noch nicht mal in aller Deutlichkeit bewusst, dass ich dazu nicht in der Lage war!

Auch musste und wollte ich lernen, noch mehr die Verantwortung für mich und mein Tun zu übernehmen. Das fiel mir gar nicht leicht.

Dies alles zu erkennen und zu lernen war ein Prozess, manchmal ein sehr anstrengender und mühsamer, der Zeit brauchte und der diverse Veränderungen in meinem Leben mit sich brachte. Ein Prozess, der mich aber auch stärkte und mir wiederum half, weiter an mir zu arbeiten, mich mit meiner Geschichte auseinanderzusetzen, auch mit meinen Eltern, besonders mit meinem Vater.

Eine weitere wesentliche Veränderung ergab sich dadurch, dass es mir möglich wurde, wieder ins Berufsleben zurückzukehren. Ich konnte jahrelang aufgrund meiner schlechten psychischen Verfassung nicht arbeiten. Das veränderte sich nun. Als Erstes übernahm ich die Buchhaltung für einen Verein. Dies zu schaffen war ein guter Start, der erste Schritt. Auch im Hinblick darauf, was sich bald als Nächstes anbot: Freiberuflich in meinem studierten Beruf tätig zu werden. Ich begann mit wenigen Stunden pro Woche, so dass ich weiter drei Stunden Therapie beibehalten konnte. Nach und nach konnte ich mein Arbeitsvolumen ausbauen, weil es mir immer besser ging. Gleichzeitig reduzierte ich die wöchentlichen Therapiestunden, was wiederum eine weitere Ausweitung meines Arbeitens begünstigte.

Wenn anfangs auch nur für wenige Stunden, hatte das Arbeiten dennoch sehr positive Auswirkungen. Einerseits lenkte es mich meist gut ab, andererseits stärkte es mich nicht unwesentlich. Die Therapeutin hatte es auch aus therapeutischer Sicht sehr begrüßt und für unbedingt notwendig erachtet, dass ich mich wieder beruflich engagierte, wieder mit dem Arbeiten begann. Es war wahrlich nicht leicht, brachte mich auch in Kollision mit meinen Wünschen nach Zeit für mich allein, aber ich profitierte sehr davon. Auch das war wie ein zu mir Zurückkehren.

Man kehrt zu etwas zurück, was schon geschehen ist,
und wiederholt es,
aber mit einem neuen Schlüssel.
Elisabeth Bronfen

Wenn Gefühle auf Worte treffen
Siri Hustvedt im Gespräch mit Elisabeth Bronfen

9. Die Therapie und die Therapeutin

9.1 Der äußere Rahmen der Therapie und seine Wirkung auf mich

Ich beginne mit ein paar Bemerkungen zu der *Art der Therapie.* Soweit ich das beurteilen kann, handelte es sich um eine tiefenpsychologisch fundierte Gesprächstherapie mit körpertherapeutischen Anteilen; die Traumdeutung war ein weiterer wichtiger Bestandteil. Im Laufe der Therapie ging unsere Arbeit immer mehr in eine Psychoanalyse über, was für mich auch deshalb besonders passend war, weil ich bis in die tiefsten Tiefen meiner Seele, meiner Kindheit schauen wollte. Ich wollte unbedingt wissen, was mit mir los ist und Vieles aus meinem aktuellen Leben überprüfen und vor allem ändern.

Nicht nur im Zusammenhang mit der Therapieform stehen die *verschiedenen Formen des therapeutischen Settings.* Zu Anfang saßen wir uns gegenüber mit einem für mich guten räumlichen Abstand. Weil es mir so schwerfiel, sie anzuschauen, schlug die Therapeutin bald vor, dass wir uns so setzen, dass ich sie angucken kann, aber nicht muss, also „über Eck". Darauf ging ich ein und fühlte mich danach viel wohler. Ihren Vorschlag, mich hinzulegen, um besser bei mir sein zu können, lehnte ich sofort ab mit der Begründung, mich dann nicht wehren zu können.

Später setzte ich mich auf Matratzen auf den Boden mit Blick zur Wand gegenüber. An dieser Wand war ein Spiegel angebracht, über den ich die Therapeutin ansehen konnte, denn das direkte Anschauen war weiter ein Problem für mich. So fiel es mir erheblich leichter. Und wenn ich wollte, schaute ich zur Seite und sie damit direkt an. Ich wandte mich immer dann zu ihr hin, wenn ich es brauchte. Diese verschiedenen Varianten waren auch deshalb hilfreich, weil ich, wie sie es nannte, nicht bei mir blieb, sondern mich oft von ihrem Gesichtsausdruck beeinflussen und ablenken ließ. In der intensiven, harten Phase mit Erinnerungen an die Taten meines Vaters setzte sich die

Therapeutin ebenfalls auf den Boden, blieb also nicht in ihrem Sessel sitzen. Als ich das nach einer Weile registrierte, nahm ich an, es ist so bequemer für sie. Erst nach mehreren Stunden verstand ich den Zusammenhang. Sie wollte nicht auf mich herunter schauen, wie ich es mit meinem Vater so oft erlebt hatte. Ich registrierte es sehr dankbar und berührt. Bald saß sie immer mit mir auf einer Ebene.

Erst einige Jahre später war es mir möglich, auch weil ich es selbst wollte, mich auf ihre Couch zu legen. Dabei entdeckte ich, welch große Vorteile das hat. Ich konnte meine Gefühle viel besser zulassen und mich intensiver darauf einlassen. Und ich konnte bei mir bleiben! Letzteres empfand ich wie ganz in mir drin sein.

Sie setzte sich oft an das Kopfende der Couch, also für mich nicht sichtbar. Oder, wenn ich es wünschte und brauchte, sie anschauen zu können, setzte sie sich entsprechend anders hin. Es gab auch zu dieser Zeit Stunden, in denen ich lieber sitzen als liegen wollte. Ich konnte also immer danach gehen, was mir entsprach.

Meist waren wir in einem großen, wunderschönen *Raum,* manchmal aber auch in einem kleineren. Auch das konnte ich im Verlauf der Therapie, so der jeweils andere Raum frei war, mit bestimmen. Manchmal brauchte ich viel Raum um mich, und manchmal wollte ich es irgendwie näher haben. Zeitweise nannte ich den schönen großen Raum „mein Kinderzimmer“! Ich fühlte mich dort, in ihrem Raum, geborgen, aufgehoben und beschützt, was für mich etwas sehr Wertvolles war.

Und ich durfte nach dem Ende einer Therapiestunde in „unserem“ Raum bleiben! Dies bot mir die Therapeutin nach heftigen und intensiven Stunden an. Ich traute mich bald, sie zu bitten, noch ein bisschen bleiben zu dürfen. Später gestattete sie mir das, auch wenn niemand mehr in der Praxis war. Sie muss wohl Vertrauen zu mir haben, hatte ich gedacht und mich sehr gefreut. Irgendwann vereinbarten wir, dass ich jeden Dienstag nach der Stunde noch in unserem Raum bleiben durfte, wenn ich das wollte. Dieses Angebot nahm ich natürlich sehr gern an. Manchmal weinte ich dann lange und mehr oder weniger heftig, oder ich versuchte, eine schwere Stunde zu verdauen. Ein anderes Mal genoss ich es einfach nur, in dieser vertrauten Umgebung sein zu dürfen. Ich kuschelte mich in meine Ecke, auf die Matratzen, und ab und zu schlief ich sogar ein. Das war für mich wie ganz viel Liebes von ihr zu bekommen. Ich erinnere noch heute freudig und dankbar, wie schön es war, sie zu fragen, ob ich noch bleiben dürfe, und dann diese Antwort zu bekommen: Aber natürlich! –

Es kam aber auch vor, dass die Therapeutin meinte, ich solle nicht bleiben, damit ich mit den schlimmen Erinnerungen nicht mehr dort saß, sondern wieder besser ins Jetzt zurück fand. Das formulierte sie ungefähr so: Ich möchte Sie heute hier nicht alleine lassen! – Oder ich blieb nach einer schweren Stunde, und sie kam nochmals in unseren Raum und schaute nach mir. Zeigte mir auch so, ich bin nicht mehr alleine mit allem.

Zu Beginn der Therapie hatte ich *jede Woche eine Stunde.* Wenn es nötig war und die Therapeutin die Möglichkeit hatte, bot sie mir eine weitere Stunde an. Nach und nach erhöhten wir auf zwei und später sogar auf drei Stunden pro Woche. Es fiel mir sehr schwer, nach außen und vor mir selbst dazu zu stehen, dass ich das brauchte. Selbst das war mir manchmal noch zu wenig.

Ihr war es außerdem sehr wichtig, dass wir immer an den gleichen Tagen und immer zur selben Uhrzeit, mit wirklich nur wenigen Ausnahmen, unsere Stunden abhielten. Diese Kontinuität war immens wichtig, ich empfand sie wie ein Netz, das mich auffängt, hält und trägt. Es gab Zeiten, in denen ich große Schwierigkeiten damit hatte, nicht alle meine Lieben zu sehr mit dem furchtbaren Thema zu belasten. Andererseits musste ich darüber reden, mir Vieles von der Seele reden, benötigte ich Zuspruch, Verständnis und Verlässlichkeit – auch deshalb schätzte ich diese Kontinuität sehr und die Möglichkeit von wöchentlich drei Stunden Therapie.

Ihre *Therapieferien* waren folglich für mich ein riesiges Problem. Grundsätzlich ging sie mit meinen Schwierigkeiten bezüglich der Ferien sehr verständnisvoll um. Sie bot mir immer dann, wenn es mir sehr schlecht ging, wir uns in einer intensiven Therapiephase befanden, doch eine oder auch mehrere Stunden während dieser Zeit an. Sie erklärte dazu, sie habe Büroarbeiten oder ähnliches zu erledigen, sei also sowieso in der Praxis oder sie kam deshalb extra in die Praxis. Dann fragte ich mich, manchmal aber auch sie, warum sie das wohl macht. Sie erklärte mir ihr Handeln beispielsweise damit, dass sie im Moment einfach die Notwendigkeit zusätzlicher Stunden sähe.

..., die die Frau, die ich heute bin,
mit dem Kind, das ich gewesen bin, verbunden (hat)
„Die Frau"
Annie Ernaux (über ihre Mutter)

9.2 Wie ich die Therapeutin erlebte

In allen bisherigen Kapiteln wird das Verhalten der Therapeutin, ihre Arbeitsweise und ihr Umgang mit mir bereits sehr deutlich, aber ich kommentiere oder erwähne dabei nicht immer explizit, wie das jeweilige Verhalten auf mich wirkte. Außerdem sind in diesen Kapiteln viele relevante Aspekte nicht erwähnt. Auch deshalb möchte ich nun beschreiben, wie ich die Therapeutin und ihr Verhalten wahrnahm, was mir besonders gut tat und half oder womit ich Probleme hatte. Ich berichte davon auch deshalb, weil ich es sehr spannend finde, mit dem großen Abstand von heute, mir ihre Art, ihre Arbeitsweise, ihr Engagement und all die unterstützenden bis hin zu den damals für mich sehr schwierigen Aspekten genauer anzuschauen.

Im Mittelpunkt der Betrachtung steht für mich im Folgenden, wie ich die Therapeutin bezüglich des Bearbeitens der sexuellen Gewalt erlebte; ganz besonders ihr Umgang mit mir im Zusammenhang mit dem Auftauchen von Erinnerungen und Bildern, ohne diese jeweils noch einmal zu schildern. Der Fokus liegt auf meinem Erleben, darauf, wie sie sich verhielt und wie es mir damit erging. Aber ich werde auch versuchen, ihr Spektrum an therapeutischen Möglichkeiten und die persönlichen, menschlichen Anteile aufzuzeigen.

Als ich mich diesem Kapitel näherte, fiel mir folgendes spontan zu ihr ein:

Streng ... und so liebevoll! Und: Abgegrenzt ... und so sehr für mich da! ... Das kam direkt und unmittelbar aus mir heraus. Streng? Ja, streng. Vielleicht würde „ernsthaft und sehr konzentriert" sie noch besser beschreiben, denke ich jetzt. Mir fiel jedoch „streng" ein, denn so habe ich sie damals manchmal empfunden. Aber sie hat mir auch so viel gegeben. Auch viel Liebe. Ja, so fühle ich es. Und dennoch nahm ich sie öfter sehr abgegrenzt wahr. Und ich fühlte wieder meine große Dankbarkeit ihr gegenüber.

Außerdem fällt mir gleich noch ein, wie aufmerksam sie war. Ich denke dabei an die, bei einer Berührung von ihr, aufgetauchten erotischen Gefühle und die mit großer Bestimmtheit ausgesprochene Folge, dass es deshalb erst einmal keine Berührungen mehr geben dürfe. Wie schlimm das für mich war!

Auch das möchte ich nicht unerwähnt lassen, dass ich bereits bei den Vorarbeiten zu diesem Kapitel große Freude spürte – und strahlte. Ich freute mich darauf, davon zu berichten.

Ich beginne jedoch mit einigen für mich schwer anzunehmenden Aspekten, die ich zuerst der Therapeutin anlastete. Mit einigem Abstand wurde mir zunehmend klarer, dass es sich meist um Übertragungen handelte.

Für mich wäre es beispielsweise hilfreich gewesen, hätte die Therapeutin schon zu Beginn der Therapie darauf hingewiesen und erläutert, warum es vorkommen kann, dass sie während einer Therapiestunde einmal *lange Zeit nichts sagt,* auch wenn ich ganz viel sage, von früher erzähle oder was auch immer. Ich verstand nämlich oft nicht, warum sie das weder kommentierte noch in irgendeinen Zusammenhang brachte. Zeitweise nahm ich sie nämlich genau so wahr, dass sie ganz oft nichts oder nur wenig sagte. Und das war besonders heftig für mich, wenn ich von erinnerten Gefühlen und Bildern sprach. Dann verunsicherte mich ihr Schweigen sehr. Es war doch sowieso schon so viel Scham und Pein und Zweifel und was weiß ich noch alles bei mir vorhanden! Und dann sagte sie nichts! Das brauchte ich wirklich nicht auch noch!

Sie widersprach auch hierbei manchmal meiner Wahrnehmung, weil sie sich anders erlebt hatte. Ich erinnere auch, dass sie ihr Schweigen oder ihre wenigen Worte einmal erläuternd kommentierte, doch ich glaube, ich wollte es nicht hören. Erst später verstand ich, dass es dann einfach noch nichts zu kommentieren oder zu sagen gab, ihr Schweigen also nicht gegen mich gerichtet war. Es schon gar nicht bedeutete, sie redet nicht mit mir, wie ich das von meinem Vater kannte, der mich oder meine Geschwister wegen einer Kleinigkeit wochenlang ignorierte, einfach nicht mehr mit uns sprach! Uns sogar übersah, als gäbe es uns gar nicht. Von dieser Übertragung musste ich mich immer wieder freimachen.

Ich fragte mich auch des Öfteren, hätte es gerne von ihr gewusst, *wovon es abhängt,* ob sie beim Begrüßen oder auch während der Therapiestunde freundlich, locker oder ernst ist oder gar wie abwesend oder abweisend ausschaut oder schaut. Ich fühlte mich ohne jeden Einfluss darauf. (Ich erkannte erst viel später, dass es mir mit meiner Mutter früher ähnlich erging; es sich hier um Übertragungsvorgänge handelte.) Aber ich schaffte es immerhin recht bald, mit der Therapeutin darüber zu reden, was sehr hilfreich war. Sie forderte mich danach immer wieder auf, unbedingt und gleich zu sagen, wie ich sie sehe oder wahrnehme, wenn es mich irritiere. Dann könnten wir das überprüfen. Überhaupt sei alles von Bedeutung, so die K. an anderer Stelle, was in der Therapie auftauche. Auch deshalb wäre es sehr gut, es anzusprechen. Meines Erachtens war es für das Gelingen meiner Therapie bzw. Analyse eine immens wichtige Voraussetzung, dies umgesetzt zu haben!

Es kam auch vor, dass ich sie völlig erschöpft oder gar krank wahrnahm und dann dachte, das hat mit mir zu tun. Ich bin ihr zu viel oder zu anstrengend! Wenn ich genau das in Worte fasste, geschah das Verblüffende meist sofort. Meine Wahrnehmung veränderte sich; manchmal sogar dann, wenn die K. bestätigte, dass sie erschöpft sei oder etwas Anderes sie beeinträchtigte. Dann war das Störende, mich Ablenkende auch deshalb beseitigt, weil ich wusste, es hat nichts mit mir zu tun!

Für mich war es noch schlimmer, wenn ich sie so wahrnahm: *Sie sagt nichts und guckt völlig kalt und von oben herab auf mich.* Das war so furchtbar! Dagegen half ebenfalls nur, es anzusprechen und zu klären. Aber wie schwer fiel mir das! Sobald ich es ausgesprochen hatte, veränderte sich meine Wahrnehmung ganz rasch. Und irgendwann erkannte ich, dass ich sie immer dann in dieser Weise erlebte, wenn es sich um Erinnerungen und Bilder mit meinem Vater handelte. Dann sah ich sie nämlich genauso, wie ich es von meinem Vater kannte. Grausam war das! Gerade sie, die ich doch lieb hatte, deren Unterstützung ich dringend benötigte, gerade sie erlebte ich so! Nachdem ich mit ihrer Hilfe erkannte, dass es sich auch hier um eine Übertragung handelte, ich in ihr meinen Vater sah, konnte ich mir genau das in entsprechenden Situationen selbst sagen, und das Bedrückende löste sich auf.

Ich will gleich noch etwas notieren, was mir gar nicht gefiel, womit ich öfter haderte. Selbst nach ganz schweren Stunden musste ich *nach Ablauf der 50 Minuten gehen!* Manchmal machte sie noch nicht einmal eine Minute länger! Das fand ich hart und grausam. Zeitweise erschien es mir so, als würde sie die Stunde ganz oft überpünktlich beenden. Ich betone „erschien es mir", denn wenn ich mich darüber beschwerte, hatte sie meist eine andere Wahrnehmung. Gerade in letzter Zeit hätten wir die Zeit mehrmals überschritten oder ähnlich ihre Aussagen. Sie warf mir wohl nicht nur einmal vor – sie würde sicher sagen, sie habe mich darauf aufmerksam gemacht! – , ich würde des öfteren erst zum Ende der Stunde mit ganz schweren Themen oder Erinnerungen kommen, so dass wir fast zwangsläufig die Zeit überschreiten müssten. Das war auch mir aufgefallen, ich widersprach deshalb nicht lange. Auch deshalb nicht, weil sie meist gleich erklärte, dass das nichts Ungewöhnliches sei, sie es verstehe und weil ich wusste, dass sie das Überziehen der Stunde auch duldete. Letzteres besonders dann, wenn wir noch am Arbeiten oder ich noch am Verarbeiten von heftigen Bildern war. Und wenn es ihr zeitlich möglich war. Manchmal sagte sie allerdings mit Bedauern, sie müsse los, weil ... Ich glaube im Nachhinein, dass ich das Überziehen als Beweis dafür brauchte oder als Ausdruck dafür ansah, sie mag mich, sie ist sehr für mich da; am liebsten wäre es mir gewesen, sie ist nur für mich da! Aber das stimmt nicht ganz; das hätte mir nämlich eher wieder Angst bereitet ...

Ihr Verhalten im Zusammenhang mit meinen Erinnerungen

Dazu gäbe es außerordentlich viel zu berichten. Ich will versuchen, mich zu beschränken, auch wenn es mir schwer fallen wird, weil so Vieles wichtig und besonders für mich war.

Wie sie mich bestärkte und bestätigte
Gleich an erster Stelle will ich *ihre Bestätigungen oder Bestärkungen* in Form des nicht selten tief aus ihrem Innern kommenden *„Ehem"* nennen. In den übrigen Kapiteln wird dieses oft von mir zitiert. Ich kann kaum in Worte fassen, wie diese „Äußerung" auf mich gerade dann wirkte, wenn ich Unterstützung oder Ermutigung und besonders ihr Mitfühlen benötigte; wie tief berührt ich oft davon war. Es brauchte Monate, bis ich davon überhaupt Notiz nahm. Ich glaube, sie fragte mich einmal, ich hatte sie wohl kritisiert, weil sie nichts sagte, ob ich denn nicht diese ihre Bestärkungen (oder Bestätigungen) gehört hätte. Ganz sicher nicht! Aber ab diesem Zeitpunkt nahm ich sie zunehmend wahr! Ihr „Ehem" konnte in normalem Tonfall intoniert und relativ kurz sein, als kurze Bestätigung, oder, was oft der Fall war, sehr tief aus ihrem Innern kommend, langgezogen und wiederholt werden: Ehemmm! Ehemmm! – Blieben diese Bestärkungen aus, fragte ich mich nicht selten verunsichert, ob sie mir nicht glaubt, was ich erinnerte oder erzählt hatte. Wenn dann doch ihr „Ehem" kam, vielleicht erst nach einer längeren Pause, fühlte ich mich verstanden, bestärkt, aufgehoben oder angenommen. Eine so kleine Äußerung mit solch großer, besonderer Wirkung!

Eine andere, längere Äußerung, nämlich eine von ihr gestellte Frage, schließt sich daran gleich an. Diese stellte sie zum Beispiel in Situationen, in denen wir intensiv gearbeitet und ich anschließend eine ganze Weile geschwiegen hatte. Dann fragte sie nach einer Weile des gemeinsamen Schweigens mit lieber Stimme: *Wo sind sie denn jetzt?,* mit Betonung auf dem „wo". Vom Inhalt der Frage einmal ganz abgesehen, alleine so gefragt zu werden! Das vermittelte mir das Gefühl, sie ist ganz bei mir, es ist ihr wichtig, was mit mir ist, was mich innerlich bewegt oder beschäftigt. Es gab noch andere ähnliche Fragen oder Bemerkungen, aber diese erinnere ich besonders gerne.

Überhaupt sei hier ihr liebes Lächeln, ihr liebevoller, fürsorglicher Blick, ihre warmherzig mich anschauenden Augen und ihre von mir so gerne gehörte Stimme genannt. Nur um dieser zu lauschen hörte ich mir während ihrer Ferien immer wieder ihre auf meinem Anrufbeantworter gespeicherten Ansagen wegen Terminverschiebungen an!

Wenn ich aus Angst große Schwierigkeiten hatte, Erinnerungen an Gefühle und Bilder von den Übergriffen zuzulassen, mich überhaupt auf meine Erinnerungen einzulassen, dann *ermutigte sie mich,* bestärkte mich, dass es mir doch guttun würde, wenn das von meiner Seele runterkäme. Ich würde das doch schon lange genug mit mir herumtragen. Später wies sie mich darauf hin, ich hätte doch schon erlebt, dass es zuerst sehr hart und schmerzhaft sei, mich zu erinnern, es mir aber danach doch so viel besser gehe.

Es kam vor, dass ich sagte, es sei noch etwas in mir, aber ich könne es nicht sagen, könne einfach nicht darüber reden! Ich quälte mich öfter lange damit herum, aber ich vermochte nicht auszusprechen, was ich erinnerte oder fühlte. Die lockere, liebe und geduldige Art der K. brachte mich am Ende doch noch dazu, genau das zu tun, auch wenn ich erst in einer der folgenden Stunden dazu fähig war.

Für mich strahlte sie zudem *große Lebensfreude* aus! Ich erlebte sie außerordentlich lebensbejahend. Das liebte ich an ihr! Auch daran richtete ich mich immer wieder auf; das kam mir vor wie Energiezufuhr.

Ihre hilfreichen guten Ideen

Ich nähere mich noch mehr meinem damaligen Erinnern an, wie die Therapeutin damit umging und wie sich das wiederum für mich anfühlte. Ich erwähnte bereits, dass ich manchmal nicht aussprechen konnte, was ich erinnerte und wie mir die K. dabei half, indem sie mich geduldig ermutigte. Aber manchmal konnte ich dennoch nichts sagen, es nicht in Worten fassen, was an Bildern und Gefühlen auftauchte. Und dazu fallen mir ihre für mich erst einmal etwas kindlich anmutenden, beim Umsetzen jedoch sehr hilfreichen Ideen und Vorschläge ein:

Wie sie mir vorschlug, meine Gefühle zu malen, und ich dem nachging. Am Beeindruckendsten fand ich es, mit Fingerfarben zu malen, weil ich damit noch näher an meinen Gefühlen war und sie auf diese Weise ganz unmittelbar ausdrücken konnte.

Oder wie ich bald darauf sogar selbst vorschlug zu malen, was ich mit meinem Vater erinnerte. Wie ich ihr dies auch mittels Knete „zeigen" konnte. In einer anderen Stunde hatte sie verschiedene Gegenstände vor mir ausgebreitet und mich aufgefordert, meine Augen zu schließen und mich meinen Gefühlen und Bildern zu überlassen. Auch so konnte ich meine Erinnerungen „darstellen".

Ich bin jetzt noch fasziniert davon, was alles in dieser Therapie möglich war, welche verschiedenen Angebote sie mir unterbreitete und wie sehr mir das half. Aber auch davon, wie sie dann mit dem Ergebnis umging. Ein gemaltes Bild zum Beispiel bewahrte sie auf, nachdem sie gesagt hatte, es sei besser, es dort zu lassen, als dass ich es mit nach Hause nähme. All die von mir auf die beschriebenen Weisen ausgedrückten Gefühle und Bilder nahm sie sehr ernst. Zweifelte nichts an, kommentierte aber auch nichts. Fragte höchstens einmal nach, wenn ihr nicht klar war, was ich gemalt oder welche Bedeutung das Gemalte hatte. Ganz, ganz ernst nahm sie es, was für mich immens wichtig war. Aber ich bin auch sehr froh, dass ich mich auf all das eingelassen habe, wenn auch erst mit dem Abwehrversuch, das sei doch kindisch.

Sie erklärte mich mir

Damit meine ich, dass die Therapeutin mir immer wieder versuchte zu verdeutlichen, *warum die kleine Ann, oder auch die große Ann, so oder so fühlte, reagierte und handelte oder auch nicht handelte.* Ich lernte dadurch, mich zu verstehen und anzunehmen, insbesondere als das kleine Mädchen, das ich damals war.

Ich denke dabei an zentrale Themen wie die Frage nach der Schuld. Wie bedeutsam und entlastend war es, von ihr eines Tages die mit fester Stimme formulierte Aussage zu hören: Sie sind nicht schuld! –

Ebenso entlastend empfand ich ihre Erläuterungen zu der Tatsache, dass Adolf mir nach seinen Attacken Geld zusteckte. Dies bedeute oft, wie auch die Schuldübernahme, einen gewissen Schutz für das kleine Mädchen, so die K., denn damit phantasiere es sich als Handelnde. Diese Phantasie befreie das Kind von den eigentlich nicht auszuhaltenden Gefühlen der absoluten Hilflosigkeit und dem völligen Ausgeliefertsein.

Im Zusammenhang mit der Gewalt durch meinen Vater erinnerte ich sehr oft, mich dabei *wie tot gefühlt* zu haben. Das schütze die Psyche, hatte die K. damals erklärt. Außerdem sei es manchmal besser, sich tot zu fühlen oder zu stellen, also besser nichts zu fühlen als „das". Das sei eine typische Reaktion bei traumatischen Erlebnissen. An anderer Stelle hatte sie gemeint: Die kleine Ann dachte damals doch, das geht das ganze Leben so weiter! Dann wollte sie lieber tot sein … –

Ich denke auch an ihre hilfreiche Demonstration, sie hatte ihre flache Hand auf meine ausgestreckte gelegt, mit der sie mir bewies, dass ich bei einer *Berührung immer etwas fühle,* ganz egal, ob ich das will oder nicht.

Dies stand im Zusammenhang mit der Situation mit Bernd, bei der ich mich auch an schöne Gefühle erinnerte, aber auch in Bezug auf meinen Vater. Damals hatte ich gesagt, das sei doch einfach nicht zu verkraften, dass *„dabei" auch schöne Gefühle* spürbar waren. Das sei ja gerade das Schlimme, so die K., Ohnmacht und Hilflosigkeit wären schon schlimm, doch dabei könne das Kind Wut entwickeln, was bei den schönen Gefühlen aber nicht möglich sei. Deshalb sei das noch schlimmer für ein kleines Mädchen.

Einige Stunden zuvor hatte ich mich so zu meinem Vater geäußert: Ich bekomme sein Lächeln und das, was er mit mir gemacht hat, einfach nicht zusammen! Der hat mich manchmal so lieb angelächelt! – Das könne frau auch nicht zusammenbekommen, aber das sei eine Spaltung in ihm! Das habe nichts mit mir zu tun!

Als ich davon sprach, unter allergrößten Mühen tatsächlich sagte, ich könnte sie vergewaltigen, was ich doch auf keinen Fall tun würde, aber dieser Gedanke sei in mir aufgetaucht, reagierte sie mit großem *Verständnis:* Sie wissen es! Und ich weiß es! Wir wissen, dass Sie das nicht wollen! Dass das aus Ihrem Unbewussten kommt. – Ich hatte meine Augen geschlossen. Sie sprach ruhig, ganz weich, aber auch fest, sachlich. Unbeschreiblich, was es mir bedeutete, was sie sagte. Die K. fuhr fort: ... Dass es nichts ist, was Sie wollen, sondern etwas, was Sie selbst erlebt haben! – Es sei sicher kein Zufall, dass das gerade jetzt auftauche, jetzt, nach ihren Ferien: Da haben Sie sich ja ausgeliefert gefühlt, waren abhängig von meinen Plänen. – Und das würde ich ja kennen. Darum hätte ich das mit ihr machen wollen, was ich kennen würde. Deshalb meine Phantasie. Ich war so berührt und beeindruckt. – Also wollte ich Ihnen damit sagen, zeigen, was ich erlebt habe? – Ja! –

Nachdem sie an anderer Stelle meinte, ich würde mit ihr kokettieren, was ich gar nicht gerne hörte, ergänzte sie, ich wolle jetzt erleben, dass nichts passiert! Schweigend und sehr dankbar nahm ich diese Erläuterung damals auf.

Dies waren nur einige Beispiele, dass und wie die Therapeutin mir meine Reaktionen, meine Gefühle, mein Handeln und Ähnliches erklärte. Aber manchmal entzog sie sich auch solchen Erklärungen. Es kam ein paarmal vor, dass ich sie darum bat und sie sich dem verweigerte mit der Begründung, die ich heute noch besser verstehe als damals, sie wolle nicht theoretisch werden, denn damit kämen wir von meinen Gefühlen weg und das wäre nicht gut.

Ihr Umgang mit meinen Erinnerungen

Erst mit einem gewissen Abstand ist mir die volle Dimension der Wichtigkeit ihrer Einstellung und ihres Verhaltens bezüglich der Deutung meiner Erinnerungen und Bilder bewusst geworden. Besonders während der Jahre mit den zunehmend auftretenden Bildern mit meinem Vater erlebte ich ihre Abgrenzung und absolute Zurückhaltung als irritierend bis störend. Manchmal ärgerte es mich, dass sie nichts sagte, nichts kommentierte, nur zuhörte oder gerade einmal ein paar Fragen an mich richtete! Viel schwerer aber wog die sich für mich durch ihre Zurückhaltung ergebende Verunsicherung, die meine eigenen Zweifel noch verstärkte. Glaubt sie mir nicht? Denkt sie, ich spinne mir das zusammen? Auf entsprechende Bemerkungen von mir reagierte sie meist ruhig und gelassen.

Jetzt kann ich mit großer Achtung ihr gegenüber festhalten, dass sie, die Therapeutin, *mir nie meine Gefühle, meine Erinnerungen und Bilder* im Zusammenhang mit der erlebten sexuellen Gewalt *bestätigt hat,* diese nicht in Frage stellte, nicht ausreden wollte *oder gar ausredete.* Sie nahm das Erinnerte, alle meine dazu gemachten Äußerungen einfach auf und an und vor allem sehr, sehr ernst. Sie ließ mir den Raum, die Zeit und die Freiheit, alles selbst zu erkennen und einzuordnen. Natürlich benötigte ich dabei ihre Unterstützung, aber sie hielt sich mit Deutungen völlig zurück.

Sie formulierte ihre Haltung einmal mit den folgenden Worten: Ich würde nach Antworten suchen und hätte die gerne von ihr. Aber das müssten dann nicht meine Antworten sein. Ich müsse die Antworten selbst finden. Bei einer anderen Gelegenheit hatte sie betont, *es gehe darum, dass ich meinen Gefühlen glaube, dem, was ich erinnerte, und welche Bedeutung ich meinen Erinnerungen und Bildern beimesse.* Welche Bedeutung *ich* darin sähe! Es sei wichtig, zu klaren Gefühlen zu kommen, aber das könne nur ich leisten.

Und es steht nicht im Widerspruch zu dieser ihrer Haltung, dass *sie immer wissen wollte, was war!* Was „in mir war", so drückte ich das damals oft aus. Sie wollte von meinen Erlebnissen hören!

Ich hatte mich einmal auf die vorhergehende Stunde bezogen und gesagt, mein Gefühl sei gewesen, ich sage nichts mehr! Sie würde mir das ja sowieso nicht glauben. Es fiel mir schwer, das so offen auszusprechen. Was hinter dem „ich sage nichts mehr" stehe, wisse ich doch sicherlich, entgegnete sie. Ich verneinte ihre Frage. – Heißt das nicht, die Mama soll fragen? – Was soll die denn fragen? – Mit weicher, lieber Stimme sagte sie: Vielleicht sollten Sie das sagen, was sie fragen soll! – Nachdem ich davon sprach, sie soll nach meinen Schmerzen fragen und nach dem Grund meiner Traurigkeit, hatte sie gefragt: Und, willst Du mir das mal alles sagen? – Das kann man doch nicht sagen! – Doch! Mir kannst Du es sagen! – ... Ob die Mama das wirklich wissen will? – Ja, ganz fest von ihr, die will das wissen! – Damals fragte mich die K. ganz viel. Zum Schluss wandte ich mich dann wieder an sie: Wolltest Du das wirklich wissen? – Ja! ... Und ich bin so froh, dass Du mir das gesagt hast. Dass das endlich mal raus ist! – Ja, ich bin auch froh ... – Und ich finde es toll, dass Du die Kraft hattest, es mir zu sagen! –

Einige Stunden zuvor hatte ich wieder einmal erwähnt, da war noch etwas mit meinem Vater, und dass ich glaubte, sie daran erinnern zu müssen, damit wir das nicht vergessen. Auf meine Bitte hin „Mama, vergiss es nicht!" hatte sie zu meiner wirklich großen Beruhigung betont, nein, sie würde das nicht vergessen.

Ich habe mich und sie damals nicht nur einmal gefragt, ob sie das überhaupt alles hören und wissen will. Erkannte aber, dass es darum geht, dass ich es doch wissen will. Ja, das sei das Wichtigste, bestätigte sie damals. Mit eindringlichem Blick hatte sie ergänzt: Und ich will es auch wissen! – Zum Schluss dieser Stunde musste ich aber doch noch einmal fragen, mit kindlicher Stimme: Wollen Sie es wirklich wissen? – Ja! –

Und auch das steht nicht im Widerspruch zu ihrer Zurückhaltung. Es war ihr nicht zu viel! Auch wenn ich das oft befürchtete und es auch absolut hätte verstehen können, immer wieder vermittelte oder sagte sie mir, es ist ihr nicht zu viel!

Ich füge eine Erklärung an, mich damit dem Abschluss dieses Abschnittes nähernd, die auch mit mir, aber sehr viel mehr mit der Therapeutin zu tun hat. Ich hatte ihr von einem Gespräch mit meiner Schwester Lisa berichtet, in dem es darum ging, ob und warum sie Probleme damit hat, sich noch eingehender auf ihre Erlebnisse mit unserem Vater einzulassen. Ich sei sehr behutsam mit ihr umgegangen, um sie nicht zu beeinflussen. Daraufhin sagte die K., gerade weil es ein sensibles Thema und die Andere unsicher sei, wäre die Gefahr der Beeinflussung besonders groß. Je größer die Unsicherheit, um so leichter ließe frau sich beeinflussen. Sie würde mich ebenfalls beeinflussen, da dürften wir uns gar nichts vormachen! Es sei wichtig, besonders beim Thema sexuelle Gewalt, sich darüber im klaren zu sein. Deshalb sei es auch nötig, sich Zeit zu nehmen. Es brauche viel Zeit und Geduld, um zu spüren und zu wissen, was war, was für mich stimme. Ich dachte damals, sicher hat sie auch deshalb oft nichts gesagt, was mich ja so irritierte, um mich nicht zu beeinflussen!

Wie bedeutsam die Kombination von Zurückhaltung bei der Deutung der Erinnerungen und die gleichzeitige Bereitschaft, alles hören und wissen zu wollen, für das Bearbeiten traumatischer Erlebnisse ist, geht, wie ich meine, aus all meinen Schilderungen deutlich hervor.

Ihr Umgang mit der kleinen Ann

Nicht nur aus heutiger Sicht ist dieser Aspekt mit am Berührendsten. Eigentlich hätte ich dies gleich zu Anfang des Kapitels schreiben müssen: Wie die Therapeutin mit der kleinen Ann spricht, wie liebevoll sie mit ihr spricht, wie sie sie fragt und bei den Erinnerungen behutsam begleitet. Und besonders wundervoll, dass *sie die kleine Ann* sogar *mit DU ansprach:* Kannst Du Dich sehen? (...) Hast Du ein Bild, wie alt Du bist? – Sie sprach auch mit der großen Ann auf diese Weise: Jetzt bist Du doch nicht mehr alleine! (...) Nun verstehst Du die kleine Ann doch besser und kannst zusammen mit ihr weinen ... – Und die kleine Ann erinnert sich, fühlt sich aufgehoben und verstanden. Fragt sogar die Therapeutin in dieser Form: Willst Du das wirklich hören? – Oder an anderer Stelle: Kannst Du das (Furchtbare) wirklich aushalten? – Und die antwortet: Ja, ich kann das aushalten! –

Aus heutiger Sicht hat diese wundervolle Art, die kleine Ann mit DU anzusprechen, für mich auch diese Bedeutung: Die schlimmen Taten meines Vaters hatte ich abgespalten, verdrängt; die waren wie außerhalb von mir. Erst in der Therapie traute ich mich nach und nach, die Erinnerungen daran zuzulassen. Ich erinnerte ... und die Therapeutin war so nahe und behutsam dabei, nahm die kleine Ann und ihre Bilder und Gefühle von damals liebevoll auf. Sprach die kleine Ann wie eine liebevolle Mama mit Du an, sprach wie mit ihrem Kind. Und damit konnte ich die kleine Ann, all das, was mir passiert war, in mich hinein nehmen. Die kleine Ann war und ist innerlich nicht mehr alleine, kann sich auch in mir geborgen fühlen. Bei der Therapeutin und in mir.

Ich erinnere gerade, wie ich zum Ende einer sehr intensiven Stunde zur K. sagte: Wir vergessen die kleine Ann doch nicht? – Nein … – Und auch das sprach ich einmal aus: … die kleine Ann ist jetzt hier bei uns! Nicht mehr dort! Ich hab' sie gestern mit hierher genommen! – Ja … –

Das Du der Therapeutin hatte auch noch eine andere wundervolle Auswirkung. Weil sie mich so ansprach während dieser intensiven, nahen und schweren Sequenzen, traute ich mich, wie oben bereits erwähnt, *sie ebenfalls mit Du anzusprechen und sogar „Mama" zu ihr zu sagen* – wie ich es fühlte und wie es für mich passte. So hatte ich einmal erwähnt, in mir wäre dauernd: Mama, Mama. … Ich muss Dir noch was sagen! – Sie nahm das nicht nur in diesem Fall ganz selbstverständlich an. Manchmal sagte ich einfach Mama, nur so vor mich hin, Mama, Mama; manchmal kaum hörbar, andere Male sehr wohl hörbar. Es fühlte sich so wundervoll an, Mama zu fühlen und auch noch zu sagen!

Und das erinnere ich noch sehr: Ich hatte von dem Anruf bei meinem Vater berichtet, bei dem er zum Ende des Gesprächs fragte: Und was ist, wenn ich sage, ja, ich habe das gemacht? – Während der gesamten Therapiestunde hatte ich ganz starkes Herzflattern, was ich am Ende meines Berichtes nochmals erwähnte. Die Reaktion der Therapeutin: Ich habe vorhin schon gedacht, man müsste Ihr kleines Herz beschützen. … Beschützend in beide Hände nehmen! – Sie guckte so lieb, so innig. Später fragte sie, was ich denn jetzt fühlen würde. – Mama …, antwortete ich leise, beschämt. – Was könnte Ihr Herz denn jetzt gebrauchen? Was würde Ihnen gut tun? – Die Mama soll es in den Arm nehmen, mich in den Arm nehmen! – Ja! Ja, das machen wir! Das ist gut!, und sie kam fröhlich auf mich zu, streckte mir ihre Arme entgegen.

In einer anderen Stunde hatte ich ausführlich davon berichtete, wie schlecht es mir während ihres nur einwöchigen Urlaubs ergangen war. Dies insbesondere deshalb, weil die Taten meines Vaters immer klarer wurden, ich das nicht mehr wirklich anzweifeln konnte. Ich hatte die Urlaubswoche irgendwie überstanden, wenn auch wieder mit Gedanken daran, nicht mehr leben zu wollen. Später sprach ich von meinem Gefühl, dass sie, die Therapeutin, das, was ich erzähle, was ich erlebte, oft ganz anders sehen würde als ich. – Das ist auch gut so!, hatte sie ganz fest geantwortet. Ich zweifelte auch an, ob das, was ich ihr berichten würde, überhaupt so bei ihr ankäme, wie ich es erlebte und fühlte. Mit liebem und verständnisvollem Blick sagte sie: Doch! Das kommt an! Ganz bestimmt! – Ich dachte nur, hoffentlich! Nun sah sie mich an, beugte sich ein wenig vor und sagte: Ich habe jetzt ein Bild … als würde ich Sie auf meinen Schoß nehmen, Sie halten und sagen: Es wird alles gut! – Was war ich überrascht und berührt von ihrer Aussage und diesem schönen Bild! Ich sagte aber nur matt: Ich hoffe es! Hoffe es so sehr! –

Als ich einmal davon sprach, wie sehr ich sie vermisst hatte, fügte ich hinzu, ich könne aber nicht in Worte fassen, was ich vermissen würde, wenn sie nicht da sei. – Vielleicht, weil es nichts Verbales ist ..., meinte sie mit weicher Stimme. Ja, dachte ich und sagte, sehr berührt: Ich fühle es so: Jetzt, wo Sie wieder da sind, spüre ich wieder Energie. Wie wenn die Sonne auf eine Blume scheint... –

Vielleicht erwähnte ich bereits an anderer Stelle, dass ich zu ihr sagte, ich würde sie richtig körperlich vermissen. Das sollte heißen, mit meinem ganzen Körper. Was aber nichts mit Berührungen im eigentlichen Sinn zu tun hatte, sondern damit, dass sie mich tief berührte, mich ganz und komplett „erreichte". Ich nehme an, so ergeht es kleinen Kindern, wenn sie die Mama oder den Papa vermissen.

In diesem Zusammenhang passt es, kurz auf das Thema *Abhängigkeit* zu sprechen zu kommen, denn dieses begleitete mich ebenfalls einige Jahre. Wie viele Stunden plagte ich mich damit ab: Kann ich jemals wieder von ihr weggehen? Sie ist mir doch so wichtig! Ich fragte mich sogar, ob sie mich später gehen lassen wird, was vor dem Hintergrund des Verhaltens meiner Eltern absolut verständlich ist. Einerseits wollte ich es nahe haben mit der Therapeutin und gleichzeitig hatte ich Angst vor zu viel Nähe und vor Abhängigkeit.

Irgendwann spürte ich meine Angst vor Abhängigkeit, aber bald darauf, dass es auch sehr schön ist, mich (von ihr) abhängig zu fühlen. Hier sei noch hinzugefügt: Und *nicht* ausgenutzt zu werden!

Ich muss noch von weiterem, sehr Berührenden berichten. Nein, nicht „ich muss"! Ich *will!*

Um mich mir als Kind begreiflicher zu machen, mein Verständnis für die kleine Ann, nicht nur im Zusammenhang mit den schlimmen Erlebnissen, zu wecken und zu erweitern, hatte die K. des öfteren beispielhaft *von ihren Kinder berichtet.* Sie wollte mir damit helfen, mich mit meinen kindlichen Bedürfnissen besser annehmen zu können. Irgendwann kam sie wieder auf ihre Kinder zu sprechen und fragte mich anschließend, wie das für mich sei, wenn sie von ihnen berichte. Mal fände ich es ganz süß, manchmal wäre es eher neutral, aber es käme auch vor, dass ich von ihrer Welt nichts hören wolle, lautete meine etwas verschämt geäußerte Antwort, und ich hatte hinzugefügt: Dann will ich Sie nur wie Sie hier sind! – Nach einer kleinen Pause sagte sie: Haben Sie denn auch schon einmal gespürt, dass Sie mit in meiner Kinderwelt sind? Dass ich Sie mit in meiner Kinderwelt habe? – So oder ähnlich ihre Worte. Vielleicht hatte sie auch gesagt, sie habe mich mit in ihrem Kinderherz. Das kann aber auch bei einer anderen Gelegenheit von ihr hinzugefügt worden sein, oder ich habe es so gefühlt. Völlig überrascht und berührt hatte ich mit kindlicher Stimme geantwortet: Ja, manchmal habe ich das schon so gespürt ... – Wie viel mir das damals bedeutete ist kaum zu beschreiben, aber sicher nachzufühlen.

Sie war für mich da

Verlässlich, geduldig, verständnisvoll, mitfühlend und mit so viel Liebe – sie war einfach immer sehr für mich da! Auch wenn es mir früher manchmal nicht genug erschien oder ich sie zeitweise nicht wie gerade beschrieben erlebte, nahm ich grundsätzlich sehr wohl ihr großes Engagement dankbar wahr. Sehr bedeutsam und vor allem beruhigend zudem zu wissen, sie immer anrufen zu können, wenn ich es brauchte. Das hatte sie schon bald zu Anfang der Therapie so oder ähnlich ausgedrückt: Wenn Sie mich brauchen, rufen Sie mich an! – Oder sie sagte nur knapp, aber eindringlich zum Ende der Stunde: Ich bin da! – Ich traute mich jedoch nur selten, dieses Angebot anzunehmen, bin mir heute aber absolut sicher, dass sie es auch öfter akzeptiert hätte. Ich erinnere, dass ich nicht nur einmal sagte, ich hätte sie gerne angerufen, aber hätte mich nicht getraut oder ähnlich, und dass sie dann mit weicher Stimme fragte: Und, warum haben Sie das nicht getan? – Oder dass sie auf ähnliche Weise ihr Bedauern darüber ausdrückte. Aber schon alleine zu wissen, ich kann sie anrufen, hatte für mich oft eine beruhigende und stabilisierende Wirkung.

Ich erinnere hier noch einmal an den bereits an anderen Stellen erwähnten Italienurlaub und meine große Angst davor sowie an das Angebot der K., sie jederzeit aus dem Urlaub anrufen zu dürfen. Das mache sie doch mit ihren Kindern auch so, wenn die alleine wegfahren. Die würden den Kontakt mit zuhause brauchen, könnten jederzeit anrufen. – Sie können mich ja auch anrufen! Es ist doch wichtig, dass diese Sicherheit da ist!, so oder ähnlich damals ihre Worte.

Und was es für mich bedeutete, mit ihr nach der ersten furchtbaren Nacht telefonieren zu können. Sie ist für mich da! Es war so wundervoll, das zu erleben! Und noch dazu ihre tolle Idee, auf die Entfernung eine Therapiestunde zu halten, ich in Italien, sie in ihrer Praxis sitzend, in unserem Raum, zu einer vorher festgelegten Zeit. Noch immer berührt es mich ganz tief in meinem Innern, wenn ich daran denke. Das ist wieder etwas, was ich gar nicht recht in Worte fassen kann, was aber vielleicht die damaligen Worte der Therapeutin gut verdeutlichen: Es gehe doch jetzt um das innere Kind! Darum, dass es dem gut gehe!

Im Laufe der Jahre konnte ich immer mehr sehen, also immer mehr annehmen, wie sehr sie für mich da ist! Es gab Zeiten, in denen ich damit haderte, dass sie auch andere KlientInnen therapierte. Aber später hatte ich oft das Gefühl, als sei sie nur für mich da! Wohl wissend, dass das nicht der Fall war. Ich denke, ich konnte das alles erst nach und nach wahr- und annehmen. Damals fiel es mir nämlich sehr schwer, so viel Gutes und Liebes anzunehmen. Auch das will gelernt sein! Bei mir jedenfalls war es so.

Sie ließ sich von mir anschauen
Ich denke daran, wie sie sich von mir anschauen ließ ..., einfach nur anschauen! Ohne Worte. Wie beglückend für mich, dass sie das zuließ!

Ich sah sie an wie es kleine Kinder tun, die uns von oben bis unten genau anschauen. Einfach so. Ich nahm die Therapeutin damit in mich „hinein“, in mich auf, so notierte ich es damals. Ich hatte sie nach den entsprechenden Stunden dann „in mir“, konnte mich noch besser innerlich mit ihr verbunden fühlen, was in den ganz schlimmen Phasen der Therapie extrem bedeutsam war und was mir dazu verhalf, mich nicht (so) alleine zu fühlen.

Nach vielen Stunden hatte ich sie auf diese Weise noch vor Augen, konnte daran denken, wie sie dort saß, wie sie gekleidet war – ja, auch das erinnerte ich oft! – und wie sie mich angesehen hatte. Besonders ihre Haare und ihre Hände hatten mich fasziniert, die musste ich wieder und wieder anschauen, aber nicht nur die ...

Ihr in ihre Augen zu schauen, nur in die Augen, aber auch von ihr ebenso angesehen zu werden – das war einfach wunderschön! Nicht nur nach sehr intensiven Stunden kam es vor, dass ich sie so eine ganze Weile ansah, sie mich meist auch ... Und dann sagte sie, halb fragend, halb die Stunde damit beendend: Gut?! – Das U dabei sehr betont und ich antwortete meist mit einem zögernden Ja. Zögernd oder verhalten, weil es eigentlich nicht gut, nämlich immer noch nicht genug war. Beim Schreiben muss ich über mich selbst lächeln, auch wenn ich das noch immer gut verstehe!

Irgendwann fiel mir auf, dass sie ihre Augen für eine kleine Weile ganz sanft, so schien es mir, schloss und auf die gleiche bezaubernde Weise öffnete und mich wieder ansah ... Dann fühlte ich mich wie von ihr in den Arm genommen und gehalten. Eine wunderbare Alternative zu Berührungen, wenn die nicht möglich waren oder nicht passten. Und dies gab sie mir in den sehr intensiven Jahren ganz oft! Auch ohne dass es mir schlecht ging! Auch das nahm ich dann mit nach Hause.

Sie hatte mich zuvor einmal darauf aufmerksam gemacht, als ich wohl damit haderte, dass sie mich nicht in den Arm nehme konnte, weil Berührungen vorübergehend nicht möglich waren. Damals fragte sie, ob ich denn nicht wahrgenommen hätte, dass sie mich mit ihren Augen gehalten habe. Nein, das hatte ich bis dahin nicht; aber danach! Und wie!

Es kam aber auch einmal vor, da war es ihr wohl zu viel. Sie sprach davon, ich sähe sie an, als wolle ich sie mit meinen Augen verschlingen. Das wollte ich nun gar nicht hören, schämte mich und fühlte mich zurückgestoßen. Aber ich wusste ganz genau, was sie meinte ...

Sie freute sich mit mir und für mich
Auch das gehört für mich mit zu dem, was die Therapeutin mir gab: Dass sie sich für mich und mit mir freute; ich glaube, dass sie manchmal sogar ein bisschen stolz auf mich war. Sei es, weil ich mich nach längerem Quälen doch auf meine Erinnerungen einließ, mich danach auch leichter und freier fühlte, oder sei es nach den Konfrontationen mit meiner Mutter und besonders mit meinem Vater. Egal, bei welchen dieser schwierigen und wichtigen, kleinen oder größeren Schritte, ich ahnte oder sah ihre Freude. Und wenn ich Zeit dafür benötigte, für all diese Schritte, dann zeigte sie großes Verständnis. Mir ist klar, dass das für die Arbeit einer Therapeutin mehr oder weniger selbstverständlich ist. Dennoch erwähne ich es, weil es für mich nicht selbstverständlich und so wertvoll war.

Was sie mir von sich gab, mir schenkte

Ich beginne mit einer Begebenheit aus dem zweiten Jahr meiner Therapie. Es ging nur indirekt um das Thema sexuelle Gewalt. Vordergründig ging es, neben anderen Themen, um meine lieben Gefühle für die Therapeutin und um das, was ich befürchtete und ihr gegenüber so ausdrückte: Bei den starken Gefühlen für sie könne das nur in einer Katastrophe enden!, hatte ich traurig und erschüttert geäußert. Als diese Stunde zu Ende ging, sagte ich, ich wolle nicht gehen. – Ich gehe einfach nicht!, hatte ich fest und trotzig hinzugefügt. Entgegen meiner Erwartung reagierte die Therapeutin ganz geduldig, fragte, was wir denn nun tun sollten. Und, zu meiner Überraschung: Was muss denn die Mama tun, damit sie gehen darf? – Ihrer Tochter würde sie manchmal etwas von sich dalassen, wenn sie wegfahre, ihre Tochter das gar nicht möge oder deshalb traurig sei. – Dann gebe ich ihr etwas von mir, das sie behalten darf, bis ich wiederkomme. – Ich war so glücklich über diese liebe Reaktion, sagte, das wäre eine gute Idee. Sie überlegte einen Moment, streifte kurz ihren Armreif vom Arm ..., zog ihn jedoch wieder an und äußerte diese Idee: Was halten Sie davon, wenn Sie *einen Satz mitnehmen?* ... Ich komme wieder, sie kommen wieder..., wir sehen uns wieder! – Eigentlich hätte ich gerne ihren Armreif mitgenommen und ihn Tag und Nacht getragen. Aber das sagte ich natürlich nicht, dafür schämte ich mich viel zu sehr. Ich nahm stattdessen und auch freudig den Satz mit nach Hause und sagte ihn immer wieder vor mich hin und auch dies: Die Mama hat mir einen Satz geschenkt! –

Ich hatte der K. einmal von dem Buch „Soulkiss“ der amerikanischen Schriftstellerin Shay Youngblood (Piper Verlag, München; 1998) erzählt, worin ein kleines Mädchen von seiner Mama jeden Tag ein Wort auf den Weg zur Schule mitbekam, was mich damals sehr rührte. Deshalb hatte ich davon berichtet. Die Kleine wiederholte es wie ein Gebet, wenn ihr die Mama fehlte, die ihr gesagt hatte, sie werde den ganzen Tag an dasselbe Wort denken. Diese Vorstellung hatte dem kleinen Mädchen das Getrenntsein erträglich gemacht. ... Nun hatte mir meine Mama einen Satz mitgegeben.

Und ihren *Armreif* bekam ich dann doch noch, rund ein Jahr später. Das geschah so: Ich war mit starkem Herzklopfen in die Stunde gekommen, hatte von meiner totalen Unsicherheit und dem Gefühl von Hilflosigkeit gesprochen. Was denn mein Herz brauchen, was ihm guttun könnte, fragte sie. Ich dachte: Liebe! Und bekam noch stärkere Herzprobleme, Angst, und das Gefühl, ich muss von ihr wegrücken. Sie sagte, sie würde mich so wahrnehmen, ich wolle von ihr gehalten werden; sie solle mein Herz halten. Ich gestand ihr, dass ich an Liebe dachte und gerne in den Arm genommen werden würde, ihre Hand in meiner zu spüren wünschte. Bei alldem würde ich aber Angst fühlen. Ob sie mit ihrer Hand kommen solle?, hatte sie gefragt. – Nein!, sagte ich ganz fest. Ich hätte Angst davor, dass ihre Hand was an meinem Körper machen könnte ... Das wäre aber nicht das, was ich wolle! Nein, das sei klar, entgegnete sie, aber ich würde befürchten, sie tue sowas! Was war ich froh über diese Reaktion! Nachdem sie mich fragte, ob ich mir vielleicht etwas Anderes bei ihr holen wolle oder könne, hatte ich sofort an ihren Armreif gedacht. Nur gedacht, nichts gesagt! Und schon streifte sie ihren Armreif vom Arm und legte ihn neben mich auf meine Matratze. Das war so wundervoll! Ich nahm ihn berührt in beide Hände, legte mich abgewandt von ihr hin, hielt ihn weiter in meinen Händen. Mama, Mama, fühlte ich immer wieder. Sagte es ihr. Richtete mich wieder auf, auch um sie sehen und ansehen zu können. Hielt ihren Armreif weiterhin zärtlich in meinen Händen, zwischendrin drückte ich ihn an mich.

Dann sagte ich: Jetzt würde ich doch gerne ihre Hand spüren ... – Sie kam langsam zu mir, kniete sich vorsichtig auf den Rand meiner Matratze und legte ihre rechte Hand flach hin. Ich schaffte es nach einer kleinen Weile, meine zur leichten Faust geformte Hand in ihre geöffnete zu legen. Ich konnte schon damals kaum beschreiben, wie sich das anfühlte. Nur schön! Und dann legte sie ihre andere Hand obenauf, so dass meine geschlossene Hand von ihren Händen zart umhüllt wurde. In meiner rechten Hand hielt ich weiter ihren Armreif. Drückte ihn auch einmal an mein Herz.

Wir schauten uns eine Weile an. Sie schloss zwischendrin ihre Augen, öffnete sie wieder und sah mich an. Ganz lieb und zärtlich wirkte das. Drückte auch einmal sanft meine Hand. Dann fragte sie, was ich davon mitnehmen wolle, wenn wir jetzt auseinander gingen. – Haben Sie ein Bild oder gibt es etwas, was Sie mitnehmen wollen? – Ich dachte natürlich sofort an den Armreif, zuckte aber nur mit den Schultern. – Dann nehmen Sie den Armreif mit!, sagte sie doch tatsächlich. Ich gestand, dass das auch mein Wunsch war, ich mich nicht traute zu fragen. Das beantwortete sie mit einem lieben

Schmunzeln. Ich beendete meine Tagebuchnotizen zu dieser Stunde mit folgendem Satz: Ich habe mich heute wirklich ganz und gar meinen Gefühlen überlassen, und sie ist so toll damit umgegangen!

Den Armreif nahm ich mit nach Hause, trug ihn in der Folgezeit wenn möglich mit mir herum oder zog ihn sogar an und nachts hielt ich ihn in meiner Hand, die auf meinem Herzen lag. Es war so wunderbar, etwas von der K. für mich zu haben! Ich notierte dazu in meinem Tagebuch: Wenn sie so lieb zu mir ist, wird mir noch mehr bewusst, was mir früher fehlte, und das ist so traurig!

In der darauf folgenden Stunde brachte ich ihren Armreif natürlich wieder mit. Ich sagte gleich, ich würde ihn gerne während der Stunde noch in Händen halten, ihn ihr aber dann ... Ich könne ihn noch behalten, wenn ich wolle, sagte sie gleich darauf; so lange, wie ich ihn brauchen würde. Wir könnten ihn dann gegen etwas Anderes austauschen. Im Laufe der Stunde fühlte ich mich ganz klein, ließ Erinnerungen zu ... Leider ging es mir wieder so, dass ich die K. nicht in meiner Nähe haben konnte aus Angst, was mich sehr traurig stimmte. Sie fragte, mit ihrer lieben Stimme: Können Sie sich vorstellen, wenn Sie den Armreif von mir haben, sich zu phantasieren, dass ich Sie halte? –

In der nächsten Stunde erklärte sie mir dann, wenn sie mich anschaue, lieb anschaue, das sei doch wie in den Arm nehmen, stehe doch dafür! Erst dann konnte ich das so annehmen und genießen.

Ich brachte ihren Armreif danach, wie ich meine mich zu erinnern, in jede Stunde mit, hielt ihn dann in meinen Händen oder legte ihn nur neben mich. Irgendwann kam, was kommen musste, sie wollte ihn zurückhaben. Sie habe schon eine Idee, was sie mir Anderes anbieten könnte. Ich wolle ihn aber nicht wieder hergeben, sagte ich. Sie schaute mich nur lieb an ... In der nächsten Stunde sprach ich von meiner Traurigkeit wegen der anstehenden Rückgabe. Sie habe mir noch nichts Anderes mitgebracht, denke aber an einen glänzenden Stein. Ich hätte doch einmal von einem solchen geträumt ... Ich fand's toll, dass sie sich daran erinnerte, sagte aber auf ihre Frage hin, ob ich den Armreif denn entbehren könne, nein, ich würde ihn am liebsten gar nicht mehr zurückgeben. Währenddessen hielt ich ihn liebevoll in meinen Händen.

Einige Stunden später holte sie ein kleines Säckchen hervor, entnahm diesem fünf Steine. Wir betrachteten sie. Die K. hatte sich in meine Nähe gehockt. Ich entschied mich für den wie eine kleine Schildkröte aussehenden grünen Stein, der am hinteren Ende einen kleinen Schaden hatte. Ich müsse dabei an eine Verletzung denken. ... Das würde doch passen, hatte ich ergänzt. Obwohl mir eigentlich keiner gefiel, nahm ich diesen Stein. Ich wollte doch lieber ihren Armreif behalten! Den ich ihr dennoch mit den Worten aushändigte: Ich kann Ihnen gar nicht sagen, was mir es bedeutete, dass ich ihn so lange behalten durfte! –

Nun hatte ich also einen *schönen, schimmernden Stein* von ihr. Erst zuhause hatte ich entdeckt, dass er glitzert! Und den verlangte sie nicht zurück, aber das wusste ich damals noch nicht. Ich trug ihn meistens bei mir, hielt ihn nachts in meiner Hand, nahm ihn mit, egal, wohin ich ging oder fuhr. Immer hatte ich etwas von meiner Mama bei mir. So erlebte und fühlte ich es. Einige Monate später fragte ich sie, ob sie ihn denn nicht wieder haben wolle. – Im Moment nicht!, hatte sie entgegnet und später hinzugefügt: Ich denke, Sie brauchen ihn noch! – Ja, das ist wie ein Schutz!, war meine spontane Antwort. Danach erinnerte ich immer wieder ihre liebe Aussage, sie denke, ich brauche ihn noch ...

Damals dachte ich an das *Sonnenblumenbild* und dass ich das auch noch brauche! Das hatte sie mir gleich richtig geschenkt. Irgendwann bereits von ihr erwähnt, schenkte sie es mir nur einige Stunden nach der Stunde mit dem Stein. Wieder war es sehr heftig zugegangen, ich von großer Angst geplagt. Und dann holte sie dieses Foto hervor... – Dass Sie etwas für mich haben, dass das für mich sein soll ..., das kann ich gar nicht glauben!, so meine erste Reaktion, begleitet von Schwindelgefühlen.

Ein Foto mit einer Sonnenblume, die in einer Umgebung ganz alleine steht, in der ansonsten nur Sand und ein wenig niederes Grün zu sehen sind. Wie es dazu kam, dass sie mir gerade dieses Foto schenkte, führe ich später noch näher aus. Aber was sie beim Überreichen des Fotos ungefähr geäußert hatte, das sei bereits hier erwähnt: Auch in einer kargen Umgebung kann eine Blume wachsen! –

Indem die Therapeutin mir etwas schenkte oder mir etwas Persönliches von sich gab, konnte ich noch besser und noch einmal anders die innere Verbindung zu ihr halten. Fühlte ich mich ihr noch näher. Aber das beschreibt nur sehr unzureichend, welch tiefe Bedeutung diese Geschenke für mich hatten und noch immer haben. Und ja, sie alle und ihre große und wundervolle Bedeutung stehen auch im Zusammenhang mit dem Thema dieses Buches!

Bei dem *kleinen roten Kissen,* von dem ich nun berichte, handelt es sich allerdings um kein Geschenk, sondern um etwas, das auch mit Nähe zur Therapeutin und dem Umgang damit zu tun hat. Irgendwann hatte sie es aus dem Büro geholt. Ich glaube, sie wollte mich zum Ende einer harten Stunde in eine Decke packen, holte das kleine Kissen noch hinzu, packte es in meinen Rücken. In der Folge lag es fast immer an meinem Platz, und ich gewann es richtig lieb; ich schmiegte es ganz oft an mich. Manchmal legte ich meinen Kopf darauf, besonders ab dem Zeitpunkt, als ich auf der Couch lag. Oder ich hielt es nur in meinen Händen oder legte es neben mich. Nein, ich nahm es nicht mit nach Hause. Es erfüllte seinen Sinn gerade dadurch, dass es dort für mich da war.

Eine weitere Bedeutung dieses, wie ich es irgendwann einmal nannte „Mamakissens“ tat sich mir allerdings erst einige Jahre später auf. Und das kam so: Ich hatte in der entsprechenden Therapiestunde begeistert von meinem glücklichen Gefühl der letzten Tage und Wochen gesprochen. Ich würde die Therapie jetzt richtig genießen, weil nicht mehr so Schweres und Hartes zu bearbeiten wäre. Eine Freundin habe am Tag zuvor viel Positives über meine Veränderungen gesagt. Auch das zu meiner Freude. Und gestern auf der Rückfahrt von einem Kletterausflug hätte ich tatsächlich gedacht: Ich bin ein Glückspilz! – Ich strahlte die Therapeutin an und sie mich: Das hab' ich bestimmt noch nie von mir gedacht. Aber so fühlte ich es. – Während ich darüber redete, spürte ich plötzlich Angst. Die K. nannte es Glücksangst. Aus einem Impuls heraus nahm ich das kleine rote Kissen, schmiegte es an meinen Oberkörper. Ich wolle gerne noch etwas sagen, aber wieder hätte ich Angst. Ich zitiere aus meinen Aufzeichnungen von dieser Stunde:

– Ich will es dennoch sagen. – Ja! – Ich finde es so toll, dass es mir so gut geht! Dass ich so viel für mich geschafft habe! Und ich weiß, dass das sehr viel mit mir selbst zu tun hat. Hab' dafür ja ganz viel gearbeitet. Aber ich weiß, es hat auch mit ihnen zu tun! Ich finde, und ich sage jetzt „wir“, ich finde, WIR haben das geschafft! Und ich bin Ihnen so dankbar dafür! – Wie liebevoll sie mich anlächelte! Ich schloss meine augen, wieder ganz bei mir.

Von wegen schwer, ihr das zu sagen, dazu meinte sie bald lächelnd, fröhlich: Das haben sie mir doch eigentlich schon vor einer halben stunde gesagt! – Ich stutzte; sie fuhr fort: Als sie sagten, ich bin ein glückspilz! Das hieß doch, was bin ich für ein glückspilz, dass ich bei dir therapie mache! – Ich lächelte. Das stimme; gestern hätte ich genau das gedacht.

Bezüglich meiner Angst fragte sie, ob es mir mit dem aussprechen zu nahe geworden wäre? Nein, ich hätte uns eher getrennt erlebt, das aber als angenehm empfunden. – Also ich vermute, dass sie diese ängste vorschieben, weil es ihnen doch zu nahe wird. Ihre ängste schaffen dann was trennendes, wirken trennend. – Das auszusprechen bedeute doch viel nähe! (...)

Die K.: Da haben wir ja wohl etwas gemeinsam! Das kissen hat doch für sie was symbolisches (o. ä.), da projizieren sie ja etwas drauf. Da kommt doch etwas drittes für uns beide hinzu, worauf wir unsere ängste vor verschmelzung projizieren. Ist doch klar, dass unsere arbeit hier viel nähe schafft. Wir haben das zusammen erreicht! Es ist etwas gemeinsames und dadurch entsteht sehr das gefühl von nähe. ... Und das rote kissen wirkt entlastend. Als schutz vor verschmelzung! –

Später fügte sie hinzu, die distanz wäre dann auch wieder notwendig! (Alles nur so ungefähr!)

Ich ging so glücklich nach hause, fühlte mich sehr mit ihr verbunden. Ein tolles und freies gefühl dabei! Sie wird das schöne sicher auch noch in sich haben.

Meine Wünsche an sie

Im zweiten Jahr meiner Therapie listete ich einmal auf, was ich mir alles von der Therapeutin wünschte. (An dieser Stelle stehen die Wünsche im Fokus, die in Bezug auf das Bearbeiten der sexuellen Gewalt von besonderer Bedeutung waren.) Es fiel mir damals sehr schwer, meine Wünsche auszusprechen. Ich fragte mich während der Stunde, wohin das führen mag, wenn ich die ausspreche. Will ich ihr das wirklich sagen? Ich sprach dann meine Gedanken aus: Was ändert sich eigentlich, wenn ich Ihnen sage, was ich gerne von Ihnen hätte? Ausgeliefert fühle ich mich ja jetzt schon! Dass ich sagte, ich brauche Sie, das ist doch schon gefährlich. … Deshalb hatte ich zuerst die Form gewählt, ich brauche *das* doch so, das Hiersein, die Therapie! –

Daraufhin sprach ich die Therapeutin direkt an, sagte, was ich mir von ihr wünschen würde. Ich würde Ermutigung brauchen, das Thema anzugehen; dass sie mal sagt, toll, dass du das machst! Brauchte ganz viel Verständnis und ganz viel Vertrauen. Und Trost und Unterstützung, wenn es ganz schlimm und arg zugeht. Auch Unterstützung dabei, dass ich nicht mehr denke, ich sei schuld!

Nachdem ich dies ausgesprochen hatte, war ich froh; setzte mich aufrechter hin. Ihre lächelnde und für mich überraschende Reaktion: Ich finde es sehr gut, dass Sie es gesagt haben! – Sie freue sich schon auf die nächste Stunde, weil diese nun beendet sei. Zu meinem Erstaunen fügte sie hinzu: Ich finde, Sie haben auf manches sogar ein Recht! – Ein Recht?, fragte ich erstaunt, können sie mir ein Beispiel dazu geben? – Sie haben beispielsweise ein Recht darauf, dass ich Sie verstehe! Mir fällt noch sehr viel mehr dazu ein, aber wir sind schon über die Zeit. Aber dennoch, um es festzuhalten, dies noch: Ich denke, dass Sie manches von dem von Ihnen Angesprochenen nicht bemerken, dass ich Ihnen das gebe. Außerdem haben Sie bestimmte Vorstellungen, wie das aussehen soll, was Sie wollen. Darüber sollten wir noch reden. Auch darüber, dass Sie mich mit meinen anderen KlientInnen teilen müssen. Dabei denke ich daran, dass Sie bei so vielen Geschwistern auch immer teilen mussten. – Ich war richtig erleichtert darüber, wie sie meine Wünsche aufgenommen hatte.

Grundsätzlich nahm die Therapeutin meine an sie gerichteten Wünsche an, ohne sie groß zu kommentieren. Was mich besonders berührte war diese Aussage von ihr: *Sie sind mit allen Ihren Wünschen und Gefühlen willkommen!* – Das klang noch lange in mir nach. Danach wiederholte ich diese lieben Worte immer mal wieder, indem ich sie vor mich hinsagte.

Sie verstehe Therapie so, sagte sie an anderer Stelle, dass hier alle Gefühle einen Platz haben, also auch kindliche Wünsche. Diese gehörten erst recht in die Therapie! Unbeschreiblich, wie auf- und angenommen ich mich gerade mit meinen kindlichen Gefühlen und Wünschen fühlte.

Ich habe mir bei ihr Vieles von dem holen und erfüllen können, was ich ersehnt hatte. Gerade weil ich dies mit meiner Mutter nicht erlebt hatte, registrierte ich genau und genoss es ausgiebig, wie sorgsam und liebevoll die K. besonders mit meiner Bedürftigkeit umging. Auch wenn sie mir nicht alle meine Wünsche erfüllte, erfüllen konnte, erlebte ich es so: Ich habe sehr viel von ihr bekommen!

Ich hatte während der Zeit der Therapie zahlreiche Male betont, mir bei und mit ihr so viel Schönes nachgeholt zu haben. Ich sprach sogar davon, es fühle sich wie *Wiedergutmachung* an. Bei einem vor kurzem geführten Gespräch mit der Therapeutin meinte sie, ich könne nichts nachholen und das Fehlende, die Schäden und das Schlimme könnten auch nicht wieder gut gemacht werden. Das sei einfach nicht möglich! Es sei aber sehr wichtig gewesen, dass ich mit all meinen Wünschen und Gefühlen da sein konnte und bei ihr gut damit aufgehoben war. Ich stimme ihr in gewissem Maße zu und dennoch ist weiter das Gefühl vorhanden, etwas nachgeholt und für mich wieder gut gemacht zu haben.

Zum Schluss dieses Abschnitts komme ich auf das weiter oben erwähnte Sonnenblumenbild zurück. Dieses Bild versinnbildlicht meine an die Therapeutin gerichteten *Symbiosewünsche.* Das will ich gerne erläutern.

Ich habe bereits davon berichtet, als es um die Therapieferien ging und darum, wie sehr ich die Therapeutin vermisste, dass es mir nach ihrer Rückkehr so erging: Jetzt, wo sie wieder da sei, würde ich Energie spüren, so wie wenn die Sonne auf eine Blume scheint. Ähnlich äußerte ich mich einige Monate später: Wenn Sie mich anschauen, ich in Ihre Augen schaue, das ist für mich wie Energie! Wie ich schon einmal sagte: Wie wenn die Sonne auf eine Blume scheint! ... Wie Leben! – Noch heute fühle ich, wie begeistert ich damals von meinem Vergleich war. Irgendwann danach, von einer Frankreichreise zurückgekehrt, hatte ich erwähnt, ich würde doch Sonnenblumen so sehr lieben; hätte dort einige wunderschöne Felder gesehen. Ich fügte hinzu, wiederum mit weicher Stimme: Sonne ... Blume ... Sie verstehen? – Sie zeigte mir mit liebem Blick, dass sie verstand.

Darauf kam sie eines Tages indirekt zurück. Wir befanden uns in der letzten Stunde vor einer ihrer von mir gefürchteten Therapieferien. In meinem Kummer hatte ich gesagt, ich wolle jeden Tag eine Kerbe in die Tür machen, bis sie wiederkomme, weil ich sie doch so sehr vermissen werde. Daraufhin sagte sie: Sie können doch auch jeden Tag eine Sonnenblume malen... – Nach diesen Worten fühlte ich mich wie verzaubert.

Auch diesen Satz von ihr sagte ich ganz oft vor mich hin. Und malte während der Ferien jeden Abend eine wunderschöne Sonnenblume. Zum Schluss ergab es ein Feld von Sonnenblumen; dabei auch größere mit einigen ganz kleinen Blüten. Dieses Bild habe ich noch heute.

Vor diesem Hintergrund schenkte sie mir das Sonnenblumenfoto, das sie während eines Urlaubes an der Ostsee aufgenommen hatte. Auch dieses Geschenk nahm ich jahrelang mit auf alle meine Reisen. Es lag immer in der Nähe meines Bettes, weil das Übernachten doch so schwierig für mich war.

Meine lieben Gefühle für sie

Ich erlebte mich auch mit meinen lieben Gefühlen, mit meiner Liebe für sie bei ihr willkommen, aufgehoben und geborgen. Ich staunte immer wieder aufs Neue, wie liebevoll und selbstverständlich sie meine Liebe annahm und „aushielt“, wie ich es früher nannte, und dass es ihr nicht zu viel wurde. Unbeschreiblich, was mir das bedeutete! Natürlich hatte ich Angst, weil ich sie liebte, auch sie „macht was“, dass sie meine Liebe ausnutzt. Wie wichtig und wertvoll zu erleben, dass sie mich und meine Liebe nie ausnutzte!

Aber steht meine Liebe zu der Therapeutin tatsächlich im Zusammenhang mit dem Thema dieses Buches, der sexuellen Gewalt?, fragte ich mich zu Beginn dieses Abschnitts. Scheinbar nicht. Aber nur scheinbar. Diese Liebe war für das Bearbeiten der furchtbaren Erlebnisse und ganz besonders für mein Überleben extrem wichtig. Überlebenswichtig! Die lieben Gefühle gaben mir inneren Halt. Ohne diese lieben Gefühle hätte ich das alles nicht durchgestanden. Da bin ich mir absolut sicher.

Doch sie bedeuteten für mich noch viel mehr, nämlich ein Zurückkommen zu meinen Gefühlen.

Wenn ich mich der K. sehr nahe fühlte, dann fühlte ich mich auch mir sehr nahe. So bekam ich langsam Zugang zu meinen eigenen Gefühlen und eine Ahnung davon, wie es ist, ganz bei mir zu sein. Dazu kam eine weitere Entdeckung, mit der ich überhaupt nicht gerechnet hatte: Ich fühlte Liebe für sie... und irgendwann spürte ich auch Liebe für mich. Welch eine wunderbare Überraschung! Liebe für mich selbst hatte ich zuvor noch nie empfunden. Das wiederholte ich daraufhin bewusst des Öfteren: Ich dachte an die K., fühlte Liebe für sie, und wandte mich daraufhin innerlich mir zu ... und fühlte Liebe für mich! Damals brauchte ich noch den Umweg über sie.

Ich deutete bereits an, wie schwer es mir fiel, die Gefühle für die K. anzunehmen. Immer wieder war das Thema in den Therapiesitzungen. Oder ich traktierte mich mit den Worten: Das geht doch nicht! Das ist doch nur deine Therapeutin! – Aber die lieben Gefühle waren da! Mit Wut auf sie tat ich mich leichter, wesentlich leichter. Obwohl das auch nicht leicht war. Und dann las ich eines Tages in einem Sachbuch von einer Analyse mit einer Patientin („Stumme Liebe", herausgegeben von Eva Maria Alves). Diesen Bericht hatte ihre Analytikerin verfasst. Ich las, welch große Bedeutung und Erleichterung es für die Klientin darstellte, dass ihre Therapeutin sich von ihr lieben ließ. Das war der Satz, den ich brauchte: *Dass sie sich von ihr lieben ließ!* – Es ist mir noch heute so nah, welche Wirkung dieser Satz, diese Aussage für mich hatte. Wie oft sagte ich mir diesen Satz vor mich hin: Dass sie sich von ihr lieben ließ! Natürlich sprach ich darüber mit meiner Therapeutin. Sie freute sich sichtlich mit mir, weil es mir dadurch gelang, besser zu meinen Gefühlen zu stehen.

Sie hatte mir immer wieder zu verstehen gegeben, dass die lieben Gefühle doch nicht wirklich etwas mit ihr, der Therapeutin, zu tun hätten. Klar habe sie eine Wirkung auf mich, wie sie für mich da sei und mit mir umgehe, aber letztendlich würde ich sie doch nicht kennen. Es sei *mein* Vermögen, diese Gefühle zu haben. Und das sei doch gut!

Es dauerte ziemlich lange, bis ich das nachvollziehen konnte.

Hilfreich war, dass sie in diesem Kontext auf meine Kindheit verwies. Meine Schwierigkeiten mit diesen Gefühlen hätten mit meinen Eltern zu tun. Meine kindlichen Gefühle wären damals abrupt beendet worden, wodurch ich keinen Übergang von den kindlichen Gefühlen, der kindlichen Beziehung zu den Eltern hin zu erwachsenen Gefühlen für sie erlebt hätte. Sie denke, meine Schwierigkeiten in der Beziehung zu ihr rührten daher, dass ich beide Gefühle, die kindlichen und erwachsenen, für sie spüren würde.

Und sie stellte meine Liebe für sie auch in Zusammenhang mit der für meinen Vater. Ich hätte diese in der Therapie re-inszeniert, um wieder gut zu machen, was ich mit meinem Vater erlebte. (An dieser Stelle sprach sie tatsächlich von Wiedergutmachung!)

Ich kann dazu nur sagen: Die lieben Gefühle und die Mamagefühle für die Therapeutin haben mich nicht nur gerettet und am Leben gehalten, sie haben eine Wunde in mir geheilt!

Und dies kommt noch hinzu: Ich bekam in all den Jahren *sehr viel Liebe von ihr!* Genau das erinnerte und fühlte ich vor langer Zeit bei Gedanken an die Therapeutin: Ich habe ganz viel Liebe von ihr bekommen! Ich stutzte, dachte, nein, das stimmt doch nicht! Die hat mich doch nicht lieb gehabt! Natürlich war sie empathisch und sehr dabei und engagiert und liebevoll im Umgang mit mir, aber sie hat mir doch keine Liebe gegeben!

Aber das Gefühl blieb und ist noch heute da. Ich habe ganz viel Liebe und viel Liebes von ihr bekommen!

Zum Schluss ...

... dieses Kapitels zitiere ich eine Sequenz mit der Therapeutin, die auch sehr viel über sie aussagt und darüber, wie ich sie erlebte: Ich sprach davon, dass sie doch sehr viel Müll von mir und sicher auch von ihren übrigen KlientInnen abbekäme. Sie erklärte ganz fröhlich, daraus hätten wir aber doch immer etwas gemacht, das sei doch gerade das Tolle daran! Sie würde es auch gar nicht als Müll bezeichnen: Das ist doch unser Material, aus dem wir dann etwas machen! Das bleibt ja kein, wie sie es nannten „Müll"! – Und gerade dies sei es, was sie an ihrer Arbeit so begeistere. Meine Frage, wenn sie aus der Stunde wegginge, ob es nicht mal vorkäme, dass sie denke, heute habe ich genug von eurem Müll, beantwortete sie sehr ausführlich im Sinne von, wäre das so, dann habe sie ihre Grenzen nicht bemerkt, nicht beachtet. Das bedeute, sie könne aus dem „Müll" nichts mehr gestalten. Also das sei dann ihr Fehler.

10. Der Schreibprozess

In diesem Kapitel berichte ich davon, was der Schreibprozess bei mir bewirkte und welche Bedeutung das Schreiben dieses Buches für mich hatte. Zu diesem Prozess gehören nicht nur das Schreiben an sich, sondern auch alle Vorarbeiten einschließlich des Lesens in alten Aufzeichnungen, das Reden über mein Schreiben während dieser Zeit und die dazu wieder aufgenommenen Gespräche mit meiner ehemaligen Therapeutin.

Beim Verfassen des Buches vertieften sich nicht wenige meiner während der Therapie gewonnenen Erkenntnisse. Manches wurde mir noch klarer, oder ich konnte es nun neu oder richtig einordnen, Anderes überhaupt erst in der ganzen Dimension erfassen. Bei Letzterem denke ich besonders an die Taten meines Vaters, das Verhalten meiner Mutter und an die Folgen der sexuellen Gewalt. Ich erlebte den Schreibprozess als ein Vertiefen und Abrunden meiner damaligen Therapie.

10.1 Einige allgemeinere Bemerkungen zum Schreiben

Es kam recht oft vor, dass ich zum Schreiben meine Wohnung verlassen musste. Dann setzte ich mich, meist am Vormittag, in ein Bistro, und selbst wenn es zunehmend lauter wurde, blieb ich dort. Ich brauchte den räumlichen Abstand und das Betriebsame, um schreiben zu können.

Besonders als ich an den Kapiteln zum Vater und zur Mutter arbeitete, unterbrach ich mein Schreiben und legte kürzere oder längere Pausen ein – entweder weil es zu heftig war oder weil ich nicht weiterschreiben konnte und die Gründe dafür herausfinden wollte.

Das Schreiben löste manchmal Reaktionen aus, die ich aus der Zeit vor und während der Therapie sehr gut kannte: Schmerzhafte Rückenprobleme, Wut, Herzklopfen vor Wut, überhaupt Herzprobleme. Ein paarmal sogar die so unangenehmen Tachykardien. Und die von mir gefürchteten tiefen Ängste, die ich meist, aber leider nicht immer gut einordnen und annehmen konnte. Es berührte oder erschütterte mich, was ich von früher las, manchmal begleitet von tiefer Traurigkeit mit zum Teil heftigem Weinen. Oder ich spürte, wie früher, „Mamahunger"; auch dies erwähnte ich bei den Gesprächen mit der Therapeutin.

Meist empfand ich jedoch große Freude über das Erreichte. Und über das Schreiben und Formulieren an sich, und dies trotz des heftigen Themas.

Beim Lesen in den Therapie-Aufzeichnungen lächelte oder lachte ich auch, schüttelte den Kopf – entweder über die Therapeutin oder über mich oder auch über uns beide. Es war einfach köstlich, den Humor der K. wieder mitzuerleben. Wir haben damals viel gelacht, wenn auch nicht gerade in den ersten Therapiejahren. Humor ist sehr hilfreich und entlastend. Ich staunte auch über meine Kühnheit und meinen Mut, darüber, was ich mich alles traute, (zu) ihr zu sagen. Ich denke dabei an meine Offenheit bezüglich meiner Gefühle für sie oder wenn ich ihr sagte, wie ich sie wahrnahm, was manchmal sehr, sehr schwierig anzusprechen war.

Auch die Arbeit der Therapeutin, ihr Vorgehen und viele interessante Details rückten beim Lesen und darüber Schreiben nochmal mehr und anders in meinen Blick.

10.2 Unterstützung beim Schreiben

Zuerst sei dankbar und mit viel Herz erwähnt, wie sehr mich alle meine Lieben ermutigten, bestärkten und unterstützten. Dabei denke ich nicht nur an die von ihnen gesponserten Schreiburlaube, aber besonders gerne gerade daran.

Und nun komme ich zu den wieder aufgenommenen Gesprächen mit meiner damaligen Therapeutin, deren Unterstützung insbesondere im Zusammenhang mit dem Schreiben über Vater und Mutter, aber auch mit dem Veröffentlichen meines Buches notwendig wurde.

Von Anfang an benötigte ich ihre Ermunterung und Bestärkung allein dafür, meinen Wunsch zu schreiben ernst zu nehmen. Immer wieder rang ich um Zeit und Raum dafür. Ich würde mich fragen, ob ich mein Schreiben nicht zu wichtig nähme? Das *müsse* frau sogar zu wichtig nehmen, hatte die K. entgegnet. Dass ich mich das fragen würde stehe im Zusammenhang mit dem Thema, der sexuellen Gewalt. Ein anderes Mal quälten mich sogar Schuldgefühle, weil ich schreibe und mir das überaus wichtig ist.

Kam ich an einem Punkt oder Thema nicht weiter oder konnte einfach nicht schreiben, besprach ich das, wenn notwendig, mit der Therapeutin. Nicht nur einmal bemerkten wir, dass mir der Abstand zu dem damaligen Geschehen verloren gegangen war. Also pausierte ich, um wieder zu einem guten Abstand zu kommen. Manchmal musste ich nur kurz erwähnen, was mich gerade beschäftigte oder behinderte, dann kam ich zusammen mit ihr rasch zu den Gründen dafür – und schon konnte ich weiter arbeiten. Oder sie unterstützte mich einfach nur durch ihre lockere und ermutigende Art. Als ich sie jedoch einmal um einen Kommentar zu einem Text oder Ausschnitt bat, kam kurz und knapp ihre Aussage, das sei mein Buch. Sie wolle das nicht kommentieren. Lächelnd hatte ich gedacht: Ja, so kenne ich sie!

Ich las ihr Auszüge vor oder fasste zusammen, was ich zu einem bestimmten Thema geschrieben hatte, erwähnte meine Zweifel oder Fragen. Wir sprachen darüber, sie gab hilfreiche Erläuterungen oder stellte Fragen, was mir zu Klarheit über Formulierungen oder Inhalte verhalf.

Durch die Gespräche mit der Therapeutin vertieften sich manche Erkenntnisse oder stellten sich überhaupt erst ein; auch erschlossen sich mir dadurch einige theoretische Hintergründe. Als Beispiel sei hier nur kurz, später mehr dazu, die Erkenntnis genannt, dass es sich bei meinen Herzproblemen um verkörperte Erinnerungen handelt.

Beim Lesen in den Therapie-Aufzeichnungen im Zuge der Vorbereitungen fürs Schreiben bemerkte ich verwundert, dass die Therapeutin manchmal, eher zum Ende einer Stunde hin, eine Aussage machte oder eine Frage stellte zu einem Thema, das ich gar nicht angesprochen hatte, das mir nur diffus zugänglich war. Wie kam sie denn darauf?, fragte ich mich und dann auch sie, da ich mich nun erinnerte, dass das mehrmals der Fall war. Sie sei Analytikerin und als solche versuche sie, das Unbewusste zur Sprache zu bringen, in Sprache zu verwandeln.

Ich diskutierte mit ihr eingehend die Frage, ob und wie ich darüber schreibe, dass ich Liebe für sie fühlte und sogar erotische Gefühle im Zusammenhang mit ihr aufgetaucht waren. Eigentlich würde ich es bevorzugen, das mehr oder weniger für mich zu behalten. Ihre Reaktion: Wenn Sie das nicht schreiben, blenden Sie das Leben aus. Das gehört doch beides zum Leben! Die Liebe und die Erotik. Und beides gehört unbedingt auch in die Therapie. Stellen Sie sich einmal vor: Ein Leben ohne Erotik! Das wäre doch nix! – Bei den letzten Worten lächelte sie. Und ich stimmte ihr lächelnd zu. Sie erinnerte mich anschließend daran, dass ich oft ihren Gesichtsausdruck ansprach und damit zum Ausdruck brachte, wie ich sie wahrnahm. Sie betonte, wie wichtig es gewesen sei, dass wir das überprüften und klärten. In einer Therapie spiegele die Therapeutin der Klientin, was in dieser selbst vorhanden sei. Das anzusprechen und zu klären sei gerade bei der Bearbeitung sexueller Gewalterlebnisse von großer Bedeutung.

10.3 Zum Kapitel „Die Mutter"

Die Bedeutung und die Auswirkungen des Schreibens wurden besonders deutlich beim Abfassen der Kapitel zu den Eltern.

Im Kapitel zur Mutter lege ich ausführlich dar, warum solch große Schwierigkeiten speziell bei diesem Kapitel auftraten und dass mir das Schreiben zu einer ganz neuen Sicht bezüglich ihrer Verantwortung verhalf. Ich benötigte einige Zeit, bis ich herausfand, dass das Verhalten meiner Mutter, ihr Wegschauen und mich nicht Beschützen, für mich noch schlimmer war, als ich es beim Bearbeiten während der Therapie zulassen konnte oder wollte. Erst durch das Schreiben wurde mir die Dimension ihres Anteils an dem Furchtbaren bewusst, konnte ich das Ausmaß ihrer Verantwortung in vollem Umfang an mich herankommen lassen.

10.4 Zum Kapitel „Der Vater"

Zum Schreiben über den Vater gäbe es Vieles zu berichten; ich beschränke mich auf das Wesentliche. Am erschütterndsten war, dass mir das Ausmaß dessen, was mein Vater mir angetan hatte, erst beim Schreiben in vollem Umfang bewusst wurde. Es stellte sich heraus, dass ich die während der Therapiejahre erinnerten Übergriffe zum Teil wieder verdrängt oder vergessen hatte, sicher um besser damit leben zu können. Umso heftiger traf es mich beim Zusammenstellen dessen, worüber ich berichten wollte.

Ich ließ die erste, noch sehr unvollständige Fassung des Kapitels nach einiger Zeit für mehr als ein Jahr liegen, schrieb viele der übrigen Kapitel, bevor ich mich diesem Text wieder näherte. Nun las ich die erste Fassung ... Nein, ich hatte nichts wieder vergessen. Aber das Schreiben fiel mir sehr, sehr schwer. Wie früher, wenn es um meinen Vater ging, plagten mich heftige Rückenschmerzen – wie eine frisch zugefügte Wunde. Manchmal weinte ich lange, ohne mich beruhigen zu können. Erst als ich während des Weinens unwillkürlich meine Sehnsucht nach der Therapeutin als Mama aussprach, konnte ich die Trostlosigkeit bannen. Ich gestand mir ein, über das Geschehene in aller Offenheit berichten zu wollen. Dafür bedurfte ich der Rückversicherung derjenigen, die mein Leid in vollem Umfang kannte, mir Glauben schenkte und in den schweren Zeiten als Mama lebenswichtig gewesen war: Der Therapeutin.

In einem der Gespräche mit ihr sagte ich, es gäbe wirklich keine Zweifel mehr an dem, was mein Vater mit mir gemacht habe. In der ersten Fassung des Kapitel's hatte ich noch offenere Formulierungen für seine Übergriffe verwendet, wie beispielsweise „ich gehe davon aus, dass er ..." dies oder jenes gemacht hat. Aber ich beließ es bei Andeutungen. Nun, beim Korrekturlesen und Ergänzen des Kapitels, wurde mir immer klarer, nein, konnte ich es zulassen, was meine erinnerten Bilder und Gefühle bedeuteten. Ich brauchte die Zweifel nicht mehr! Ich las nun die erste Fassung mit großem Abstand, fast wie eine unbeteiligte Dritte, und dachte, ja, hier ist alles beschrieben. Dabei entstand ein Gefühl von: Jetzt ist es wie amtlich bestätigt! Nun kann ich es endgültig nicht wieder verdrängen!

Aber es fehlte noch etwas. Dies wurde mir besonders durch die Erkenntnis deutlich, dass es wichtig ist, exakt in Worte zu fassen, was ich erinnerte. Also nicht nur Bilder oder Gefühle wiederzugeben oder das, was ich gemalt oder mit Knete ausgedrückt hatte, sondern die jeweilige Bedeutung klar zu benennen. Darum gehe es doch in der Therapie, dass die erinnerten Bilder und Gefühle in Sprache umgewandelt werden, so die Therapeutin, also das, was geschah, zu versprachlichen. Das war mir zuvor gar nicht bewusst. Vielleicht weil ich jemand oder auch mich noch beschützen wollte? Die Therapeutin sprach in diesem Zusammenhang von Taburesten. Mir erging es dann so, dass ich zunehmend freudig strahlte, richtig begeistert war. Die Idee des Versprachlichens fand ich befreiend. Ich setzte sie gleich bei den Kapiteln zu den Brüdern um. Ergänzte und verdeutlichte!

Und dennoch zögerte ich beim Vater-Kapitel, überlegte, nur in meiner Version alle Taten genau zu benennen, in der offiziellen jedoch nicht. Doch davon kam ich rasch wieder ab. Ich wollte keinen Spielraum lassen für Interpretationen.

Andererseits dachte ich, wenn ich es genau benenne, kann ich es selbst auch nicht wieder vergessen und verdrängen! Aber will ich das überhaupt? So ging es hin und her. Es sei einfach ein großes Problem, so etwas über die eigene Familie, den eigenen Vater und die Brüder zu schreiben, betonte die K.

Eine befreundete Therapeutin meinte, mit dem Versprachlichen wäre ich auf Augenhöhe mit dem Vater. Aber nicht nur deshalb entschied ich mich für das Versprachlichen meiner erinnerten Gefühle und Bilder, sondern besonders deshalb: Das bin ich der kleinen Ann schuldig, dass ich exakt schreibe, was war!

Wie bereits im Kapitel Der Vater angedeutet, löste das Versprachlichen einen bemerkenswerten Traum aus, den ich leicht gekürzt wiedergebe:

> *Ich bin in meiner wohnung, 2. stock, normale fenster, als mehrere männer eine scheibe einschlagen. Unvorstellbar, da 2. stock, aber es gelang. (...) Sie versuchten es auch bei der balkontüre, das gelang nicht. Plötzlich einer an einem fenster, steckt den kopf rein, verhöhnt mich. Ich versuche, ihn zu verscheuchen, ihn anzugreifen, aber es gelingt mir nicht richtig. Dann liegen 2 dieser männer auf dem boden, wirken ganz klein, und ich steche mit gabeln und messern auf sie ein! In den bauch. Ziehe sie genussvoll wieder raus. Die leiden. Ich bleibe locker, mache so 2 – 3 schnitte oder gabelstiche pro person!*
>
> *Und dann kommt der saukerl, der zum fenster reinschaute, fast reinkam. Ich hab ein sehr langes, breites messer ... Und das steche ich ihm in den bauch, in den unterleib! Wunderbar war das! Er leidet, blutet, halb sitzt er, halb liegt er. Und ich genieße es!! Alle drei habe ich gepiesackt! Aber wie! (...) Die beiden ersteren eher klein, der dritte groß, also normale größe.*

Beim nächsten Treffen mit der Therapeutin sagte ich, ich müsse ihr mal wieder einen Traum schildern. Sie schaute gespannt und ich erzählte. Besonders betont und ausführlich diese meine Worte: Ich habe ein langes, breites Messer und steche dem Größeren in den Bauch, den Unterleib. Spüre geradezu, wie das Messer in den weichen Leib eindringt … Und das hat mir richtig Freude bereitet! – Ja! – Aber beim Erwachen sei ich dann sehr aufgeregt gewesen. Das sei doch verständlich, meinte sie. Ich berichtete von meinen Gedanken beim Aufschreiben des Traumes: Die drei sind die drei Täter. Der große ist mein Vater. Und für mich steht der Traum im Zusammenhang mit dem Versprachlichen. – Toll! Toll! – Der passt doch, der Traum, oder? – Ja! Absolut! – Der muss unbedingt mit in mein Buch rein!, stellten wir beide fest, denn er zeigt das nochmalige Bearbeiten durch das Schreiben! Deutlicher kann ich das Befreiende des Schreibens kaum darstellen als mit diesem Traum.

Auch was einige theoretische Hintergründe anbetrifft war das Verfassen des Vater-Kapitels sehr erhellend. Wie ich dort ausführlich dargelegt habe, wurden mir die verschiedenen „Formen des Erinnerns“ erst beim Schreiben bewusst.

Hier will ich sie nur kurz noch einmal für die Lesenden erwähnen, die das Kapitel zum Vater nicht lesen können oder wollen.

Als *körperliche und psychische Erinnerungen* verstehe ich die Angst- und Panikattacken sowie die Schlafstörungen, aber auch die Herzsymptome. Letztere habe ich erst beim Schreiben des Vater-Kapitels als solche erkannt; darauf gehe ich gleich näher ein.

Meine *Träume* spiegelten Erlebtes und Erinnertes wieder und unterstützten so die therapeutische Arbeit.

Als *indirekte Erinnerungen* bezeichne ich Situationen meines damaligen Alltags, die mich an Situationen mit meinem Vater erinnerten.

Außerdem ließ ich *meine Hände „sprechen“*, drückte meine Erinnerungen beispielsweise durch Malen und mittels Knete aus.

All das konnte wiederum Erinnerungen an Situationen, einhergehend mit körperlichen Reaktionen, auslösen.

Ich arbeitete also am Kapitel zum Vater und erwähnte im Gespräch mit der K. die Formen des Erinnerns, dabei unter anderem die körperlichen Erinnerungen. Hier müsse ich auch meine Herzsymptomatik mit hineinnehmen, sagte sie zu meinem Erstaunen.

In einem weiteren Gespräch wollte ich sie fragen, wieso die Herzattacken auch körperliche Erinnerungen seien, wie sie denn darauf komme. Bevor ich jedoch die Frage stellte, fiel mir das Buch „Neujahr“ von Juli Zeh ein, das ich am Wochenende gelesen hatte und dessen Inhalt ich kurz wiedergab. Der Protagonist sei von Herzattacken (und Angst und Panik) geplagt worden, was mich sehr an mich erinnert habe. Das zu Grunde liegende Trauma würde später im Buch geschildert.

Nun kam ich auf ihre Aussage zurück, meine Herzprobleme seien körperliche Erinnerungen. Die K. lächelte vielsagend, als ich meinte, ich könne damit nichts anfangen: Wenn ich versuchen will, das zu verstehen, irgendwie nachzuvollziehen, ist da eine Wand. Ich kann einfach nicht darüber nachdenken. – Ich erinnerte sie daran, dass es mir bei unserem letzten Treffen doch so ergangen sei: Als Sie davon sprachen, blieb mir richtig die Luft weg. Wurde es mir ganz eng. Mir ist aber nichts dazu eingefallen. Da fehlt mir völlig das Verständnis. – Daran können sie sehen, dass das immer noch in Ihrem Körper ist! – Ich sei doch mit den Herzproblemen zu ihr in die Therapie gekommen: Und das, was Sie von dem Buch erzählten, er habe die geschilderten Symptome wegen eines vor vielen Jahren erlebten Traumas, das zeigt es doch! –

Erst nach und nach wurde mir klar: Ich habe mir mit der Schilderung aus diesem Buch ja selbst die Antwort gegeben und hab's noch nicht mal gemerkt! Unglaublich! – Lächelndes, zustimmendes Nicken der K. Ich erkannte: Er hatte die Herzsymptome wegen des erlittenen Traumas. Sein Körper hat ihn so darauf aufmerksam gemacht, auch wenn erst einmal nicht klar war, auf was der Körper ihn hinwies. –

Etwas später sagte ich, da sei bei mir aber immer noch ein kleiner Widerstand. Das zeige, so die K., wie tief das sitze, wie sehr das Verdrängen in den Körper eingegangen sei. Ich fügte hinzu: Ich hab' irgendwo mal gelesen „verkörperter Schrecken". – Ja… – Mir fiel die Übernachtung von vor vielen Jahren ein, meine Eltern nebenan in ihrem Campingbus. Und am nächsten Tag begannen die Herzattacken…

Bei den Notizen zu dem Gespräch mit der Therapeutin erinnerte ich, dass sie schon vor Jahren gesagt hatte: Der Verstand vergisst, der Körper nicht! – Nun staunte ich nochmals sehr, wie schwer es mir fiel zu verstehen, mein Körper hat sich über mein Herz erinnert. Die Herzattacken wiesen auf die traumatischen Kindheitserlebnisse hin, sind körperliche Erinnerungen an die sexuelle Gewalt.

10.5 Über das Veröffentlichen – Erster Teil

Wie schwierig das Schreiben war habe ich ausreichend dargelegt. Aber die Texte zu veröffentlichen? Unvorstellbar! Nun, dabei ist es nicht geblieben, sonst gäbe es hier nichts zu lesen. Auch dies war ein Prozess, und davon will ich ebenfalls berichten.

Ursprünglich wollte ich den Therapieprozess nur für mich in Worte fassen. Das hat sich im Laufe des daran Arbeitens verändert, da nicht nur ich zu der Überzeugung kam, dass meine Schilderungen für andere Betroffene und auch für Fachkräfte von Relevanz sein können. Der durch das Schreiben und immer wieder darüber Reden gewonnene größere Abstand war für dieses Vorhaben Voraussetzung. Er erst ermöglichte es mir, überhaupt an ein Veröffentlichen denken zu können. Doch fürs Umsetzen benötigte ich noch größeren Abstand, wie sich bald zeigte.

Ich hatte mit der K. über meinen Wunsch gesprochen, meine Geschichte als Buch veröffentlichen zu wollen. Das sei aber nicht leicht. Ich wolle die Texte zunächst als Blog im Internet veröffentlichen, als blog'buch, wie ich es nennen würde. Dennoch bliebe es unvorstellbar. Das sei doch auch ein großer Schritt, bestätigte die K.

Daraufhin entschied ich mich tatsächlich, meine Geschichte als Blog ins Internet zu stellen, mit der Idee, dass mir das ein Veröffentlichen als Buch erleichtern wird. Nachdem ich die ersten Kapitel im Netz veröffentlicht hatte, überlegte ich ganz genau, welchen Personen ich den Link zur Verfügung stellen wollte. Das war zunächst ein sehr enger Kreis meiner Freundinnen sowie die Therapeutin. Diesen Kreis erweiterte ich erst nach und nach.

Ich bekam viele wunderbare Kommentare zu meinem blog'buch, von denen mich manche sehr bewegten und berührten.

Hier gehe ich jedoch nur auf die Reaktionen meines Bruders Heiner und meiner Nichte Lea ein.

Heiner war der erste aus dem Kreis der Familie, der mein blog'buch komplett las und das in sehr kurzer Zeit. Er sagte dazu, mit Tränen erstickter Stimme, dass er nicht wusste, *wie* schlimm es mir als kleinem Mädchen ergangen war, er erst jetzt den vollen Umfang erfahren habe und dass er mich leider nicht habe beschützen können, was für ihn kaum auszuhalten wäre. Aber er sei doch nur zwei Jahre älter, wie hätte das gehen sollen?, wandte ich ein. Ja, das wisse er, dennoch hätte er mich so gerne beschützt. Er weinte. Auch bei seiner Erinnerung an das Bild, ich auf dem Schoß des Vaters... Damals habe er mich beneidet. Jetzt aber wohl nicht mehr?, fragte ich. Nein. Er ärgere sich sehr, dass er unseren Vater zu dessen Lebzeiten nicht deutlich zur Rede gestellt habe. Außerdem habe er Vieles von früher vergessen, speziell beim Lesen des Kapitels über die Mutter wieder erinnert, auch weil dort die Atmosphäre von zuhause sehr gut wiedergeben sei.

Ein solches Mitfühlen hatte ich mir von Seiten meiner Familie und speziell von Heiner während der Therapiezeit sehr gewünscht. Umso mehr freute und berührte es mich nun.

Er hatte sich noch verwundert darüber geäußert, dass ich für die Therapeutin, obwohl sie einige Jahre jünger ist als ich, Mamagefühle empfinden konnte. In den ersten Jahren der Therapie wusste ich das nicht, ahnte es aber. Dennoch beeinflusste mich das in meinen Gefühlen für sie und auch sonst überhaupt nicht.

Meine Nichte Lea, die Tochter meines Bruders Bernd, las ebenfalls mein blog'buch. Ich lernte sie kennen, nachdem fast alle Kapitel geschrieben und in den Blog eingestellt waren. Ich hatte sie als kleines Mädchen bei Familienfeiern vor sehr langer Zeit erlebt, aber danach nicht mehr.

Nun begegneten wir uns anlässlich eines Geburtstages, redeten recht schnell ganz offen miteinander. Dabei deutete sie schon an, dass sie vom „sexuellen Missbrauch" durch meinen Vater „gehört habe". Für mich stellte sich natürlich die Frage, sage ich ihr das offen von ihrem Vater, meinem Bruder Bernd? Mit ihm hat sie seit vielen Jahren keinen Kontakt. Bei einem Treffen zu zweit spürte ich, ich kann es ihr sagen. Auch weil ich sie so erlebte: Sie will wissen! Wissen, was in ihrer Familie war, wissen, was mit ihr los ist. Und ich sagte es und das war gut. Ich sei jederzeit für sie da, wenn sie darüber reden wolle, wenn es arg für sie werden würde oder wenn sie mich brauche. Sie fing noch in der Nacht, wie sie mir am nächsten Tag mitteilte und wie ich es vermutet hatte, damit an, in meinem Blog zu lesen – bis zum Kapitel zu ihrem Vater, also zu Bernd. Die übrigen Kapitel las sie in den nächsten Tagen. Sie reagierte erschüttert, besonders über die Taten meines Vaters. Wir waren beide sehr berührt von unserer Begegnung und den Folgen und freuen uns nun, in der jeweils Anderen eine Person aus der Familie zu haben, die genau hinschaut und auch wissen will.

10.6 Über das Veröffentlichen – Zweiter Teil

Das Veröffentlichen als blog'buch machte es wirklich leichter, ans Veröffentlichen als gedrucktes Buch zu denken. Leichter ja, aber nicht leicht.

Irgendwann war ich soweit, mein Buchprojekt Verlagen vorstellen zu wollen. Die Vorarbeiten dafür, das Heraussuchen von in Frage kommenden Verlagen, das Erstellen von Exposé, Inhaltsangabe und die Auswahl von Leseproben nahmen viel Kraft und Zeit in Anspruch. Außerdem hatte ich das letzte Kapitel über den Schreibprozess vor mir hergeschoben. Nun wollte ich auch damit beginnen.

Plötzlich auftretende Schlafprobleme irritierten mich. Ich vermutete bald, dass sie mit dem Schreiben des letzten Kapitels oder mit dem Veröffentlichen als Buch zu tun haben könnten oder gar mit beidem. Auch darüber sprach ich mit der K. Das letzte Kapitel zu schreiben bedeute, das Buch zu beenden und abzuschließen. Deshalb vielleicht meine Schlafprobleme? Das könne gut sein, bestätigte sie, da ich mich damit noch einmal und nun endgültig von dieser Thematik verabschieden würde.

In der Folge wurden die Schlafstörungen immer heftiger, und es kamen Ängste hinzu. Ängste so wie früher, von unten aus dem Bauch heraus. Furchtbar! Was ist denn jetzt los? Irgendwann hatte ich sogar wieder Angst vor der Angst – auch wie früher! Hat das wirklich mit dem Willen zu veröffentlichen zu tun? Ich konnte es nicht glauben.

Auch nicht, dass es mich wieder und nochmal so heftig erwischen würde. Richtige Angstattacken! Schlotternde Beine, Kälte, Herzklopfen. Einschlafen und erschrecktes Erwachen mit starkem Herzklopfen. Alles wie gehabt, wie zu Beginn meines Absturzes und in den Jahren danach. Dass es mir noch einmal so schlecht ergeht! Das gibt's doch nicht! Ich war schon ein bisschen enttäuscht von mir, aber mehr noch erschüttert. Ich rang um meine Fassung.

Ich berichtete der Therapeutin ausführlich von meinen Ängsten und meinem schlechten Befinden und kam dabei zu dem Schluss, für eine Veröffentlichung in Form eines Buches wohl immer noch nicht bereit zu sein. Vom Kopf her ja, aber nicht vom Herz. Die Ängste sollten mich am Veröffentlichen hindern, also daran hindern, nun wirklich mit meiner Geschichte an die Öffentlichkeit zu gehen. Die Therapeutin betonte, ich würde mich durch das Anschreiben der Verlage verletzbar und angreifbar machen. Dies alles habe mit der sexuellen Gewalt zu tun.

Außerdem sei es wichtig, die verschiedenen Ebenen auseinander zu halten. – Sie meinen, dass ich mein Buch veröffentlichen will, dass es andererseits an alte, basale Ängste andockt? – Ja! ... Auch an den Schmerz... –

Auf ihre Anregung hin nahm ich die kleine Ann innerlich „an die Hand"... Und redete mit meiner Angst. Sie habe mich darauf aufmerksam gemacht, dass ich fürs Veröffentlichen noch nicht bereit sei. Das sei auch in Ordnung. Aber nun wolle ich mich nicht mehr davon abhalten lassen, solle sie keine Macht mehr über mich haben. Eher so: *Die Täter* dürfen jetzt keine Macht mehr über mich haben! – Und wenn die Angst nur in Anflügen auftauchte, sagte ich mir: Ich mache weiter! Ich lasse mich nicht beirren! –

Ich brauchte noch ein bisschen Zeit, dann schrieb ich nach und nach viele Verlage an und empfand zunehmend Freude dabei. Sogar Glücksgefühle stellten sich ein. Aber es war auch mal wieder schwierig.

Als ich nun am Ende des Kapitels zum Schreibprozess hinzufügen wollte, das Schreiben und das Veröffentlichen sei für mich der letzte Schritt der Abgrenzung gegenüber meinem Vater, fragte ich mich, ob da nicht ein Widerspruch besteht? Zum einen war das Schreiben ein Vertiefen und Abrunden meines Therapieprozesses, hatte es mir weitere Erkenntnisse und neue Sichtweisen gebracht und zu noch mehr Abstand verholfen. Andererseits erlebte ich solche Reaktionen wie die Herzattacken und heftigen Ängste. Aber besteht da tatsächlich ein Gegensatz? Oder ist es nicht eher etwas, was genau zu dieser Thematik passt? Ich zeige auf, dass die Folgen solch furchtbarer Taten nie ganz verschwinden, dass frau dennoch ihr Leben befreit und glücklich führen und gestalten kann.

Also kann der Satz doch stehen bleiben: Für mich stellt das Schreiben und das Veröffentlichen den letzten Schritt der Abgrenzung gegenüber meinem Vater dar.

Nachwort von Prof. Dr. Helga Krüger-Kirn

In schwarzen Farben gekleidet sitzt Frau Salandre Stunde um Stunde vor mir und berichtet von ihren Herzattacken, den mehrmaligen Notarzteinsätzen bei Verdacht auf Herzinfarkt und ihrer Suche nach den Ursachen für die Herzanfälle. Sie vermutet einen Zusammenhang mit der erlebten sexuellen Gewalt durch ihre Brüder oder auch mehr…

Was verbirgt sie hinter der sorgsam auf den Abstand achtenden schwarzen Fassade? Und was hat dazu geführt, dass ihr Körper wie ein Schutzwall wirkt, hinter den sie sich zurückgezogen hat? Wird es mir gelingen, eine Beziehung zu ihr aufzubauen, die es ihr ermöglicht, über ihre Beweggründe zu sprechen und dem bis dato Unaussprechlichen Raum zu geben? Diese und ähnliche Fragen begleiteten meine ersten Begegnungen mit Frau Salandre.

Das Buch vermittelt einen tiefgehenden Eindruck davon, mit welchen Herausforderungen, Ängsten und Hoffnungen ihr therapeutischer Entwicklungsweg verbunden war. Im Rahmen eines Nachwortes kann ich dieser Komplexität nicht gerecht werden. Meine nachfolgenden Überlegungen betrachte ich daher nicht als Analyse des Therapieprozesses, geschweige denn als Einordnung in diagnostisch sinnvolle und behandlungsrelevante Interventionen. Stattdessen möchte ich mich auf einige Aspekte konzentrieren, die mich in diesem Therapieprozess sehr herausgefordert und dazu angeregt haben, darüber hinaus auf theoretischer Ebene nachzudenken. Im Fokus steht die im Diskurs der therapeutischen Wirksamkeitsstudien über alle Therapieschulen hinweg als zentral erachtete therapeutische Beziehung.

Mit dieser Herangehensweise möchte und kann ich mich nicht von der persönlichen Betroffenheit distanzieren, die mit dem Leid traumatischer Erfahrungen einhergeht und dann besonders heftig ist, wenn emotionale, physische und/oder sexuelle Gewalt die kindliche Beziehungswelt prägen.

Dass das traumatische Geschehen in der Kindheit alle Lebensbereiche bis in die Gegenwart überlagert, hat auch Frau Salandre so empfunden. Die nachhaltigen Spuren in der psychischen Entwicklung und dem Urvertrauen kamen nicht nur in der persönlichen Beziehungsgestaltung, sondern auch im Rahmen des psychoanalytischen Prozesses und der Übertragungsbeziehung auf vielfältige Weise zum Ausdruck. Die emotionale Herausforderung für mich als Therapeutin bestand besonders darin, trotz der Anteilnahme an der traumatischen Lebensgeschichte eine therapeutische Position aufrechtzuerhalten, die Frau Salandre dabei unterstützte, ihre traumatischen Erfahrungen zu verarbeiten. Die beständige Herausforderung während des therapeutischen Prozesses lässt sich am ehesten als ein Oszillieren zwischen Mitgefühl und Verständnis für ihre Sehnsucht nach Nähe auf der einen Seite und einer notwendigen abgegrenzten Haltung als Therapeutin auf der anderen Seite beschreiben. Die im psychoanalytischen Diskurs damit verbundenen Themen wie Übertragungsliebe, erotisierende Übertragung und Gegenübertragung haben hier einen ebenso festen Platz wie Scham, Aggression und Hass sowie der Umgang mit therapeutischer Abstinenz.

Weniger diskutiert wird bisher, dass die Analyse der Übertragungs- und Gegenübertragungsdynamiken auch durch die gemeinsame geschlechtsspezifische Eingebundenheit in geschlechtliche Normen beeinflusst wird. So betrachtet, kommt in unseren Deutungs- und Verarbeitungsangeboten nicht nur das therapeutische Anliegen zum Tragen, Einsicht in und Veränderung der mit den traumatischen Erfahrungen entwickelten psychischen Strukturen zu unterstützen. Mit der Behandlung des subjektiven Leids ist auch die Aufgabe verbunden, jene Betroffenheit zu reflektieren, die sich über die gemeinsame Eingebundenheit in die symbolische Ordnung der Geschlechter in unserer Gesellschaft, in unseren affektiven und kognitiven Fühl- und Denkweisen widerspiegelt. Dass sich darin bis heute nicht nur eine auf patriarchaler Hierarchie beruhende Geschlechterdifferenz artikuliert, sondern auch eine kollektive Abwehr gegenüber sexueller Gewalt/sexuellem Missbrauch, hat mich dazu bewogen, während des Therapieprozesses über die Grenzen des Therapieraums hinaus nachzudenken und den Therapieort als politischen zu reflektieren.

Freuds Aussage, dass „[j]ede psychoanalytische Behandlung ein Versuch [ist], verdrängte Liebe zu befreien, die in einem Symptom einen kümmerlichen Kompromiss-Ausweg gefunden hat […]“ (Freud 1907a, S. 118), artikuliert auf wunderbare Weise die Sehnsucht nach einem ‚besseren Leben‘ und zielt im Umkehrschluss auf eine Wiedergewinnung und (Re-)Integration verletzter und blockierter Selbstaspekte. Damit lässt sich der Therapieprozess über den Versuch, den je eigenen Sinn der Symptome zu verstehen und die subjektiven Leidenserfahrungen zu überwinden, als Raum der Selbsterkenntnis charakterisieren, als Raum für die Entdeckung eigener Potenziale, Fähigkeiten und Schwächen. Man könnte es auch so beschreiben: Dem Therapieprozess liegt ein Erkenntnisinteresse nach bisher verunmöglichten und unerhörten Begehrensweisen einer selbstbestimmten Individuation zugrunde. Damit eng verbunden ist ein ganzheitlicher Fokus auf Psychosomatik als Verbindung von Leib-Seelischem, in der sich die gesamte Lebenssituation spiegelt und zwar sowohl mit ihren Bezügen zu den familiären Beziehungen sowie darüber hinaus mit deren mit sozialem und kulturellem Sinn versehenen Strukturen. Demzufolge lassen sich die manifesten Symptome von Frau Salandre wie Schlafstörungen und Angstzustände begleitet von herzsymptomatischen Panikattacken als Traumafolgestörungen verstehen. Als Krankheit verkleidet werden die psychosomatischen Symptome nicht nur als entwicklungshemmende Störungen, sondern als verborgene, auf den Körper zurückgezogene Ausdrucksversuche von Gefühlen, Wahrnehmungen und Interaktionen verstanden, die ehemals in der Beziehung mit den Bezugspersonen nicht realisierbar waren. Das heißt, dass die Symptomatik nicht nur auf die betroffene Person zurückweist, sondern auf die gesamte Herkunfts- und gegenwärtige Familiensituation. Daran schließen sich unmittelbar Fragen danach an, wem die Entwicklung dieser Symptomatik noch genützt hat. Bei Frau Salandre wirkten diese Abwehrbewegungen über die Kardiologentermine und nächtlichen Notaufnahmen hinein in die Aufrechterhaltung familiärer Verpflichtungen, einer harmonischen Ehe und einem erfolgreichen Berufsleben. Bis dann das Schlafen immer schwieriger wurde. Nachts machten sich die Geister bemerkbar, vor denen sie tagsüber auf der Flucht war und die sie nicht in Worte fassen konnte. Welche Gespenster raubten ihr den Schlaf oder psychoanalytisch gesprochen, bahnten sich auf diese Weise einen Weg aus der Verdrängung?

Über das Schweigen Worte finden

Im Therapieraum stellte mich die spezifische Konflikt- und Traumageschichte von Frau Salandre vor die Aufgabe, die durch Zurückweisung, Angst, Schmerz, Verzweiflung und Einsamkeit mitbedingten (unbewussten) Sinnstrukturen in ihrem Denken, Fühlen und Handeln verstehen zu lernen. Wie lässt sich das Sprachlose, wo die Erfahrung von sexualisierter Gewalt keine Worte, sondern nur Angst und Einsamkeit zurückließ, das Nicht-Bewusste, aber doch potentielle Wissen um die traumatisierenden Erfahrungen in Worte fassen? Wir lernten, dass sich dort, wo sich Erfahrungen (noch) nicht in Worte fassen lassen, auch das Schweigen mehrere Gesichter hat. Schweigen als geteilte Erfahrung kann mehr als (erneutes) wegschauendes Schweigen bedeuten, mehr als die Angst vor Überforderung oder sich (erneut) als ohnmächtiges, hilfloses, ausgeliefertes Opfer oder sich falsch verstanden oder zurückgewiesen zu fühlen. Verbunden im gemeinsamen Schweigen entfaltet sich eine Zuversicht; eine Zuversicht, wahrgenommen und sich "gehört" zu fühlen. Ebenso wie das Schweigen hat auch das Sprechen mehrere Facetten und besteht nicht nur aus Wörtern, sondern ist zugleich gemeinsame interaktive Praxis. Das gemeinsame tastende Suchen nach (Sprach)Bildern und Worten half, der quälenden Sprachlosigkeit etwas entgegenzusetzen. So gelang es allmählich, eine Sinnhaftigkeit in der Sprachlosigkeit zu entdecken und einen Zugang zu der Erkenntnis zu entwickeln, dass sich diese auf unbewusster, unwillkürlicher Ebene als Rüstung gegen den Schmerz auf den Körper und in besonderer Weise um das Herz gelegt hatte. So gesehen lässt sich die spezifische Körpersymptomatik als Hinweis auf verdrängte Erfahrungen übersetzen und als praesymbolischer Ausdruck von nicht verarbeiteten Erfahrungen mit den dazugehörenden Affekten. Die leibliche Körpermetaphorik, in der sich Vor- und Unbewusstes artikuliert, rückte als eine zentrale Erkenntnismodalität mit in den Fokus. Dieser Prozess lässt sich als vielfältiger und komplexer Austausch verbaler und averbaler Mitteilungen verstehen.

Denn wenn Sprechen zu einer Sprache zurückfinden soll, die an die ehemals unterdrückten Erfahrungen anknüpfen kann, benötigt das Sprechen bzw. Deuten eine sinnlich spürbare Bezugnahme auf den eigenen und fremden Körper. Eine Veränderung des Tonfalls, Verschiebung des Atemflusses oder eine unerwartete körperliche Bewegung im Kontext des Sprechens wurden als Zeichen gelesen, die in der Regel eine innere Mobilisierung und Annäherung an verdrängte Themenkomplexe anzeigten. Körpererleben und Körperausdruck gelten daher ebenso wie Worte als Instrumente des Verstehens und ermöglichen eine Annäherung an die in der Szene verborgenen unbewussten Anteile. So werden alle Ebenen der Mitteilungen im Rahmen der therapeutischen Beziehung nicht nur auf der psychischen und symbolhaften Ebene erforscht, sondern auch in ihrer emotionalen, affektiven und körperlich-vegetativen Gegenwärtigkeit, um eine treffende Deutung zu finden.

Frau Salandre hat meine Bemühungen, die Vielgestaltigkeit des Sprechens und die gemeinsame Suche nach einer Sprache, die im und mit dem Trauma verloren gegangen ist, als intensives Bearbeiten ihrer psychischen Lebenssituation erlebt. Im Rahmen dieses Prozesses konnte sie nach und nach Vertrauen entwickeln, aber auch ein Begreifenwollen ihrer eigenen Situation als Kind und Jugendliche. Rückblickend hat die Erfahrung, die wiederentdeckte und schmerzlich vermisste Zuversicht, mir bzw. den anderen vertrauen zu können, eine nicht zu unterschätzende Bedeutung für ihre zunehmende psychische Stabilisierung. Die Kehrseite davon waren Gefühle von Misstrauen, die bestenfalls dazu dienten, das sich neu entfaltende Vertrauen zu überprüfen. Solche Phasen der Therapie waren für uns beide dann besonders anstrengend und ängstigend, wenn angesichts der dominierenden Erfahrung von Angst, innerer Bedrohung, Einsamkeit und Schuld das Vertrauen drohte, erneut so fragil zu werden, dass es an den Rand der Verzweiflung und des Abbruchs führte.

Mit weiteren Herausforderungen an unsere Beziehungsarbeit waren auch Enttäuschungen von Wünschen verbunden, die sich bei mir in der Gegenübertragung als Forderungen anfühlten. Erst die Bearbeitung dieser von Frau Salandre als versagend erlebten Zurückweisungen konnte einen Zugang eröffnen, diese als Reinszenierungen von Ohnmacht mit umgekehrten Positionen zu verstehen. Trotz der Herausforderungen und Erschütterungen der Übertragungs-Gegenübertragungsbeziehung bot die Verlässlichkeit des Settings einen Rahmen und Möglichkeitsraum, die Enttäuschungen durch die Therapeutin zu verarbeiten und mit dem eigenen Erleben und der eigenen psychischen Innenwelt in Bezug zu setzen.

Das Durcharbeiten der mit diesen Erwartungen verbundenen Konflikte war ein Wendepunkt und hat wichtige Voraussetzungen geschaffen: Nicht nur dass Frau Salandre ihre Wünsche selbstbestimmter und mit weniger Schuldgefühlen formulieren konnte, sondern auch, dass sie mich als Subjekt mit eigenen Gefühlen wahrnehmen konnte und nicht als Bedrohung.

Gleichwohl gab es immer wieder auch Situationen, die dazu verführten, sich aus Hilflosigkeit auf je unterschiedliche Weise in ein Netz von Erklärungen und Theorien zu begeben. Diese Dynamik zu erfassen und als Überforderung der emotionalen Ressourcen zu verstehen, war für das Gelingen des Therapieprozesses hilfreich. Auf TherapeutInnenseite war hier ein Innehalten und aufmerksames Nachdenken nicht nur sinnvoll, sondern auch notwendig, um zu überprüfen, ob und welche Strukturierungsfunktion theoretische Erklärungen übernahmen, wenn ich den Halt in der Ohnmacht gegenüber dem Ausmaß der sexuellen Gewalt zu verlieren drohte. Die unabdingbare Verpflichtung zur Supervision übernimmt hier wichtige Funktionen: Erst darüber kann die Abwehr der eigenen Verwobenheit in die kollektive Abwehr gegenüber Ohnmacht und sexueller Gewalt verstehbar werden und dazu beitragen, wieder zu jener psychoanalytischen Grundhaltung zurückzufinden, in der das Beziehungsgeschehen in der Therapie reflektiert werden kann.

Sehnsucht und Begehren

Welches Maß an Nähe und libidinöser Zuwendung oder gar körperlicher Berührung ist einem psychischen Entwicklungsprozess förderlich und sogar notwendig? Wann eröffnen sie einen Raum für psychische Verarbeitung/symbolisches Denken und wann sind sie überwiegend als Abwehr von konflikthaften und/oder ängstigenden Beziehungserfahrungen und Begehrensweisen zu sehen? Wenn wir uns an die eingangs zitierten Sätze von Freud erinnern, dass in jedem Symptom letztlich missglückte Interaktionserfahrungen und Spuren unerhörter Liebe schlummern, und diese mit unserem Wissen über die psychische Entwicklung in Bezug setzen, macht es Sinn, im Rahmen nonverbaler und leiblich basierter Interaktionsmuster körperliche Berührung in das Deutungsrepertoire einzubeziehen. Dieser Perspektive steht in einer von Freud geprägten psychoanalytischen Perspektive ein Abstinenzbegriff gegenüber, der mit Neutralität in Verbindung steht und das Sprechen der Analytikerin und ihre Deutungen im Wesentlichen auf verbale Interventionen begrenzt. Die Angst, dass Berührungen missbräuchlichen Charakter annehmen oder die Patientin hindern könnten, analytisch, sprich selbstreflexiv zu arbeiten, sind häufig vorgetragene und ernst zu nehmende Einwände. Gleichwohl gibt es im Verlauf der Analyse immer wieder auch Situationen, wo Worte die Psyche/Seele nicht mehr erreichen bzw. ausreichend halten können. Besonders wenn zentrale Verarbeitungs- und Symbolisierungsprozesse im Verlauf der psychischen Entwicklung frühzeitig blockiert wurden – wie dies bei komplexen traumatischen Erfahrungen sehr häufig vorkommt – ist es angezeigt, über Berührungen nachzudenken. Was die Psychoanalyse der Sehnsucht hinzufügen kann, ist eine Lesart, in der der Wunsch nach Nähe und Berührung in der Psychotherapie übersetzungsbedürftig ist und auf frühere Erfahrungen, sowohl reale als auch gewünschte, verweist. Solche Begehrensweisen als Übertragungsphänomene zu übersetzen, stellt das Fundament eines professionellen Umgangs mit der zwischenmenschlichen Nähe und einer vertrauensvollen Beziehung dar und bietet nicht nur einen notwendigen Schutz vor Übergriffen und damit einer Retraumatisierung, sondern ist unabdingbare Voraussetzung für einen gelingenden Therapieprozess.

In meinem Verständnis schließt die Frage, wann welche Art von Deutung wirksam und sinnvoll ist, auch körpertherapeutische Interventions- und Deutungsmuster mit ein. Gleichwohl sei hier einschränkend angemerkt, dass ein professioneller Umgang mit körperorientierten Interventionen genauso gelernt und gleichermaßen reflektiert werden muss wie verbale Übertragungs- und Gegenübertragungsdynamiken. Mit anderen Worten: Auch körpertherapeutische Interaktionen benötigen eine selbstreflexive Erforschung der Motive/Bewegungen der Analytikerin. Für mich bedeutete dies, das Anliegen von Frau Salandre einzuordnen und anzunehmen, um dabei einerseits den Übergang zwischen körperlich-affektiven Zuständen und sprachlichen Repräsentationen zu unterstützen und sie anderseits dazu zu ermutigen, sich mit ihrem Begehren anzuvertrauen und vertraut zu machen.

Blick und Berührung im Therapieraum

Blickkontakt und ebenso körperliche Berührungen, selbst wenn sie sich auf alltägliche Begrüßungsrituale beziehen, stellen für Menschen mit sexueller Gewalterfahrung eine besondere Herausforderung dar. Zu groß ist die Angst vor intrusiven Erfahrungen, die als unverarbeitete Fragmente in das Bewusstsein einbrechen können und die Gegenwart zu überwältigen drohen. Hier nun rückt der Körper erneut und in besonderer Weise in den Fokus, denn im Zentrum stehen Spaltungsvorgänge, in die auch der Körper involviert ist. Der Therapeutin in die Augen zu schauen war für Frau Salandre ein wiederkehrender Wunsch, den sie auf verschiedene Weise umkreiste und der verschiedene Bedeutungsebenen hatte. Im therapeutischen Beziehungsraum wurde spürbar, dass der ersehnte Blick ebenso wie das Ringen um Worte eine zaghafte Annäherung an ein verborgenes, noch nicht artikulierbares Unbewusstes symbolisierte. Im Folgenden gehe ich auf die Szene des visuellen Kontakts ein, die Frau Salandre auf Seite 36 beschrieben hat.[1] Ausgehend von meiner Gegenübertragung folgte ich dem ersehnten Blickkontakt von Frau Salandre und versuchte, die spürbare Angst vor Zurückweisung und Beschämung abzumildern, indem ich die latent spürbare Sehnsucht nach einfühlend-sinnlicher Spiegelung in Worte fasste. Um Frau Salandre nicht zu überfordern, machte ich ihr den Vorschlag, sich dem Blickkontakt kleinschrittig anzunähern, zunächst die Augen zu schließen und sich vorzustellen, was passiert, wenn sie mich ansehen würde. Erst in einem weiteren Schritt näherten wir uns konkret dem realen Blickkontakt. Um der Abwehr der Sehnsucht nach einem zugewandten Gegenüber nicht erneut durch Beschämungs- und Bestrafungsangst Raum zu geben, habe ich auch den zweiten Schritt hin zu einer Begegnung von Antlitz zu Antlitz modifiziert, indem ich die Augen schloss und Frau Salandre die Entscheidungskompetenz übertrug, zu bestimmen, wann sie bereit dazu war, mir im Blick zu begegnen und ich die Augen öffnen sollte. Diese Szene deutete eine Wende an zwischen ihrer bisherigen Abwehr dieser Wünsche und einem langsamen, steten Zulassen ihrer Sehnsucht, ihren Bedürfnissen mehr Raum zu geben und sich mit den darin verborgenen Ängsten und traumatischen Erfahrungen auseinanderzusetzen. Immer wieder konnten wir im weiteren Verlauf aus unterschiedlichen Perspektiven auf diese gemeinsam inszenierte Szene Bezug nehmen und die Ängste, sich psychisch mehr mit ihrem Begehren zu verbinden, bearbeiten. Was sich in der Szene zunächst nur ahnungsweise und unbewusst andeutete und welche Sehnsüchte nach sinnlicher Nähe bei Frau Salandre geweckt wurden, konnte sie im weiteren Verlauf der Therapie entfalten.

An dieser Stelle erscheint es mir sinnvoll, kurz innezuhalten und darauf aufmerksam zu machen, dass die psychische Entwicklung im therapeutischen Raum einer Logik folgt, die dem Prozess der (idealtypischen) kindlichen Entwicklung in gewisser Weise ähnlich und doch diametral gegenläufig ist. Denn während sich in der kindlichen Entwicklung via elterlicher Spiegelung der affektiven Körperäußerungen und Impulse allmählich ein psychischer Raum entwickelt, führt in der Therapie die Erforschung der psychischen Strukturen (respektive der Abwehrmechanismen) zu einer Wiederentdeckung ehemals

[1] Eine ausführliche Beschreibung und Diskussion finden Sie in dem Aufsatz: Krüger-Kirn, Helga (2013): Der unerhörte Blick im Körper der Frau. Überlegungen zum körpersprachlichen Übertragungsraum in Frau-Frau-Therapien. In: Körper-Tanz-Bewegung. Zeitschrift für Körperpsychotherapie und Kreativtherapie Nr. 3,103–110+Nr. 4, 146–152. Ernst-Reinhard-Verlag.

unterdrückter emotionaler und körperbasierter Erfahrungen und damit zu einer (Re)Integration und Etablierung eines bewussten somato-psychischen Raums. Für das eigene Begehren einen Ausdruck zu finden und damit erkannt und verstanden zu werden, trägt zu einer Verankerung im eigenen Körper-Selbst bei und gilt als vitaler Bestandteil eines psychischen Entwicklungsprozesses. Damit rückt auch die Frage in den Fokus, inwiefern gerade die leibhaftige Präsenz eines Gegenübers von zentraler Bedeutung ist, um die abgewehrten und/oder verheimlichten Anteile bearbeiten und integrieren zu können. Wie in der kindlichen Entwicklung gewinnt auch im Therapieprozess an der Schwelle zur Symbolisierung/bewussten Verarbeitung die Anerkennung durch die Andere eine zentrale Bedeutung und zwar nicht nur auf verbaler Ebene, sondern durch deren leibhaftige und visuelle Präsenz. Blickkontakt symbolisiert hierbei eine existentielle Erfahrungsdimension für die eigene psychische Subjektivität. Hier kommt in der Übertragung eine Mutterfigur ins Spiel, die die Sehnsucht nach anerkennender Resonanz erkennen und beantworten soll. Dabei geht es um einen Raum, in dem Frau Salandre eine Erfahrung machen kann, an der das eigene Begehren beteiligt ist. Dieser Ort, an dem sie ihre Bedürfnisse und Begehrensweisen verankern kann, musste sich in der Interaktion mit mir erst entwickeln. Denn der Ort, der dafür in Frage gekommen wäre, war bisher zu sehr mit den sexuellen Gewalterfahrungen und den Zurückweisungen bzw. der Ignoranz durch die Mutter assoziiert.

In gewisser Weise ist Frau Salandre ‚körperlos' aufgewachsen, denn eine dem Kind angemessene emotionale Resonanz ihrer leiblich affektiven Bedürfnisse und -erfahrungen durch Mutter/Vater fehlten ihr weitestgehend. Zudem haben die sexuellen Gewalterfahrungen und deren induzierte Verheimlichung nicht nur ihre Entwicklung insgesamt, sondern in besonderem Maße ihr körperliches Selbstkonzept beschädigt. Hier macht es Sinn, die fehlende Nähe zur Mutter auch dahingehend zu lesen, dass die Mutter nicht nur ihr Selbstvertrauen behindert hat, sondern auch ihr geschlechtliches Selbstbewusstsein. Die in der Übertragung ersehnte Nähe konnte auf diese Weise als Ausdruck ihrer innerseelischen Not übersetzt werden, sich, neben dem unterbrochenen Dialog mit dem Mutterkörper, dem eigenen Begehren nach Selbstbestimmung und -gestaltung anzunähern. Der Beziehungskontext mit der Mutter scheint in dieser Perspektive in zweifacher Hinsicht bedeutungsvoll: Als Ort der Anerkennung der traumatischen Erfahrungen sowie als Ort der Sehnsucht nach Identifizierung und Entwicklung. Denn über den Generationenunterschied hinweg gibt es eine körperliche Gleichheit, die den gesamten sexuellen Körper betrifft. Vielfältige Teilrepräsentanzen wie Haare, Gesicht, Brust, Unterleib und Geschlechtsorgane sind untrennbar mit den sexuellen Gewalterfahrungen in einer bestimmten Weise weiblich besetzt und von gesellschaftlichen Idealen von weiblicher Körperlichkeit und Weiblichkeit nicht zu trennen.

Um im therapeutischen Raum das Begehren nach Berührung in all seinen Ausgestaltungen explorieren zu können, muss in der Übertragungsfigur die konflikthafte Gleichzeitigkeit von erwachsenem Frauenkörper und kindlich-regressiven Entwicklungsstufen und Bedürfnissen bedacht werden. Gerade aufgrund von sexuellen Gewalterfahrungen lassen sich induzierte und selbstbestimmte Erregungszustände kaum auseinanderhalten. Deshalb müssen mit Blick auf eine Sehnsucht nach Berührung die verschiedenen Übertragungsebenen differenziert und getrennt werden. Dies lässt sich nur bedingt auf der Ebene von Worten erforschen. Wie von Frau Salandre beschrieben, haben wir uns dem sinnlichen Erfahrungs- und Differenzierungsraum der Berührung über die Hände angenähert.

Grundsätzlich ist es im Rahmen der Differenzierungsarbeit zwischen induzierten und selbstbestimmten Erfahrungen und einer damit einhergehenden Unterscheidung von Heute und Damals nahe liegend, aber auch notwendig, die Therapeutin in der Mutterposition zu denken und sich ausgehend von einer ersehnten Nähe zum Mutterkörper voranzutasten. In dieser Übertragungsszene erscheint Frau Salandre als bedürftiges Mädchen, die Therapeutin als dem Kind zugewandte Mutter, die versucht, die Bedürfnisse und affektiven Äußerungen des Kindes zu verstehen, in Worte zu fassen und zu beantworten. Dieser Prozess wäre ohne die gemeinsame Bereitschaft zu immer neuen Versuchen, die berührungsnahen Sensationen in der Übertragung und Gegenübertragung, sowohl vertikal im Kontext der Gegenwart wie horizontal im Kontext der Vergangenheit, zu erforschen und einzuordnen, nicht möglich gewesen. Darüber hinaus mussten die Auswirkungen der sexuellen Gewalterfahrungen mit bedacht werden, die zwar die Mutter nicht unmittelbar betrafen, aber sie unbewusst zu einer Komplizin machten. Deren Ausblendung hätte zu Fehldeutungen, misstrauischen Übertragungen und entsprechenden Gegenübertragungen beigetragen.

Bleibt die sinnliche Sehnsucht nach Berührung oder nach Blickkontakt aber ausschließlich auf den frühkindlichen Erfahrungsraum bezogen, können nur bestimmte Übertragungs-Gegenübertragungsfiguren zugelassen werden. Eine nicht zu unterschätzende Rolle spielt hier, eingebettet in die Mutter – Tochter – Beziehung, das kollektive Homosexualitätstabu. Hier realisiert sich ein Tabu, indem die (homo-)erotischen Übertragungen und sinnlich-körperlichen Berührungswünsche sexualisiert und dementsprechend vermieden und erneut tabuisiert werden. So verstanden kann nicht nur die Sehnsucht nach Berührung in ein anderes Licht rücken, sondern auch in einen interaktionellen Raum, in dem das sinnliche Begehren nicht mehr ‚unerhört' bleiben muss. In diesem Sinne konnten die Berührungsinteraktionen sowohl als Reaktion auf das Verlorensein im kaum nachvollziehbaren Leiden sowie als ein Vermissen nach mütterlicher Resonanz eingeordnet und übersetzt werden. Derartige Erfahrungen machten Frau

Salandre Mut, sich weiter mit ihren Bedürfnissen und Sehnsüchten zu beschäftigen. Sie schafften zudem eine wichtige Voraussetzung für eine anerkennende Liebe zu sich selbst. Auch für mich war es ein zutiefst berührender Prozess, mit Frau Salandre an der Entwicklung stimmiger körpersprachlicher Differenzierungen und Berührungen zu arbeiten. Diese besondere Übertragungsdynamik beschreibt Daniel Stern (2005) als *now moment,* eine aus dem Moment der Begegnung heraus besondere Form von Verstehen. An dieser Stelle sei unterstrichen, dass sich now moments auf die gesamte Spannbreite affektiver Erfahrungen, also auch aggressive und zurückweisende, beziehen.

Während auf der konzeptionellen Ebene die Bedeutung nonverbaler Interaktionsmuster für die subjektive Entwicklung und das Übertragungs-Gegenübertragungsgeschehen für die meisten therapeutischen Schulen konsensfähig ist, spannt sich an der Debatte bezüglich des Umgangs mit körpernahen Dimensionen eine Kontroverse auf. Neben einer sicherlich bedeutsamen Technikdebatte verhindert vor allem die normative Einengung des sinnlichen Begehrens auf genitale Sexualität ein produktives Weiterdenken. Wiewohl die Übertragungsliebe als zentrales Moment in der Behandlung gilt, führt gerade die sexualisierte Fehldeutung des libidinösen Interesses am Anderen zu einer Tabuisierung von Körperlichkeit, sodass die Beziehungsdynamik in der Therapie zunehmend körperlos zu werden droht. Weil sich bis heute im therapeutischen Diskurs das Deutungsmuster der versorgenden, beruhigenden Mutter als kollektive Basis einer ‚weiblich-therapeutischen Verkehrsform' (frei nach Poluda-Korte, 1999) etabliert hat, rückt, vor dem Hintergrund der bisherigen Überlegungen, ein kollektiv wirksames ‚Homosexualitätstabu' in uns Allen in den Fokus. Nur so lässt sich erklären, dass eine differenzierende und vertiefte Erforschung des therapeutischen Begegnungsraumes in haltende und erotisch-sinnliche Begehrensweisen bis heute im theoretischen Diskurs noch kaum erfolgt ist. Sowohl klinisch wie auch politisch fälschlicherweise als Sexualität missgedeutet, steht sie einer produktiven weiblichen Entwicklung im Wege (vgl. Krüger-Kirn 2012). Hier zeigt der therapeutische Prozess von Frau Salandre, wie sehr ‚das Erleben von und Nachdenken über Sexualisierung und deren Tabuisierung immer wieder aufs Neue dazu herausgefordert haben, über eigene Verwicklungen in der Übertragungs-Gegenübertragungsdynamik nachzudenken. Auf diese Weise konnte ein Perspektivenwechsel neue Räume eröffnen, in denen unbewusste Bedeutungszusammenhänge spürbar, sichtbar und aus der bisher ‚unerhörten' Abwehr befreit werden konnten. So betrachtet überschneiden sich konkrete Behandlungsanliegen mit prinzipiellen Fragen zur weiblichen Identitätsentwicklung. Dabei geht es grundsätzlich um tabuisierte Erfahrungen, die einer weiblichen Selbstermächtigung im Wege stehen. Eine solche Perspektiverweiterung trägt dazu bei, über den subjektiven Schauplatz der psychischen Konflikte hinauszugehen und die psychischen Verarbeitungs- und Verinnerlichungsstrukturen nicht unabhängig von der Wirkmacht hegemonialer Körper- und Weiblichkeitsbilder respektive der kollektiven Abwehr der Homosexualität zu betrachten.

Die politische Seite der Therapie

Für mich als Therapeutin ist dieser Zusammenhang gleichsam Anliegen und Auftrag, die bisher in den Blick genommenen entwicklungspsychologischen und therapeutischen Reflexionen auf geschlechtliche Aneignungsprozesse von Körper und Sexualität nicht nur in einen Zusammenhang mit gesellschaftlichen Sexualitäts- und Weiblichkeitsvorstellungen, sondern in einen Bezug zu sexueller Gewalt zu stellen.

In familiären Strukturen, sofern sie wie bei Frau Salandre patriarchal geprägt sind, rückt die Verwobenheit von Weiblichkeitsvorstellungen und geschlechtlichen Machtverhältnissen augenscheinlich in den Fokus. Der Blick auf sexualisierte Machtansprüche in familiären Strukturen trägt dazu bei, über die Ebene biographischer Selbstverhältnisse hinauszugehen. Für die therapeutische Behandlung wie für das Nachdenken über sexualisierte Gewalt geht es daher auch darum, die in den sexuellen Gewaltpraktiken repräsentierte Macht in familiären (und darüber hinaus in sozialen und institutionellen) Strukturen nicht zu ignorieren und auf diese Weise weiter zu verschleiern. Subjektive Erfahrungen im Geschlechterverhältnis zu kollektivieren und individuelles Schweigen ins Verhältnis zu gesellschaftlichem Schweigen zu stellen, ist in zweierlei Hinsicht hilfreich und notwendig: Zum einen, um sich aus der individuellen Schuld- und Betroffenheitszuschreibung zu befreien, und zum anderen, um sich die psychosozialen Folgen normativer sozialer Ordnungen bewusst zu machen. Es ist und bleibt daher eine zentrale Aufgabe, sexuelle Gewalt immer auch als Form der Herrschaft zu untersuchen. Eben diese Zusammenhänge zwischen Machtmissbrauch und gesellschaftlichen Geschlechterstrukturen rücken die politische Tragweite von Psychotherapie im Allgemeinen und mit Blick auf sexualisierte Gewalt im Besonderen in den Fokus (vgl. hierzu Krüger-Kirn 2019). Die hier vorliegende Versprachlichung von geschlechterbasierter Unterdrückung und Gewalt ist daher nicht nur auf der persönlichen Ebene als ein mutiger Schritt zu verstehen, sondern als ein wichtiger Beitrag bzw. eine Voraussetzung für politisches Handeln. Es sind die niedergeschriebenen Zwiegespräche, die in diesem Buch erneut zu Wort kommen und auf diese Weise Zeugnis ablegen – nicht nur von Frau Salandres persönlichem Entwicklungsprozess, sondern auch von der Sinnhaftigkeit, sich den Traumata zu stellen, die weit über die sexuellen Gewalterfahrungen hinausgehen, um eine Gestaltungs- und Deutungshoheit über sich und die eigenen Lebensvorstellungen (wieder) zu gewinnen.

Eigenes Denken, Erinnern und Erleben im Rahmen von Therapie zu Wort kommen zu lassen, kann zu (Selbst)Liebe und Selbstbestimmung beitragen. Einen nicht zu unterschätzenden Beitrag dazu hat bei Frau Salandre auch das Tagebuch schreiben geleistet: Im Schreiben konnte sich Frau Salandre selbst zur Sprache bringen und die im therapeutischen Kontakt erfahrene emotionale Resonanz und Selbstreflexion fortsetzen. Auf diese Weise konnte sie nicht nur ihr Erleben weiter erkunden, sondern sich dem Bann der Angst und der quälenden sprachlosen Leere entgegenstellen.

Als die gewonnen Einsichten und Erfahrungen das Selbsterleben im Alltag zunehmend selbstbestimmter, heller und lustvoller machten, war auch die Zeit gekommen, dieses Buch zu schreiben.

Prof. Dr. phil. Dipl. Psych. Helga Krüger-Kirn April 2022

Literatur

Freud, Sigmund (1907a): Der Wahn und die Träume. In: W. Jensens „Gradiva“. Gesammelte Werke VII, 29–125. Fischer, Frankfurt / M.

Krüger-Kirn, Helga (2012): Perspektiven auf emanzipatorische Praxen in subjektiven Selbst- und Körperverhältnissen. In: Carmen Birkle, Ramona Kahl, Gundula Ludwig, Susanne Maurer (Hg): Emanzipation und feministische Politiken: Verwicklungen, Verwandlungen, Verwerfungen. Helmer Verlag

Krüger-Kirn, Helga (2013): Der unerhörte Blick im Körper der Frau. Überlegungen zum körpersprachlichen Übertragungsraum in Frau-Frau-Therapien. In: Körper-Tanz-Bewegung. Zeitschrift für Körperpsychotherapie und Kreativtherapie Nr. 3, 103–110 und Nr. 4, 146–152. Ernst-Reinhard-Verlag

Krüger-Kirn, Helga (2019): #MeToo und die Debatte um geschlechtliche Gleichstellung. Psychoanalytische Annäherungen. In: Psychoanalyse im Widerspruch, Nr. 62, 32 (2) 2019, 3–25. Psychosozial-Verlag

Poluda-Korte, Eva (1999): Die psychosexuelle Entwicklung der Geschlechter im Vergleich. Forum der Psychoanalyse 15, 101–115. Springer-Verlag

Stern, Daniel N. (2005): Das Gegenwartsmoment. Veränderungsprozesse in Psychoanalyse, Psychotherapie und Alltag. Brandes & Apsel, Frankfurt/M.